ANNUAL REVIEW OF PLANT PHYSIOLOGY

ANNUAL REVIEW OF PLANT PHYSIOLOGY

WINSLOW R. BRIGGS, *Editor*
Carnegie Institution of Washington, Stanford, California

PAUL B. GREEN, *Associate Editor*
Stanford University

RUSSELL L. JONES, *Associate Editor*
University of California, Berkeley

VOLUME 32

1981

ANNUAL REVIEWS INC. 4139 EL CAMINO WAY PALO ALTO, CALIFORNIA 94306 USA

ANNUAL REVIEWS INC.
Palo Alto, California, USA

REPRINTS The conspicuous number aligned in the margin with the title of each article in this volume is a key for use in ordering reprints. Available reprints are priced at the uniform rate of $2.00 each postpaid. The minimum acceptable reprint order is 5 reprints and/or $10.00 prepaid. A quantity discount is available.

International Standard Serial Number: 0066-4294
International Standard Book Number: 0-8243-0632-5
Library of Congress Catalog Card Number: A-51-1660

PRINTED AND BOUND IN THE UNITED STATES OF AMERICA

PREFACE

The publication in July 1981 of Volume 50 of the *Annual Review of Biochemistry* marks the 50th Anniversary of the founding of one of the world's most respected publishers of scientific reviews — ANNUAL REVIEWS INC. of Palo Alto, California.

Annual Reviews Inc. was established by Dr. J. Murray Luck, Professor of Biochemistry, Stanford University, who initiated publication of the *Annual Review of Biochemistry* in 1932 to help keep biochemists abreast of the proliferating research literature in their field. His plan was to publish one compact volume each year containing critical reviews of the latest research findings in Biochemistry. The growth of the scientific literature was so rapid that Dr. Luck's successful formula was soon applied to other disciplines by the addition of new annual publications in various biological, medical, physical, and social sciences over the years. The *Annual Review of Plant Physiology*—seventh in the series—made its debut in 1950, and today the Annual Reviews family of publications consists of 24 series that regularly serve hundreds of thousands of scholars, students, and professionals worldwide.

The Editors and members of the Editorial Committee for the *Annual Review of Plant Physiology* are proud to add their congratulations to those of scientists everywhere for the valuable services of Annual Reviews Inc. to all branches of the scientific community.

THE EDITORS

Annual Review of Plant Physiology
Volume 32, 1981

CONTENTS

ARTICLES IN OTHER *ANNUAL REVIEWS* OF INTEREST TO PLANT PHYSIOLOGISTS

From the *Annual Review of Phytopathology,* Volume 18 (1980):

Lignification as a Mechanism of Disease Resistance, *C. P. Vance, T. K. Kirk, and R. T. Sherwood*

Theory of Genetic Interactions Among Populations of Plants and Their Pathogens, *K. J. Leonard and R. J. Czochor*

Germplasm Resources of Plants: Their Preservation and Use, *G. F. Sprague*

From the *Annual Review of Microbiology,* Volume 34 (1980):

Oxygen and Hydrogen in Biological Nitrogen Fixation, *Robert L. Robson and John R. Postgate*

Some Aspects of Structure and Function in N_2-Fixing Cyanobacteria, *W. D. P. Stewart*

From the *Annual Review of Genetics,* Volume 14 (1980):

Molecular Arrangement and Evolution of Heterochromatic DNA, *Douglas L. Brutlag*

From the *Annual Review of Ecology and Systematics,* Volume 11 (1980):

Herbivory in Relation to Plant Nitrogen Content, *William J. Mattson, Jr.*

• The Mineral Nutrition of Wild Plants, *F. Stuart Chapin, III*

Physiological Ecology of Tropical Succession: A Comparative Review, *F. A. Bazzaz and S. T. A. Pickett*

Aquatic Primary Productivity and the ^{14}C-CO_2 Method: A History of the Productivity Problem, *Bruce J. Peterson*

From the *Annual Review of Nutrition,* Volume 1 (1981):

Formation and Mode of Action of Flavoproteins, *A. H. Merrill, Jr., J. D. Lambeth, D. E. Edmondson, and D. B. McCormick*

From the *Annual Review of Biochemistry,* Volume 50 (1981):

• Lectins: Their Multiple Endogenous Cellular Functions, *S. H. Barondes*

Organization and Expression of Eucaryotic Split Genes Coding for Proteins, *Richard Breathnach and Pierre Chambon*

The Mechanism and Regulation of ATP Synthesis by F_1-ATPases, *Richard L. Cross*

Proton-Translocating Cytochrome Complexes, *Mårten Wikström, Klaas Krab, and Matti Saraste*

Birgit Vennesland

Ann. Rev. Plant Physiol. 1981. 32:1–20

RECOLLECTIONS AND SMALL CONFESSIONS

❖7704

Birgit Vennesland

Forschungsstelle Vennesland der Max-Planck-Gesellschaft,
1 Berlin 33, Germany

CONTENTS

The Preacher sought to find out acceptable words - even words of truth

Ecclesiastes 12, 10

On seeking inspiration from the prefatory chapters of my distinguished predecessors, I began to wonder what I had done to deserve inclusion in this series, beyond surviving until retirement. I represent a scientific generation that lived through a developmental explosion when scientific research came of age. Maybe I am typical of those biochemists with plant physiological interests who trained in the 1930s, when refrigerated centrifuges and spectrophotometers and isotopic tracers were only available in homemade versions, fractions were hand-collected, and jobs were rare. My generation has seen greater scientific change in our lifetime than any generation before us. All I can do is describe what it was like.

0066-4294/81/0601-0001$01.00

LIFE HISTORY

I was born together with my sister Kirsten on a stormy November day in 1913 in Kristiansand, a small town on the southern coast of Norway. Ever since, when I return to Norway and look at it, my nose tickles a warning of impending tears. I had forgotten how beautiful it was. This was one of the reasons that I moved, in 1968, from Chicago to Berlin, but that is a later part of the story.

My ancestors, so far as I know, were all Norwegian farmers. My father grew up a country boy, but my maternal grandparents settled in Kristiansand, a town of about 20,000 population. Both my parents attended the normal school in Kristiansand with the object of becoming school teachers, but my father made frequent trips to the USA and Canada, where he functioned variously as a school teacher (in a small Norwegian settlement in Iowa), a bookkeeper, a lumber dealer, a real-estate agent, and finally a DDS, graduate of the Chicago School of Dental Surgery.

In May 1917 my parents got tired of waiting for the end of World War I, and my mother bundled my twin sister and me onto a ship to rejoin my father, who had been caught by the war on the other side of the Atlantic. I grew up on the north side of Chicago and attended Waters Grammar School and Roosevelt High School, before entering the University of Chicago in the fall of 1930. Except for two exciting years in the Department of Biochemistry at the Harvard Medical School (in 1939–41), I lived continuously in Chicago until I returned to Europe in 1968.

My parents were liberal-minded middle class citizens, glad to be in the United States—Europe was a rather unhappy part of the world in the 1920s, but life in the USA was comfortable during this decade. We grew up bilingual, speaking Norwegian at home, and our house was always overflowing with books, both English and Norwegian. My mother had an intense interest in women's rights, without any urge to participate in politics directly. She maintained an emotional attachment to Norway which was apolitical, but ensured that my sister and I returned to Norway for three extensive visits, in 1921-22, in 1930, and in 1935, so we would not "forget."

I am a product of the public schools of Chicago. In my recollection, they were not so bad. Standards may not have been very high, but we were encouraged to read extensively and to try our hand at writing or debating or acting or singing, or whatever other hobby might interest us. For a period, I was an ardent girl scout, and spent quite happy summers outside the city earning badges for flower, bird, and star finding. The science involved was nonexistent. One acquired handbooks that enabled one to attach a name to what one could see: that was all. But I enjoyed these activities very much and began to think about being some kind of "naturalist." My picture of a naturalist was of a person who spent most of his time collecting things that grew or flew and learning all about them.

The museum of natural history (then the Field Museum) strengthened this interest. For years I spent most of my Saturdays haunting the Field Museum, the Art Institute, and the Public Library in downtown Chicago. I read hungrily and quite indiscriminately. My first favorite fictional characters were probably "Tarzan of the Apes" followed by Jim of "Treasure Island." Robert Louis Stevenson was a little too realistic. I found it cruel of him to give a sore throat to the hero of "Kidnapped."

My father had been prevented by strong pecuniary pressure from entering the occupation he thought he would have liked best, which was civil engineering: bridge building to be exact. In the eyes of the small boy in a wild and mountainous countryside, the man who designed and built bridges was the great hero-figure. My mother had become a school teacher by passionate choice and looked for the same inclinations in her offspring. The result was that when I was asked what I wanted to be, at an age when most boys opted for fireman or locomotive driver and most girls expressed a preference for nursing or acting, I lisped that I wanted to be a teacher. This conviction never left me, but it is hard to say how much was instilled from outside. My mother was a rather compelling educator of the very young. If I search my heart in this matter, I would say that it was the subject matter to be taught that interested me. But what subject matter? Here choice was a little difficult because I was interested in almost everything.

I suppose that under other circumstances I could have ended up in any of a large variety of occupations. But fate took a hand. The University of Chicago offered tuition scholarships to graduating high school students on the basis of competitive examinations. In those years the examination was given in a considerable variety of standard high school subjects, but the examinee picked only one. When invited to take the exam in physics, I was only too glad to say yes. I placed first, with the result that I entered the University of Chicago a little before my seventeenth birthday in the fall of 1930.

From this course of events you might infer that I had a special interest in physics, but that wasn't true. I preferred to take an examination in physics because our high school physics course was taught with a little rigor and was therefore considered "hard." Even in grammar school I had learned that it was much easier to "shine" in a "hard" subject. Later I learned that it had been my great luck that our high school physics course had been a very simplified version of the first year physics sequence taught at the University of Chicago. Had I taken the examination in chemistry, I would have laid an egg. Our high school chemistry course was not taught as a science but as an odd mixture of facts about chemical technology.

In high school I had taken what was then known as a general language major, which was considered, probably rightly, the best college preparatory course. I entered the University of Chicago in the very early days of the reign of R. M. Hutchins, President, then Chancellor, of the University. Hutchins was planning a big reorganization of college education. The class in which I entered was the last class subject to the much-maligned "old" plan. Our successors flourished under the "new" plan, which underwent so many subsequent revisions that one quite lost track of it. But I don't want to get involved in the subject of specialist (old) vs general (new) education. For myself the main point was that I had, so to speak, the benefits of the best of both systems. I was enrolled for the first two quarters in a general science course entitled "The Nature of the World and of Man." This course was staffed by top science professors, each of whom gave a few lectures in his specialty. Most of them were good lecturers. I was spellbound. Since the subject organization was partly alphabetical, we began, I believe, with astronomy and ended with zoology, and I fell in love with each subject in turn, which meant that I began my college career with a decision to become an astronomer and shifted to a committment to zoology, having changed my mind repeatedly over the course of six months. The result of this series of changes was that I chose to follow a premedical program, since it provided for a maximal mix of both physical and biological sciences, at least at the beginning.

I had no real commitment to medicine. Though I dreamed a little of medical research, it was in terms of *Arrowsmith* and *Microbe Hunters* (my major as a premed was bacteriology). I took the first course in biochemistry in the beginning of my junior year, and found myself in a dilemma. From now on it would all be biology. I was finished with the physical sciences, if I continued with my original plan. Though committed to biology, I felt that I needed much more chemistry to think successfully about biological problems. The obvious compromise was biochemistry. There were few BS degrees given at that time with a major in biochemistry, mainly because all of the necessary prerequisites plus a year of biochemistry could not easily be crowded into a four-year program.

The credits on my college transcript showed an appalling concentration on science. As a horrifying example of the uneducated specialist, I was exhibit A. Presumably, I had never read a book except for science texts. This presumption was not true. I had actually done all the assigned reading for the "survey" courses in the humanities and social sciences plus most of the assignments that Hutchins and Adler gave in their "Great Books" course. It was my exceeding good fortune that I went through college with my physical wants provided for in the University dormitories and no need to earn my way by taking part-time jobs. Thus, I had plenty of free time to indulge in hobbies, and I have always been a voracious reader with rather catholic tastes. It only enhanced my pleasure that I could read St. Augustine and Dostoevsky and Democritus and Hume for fun, without subsequently being subjected to the indignity of a multiple choice examination. The use of such examinations as a measure of the well-rounded human being is in my opinion a barbarism. The important thing in the humanities is the exposure to the works of great thinkers and artists. For this the University of Chicago in the early thirties was an excellent place. One could attend or cut any lectures one pleased, and the libraries were open til ten in the evening.

Becoming a Biochemist

I took my BS in biochemistry in the spring of 1934 and spent five subsequent months working as a chemical technician in the department of medicine of the University of Illinois Medical School on Chicago's West Side. That was my first taste of earning my own living. The work was not uninteresting, because I was attached to a research project, but I realized that I had still only scratched the surface of what I needed and wanted to know if I was going to do research in biology. In the fall I returned to the University graduate school in the same department of biochemistry.

The decade of the thirties was not a happy time in the graduate schools. It was the decade of the great depression. Most students had to live on a shoestring. The department of biochemistry happened to be better off than most. It provided its graduate students with teaching assistantships: $ 200 for each 12-week quarter = $ 800 per year, from which a minimum of 4 X $ 33 = 132 was repaid as tuition. A single person could almost live on the remainder, though not with any frills. We took outside jobs wherever we found them. I remember doing pH determinations for the meat packers, with a homemade glass electrode. The necessary special glass had just become available commercially.

In *The Adventures of Augie March,* Saul Bellow has described the milieu around the Midway in the thirties. His time there coincided apparently in part with mine and our paths probably crossed. Though I have no clear recollection of having met him, I have suspicions. But the picturesque squalor described by Bellow wasn't so picturesque (or squalid either) for the graduate science students. We spent most of our time, morning, noon, and night, in the laboratory. Meals were cooked there, and romance sometimes flourished there.

Throughout my years as a student of biochemistry at the University of Chicago, Fred C. Koch was Departmental Chairman. Koch's research interests were in steroids, both in the form of vitamins and of sex hormones —a very ambitious program. In addition, he was interested in the development of analytical methods suitable for use in the clinical laboratories attached to the University hospital. In the student training there was a heavy emphasis on blood and urine analysis. This had good practical reasons. Hospital laboratories provided jobs for biochemists, and jobs were a very precious commodity.

The biochemistry of the thirties was mainly concerned with the chemistry of small molecules. Vitamins, hormones, and growth factors were being identified. From the chemistry of carbohydrates, fats, and amino acids, we made a conceptual leap to the physiological aspects of diabetis, ketogenesis, and glyconeogenesis. In between there was the very mysterious process of metabolism which took place in something equally mysterious called protoplasm and was catalyzed by agents called enzymes.

Enzymes, we were taught, were of unknown chemical nature. This may seem surprising, since Sumner had long since identified urease as a crystalline protein, and Northrop and Kunitz were crystallizing the proteolytic enzymes of the digestive tract. But Fred Koch was a great admirer of Willstätter, and adhered to Willstätter's view that peroxidase was not a protein. It has been pointed out that Willstätter, who was a first-rate chemist, had the sheer bad luck of selecting for purification an enzyme with a very high turnover number. He purified it to a point where the available protein tests were negative and the material left no ash, though it still gave a good enzyme test. (This business about the ash was the origin of a certain friction between Willstätter and Otto Warburg, as described in a later section.)

The opposition to the conclusion that enzymes were proteins was not irrational. The prevailing view was that "proteins are big polypeptides. Such molecules do not have catalytic properties. They can absorb other substances, however. The isolation of a few crystalline proteins with enzymatic properties is the result of the absorption of the enzyme on the protein." What was actually demanded was the isolation of the active center of the enzyme. Later the identification of special prosthetic groups for various reaction types made the notion that enzymes are proteins easier to accept. But what really clinched the point was not one single discovery but a massive accumulation of findings with many different enzyme reactions, showing that no enzymatic activity could be separated from protein.

For my PhD I picked my own problem with Martin Hanke's sponsorship and help. By 1938 I had completed my thesis on the *Oxidation-reduction requirements of an obligate anaerobe.* This thesis didn't, in my later opinion, amount to much. I was trying to approach the problem of what it was that distinguished obligate anaerobes from other bacteria, and I didn't select a very good approach. But the work taught me two things: It taught me that bacteria require CO_2 for growth, and it taught me the importance of collaborators. If you have to do absolutely everything all by yourself, progress can be very slow, especially if you have no money.

Boston and Radioactive Carbon

After working for a year as research assistant for E. A. Evans, Jr., I had the colossal good luck in 1939 to get a fellowship from the International Federation of University Women. This fellowship carried the stipulation that it had to be used in a foreign country. Arrangements were made for me to join Meyerhof, who was at that time in Paris. A summary and evaluation of Meyerhof's scientific achievements can be found in Nachmansohn's autobiographical chapter (5), in which the latter also outlines the problems faced by Meyerhof in 1939.

I was to leave for Paris in the middle of September. On the outbreak of World War II, all passports of American citizens were cancelled. You had to reapply for a new one in order to go to Europe, and you had better have a good reason. This was at first a bitter blow. In retrospect, however, it is clear that the fates had been kind, because I managed to find sanctuary in Baird Hastings's department of biochemistry at the Harvard Medical School. When I arrived, he assigned me to his ^{11}C-glycogen project, which was just getting organized. As a departmental chairman and group leader of the ^{11}C project, Baird Hastings was one of the kindest and most considerate human beings that I have ever encountered. He ran a happy department in an unhappy time. The following two years were for me depressing in terms of the continual escalations of the war; and at the same time they were intensely stimulating and exciting as far as scientific work was concerned.

The utility of deuterium for studying metabolic processes or paths had recently been demonstrated at Columbia by Schoenheimer and his coworkers, but the usefulness of D was limited. ^{11}C was the first radioactive isotope of carbon made available to biochemists. It had a half-life of about 20 minutes—a severe limitation which required speed in all operations. There was a substantial group involved. To achieve the necessary velocity, four or five people actively participated in each labeling experiment. We studied the synthesis of glycogen by rat liver in vivo, with the object of learning something about how glycogen was made from precursors of low molecular weight. The most interesting discovery was that the carbon of CO_2 got into the carbon of the glycogen. There were several laboratories working on related reactions at this time, more or less unaware of each other. The full story has been told elsewhere (2) and calls for no recapitulation. Earl Evans, who had been working with Hans Krebs in Sheffield, came back to the University of Chicago in 1940, got hold of some ^{11}C from Louis Slotin, and showed that the α-ketoglutarate, made during the Krebs cycle by pigeon liver brei, was labeled by $^{11}CO_2$. In the fall of 1941, I returned to the University of Chicago as an Instructor of Biochemistry with the intention of continuing ^{11}C experiments with Evans and Slotin, but we didn't get much accomplished before the United States was also involved in the war. During the following year or so, the chemists who could make and handle isotopes were swallowed by war projects, and research unrelated to the war effort ceased almost completely. I was attached to a Malaria project, but my teaching assignments were so heavy that there was time for little else.

In retrospect, when I revive my recollections of my two years at Harvard, it seems to me that there was an electrical intellectual excitement in the Boston air. This was partly due to ^{11}C, but there was also something else. It was the time of the great migration of scientists from Europe to America. They came first to the eastern seaboard and often gathered in Woods Hole, which was within easy weekend reach of Boston. Through Gertrude Perlmann, I had an opportunity to read Fritz Lipmann's paper, "Metabolic Generation and Utilization of Phosphate Bond Energy," when it was still in press in Volume 1 of *Advances in Enzymology.* This manuscript seemed to have the power of a revelation. Now I could understand why Meyerhof had suggested that I work on the "oxidizing enzyme" of glycolysis. Now I felt that I understood glycolysis for the first time. Of course it wasn't until the postwar years that the biochemical seeds sown by the immigrants grew and bore fruit a hundredfold.

β-Carboxylations and Stereospecificity

When things began returning to normal after the war was over, I turned my attention to plants, with the object of looking for dark fixation of CO_2, Krebs cycle, and other enzyme reactions. The known facts about the metabolism of *Crassulacean* plants suggested strongly that a set of reactions similar to those of the Krebs cycle plus the dark carboxylations known as Wood-Werkman reactions, would probably be found in plants also. I believe that what attracted me to higher plants as experimental material was the paucity of information available about their metabolism, plus the fact that they were the seat of one of the most fascinating processes in nature: photosynthesis.

James Franck and Hans Gaffron were located in a neighboring building, and I often went over to listen to their seminars and discussion groups, in this way learning also a good deal about the history of the photosynthesis problem and the development of current experimentation. Since different research groups at the same institution were not supposed to compete, it was agreed that I would limit myself to dark reactions, unless it was possible to find an area for collaboration. My group had invested much time and effort in the preparation of the pyridine nucleotides, and Eric Conn gave some of our TPN (NADP) to one of Franck's students (Tolmach), who was overjoyed when it worked as a reagent for eliciting O_2 evolution. But no real collaboration developed here, mainly because James Franck didn't like enzymes. I suspect that the root of this antipathy was similar to the root of Otto Warburg's unreasonable antipathy for radioactive isotopes. After a certain age, it becomes difficult if not impossible to master a new technique. It happens to all of us.

It was about 1949–50, as I recall, that Frank Westheimer of the Chemistry Department and I discovered that we were both interested in β-carboxylations, except that he, of course, was interested in reaction mechanism, whereas I was looking for enzymes that catalyzed such reactions. When Harvey Fisher asked if he could do a joint PhD with Westheimer and me, I rather expected that what would emerge would be a carboxylation project. But no, what Westheimer wanted to work with was pyridine nucleotides. He had been using deuterium to study the mechanism of oxidation of alcohols by chromates. (That problem, Harvey informed me, was known as "how to make better cleaning solution.") Now Westheimer wanted to know how diphosphopyridine nucleotide (NAD) and alcohol dehydrogenase oxidized ethanol. Hydrogen must either be transferred directly, or the extra hydrogen in the reduced pyridine nucleotide must have its origin in water. The problem was beautifully formulated from the very beginning. By great good luck, Pabst put diphosphopyridine nucleotide on the market at just the right time, so that the preparative work required was simplified. The first serious experiment was successful. We soon knew that there was direct transfer of one hydrogen atom. Since Ogston's celebrated paper (1) had appeared, and we had sweated out the significance of Ogston's paper in connection with citrate synthesis, as told elsewhere (7), the possibility of using deuterium to do stereospecificity studies was apparent. All in all, this was a happy and very successful collaboration, perhaps because Westheimer and I brought different past experience and skills to a problem of common interest. What little I could contribute to mechanism studies rests on the tutoring I received from Westheimer.

A summary of known information about the stereospecificity of hydrogen transfer in pyridine nucleotide dehydrogenase reactions has appeared (8), and there the reviewers give me a little more credit than is my due for getting this work started. That isn't quite fair to Westheimer, though I feel a flush of gratitude to the reviewers. But when I think of the long line of students and postdocs who worked on this project and contributed all the hard labor, I feel very humble indeed. I do deserve credit for durability. Perhaps I stuck with it longest.

Photosynthesis and Otto Warburg

After Gaffron left Chicago, I began to work seriously with chloroplast preparations, trying to learn something about photophosphorylation and the Hill reaction. A fact that particularly caught my fancy was the catalytic effect of CO_2 on the latter. Warburg was putting a strong emphasis on this point because it fit his theory of photosynthesis. Gaffron said Warburg's experiments were irreproducible. I decided to try it myself, and found that one could get quite good stimulation of Hill reactions with CO_2, provided one picked the right conditions. This was the background for my initial visit to Warburg's laboratory in 1961.

A succession of visits ensued and culminated in my accepting a position as a director at Warburg's institute in West Berlin in 1968. There I began to work on nitrate reduction by *Chlorella.* There were complex reasons for the selection of this problem. One was that Warburg regarded nitrate as the "natural" Hill reagent. Later I gradually developed a strong suspicion that the reason Warburg got such fantastically low values for the overall quantum requirement of photosynthesis was mainly that he had nitrate in the medium and excess carbohydrate in the cells. Better methods have long since superseded those used by Warburg, and the problem of the quantum requirement is no longer cogent.

The editors made it clear that I was not to write a scientific review, so I will exercise heroic self-restraint and refrain from saying much more about nitrate reduction.

Warburg's theory of photosynthesis was sensible and internally self-consistent, though it paid no attention to any experiments that he hadn't done himself. There were some assumptions in his theory and he was quite aware of them. I asked him once whether he didn't think there might be more than one kind of light reaction. "Of course, there might be. But you should keep everything as simple as you possibly can. Never introduce any more detail into a theory than you must." Me: "But, Herr Professor, I think there is evidence that there are two different kinds of light reactions." Warburg: "What is your evidence? Have you seen it?" Me: "Well, not exactly—uh. There is lots of evidence from different labs." Warburg: "Don't tell me about anybody else's experiments. I want to see yours. Show me. That's the way one should argue. I'll show you my experiments. Then you show me your experiments." Me: "All right."

When I last looked at Warburg's desk, it looked as though he was doing energy calculations for an equation which I had suggested to him:

$$H_2 + HNO_2 \xrightarrow{h\nu} NH_3 + O_2 \quad 1.$$

I had thought of this formulation when I was reading his earlier work on nitrate reduction. The advantage of Equation 1 over Warburg's earlier formulation was that the new equation is really a one-quantum reaction, almost. Now one didn't need a back reaction of O_2 to make up the energy deficit, and CO_2 was pushed over to the Calvin cycle where it belonged. Had it been possible to convince him that there was a second light reaction that caused dismutation but led to no O_2 evolution, he might have added:

$$NH_3 + HOH \xrightarrow{h\nu} NH_2OH + H_2 \quad 2.$$

and

$$NH_2OH + HOH \xrightarrow{h\nu} HNO_2 + 2H_2 \quad 3.$$

All three equations add up to look like water-splitting, but in principle they represent an attempt to divide the overall process into three chemical steps, each of which might be energized by one light quantum.

In my opinion, the apparent naivete in Warburg's theories was studied and intentional. The rules seemed to be: keep maximal simplicity and stick to minimal numbers. Make changes only when you must. The advantage of writing balanced equations is mainly that it is an indispensable approach to correct energy calculations. The chemical identity of the components need not be taken literally. One knows that the entire process is far more complicated than the symbols suggest. The very starkness of the symbols protects one from the easy semantic error of confusing the picture with the phenomena that the picture represents. To me, this approach to theory building seems preferable to the currently accepted practice of beginning with a diagram of an hypothetical electron-transport chain. This latter term, invented to represent a sequence of reactions in time, has been converted in the minds of many to a real chain along which electrons flow. Arguing about the properties of such chains is a little like counting angels on the head of a pin, though it seems to pay.

Lest the previous paragraphs give the impression that my relations with Warburg were totally harmonious, I hasten to add that the harmony was mainly confined to scientific questions at a fairly elementary level. Warburg was a man of violent emotions. Anger was predominant, and he wasn't very rational when his emotions took control. We all have a rational self and an irrational, emotional self. Mostly, the rational self is in control. Warburg's emotional self was unusually powerful. Science was his recipe for staying rational.

A mutual acquaintance—not a scientist—was reported to be so disappointed in an emotional relationship that he got sick. Warburg's advice was, "Tell him not to think about anything but science—think about absolutely nothing else—only science." Since the sick man was not a scientist, this prescription probably didn't help him much. But I regard it as a clue to the reason that science, for Warburg, was such an obsession.

On a visit to Warburg's laboratory, I had asked him once for his opinion about the claim in an earlier biographical sketch that Warburg had never made any mistakes. He pondered awhile and said, "Of course, I have made mistakes—many of them. The only way to avoid making any mistakes is never to do anything at all. My biggest mistake—", here he paused a long time before continuing, "my biggest mistake was to get much too much involved in controversy. Never get involved in controversy. It's a waste of time. It isn't that controversy in itself is wrong. No, it can even be stimulating. But controversy takes too much time and energy. That's what's wrong about it. I have wasted my time and my energy in controversy, when I should have been going on doing new experiments, and now—". In the course of the conversation, Warburg also tried to explain to me why he got so mad at James Franck. Warburg: "He said—he said—I couldn't measure light. He was a theoretician. By himself he couldn't measure anything, and he said I—I couldn't—measure—." Something curious was happening. Warburg was getting a little incoherent. In the course of telling me about why he got angry, Otto Warburg got angry all over again. He got as angry as he would have been if James Franck had been sitting there in the same room, now, telling him how sorry he (Franck) was, that his (Warburg's) measurements couldn't possibly be right. At that time, however, I hadn't learned to recognize Warburg's anger. He didn't usually signal it the way most of us do, by raising the pitch and/or the volume of his voice. He had perfect control over his behavior.

Warburg's views on photosynthesis could be considered to be a further development of Willstätter's views. The well-selected library of the Institute for Cell Physiology included Willstätter's works on chlorophyll and photosynthesis. But there was an antagonism to Willstätter on Warburg's part. What was its origin? Warburg told me about this as follows:

> Willstätter gave a lecture about peroxidase and the audience was very big. When he had finished and it was time for questions, I asked him if there was any ash. That wasn't such a stupid question for a young man to ask, was it? Willstätter said, "No, Herr Warburg, there isn't any ash." It wasn't what he said, but the way he said it. He was a good speaker. Because of the way he said it, five hundred people laughed.

In his fine biography of Warburg (3), Hans Krebs cites two sentences concerning the 46-year war over bioenergetics as an example of Warburg's often very aggressive polemical style. I translated the original German version of that manuscript into English. Warburg asked me to do this for him. Now I know perfectly well that the translator isn't supposed to try to improve on the original. He is supposed to be accurate. I did the translation as conscientiously as though I were taking an examination for a translator's certificate, except at one place. After pondering carefully, I translated the "war" into "argument," knowing that this would flunk me. Except that I thought the examiner wouldn't notice. The examiner, I assumed, would be Warburg himself. At first, this guess seemed right. Warburg concentrated on the science and didn't notice the omitted war. He was pleased and complimented me on how well I understood his experiments and theories. Since he was never prodigal with compliments, I felt pretty good. But there was a second examiner, a Warburg No. 2, who acted as a censor. I hadn't known about him. The censor didn't know any science, but he could read English. So the day after getting the nice A, I was told my grade had been changed to an F. "I won't use any of it," said Warburg, in anger. The baleful glare of his blue-gray eyes could be uncannily frightening. Of course it didn't help to try to point out that I was only giving his own advice back to him. Advice is a commodity that can readily be transmitted from senior to junior, but seldom in the other direction.

Warburg was a stubborn man, and apparently something of a rebel in his youth. In German universities there is a procedure called habilitation, which qualifies a person to hold a professorship. This procedure is a kind of final examination, in which the candidate demonstrates his teaching competence as well as his mastery of a research area. Warburg was proud of the fact that he had never submitted to the habilitation process.

Krebs has told how Warburg became a professor by sending lamb chops to Fisher (3). Warburg told me once how he first managed not to become a professor. Warburg: "I have never given a lecture to students. Never. Not one. When I was young, my Professor said, 'The time has come. You have to give some lectures.'
I said, 'I won't.'
He said, 'You must.'
I said, 'I won't.'
He said, 'Oh come, just a few, not many, you can easily do it.'
I said, 'I won't.'
So they held a meeting and decided it must be schizoid-paranoia." Warburg smiled when he got to the last sentence.

In a book review, I once wrote a kind of sketch of Warburg (6). His comment, on reading this review was, "How nice, how very nice. But I am not like that. Really not. You are mistaken." Indeed I was mistaken, as I later discovered. Up to that time, I had seen only the sunny side of a very complex character. The brightest sun casts the darkest shadow.

I hope Warburg approves of the present manuscript, as he once did of another one. This latter paper described some work that had been done in his laboratory, and was dutifully mailed to him for his approval. That manuscript crossed the Atlantic six times. After two revisions I finally got the hoped-for letter: Dear Dr. Vennesland: Imprimatur! Warburg.

SOME PREJUDICES AND OPINIONS

The editors have given me laissez-faire. I can express my prejudices and opinions without editorial censorship. After a lifetime subjected to the discipline of science editors, I find this new freedom heady stuff. Here we go! Let me apologize for the next section by explaining that it is not completely factual; but it contains some grains of truth, nevertheless.

The Problems of Scientific Publishing

These are so great that I am going to have to switch to parables. The journals, some of them, are making brevity such a virtue that clarity has become a sin. It isn't that I have anything against editors. Editors work hard for little or nothing but the glory of having their names appear in public; and the job is usually sheer drudgery. Editors deserve our gratitude for the time and effort they invest in preserving high standards. OK. But there's more to the job of an editor than that. I demand that the editor use his head before I'm going to give him any gratitude.

Here is parable 1. I once got a rejection slip from a well-known journal saying that there might have been some interest in this material if the work had been done with a purified enzyme, but since all of the work had been done with nothing but crude extracts, it was totally unsuitable - oh - you know the usual lines. They were accompanied by a more soothing suggestion that I try another journal. Now, Ladies and Gentlemen, imagine my feelings! I know you can. Half of that paper dealt with experiments with purified enzyme! Imagine my feelings - pardon - my becoming a little incoherent. That bastard hasn't even taken the trouble to read my paper. Imagine? *My* paper! That's what my feelings were. Now it turned out all to the good, because the paper probably *was* more suitable for the second journal, and it was also revised for the second journal and became a better paper. This time we put the last half in front and the front half in the back. I couldn't resist that final slap at the anonymous referee who wasn't interested in my paper.

My second parable is inspired by a horrifying suggestion that I recently read somewhere—perhaps in TIBS—that referees should be named and authors should be anonymous. When I act as referee, I try to make my hands soft as silk. I am not an intemperate critic when I am writing anonymously to a defenseless author. I know who he is, but he doesn't know me. He can't hit me back because he doesn't know whom to hit; and I have never enjoyed slapping defenseless children. But an editor once had the nerve to write to me, to ask me for my opinion of a paper by an unidentified author. This isn't the normal kind of request made to a referee. The referee is usually told who the author is. Now, I'm not to be told who the author is, but of course the editor knows who he is, and the editor wants my honest opinion, so the wise editor can protect the author from injustice, can't he? So what did I do? I blasted that author—oops that was a Freudian slip—I mean editor—no—I mean author—who do I mean anyway?—with the roar of all the cannon I have in my artillery! "I don't know who the author is, but in my opinion the paper shows that he doesn't know what he is talking about. The following two examples should suffice, etc. I hope that these are written in a form suitable for transmission to the author." I never received the reply of the outraged author. It was probably unprintable.

Women in the Universities

I have already mentioned the debt I owe to the International Federation of University Women. It seems obligatory to comment on the women's rights movement with its attendant publicity, which has been louder in the USA perhaps than in Europe. First of all, I feel a genuine sisterly sympathy for any young woman who is determined to make her way in a research career, and I am glad that it has been recognized in principle, at least, that women should be considered for assistant professorships and promotion on the academic ladder on the same basis as men. Reverse discrimination, however, bothers me very much indeed. I do not approve of it in any form. In the long run, such a practice will guarantee a lower average quality in academia. In other words, I don't think that one should try to establish a particular sex ratio for academic appointments in too short a time. And I don't even think we should try to calculate what an appropriate "fair" ratio of sexes should be. Maybe the men really do have brains better suited to perform best in hardware subjects like physics. Maybe the girls have brains better suited to perform best in biology? Let's let the optimal mix demonstrate itself.

The entry of women into public life in more than occasional numbers is a fairly recent phenomenon in human history. The cause undoubtedly lies in underlying economic changes, the same ones that have caused a large population shift from country to city. If we go back several hundred years to colonial America or to Europe in the eighteenth century, most of our female ancestors were living on farms and engaged in a form of housekeeping that was a full-time job. The raw materials often came straight from the field. It was the man's job to provide them and the woman's job to convert them into daily meals and clothing. The upper classes had servants, true, but these rulers were a tiny minority. The great bulk of the population lived like Adam and Eve, delving and spinning, and no one questioned the importance of the woman's job for the family prosperity. What has housekeeping become now? Every single task has been lightened with labor-saving devices to such a point that fewer and fewer able-bodied women can take pride in their housekeeping performance as a difficult job well done. With the exception of the care of small children, which I grant is an all-consuming task, women are left with no satisfying occupation except for the period of child-rearing. The consequence is that women move into the market place in increasing numbers, demanding a share of the more interesting positions and society is adapting to this change. My impression is that most men view this movement with sympathy and understanding. Their wives and daughters are involved. And take note, you blacks and women, who demand instant action. Take note: If you get all the reverse discrimination you want, this will provide what some will regard as proof that you really are "inferior." And you ought to be able to figure out what that proof will be. Most of the intellectual lightweights on the staff *will be* blacks and women.

Research Financing

The support provided for the individual scientist by the Max-Planck-Society is ideal. I have nothing but gratitude for this support and for the manner in which it is administered. Since my move to Germany preceded the crunch in research funding in the USA, my experience of research poverty is limited to the thirties, which is a long way back. Nevertheless, one's earlier recollections are often the most vivid. My generation could have accomplished much more if we hadn't been forced to spend so much of our time on do-it-yourself activities. Though this was not totally bad training, it seems reasonably clear that one individual cannot aspire to becoming an expert glass blower and instrument maker, bench chemist and lecturer, as well as an authority on the theoretical aspects of his subject. Even if the talents for these occupations are all there, human life is too short to acquire the necessary expertise in all crafts and disciplines. The provision of positions with other positions attached to make a unit recognizes this need for cooperative productivity. The problem of hierarchy (who is the boss?) remains, and is solved in the Max-Planck-Society by providing real tenure only for the top positions. Isn't there room here for an adjustment?

In some areas, seven independent groups of three or four scientists may be more productive than one group of 25. This is a plea for more intermediary tenure positions at an independent level within the Max-Planck-Society. I grant that there are problems which can only be handled by very large units, but only a fraction of biological research is of such a nature.

Science: A Personal View

I have never before attempted to articulate my view of the natural sciences. When I was still quite young, formal philosophy appealed to me, but its attractions gradually vanished as I immersed myself in the study of living things. The contents of this section reflect this transition, and are personal in the sense that they express tastes or preferences which are bound to vary with the individual. But I have attempted also to express as clearly as possible my own rational view of the relationship of a self-conscious human being to the universe of changing things.

Scientific thinking is thought about the real world. A mind trained (or untrained) in classical logic has a tendency to confuse scientific thinking with the thinking of classical logic because the logic is the same. But scientific thinking does not permit the use of symbols, concepts—call them what you will—unless they stand for things that are real. Therefore, it is very important to be especially careful to examine the real meaning of the words we are using.

Logic is an internally consistent system of thinking. Classical logic is expressed in terms of symbols and speech. These entities need have no correspondence to the things of reality. Classical logic is attractive to the young mind which knows speech but doesn't know much more. The young mind likes to dance. Logic is the dance of the mind. Classical logic is the art of reasoning about nothing. As symbols are not real things, they are no-thing. There is no progress in logic. It is the process the mind uses; therefore it need not be taught. It is there in the mind as the process the mind uses to think.

I prefer to work with plants and to think about humans, which seems foolish, but that's the way it is. For example, one thing I have often wondered about is why most (not all) of the smells emanating from the plant world are experienced as a pleasant feeling, whereas the opposite is true for the smells emanating from the animal world. But "why" is not the right word with which to question nature. The small child often uses "why" in the sense of "tell me more." The adult uses "why" in the sense of "tell me your motives." "Why" is a word that applies only when two brains are communicating with the language of human speech. "Why" is a nonsense syllable in the system man uses to communicate with nature. This system is the scientific system. Don't ask why? Ask what it is? And think of it as things like chemical compounds or as a process involving molecules.

The word revolution is being applied right and left to the development of modern biological sciences—without justification in my opinion. Kuhn apparently started this custom with *The Nature of Scientific Revolutions,* which I once read with real pleasure. Kuhn has applied himself to the history of science by applying the scientific method of observation to the process whereby scientists arrive at concensus (4). His method is a welcome change from that of the philosopher-historians. But I have been surprised by the fact that many of my fellow scientists didn't seem to read Kuhn the way I did. Some of them wanted to claim kinship with Galileo every time they made a new discovery.

New discoveries are very nice things, everybody is happy, progress is great, business is good, we are going to heaven fast. That's right. It is fun to make discoveries. But that is not the nature of revolutions. Revolutions hurt, because something we believed in has to be rejected. That's how you can tell the difference between a discovery and a revolution. If the molecular biologists want a big word to describe a big thing—the surge of new information in the field of genetics and cell differentiation—let them use the word explosion.

The basic premise of biology—that it can be understood in the terms of the physical sciences—has in no way been altered by the molecular biologists. Nevertheless, modern biology has given us a deep sense of kinship with all living things. Molecular biology is the Saint Francis of the sciences. I used to think that speculation about the course of evolution was pointless because it was impossible to duplicate the process. But if the nucleic acids carry within themselves the records of their origins, as Eigen suggests, then we may in fact be able to deduce the chemical course of our origins from dust. The prospect is wonderfully exciting. I regret being born too soon.

But this basic assumption of the biological sciences that all biological phenomena can be explained in terms of the laws of physics and chemistry doesn't always seem to apply. The mind-body problem defies such an explanation. Our awareness, our consciousness, our capacity for thought reflect complex chemical processes occurring in our brains. Free will implies a capacity to control these chemical events—albeit to a limited extent. We cannot deduce the experience of consciousness from physical and chemical laws. At this level, biology endows chemistry and physics with a new set of properties.

Mind is not a real thing, but a different entity, associated in a curious relationship with processes in which only things are involved. The natural scientist can examine the things involved and note their correlation with subjective phenomena.

The scientific method cannot answer all our questions. We insist on raising senseless questions. We want purpose and design in the universe. Our notions of purpose and design are based only on our own built-in natural drives.

THE SERIOUS AND THE NOT-SO-SERIOUS

I appreciate the logic of the theologians who said, "God's name is secret." But they weren't theologians as defenders of dogma. They were searchers for truth. I cannot subscribe to any dogma. I consider myself religious, but many religious people wouldn't. Man is an animal, endowed with a brain that tells him he is going to end in Nirvana, and the same brain gives man a set of drives that makes him fight his own extinction with all his might. This is the human dilemma. The great religions provide comfort by either telling us there isn't such a place as Nirvana, there is a Heaven and a Hell, or by telling us that Nirvana is really very pleasant, which comes out in a curious sense to the same thing. The "Hell" of the Christian is the "Rebirth" of Buddhism.

I believe that the best thing a human being can do is to try to increase understanding of the entire reality—the all—around us. This can only be done by the techniques of the natural sciences. One can't figure it out by meditation. One must observe nature. In this sense, the natural scientist isn't creative at all. Perhaps this word should be applied only to the artist. Still, there is an aspect of the work of the scientist that is very similar to the work of the artist. The scientist puts questions to nature and nature answers. No speck of imagination here. The art comes in the process of figuring out how to put the questions, and the thrill comes when you get the answer—the thrill is the *signal* that tells you you have the right answer.

I enjoy great literature. There are passages in Shakespeare and the Bible that have moved me in a way impossible to describe. One has to be in the mood for that kind of reading, and one's taste changes with circumstances, so I can't tell what my favorite passages are. It might be this one this week and next month it will be another one. I like fiction that can be read with several layers of meaning, but I have never been able to wade through Joyce's monumental later work. I don't like nonsense. I cannot understand that anyone can be interested in the meaning of symbols that represent nonsense.

Ariel's song is a sample of the kind of poetry that gives me the same kind of thrill as the thrill of scientific discovery.

Full fathom five thy father lies;
  Of his bones are coral made;
Those are pearls that were his eyes;
  Nothing of him that doth fade
But does suffer a sea-change
Into something rich and strange.
Sea-nymphs hourly ring his knell:
      Burden. Ding-dong.
Hark! now I hear them—Ding-dong
    bell.

Ariel's song—The Tempest (Shakespeare)

The above lines may have been inspired by the exhilarating excitement of the naval explorations of Elizabethan England, but they can be applied equally to describe the excitement of scientific discovery. The scientists aren't taking the mystery out of life. Even if they wanted to, they couldn't do it.

I have borrowed the title of this article from a book by Barbra Ring, a Norwegian novelist and newspaper woman who was born in my home town, Kristiansand, a good many years before I was. After getting a divorce from an insanely jealous husband, she made her living by writing. She provided the original version of my favorite Kjutta joke. Kjutta, I must explain, was once a real character who lived in Kristiansand, but he has become a legend there in the form of a hundred varieties of Kjutta stories.

Kjutta was in a fight and was beaten up. He is describing the fight: "First, I hauled off and slammed him on the nose and didn't hit him, so I slammed him again the same way. Then he began to run, but I got ahead of him."

Literature Cited

1. Bentley, R. 1978. Ogston and the development of prochirality theory. *Nature* 276:673–76
2. Krebs, H. A. 1974. The discovery of carbon dioxide fixation in mammalian tissues. *Mol. Cell. Biochem.* 5:79–94
3. Krebs, H., Mitarbeit von Schmid, R. 1979. *Otto Warburg: Zellphysiologe—Biochemiker—Mediziner,* 1883–1970. Stuttgart: Wissenschaftliche Verlagsgesellschaft. 168 pp.
4. Kuhn, T. S. 1977. *The Essential Tension.* Chicago/London: Univ. Chicago Press. 366 pp.
5. Nachmansohn, D. 1972. Biochemistry as part of my life. *Ann. Rev. Biochem.* 41:1–28
6. Vennesland, B. 1963. Review of *New Methods of Cell Physiology. Perspect. Biol. Med.* 6:385–88
7. Vennesland, B. 1974. Stereospecificity in biology. *Top. Curr. Chem.* 48:39–65
8. You, K. S., Arnold, L. J. Jr., Allison, W. S., Kaplan, N. O. 1978. Enzyme stereospecificities for nicotinamide nucleotides. *Trends Biochem. Sci.* 3:265–68

Ann. Rev. Plant Physiol. 1981. 32:21–81

BIOCHEMICAL MECHANISMS OF DISEASE RESISTANCE[1]

◆7705

Alois A. Bell

U.S.D.A., SEA-AR, National Cotton Pathology Research Laboratory, College Station, Texas 77841

CONTENTS

[1]The US Government has the right to retain a nonexclusive royalty-free license in and to any copyright covering this paper.

INTRODUCTION

In the first *Annual Review of Phytopathology,* Cruickshank (71) reviewed a small but exciting group of papers which indicated that plants, like animals, use postinfectional biochemical responses to resist diseases. Since this 1963 review, research on the biochemical responses of plants to diseases and pests has increased rapidly, now providing hundreds of articles each year. Many of these studies have been discussed in recent books (78, 82, 116, 160, 161, 171, 354) and reviews (8, 31–34, 38, 72, 146, 237, 281, 360, 367, 394, 404). In this review, selected references are used to illustrate both newly emerging concepts and future research needs.

APPROACHES TO RESISTANCE STUDIES

Disease resistance is the ability of a plant to prevent, restrict, or retard disease development, and occurs at high, moderate, or low levels. In contrast, virulence is the ability (low, moderate, or high) of an infectious agent to overcome resistance. Both resistance and virulence are each the combined result of multiple biochemical components (171). Consequently, any single putative component of resistance should be evaluated with many experimental approaches to determine its role in the total defense system. Some of the experimental approaches used to compare biochemical responses in resistant and susceptible plants include: genetically different plants, genetically different pathogens, environmentally altered resistance, ontogenic changes in resistance, induced resistance, and induced susceptibility.

Genetically Different Plants

Cultivars within a plant species, and especially species within a genus, generally show a wide variation in their levels of resistance to any given pathogen. Such resistance is usually divided into two categories: race-specific resistance, which is expressed against some but not other isolates of a pathogen, and race-general resistance, which is expressed against all isolates. The genetics of these types of resistance and breeding methods for transferring them have been reviewed (79, 277, 324). Most studies of genetically different plants have concentrated on race-specific resistance, especially that expressed against biotrophic or partially biotrophic pathogens.

Strict biotrophic pathogens live and grow in contact with cells of the susceptible host without eliciting cell death and defense responses, whereas necrotrophic pathogens quickly cause cell death (and probably defense reactions) in advance of the penetrating pathogen. Partially biotrophic pathogens, such as the fungi *Colletotrichum, Cladosporium,* and *Phytoph-*

thora and the bacteria *Xanthomonas* and *Pseudomonas,* normally maintain their biotrophic relationship with their host for a few days and then eventually give rise to necrotic cankers and leaf spots or blight. Race-specific resistance against completely or partially biotrophic pathogens is usually controlled by single dominant genes (79). Resistance results when the biotrophic relationship is not established or maintained, resulting in rapid induction of defense reactions including necrosis of the invaded cell(s). This resistance response to biotrophic pathogens has been referred to as hypersensitivity, incompatibility, or local lesions; I will refer to it as necrogenic resistance.

High levels of resistance against necrotrophic pathogens appear much the same as those against biotrophic ones, but susceptibility has opposite qualities. Susceptibility, rather than resistance, is usually controlled by a single dominant gene or the cytoplasm [possibly a mitochondrial gene (79)], and similar but more intense biochemical responses to the pathogen occur in the susceptible compared to resistant tissues (416).

Comparisons of resistant and susceptible cultivars have involved whole plants, excised plant parts, tissue slices, callus cultures, and free protoplasts. Detached plant parts generally show less resistance than the whole plants, but have been extremely useful for studying responses to foliar pathogens. Tissue slices from storage organs generally undergo extensive biochemical changes during the first 12 hr after wounding (199), and must age for 12 to 20 hr before they show rapid race-specific reactions typical of the whole plant (121).

Race-specific resistance has been demonstrated in callus cultures from leaves or stems of cotton (325), tobacco (25, 175, 256), and soybeans (168). Resistance in callus cultures depends upon critical concentrations of 2,4-D, naphthalene acetic acid, and coconut milk, and on temperature (25, 168). Even under optimal conditions, the level of disease resistance in soybean callus cultures is less than in the whole plant (168). At present, tissue cultures have offered little advantage over whole plants or plant parts for biochemical studies.

In the future, haploid cell cultures could become extremely valuable for the step-wise mutational removal and evaluation of enzymes involved in the synthesis of secondary products such as the phenols involved in melanization. We (37, 411, 412) have used UV-induced mutation for the step-wise removal of enzymes and evaluation of melanin biosynthesis in pathogenic fungi. Our techniques may be adaptable to haploid plant cells, provided that whole plants can be regenerated from mutant protoplasts. Mutant potato plants with resistance to *Phytophthora infestans* and *Alternaria solani* have already been found among plant populations regenerated from random protoplasts of a susceptible cultivar (340).

Mesophyll protoplasts have been used extensively to study the nature of resistance to virus diseases. Resistance that causes greatly reduced virus synthesis in plants is expressed in isolated protoplasts. However, necrogenic resistance frequently is not expressed by protoplasts, indicating that whole cells or mass cell action is necessary for such resistance (69, 272). Alternatively, the 2,4-D normally included in the medium may alter the resistance response (242). Protoplasts have also been used to study the nature of cell death associated with necrogenic resistance to biotrophic pathogens (92, 169) and with susceptibility to phytotoxins (294, 363, 364).

Genetically Different Pathogens

Isolates of a pathogen that can be distinguished by the differential expressions of resistance by a group of host cultivars or species are called races (bacteria and fungi) or strains (viruses). Some studies of resistance have included different races as well as several cultivars with different race-specific genes for resistance (92, 209). Likewise, races, species, or isolates with variable virulence levels have been used to study resistance mechanisms in single cultivars (355, 403) and in race-general resistant cultivars (269). Isolates with variable virulence have been compared for their ability to degrade host-produced antibiotics (381, 385) and produce phytotoxins (422) to show the importance of these characters for virulence. In other studies mutants that lack ability to produce catalase (2) or pectinase enzymes (172) have been used to evaluate these enzymes.

Environmentally Altered Resistance

Resistance levels controlled by certain genes are strongly influenced by temperature. For example, "Acala 4-42-77" cotton has high resistance to *Verticillium* wilt at 29°C but is susceptible at 25°C (35). In contrast, wheat cultivars with the *Sr6* allele for resistance to *Puccinia graminis* have high resistance at 20°C but are susceptible at 25°C (261). Thus, increases in temperature may either increase or decrease plant resistance to fungal pathogens. Resistance to viruses (162, 299), nematodes (97, 187), and bacteria (225) frequently decreases or disappears as temperatures are increased from 20–28°C to 30–35°C.

Heat shock (e.g. exposure to 44°C for 1 hr or 50°C for 1 min) administered to plants before inoculation generally causes a temporary loss of resistance against biotrophic or partially biotrophic pathogens (60, 158), but greatly increases resistance to many necrotrophic pathogens, particularly those producing host-specific toxins (364). Heat shock applied 12–48 hr after inoculation with biotrophic pathogens has the opposite effect as treatments applied before inoculation, stimulating strong resistance reactions even in normally suceptible hosts (215, 321). Comparisons of cultivars under temperature regimes that give susceptible or resistant reactions have been used increasingly for biochemical studies of resistance.

Other factors that influence resistance include: the intensity, duration, and quality of light; moisture levels; nutrient levels; and agricultural and industrial chemicals. Low light intensity generally decreases resistance (144, 192, 327), but also may increase it (41), depending on the specific host-pathogen combination. Long photoperiods generally result in higher levels of resistance than short ones (196). Both light filters and chemical inhibitors of photosynthesis [e.g. 3-(3,4-dichlorophenyl)-1,1-dimethyl urea] have also been used to manipulate resistance and biochemical responses (56, 279). Keeping leaves watersoaked negates resistance to bacteria, but resistance is resumed 2–3 hr after the excess water is allowed to dissipate (67, 426). In regard to nutrients, increasing concentrations of potassium and calcium most often increase resistance, whereas excess nitrogen decreases resistance, and phosphorus has variable affects (184). These and other effects of environment on resistance have been reviewed (32).

Ontogenic Changes in Resistance

Resistance levels vary among plant parts and tissues (183), and among genetically identical plants of different age (197, 255). Each plant part also changes in resistance with age. Resistance levels in stems and roots generally increase rapidly during the first two weeks of seedling or new shoot growth, and slowly thereafter (177, 179), whereas resistance levels in leaves and fruits frequently decline with age (197, 279, 409). Leaves of any given developmental age are usually most resistant on plants at about the time of flowering. These ontogenic changes and their possible biochemical explanations have been reviewed (33).

Induced Resistance

Plants frequently show enhanced levels of resistance to disease following inoculation with pathogens or treatment with chemicals that cause local necrotic lesions or numerous scattered necrotic cells. This phenomenon is referred to as induced resistance (259). Several reviews (cited in 171) give an update on the subject; specialized reviews have been concerned with induced resistance to bacterial and fungal infection (227), fungal diseases (194), vascular wilt fungi (34), and viruses (241); the phenomenon has also been observed with nematodes (285).

Induced resistance has been studied most extensively in solanaceous plants (particularly tobacco), in cucurbits, in rosaceous plants, and in cereals. However, induced resistance apparently occurs in many, if not all, plant families. Heat-killed cells, cell walls, cell-free sonicates, and culture filtrates of pathogens, as well as live pathogens, are effective inducers of resistance in some hosts (36, 38, 156, 243, 266, 306, 335, 350). The common feature among all inciting agents is their ability to cause chronic cell injury, evidenced by increased electrolyte leakage, ethylene evolution, peroxidase activity, and usually cellular necrosis.

Induced resistance is generally observed as a reduction in both the size and number of visible lesions produced by the target pathogen. However, microscopic observations (21) indicate that the apparent reduction in number, at least in some cases, may not be real; rather, certain necrotic reactions in the induced-resistant plant are limited to single cells or a few cells, and are not visible without aid of staining procedures and a microscope. Thus, reduction in lesion size probably is the major result of induced resistance.

Induced resistance develops similarly in all plants with only minor variations due to effects of plant species and inducing agents. A typical example of induced resistance is that developed against *Colletotrichum lagenarium* in cucumber. Resistance is induced by inoculation of the first true leaf with virulent or avirulent races of *C. lagenarium* (189, 227, 313), tobacco necrosis virus (189, 227), avirulent *Cladosporium cucumerinum* (227), or the virulent bacterial pathogen *Pseudomonas lachrymans* (58). The extent and duration of induced resistance is directly proportional to the inoculum concentrations of the inducing pathogen. Systemic resistance first appears in the second true leaf about 72–96 hr after the inducing inoculation, reaches a maximum at about 7 days, and persists for 4 to 7 weeks. A "booster inoculation" on a higher leaf extends induced resistance until fruiting. As few as 2 to 10 lesions per plant induce resistance; removing the inoculated leaf after 96 hr does not diminish it. Cucumber, watermelon, or muskmelon scions become resistant when grafted onto cucumber roots that received an inducing inoculation (189). Thus, the "signal" for induced systemic resistance is not species- or genus-specific.

The magnitude of induced resistance may vary among tissues, depending on the host species, cultivar, and the inducing agent. Removal of the epidermis before inoculation, for example, causes a concomitant loss of induced resistance to *C. lagenarium* in susceptible, but not resistant, cucumber cultivars (313). Apparently, induced resistance occurs only in the epidermis in one cultivar but throughout leaf tissues in the other. A similar situation has been observed with kinetin-induced resistance to tobacco mosaic virus (TMV) (203). Kinetin induces marked reduction of local lesion size in whole tobacco leaves, but not in leaves with the lower epidermis removed. Accordingly, kinetin treatment markedly decreases TMV multiplication in the epidermis but actually stimulates TMV multiplication in the mesophyll cells. These results may explain the puzzling observation (21, 113, 369) that virus multiplication in whole leaves frequently is not reduced by surface-applied treatments that induce high levels of resistance to local lesion expansion.

The similarities between induced resistance in plants and immunization by vaccination in mammals have intrigued many researchers, leading to extensive biochemical studies of induced resistance. The practicality of this approach has been demonstrated recently by the development of two groups of disease control chemicals, the dichlorocyclopropanes (231) and tricyclazole-related compounds (414); both apparently work by enhancing the resistance of rice to *Piricularia oryzae* rather than by directly killing the fungus. Many of the herbicides and growth regulators also enhance resistance, especially when applied at relatively low concentrations before infection (34, 47, 217, 242, 390). Understanding these interactions may allow us to develop herbicides that kill weeds and, at the same time, increase pest resistance in crop plants.

Induced Susceptibility

Inoculation of a plant with a virulent pathogen prior to or simultaneously with an avirulent one often causes induced susceptibility (or loss of resistance) to the avirulent pathogen. For example, inoculation of beans with virulent *Xanthomonas phaseoli,* either 4 days prior to or simultaneously with the avirulent *Xanthomonas vesicatoria* (virulent to tomato), increased populations of *X. vesicatoria* in diseased tissues (174). Likewise, simultaneous inoculation of pepper leaves with virulent and avirulent strains of *X. vesicatoria* reduced resistance and the early intense electrolyte leakage normally associated with the resistance response (356). Heat-killed cells of the virulent strain also effectively reduced the resistance response, indicating that some component, possibly in the cell wall, may interfere with the resistance response.

Virulent biotrophic fungal pathogens, such as *Puccinia graminis, Phytophthora infestans,* and *Erysiphe graminis,* or biotrophic nematodes, may also induce susceptibility. Compatible interactions established between *P. graminis* and invaded wheat cells at 26°C are maintained at 20°C; however, the cells around the colony borders give a normal resistant reaction, surrounding the colonized "susceptible" cells with a necrotic halo (149). This process indicates a highly localized permanent induction of susceptibility in cells where haustoria of the pathogen are established. Giant cells established by nematodes at 32°C for 2 to 3 days likewise are maintained at 27°C, giving susceptibility at the lower temperature where resistance is normally expressed (97). Susceptibility induced by *Erysiphe graminis* to *Sphaerotheca fuliginea* in barley leaves is also highly localized. *S. fuliginea* establishes infection in 75% of the cells that harbor haustoria of *E. graminis,* but a very low frequency of infection occurs four cell-rows away. Attempts to identify "suppressors" responsible for induced susceptibility to fungi will be discussed in the section on recognition.

Nematodes and mycorrhizal fungi may enhance susceptibility to many soil-borne pathogens. The ability of nematodes to break resistance depends on the source and level of resistance to the other pathogen (42) and on levels of resistance to the nematode (287). During feeding, nematodes disrupt plant cells and increase leakage of metabolites that may serve as nutrients for other pathogens (388). This and other mechanisms by which nematodes enhance susceptibility have been reviewed (32, 34). Mycorrhizal fungi probably enhance susceptibility by increasing phosphorus uptake by the plant, since phosphorus fertilization often has a parallel effect (330).

Various growth regulators have been used to induce susceptibility. For example, susceptibility has been induced in tomato to the nematode *Meloidogyne incognita* by cytokinins (98) and NAA plus cytokinin (329), in tobacco to tobacco mosaic virus by 2,4-D (348) and ethylene (301), in corn to *Helminthosporium* leaf spot by IAA (166), and in wheat to *Puccinia graminis* by ethylene (261). To induce susceptibility, growth regulators usually have been supplied at relatively high concentrations after inoculation.

Metabolic inhibitors have been used to induce susceptibility, and thus they implicate various metabolic pathways in resistance. The protein synthesis inhibitors cycloheximide and blasticidin S and the RNA synthesis inhibitor actinomycin D have been used extensively to show that postinfectional synthesis of RNA and protein apparently are essential for resistance (22, 120, 158, 380, 424). The results from such experiments need to be interpreted with caution, however, since most inhibitors of protein and RNA synthesis have high affinities for cell membranes and may be causing additional effects by altering membranes. Actinomycin D and cycloheximide at certain concentrations actually stimulate some defense reactions such as phytoalexin synthesis (170). Other metabolic inhibitors shown to induce susceptibility include: the respiratory inhibitors 2,4-dinitrophenol, sodium azide, and 2-thiouracil (228, 284); the photosynthesis inhibitor 3-(*p*-chlorophenyl)-1,1-dimethylurea (327); and the ascorbic acid synthesis inhibitor lycorine (14).

HISTOLOGY AND CYTOLOGY OF RESISTANCE

Improved technology for ultrastructural studies has led to a tremendous number of histological studies during the past 10 years. I will discuss only selected studies of resistance responses to viruses, bacteria, fungi, and nematodes.

Viruses

High levels of race-general resistance greatly curtail virus multiplication, frequently to less than 5% of that in the susceptible cultivar (419). Slow, progressive increases in virus levels may occur for many weeks, but with no apparent effect on the host tissue. With race-specific resistance, virus multiplication occurs readily in the resistant plant, similar to that in the susceptible plant, until about 36–96 hr after inoculation; then virus multiplication drops markedly, concurrent with the formation of local chlorotic or necrotic lesions (286). High levels of race-specific resistance result in very rapid formation of local lesions that show intense melanization and remain small and highly confined, whereas progressively lower levels of resistance result in slower developing, progressively larger lesions, until low resistance may lead to a systemic necrosis (61, 162). The extensive systemic necrosis resulting from low resistance often causes greater damage to the plant than the high levels of virus multiplication in the "susceptible," but biotrophic, plant-parasite relationship. The tissues around confined local lesions develop high levels of induced resistance and apparently stop the spread of the virus.

The histology of local lesion formation has been studied extensively in *Gomphrena* leaves infected with tomato bushy stunt virus (12, 297). The first evidence of a resistance response is an increase in electrolyte leakage starting 5 hr after inoculation (297). This leakage increases slowly but progressively until a large increase occurs after 25 hr. At 30 hr, about 8 hr before necrotic lesions are visible, the lesions appear as yellow fluorescent spots under 365 nm UV light. Alterations of chloroplasts, mitochondria, cell walls, and plasma membranes and occasional severely degenerated cells in the lesion area are also seen at 30 hr. By 48 hr, small callose deposits are found in pit areas and cell walls at the periphery of the necrotic spot, and live cells around the lesion show starch accumulation, particularly in chloroplasts of bundle sheath cells. Membrane-bound bundles of virus particles embedded in callose also occur outside the protoplasts in the necrotic area (12). The cells encircling the lesion develop thickened walls, and massive depositions of materials containing lignin and suberin are infused into the intercellular spaces.

Sequential breakdown of organelles has been reported for tobacco undergoing local lesion formation on infection with tobacco mosaic virus (76, 77, 108). First, chloroplasts swell, display vesicular bodies and swelling of thylakoids, lose surrounding envelopes, and burst. Then the tonoplast ruptures, but mitochondria and nuclei may remain intact until cells are otherwise severely disorganized. The nuclei assume a dark granular appearance and nucleoli become enlarged. As in infected *Gomphrena,* callose deposits are formed in sieve plate pores, plasmodesmata, and cell walls; electron-dense materials derived from phenylalanine plug tracheids and occur in intercellular spaces. Crystalline packets of virus particles appear in completely collapsed cells, whereas particles remain dispersed in the few intact cells containing virus.

Bacteria

High levels of race-general resistance to bacteria greatly restrict the multiplication of bacteria without involvement of necrosis, whereas race-specific resistance involves the rapid formation of necrotic cells or lesions similar to the local lesions that restrict virus multiplication and spreading. Both virulent and avirulent strains multiply exponentially until necrosis occurs and then populations decline. This decline also occurs in intercellular spaces between lesions, indicating accumulation of antibiotics (105, 335, 403). Because of an earlier onset of necrosis and, occasionally, slower rates of multiplication in the resistant plant, the avirulent strains normally reach population peaks only 1.0 to 0.1% as great as those of virulent strains (58, 105, 403).

The pattern of necrosis associated with race-specific resistance depends on inoculum concentrations injected into leaves: 10^4-10^5 cells per milliliter cause nonapparent necrosis of individual tobacco cells (375) and clusters of a few cells in cotton (105) and bean (403); 10^6-10^7 cells cause visible necrotic lesions of variable size depending on the host and pathogen; 10^8 or more cells cause confluent necrosis, and the entire inoculated area collapses and desiccates, Necrosis typically first appears at 18–24 hr, and collapse and desiccation proceed through 72 hr after inoculation of bean (319, 403), cotton (59, 105), and pepper (65). In tobacco, necrosis develops very rapidly and may be apparent in 6–12 hr. Necrosis in susceptible cultivars develops 3–5 days after inoculation in bean, pepper, and cotton and 18–48 hr in tobacco.

Disruption of plasma membranes is indicated by a large increase in electrolyte leakage slightly before or concurrent with necrosis in both susceptible and resistant cultivars of pepper (65) and bean (429). Avirulent, but not virulent, strains of bacteria also may increase electrolyte leakage of pepper as much as 16 hr before necrosis. The degree of increase varies considerably for different avirulent strains. Treatments that induce resistance result in a marked increases in both the quickness and extent of electrolyte leakage when bacteria or water are infiltrated into tissue 6–42 hr later (66).

The sequence of ultrastructural changes associated with necrotic resistance have been described for tobacco (134, 333), cotton (59, 105), and bean (319, 349) and are similar, except for the more rapid development in tobacco. The resistance of plants to both saprophytic bacteria and avirulent bacteria apparently begins with attachment of bacteria to cell walls (134, 333, 349, 357). Subsequently, saprophytes may swell and burst indicating toxic action (175), or they become engulfed on the cell wall by a fibrillar structure (333, 349) resembling cutin or suberin (59, 134); no necrosis or degeneration occurs within the host cell. Avirulent bacteria also attach to plant cell walls and cause visible erosion of the outer wall, probably the cuticle, by 20 min to 4 hr after inoculation. Extrusions of fibrillar and granular materials from the host cell wall then envelop the attached bacterium. Next to the sites of bacterial attachment, the plasmalemma becomes convoluted, and numerous membrane-bound vesicles accumulate between it and the cell wall; the cell wall and "cuticle" also thicken and become more electron dense. Cytoplasm accumulates and aggregates against the plasmalemma closest to the bacteria and exhibits vesiculation and accumulation of osmiophilic droplets. Finally, there is extensive degeneration of all membrane systems and organelles leading to final collapse and apparent necrosis of the responding cell. Some adjacent cells also may undergo similar necrotic changes, whereas other adjoining nonnecrotic cells appear normal. Attachment, cell wall erosion, and envelopment of bacteria do not occur in the susceptible reaction; membrane, organelle, and cell disruption usually occur several days later than in the resistant response.

Fungi

Fungal spores generally germinate, and germ tubes grow equally well on the surface of both resistant and susceptible plants. High levels of race-general resistance often prevent penetration of fungi into plants, whereas race-specific resistance usually inhibits later fungal growth stages. Resistance to penetration occurs as cell wall appositions that are laid down between the convoluted plasmalemma and the inside of the wall (342). These appositions vary considerably in morphology and color depending on the plant species. Successful resistance by appositions is not normally accompanied by cell death. In contrast, race-specific resistance to fungi, like that to viruses and bacteria, almost always is associated with a rapid necrosis of the approached or penetrated host cell, accompanied by inhibition of fungal growth.

High levels of resistance result in fewer, smaller, and more heavily melanized lesions on leaves, stems, and roots, or in the vascular system, than low levels of resistance (85, 207, 226, 233, 240, 246, 251, 267, 269). This restriction of lesions apparently happens because resistance responses, including necrosis, occur either before or within a few hours after penetration of the cell wall. As levels of resistance decline among different plants, there is usually a progressive decrease in the speed of response to penetration, an increase in the number and size of necrotic lesions, and an increase in the extent of fungal development (226, 233, 240, 251, 267). Low levels of resistance allow limited but progressive development and sporulation of the fungus in spite of the necrogenic resistance that occurs too late to completely prevent secondary spread of the fungus to new cells that are not yet exhibiting resistance.

The time sequence between plant cell necrosis and inhibition of fungal growth varies for different cultivars and races; in some host-parasite systems fungal inhibition occurs several hours before necrosis (148, 226, 255, 267), in other systems the two events are concurrent (226, 267, 320), and in a few systems fungal inhibition may occur several hours after necrosis (226, 251). This variation has led to the popular, pedantic debate as to whether necrosis (hypersensitivity) is the cause or result of resistance. In fact, such variations should be expected, because of the variation in sensitivity of fungal strains to plant antibiotics and the diverse mechanisms involved in antibiotic synthesis and release. Some antibiotics are secreted by live protoplasts, some are stored in vacuoles and released by rupture of the vacuolar membranes, and still others are formed only when substrates released from vacuoles are mixed with enzymes in other parts of the protoplast or cell wall.

A much neglected fact is that different tissues and even cells may have different levels of resistance to a given fungus. For example, the *Sr6* temperature-sensitive allele in wheat confers resistance to the stem rust fungus only in mesophyll cells; the epidermal cells are susceptible and do not become necrotic (148, 149, 320). At 19°C, wheat (*Sr6*) mesophyll cells undergo a necrotic reaction in response to avirulent strains and resistance of the whole leaf results; at 26°C, mesophyll cells do not undergo necrosis and susceptibility results. These observations may explain why biochemical studies based on whole leaves of wheats containing the *Sr6* allele have given results disparate from those obtained with other cultivar-race systems.

Another example of cellular variation occurs among epidermal cells of barley (198). Only about 0.7% of germinated conidia of *Erysiphe graminis* form primary haustoria in resistant cultivars compared to 60% in susceptible cultivars. In the resistant cultivar, most haustoria are initially located in subsidiary cells next to stomatal guard cells, indicating a lower level of resistance in these cells. Once primary haustoria are formed in the subsidiary cells, nearly 36% of the secondary hyphae then form secondary haustoria in the adjacent epidermal cells, indicating subsequent induction of susceptibility. Future histological studies of plant-parasite interactions need to be directed more toward cellular specialization.

The specific sequence of cellular events associated with necrogenic resistance in response to fungi (48, 148, 167, 234, 255, 300) is similar to that described in the previous sections for viruses and bacteria. For example, the early ultrastructural symptoms of resistance in wheat to rust fungi include an increase of electron density, perforation and invagination of the plasmalemma, disruptions of noninvaginated plasmalemma, vacuolation of the cytoplasm, and accumulation of electron-dense material along the vacuolar membranes. Later, the electron-dense accumulations increase in size and also occur along the chloroplast and mitochondrial membranes. Finally, entire protoplasts become electron dense and collapse. In fungal haustoria in resistant plants there is a uniform increase in electron density of the cytoplasm. Many of the same ultrastructural symptoms in plant cells are caused by the release of *cis*-dihydroxyphenols from vacuoles into the cytoplasm (275).

Nematodes

In susceptible roots, larvae of the nematodes *Meloidogyne, Rotylenchus,* and *Heterodera* usually penetrate to the pericycle and then cause major changes in the plant cells surrounding the head of the nematode. Normal vascular tissue is replaced by a large multinucleated "giant cell" that is rich in organic nutrients used by the nematode. Hypertrophy and cell proliferation accompany giant cell formation causing galls to form on the roots. Once feeding begins in the giant cell, larvae become sedentary, develop into adult females, and after about 20–35 days eggs masses are laid. Thus, the welfare of the animal depends on the establishment and maintenance of the giant cell.

The frequency and path of penetration by nematodes is generally similar in resistant and susceptible plants. In highly resistant alfalfa, after 3 to 4 days, the number of animals in roots declines sharply, presumably a consequence of failure to establish a feeding relationship. After a week, very few animals are found, and galls fail to develop subsequently (137, 311). In other crops, animal numbers initially decrease very little in resistant cultivars; but their development rate is slowed, fewer females lay egg masses, and each mass contains fewer eggs; also, fewer and smaller galls are formed (97, 188, 264, 265, 285). After sufficient time for "giant cell" formation (3–4 days), necrotic cells are frequently found around the area of the animal's head in the resistant plant (97, 188, 309, 310, 395, 402). After inducing necrosis, however, the nematode may migrate to nonnecrotic cells and recommence feeding; these cells subsequently become necrotic. The cellular events associated with the necrosis closely resemble those associated with the host-specific necrogenic resistance to viruses, bacteria, and fungi.

MODIFICATION OF CELL WALLS FOR DEFENSE

The Cuticle

The surfaces of plant cell walls exposed directly to the air are covered by the biopolyester membrane, cutin. On the surface of epidermal cells, the thickness of cutin varies from 0.5 to 14 μm, whereas the cutin layer over cells in the substomatal cavities or in the intercellular spaces of the leaf mesophyll is very thin (0.15–1.0 μm) (224). Cutin is generally attached to the pectinaceous layer of the cell wall and is often imbedded under waxes. Cutin and waxes together comprise the cuticle.

The chemistry and biochemistry of cutin and the related polyester suberin were reviewed recently (224). Cutin is composed primarily of hydroxy and epoxy derivatives of palmitic and oleic acids. The unsubstituted acids and various cinnamic acid derivatives also may be esterified into the polymer. Most ester linkages occur between C-1 carboxyl groups and the C-16 or C-18 hydroxyl groups, although linkage with the secondary alcohol groups also occurs.

The cuticle is a hydrophobic surface and thus prevents water from accumulating as a film on cell surfaces; it also restricts the flow of nutrients to the cell surface. Both traits prevent certain microorganisms from becoming established on plant surfaces.

The cuticle also acts as a chemical and physical barrier to the germination and penetration of fungi into leaves. Components of wax, as well as cutin acids, inhibit germination of certain pathogens (397), but other pathogens use cutin as a sole carbon source for growth in artificial media (224). Cuticle thickness has been correlated with levels of resistance to fungi that penetrate directly into tissues (30, 397).

Conclusive proof that cutin is an effective physical barrier to pathogens was provided recently when pure cutinase, a unique glycoprotein, was obtained from several plant pathogens (224, 238, 252, 337). First, antibody to cutinase from *Fusarium solani* f. sp. *pisi* was prepared and conjugated to ferritin; and the conjugate was used to show that germinating spores of the fungus excreted cutinase during the penetration of pea cuticle (337). Next, the antibody and diisopropylfluorophosphate, a potent cutinase inhibitor, were used to prevent infection of the host, even though neither treatment had any effect on the germination or growth of the spores (252). Thus, cutin is an effective physical barrier, and cutinase or a wound through cutin is necessary to breach this barrier.

The observations that avirulent bacteria apparently attach to the cuticle of mesophyll cells and later become engulfed by an apparently modified cuticle (59, 134) indicate a postinfectional role for the cuticle that needs further investigation.

Carbohydrate Appositions

As fungi begin to penetrate the cell wall, either with infectious hyphae or haustoria, the resistant host responds by synthesizing new carbohydrates, particularly callose and cellulose, which are added to the inside of the cell wall just outside the plasmalemma. These appositions may continue even after the fungus penetrates the original wall, until they become dome shaped or elongate and are called papillae (4). Cells adjoining or near those invaded by pathogens also may deposit new carbohydrates onto thickening secondary walls. In susceptible hosts, papillae and secondary wall thickening are often poorly developed or missing (5, 167).

The thickening of walls and deposition of callose in and around viral local lesions have been studied in bean (345, 374), tobacco (108, 345, 359), and *Gomphrena* (10). In bean, thickening to as much as five times the normal wall width occurs by additions to the walls of cells surrounding the local lesion. Callose deposition, based on enhanced natural yellow fluorescence or fluorescence following aniline blue staining (345), can be detected several hours before local lesions appear in most hosts. The fluorescent spots (callose) enlarge with time, covering the lesion area, and then gradually disappear as necrosis and browning of cells progress. Even after complete browning, callose can be seen in walls of live cells around lesions (359), in plasmodesmata and sieve plate pores (108), and around membrane-bound bundles of virus particles in necrotic cells (10). Inhibition of necrosis in bean by either ascorbate infiltration or elevated temperatures results in the absence of callose deposition, as well as other resistance responses (359). Thus, the early deposition of callose may be an important component of the defense against viruses, possibly by plugging plasmodesmata and thereby preventing virus movement.

Cell wall thickening and callose deposition also occur in healthy cells surrounding necrotic lesions caused by fungi, regardless of whether the necrotic reaction is considered as one of resistance or susceptibility (123). However, more massive callose deposition and greater wall thickening have been associated with the resistant than the susceptible reaction of tomato to *Cladosporium* (85, 233, 234). In potato susceptible to *Phytophthora,* wall appositions are poorly developed, consisting mostly of callose, whereas haustoria in the resistant host are surrounded by an electron-dense matrix that in turn is completely encased in a callose-like material (167). Similar electron-dense materials occur within other carbohydrate appositions, suggesting that phenolic materials (lignin, tannin, or melanin) may have to be combined with the carbohydrates to form an effective barrier.

The effectiveness of papillae in blocking fungal penetration of host cells has been shown in several ways. Papillae in susceptible barley cultivars were enhanced in size in cytoplasm-enriched ends of centrifuged cells (402) and by calcium phosphate treatments (6), and were more effective than normal papillae in resisting penetration by *Erysiphe.* Likewise, when papillae formation was blocked by cycloheximide in 12 species of Gramineae, three avirulent pathogens normally excluded by papillae were able to penetrate epidermal cells of all species (342). In other cases, papillae delayed fungal penetrations of cells until necrogenic resistance led to death of both the host and fungal cells (300). Thus, papillae are an important component of resistance, but they probably must contain other materials in addition to callose and cellulose to be effective.

Gels and Tyloses

Vascular gels that coat the walls and fill the lumina of infected vessels have been found in numerous plant species infected with fungi (382). These gels generally form within 48 hr in both resistant and susceptible cultivars, but often are dissolved subsequently, presumably by fungal enzymes, in susceptible cultivars. Two theories have been presented on the origin of gels. VanderMolen et al (382) believe that the gels arise from the perforation plates, end walls, and pit membranes by a process of distention of primary wall and middle lamella constituents—probably pectinaceous materials, hemicellulose, and other carbohydrates. Moreau et al (270), noting extensive development of golgi and other secretory organelles in paravascular parenchyma, suggest that new carbohydrates and other materials may be synthesized and secreted into the vessels to form gels. Further studies are needed to clarify the relative contributions of "old" and "new" carbohydrates to the gels.

Tyloses are formed by protrusions of cell walls of paravascular cells through pits into the xylem vessels. Once a tylose is formed it may undergo secondary wall thickening. Tyloses frequently develop more rapidly or abundantly in resistant than in susceptible cultivars, such as in cotton (246), elm (103), tomato (26), and sweet potato (64) infected with wilt fungi. In cotton, phytoalexins are formed almost immediately after the tyloses and coat them, possibly offering protection from microbial attack (246). Without tyloses, the phytoalexins would be diluted in the xylem stream, probably making them ineffective. Thus, the two components working together are needed for effective resistance. The roles of gels and tyloses in resistance to vascular wilt fungi have been reviewed (34).

Structural Proteins

The importance of structural proteins in cell walls has become increasingly apparent in recent years. Such proteins also might be extremely important in processes such as gel and tylose formation and the binding of appositional material to cell walls. Lignin complexes formed with structural proteins are much more resistant to acid hydrolysis than comparable complexes with carbohydrates (238). Many wall-bound proteins also have enzymatic activities and probably are important for chemical reactions in cell walls.

A few studies indicate that structural proteins may be involved in resistance. Most of these studies involve the so-called b-proteins of tobacco (127, 204). Five different b-proteins have been isolated from tobacco leaves in which systemic resistance has been induced by various local lesion-producing viruses (204), polyacrylic acid (204), or the fungus *Thielaviopsis basicola* (127). The concentrations of b-proteins, which are extracted with acidic (pH 2.8) buffer, are generally proportional to the amount of necrosis caused by the resistance-inducing treatment. The b-proteins appear in nontreated leaves that show systemically induced resistance, but are missing from those which lack induced resistance. They are the most acidic proteins in the crude protein extracts, based on electrophoretic behavior, and have molecular weights of 15,500 to 15,800 (127). Although their appearance is correlated with induced resistance in tobacco, the cellular localization and biological function of the b-proteins is uncertain.

Wounding or inoculation with local lesion viruses causes accumulation of two base-extractable glycoproteins in bean (213). The first glycoprotein occurs maximally about 36 hr after wounding and contains both β-(1,3)- and β-(1,4)-linked glucopyranosyl residues; during the next 24 hr this protein largely disappears. The second glycoprotein, which contains only β-(1,4)-linked glucose, along with other sugars, begins appearing after 96 hr and accumulates to at least 144 hr. Both of these glycoproteins closely resemble structural proteins of the cell wall.

Cytochemical tests indicate that a proteinaceous intercellular matrix is formed at the juncture of necrotic and healthy cells in resistant corn infected with *Helminthosporium,* but this "protein" is not formed in the susceptible reaction (48). Thus, structural proteins may be involved in modifications of the outer cell wall. More attention should be given to the cell wall proteins and their roles in resistance.

Lignin and Phenolic Acid Complexes

Lignin is a phenolic polymer formed mostly by the free radical condensation of hydroxycinnamyl alcohols. The alcohols are derived from phenylalanine as shown in Figure 1. A few hydroxycinnamic acid residues may be esterified directly into the alcoholic lignin polymer, and residues of hydroxycinnamyl aldehydes apparently occur in some lignins, since the lignins react with acidic phloroglucinol, a chromogenic reagent for aldehydes. Lignin forms covalent bonds with cellulose, pectates, and structural proteins when synthesized in the presence of these compounds (238). Likewise, it forms ester linkages with fatty acid polyesters, similar to those in cutin, to yield suberin (224); suberized cells are rarely penetrated by pathogens. The hydroxycinnamic acids also form complexes with polysaccharides, proteins, suberin, and cutin by esterification (114, 224). Thus, both lignin and cinnamic acids cause modifications of cell walls that may contribute to disease resistance.

The occurrence, speed, and extent of lignification have been related to resistance in many host-parasite combinations (15, 315, 413). For example,

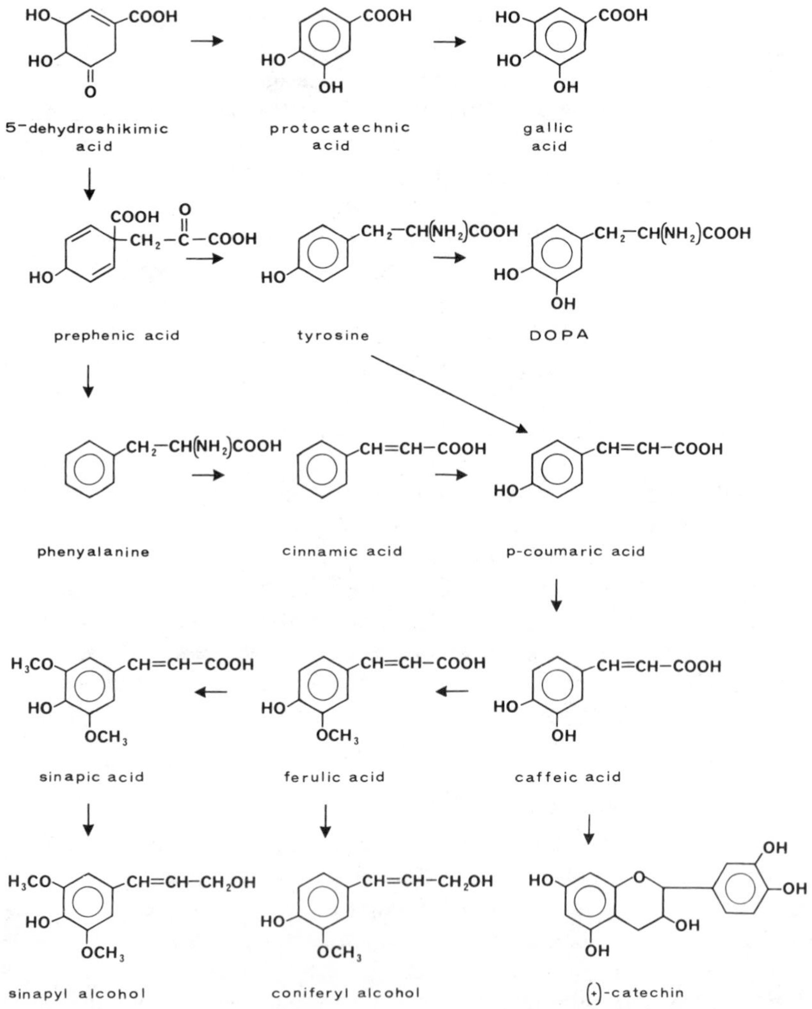

Figure 1 Relationships and biosynthetic derivation of various lignin precursors and dihydroxyphenols from the shikimic acid pathway.

lignification of cell walls is stimulated in and around viral local lesions (12, 214) and in necrotic cells formed near nematodes in resistant cultivars (128, 310); no stimulation occurs in susceptible cultivars. In virus-inoculated bean, lignin begins increasing at about 24 hr after inoculation and continues to 120 hr concurrent with increased secondary wall synthesis. Cells surrounding local lesions in bean and *Gomphrena* show extensive deposition of lignin and suberin in both the outer cell wall and the intercellular spaces.

Virus spread usually does not occur beyond the suberized cells. Cultivar resistance of potato to fungi also has been related to lignification and formation of phenolic acid esters (413).

Coniferyl alcohol, coniferyl aldehyde, and apparently lignin accumulate rapidly in highly resistant flax cultivars and more slowly in cultivars with moderate resistance to *Melampsora lini* (209). These compounds are not accumulated in susceptible cultivars. Both the aldehyde and alcohol are highly toxic to several fungi and appear to function as phytoalexins in flax. Lignin intermediates may play similar roles in other host-parasite relationships. The instability of the extracted alcohols and aldehydes could easily cause them to be overlooked.

In many trees, the cinnamyl alcohols are converted into dimers called lignans. Many of these, e.g. hydroxymatairesinol (336), accumulate in infected sapwood and apparently prevent decay.

Treatments that induce resistance may enhance lignification in response to infection. Tangerines treated with ethylene are more resistant to *Colletotrichum* and accumulate phenolic compounds and lignin more intensely than untreated fruit (47). Likewise, slices of carrot treated with heat-killed conidia of *Botrytis cinerea* are more resistant to it and have greater peroxidase activity and lignin and suberin synthesis than untreated slices (156).

The formation of lignified appositions (papillae) between the cell wall and plasma membrane, as well as lignification of the outer wall, appears important for resistance of species in the Gramineae to leaf-spotting fungi (293, 314, 316, 342, 379). The papillae and altered walls in wheat (317) and reed canarygrass (341) contain lignin as well as callose and cellulose. The lignin apparently is concentrated toward the center of the papilla, next to the fungus, while the cellulose is mostly positioned next to the plasma membrane—possibly to insulate the membrane from the lignin. The infection-induced lignin in wheat is different than that in healthy plants, because it does not react with acidic phloroglucinol and has a high sinapyl alcohol content (314). The latter compound is fluorescent and may contribute to the yellow to yellow-green fluorescence of the lignified papillae and outer cell walls which appear as "haloes" around infection sites, although callose also could be involved. Correlations between the development of fluorescent papilla or cells and cultivar resistance of barley to *Erysiphe graminis* (219) and wheat to *Puccinia graminis* (320) have been reported. Lignin synthesis has also been shown to occur more rapidly in resistant than in susceptible wheat infected with *E. graminis* (136).

Key enzymes in the synthesis of lignin are: phenylalanine ammonia lyase (PAL) and tyrosine ammonia lyase (TAL), which convert phenylalanine and tyrosine to cinnamic acid and 4-hydroxycinnamic acid, respectively; cinnamic acid-4-hydroxylase (CAH), which converts cinnamic acid to 4-

hydroxycinnamic acid; caffeic acid O-methyl transferase and 5-hydroxyferulic acid O-methyl transferase which add methyl groups to the C-3 and C-5 hydroxyls of caffeic and 5-hydroxyferulic acid, respectively; and peroxidase (PO) which polymerizes the hydroxycinnamyl alcohols. In general, increased activity of all of these enzymes slightly precedes accumulation of lignin (115, 128, 136, 236, 260, 378, 379), and such increases are correlated with disease resistance (115, 128, 136, 378, 379). In tobacco and wheat, the peak of increase in PAL or CAH activity precedes by 6–12 hr the peak for methyl transferase activity, which in turn precedes the peak in PO activity by about 24 hr (236, 260).

In reed canarygrass (378) and Japanese radish (15), the increases of PO activity are associated primarily with the most cathodic isozymes, which also are the major isozymes in the ionically and covalently wall-bound fractions from canarygrass. Increases of PO activity in cell walls of canarygrass are concentrated near sites of attempted penetration where lignification is greatest, and inhibition of lignin synthesis by cycloheximide also causes inhibition of new synthesis of the cathodic isozymes of PO (378). In radish, one of the cathodic isozymes is also the most efficient in forming lignins that include sinapyl alcohol residues. Thus, specific cathodic isozymes of PO appear to be responsible for lignin synthesis in the walls of these two hosts. Both of these studies are excellent examples of the increased attention which must be given to the natural substrate specificities of PO isozymes.

Recent evidence clearly shows that lignification protects cell walls and papillae from digestion by fungal enzymes. Lignified papillae and cell wall "haloes" from wheat were highly resistant to degradation by mixed enzymes from each of 14 pathogenic fungal species (316). Such resistance was apparent by 12 hr after inoculation, showing the importance of even small amounts of lignin. Extraction of lignin from papillae or walls allowed their digestion by enzymes (317); however, removal of only phenolic esters did not destroy the resistance to enzyme degradation. Commercial cellulase also macerated wheat leaves except in cells where lignification had been induced by inoculation of wounds with fungi (293).

The induction of lignification, at least in wheat, appears to involve greater specificity than has been found for induction of phytoalexins in other species. Of 15 chemicals known to elicit phytoalexin synthesis in other crops, only mecuric chloride elicited lignin synthesis in wheat (293). Likewise, three different yeasts or bacteria induced only small amounts of lignin, whereas 20 of 22 filamentous fungi ranging from saprophytes to virulent necrotrophic pathogens caused moderate to strong responses. Thus, the filamentous fungi appear somewhat specific in their ability to induce lignification at least in wheat. Hence fungal cutinase or its products may be involved, since this enzyme has been found mostly in fungi that penetrate plant cells directly.

PRODUCTION OF TANNINS AND MELANINS FOR DEFENSE

Necrosis associated with race-specific resistance normally is characterized by the formation of brown to black pigments (melanin) throughout the cell walls and the collapsed protoplasts. Walls of adjoining live cells also may become melanized. The intensity of melanin formation often is greatest in highly resistant plants, suggesting that melanins or their precursors contribute to resistance. Melanins in plants are formed principally from various *ortho*-dihydroxyphenolic compounds (27, 249, 288). The enzymes polyphenoloxidase (PPO) and peroxidase (PO) oxidize the colorless dihydroxyphenols to give the colored *ortho*-quinones (262). Certain dihydroxyphenols are conjugated with each other or with glucose hydroxyl groups to form polydihydroxyphenolic oligomers and polymers called tannins. Phenolic residues in the originally colorless tannin molecule also may oxidize to quinone residues. The colored condensation products of quinones and tannins constitute the plant melanins.

Two books (368, 405) deal with the biochemistry of the dihydroxyphenols and related compounds. All *ortho*-dihydroxyphenols are derived from the shikimic acid pathway as shown in Figure 1. Based on their biosynthesis they may be divided into four groups: gallotannins, 3-hydroxytyramine, caffeic acid esters, and flavonoids. The gallotannins (syn., tannic acid, hydrolyzable tannin) are derived from gallic acid and glucose. Several residues of gallic acid, or its oligomers, are esterified to single glucose molecules to form the various tannin molecules. In ellagitannins, some of the gallic acid residues attached to glucose are dimerized, apparently by PO (200). Thus, the hydrolyzed tannin yields ellagic acid. 3-Hydroxytyramine is formed by decarboxylation of 3,4-dihydroxyphenylalanine (DOPA) which is formed from tyrosine by PPO. Caffeic acid likewise is formed by the action of PPO on 4-hydroxycinnamic acid; it commonly forms esters with quinic acid to form chlorogenic, isochlorogenic, and neochlorogenic acid; but it also may form esters with glucose and shikimic acid. The dihydroxyflavonoids are derived by condensation of caffeic acid with a triacetate unit. The flavan-3-ols, catechin and gallocatechin, undergo C4-C8 polymerization to form the condensed proanthocyanidins (syn., condensed tannins, flavolans). Because of its key role in the formation of cinnamic acid, PAL is a critical enzyme for the formation of caffeic acid esters and flavonoids but not for gallotannins and 3-hydroxytyramine (55).

The gallotannins and condensed proanthocyanidins have long been used in the leather industry to denature animal proteins in skin and preserve it against microbial degradation. Hydroxyl groups of tannin molecules form hydrogen bonds with carbonyl groups of the protein amide linkages; and quinone residues form covalent bonds with free amino groups in lysine (185) and other amino acids. Both reactions modify protein structure. PO, besides forming quinones, converts dihydroxyphenols to free radicals that may undergo various reactions with cellular constituents. Thus, as expected, the tannins and the *ortho*-quinones have some toxicity to most microorganisms and viruses; but more important, they inactivate extracellular enzymes produced by microorganisms (13, 27, 154, 155, 177). The monomeric dihydroxyphenols also have some antibiotic and enzyme denaturing activity (276), but this may be partially due to products formed by autoxidation, or oxidation by PPO or PO from the microorganism. Mixing the dihydroxyphenols with PO or PPO generally increases both their antibiotic and enzyme-denaturing activities (377), indicating that the quinones and free radicals are much more active than the phenols. Because the quinones and free radicals have very short half-lives, it is important that the dihydroxyphenols and PO, or PPO, be mixed in the presence of the pathogen for maximum effectiveness as antibiotics or enzyme denaturants. However, once plant cell walls are melanized, they probably remain resistant to hydrolysis for extended periods of time.

Studies on the subcellular localization and on timing of synthesis, storage, and release of the dihydroxyphenols are necessary to evaluate the importance of these compounds in resistance. Many studies of this type have been attempted, but their results must be interpreted with caution. Techniques commonly used for fixation in electronmicroscopy may rupture large vacuoles containing dihydroxyphenols, resulting in artifactual staining of the cytoplasm, organelles, and membranes (275); these symptoms are often reported for cells expressing resistance. Caffeine can be used to retain 3-hydroxytyramine in banana and flavanols in cotton inside the vacuoles of phenol-storing cells prepared for electron microscopy (275). This technique should be of value in studies of dihydroxyphenol localization in diseased tissues.

Another problem in ultrastructure studies is the interpretation of electron dense areas in cell walls. Heath (157) has shown in cereals that such areas contain large amounts of silicates, which instead of phenolic melanins might be responsible for the electron density. It is possible, however, that melanins or tannins are deposited in the cell walls and then serve as scavengers for the silicates. Wheeler et al (412) found that DOPA and catechol were readily converted to melanins in the outer cell wall and matrix around the wall when fed to microsclerotia of mutant albino fungi. However, both melanins showed extremely weak electron density and were barely discernible, even though the parent dihydroxyphenols appear as extremely electron dense areas. This observation suggests that melanin concentrations in walls or other cell parts of plants may be much greater than has generally been thought. Improved methods for determining melanin localization and estimating melanin concentrations are critically needed.

The cellular and subcellular localization of the flavanol dihydroxyphenols (catechin, gallocatechin, and their condensed proanthocyanidin derivatives) have been studied in cotton with the aid of a specific histochemical reagent (248) and with improved electron microscopy methods (248, 273, 275). In roots, the flavanols first appear in the cap about 0.5 mm back from the tip and then in the endodermis about 1 mm back from the tip; flavanols in cells first appear in numerous small vacuoles. At 3 mm back from the tip, endodermal cells have completed synthesis and accumulation of flavanols in large vacuoles that fill most of the cell. Flavanol accumulation is complete in as little as 30 min at 35°C. Further back from the root tip, flavanols also appear in healthy hypodermis, but not in the epidermis. Stele tissue remains largely free of flavanols unless the xylem becomes infected. Then the xylem ray cells synthesize flavanols even at great distances from the pathogen, whereas paravascular cells synthesize terpenoid phytoalexins that are exuded into the xylem vessels (38). In shoots, flavanols occur in the epidermis and randomly scattered cells of the cortex, pith, and mesophyll.

Attempts to evaluate the contribution of melanin in resistance have two inherent difficulties. First, the most active intermediates, the quinones and free radicals, are too transient and unstable for meaningful quantitative measurements. Second, the terminal product melanin is highly resistant to enzyme or acid hydrolysis and consequently defies quantitation. To estimate the importance of phenol oxidation and melanin generation, one is then forced to measure: 1. activities of enzymes involved in dihydroxyphenol synthesis, such as PAL, that also feed intermediates into lignins, and phytoalexins; 2. concentrations of dihydroxyphenols that are transient intermediates in the synthesis of melanin; or 3. activities of PO and PPO, both of which can be destroyed by quinones and melanin products. Obviously, none of these criteria alone is an adequate measure of the total system. Perhaps only melanin-deficient mutants, such as have been developed in fungi (37, 411, 412), can eventually give unequivocal answers as to the importance of dihydroxyphenols and melanins. Mutant callus cells of sycamore (126) and tobacco (43) with greatly enhanced phenol synthesis have been obtained with *p*-fluorophenylalanine treatments, but these have not been used in host-parasite studies.

The biochemistry of PAL and its importance in biosynthesis of dihydroxyphenols have been reviewed by Camm & Towers (55). Increases in this enzyme often are greater in resistant than in susceptible cultivars in response to various infectious agents (258, 303, 421), particularly in plant species that form caffeic acid esters as their major dihydroxyphenols. The exceptions have usually resulted when very young leaves (421) or sliced tissues (68) were used in experiments. These tissues have high levels of PAL activity before inoculation, so that enzyme concentration is probably not a limiting factor.

Treatments that inhibit PAL activity or synthesis may inhibit resistance. α-Aminooxyacetate (100 μM), an inhibitor of PAL activity but not its synthesis, decreased chlorogenic acid synthesis by about 75% in TMV-infected tobacco, reduced melanin production, and increased lesion size two- to fourfold; virus particles in treated leaves occurred beyond the lesion boundary (258). Likewise, placing tomatoes at 32°C, compared to 27°C, caused a marked drop of PAL activity in both healthy and diseased plants and induced susceptibility to nematodes (50). Apparently PAL was partially inactivated at 32°C, which may have contributed to the susceptibility.

Concentrations of various dihydroxyphenols have been related to cultivar resistance (233). DOPA concentrations in three *Cercospora*-resistant sugar beet cultivars were consistently higher than those in three susceptible cultivars throughout the growing season (159). Also, both concentrations of DOPA and resistance declined with age. Concentrations of caffeic acid and its esters have been related to cultivar resistance in watermelon to *Alternaria* (62), in chili pepper to *Colletotrichum* (44), in tomato to root knot nematode (176), and in tobacco to tobacco mosaic virus (370). In rice inoculated with avirulent strains of *Piricularia oryzae*, dihydroxyphenols and PO increased and ascorbic acid decreased more rapidly than in rice inoculated with virulent strains (355). Flavonoids also accumulate more rapidly in resistant than susceptible corn cultivars in response to infection by *Colletotrichum* (143). Cotton cultivars resistant to *Verticillium* wilt have high constitutive levels of condensed proanthocyanidins in leaves (38), and concentrations of the proanthocyanidins decrease with age in leaves (38, 173) and increase with age in root and stem bark (177), in each case concurrent with similar changes in resistance to fungal diseases. Collectively, these observations indicate that high levels of dihydroxyphenols are desirable for resistance, but the phenols still must be released and oxidized to be effective.

Few studies have concentrated on the transfer of dihydroxyphenols from vacuoles into cytoplasm, walls, and intercellular spaces where they can be oxidized by PO or PPO. A study in apple, however, indicates that this approach may be promising. Resistant apple leaves "leaked" dihydroxyphenols more rapidly than susceptible ones when infiltrated with the bacterium *Erwinia amylovora* (3). Further, the leakage of phenols preceded electrolyte leakage by about 1 hr in the resistant cultivar, but the two events occurred simultaneously in the susceptible cultivar. These differences could allow earlier and more effective melanization in the resistant cell walls.

Chemicals that enhance melanin concentrations should enhance resistance. In fact, the dichlorocyclopropanes that induce resistance in rice to *Piricularia oryzae* also stimulate melanin formation in response to the fungus (231). Tricyclazole similarly induces resistance in rice and at 1 ppm blocks the reduction of polyhydroxynaphthalenes to tetralones by *Piricularia,* resulting in accumulation of naphthoquinones (414). It is tempting to speculate that some pathogenic fungi may naturally reduce plant dihydroxyphenols to tetralones as a mechanism of inhibiting melanization and overcoming resistance. Dichlorocyclopropanes or tricyclazole then might restore normal melanization and resistance by blocking the reduction.

Attempts to correlate PO or PPO with resistance have been inconsistent (24, 41, 52, 193, 280, 307, 331, 376, 389, 420), leading some investigators to conclude that increases of these enzymes must be the result rather than the cause of resistance. Undoubtedly, these enzymes are not the sole cause of resistance; the dihydroxyphenols must also be made and mixed with the enzymes for an effective antibiotic and melanizing system. Few studies give proper consideration to the total system or time-space relationships. More attention also must be paid to the reaction specificities of different isozymes for natural products. Sheen (338) found that caffeic acid, chlorogenic acid, rutin, and quercetin were oxidized most rapidly by the most anodic isozymes of peroxidase from tobacco, and least rapidly by the cathodic ones.

In cotton, PPO apparently is most important for adding hydroxyl groups during the synthesis of dihydroxyphenols in plastids, whereas PO is the primary enzyme responsible for the oxidation of phenols to melanins in cell walls (274). Infection-induced increases of PO in cotton (392), and PO and PPO in tomato (80), are localized in cell walls and not in cytoplasmic organelles. Hydrolytic enzymes release PO from washed cell walls of soybean (23) and cotton (362), indicating that considerable PO may be covalently bound to walls. Unfortunately, many studies of PO and PPO have examined only the readily extractable enzymes. Studies of PO and PPO in diseased tissue are also complicated by the synthesis of at least PO by bacteria, fungi, and nematodes.

Viruses offer a unique opportunity to evaluate host PO and PPO because the enzymes are not present in the viruses. Several lines of evidence indicate the importance of PPO and PO in local lesion formation. First, supplying exogenous 3,4-dihydroxyphenylalanine, caffeic acid, chlorogenic acid, or gallic acid to resistant tobacco leaves infected with TMV causes lesion areas, but not other leaf areas, to darken markedly within a few hours, 6 to 18 hr before necrosis normally appears. Thus, an increase of oxidase activity in cell walls precedes necrosis. Equally important, the artificially melanized cells do not collapse, indicating that melanization can occur without cell death. Second, extractable PO and PPO increase in local lesion hosts, but not in systemic hosts, just before lesions are apparent (52). Third, treatment of leaves with ascorbate inhibits melanization and allows lesion size to increase (288). Fourth, natural aging of leaves results in increases in PO and reductions in lesion size (389, 406). Fifth, systemically induced resistance to viruses is associated with systemic increases in PO, PPO, and catalase activities (24, 389). Sixth, inhibition of systemic increase in PPO by 2-thiouracil also prevents induced resistance (24). When leaves are excised (389), or plants are held at high temperatures before returning them to low temperatures (52), resistance and PO and PPO activity are often not correlated. However, both treatments alone and other stresses that induce ethylene formation (139) cause appreciable increases in PO. Thus, the importance of PO and PPO activity for the resistance response is not negated, since activity levels are already high before infection.

PRODUCTION OF ANTIBIOTICS FOR DEFENSE

In addition to lignins, tannins, and melanins, plants produce lytic enzymes, proteinaceous inhibitors of enzymes and viruses, and a great variety of natural products that may function as antibiotics in resistance. The natural products are further subdivided into constitutive antibiotic systems and phytoalexins (antibiotics that are synthesized de novo after infection).

Lytic Enzymes

Various plants produce β-(1,3)-glucanase and chitinase (1, 278, 401), which together lyse and dissolve walls of nonmelanized fungal cells (401). Ethylene stimulates de novo synthesis of these enzymes (1), indicating that the enzymes may play a role in postinfectional disease resistance. The activity of β-(1,3)-glucanase increases more rapidly and to higher levels in resistant than susceptible muskmelon in response to *Fusarium oxysporum* (278). Pretreatment of susceptible muskmelon with laminarin (a β-1,3 polyglucan) prior to inoculation increased β-(1,3)-glucanase activity and markedly reduced disease. Fungal laminarins introduced into the xylem of avocado also are destroyed quickly (210), indicating that β-(1,3)-glucanase activity may be stimulated by various fungi. However, since plant callose [also a β-(1,3)-polyglucan] first accumulates and only later disappears in resistance responses to various pathogens, the roles of β-(1,3)-glucanases are unclear. Pathogenic fungi, as well as plants, produce chitinase, also complicating its evaluation in resistance (99). Nevertheless, the frequent lysis and disappearance of fungal structures in host tissues suggests that chitinase may be important for resistance.

Plants produce proteolytic enzymes that may inactivate viruses. Proteases that can degrade the potato virus X protein coat from either the N-terminus (a trypsin reaction) or the C-terminus have been shown in crude plant sap (218, 373). However, the special relationship of the proteases and the virus in intact cells and the importance of proteases to resistance are unknown.

Protein Inhibitors of Enzymes and Virus

Numerous protein inhibitors of trypsin and chymotrypsin occur in legume seeds and potato tubers (99, 125). These inhibitors are thought to have a role in the ecological relationships between herbivores and plants (237, 281, 367), but they may also contribute to disease resistance. Proteinase inhibitor activity increased rapidly and remained high in a resistant tomato infected with *Phytophthora infestans,* whereas slight or no increase, followed by a progressive decline in activity, occurred in three susceptible cultivar-race combinations (296). Inoculation of a single leaf systemically induced a twofold increase in the inhibitor along with resistance. However, the nature of the inhibitor was not established.

Various proteinaceous inhibitors of virus infection have been reported. For example, four such inhibitors were found among 29 species of the Chenopodiaceae (353); similar inhibitors occur in the Amaranthaceae. These inhibitors prevent transfer of virus from plants in these families to plants of the Solanaceae or Leguminosae, but not to plants of the original species. The proteins apparently are basic glycoproteins with molecular weights of 25,000 to 38,000. These observations suggest that these infection inhibitors may elicit defense responses, since foreign glycoproteins often act as defense elicitors (See section on Recognition).

Constitutive Antibiotic Systems

Plants produce thousands of natural products, many of which are unique to specific taxonomic groups and are toxic to pests. Several reviews (30, 31, 34, 146, 237, 281, 367, 404) are available on the probable importance of these constitutive antibiotics in plant-pest interactions. The compounds may contribute to disease resistance, but appear particularly important for resistance to insects and herbivores (237, 281, 367).

Constitutive antibiotics may be completely synthesized and stored in vacuoles, lysigenous glands, ducts, heartwood, and periderm. Alternatively, less toxic precursors to the antibiotics may be stored in vacuoles, with the final antibiotic being formed rapidly by the action of hydrolases and oxidases located in other parts of the cells. I prefer to distinguish the second group as wound antibiotics, although the first group, obviously, also has to be released from vacuoles or glands to contact a pathogen or pest.

Seeds and seedlings may contain high levels of unusual metabolites that are antibiotic. For example, oil from American elm seed contains 55% capric acid (94), and homoserine constitutes 70% of the free amino acids in 5-day-old pea seedlings (165). Capric acid is toxic to *Ceratocystis ulmi,* and might be involved in seedling resistance to this fungus. All 13 isolates of *Pseudomonas pisi,* a bacterial pathogen of pea, could utilize homoserine as a sole carbon source for growth, whereas 116 of 117 avirulent isolates from 14 other *Pseudomonas* species could not (165). Homoserine is an intermediate in the synthesis of methionine, threonine, and lysine, and does not inhibit growth in the presence of these compounds (326). Other amino acids that are similarly intermediates in biosynthetic pathways also inhibit growth of *Pseudomonas syringae* when used as sole nitrogen sources, indicating the possible general importance of imbalanced amino acid ratios for suppressing bacterial growth.

Roots of plants may be protected by antibiotics produced in the epidermis and periderm and exuded into soil. For example, the epidermal cells of cotton plants become filled with toxic terpenoid aldehydes after they are a few days old, and these cells rarely are penetrated by pathogens (178). Infection of hypocotyls increases the quantities of terpenoids that are exuded by roots.

More than one-third of the species in the Compositae suppress nematode populations in the surrounding soil (132). Marigold (*Tagetes erecta* L.) and margosa (*Azadirachta indica* Juss.) root exudates are particularly toxic to nematodes, and these plants have been used for biological control (7). The active antibiotics from marigold are the polyenes, α-terthienyl and derivatives of bithienyl, whereas those in safflower are polyacetylenes (220). Carrot roots also accumulate the antibiotic polyacetylene falcarindiol in the periderm and pericycle (124). This compound may account for the higher levels of resistance in these tissues to pathogens. A natural nematocide from *Daphne odora* Thunb, has been identified as a complex diterpenoid derivative in which a cyclopentenone ring is essential for activity (221). Similar diterpenoids from various other plants are toxic to gram-positive bacteria.

Numerous constitutive antibiotics in wood inhibit fungal growth (16). For example, wood-rotting fungi are inhibited in wood of ironbark by stilbenes and ellagitannins (155). Concentrations of β-phellandrene in the stem cortex of pine are closely correlated with resistance to fusiform rust, and genetic manipulation of this compound has been suggested for improving resistance (318). Borbonol occurs throughout the wood and leaves of avocado and may contribute to resistance against *Phytophthora* root rot (428).

The surfaces of many plant leaves are covered by glandular trichomes that exude a variety of constitutive antibiotics that are probably most important for insect resistance (237, 281). Surface washings from *Acer* leaves contain gallic acid and other compounds (gallotannins?) (186), and those from cotton contain catechins and condensed proanthocyanidins that originate from the glandular trichomes (unpublished). Both groups of compounds are weak antibiotics against phylloplane (leaf surface) microorganisms. Other antibiotics recovered from leaf surfaces include: luteone, the isopentenyl isoflavone from *Lupinus* spp. (147); parthenolide, the sesquiterpene lactone from glands of *Chrysanthemum parthenium* (45); and sclareol, the diterpenoid from tobacco (19). Each of these is toxic to various fungi and is a major component of leaf exudates. Spraying leaves with sclareol provides effective control of rust diseases, and parthenolide, added to inoculum droplets containing spores, reduces infection by two fungal pathogens on *Chrysanthemum* and bean.

More than 20 different antibiotic terpenoid aldehydes and quinones have been identified from the lysigenous pigment glands in leaves of *Gossypium* species (40). Gland contents differ among cytogenic groups, and differences apparently result from the variable occurrence of four key enzymes that regulate terminal parts of the terpenoid pathway. The importance and potential genetic manipulation of these compounds for pest resistance has been reviewed (38).

The saponins and alkaloids are examples of compounds normally stored in vacuoles of cells in various tissues. These compounds are apparently most important for insect resistance. Consequently, cultivars with high alkaloid contents, for example, may not have any more resistance to certain diseases than those with low alkaloid content (112, 343). Sometimes, however, saponins and alkaloids do appear to be important for disease resistance.

Roots of oats contain the antibiotic spirostanol saponins, avenacin A and B, in a ratio of 10:1, in vacuoles of cells (244, 396). The steroid aglycone of avenacin A has a C-3 sugar side chain composed of 2 moles of glucose and 1 mole of arabinose. Fungal invasion of oat roots triggers release of avenacins from protoplasts simultaneously with electrolyte and nutrient leakage. The virulent pathogen *Fusarium avenaceum,* however, is able to detoxify low concentrations of avenacin A by producing a cell wall-bound β-glucosidase and α-arabinosidase, which convert it to the nontoxic avenamin A (244, 396). High concentrations of avenacin A are toxic to *F. avenaceum.* In oat leaves, avenacin A occurs in vacuoles as a nontoxic 26-desglucoside that can be degraded by a specific β-glucosidase located in another part of the plant cell. Thus, avenacin A exists as a constitutive antibiotic in roots but as a wound antibiotic in foliage.

The alkaloids in solanaceous plants also have been implicated in disease resistance. The skins and healed wound surfaces of potato tubers contain high concentrations of fungitoxic alkaloids which apparently prevent infection of healthy tissue by most fungi. However, infections of fresh wounds on potato often suppress normal alkaloid synthesis, diverting mevalonate to synthesis of terpenoid phytoalexins (75, 344). Infection of resistant, but not susceptible, tobacco roots by nematodes causes content of the alkaloid nicotine to double in leaves (145), possibly inducing resistance to other pests.

Concentrations of the alkaloid tomatine are greater in extracts from *Fusarium*-resistant tomatoes than in those from susceptible tomatoes (142). This alkaloid also is toxic to the pathogen *Cladosporium fulvum,* suggesting that specific leakage of α-tomatine from vacuoles may be responsible for race-specific resistance (95). Accelerated leakage of ^{32}P-labeled compounds is constantly associated with resistance in several cultivar-race combinations (384), but tomatine levels in the exudates are not known. Future research on constitutive antibiotics needs to concentrate on time-space relationships between leakage of antibiotics and microbial ingress into cells.

Wound antibiotics are generated from: allyl sulfoxides in onions and garlic; glucosinolates in cabbage; cyanogenic glucosides in sorghum, lima beans, and peach; *para*-hydroquinone glucosides in pear and walnut; and benzoxazinones in corn, wheat, and rye. Phytoalexins have not yet been identified or clearly related to resistance in these crops. Thus, the wound antibiotics may be important for resistance.

Most of the meager data comparing wound antibiotics in resistant and susceptible plants were reviewed previously (30). The correlation between the accumulation of *para*-benzoquinone (and an oxidation product) and cultivar resistance in barley to *Selenophoma* leaf spot may be due to a wound antibiotic system (107). Benzoquinones are generated from glucosides in pear and wheat, but their origin in barley is not known. Development of corn cultivars with a genetic lesion for production of 2,4-dihydroxy-7-methoxy-2H-1,4-benzoxazin-3(4H)-one (DIMBOA) has allowed evaluation of this wound antibiotic, indicating that it is important for resistance to certain insects, fungal pathogens, and bacterial pathogens. However, cultivar resistance to strains of *Erwinia chrysanthemi* apparently depends on factors besides DIMBOA (230).

Virulent and avirulent pathogens may vary in their sensitivity to wound antibiotics. In birdsfoot trefoil, cyanide is generated from the cyanogenic glucosides limamarin and lotaustralian. *Stemphylium loti,* a virulent pathogen of trefoil, was able to adapt to cyanide in culture much more readily than three *Stemphyllium* spp. that only attack other crops (117). Protein synthesis was necessary for the adaptation, which resulted in the metabolism of the cyanide (118). Likewise, 80% of 34 *E. chrysanthemi* isolates from corn were relatively resistant to DIMBOA, while only 1 of 25 isolates from other crops had similar resistance (230). Thus, DIMBOA tolerance appears to be a definite ecological advantage to the pathogen in establishing a successful host-parasite relationship.

Phytoalexins

These antibiotics, produced after infection, include a variety of natural compounds: isoflavonoids, flavonoids, dihydrophenanthrenes, stilbenes, coumarins, isocoumarins, terpenoids, furanoacetylenes, polyacetylenes, and polyenes. Different phytoalexins are often unique to a few genera or closely related species (39, 181, 182), and thus may be useful in chemotaxonomic studies of plants. Several reviews tabulate and give chemical structures for most known phytoalexins and related toxic compounds (138, 146). Cruickshank (72; see also 171) thoroughly reviewed the importance of phytoalexins in disease resistance. Detailed reviews on the biochemistry and biology have been published for isoflavonoid phytoalexins of the Leguminosae (386), sesquiterpenoid phytoalexins of the Solanaceae (360), and sesquiterpenoid phytoalexins of the Malvaceae (38).

Phytoalexin biosynthesis has been studied most extensively in the Leguminosae. Structures of typical phytoalexins found in this plant family are shown in Figure 2; formonetin is the isoflavone precursor of the phytoalexins vesitol, sativan, and medicarpin, and daidzein is the precursor of the phytoalexins glyceollin and phaseollin. The probable biosynthetic pathways have been determined for: maackiain in clover (84); vesitol, sativan, and medicarpin in alfalfa (83); pisatin in pea (386); vignafuran in cowpea (257); glyceollin in soybean (291, 425); phaseollin in bean (418); and wyerone in broad bean (53). All of these phytoalexins except wyerone are derived from the condensation of a cinnamyl-CoA and three acetyl-CoA units; the right-hand ring (Figure 2) in each case comes from the cinnamyl unit. The immediate products of acetate and cinnamate condensation are chalcones which first isomerize to flavonones and eventually give rise to isoflavones such as daidzien and formonetin. Most of the additions of hydroxyl, methyl, and isopentenyl groups to the aromatic rings apparently occur at the isoflavone level. Subsequently, two reductive reactions and a step involving formation of carbonium ions gives rise to the pterocarpan or isoflavan ring system. The original C-3 from cinnamic acid is eliminated from the isoflavone ring to give rise to the benzofuran ring in vignafuran (257). Acetate, malonate, or oleate are readily incorporated into wyerone, suggesting that it is derived by dehydrogenation and oxidation of fatty acids (53).

Examples of phytoalexins from other plant families are shown in Figure 3. 2,7-Dihydroxyflavan in *Narcissus* (70) is probably derived from acetate and cinnate via the chalcone, as is the case with other flavones. Orchinol from orchids (109) and viniferin from grapes (302) are apparently derived from stilbenes such as resveratrol (Figure 2), which is also a phytoalexin

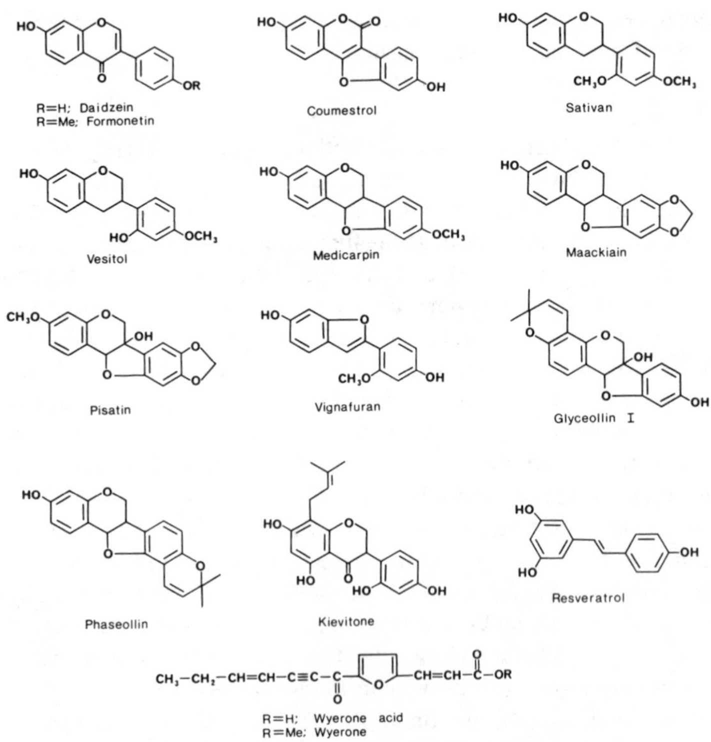

Figure 2 Phytoalexins and isoflavone precursors in the Leguminosae.

in peanut (146). The stilbenes are derived from cinnamate and acetate, but one of the acetate carboxyls is eliminated so that the carboxyl group of cinnamate is incorporated directly into the left-hand ring (Figure 3). Xanthotoxin (190) is typical of furanocoumarins formed from acetate and cinnamate. Momilactones from rice (57), casbene from castor bean (351), rhisitin and hydroxylubimin from potato (328), capsidiol from pepper (361), desoxyhemigossypol, hemigossypol, and gossypol from cotton (38), and ipomeamarone from sweet potato (205) are terpenoids derived from mevalonic acid. The structure of safynol from safflower (9) is consistent with an origination from fatty acids, although this has not been determined. Isocoumarins, such 6-methoxymellein from carrot, are normally derived from pentaketides synthesized from acetate units and are also commonly found in fungi.

Nearly all species in the Leguminosae, Malvaceae, and Solanaceae synthesize phytoalexins in association with the necrogenic resistance response to pathogens. Other defense responses, such as lignin and tannin synthesis and cell wall appositions, are consistently associated with the synthesis of

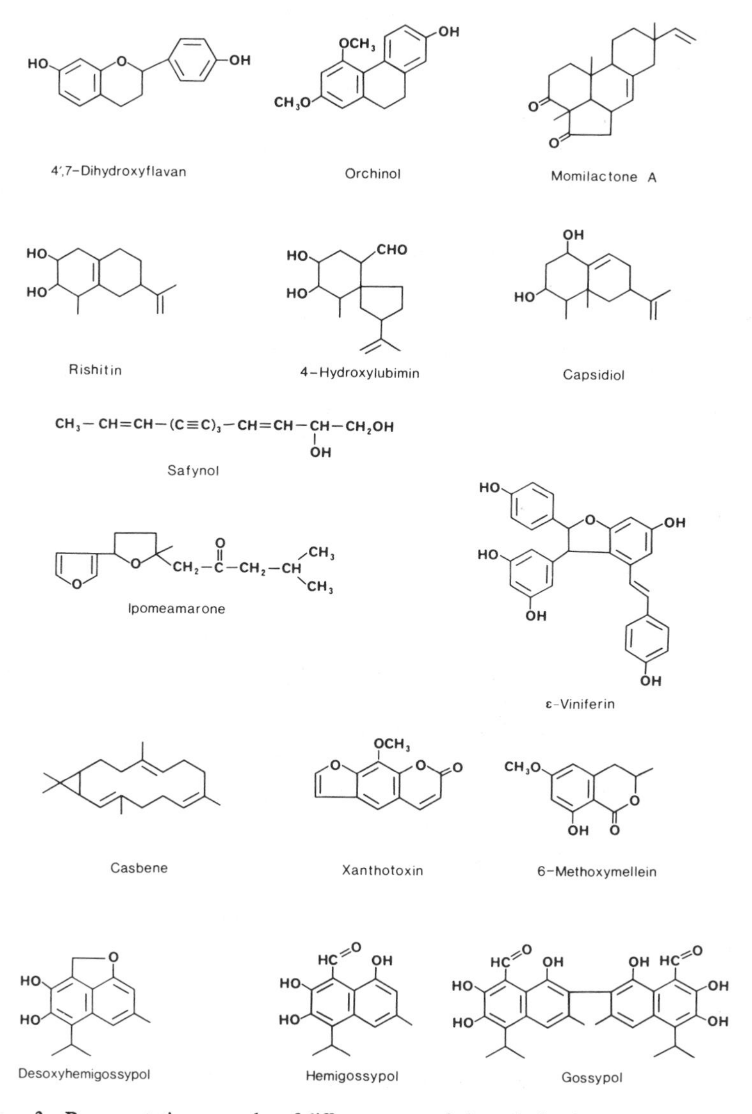

Figure 3 Representative examples of different types of phytoalexins formed by various plant families.

phytoalexins (38, 88). In cotton, the accumulation of tannins follows the accumulation of terpenoid phytoalexins by about 24 hrs (38). Phytoalexins are apparently synthesized in live cells (104, 253), but cell necrosis follows very quickly, possibly within minutes. Thus, peak concentrations of phytoalexins in tissues almost always coincide with the onset of visible necrosis.

The phytoalexins glyceollin (202), phaseollin (129), pisatin (346), rishitin (245), and phytuberin (245) are toxic to plant cells as well as microorganisms, and can cause disruption of the plasmalemma and vacuolar membranes and inhibition of mitochondrial respiration even in the plant of origin. Thus, accumulation of critical phytoalexin concentrations in vacuoles possibly leads to rupture of vacuoles (151) and then to the total disorganization of membranes characteristic of necrogenic resistance. Injury in adjoining plant cells may be prevented by plant detoxification; metabolism by healthy tissues of rhisitin in potato (62), glyceollin in soybean (425), and phaseollin in bean (129) has been demonstrated.

It would be interesting to know whether enzymes, e.g. peroxidase, are responsible for the plant detoxifications of phytoalexins. If so, the inactivation of peroxidase by tannins and melanins might then allow the observed accumulation and persistance of the phytoalexins in the dead cells but not in nonmelanized live cells. The tannins, or melanins, also might prevent detoxification of the phytoalexins by extracellular enzymes (229) or wall-bound enzymes (244, 396) of the pathogens. The accumulation of phytoalexins in soybean is, in fact, as much a function of inhibited degradation as of facilitated synthesis (425).

Patterns of phytoalexin accumulation in resistant plants mimic those of necrogenic resistance described in the histology section. Biotrophic pathogens elicit rapid, intense accumulation of phytoalexins in association with viral local lesions (20, 216), necrotic flecks caused by rust fungi (104), and necrosis at feeding sites of nematodes (201, 312, 393). Little, if any, phytoalexin forms in the susceptible hosts infected by biotrophs.

The terpenoid aldehydes formed in response to the nematode *Meloidogyne incognita* in the pericycle of cotton roots are related to resistance, but the constritutive terpenoid aldehydes formed in the older root epidermis away from the nematode are not (390). Consequently, misleading data are obtained if terpenoid aldehydes are measured only in the whole root. This example clearly emphasizes the need to study phytoalexins histochemically at the tissue or cellular level where the host-parasite interaction occurs.

Resistance to partially biotrophic organisms is characterized by quicker initiation of phytoalexin accumulation in resistant than in susceptible cultivars. Considerable amounts of phytoalexins may eventually accumulate in the susceptible cultivar as necrosis becomes prominent. This pattern has been shown in resistance to: bacterial diseases in bean (130, 403) and cotton (106); *Phytophora* rots of soybean (423) and cowpeas (290); *Cladosporium* leaf spot of tomato (86); and fungal wilt diseases of cotton (246), tomato (365), and alfalfa (212). The phytoalexin effective against the bacterium *Pseudomonas syringe* in bean apparently is not phaseollin, although large amounts of this antibiotic accumulate in the resistant host (63). Likewise, the most potent antibacterial agent in cotton apparently is 2,7-dihydroxy-cadalene (M. Essenerg, personal communication) rather than the terpenoid aldehydes which are highly toxic to fungi. Thus, different phytoalexins may be involved with different pathogens, making multiple phytoalexin responses the most desirable. Resistant cultivars of both cotton and tomato produce a greater variety of phytoalexins than susceptible cultivars against several pathogens (38, 86, 365).

With necrotrophic pathogens, phytoalexin synthesis is often initiated quickly and at about the same time in both susceptible and resistant hosts in response either to virulent or to avirulent pathogens (63, 110, 191, 195, 211, 223, 352). By 48 to 96 hr after inoculation, however, the greatest concentrations usually exist in lesions on the resistant plant. This difference may be due to greater capacity for phytoalexin synthesis in resistant plants. Or the pathogen in the susceptible plant may metabolize the phytoalexin, slowing its accumulation. Eventually, phytoalexins may accumulate to toxic levels even in susceptible plants, causing confinement of the pathogen after lesions become relatively large (352).

Virulent and avirulent necrotrophic pathogens and saprophytes may metabolize phytoalexins in a variety of ways. In some cases, the product is as toxic as the original phytoalexin, offering no apparent advantage for the change. In other cases, the metabolic product is considerably less toxic. These variations and anomalies are most evident from metabolic studies of the bean phytoalexin phaseollin.

Stemphylium botryosum, virulent to alfalfa but not bean, converts phaseollin to the equally toxic phaseollinisoflavan (164), whereas *Septoria nodorum,* virulent only to wheat, converts phaseollin to 12,13-dihydrodihydroxyphaseollin which is much less fungitoxic (18). Races of *Collectotrichum lindemuthianum,* virulent or avirulent to bean, and moderately virulent *Botrytis cinerea* convert phaseollin to 6a-hydroxyphaseollin, which is less toxic to *Botrytis* but not to *Colletotrichum* (51, 381). However, *Colletotrichum* further converts this product to the less toxic 6a, 7-dihydroxyphaseollin (51). The virulent *Fusarium solani* f. sp. *phaseoli* (387) converts phaseollin to the less toxic 1a-hydroxyphaseollin, but so does the saprophyte *Cladosporium herbarum* (381). In contrast, neither virulent *Thielaviopsis basicola* nor *Fusarium oxysporum* f. sp. *phaseoli* are able to metabolize phaseollin (381). Thus, while phaseollin can be metabolized in many ways, these abilities are not consistently related to virulence.

Some of the apparent inconsistencies between phaseollin metabolism and virulence may be due to the fact that bean produces multiple phytoalexins of which kievitone and phaseollidin, as well as phaseollin, are often prominent early in resistance responses. Lesions resulting from *Fusarium solani*

f. sp. *phaseoli,* however, contain only traces of these phytoalexins because of detoxification (387). Water is added across the double bond of the isopentenyl groups in these phytoalexins to form the less toxic hydrated derivatives. Kuhn & Smith (229) recently isolated the responsible enzyme, kievitone hydratase, from *F. solani* f. sp. *phaseoli,* and concluded that it is an adaptive extracellular enzyme.

Solanaceous plants like beans have multiple phytoalexins, and attempts to correlate virulence with detoxification of individual phytoalexins have been similarly frustrated (398, 399).

Most pea cultivars produce only one major phytoalexin, pisatin, and the ability to detoxify pisatin has correlated well with virulence. *Fusarium solani* f. sp. *pisi* demethylates pisatin to the less toxic 6a-hydroxymaackiain, which is further degraded to 3-hydroxymaackiain isoflavane and finally to smaller products including CO_2 (119). In an extensive survey of more than 50 isolates of *F. solani,* Van Etten et al (385) found that all isolates highly virulent to pea were tolerant of pisatin and able to demethylate it, whereas all isolates highly sensitive to pisatin could not demethylate it and were avirulent. Thus, the ability to metabolize pisatin appears essential for high virulence. *Ascochyta pisi,* another pea pathogen, also demethylates pisatin (cited in 119).

The clover phytoalexins, medicarpin and maackiain, undergo various fungal-mediated transformations involving hydroxylation, methylation, demethylation, and occasionally ring fission (180). Most of these conversions cause detoxification and generally are correlated with virulence. Highly virulent *Sclerotinia trifoliorum* rapidly detoxifies both phytoalexins in vitro and in vivo (250). Moderately virulent *Botrytis cinerea, Botrytis fabae,* and *Colletotrichum lindemuthianum* also detoxify the phytoalexins but at somewhat slower rates, particularly in vivo (180, 250). *Helminthosporium carbonum,* a pathogen of corn but not clover, cannot metabolize these phytoalexins, but the avirulent *C. coffeanum* can.

Broad bean also forms the phytoalexin maackiain, but the most prominent phytoalexins are the furanoacetylenes wyerone, wyerone acid, and wyerone epoxide. In this plant, unlike clover, *B. fabae* is more virulent than *B. cinerea.* Likewise, *B. fabae* is less sensitive to the furanoacetylenes (322) and detoxifies these compounds more rapidly than *B. cinerea* (153, 323).

Various ontogenic changes in resistance have been related to phytoalexins. Potato sprouts from both susceptible and race-specific resistant cultivars are highly resistant to *Phytophthora infestans.* Likewise, a rapid accumulation of rishitin and related compounds occurs in sprouts of all cultivars. High alkaloid content in sprouts may cause lysis of *P. infestans* hyphae, thus triggering a phytoalexin response even in the susceptible cultivars; the fungal cytoplasm is a potent elicitor of phytoalexin synthesis (239).

In 6-day-old (susceptible) and in 12-day-old (resistant) cotton seedlings, hemigossypol accumulation in response to *Rhizoctonia* begins at the same time, but it does not persist in young tissue beyond 48 hr when tissues become extensively rotted (197). In old cotton hypocotyls, which are resistant, the tissues are less damaged, probably because of much greater tannin content, and phytoalexin accumulation then continues until *Rhizoctonia* is contained. Aging carrot roots (133) or pea leaves (17) progressively show diminished resistance and less accumulation of 6-methoxymellein and pisatin, respectively, in response to fungal pathogens. Treatments of pea leaves with benzyladenine delays both senescence and loss of resistance while maintaining high levels of pisatin production.

Environmental variables that alter resistance often have a predictable effect on phytoalexin synthesis. High temperatures that increase resistance also hasten the onset and accelerate rate of synthesis of hemigossypol and phaseollin in cotton (35) or bean (304), respectively. Likewise, high light intensity and long as compared to short photoperiods enhance both resistance and accumulation of safynol in safflower (372) and medicarpin in clover (74), respectively. High compared to low levels of nitrogen reduce quantities of phytoalexins accumulated in clover and have similar effects on resistance (73). Many other environmental effects on disease resistance also may be mediated through quantitative changes in phytoalexin production.

Treatments that induce resistance frequently induce phytoalexin synthesis at least in the treated tissue. The degree of resistance induced in cotton to *Verticillium* wilt by heat-killed and heat-inhibited conidia is directly related to amounts of terpenoid aldehydes produced in xylem tissue (36). Likewise, resistance induced in soybean by avirulent *Phytophthora cactorum* against *P. megasperma* is correlated with glyceollin synthesis (366), and induced resistance in tobacco is associated with antibiotic production in treated leaves (306, 335). Spraying tomatoes with the pepper phytoalexin capsidiol also protects plants against *P. infestans* (400). Quickness or intensity of phytoalexin synthesis might also be increased in leaves with induced systemic resistance; however, this possibility has largely been overlooked.

Induced susceptibility has been associated with reduction of phytoalexin synthesis. For example, inoculation of potato tubers with a virulent race of *Phytophthora infestans* 12 hr before inoculating with an avirulent race induces susceptibility to the latter. Corresponding reductions in rishitin and phytuberin accumulations also occur (391). Growth regulator treatments that reduce resistance may strongly inhibit phytoalexin synthesis. Concentrations of 2,4-D as low as 2×10^{-6}M completely inhibit phaseollin accumulation by suspension cultures of bean cells (90). Various inhibitors of protein and RNA synthesis and respiration reduce resistance and similarly inhibit phytoalexin synthesis.

The accumulated evidence overwelmingly shows the importance of phytoalexins in resistance.

HOST RECOGNITION OF PATHOGENS

Postinfectional resistance responses (lignification, melanization, phytoalexin synthesis, etc) are not normally expressed against virulent races of biotrophic pathogens. There is also a delay in these responses against virulent races of partially biotrophic organisms. Genes for race-specific resistance simply restore the normal pattern of rapid postinfectional resistance (recognition) expressed against most avirulent pathogens. Thus, virulent races either produce suppressors that inhibit or delay active resistance or fail to produce elicitors of defense reactions. Several reviews (8, 54, 81, 332) and books (78, 354) have dealt with the nature of host recognition of pathogens.

Suppressors

In animal host-parasite systems, susceptibility is often dependent on antigenic similarity of the host and pathogen, which results in a delayed or attenuated immune response. Such antigens apparently reside on cell surfaces. Common antigens also have been demonstrated between plants and biotrophic plant pathogens such as rust fungi, smut fungi, and nematodes, or partially biotrophic pathogens, including both fungi and bacteria (81). However, the same common antigens are usually found in different races of the pathogen and in both susceptible and resistant cultivars (287). Common antigens might function in suppressing resistance, but their precise identity, cellular location, and role is still unknown.

Plant pathogenic bacteria produce an extracellular polysaccharide slime that appears essential for virulence. Mutants lacking slime invariably are avirulent and trigger immediate host resistance responses. The slime is a peptidoglycan composed mostly of a complex highly branched polysaccharide, but also contains 2–5% protein (101, 102). A pure polysaccharide fraction, xanthan, has been prepared and identified from slime of *Xanthomonas campestris* (282). Xanthan has an extremely high affinity for water, and readily forms mixed gels with various plant cell wall polysaccharides (101). Accordingly, extracellular polysaccharide slime preparations from various bacteria induce persistant water soaking of leaves, specifically in susceptible cultivars of the host of the bacterium from which the slime was derived (101, 102). This water soaking by such preparations probably suppresses resistance, because artificial water-soaking (67) or suspending of bacteria in agar gels within intercellular spaces (357) suppresses resistance.

In resistant cotton cultivars, the slime preparation from *Xanthomonas malvacearum* also elicits a necrogenic resistance response (101). Thus, the same compound may be a suppressor or elicitor of resistance depending on the host cultivar.

Phycomycetous fungi such as *Phytophthora* and *Pythium* contain branched polyglucans called mycolaminarins. These polysaccharides may make up as much as one-third of the fungal dry weight (415), and are thought to act primarily as storage carbohydrates in the cytoplasm. The mycolaminarins usually contain mostly β-(1,3) linkages with fewer β-(1,6) linkages; the ratios of the two linkages and size of the polymers vary with different genera and species. The mycolaminarins have numerous interesting effects on host-parasite relationships, including: complete suppression of virus infection in potato and tobacco (415), agglutination of potato protoplasts (298), wilt-phytotoxin activity (210, 417), elicitation of phytoalexins in the Leguminosae (8), suppression of resistance in potato tubers (91, 122), and suppression of lysis of potato protoplasts by elicitors (93). Apparently the mycolaminarins bind with the protoplast membrane of the plant cell (298). The binding of mycolaminarin in tobacco and potato may prevent binding of virus particles and elicitors, whereas in other plants it may accelerate water loss or elicit defense reactions because of stresses in the membranes that cause leakage. Laminaribiose and methyl-β-D glucopyranoside, structural analogs of β-(1–3) glucans, also inhibit resistance responses elicited by mycelial homogenates of *Phytophthora infestans* in potato (254).

Doke et al (91, 93) and Garas et al (122) suggested that some of the specificity of the interactions between races of *P. infestans* and cultivars of potato is the result of selective suppression of resistance by the mycolaminarins. Mycolaminarins prepared from seven fungal races were incubated with protoplasts from 9 cultivars that varied in race-specific resistance. After 15 minutes, a nonspecific elicitor preparation, extracted from *P. infestans* cell walls by autoclaving, was added to the protoplasts, which were observed for lysis. Particularly within cultivars, there was a correlation between susceptibility and suppression of protoplast lysis by the mycolaminarin preparations. However, mycolaminarins were ineffective if potato tissue or protoplasts were previously or simultaneously exposed to elicitors. Thus, there is the difficulty of proving that the mycolaminarins, which are cytoplasmic constituents, reach plant cell membranes appreciably before elicitors from fungal cell walls. The suppressive mycolaminarins, but not elicitors, apparently are released into germination fluids of cystospores (91). However, intercellular metabolites commonly leak for a few hours during spore germination (347). Thus, there still is the problem of proving the continued secretion of laminarins by hyphae and haustoria.

Extracellular microbial metabolites also may suppress resistance. *Pseudomonas phaseolicola* produces the extracellular metabolite phaseolotoxin, which produces chlorosis in plant tissue. Treatment of resistant bean cultivars with phaseolotoxin before inoculation suppressed the necrogenic resistance response concomitant with reduced phytoalexin production (131). Many other microbial metabolites also are known to cause chlorosis (31), but their roles in resistance have not been studied.

Elicitors

The disease resistance responses of plants may be "triggered" by a variety of physical, chemical, biotic, and even genetic elicitors. UV irradiation, chilling, and drought stress (100) are among the physical elicitors. Heavy metal ions, particularly Cu^{2+} and Hg^{2+} (150), are extremely good chemical elicitors. Other chemical elicitors include organic amines, antibiotics, fungicides, dyes, respiratory inhibitors, polycations, RNA, DNA, basic proteins, and surfactants (163, 170). Cytoplasmic constituents of fungi (254, 268, 271) and bacteria (49), lipopolysaccharides from the bacterial outer wall membrane (135, 263), or rigorously extracted wall components from fungi (16, 92, 371) are nonspecific elicitors. Likewise, polysaccharides (11), protein-lipopolysaccharides (427), glycoproteins (87, 89, 96, 235, 358), and hydrolytic enzymes (89, 170) purified from culture filtrates may elicit defense responses. Genetic lethal responses in cotton (247), genetic dwarfing in pea (140), and genetic tumors in tobacco (339) are accompanied by "spontaneous" induction of defense responses. Likewise, plant extracts or diffusates may act as elicitors even in the host of origin (152). The common feature among all elicitors, at effective doses, appears to be that they cause chronic injury of protoplast membranes, resulting in slow leakage of cytoplasmic ions and molecules (96, 131, 150, 283, 295, 383, 407). Higher doses of elicitors or phytotoxins may cause protoplast disruption too quickly for completion of the energy-dependent postinfectional responses.

Heteropolymers on the outer walls of pathogens may specifically elicit resistance responses in resistant cultivars. Conidia of *Verticillium dahliae* killed with mild hot water treatments elicit resistance responses in xylem vessels of highly resistant *Gossypium barbadense* but not in those of susceptible or low resistant cultivars of *G. hirsutum* (36, 38). This specificity is not shown by crude culture filtrates, cell homogenates, or purified cell walls, indicating that specificity resides with the outer surface of the intact cell wall. Both heat-killed cells and the extracellular polysaccharide of *Xanthomonas phaseoli* also induce greater resistance responses in resistant than susceptible cotton cultivars (38, 101).

An elicitor was produced by or released from cells of *Phytophthora megasperma* var *sojae* race 1 at an accelerated rate when the fungus was placed in contact with tissue, or in filtered juice, of the resistant 'Harosoy 63' soybean for 8 hr; no elicitor, however, accumulated when the susceptible cultivar or its juice was used (111). A specific elicitor for 'Harosoy 63' was subsequently isolated and shown to occur in race 1 but not race 3 which is virulent to 'Harosoy 63' (208). The elicitor apparently is a surface glycoprotein. Tightly bound specific elicitors of resistance in soybean cultivars also were extracted from the cellular envelopes of *Pseudomonas glycinea* (49). These materials could not be extracted by the usual methods used for extracellular polysaccharides or lipopolysaccharides, but because of incomplete characterization, their identity or location in the cell wall is uncertain.

Specific elicitation of membrane leakage in resistant tomato leaves by high molecular weight products from culture filtrates of races of *Cladosporium fulvum* has also been reported (383). However, subsequent studies (87, 96, 235) failed to confirm this report. Rather, several glycoproteins that elicited electrolyte leakage in all cultivars were isolated from the filtrates (232). These glycoprotein elicitors quickly bound to isolated cells, but not isolated cell walls, suggesting that their site of action was the plasma membrane (96).

Phytotoxins As Hyperelicitors

Many necrotrophic pathogens secrete diffusable phytotoxins that cause necrosis and probably activate many of the resistance responses well in advance of the pathogen. Such phytotoxins can be regarded as hyperelicitors that activate and allow dissipation of resistance responses in advance of the fungus. In support of this concept, plants and tissues vulnerable to host-specific toxins are not known to contain or form persistant antibiotics, although they do form labile antibiotics. Phytotoxins that are host-specific also have the following characteristics in common with other elicitors and necrogenic resistance: 1. host sensitivity to give the necrogenic response in each case is usually related to single dominant genes (79, 277, 324); 2. heat shock (40–50°C) prior to exposure negates the necrogenic response for a few days until normal sensitivity is recovered (46, 364); 3. pretreatment with cycloheximide may block the necrogenic response (305); 4. rapid loss of membrane potential (283) and rapid increases in oxygen uptake (28, 308) occur concurrent with necrosis; 5. protoplast membranes leak electrolytes prior to necrosis (206, 222); *H. maydis* toxin initially affects mitochondria (292), but protoplast membrane distortion and leakage soon follow (206, 294); 6. ethylene evolution is enhanced in association with necrosis (410); and 7. ultrastructural effects on the host initially involve invagination and rupture of the plasmalemma and later disruption of all cell membranes (289, 408). Several glycopeptide phytotoxins already are known to elicit phytoalexin synthesis, and many wilt phytotoxins elicit melanization in xylem vessels, which characteristically is associated with multiple resistance responses (34). The effects of other phytotoxins on resistance responses need to be studied.

One difference between what we now call elicitors and host-specific phytotoxins might be the slightly greater potency of the phytotoxins. However, the *Rhizopus* elicitor causes necrogenic resistance at 2×10^{-8}M (358), which exceeds the activity threshold of many phytotoxins for inducing necrosis. Some phytotoxins are smaller molecules than microbial elicitors and thus diffuse more readily in tissue. For example, helminthosporoside is apparently a sesquiterpenoid digalactoside and rynchosporoside is a mixture of mono-, di-, and triglucosides of propanediol (29). The small size may allow the phytotoxins to diffuse and trigger defense reactions extensively before contact with the pathogen. The host might then overreact in the production of lignins, melanins, and phytoalexins to its own detriment. Spontaneous defense responses are associated with genetic lethality in middle-aged hybrid cotton plants (247). Genetic stunting of peas is also associated with relatively low levels (50 μg/g) of spontaneous pisatin production in the mutant plants (140). Thus some disease symptoms induced by necrotrophic pathogens may be the result of hyperactive resistance responses, rather than direct killing of the host by the pathogen.

The binding of phytotoxins and elicitors to plant cells also appears to be similar. Partially purified helminthosporoside from *Helminthosporium sacchari* apparently binds more readily and in greater quantities to plasma membranes of susceptible compared to resistant sugar cane cultivars (363). Binding is intermediate in cultivars with intermediate levels of resistance. The binding and toxic activities are blocked by methyl α-D galactoside and di- and trisaccharide sugars containing α-linked galactose. These results suggest that galactan-binding lectins on the plasmalemma might be the binding site. Lectin binding also may be necessary for necrogenic resistance. Lectins from potato agglutinate avirulent mutants but not virulent strains of *Pseudomonas solanacearum,* and binding appears essential as an initial step for the resistance response (334). Potato lectin also agglutinates germinated cystospores of *P. infestans;* virulent and avirulent races are equally affected (121). However, N,N'-diacetylcitobiose (the hapten of potato lectin) prevents the agglutination and inhibits necrogenic resistance (cited in 121). Thus, the effects of lectin binding on membrane permeability, rather than simply the presence or absence of lectin binding, may be critical to determining necrogenic responses to elicitors or phytotoxins. Differential binding to membranes, however, does not necessarily involve only lectins. Interactions with other proteins or lipid bilayers might also be important.

REFLECTIONS

Tremendous progress has been made over the past 10 years in our understanding of the biochemistry of disease resistance. Many of the natural products of plants, once regarded as "secondary," are obviously *essential* to the plants' abilities to resist diseases, pests, and environmental adversities.

The extreme diversity in the structures of the natural plant antibiotics suggests that biochemical geneticists may be able to regulate the structures of antibiotics through interspecific hybridization and genetic engineering. For example, only a few gene changes should be necessary to have a pea plant that makes the more toxic phaseollin or glyceollin instead of pisatin. Obviously, changes in membrane systems or detoxification systems may also be required, but the idea appears feasible. In fact, we have already been able to make two such changes in the terpenoid biosynthetic pathway of cotton by interspecific hybridization (38).

It also appears feasible to change the host-parasite interactions with the intervention of nontoxic chemicals. The chemical enhancement of natural resistance (melanization and phytoalexin synthesis) in rice to *Piricularia oryzae* by the cyclopropanes indicates the feasibility of this approach (57).

The bases of plant recognition of pathogens continue to baffle us, but evidence increasingly points to the interaction of wall components or exuded chemicals of the pathogen with the wall and plasmalemma of the host as being critical. Much more needs to be known about lectins and other membrane proteins that interact with elicitors and suppressors. At the same time we must not ignore the lipid portion of the membrane. Some elicitors may change the conformational structure of the lipid bilayer in the same manner as drought (347) and in this way excite defense responses.

Resistance depends on many biochemical components. Elimination of a single component may weaken but will not necessarily destroy resistance. Accordingly, some essential components will occur similarly in both susceptible and resistant plants. In order to avoid mistaken conclusions, we should not dwell on one or two differences between susceptible and resistant plants to the exclusion of other defense components. Future studies should regard resistance as a multicomponent dynamic system, paying particular attention to time sequences and cellular specialization.

Literature Cited

1. Abeles, F. B., Bosshart, R. P., Forrence, L. E., Habig, W. H. 1970. Preparation and purification of glucanase and chitinase from bean leaves. *Plant Physiol.* 47:129–34
2. Abo-El-Dahab, M. K., El-Goorani, M. A. 1972. Catalase activity of virulent and avirulent strains of *Pseudomonas solanacearum*. *Phytopathology* 62:294–95
3. Addy, S. K. 1976. Leakage of electrolytes and phenols from apple leaves caused by virulent and avirulent strains of *Erwinia amylovora*. *Phytopathology* 66:1403–5
4. Aist, J. R. 1976. Papillae and related wound plugs of plant cells. *Ann. Rev. Phytopathol.* 14:145–63
5. Aist, J. R., Israel, H. W. 1977. Effects of heat-shock inhibition of papilla formation on compatible host penetration by two obligate parasites. *Physiol. Plant Pathol.* 10:13–20
6. Aist, J. R., Kunoh, H., Israel, H. W. 1979. Challenge appressoria of *Erysiphe graminis* fail to breach preformed papillae of compatible barley cultivar. *Phytopathology* 69:1245–50
7. Alam, M. M., Masood, A., Husain, S. I. 1975. Effect of margosa and marigold root-exudates on mortality and larval hatch of certain nematodes. *Indian J. Exp. Biol.* 13:412–14
8. Albersheim, P., Valent, B. S. 1978. Host-pathogen interactions in plants. *J. Cell Biol.* 78:627–43
9. Allen, E. H., Thomas, C. A. 1971. A second antifungal polyacetylene compound from *Phytophthora*-infected safflower. *Phytopathology* 61:1107–9
10. Allison, A. V., Shalla, T. A. 1974. The ultrastructure of local lesions induced by potato virus X: A sequence of cytological events in the course of infection. *Phytopathology* 64:784–93
11. Anderson, A. J. 1978. Isolation from three species of *Colletotrichum* of glucan-containing polysaccharides that elicit browning and phytoalexin production in bean. *Phytopathology* 68: 189–94
12. Appiano, A., Pennazio, S., D'Agostino, G., Redolfi, P. 1977. Fine structure of necrotic local lesions induced by tomato bushy stunt virus in *Gomphrena globosa* leaves. *Physiol. Plant Pathol.* 11:327–32
13. Arinze, A. E., Smith, I. M. 1979. Production of a polygalacturonase complex by *Botryodiplodia theobromae* and its involvement in the rot of sweet potato. *Physiol. Plant Pathol.* 14:141–52
14. Arrigoni, O., Zacheo, G., Arrigoni-Liso, R., Gleve-Zacheo, T., Lamberti, F. 1979. Relationship between ascorbic acid and resistance in tomato plants to *Meloidogyne incognita*. *Phytopathology* 69:579–81
15. Asada, Y., Ohguchi, T., Matsumoto, I. 1979. Induction of lignification in response to fungal infection. See Ref. 78, pp. 99–112
16. Ayers, A. R., Ebel, J., Valent, B., Albersheim, P. 1976. Host-pathogen interactions. X. Fractionation and biological activity of an elicitor isolated from the mycelial walls of *Phytophthora megasperma* var. *sojae*. *Plant Physiol.* 57:760–65
17. Bailey, J. A. 1969. Phytoalexin production by leaves of *Pisum sativum* in relation to senescence. *Ann. Appl. Biol.* 64:315–24
18. Bailey, J. A., Burden, R. S., Mynett, A., Brown, C. 1977. Metabolism of phaseollin by *Septoria nodorum* and other non-pathogens of *Phaseolus vulgaris*. *Phytochemistry* 16:1541–44
19. Bailey, J. A., Carter, G. A., Burden, R. S., Wain, R. L. 1975. Control of rust diseases by diterpenes from *Nicotiana glutinosa*. *Nature* 225:328–29
20. Bailey, J. A., Vincent, G. G., Burden, R. S. 1976. The antifungal activity of glutinosone and capsidiol and their accumulation in virus-infected tobacco species. *Physiol. Plant Pathol.* 8:35–41
21. Balazs, E., Barna, B., Kiraly, Z. 1976. Effect of kinetin on lesion development and infection sites in Xanthi-nc tobacco infected by TMV: single-cell local lesions. *Acta Phytopathol. Acad. Sci. Hung.* 11:1–9
22. Barnett, A., Wood, K. R. 1978. Influence of actinomycin D, ethephon, cycloheximide and chloramphenicol on the infection of a resistant and a susceptible cucumber cultivar with cucumber mosaic virus (Price's No. 6 strain). *Physiol. Plant Pathol.* 12:257–77
23. Barnett, N. M. 1973. Release of peroxidase from soybean hypocotyl cell walls by *Sclerotium rolfsii* culture filtrates. *Can. J. Bot.* 52:265–71
24. Batra, G. K., Kuhn, C. W. 1975. Polyphenoloxidase and peroxidase activities associated with acquired resistance and its inhibition by 2-thiouracil in virus-infected soybean. *Physiol. Plant Pathol.* 5:239–48
25. Beachy, R. N., Murakishi, H. H. 1971. Local lesion formation in tobacco tissue culture. *Phytopathology* 61:877–78

26. Beckman, C. H., Elgersma, D. M., MacHardy, W. E. 1972. The localization of fusarial infections in the vascular tissue of single dominant-gene resistant tomatoes. *Phytopathology* 62:1256–60
27. Beckman, C. H., Mueller, W. C., Mace, M. E. 1974. The stabilization of artificial and natural cell wall membranes by phenolic infusion and its relation to wilt disease resistance. *Phytopathology* 64:1214–20
28. Bednarski, M. A., Scheffer, R. P., Izawa, S. 1977. Effects of toxin from *Helminthosporium maydis* T on respiration and associated activities in maize tissue. *Physiol. Plant Pathol.* 11:129–41
29. Beier, R. C. 1980. *Carbohydrate chemistry. Synthetic and structural investigation of the phytotoxins found in Helminthosporium sacchari, and Rhynchosporium secalis.* PhD thesis. Montana State Univ., Bozeman. 335 pp.
30. Bell, A. A. 1974. Biochemical bases of resistance of plants to pathogens. In *Proceedings of the Summer Institute on Biological Control of Plant Insects and Diseases,* ed. F. G. Maxwell, F. A. Harris, pp. 403–62. Jackson: Univ. Press Miss. 647 pp.
31. Bell, A. A. 1977. Plant pathology as influenced by allelopathy. In *The Role of Secondary Compounds in Plant Interactions (Allelopathy),* pp. 64–99. U.S. Dept. Agric., Agric. Res. Serv., Miss. State
32. Bell, A. A. 1980. Plant pest interaction with environmental stress and breeding for pest resistance: plant diseases. In *Breeding Plants for Marginal Environments.* New York: Wiley. In press
33. Bell, A. A. 1980. The time sequence of defense. See Ref. 171, pp. 53–73
34. Bell, A. A., Mace, M. E. 1980. Biochemistry and physiology of resistance. In *Fungal Wilt Diseases of Plants,* ed. M. E. Mace, A. A. Bell, C. H. Beckman. New York: Academic. In press
35. Bell, A. A., Presley, J. T. 1969. Temperature effects upon resistance and phytoalexin synthesis in cotton inoculated with *Verticillium albo-atrum. Phytopathology* 59:1141–46
36. Bell, A. A., Presley, J. T. 1969. Heat-inhibited or heat-killed conidia of *Verticillium albo-atrum* induce disease resistance and phytoalexin synthesis in cotton. *Phytopathology* 59:1147–51
37. Bell, A. A., Puhalla, J. E., Tolmsoff, W. J., Stipanovic, R. D. 1976. Use of mutants to establish (+)-scytalone as an intermediate in melanin biosynthesis by *Verticillium dahliae. Can. J. Microbiol.* 22:787–99
38. Bell, A. A., Stipanovic, R. D. 1978. Biochemistry of disease and pest resistance in cotton. *Mycopathologia* 65:91–106
39. Bell, A. A., Stipanovic, R. D., Howell, C. R., Fryxell, P. A. 1975. Antimicrobial terpenoids of *Gossypium:* hemigossypol, 6-methoxyhemigossypol and 6-deoxyhemigossypol. *Phytochemistry* 14:225–31
40. Bell, A. A., Stipanovic, R. D., O'Brien, D. H., Fryxell, P. A. 1978. Sesquiterpenoid aldehyde quinones and derivatives in pigment glands of *Gossypium. Phytochemistry* 17:1297–1305
41. Benedict, W. G. 1972. Influence of light on peroxidase activity associated with resistance of tomato cultivars to *Septoria lycopersici. Can. J. Bot.* 50: 1931–36
42. Bergeson, G. B. 1975. The effect of *Meloidogyne incognita* on the resistance of four muskmelon varieties to *Fusarium* wilt. *Plant Dis. Rep.* 59:410–13
43. Berlin, J., Widholm, J. M. 1978. Metabolism of phenylalanine and tyrosine in tobacco cell lines resistant and sensitive to *p*-fluorophenylalanine. *Phytochemistry* 17:65–68
44. Bhullar, B. S., Bajaj, K. L., Bhatia, I. S. 1972. Studies on the phenols of resistant and susceptible varieties of chillies in relation to anthracnose disease. *Phytopathol. Z.* 75:236–40
45. Blakeman, J. P., Atkinson, P. 1979. Antimicrobial properties and possible role in host-pathogen interactions of parthenolide, a sesquiterpene lactone isolated from glands of *Chrysanthemum parthenium. Physiol. Plant Pathol.* 15:183–92
46. Bronson, C. R., Scheffer, R. P. 1977. Heat- and aging-induced tolerance of sorghum and oat tissues to host-selective toxins. *Phytopathology* 67:1232–38
47. Brown, G. E. 1978. Hypersensitive response of orange-colored Robinson tangerines to *Colletotrichum gloeosporioides* after ethylene treatment. *Phytopathology* 68:700–6
48. Brotzman, H. C., Calvert, O. H., White, J. A., Brown, M. F. 1975. Southern corn leaf blight: ultrastructure of host-pathogen association. *Physiol. Plant Pathol.* 7:209–11
49. Bruegger, B. B., Keen, N. T. 1979. Specific elicitors of glyceollin accumulation in the *Pseudomonas glycinea*-soybean host-parasite system. *Physiol. Plant Pathol.* 15:43–51

50. Brueske, C. H. 1980. Phenylalanine ammonia lyase activity in tomato roots infected and resistant to the root-knot nematode, *Meloidogyne incognita. Physiol. Plant Pathol.* 16:409–14
51. Burden, R. S., Bailey, J. A., Vincent, G. G. 1974. Metabolism of phaseollin by *Colletotrichum lindemuthianum. Phytochemistry* 13:1789–91
52. Cabanne, F., Scalla, R., Martin, C. 1971. Oxidase activities during the hypersensitive reaction of *Nicotiana xanthi* to tobacco mosaic virus. *J. Gen. Virol.* 11:119–22
53. Cain, R. O., Porter, A. E. A. 1979. Biosynthesis of the phytoalexin wyerone in *Vicia faba. Phytochemistry* 18:322–23
54. Callow, J. A. 1977. Recognition, resistance and the role of plant lectins in host-parasite interactions. *Adv. Bot. Res.* 4:1–49
55. Camm, E. L., Towers, G. H. N. 1973. Review article: phenylalanine ammonia lyase. *Phytochemistry* 12:961–73
56. Campbell, G. K., Deverall, B. J. 1980. The effects of light and a photosynthetic inhibitor on the expression of the *Lr20* gene for resistance to leaf rust in wheat. *Physiol. Plant Pathol.* 16:415–23
57. Cartwright, D., Langcake, P., Pryce, R. J., Leworthy, D. P., Ride, J. P. 1977. Chemical activation of host defense mechanisms as a basis for crop protection. *Nature* 267:511–13
58. Caruso, F. L., Kuc, J. 1979. Induced resistance of cucumber to anthracnose and angular leaf spot by *Pseudomonas lachrymans* and *Colletotrichum lagenarium. Physiol. Plant Pathol.* 14:191–201
59. Cason, E. T. Jr., Richardson, P. E., Essenberg, M. K., Brinkerhoff, L. A., Johnson, W. M., Venere, R. J. 1978. Ultrastructural cell wall alterations in immune cotton leaves inoculated with *Xanthomonas malvacearum. Phytopathology* 68:1015–21
60. Chamberlain, D. W. 1972. Heat-induced susceptibility to nonpathogens and cross-protection against *Phytophthora megasperma* var. *sojae* in soybean. *Phytopathology* 62:645–46
61. Cho, E. K., Goodman, R. M. 1979. Strains of soybean mosaic virus: classification based on virulence in resistant soybean cultivars. *Phytopathology* 69:467–70
62. Chopra, B. L., Jhooty, J. S., Bajaj, K. L. 1974. Biochemical differences between two varieties of watermelon resistant and susceptible to *Alternaria cucumerina. Phytopathol. Z.* 79:47–52
63. Christenson, J. A., Hadwiger, L. A. 1973. Induction of pisatin formation in the pea foot region by pathogenic and nonpathogenic clones of *Fusarium solani. Phytopathology* 63:784–90
64. Collins, W. W., Nielsen, L. W. 1976. *Fusarium* wilt resistance in sweet potatoes. *Phytopathology* 66:489–93
65. Cook, A. A. 1973. Characterization of hypersensitivity in *Capsicum annuum* induced by the tomato strain of *Xanthomonas vesicatoria. Phytopathology* 63:915–18
66. Cook, A. A. 1975. Effect of low concentrations of *Xanthomonas vesicatoria* infiltrated into pepper leaves. *Phytopathology* 65:487–89
67. Cook, A. A., Stall, R. E. 1977. Effects of watersoaking on response to *Xanthomonas vesicatoria* in pepper leaves. *Phytopathology* 67:1101–3
68. Corsisi, D. L., Pavek, J. J. 1980. Phenylalanine ammonia lyase activity and fungitoxic metabolites produced by potato cultivars in response to *Fusarium* tuber rot. *Physiol. Plant Pathol.* 16:63–72
69. Coutts, R. H. A., Barnett, A., Wood, K. R. 1978. Aspects of the resistance of cucumber plants and protoplasts to cucumber mosaic virus. *Ann. Appl. Biol.* 89:336–39
70. Coxon, D. T., O'Neill, T. M., Mansfield, J. W., Porter, A. E. A. 1980. Identification of three hydroxyflavan phytoalexins from daffodil bulbs. *Phytochemistry* 19:889–91
71. Cruickshank, I. A. M. 1963. Phytoalexins. *Ann. Rev. Phytopathol.* 1:351–74
72. Cruickshank, I. A. M. 1976. A review of the role of phytoalexins in disease resistance mechanisms. *Pontif. Accad. Sci.* Scripta Varia, "Natural Products and the Protection of Plants" 41:509–69
73. Cruickshank, I. A. M., Spencer, K., Mandryk, M. 1979. Nitrogen nutrition and the net accumulation of medicarpin in infection-droplets on excised leaflets of white clover. *Physiol. Plant Pathol.* 14:71–76
74. Cruickshank, I. A. M., Veeraraghavan, J., Perrin, D. R. 1974. Some physical factors affecting the formation and/or net accumulation of medicarpin in infection droplets on white clover leaflets. *Aust. J. Plant Physiol.* 1:149–56
75. Currier, W. W., Kuc, J. 1975. Effect of temperature on rishitin and steroid glycoalkaloid accumulation in potato tuber. *Phytopathology* 65:1194–97
76. DaGraca, J. V., Martin, M. M. 1975. Ultrastructural changes in tobacco

mosaic virus-induced local lesions in *Nicotiana tobacum* L. cv. "Samsun *NN*". *Physiol. Plant Pathol.* 7:287–91
77. DaGraca, J. V., Martin, M. M. 1976. An electron microscope study of hypersensitive tobacco infected with tobacco mosaic virus at 32°C. *Physiol. Plant Pathol.* 8:215–19
78. Daly, J. M., Uritani, I., eds. 1979. *Recognition and Specificity in Plant Host-Parasite Interactions.* Tokyo: Jpn. Sci. Soc. Press. 355 pp.
79. Day, P. R. 1974. *Genetics of Host-Parasite Interaction,* ed. A. Kelman, L. Sequeira. San Francisco: Freeman. 238 pp.
80. Delon, R. 1974. Localisation d'activites polyphenoloxydasiques et peroxydasiques dans les cellules racinaires de *Lycopersicon esculentum* parasitees par *Pyrenochaeta lycopersici. Phytopathol. Z.* 80:199–208
81. DeVay, J. E., Adler, H. E. 1976. Antigens common to hosts and parasites. *Ann. Rev. Microbiol.* 30:147–68
82. Deverall, B. J. 1977. *Defense Mechanisms of Plants.* Cambridge monogr. exp. biol. No. 19, Cambridge Univ. Press. 110 pp.
83. Dewick, P. M., Martin, M. 1979. Biosynthesis of pterocarpan and isoflavan phytoalexins in *Medicago sativa:* the biochemical interconversion of pterocarpans and 2'-hydroxyisoflavans. *Phytochemistry* 18:591–96
84. Dewick, P. M., Ward, D. 1978. Isoflavone precursors of the pterocarpan phytoalexin maackiain in *Trifolium pratense. Phytochemistry* 17:1751–54
85. de Wit, P. J. G. M. 1977. A light and scanning-electron microscopic study of infection of tomato plants by virulent and avirulent races of *Cladosporium fulvum. Neth. J. Plant Pathol.* 83:109–22
86. de Wit, P. J. G. M., Flach, W. 1979. Differential accumulation of phytoalexins in tomato leaves but not in fruits after inoculation with virulent and avirulent races of *Cladosporium fulvum. Physiol. Plant Pathol.* 15:257–67
87. de Wit, P. J. G. M., Roseboom, P. H. M. 1980. Isolation, partial characterization and specificity of glycoprotein elicitors from culture filtrates, mycelium and cell walls of *Cladosporium fulvum* (syn. *Fulvia fulva*). *Physiol. Plant Pathol.* 16:391–408
88. Dixon, R. A., Bendall, D. S. 1978. Changes in the levels of enzymes of phenylpropanoid and flavonoid synthesis during phaseollin production in cell suspension cultures of *Phaseolus vulgaris. Physiol. Plant Pathol.* 13:295–306
89. Dixon, R. A., Fuller, K. W. 1977. Characterization of components from culture filtrates of *Botrytis cinerea* which stimulate phaseollin biosynthesis in *Phaseolus vulgaris* cell suspension cultures. *Physiol. Plant Pathol.* 11:287–96
90. Dixon, R. A., Fuller, K. W. 1978. Effects of growth substances on non-induced and *Botrytis cinerea* culture filtrate-induced phaseollin production in *Phaseolus vulgaris* cell suspension cultures. *Physiol. Plant Pathol.* 12: 279–88
91. Doke, N., Garas, N. A., Kuc, J. 1980. Effect on host hypersensitivity of suppressors released during the germination of *Phytophthora infestans* cytospores. *Phytopathology* 70:35–39
92. Doke, N., Tomiyama, K. 1980. Effect of hyphal wall components from *Phytophthora infestans* on protoplasts of potato tuber tissues. *Physiol. Plant Pathol.* 16:169–76
93. Doke, N., Tomiyama, K. 1980. Suppression of the hypersensitive response of potato tuber protoplasts to hyphal wall components by water soluble glucans isolated from *Phytophthora infestans. Physiol. Plant Pathol.* 16:177–86
94. Doskotch, R. W., Keely, S. L. Jr., Schreiber, L. R. 1975. Isolation and identification of an antifungal agent from seeds of American elm. *Phytopathology* 65:634–35
95. Dow, J. M., Callow, J. A. 1978. A possible role for *a*-tomatine in the varietal-specific resistance of tomato to *Cladosporium fulvum. Phytopathol. Z.* 92: 211–16
96. Dow, J. M., Callow, J. A. 1979. Leakage of electrolytes from isolated leaf mesophyll cells of tomato induced by glycopeptides from culture filtrates of *Fulvia fulva* (Cooke) Ciferri (syn. *Cladosporium fulvum*). *Physiol. Plant Pathol.* 15:27–34
97. Dropkin, V. H. 1969. The necrotic reaction of tomatoes and other hosts resistant to *Meloidogyne:* reversal by temperature. *Phytopathology* 59:1632–37
98. Dropkin, V. H., Helgeson, J. P., Upper, C. D. 1969. The hypersensitivity reaction of tomatoes resistant to *Meloidogyne incognita*: reversal by cytokinins. *J. Nematol.* 1:55–61
99. Eddy, J. L., Derr, J. E., Hass, G. M. 1980. Chymotrypsin inhibitor from potatoes: interaction with target enzymes. *Phytochemistry* 19:757–61

100. Edreva, A. 1977. Comparative biochemical studies of an infectious disease (Blue mould) and a physiological disorder of tobacco. *Physiol. Plant Pathol.* 11:149–61
101. El-Banoby, F. E., Rudolph, K. 1979. Induction of water-soaking in plant leaves by extracellular polysaccharides from phytopathogenic pseudomonads and xanthomonads. *Physiol. Plant Pathol.* 15:341–49
102. El-Banoby, F. E., Rudolph, K. 1980. Purification of extracellular polysaccharides from *Pseudomonas phaseolicola* which induce water-soaking in bean leaves. *Physiol. Plant Pathol.* 16:425–37
103. Elgersma, D. M. 1973. Tylose formation in elms after inoculation with *Ceratocystis ulmi,* a possible resistance mechanism. *Neth. J. Plant Pathol.* 79: 218–20
104. Elnaghy, M. A., Heitefuss, R. 1976. Permeability changes and production of antifungal compounds in *Phaseolus vulgaris* infected with *Uromyces phaseoli.* II. Role of phytoalexins. *Physiol. Plant Pathol.* 8:269–77
105. Essenberg, M., Cason, E. T. Jr., Hamilton, B., Brinkerhoff, L. A., Gholson, R. K., Richardson, P. E. 1979. Single cell colonies of *Xanthomonas malvacearum* in susceptible and immune cotton leaves and the local resistant response to colonies in immune leaves. *Physiol. Plant Pathol.* 15:53–68
106. Essenberg, M., Hamilton, B., Cason, E. T. Jr., Brinkerhoff, L. A., Gholson, R. K., Richardson, P. E. 1979. Localized bacteriostasis indicated by water dispersal of colonies of *Xanthomonas malvacearum* within immune cotton leaves. *Physiol. Plant Pathol.* 14:69–78
107. Evans, R. L., Pluck, D. J. 1978. Phytoalexins produced in barley in response to the halo spot fungus, *Selenophoma donacis. Ann. Appl. Biol.* 89:332–36
108. Favali, M. A., Conti, G. G., Bassi, M. 1978. Modifications of the vascular bundle untrastructure in the "resistant zone" around necrotic lesions induced by tobacco mosaic virus. *Physiol. Plant Pathol.* 13:247–51
109. Fisch, M. H., Flick, B. H., Arditti, J. 1973. Structure and antifungal activity of hircinol, loroglossol and orchinol. *Phytochemistry* 12:437–41
110. Fraile, A., Garcia-Arenal, F., Sagasta, E. M. 1980. Phytoalexin accumulation in bean (*Phaseolus vulgaris*) after infection with *Botrytis cinerea* and treatment with mercuric chloride. *Physiol. Plant Pathol.* 16:9–18
111. Frank, J. A., Paxton, J. D. 1971. An inducer of soybean phytoalexin and its role in the resistance of soybeans to *Phytophthora* rot. *Phytopathology* 61: 954–58
112. Frank, J. A., Wilson, J. M., Webb, R. E. 1975. The relationship between glycoalkaloids and disease resistance in potatoes. *Phytopathology* 65:1045–49
113. Fraser, R. S. S. 1979. Systemic consequences of the local lesion reaction to tobacco mosaic virus in a tobacco variety lacking the *N* gene for hypersensitivity. *Physiol. Plant Pathol.* 14: 383–94
114. Friend, J. 1976. Lignification in infected tissue. See Ref. 116, pp. 291–304
115. Friend, J., Thornton, J. D. 1974. Caffeic acid-O-methyl transferase, phenolase and peroxidase in potato tuber tissue inoculated with *Phytophthora infestans. Phytopathol. Z.* 81:56–64
116. Friend, J., Threlfall, D. R., eds. 1976. *Biochemical Aspects of Plant-Parasite Relationships.* London/New York/San Francisco: Academic. 354 pp.
117. Fry, W. E., Millar, R. L. 1971. Cyanide tolerance in *Stemphylium loti. Phytopathology* 61:494–500
118. Fry, W. E., Millar, R. L. 1971. Development of cyanide tolerance in *Stemphylium loti. Phytopathology* 61:501–6
119. Fuchs, A., de Vries, F. W., Platero Sanz, M. 1980. The mechanism of pisatin degradation by *Fusarium oxysporum* f. sp. *pisi. Physiol. Plant Pathol.* 16:119–33
120. Furuichi, N., Tomiyama, K., Doke, N. 1979. Hypersensitive reactivity in potato: transition from inactive to active state induced by infection with an incompatible race of *Phytophthora infestans. Phytopathology* 69:734–36
121. Furuichi, N., Tomiyama, K., Doke, N. 1980. The role of potato lectin in the binding of germ tubes of *Phytophthora infestans* to potato cell membrane. *Physiol. Plant Pathol.* 16:249–56
122. Garas, N. A., Doke, N., Kuc, J. 1979. Suppression of the hypersensitive reaction in potato tubers by mycelial components from *Phytophthora infestans. Physiol. Plant Pathol.* 15:117–26
123. Garcia-Arenal, F., Sagasta, E. M. 1977. Callose deposition and phytoalexin accumulation in *Botrytis cinerea*-infected bean (*Phaseolus vulgaris*). *Plant Sci. Lett.* 10:305–12
124. Garrod, B., Lewis, B. G., Coxon, D. T. 1978. *Cis*-heptadeca-1,9-diene-4,6-diyne-3, 8-diol, an antifungal polyacety-

lene from carrot root tissue. *Physiol. Plant Pathol.* 13:241–46

125. Gatehouse, A. M. R., Gatehouse, J. A., Boulter, D. 1980. Isolation and characterization of trypsin inhibitors from cowpea (*Vigna unguiculata*). *Phytochemistry* 19:751–56
126. Gathercole, R. W. E., Street, H. E. 1978. A *p*-fluorophenylalanine–resistant cell line of sycamore with increased contents of phenylalanine, tyrosine and phenolics. *Z. Pflanzenphysiol.* 89: 283–87
127. Gianinazzi, S., Ahl, P., Cornu, A., Scalla, R., Cassini, R. 1980. First report of host b-protein appearance in response to a fungal infection in tobacco. *Physiol. Plant Pathol.* 16:337–42
128. Giebel, J., Krenz, J., Wilski, A. 1970. The formation of lignin-like substances in roots of resistant potatoes under the influence of *Heterodera rostochiensis* larvae. *Nematologica* 16:601
129. Glazener, J. A., VanEtten, H. D. 1978. Phytotoxicity of phaseollin to, and alteration of phaseollin by, cell suspension cultures of *Phaseolus vulgaris*. *Phytopathology* 68:111–17
130. Gnanamanickam, S. S., Patil, S. S. 1977. Accumulation of antibacterial isoflavonoids in hypersensitively responding bean leaf tissues inoculated with *Pseudomonas phaseolicola*. *Physiol. Plant Pathol.* 10:159–68
131. Gnanamanickam, S. S., Patil, S. S. 1977. Phaseotoxin suppresses bacterially induced hypersensitive reaction and phytoalexin synthesis in bean cultivars. *Physiol. Plant Pathol.* 10:169–79
132. Gommers, F. J., Voorin'tholt, D. J. M. 1976. Chemotaxonomy of Compositae related to their host suitability for *Pratylenchus penetrans*. *Neth. J. Plant Pathol.* 82:1–8
133. Goodliffe, J. P., Heale, J. B. 1978. The role of 6-methoxy mellein in the resistance and susceptibility of carrot root tissue to the cold-storage pathogen *Botrytis cinerea*. *Physiol. Plant Pathol.* 12:27–43
134. Goodman, R. N., Huang, P., White, J. A. 1976. Ultrastructural evidence for immobilization of an incompatible bacterium, *Pseudomonas pisi*, in tobacco leaf tissue. *Phytopathology* 66:754–64
135. Graham, T. L., Sequeira, L., Huang, T.-S. R. 1977. Bacterial lipopolysaccharides as inducers of disease resistance in tobacco. *Appl. Environ. Microbiol.* 34:424–32
136. Green, N. E., Hadwiger, L. A., Graham, S. O. 1975. Phenylalanine ammonia-lyase, tyrosine ammonia-lyase, and lignin in wheat inoculated with *Erysiphe graminis* f. sp. *tritici*. *Phytopathology* 65:1071–74
137. Griffin, G. D., Elgin, J. H. Jr. 1979. Penetration and development of *Meloidogyne hapla* in resistant and susceptible alfalfa under differing temperatures. *J. Nematol.* 9:51–56
138. Grisebach, H., Ebel, J. 1978. Phytoalexins, chemical defense substances of higher plants? *Angew. Chem. Int. Ed. Engl.* 17:635–47
139. Haard, N. F., Marshall, M. 1976. Isoperoxidase changes in soluble and particulate fractions of sweet potato root resulting from cut injury, ethylene and black rot infection. *Physiol. Plant Pathol.* 8:195–205
140. Hadwiger, L. A., Sander, C. Eddyvean, J., Ralston, J. 1976. Sodium azide-induced mutants of peas that accumulate pisatin. *Phytopathology* 66:629–30
141. Hadwiger, L. A., Schwochau, M. E. 1970. Induction of phenylalanine ammonia-lyase and pisatin in pea pods with polylysine, spermidine or histone fractions. *Biochem. Biophys. Res. Commun.* 38:683–88
142. Hammerschlag, F., Mace, M. E. 1975. Antifungal activity of extracts from *Fusarium* wilt-susceptible and -resistant tomato plants. *Phytopathology* 65: 93–94
143. Hammerschmidt, R., Nicholson, R. L. 1977. Resistance of maize to anthracnose: changes in host phenols and pigments. *Phytopathology* 67:251–58
144. Hammerschmidt, R., Nicholson, R. L. 1977. Resistance of maize to anthracnose: effect of light intensity on lesion development. *Phytopathology* 67: 247–50
145. Hanounik, S. B., Osborne, W. W. 1974. Influence of *Meloidogyne incognita* on the content of amino acids and nicotine in tobacco grown under gnotobiotic conditions. *J. Nematol.* 7:332–36
146. Harborne, J. B., Ingham, J. L. 1978. Biochemical aspects of the coevolution of higher plants with their fungal parasites, pp. 343–405. In *Biochemical Aspects of Plant and Animal Coevolution*, ed. J. B. Harborne. London: Academic
147. Harborne, J. B., Ingham, J. L., King, L., Payne, M. 1976. The isopentenyl isoflavone luteone as a pre-infectional antifungal agent in the genus *Lupinus*. *Phytochemistry* 15:1485–87
148. Harder, D. E., Rohringer, R., Samborski, D. J., Rimmer, S. R., Kim, W. K., Chong, J. 1979. Electron microscopy of

susceptible and resistant near-isogenic (*sr6/Sr6*) lines of wheat infected by *Puccinia graminis tritici*. II. Expression of incompatibility in mesophyll and epidermal cells and the effect of temperature on host-parasite interactions in these cells. *Can. J. Bot.* 57:2617–25
149. Harder, D. E., Samborski, D. J., Rohringer, R., Rimmer, S. R., Kim, W. K., Chong, J. 1979. Electron microscopy of susceptible and resistant near-isogenic (*sr6/Sr6*) lines of wheat infected by *Puccinia graminis tritici*. III. Ultrastructure of incompatible interactions. *Can. J. Bot.* 57:2626–34
150. Hargreaves, J. A. 1979. Investigations into the mechanism of mercuric chloride stimulated phytoalexin accumulation in *Phaseolus vulgaris* and *Pisum sativum*. *Physiol. Plant Pathol.* 15: 279–87
151. Hargreaves, J. A. 1980. A possible mechanism for the phytotoxicity of the phytoalexin phaseollin. *Physiol. Plant Pathol.* 16:351–57
152. Hargreaves, J. A., Bailey, J. A. 1978. Phytoalexin production by hypocotyls of *Phaseolus vulgaris* in response to constitutive metabolites released by damaged bean cells. *Physiol. Plant Pathol.* 13:89–100
153. Hargreaves, J. A., Mansfield, J. W., Coxon, D. T., Price, K. R. 1976. Wyerone epoxide as a phytoalexin in *Vicia faba* and its metabolism by *Botrytis cinerea* and *B. fabae in vitro*. *Phytochemistry* 15:1119–21
154. Hart, J. H., Hillis, W. E. 1972. Inhibition of wood-rotting fungi by ellagitannins in the heartwood of *Quercus alba*. *Phytopathology* 62:620–26
155. Hart, J. H., Hillis, W. E. 1974. Inhibition of wood-rotting fungi by stilbenes and other polyphenols in *Eucalyptus sideroxylon*. *Phytopathology* 64:939–48
156. Heale, J. B., Sharman, S. 1977. Induced resistance to *Botrytis cinerea* in root slices and tissue cultures of carrot (*Daucus carota* L.). *Physiol. Plant Pathol.* 10:51–61
157. Heath, M. C. 1979. Partial characterization of the electron-opaque deposits formed in the non-host plant, French bean, after cowpea rust infection. *Physiol. Plant Pathol.* 15:141–48
158. Heath, M. C. 1979. Effects of heat shock, actinomycin D, cycloheximide and blasticidin S on nonhost interactions with rust fungi. *Physiol. Plant Pathol.* 15:211–18
159. Hecker, R. J., Ruppel, E. G., Maag, G. W., Rasmuson, D. M. 1975. Amino acids associated with *Cercospora* leaf spot resistance in sugarbeet. *Phytopathol. Z.* 82:175–81
160. Hedin, P. A., ed. 1977. *Host Plant Resistance to Pests,* A.C.S. Symp. Ser. 62, Am. Chem. Soc., Washington DC. 286 pp.
161. Heitefuss, R., Williams, P. H., eds. 1976. *Physiological Plant Pathology. Encyclopedia of Plant Physiology, Vol. 4,* ed. A. Pirson, M. H. Zimmermann. Berlin: Springer-Verlag. 890 pp.
162. Hendrix, J. W. 1972. Temperature-dependent resistance to tobacco ringspot virus in L8, a necrosis-prone tobacco cultivar. *Phytopathology* 62: 1376–81
163. Hess, S. L., Hadwiger, L. A. 1971. The induction of phenylalanine ammonia-lyase and phaseollin by 9-aminoacridine and other DNA-intercalating compounds. *Plant Physiol.* 48:197–202
164. Higgins, V. J., Stoessl, A., Heath, M. C. 1974. Conversion of phaseollin to phaseollinisoflavan by *Stemphylium botryosum*. *Phytopathology* 64:105–7
165. Hildebrand, D. C. 1973. Tolerance of homoserine by *Pseudomonas pisi* and implications of homoserine in plant resistance. *Phytopathology* 63:301–2
166. Hoffman, S. E., Zscheile, F. P. Jr. 1973. Leaf bioassay for *Helminthosporium carbonum* toxin-search for phytoalexin. *Phytopathology* 63:729–34
167. Hohl, H. R., Stossel, P. 1976. Host-parasite interfaces in a resistant and susceptible cultivar of *Solanum tuberosum* inoculated with *Phytophthora infestans:* tuber tissue. *Can. J. Bot.* 54:900–12
168. Holliday, M. J., Klarman, W. L. 1979. Expression of disease reaction types in soybean callus from resistant and susceptible plants. *Phytopathology* 69: 576–78
169. Hooley, R., McCarthy, D. 1980. Extracts from virus infected hypersensitive tobacco leaves are detrimental to protoplast survival. *Physiol. Plant Pathol.* 16:25–38
170. Hopper, D. G., Venere, R. J., Brinkerhoff, L. A., Gholson, R. K. 1975. Necrosis induction in cotton. *Phytopathology* 65:206–13
171. Horsfall, J. G., Cowling, E. B. 1980. *Plant Disease: An Advanced Treatise. Vol. V. How Plants Defend Themselves.* New York: Academic. 534 pp.
172. Howell, C. R. 1976. Use of enzyme-deficient mutants of *Verticillium dahliae* to assess the importance of pectolytic enzymes in symptom expression of *Ver-*

ticillium wilt of cotton. *Physiol. Plant Pathol.* 9:279–83

173. Howell, C. R., Bell, A. A., Stipanovic, R. D. 1976. Effect of aging on flavonoid content and resistance of cotton leaves to *Verticillium* wilt. *Physiol. Plant Pathol.* 8:181–88
174. Hsu, S.-T., Dickey, R. S. 1972. Interaction between *Xanthomonas phaseoli, Xanthomonas vesicatoria, Xanthomonas campestris,* and *Pseudomonas fluorescens* in bean and tomato leaves. *Phytopathology* 62:1120–26
175. Huang, J., Van Dyke, C. G. 1978. Interaction of tobacco callus tissue with *Pseudomonas tabaci, P. pisi* and *P. fluroescens. Physiol. Plant Pathol.* 13: 65–72
176. Hung, C.-L., Rohde, R. A. 1973. Phenol accumulation related to resistance in tomato to infection by root-knot and lesion nematodes. *J. Nematol.* 5:253–58
177. Hunter, R. E. 1978. Effects of catechin in culture and in cotton seedlings on the growth and polygalacturonase activity of *Rhizoctonia solani. Phytopathology* 68:1032–36
178. Hunter, R. E., Halloin, J. M., Veech, J. A., Carter, W. W. 1978. Exudation of terpenoids by cotton roots. *Plant Soil* 50:237–40
179. Hunter, R. E., Halloin, J. M., Veech, J. A., Carter, W. W. 1978. Terpenoid accumulation in hypocotyls of cotton seedlings during aging and after infection by *Rhizoctonia solani. Phytopathology* 68:347–50
180. Ingham, J. L. 1976. Fungal modification of pterocarpan phytoalexins from *Melilotus alba* and *Trifolium pratense. Phytochemistry* 15:1489–95
181. Ingham, J. L. 1979. Isoflavonoid phytoalexins of the genus *Medicago. Biochem. Syst. Ecol.* 7:29–34
182. Ingham, J. L., Harborne, J. B. 1976. Phytoalexin induction as a new dynamic approach to the study of systematic relationships among higher plants. *Nature* 260:241–43
183. Innes, N. L. 1974. Resistance to bacterial blight of cotton varieties homozygous for combinations of B resistance genes. *Ann. Appl. Biol.* 78:89–98
184. International Potash Institute. 1976. *Fertilizer Use and Plant Health, Proc. 12th Colloq., Int. Potash Inst., Izmir, Turkey.* Worblaufen-Bern, Switzerland: Int. Potash Inst. 330 pp.
185. Ireland, R. J., Pierpoint, W. S. 1980. Reaction of strains of potato virus X with chlorogenoquinone in vitro, and the detection of a naturally modified form of the virus. *Physiol. Plant Pathol.* 16:81–92
186. Irvine, J. A., Dix, N. J., Warren, R. C. 1978. Inhibitory substances in *Acer platanoides* leaves: seasonal activity and effects on growth of phylloplane fungi. *Trans. Br. Mycol. Soc.* 70:363–71
187. Jatala, P., Russell, C. C. 1972. Nature of sweet potato resistance to *Meloidogyne incognita* and the effects of temperatures on parasitism. *J. Nematol.* 4:1–7
188. Jena, R. N., Rao, Y. S. 1977. Nature of resistance in rice (*Oryza sativa* L.) to the root-knot nematode (*Meloidogyne graminicola* Golden and Birchfield) II. Histopathology of nematode infection in rice varieties. *Proc. Indian Acad. Sci.* 86:87–91
189. Jenns, A. E., Kuc, J. 1979. Graft transmission of systemic resistance of cucumber to anthracnose induced by *Colletotrichum lagenarium* and tobacco necrosis virus. *Phytopathology* 69: 753–56
190. Johnson, C., Brannon, D. R., Kuc, J. 1973. Xanthotoxin: a phytoalexin of *Pastinaca sativa* root. *Phytochemistry* 12:2961–62
191. Johnson, G., Maag, D. D., Johnson, D. K., Thomas, R. D. 1976. The possible role of phytoalexins in the resistance of sugarbeet (*Beta vulgaris*) to *Cercospora beticola. Physiol. Plant Pathol.* 8:225–30
192. Johnson, L. B. 1970. Symptom development and resistance in safflower hypocotyls to *Phytophthora drechsleri. Phytopathology* 60:534–37
193. Johnson, L. B., Lee, R. F. 1978. Peroxidase changes in wheat isolines with compatible and incompatible leaf rust infections. *Physiol. Plant Pathol.* 13: 173–81
194. Johnson, R. 1978. Induced resistance to fungal diseases with special reference to yellow rust of wheat. *Ann. Appl. Biol.* 89:107–10
195. Jones, D. R., Unwin, C. H., Ward, E. W. B. 1975. The significance of capsidiol induction in pepper fruit during an incompatible interaction with *Phytophthora infestans. Phytopathology* 65: 1268–88
196. Jones, I. T. 1975. The preconditioning effect of day-length and light intensity on adult plant resistance to powdery mildew in oats. *Ann. Appl. Biol.* 80: 301–9
197. Jones, I. T., Hayes, J. D. 1971. The effect of sowing date on adult plant resistance to *Erysiphe graminis* f. sp.

avenae in oats. *Ann. Appl. Biol.* 68: 31–39

198. Jorgensen, J. H., Mortensen, K. 1977. Primary infection by *Erysiphe graminis* f. sp. *hordei* of barley mutants with resistance genes in the ml-o locus. *Phytopathology* 67:678–85
199. Kahl, G. 1974. Metabolism in plant storage tissue slices. *Bot. Rev.* 40:263–314
200. Kamel, M. Y., Saleh, N. A., Ghazy, A. M. 1977. Gallic acid oxidation by turnip peroxidase. *Phytochemistry* 16: 521–24
201. Kaplan, D. T., Keen, N. T., Thomason, I. J. 1980. Association of glyceollin with the incompatible response of soybean roots to *Meloidogyne incognita. Physiol. Plant Pathol.* 16:309–18
202. Kaplan, D. T., Keen, N. T., Thomason, I. J. 1980. Studies on the mode of action of glyceollin in soybean incompatibility to the root knot nematode, *Meloidogyne incognita. Physiol. Plant Pathol.* 16: 319–25
203. Kasamo, K., Shimomura, T. 1977. The role of the epidermis in local lesion formation and the multiplication of tobacco mosaic virus and its relation to kinetin. *Virology* 76:12–18
204. Kassanis, B., Gianinazzi, S., White, R. F. 1974. A possible explanation of the resistance of virus-infected tobacco plants to second infection. *J. Gen. Virol.* 23:11–16
205. Kato, N., Imaseki, H., Nakashima, N., Akazawa, T., Uritani, I. 1973. Isolation of a new phytoalexin-like compound, ipomeamaronol, from black-rot fungus infected sweet potato root tissue, and its structural elucidation. *Plant Cell Physiol.* 14:597–606
206. Keck, R. W., Hodges, T. K. 1973. Membrane permeability in plants: changes induced by host-specific pathotoxins. *Phytopathology* 63:226–30
207. Keeling, B. L., Banttari, E. E. 1975. Factors associated with the resistance of barley to *Helminthosporium teres. Phytopathology* 65:464–67
208. Keen, N. T. 1975. Specific elicitors of plant phytoalexin production: determinants of race specificity in pathogens? *Science* 187:74–75
209. Keen, N. T., Littlefield, L. J. 1979. The possible association of phytoalexins with resistance gene expression in flax to *Melampsora lini. Physiol. Plant Pathol.* 14:265–80
210. Keen, N. T., Wang, M. C., Bartnicki-Garcia, S., Zentmyer, G. A. 1975. Phytotoxicity of mycolaminarans—β-1,3-glucans from *Phytophthora* spp. *Physiol. Plant Pathol.* 7:91–97
211. Khan, F. Z., Milton, J. M. 1975. Phytoalexin production by lucerne (*Medicago sativa* L.) in response to infection by *Verticillium. Physiol. Plant Pathol.* 7:179–87
212. Khan, F. Z., Milton, J. M. 1978. Phytoalexin production and the resistance of lucerne (*Medicago sativa* L.) to *Verticillium albo-atrum. Physiol. Plant Pathol.* 13:215–21
213. Kimmins, W. C., Brown, R. G. 1973. Hypersensitive resistance. The role of cell wall glycoproteins in virus localization. *Can J. Bot.* 51:1923–26
214. Kimmins, W. C., Wuddah, D. 1977. Hypersensitive resistance: determination of lignin in leaves with a localized virus infection. *Phytopathology* 67: 1012–16
215. Kiraly, Z., Barna, B., Ersek, T. 1972. Hypersensitivity as a consequence, not the cause, of plant resistance to infection. *Nature* 239:456–58
216. Klarman, W. L., Hammerschlag, F. 1972. Production of the phytoalexin, hydroxyphaseollin, in soybean leaves inoculated with tobacco necrosis virus. *Phytopathology* 62:719–21
217. Kluge, S., Paunow, S., Schuster, G. 1977. On the action of some metabolically active substances on the protein content and the multiplication of viruses in leaves of *Nicotiana tabacum* L. *Phytopathol. Z.* 88:11–17
218. Koenig, R., Tremaine, J. H., Shepard, J. F. 1978. In situ degradation of the protein chain of potato virus X at the N- and C-termini. *J. Gen. Virol.* 38:329–37
219. Koga, H., Mayama, S., Shishiyama, J. 1980. Correlation between the deposition of fluorescent compounds in papillae and resistance in barley against *Erysiphe graminis hordei. Can. J. Bot.* 58:536–41
220. Kogiso, S., Wada, K., Munakata, K. 1976. Isolation of nematicidal polyacetylenes from *Carthamus tinctorius* L. *Agric. Biol. Chem.* 40:2085–89
221. Kogiso, S., Wada, K., Munakata, K. 1976. Odoracin, a nematicidal constituent from *Daphne odora. Agric. Biol. Chem.* 40:2119–20
222. Kohmoto, K., Khan, I. D., Renbutsu, Y., Taniguchi, T., Nishimura, S. 1976. Multiple host-specific toxins of *Alternaria mali* and their effect on the permeability of host cells. *Physiol. Plant Pathol.* 8:141–53
223. Kojima, M., Uritani, I. 1976. Possible involvement of furanoterpenoid

phytoalexins in establishing host-parasite specificity between sweet potato and various strains of *Ceratocystis fimbriata*. *Physiol. Plant Pathol.* 8:97–111
224. Kolattukudy, P. E. 1980. Biopolyester membranes of plants: cutin and suberin. *Science* 208:990–1000
225. Krausz, J. P., Thurston, H. D. 1975. Breakdown of resistance to *Pseudomonas solanacearum* in tomato. *Phytopathology* 65:1271–74
226. Krober, H., Petzold, H. 1972. Licht- und elektronenmikroskopische untersuchungen uber wirt-parasit-beziehungen bei anfalligen und gegen *Peronospora* spp. resistent gezuchteten sorten von tabak und spinat. *Phytopathol. Z.* 74: 296–313
227. Kuc, J., Hammerschmidt, R. 1978. Acquired resistance to bacterial and fungal infection. *Ann. Appl. Biol.* 89:313–17
228. Kuhn, C. W. 1973. 2-Thiouracil-induced changes in alfalfa mosaic virus infectivity and nucleoprotein components in hypersensitive bean. *Phytopathology* 63:1235–38
229. Kuhn, P. J., Smith, D. A. 1979. Isolation from *Fusarium solani* f. sp. *phaseoli* of an enzymatic system responsible for kievitone and phaseollidin detoxification. *Physiol. Plant Pathol.* 14:179–90
230. Lacy, G. H., Hirano, S. S., Victoria, J. I., Kelman, A., Upper, C. D. 1979. Inhibition of soft-rotting *Erwinia* spp. strains by 2,4-dihydroxy-7-methoxy-2H-1,4-benzoxazin-3(4H)-one in relation to their pathogenicity on *Zea mays*. *Phytopathology* 69:757–63
231. Langcake, P., Wickins, S. G. A. 1975. Studies on the action of the dichlorocyclopropanes on the host-parasite relationship in the rice blast disease. *Physiol. Plant Pathol.* 7:113–26
232. Lazarovits, G., Bhullar, B. S., Sugiyama, H. J., Higgins, V. J. 1979. Purification and partial characterization of a glycoprotein toxin produced by *Cladosporium fulvum*. *Phytopathology* 69:1062–68
233. Lazarovits, G., Higgins, V. J. 1976. Histological comparison of *Cladosporium fulvum* race 1 on immune, resistant, and susceptible tomato varieties. *Can. J. Bot.* 54:224–34
234. Lazarovits, G., Higgins, V. J. 1976. Ultrastructure of susceptible, resistant, and immune reactions of tomato to races of *Cladosporium fulvum*. *Can. J. Bot.* 54:235–49
235. Lazarovits, G., Higgins, V. J. 1979. Biological activity and specificity of a toxin produced by *Cladosporium fulvum*. *Phytopathology* 69:1056–61
236. Legrand, M., Fritig, B., Hirth, L. 1976. Enzymes of the phenylpropanoid pathway and the necrotic reaction of hypersensitive tobacco to tobacco mosaic virus. *Phytochemistry* 15:1353–59
237. Levin, D. A. 1976. The chemical defenses of plants to pathogens and herbivores. *Ann. Rev. Ecol. Syst.* 7:121–59
238. Lin, T. S., Kolattukudy, P. E. 1978. Induction of a biopolyester hydrolase (cutinase) by low levels of cutin monomers in *Fusarium solani* f. sp. *pisi*. *J. Bacteriol.* 133:942–51
239. Lisker, N., Kuc, J. 1978. Terpenoid accumulation and browning in potato sprouts inoculated with *Phytophthora infestans*. *Phytopathology* 68:1284–87
240. Littlefield, L. J. 1973. Histological evidence for diverse mechanisms of resistance to flax rust, *Melampsora lini* (Ehrenb.) Lev. *Physiol. Plant Pathol.* 3:241–47
241. Loebenstein, G. 1972. Localization and induced resistance in virus-infected plants. *Ann. Rev. Phytopathol.* 10:177–206
242. Loebenstein, G., Gera, A., Barnett, A., Shabtai, S., Cohen, J. 1980. Effect of 2,4-dichlorophenoxyacetic acid on multiplication of tobacco mosaic virus in protoplasts from local-lesion and systemic-responding tobaccos. *Virology* 100:110–15
243. Lozano, J. C., Sequeira, L. 1970. Prevention of the hypersensitive reaction in tobacco leaves by heat-killed bacterial cells. *Phytopathology* 60:875–79
244. Luning, H. U., Waiyaki, B. G., Schlosser, E. 1978. Role of saponins in antifungal resistance. VIII. Interactions *Avena sativa-Fusarium avenaceum*. *Phytopathol. Z.* 92:338–45
245. Lyon, G. D., Mayo, M. A. 1978. The phytoalexin rishitin affects the viability of isolated plant protoplasts. *Phytopathol. Z.* 92:298–304
246. Mace, M. E. 1978. Contributions of tyloses and terpenoid aldehyde phytoalexins to *Verticillium* wilt resistance in cotton. *Physiol. Plant Pathol.* 12:1–11
247. Mace, M. E., Bell, A. A. 1981. Flavanol and terpenoid aldehyde synthesis in tumors associated with genetic incompatibility in a *Gossypium hirsutum* x *G. gossypioides* hybird. *Can. J. Bot.* In press
248. Mace, M. E., Bell, A. A., Stipanovic, R. D. 1978. Histochemistry and identification of flavanols in *Verticillium* wilt-

resistant and -susceptible cottons. *Physiol. Plant Pathol.* 13:143–49

249. Mace, M. E., Veech, J. A., Beckman, C. H. 1972. *Fusarium* wilt of susceptible and resistant tomato isolines: histochemistry of vascular browning. *Phytopathology* 62:651–54
250. Macfoy, C. A., Smith, I. M. 1979. Phytoalexin production and degradation in relation to resistance of clover leaves to *Sclerotinia* and *Botrytis* spp. *Physiol. Plant Pathol.* 14:99–111
251. Maclean, D. J., Tommerup, I. C. 1979. Histology and physiology of compatibility and incompatibility between lettuce and the downy mildew fungus, *Bremia lactucae* Regel. *Physiol. Plant Pathol.* 14:291–312
252. Maiti, I. B., Kolattukudy, P. E. 1979. Prevention of fungal infection of plants by specific inhibition of cutinase. *Science* 205:507–8
253. Mansfield, J. W., Hargreaves, J. A., Boyle, F. C. 1974. Phytoalexin production by live cells in broad bean leaves infected with *Botrytis cinerea. Nature* 252:316–17
254. Marcan, H., Jarvis, M. C., Friend, J. 1979. Effect of methyl glycosides and oligosaccharides on cell death and browning of potato tuber discs induced by mycelial components of *Phytophthora infestans. Physiol. Plant Pathol.* 14:1–9
255. Mares, D. J. 1979. Microscopic study of the development of yellow rust (*Puccinia striiformis*) in a wheat cultivar showing adult plant resistance. *Physiol. Plant Pathol.* 15:289–96
256. Maronek, D. M., Hendrix, J. W. 1978. Resistance to race 0 of *Phytophthora parasitica* var. *nicotianae* in tissue cultures of a tobacco breeding line with black shank resistance derived from *Nicotiana longiflora. Phytopathology* 68:233–34
257. Martin, M., Dewick, P. M. 1979. Biosynthesis of the 2-arylbenzofuran phytoalexin vignafuran in *Vigna unguiculata. Phytochemistry* 18:1309–17
258. Massala, R., Legrand, M., Fritig, B. 1980. Effect of α-amino-oxyacetate, a competitive inhibitor of phenylalanine ammonia-lyase, on the hypersensitive resistance of tobacco to tobacco mosaic virus. *Physiol. Plant Pathol.* 16:213–26
259. Matta, A. 1971. Microbial penetration and immunization of uncongenial host plants. *Ann. Rev. Phytopathol.* 9:387–410
260. Maule, A. J., Ride, J. P. 1976. Ammonia-lyase and *O*-methyl transferase activities related to lignification in wheat leaves infected with *Botrytis. Phytochemistry* 15:1661–64
261. Mayama, S., Daly, J. M., Rehfeld, D. W., Daly, C. R. 1975. Hypersensitive response of near-isogenic wheat carrying the temperature-sensitive *Sr6* allele for resistance to stem rust. *Physiol. Plant Pathol.* 7:35–47
262. Mayer, A. M., Harel, E. 1979. Review: polyphenol oxidases in plants. *Phytochemistry* 18:193–215
263. Mazzucchi, U., Bazzi, C., Pupillo, P. 1979. The inhibition of susceptible and hypersensitive reactions by protein-lipopolysaccharide complexes from phytopathogenic pseudomonads: relationship to polysaccharide antigenic determinants. *Physiol. Plant Pathol.* 14:19–30
264. McClure, M. A., Ellis, K. C., Nigh, E. L. 1974. Resistance of cotton to the root-knot nematode, *Meloidogyne incognita. J. Nematol.* 6:17–20
265. McClure, M. A., Ellis, K. C. Nigh, E. L. 1974. Post-infection development and histopathology of *Meloidogyne incognita* in resistant cotton. *J. Nematol.* 6:21–26
266. McIntyre, J. L., Miller, P. M. 1978. Protection of tobacco against *Phytophthora parasitica* var. *nicotianae* by cultivar-nonpathogenic races, cell-free sonicates, and *Pratylenchus penetrans. Phytopathology* 68:235–39
267. Mendgen, K. 1978. Der Infektionsverlauf von *Uromyces phaseoli* bei anfälligen und resistenten Bohnensorten. *Phytopathol. Z.* 93:295–313
268. Metlitskii, L. V., Ozeretskovskaya, O. L., Yurganova, L. A., Savel'eva, O. N., Chalova, L. I., D'yakov, Y. T. 1976. Induction of phytoalexins in potato tubers by metabolites of the fungus *Phytophthora infestans* (Mont.) De Bary. *Dokl. Akad. Nauk. SSSR, Biochem.* 226:72–75 (Engl.)
269. Miller, H. J., Elgersma, D. M. 1976. The growth of aggressive and non-aggressive strains of *Ophiostoma ulmi* in susceptible and resistant elms, a scanning electron microscopial study. *Neth. J. Plant Pathol.* 82:51–65
270. Moreau, M., Catesson, A., Czaninski, Y., Peresse, M. 1977. Vessel associated cells. Location for the synthesis of main constituents which obstruct the vessels in cases of attack by vascular parasites. *Proc. 2nd. Int. Mycol. Congr., Tampa, Fla.*
271. Morris, A. J., Smith, D. A. 1978. Phytoalexin formation in bean hypoco-

tyls induced by cell-free mycelial extracts of *Rhizoctonia* and *Fusarium*. *Ann. Appl. Biol.* 89:344–47
272. Motoyoshi, F., Oshima, N. 1977. Expression of genetically controlled resistance to tobacco mosaic virus infection in isolated tomato leaf mesophyll protoplasts. *J. Gen. Virol.* 34:499–506
273. Mueller, W. C., Beckman, C. H. 1976. Ultrastructure and development of phenolic-storing cells in cotton roots. *Can. J. Bot.* 54:2074–82
274. Mueller, W. C., Beckman, C. H. 1978. Ultrastructural localization of polyphenoloxidase and peroxidase in roots and hypocotyls of cotton seedlings. *Can. J. Bot.* 56:1579–87
275. Mueller, W. C., Greenwood, A. D. 1978. The ultrastructure of phenolic-storing cells fixed with caffeine. *J. Exp. Bot.* 29:757–64
276. Mukherjee, N., Kundu, B. 1973. Antifungal activities of some phenolics and related compounds to three fungal plant pathogens. *Phytopathol. Z.* 78:89–92
277. Nelson, R. R., ed. 1973. *Breeding Plants For Disease Resistance.* University Park: Penn. State Univ. Press. 401 pp.
278. Netzer, D., Kritzman, G., Chet, I. 1979. β-(1,3)-Glucanase activity and quantity of fungus in relation to *Fusarium* wilt in resistant and susceptible near-isogenic lines of muskmelon. *Physiol. Plant Pathol.* 14:47–55
279. Nilsen, K. N., Hodges, C. F., Madsen, J. P. 1979. Pathogenesis of *Drechslera sorokiniana* leaf spot on progressively older leaves of *Poa pratensis* as influenced by photoperiod and light quality. *Physiol. Plant Pathol.* 15:171–76
280. Noel, G. R., McClure, M. A. 1978. Peroxidase and 6-phosphogluconate dehydrogenase in resistant and susceptible cotton infected by *Meloidogyne incognita. J. Nematol.* 10:34–39
281. Norris, D. M., Kogan, M. 1980. Biochemical and morphological bases of resistance. In *Breeding Plants Resistant to Insects,* ed. F. G. Maxwell, P. R. Jennings, pp. 23–61. New York: Wiley. 683 pp.
282. Norton, I. T., Goodall, D. M. 1980. Kinetic evidence for intramolecular conformational ordering of the extracellular polysaccharide (xanthan) from *Xanthomonas campestris. J. Chem. Soc. Chem. Commun.,* pp. 545–47
283. Novacky, A., Karr, A. L., Van Sambeek, J. W. 1976. Using electrophysiology to study plant disease development. *Bioscience* 26:499–504
284. Nozue, M., Tomiyama, K., Doke, N. 1978. Effect of adenosine 5'-triphosphate on hypersensitive death of potato tuber cells infected by *Phytophthora infestans. Phytopathology* 78:873–76
285. O'Brien, P. C., Fisher, J. M. 1978. Studies on the mechanism of resistance of wheat to *Heterodera avenae. Nematologica* 24:463–71
286. Otsuki, Y., Shimomura, T., Takebe, I. 1972. Tobacco mosaic virus multiplication and expression of the N gene in necrotic responding tobacco varieties. *Virology* 50:45–50
287. Palmerley, R. A., Callow, J. A. 1978. Common antigens in extracts of *Phytophthora infestans* in potatoes. *Physiol. Plant Pathol.* 12:241–48
288. Parish, C. L., Zaitlin, M., Siegel, A. 1965. A study of necrotic lesion formation by tobacco mosaic virus. *Virology* 26:413–18
289. Park, P., Fukutomi, M., Akai, S. 1976. Effect of the host-specific toxin from *Alternaria kikuchiana* on the ultrastructure of plasma membranes of cells in leaves of Japanese pear. *Physiol. Plant Pathol.* 9:167–74
290. Partridge, J. E., Keen, N. T. 1976. Association of the phytoalexin kievitone with single-gene resistance of cowpeas to *Phytophthora vignae. Phytopathology* 66:426–29
291. Partridge, J. E., Keen, N. T. 1977. Soybean phytoalexins: rates of synthesis are not regulated by activation of initial enzymes in flavonoid biosynthesis. *Phytopathology* 67:50–55
292. Payne, G., Kono, Y., Daly, J. M. 1980. A comparison of purified host specific toxin from *Helminthosporium maydis,* race T, and its acetate derivative on oxidation by mitochondria from susceptible and resistant plants. *Plant Physiol.* 65:785–91
293. Pearce, R. B., Ride, J. P. 1980. Specificity of induction of the lignification response in wounded wheat leaves. *Physiol. Plant Pathol.* 16:197–204
294. Pelcher, L. E., Kao, K. N., Gamborg, O. L., Yoder, O. C., Gracen, V. E. 1975. Effects of *Helminthosporium maydis* race T toxin on protoplasts of resistant and susceptible corn (*Zea mays*). *Can. J. Bot.* 53:427–31
295. Pellizzari, E. D., Kuc, J., Williams, E. B. 1970. The hypersensitive reaction in *Malus* species: changes in the leakage of electrolytes from apple leaves after inoculation with *Venturia inaequalis. Phytopathology* 60:373–76

296. Peng, J. H., Black, L. L. 1976. Increased proteinase inhibitor activity in response to infection of resistant tomato plants by *Phytophthora infestans. Phytopathology* 66:958–63
297. Pennazio, S., Appiano, A., Redolfi, P. 1979. Changes occurring in *Gomphrena globosa* leaves in advance of the appearance of tomato bushy stunt virus necrotic local lesions. *Physiol. Plant Pathol.* 15:177–82
298. Peters, B. M., Cribbs, D. H., Stelzig, D. A. 1978. Agglutination of plant protoplasts by fungal cell wall glucans. *Science* 201:364–65
299. Pfannenstiel, M. A., Niblett, C. L. 1978. The nature of the resistance of agrotricums to wheat streak mosaic virus. *Phytopathology* 68:1204–9
300. Politis, D. J. 1976. Ultrastructure of penetration by *Colletotrichum graminicola* of highly resistant oat leaves. *Physiol. Plant Pathol.* 8:117–22
301. Pritchard, D. W., Ross, A. F. 1975. The relationship of ethylene to formation of tobacco mosaic virus lesions in hypersensitive responding tobacco leaves with and without induced resistance. *Virology* 64:295–307
302. Pryce, R. J., Langcake, P. 1977. α-Viniferin: an antifungal resveratrol trimer from grapevines. *Phytochemistry* 16:1452–54
303. Purushothaman, D. 1974. Phenylalanine ammonia lyase and aromatic amino acids in rice varieties infected with *Xanthomonas oryzae. Phytopathol. Z.* 80:171–75
304. Rahe, J. E. 1973. Phytoalexin nature of heat-induced protection against bean anthracnose. *Phytopathology* 63:572–77
305. Rancillac, M., Kaur-Sawhney, R., Staskawicz, B., Galston, A. W. 1976. Effects of cycloheximide and kinetin pretreatments on responses of susceptible and resistant *Avena* leaf protoplasts to the phytotoxin victorin. *Plant Cell Physiol.* 17:987–95
306. Rathmell, W. G., Sequeira, L. 1975. Induced resistance in tobacco leaves: the role of inhibitors of bacterial growth in the intercellular fluid. *Physiol. Plant Pathol.* 5:65–73
307. Rautela, G. S., Payne, M. G. 1970. The relationship of peroxidase and orthodiphenol oxidase to resistance of sugarbeets to *Cercospora* leaf spot. *Phytopathology* 60:238–45
308. Rawn, C. D. 1977. Simultaneous changes in the rate and pathways of glucose oxidation in victorin-treated oat leaves. *Phytopathology* 67:338–43
309. Rebois, R. V., Madden, P. A., Eldridge, B. J. 1975. Some ultrastructural changes induced in resistant and susceptible soybean roots following infection by *Rotylenchulus reniformis. J. Nematol.* 7:122–39
310. Reed, B. M., Richardson, P. E., Russell, C. C. 1979. Stem nematode infection of resistant and susceptible cultivars of alfalfa. *Phytopathology* 69:993–96
311. Reynolds, H. W., Carter, W. W., O'Bannon, J. H. 1970. Symptomless resistance of alfalfa to *Meloidogyne incognita acrita. J. Nematol.* 2:131–34
312. Rich, J. R., Keen, N. T., Thomason, I. J. 1977. Association of coumestans with the hypersensitivity of Lima bean roots to *Pratylenchus scribneri. Physiol. Plant Pathol.* 10:105–16
313. Richmond, S., Kuc, J., Elliston, J. E. 1979. Penetration of cucumber leaves by *Colletotrichum lagenarium* is reduced in plants systemically protected by previous infection with the pathogen. *Physiol. Plant Pathol.* 14:329–38
314. Ride, J. P. 1975. Lignification in wounded wheat leaves in response to fungi and its possible role in resistance. *Physiol. Plant Pathol.* 5:125–34
315. Ride, J. P. 1978. The role of cell wall alterations in resistance to fungi. *Ann. Appl. Biol.* 89:302–6
316. Ride, J. P. 1980. The effect of induced lignification on the resistance of wheat cell walls to fungal degradation. *Physiol. Plant Pathol.* 16:187–96
317. Ride, J. P., Pearce, R. B. 1979. Lignification and papilla formation at sites of attempted penetration of wheat leaves by nonpathogenic fungi. *Physiol. Plant Pathol.* 15:79–92
318. Rockwood, D. L. 1974. Cortical monoterpene and fusiform rust resistance relationships in slash pine. *Phytopathology* 64:976–79
319. Roebuck, P., Sexton, R., Mansfield, J. W. 1978. Ultrastructural observations on the development of the hypersensitive reaction in leaves of *Phaseolus vulgaris* cv. Red Mexican inoculated with *Pseudomonas phaseolicola* (race 1). *Physiol. Plant Pathol.* 12:151–57
320. Rohringer, R., Kim, W. K., Samborski, D. J. 1979. A histological study of interactions between avirulent races of stem rust and wheat containing resistance genes *Sr5, Sr6, Sr8,* or *Sr22. Can. J. Bot.* 57:324–31
321. Ross, A. F., Israel, H. W. 1970. Use of heat treatments in the study of acquired resistance to tobacco mosaic virus in

hypersensitive tobacco. *Phytopathology* 60:755–69

322. Rossall, S., Mansfield, J. W., Hutson, R. A. 1980. Death of *Botrytis cinerea* and *B. fabae* following exposure to wyerone derivatives *in vitro* and during infection development in broad bean leaves. *Physiol. Plant Pathol.* 16:135–46
323. Rossall, S., Mansfield, J. W., Price, N. C. 1977. The effect of reduced wyerone acid on the antifungal activity of the phytoalexin wyerone acid against *Botrytis fabae*. *J. Gen. Microbiol.* 102:203–5
324. Russell, G. E. 1979. *Plant Breeding for Pest and Disease Control.* Boston: Butterworth
325. Ruyack, J., Downing, M. R., Chang, J. S., Mitchell, E. D. Jr. 1979. Growth of callus and suspension culture cells from cotton varieties (*Gossypium hirsutum* L.) resistant and susceptible to *Xanthomonas malvacearum* (E. F. Sm.) Dows. *In Vitro* 15:368–73
326. Sands, D. C., Zucker, M. 1976. Amino acid inhibition of pseudomonads and its reversal by biosynthetically related amino acids. *Physiol. Plant Pathol.* 9:127–33
327. Sasser, M., Andrews, A. K., Doganay, Z. U. 1974. Inhibition of photosynthesis diminishes antibacterial action of pepper plants. *Phytopathology* 64:770–72
328. Sato, K., Ishiguri, Y., Doke, N., Tomiyama, K., Yagihashi, F., Murai, A., Katsui, N., Masamune, T. 1978. Biosynthesis of the sesquiterpenoid phytoalexin rishitin from acetate via oxylubimin in potato. *Phytochemistry* 17: 1901–2
329. Sawhney, R., Webster, J. M. 1975. The role of plant growth hormones in determining the resistance of tomato plants to the root-knot nematode, *Meloidogyne incognita*. *Nematologica* 21:95–103
330. Schonbeck, F., Schinzer, U. 1972. Untersuchungen über den Einfluss der endotrophen Mycorrhiza auf die TMV-Läsionenbildung in *Nicotiana tabacum* L. var. *Xanthi*-nc. *Phytopathol. Z.* 73:78–80
331. Seevers, P. M., Daly, J. M. 1970. Studies on wheat stem rust resistance controlled at the *Sr6* locus. II. Peroxidase activities. *Phytopathology* 60:1642–47
332. Sequeira, L. 1978. Lectins and their role in host-pathogen specificity. *Ann. Rev. Phytopathol.* 16:453–81
333. Sequeira, L., Gaard, G., De Zoeten, G. A. 1977. Interaction of bacteria and host cell walls: its relation to mechanisms of induced resistance. *Physiol. Plant Pathol.* 10:43–50
334. Sequeira, L., Graham, T. L. 1977. Agglutination of avirulent strains of *Pseudomonas solanacearum* by potato lectin. *Physiol. Plant Pathol.* 11:43–54
335. Sequeira, L., Hill, L. M. 1974. Induced resistance in tobacco leaves: the growth of *Pseudomonas solanacearum* in protected tissues. *Physiol. Plant Pathol.* 4:447–55
336. Shain, L., Hillis, W. E. 1971. Phenolic extractives in Norway spruce and their effects on *Fomes annosus*. *Phytopathology* 61:841–45
337. Shaykh, M., Soliday, C., Kolattukudy, P. E. 1977. Proof for the production of cutinase by *Fusarium solani* f. *pisi* during penetration into its host, *Pisum sativum*. *Plant Physiol.* 60:170–72
338. Sheen, S. J. 1974. Polyphenol oxidation by leaf peroxidases in *Nicotiana*. *Bot. Gaz.* 135:155–61
339. Sheen, S. J., Andersen, R. A. 1974. Comparison of polyphenols and related enzymes in the capsule and nodal tumor of *Nicotiana* plants. *Can. J. Bot.* 52: 1379–85
340. Shepard, J. F., Bidney, D., Shahin, E. 1980. Potato protoplasts in crop improvement. *Science* 208:17–24
341. Sherwood, R. T., Vance, C. P. 1976. Histochemistry of papillae formed in reed canarygrass leaves in response to noninfecting pathogenic fungi. *Phytopathology* 66:503–10
342. Sherwood, R. T., Vance, C. P. 1980. Resistance to fungal penetration in Graminae. *Phytopathology* 70:273–79
343. Sherwood, R. T., Zeiders, K. E., Vance, C. P. 1978. *Helminthosporium* and *Stagonospora* leaf spot resistance are unrelated to indole alkaloid content in reed canarygrass. *Phytopathology* 68: 803–7
344. Shih, M., Kuc, J., Williams, E. B. 1973. Suppression of steroid glycoalkaloid accumulation as related to rishitin accumulation in potato tubers. *Phytopathology* 63:821–26
345. Shimomura, T., Dijkstra, J. 1975. The occurrence of callose during the process of local lesion formation. *Neth. J. Plant Pathol.* 81:107–21
346. Shiraishi, T., Oku, H., Isono, M., Ouchi, S. 1975. The injurious effect of pisatin on the plasma membrane of pea. *Plant Cell Physiol.* 16:939–42
347. Simon, E. W. 1974. Phospholipids and plant membrane permeability. *New Phytol.* 73:377–420
348. Simons, T. J., Israel, H. W., Ross, A. F. 1972. Effect of 2,4-dichlorophenoxyacetic acid on tobacco mosaic virus le-

sions in tobacco and on the fine structure of adjacent cells. *Virology* 48: 502–15

349. Sing, V. O., Schroth, M. N. 1977. Bacteria-plant cell surface interactions: active immobilization of saprophytic bacteria in plant leaves. *Science* 197: 759–61
350. Sinha, A. K., Trivedi, N. 1972. Resistance induced in rice plants against *Helminthosporium* infection. *Phytopathol. Z.* 74:182–91
351. Sitton, D., West, C. A. 1975. Casbene: an anti-fungal diterpene produced in cell-free extracts of *Ricinus communis* seedlings. *Phytochemistry* 14:1921–25
352. Smith, D. A., VanEtten, H. D., Bateman, D. F. 1975. Accumulation of phytoalexins in *Phaseolus vulgaris* hypocotyls following infection by *Rhizoctonia solani. Physiol. Plant Pathol.* 5:51–64
353. Smookler, M. M. 1971. Properties of inhibitors of plant virus infection occurring in the leaves of species in the Chenopodiales. *Ann. Appl. Biol.* 69: 157–68
354. Solheim, B., Raa, J., eds. 1977. *Cell Wall Biochemistry Related to Specificity in Host-Plant Pathogen Interactions.* Tromsø: Universitetsforlaget. 487 pp.
355. Sridhar, R., Ou, S. H. 1974. Biochemical changes associated with the development of resistant and susceptible types of rice blast lesions. *Phytopathol. Z.* 79:222–30
356. Stall, R. E., Bartz, J. A., Cook, A. A. 1974. Decreased hypersensitivity to xanthomonads in pepper after inoculations with virulent cells of *Xanthomonas vesicatoria. Phytopathology* 64:731–35
357. Stall, R. E., Cook, A. A. 1979. Evidence that bacterial contact with the plant cell is necessary for the hypersensitive reaction but not the susceptible reaction. *Physiol. Plant Pathol.* 14:77–84
358. Stekoll, M., West, C. A. 1978. Purification and properties of an elicitor of castor bean phytoalexin from culture filtrates of the fungus *Rhizopus stolonifer. Plant Physiol.* 61:38–45
359. Stobbs, L. W., Manocha, M. S. 1977. Histological changes associated with virus localization in TMV-infected Pinto bean leaves. *Physiol. Plant Pathol.* 11:87–94
360. Stoessl, A., Stothers, J. B., Ward, E. W. B. 1976. Sesquiterpenoid stress compounds of the Solanaceae. *Phytochemistry* 15:855–72
361. Stoessl, A., Unwin, C. H., Ward, E. W. B. 1972. Postinfectional inhibitors from plants. I. Capsidiol, an antifungal compound from *Capsicum frutescens. Phytopathol. Z.* 74:141–52
362. Strand, L. L., Mussell, H. 1975. Solubilization of peroxidase activity from cotton cell walls by endopolygalacturonases. *Phytopathology* 65:830–31
363. Strobel, G. A. 1975. A mechanism of disease resistance in plants. *Sci. Am.* 232:80–88
364. Strobel, G. A. 1979. The relationship between membrane ATPase activity in sugarcane and heat-induced resistance to helminthosporoside. *Biochim. Biophys. Acta* 554:460–68
365. Stromberg, E. L., Corden, M. E. 1977. Fungitoxicity of xylem extracts from tomato plants resistant or susceptible to *Fusarium* wilt. *Phytopathology* 67: 693–97
366. Svoboda, W. E., Paxton, J. D. 1972. Phytoalexin production in locally cross-protected Harosoy and Harosoy-63 soybeans. *Phytopathology* 62:1457–60
367. Swain, T. 1977. Secondary compounds as protective agents. *Ann. Rev. Plant Physiol.* 28:479–501
368. Swain, T., Harborne, J. B., Van Sumere, C. F., eds. 1979. *Recent Advances in Phytochemistry. Vol. 12. Biochemistry of Plant Phenolics.* New York: Plenum. 651 pp.
369. Sziraki, I., Balazs, E., Kiraly, Z. 1980. Role of different stresses in inducing systemic acquired resistance to TMV and increasing cytokinin level in tobacco. *Physiol. Plant Pathol.* 16: 277–84
370. Tanguy, J., Martin, C. 1972. Phenolic compounds and the hypersensitivity reaction in *Nicotiana tabacum* infected with tobacco mosaic virus. *Phytochemistry* 11:19–28
371. Theodorou, M. K., Smith, I. M. 1979. The response of French bean varieties to components isolated from races of *Colletotrichum lindemuthianum. Physiol. Plant Pathol.* 15:297–309
372. Thomas, C. A., Allen, E. H. 1971. Light and antifungal polyacetylene compounds in relation to resistance of safflower to *Phytophthora drechsleri. Phytopathology* 61:1459–61
373. Tremaine, J. H., Agrawal, H. O. 1972. Limited proteolysis of potato virus X by trypsin and plant proteases. *Virology* 49:735–44
374. Tu, J. C., Hiruki, C. 1971. Electron microscopy of cell wall thickening in local lesions of potato virus-M infected Red

Kidney bean. *Phytopathology* 61: 862–68
375. Turner, J. G., Novacky, A. 1974. The quantitative relation between plant and bacterial cells involved in the hypersensitive reaction. *Phytopathology* 64: 885–90
376. Urs, N. V. R., Dunleavy, J. M. 1974. Function of peroxidase in resistance of soybean to bacterial pustule. *Crop Sci.* 14:740–44
377. Urs, N. V. R., Dunleavy, J. M. 1975. Enhancement of the bactericidal activity of a peroxidase system by phenolic compounds. *Phytopathology* 65:686–90
378. Vance, C. P., Anderson, J. O., Sherwood, R. T. 1976. Soluble and cell wall peroxidases in reed canarygrass in relation to disease resistance and localized lignin formation. *Plant Physiol.* 57: 920–22
379. Vance, C. P., Sherwood, R. T. 1977. Lignified papilla formation as a mechanism for protection in reed canarygrass. *Physiol. Plant Pathol.* 10:247–56
380. Vance, C. P., Sherwood, R. T. 1977. Cycloheximide treatments implicate papilla formation in resistance of reed canarygrass to fungi. *Phytopathology* 66:498–502
381. Van den Heuvel, J., Glazener, J. A. 1975. Comparative abilities of fungi pathogenic and nonpathogenic to bean (*Phaseolus vulgaris*) to metabolize phaseollin. *Neth. J. Plant Pathol.* 81: 125–37
382. VanderMolen, G. E., Beckman, C. H., Rodehorst, E. 1977. Vascular gelation: a general response phenomenon following infection. *Physiol. Plant Pathol.* 11: 95–100
383. Van Dijkman, A., Sijpesteijn, A. K. 1971. A biochemical mechanism for the gene-for-gene resistance of tomato to *Cladosporium fulvum. Neth. J. Plant Pathol.* 77:14–24
384. Van Dijkman, A., Sijpesteijn, A. K. 1973. Leakage of pre-absorbed ^{32}P from tomato leaf disks infiltrated with high molecular weight products of incompatible races of *Cladosporium fulvum. Physiol. Plant Pathol.* 3:57–67
385. VanEtten, H. D., Matthews, P. S., Tegtmeier, K. J., Dietert, M. F., Stein, J. I. 1980. The association of pisatin tolerance and demethylation with virulence on pea in *Nectria haematococca. Physiol. Plant Pathol.* 16:257–68
386. VanEtten, H. D., Pueppke, S. G. 1976. Isoflavonoid phytoalexins. See Ref. 116, pp. 239–89
387. VanEtten, H. D., Smith, D. A. 1975. Accumulation of antifungal isoflavonoids and 1a-hydroxyphaseollone, a phaseollin metabolite, in bean tissue infected with *Fusarium solani* f. sp. *phaseoli. Physiol. Plant Pathol.* 5: 225–37
388. Van Gundy, S. D., Kirkpatric, J. D., Golden, J. 1977. The nature and role of metabolic leakage from root-knot nematode galls and infection by *Rhizoctonia solani. J. Nematol.* 9:113–21
389. Van Loon, L. C. 1976. Systemic acquired resistance, peroxidase activity and lesion size in tobacco reacting hypersensitively to tobacco mosaic virus. *Physiol. Plant Pathol.* 8:231–42
390. Van Loon, L. C. 1979. Effects of auxin on the localization of tobacco mosaic virus in hypersensitively reacting tobacco. *Physiol. Plant Pathol.* 14:213–26
391. Varns, J. L., Kuc, J. 1971. Suppression of rishitin and phytuberin accumulation and hypersensitive response in potato by compatible races of *Phytophthora infestans. Phytopathology* 61:178–81
392. Veech, J. A. 1977. Localization of peroxidase in *Rhizoctonia solani*-infected cotton seedlings. *Phytopathology* 66:1072–76
393. Veech, J. A. 1979. Histochemical localization and nematoxicity of terpenoid aldehydes in cotton. *J. Nematol.* 11:240–46
394. Veech, J. A. 1980. Plant resistance to nematodes. In *Plant Parasitic Nematodes.* Vol. 3, ed. B. M. Zuckerman, R. A. Rohde. New York: Academic. In press
395. Veech, J. A., Endo, B. Y. 1970. Comparative morphology and enzyme histochemistry in root-knot resistant and susceptible soybeans. *Phytopathology* 60:896–902
396. Waiyaki, B. G., Schlosser, E. 1978. Role of saponins in antifungal resistance. IX. Species-specific inactivation of avenacin by *Fusarium avenaceum. Phytopathol. Z.* 92:346–50
397. Wang, S. C., Pinckard, J. A. 1973. Cotton boll cuticle, a potential factor in boll rot resistance. *Phytopathology* 63:315–19
398. Ward, E. W. B., Stoessl, A. 1972. Postinfectional inhibitors from plants. III. Detoxification of capsidiol, an antifungal compound from peppers. *Phytopathology* 62:1186–87
399. Ward, E. W. B., Stoessl, A. 1977. Phytoalexins from potatoes: evidence for the conversion of lubimin to 15-

dihydrolubimin by fungi. *Phytopathology* 67:468–71
400. Ward, E. W. B., Unwin, C. H., Stoessl, A. 1975. Experimental control of late blight of tomatoes with capsidiol, the phytoalexin from peppers. *Phytopathology* 65:168–69
401. Wargo, P. M. 1975. Lysis of the cell wall of *Armillaria mellea* by enzymes from forest trees. *Physiol. Plant Pathol.* 5:99–105
402. Waterman, M. A., Aist, J. R., Israel, H. W. 1978. Centrifugation studies help clarify the role of papilla formation in compatible barley powdery mildew interactions. *Phytopathology* 68:797–802
403. Webster, D. M., Sequeira, L. 1977. Expression of resistance in bean pods to an incompatible isolate of *Pseudomonas syringae*. *Can. J. Bot.* 55:2043–52
404. Webster, J. M. 1975. Aspects of the host-parasite relationship of plant-parasitic nematodes. *Adv. Parasitol.* 13: 225–50
405. Weiss, U., Edwards, J. M. 1980. *The Biosynthesis of Aromatic Compounds.* New York: Wiley. 728 pp.
406. Weststeijn, E. A. 1976. Peroxidase activity in leaves of *Nicotiana tabacum* var. *Xanthi* nc. before and after infection with tobacco mosaic virus. *Physiol. Plant Pathol.* 8:63–71
407. Weststeijn, E. A. 1978. Permeability changes in the hypersensitive reaction of *Nicotiana tabacum* cv. *Xanthi* nc. after infection with tobacco mosaic virus. *Physiol. Plant Pathol.* 13:253–58
408. Wheeler, H. 1971. Pathological changes in ultrastructure: effects of victorin on resistant oat roots. *Phytopathology* 61:641–44
409. Wheeler, H. 1977. Increase with age in sensitivity of oat leaves to victorin. *Phytopathology* 67:859–61
410. Wheeler, H., Elbel, E. 1979. Timecourse and antioxidant inhibition of ethylene production by victorin-treated oat leaves. *Phytopathology* 69:32–34
411. Wheeler, M. H., Stipanovic, R. D. 1979. Melanin biosynthesis in *Thielaviopsis basicola*. *Exp. Mycol.* 3:340–50
412. Wheeler, M. H., Tolmsoff, W. J., Bell, A. A. 1978. Ultrastructural and chemical distinction of melanins formed by *Verticillium dahliae* from (+)-scytalone, 1,8-dihydroxynaphthalene, catechol, and L-3,4-dihydroxyphenylalanine. *Can. J. Microbiol.* 24:289–97
413. Whitmore, F. W. 1978. Lignin-protein complex catalyzed by peroxidase. *Plant Sci. Lett.* 13:241–45
414. Woloshuk, C. P., Wolkow, P. M., Sisler, H. D. 1980. Effect of three fungicides specific for control of rice blast disease on growth and melanin biosynthesis by *Pyricularia oryzae*. *J. Biochem. Pest.* In press
415. Wood, F. A., Singh, R. P., Hodgson, W. A. 1971. Characterization of a virus-inhibiting polysaccharide from *Phytophthora infestans*. *Phytopathology* 61: 1006–9
416. Wood, R. K. S., Ballio, A., Graniti, A., eds. 1972. *Phytotoxins in Plant Diseases.* London: Academic. 350 pp.
417. Woodward, J. R., Keane, P. J., Stone, B. A. 1980. Structures and properties of wilt-inducing polysaccharides from *Phytophthora* species. *Physiol. Plant Pathol.* 16:439–54
418. Woodward, M. D. 1980. Phaseollin formation and metabolism in *Phaseolus vulgaris*. *Phytochemistry* 19:921–27
419. Wyatt, S. D., Kuhn, C. W. 1979. Replication and properties of cowpea chlorotic mottle virus in resistant cowpeas. *Phytopathology* 69:125–29
420. Yamamoto, H., Hokin, H., Tani, T. 1978. Peroxidase and polyphenoloxidase in relation to the crown rust resistance of oat leaves. *Phytopathol. Z.* 91:193–202
421. Yamamoto, H., Hokin, H., Tani, T., Kadota, G. 1977. Phenylalanine ammonia-lyase in relation to the crown rust resistance of oat leaves. *Phytopathol. Z.* 90:203–11
422. Yoder, O. C., Gracen, V. E. 1975. Segregation of pathogenicity types and host-specific toxin production in progenies of crosses between races T and O of *Helminthosporium maydis* (*Cochliobolus heterostrophus*). *Phytopathology* 65:273–76
423. Yoshikawa, M., Yamauchi, K., Masago, H. 1978. Glyceollin: its role in restricting fungal growth in resistant soybean hypocotyls infected with *Phytophthora megasperma* var. *sojae*. *Physiol. Plant Pathol.* 12:73–82
424. Yoshikawa, M., Yamauchi, K., Masago, H. 1978. De novo messenger RNA and protein synthesis are required for phytoalexin-mediated disease resistance in soybean hypocotyls. *Plant Physiol.* 61:314–17
425. Yoshikawa, M., Yamauchi, K., Masago, H. 1979. Biosynthesis and biodegradation of glyceollin by soybean hypocotyls infected with *Phytophthora megasperma* var. *sojae*. *Physiol. Plant Pathol.* 14:157–69

426. Young, J. M. 1974. Effect of water on bacterial multiplication in plant tissue. *N. Z. J. Agric. Res.* 17:115–19
427. Zaki, A. I., Keen, N. T., Erwin, D. C. 1972. Implication of vergosin and hemigossypol in the resistance of cotton to *Verticillium albo-atrum*. *Phytopathology* 61:1402–6
428. Zaki, A. I., Zentmyer, G. A., Pettus, J., Sims, J. J., Keen, N. T., Sing, V. O. 1980. Barbonol from *Persea* spp.—chemical properties and antifungal activity against *Phytophthora cinnamomi*. *Physiol. Plant Pathol.* 16:205–12
429. Zeller, W., Rudolph, K., Fuchs, W. H. 1973. Permeabilitätsveränderungen bei resistenten und anfälligen Buschbohnensorten nach Inokulation mit *Pseudomonas phaseolicola* (Burkh.) Dowson. *Phytopathol. Z.* 77:363–72

Ann. Rev. Plant Physiol. 1981. 32:83–110

MECHANISMS OF CONTROL OF LEAF MOVEMENTS[1,2]

◆7706

Ruth L. Satter[3] and Arthur W. Galston

Department of Biology, Yale University, New Haven, Connecticut 06511

CONTENTS

"... it is easier to explain biochemistry in terms of transport than it is to explain transport in terms of biochemistry."

Peter Mitchell (70)

[1]Dedicated to Professor Erwin Bünning, whose researches and writings forced the biological world to face up to the reality of biological clocks.

[2]Abbreviations used: BAP (blue and far-red absorbing pigment); R (brief irradiation, 660 nm); FR (brief irradiation, 730 nm); DD (continuous darkness); LL (light of constant intensity); LD (light-dark cycle).

[3]Present address: Biological Sciences Group, University of Connecticut, Storrs, Connecticut 06268.

0066-4294/81/0601-0083$01.00

INTRODUCTION

The rhythmic movements of certain leguminous leaves, which continue without environmental perturbation, provide an unusual opportunity for the plant physiologist to address several problems of general significance. First, leaf movements can be viewed as convenient hands of the biological clock (14), permitting one to study the internal timekeeping mechanism whose oscillations provide the physiological signals that regulate overt rhythms. Second, since the movements can be rephased by red (17, 62, 111) or blue (62, 99) light, their study can aid in interpreting how light and the clock interact. Such interaction provides a basis for photoperiodism, thus synchronizing developmental processes in many plants with seasonal change of day length (78). Third, the reversible changes in leaf angle in plants bearing pulvini are caused by changes in the size and turgor of pulvinar motor cells; these in turn are regulated by the movement of K^+, Cl^-, and other ions into and out of such cells (reviewed in 89). Since these massive ion fluxes can be manipulated by various chemical and physical treatments, leaf movement studies can contribute to detailed molecular models of membrane-localized ion transport systems. Thus, just as the Rosetta stone helped decipher hieroglyphics by comparing them with demotic characters and Greek, so leaf movements can provide a framework to aid in deciphering the nature of the clockwork, the photobiological controls, and the ion transport mechanisms operating in plants.

The recognition that electrogenic H^+ secretion is the primary process generating the membrane potential and energizing solute fluxes through plant cell membranes (83, 112, 114) encouraged us to consider such an explanation for the ion fluxes that lead to leaf movement. Data obtained with biophysical, physiological, and electrophysiological techniques support a model in which active H^+ secretion drives the K^+ and associated Cl^- movements that regulate motor cell turgor and leaf movements. Our presentation thus includes:

1. Analysis of light-clock interaction in the control of leaf movements;
2. A review of the roles of K^+, Cl^-, and other solutes in turgor regulation;
3. Formulation of an explanation and model for K^+ and Cl^- movements based on active H^+ secretion;
4. Discussion of the implications of the model for understanding light-clock interaction and the nature of the clock.

NYCTINASTIC SPECIES

Some nyctinastic plants (e.g. *Mimosa pudica*) exhibit rapid (ca 1 sec) responses to mechanical and other stimuli, as well as slower responses to light and an internal oscillator. We omit rapid movements, which are adequately discussed in other recent reviews (3, 12, 89, 109). Slower movements, including nyctinasty, have also been reviewed more extensively elsewhere (11, 13, 89, 120).

Our analysis depends upon data obtained from a group of nyctinastic legumes including *Albizzia julibrissin, Cassia fasciculata, Mimosa pudica, Phaseolus coccineus* (also called *P. multiflorus*), *Phaseolus vulgaris, Samanea saman* (also called *Pithecolobium saman*), and *Trifolium repens.* Wherever possible, our generalizations are based on data obtained from all these plants, despite the fact that differences in their behavior complicate the generalizing process. For example, *Phaseolus* leaves continue to oscillate when plants are kept in high intensity white light of constant intensity for periods of 30 days (48), whereas movements of *Albizzia* leaflets and *Samanea* pulvini damp after 1–2 days of bright white light, but persist during several days of darkness if sucrose is available (90, 110). These results suggest that light might have different effects in different species, even though phytochrome appears to be the common primary photoreceptor for rhythmic entrainment.

Extensor and Flexor Regions

A standard terminology permits comparison between pulvini that move in different directions [see (89) for further details]. Cells whose turgor increases during leaflet opening and decreases during closure are called *extensors,* while cells that exhibit the reverse changes are called *flexors* (97). These regions are defined by function rather than by position; thus, the extensor region is on the upper sides of *Albizzia* and *Mimosa* tertiary pulvini and *Cassia* secondary pulvini, but on the lower sides of *Phaseolus* and *Samanea* secondary pulvini. Consequently, night closure movements of *Albizzia* and *Mimosa* pinnules occur in an upward direction, while those of *Phaseolus* leaves and *Samanea* pinnae occur in a downward direction.

THE CIRCADIAN OSCILLATOR

"The sensitive plant senses the sun without seeing it in any way" [translated from DeMairan, 1729 (28)].

Persistence of leaf oscillations in *Mimosa* plants transferred from natural photoperiods to dim light of constant intensity (28) provided the first evidence that organisms have internal oscillators. Continued investigations under more rigorous environmental conditions convinced Pfeffer (76), Sachs (87a), and many other nineteenth century biologists that leaf oscillations are truly endogenous. The period length under free-running (constant) conditions may be slightly shorter or longer than 24 h, depending upon species, temperature, and light conditions. Under natural conditions, the clock generating the oscillations is reset by light-dark transitions at dawn and dusk. The perturbation can cause a phase *advance* (a specific part of the cycle, for example the maximum, occurs earlier than in controls) or a phase *delay*. The net effect of these daily phase shifts is to synchronize internal time with solar time (78).

After a light stimulus or certain other environmental perturbations, the movements become transiently noncircadian. Effects of the perturbation on the clock cannot be assessed until the transients disappear and circadian oscillations have been restored. A stable shift in phase from the previous pattern would then indicate that the clock had been reset.

Effects of Light on the Clock

Light absorbed by two photosystems affects the clock. Phytochrome is one of the photoreceptors (17, 62, 111), while an unidentified blue and far-red absorbing pigment, BAP, is the other (62, 99). Phytochrome is the more effective pigment, acting only in the Pfr form. Several effects of Pfr have been reported. Irradiation of dark grown *P. vulgaris* with wavelengths between 600 and 700 nm initiates circadian leaf movements, while simultaneous irradiation with wavelengths between 700 and 800 nm cancels this effect (62). Furthermore, daily brief red irradiations, ca 660nm (R), prevent progressive diminution in the amplitude of the oscillations of *Albizzia* leaflets, while daily brief far-red irradiations, ca 730nm (FR), are ineffective (111).

Red light also phase shifts leaf movement rhythms. Prolonged irradiations are required for *P. multiflorus* (17), but 5 min pulses are sufficient for *Samanea* (111). Phase-shifting effects of R in *Samanea* can be prevented by immediately subsequent FR; thus, rhythmic phase shifting displays the R, FR reversibility expected of a phytochrome-mediated process.

There are two reports that far-red irradiation can phase shift rhythms. Such effects in *P. multiflorus* have been attributed to photoreceptors in the lamina, while phase shifting effects of red light have been localized in the pulvinus (17). In *Samanea,* repetitive FR pulses (one every 20 min for 2 h) produce a small phase shift, but the magnitude of the shift resembles that of blue light, and has been ascribed to BAP rather than phytochrome (99).

Blue light-induced phase shifts have been reported in *P. vulgaris* (62), *Albizzia* (99), and *Samanea* (99). They are smaller than those induced by R, and require longer irradiations (in *Samanea,* for exmple, 2 h of pulsed or continuous light). Blue and R probably affect different components of the clock, since their phase shift curves have different shapes and produce zero phase shift at different parts of the cycle (99).

Chemical and Physical Probes of the Mechanism

To assay the behavior of the oscillator that drives leaf movements, Bünning, W. Mayer, and colleagues at Tübingen identified several chemical and physical treatments that alter rhythmic phase and/or frequency (Table 1), presumably by interfering with the clock's metabolic machinery. Pulsed incubation of *P. coccineus* at low temperature (5°) for a few hours as leaves are opening has no effect on rhythmic phase, but a similar perturbation as leaves are closing leads to a phase delay (21). Bünning (14) therefore concluded that the oscillator has energy-independent and energy-dependent phases, the former coincident with opening and the latter with closing. Later investigations revealed that external KCl (20) or the K^+ carrier valinomycin (19) produce phase advances if supplied as leaves are opening (i.e. during the phase characterized as energy-independent), but KCl is ineffective and valinomycin produces only a small delay if supplied as leaves are closing. Transient wilting also shifts the phase of the rhythm (18); this effect could be a consequence of turgor alteration, which stimulates ion fluxes in other plants (reviewed in 126), or to changes in abscisic acid levels (reviewed in 82), since external abscisic acid alters leaflet movements in *Cassia* (6).

Mayer and colleagues have studied the effects of cAMP and compounds that affect its synthesis and degradation on leaf movement rhythms in *P. coccineus* (67, 68) and *Trifolium* (5). Several of the tested compounds alter rhythmic phase and frequency in both plants (Table 1), as well as in rats (31) and *Neurospora* (34), suggesting that regulation of cAMP levels might be part of the oscillator mechanism. However, the *Trifolium* studies do not support this model (5), since cAMP and imidazole have similar effects on rhythmic phase even though imidazole activates phosphodiesterases in ani-

Table 1 Physical and chemical treatments that alter the phase and/or frequency of leaf movement rhythms

		Continuous application		Pulsed application	
Species	Treatment	Dosage	Δ Period[a]	Dosage	Δ Phase[a]
P. coccineus	low temp.			5°, 5 h	– (21)
	wilting			6 h	+ and – (18)
	D_2O	25%	4 h ↑ (16)	99%, 6 h	– (16)
	ethanol	2.5%	4 h ↑ (54)	5–30%, 2 h	+ and – (15)
	methanol	3.0%	3 h ↑ (54)		
	theophylline	0.1%	5 h ↑ (54)	10 mM, 4 h	+ and – (67)
	theobromine	0.1%	2 h ↑ (54)		
	caffeine			10 mM, 4 h	+ and – (68)
	valinomycin			10 μg/1, 5 h	+ and – (19)
	KCl			200 mM, 5 h	+ (20)
Trifolium	cAMP	0.5 mM	1 h ↑ (5)	1 mM, 4 h	– (5)
	theophylline	2.0 mM	2 h ↑ (5)	1 mM, 4 h	+ and – (5)
	imidazole	0.5–1.0 mM	no effect (5)	10 mM, 4 h	– (5)

[a] Symbols: ↑ = increase in period length; – = phase delay; + = phase advance.

mals and microorganisms (85) and therefore would be expected to decrease endogenous cAMP levels. Furthermore theophylline, an inhibitor of phosphodiesterases (85), produces both advances and delays, while cAMP produces only delays.

Methylxanthines such as caffeine, theophylline, and theobromine promote Ca^{2+} release from the sarcoplasmic reticulum in animals (51); Bollig et al (5) have suggested that these compounds and imidazole might perturb circadian oscillations by increase in intracellular Ca^{2+}. It should be possible to test this interesting suggestion, since sensitive methods for monitoring free Ca^{2+} can be used with plant tissues (45, 119).

RAPIDLY DETECTABLE RESPONSES TO LIGHT AND TEMPERATURE CHANGES

Light and temperature alterations have rapidly detectable effects on leaf movements, in addition to effects on the clock.

Light Effects

The same two photosystems that reset the clock have rapid direct effects on the movements. Prolonged irradiation with blue or far-red light promotes opening in *Mimosa* (37), *Albizzia* (33, 50), and *Samanea* (99), while the Pfr form of phytochrome promotes closure in these same species (36, 47, 50, 98, 116). However, the effects of phytochrome conversion are apparent only when BAP is not simultaneously photoactivated. Thus, irradiation with cool white fluorescent light, which emits very little energy in wavelengths longer than 700 nm and would be expected to convert most of the phytochrome to Pfr, promotes opening rather than closure. Activation of BAP apparently masks or inhibits the action of Pfr. An interaction of phytochrome with BAP has also been noted in electrophysiological studies in corn coleoptiles, indicating a common membrane site for the two pigments (81a).

The effects of phytochrome conversion are detectable most rapidly when plants grown in LD are darkened early in the photoperiod (36, 47, 50, 95, 98). Red given immediately before darkness promotes closure while FR inhibits it; however, plants darkened toward the end of the photoperiod close irrespective of the phytochrome state. Thus sensitivity to Pfr is rhythmic; it is required for closure during certain parts of the daily cycle, but is unimportant at times when the clock specifies closure.

A similar situation occurs during an extended dark period (DD). *Albizzia* (94, 95) and *Samanea* (72) leaflets start to open rhythmically at DD=6-8 h, and start to close approximately 12 h later. Red prior to opening decreases the opening angle while FR enhances it, but light treatments before closure have a much smaller effect (95, 98). Once again, sensitivity to Pfr varies rhythmically, and R, when effective, decreases leaflet angle.

Temperature Effects

Reduction in temperature as *Albizzia* (92) or *Samanea* (97) leaflets are starting to open inhibits the movements, but similar treatment during closure is promotive. Incubation in NaN_3 (0.5 mM) also inhibits opening and promotes closure (95), indicating an alternation of energy-requiring and energy-independent phases. Furthermore, the rate of oxygen uptake by *Samanea* pulvini increases during opening and decreases prior to and during closure (100). Since leaflet movements are dependent upon solute movement into and out of pulvinar cells, energy-requiring ion transport apparently peaks during leaflet opening (96), even though this is the energy-independent phase of the clock, as assayed by Bünning and colleagues (14, 21). Thus, the energy requirements for solute transport are ca 180° out of phase with those of the oscillator that drives leaf movements.

Kinetic analysis of *Samanea* movements during a prolonged period of low temperature reveals that the immediate and delayed effects of low temperature are qualitatively different. If the treatment starts just before opening, the rate of movement decreases during the next 90 min (97). However, this inhibition is transient; after several hours at 16°, leaflets start to open slowly, continuing until they reach a wide angle (39). The oscillations then cease, as do circadian oscillations in all organisms when the temperature falls below a critical value for that species (14). Thus the *immediate* effect of low temperature is to reduce leaflet opening while the *delayed* effect is to stop the movement, locking the leaflets open. The immediate effect is interpreted as a slowing down of active transport of solutes through extensor cell membranes, with consequent inhibition of turgor changes that lead to leaflet opening, while longer range effects are interpreted as stopping the clock in its relaxation (minimal energy-requiring) phase. This interpretation emphasizes the distinction between the rhythmic solute movements responsible for turgor changes and the clock itself, whatever its nature.

TURGOR REGULATION IN THE PULVINUS

Pulvini contain all the ingredients needed for the operation of a circadian turgor-based movement capable of being phased by light. In addition to the motor cells whose turgor changes control leaf movements and the ions required for turgor regulation (101), each pulvinus contains clockwork (95, 97), the relevant pigments (58, 95, 98, 121), metabolic machinery, and substrate to keep the clock running for one circadian period or longer (110).

The persistence of free-running, light-sensitive oscillations after the selective dissection of either extensor or flexor cortical cells proves that clockwork and pigments are located on both sides of the pulvinus (73).

The pulvinus is a flexible cylinder that is usually straight during the day and curved at night. It consists of a central vascular core containing xylem and phloem surrounded by three to seven layers of collenchyma cells, which are in turn surrounded by several layers of cortical cells (23, 35, 71, 103). The *motor cells* are cortical cells whose elastic walls and position in the pulvinus permit turgor changes to be translated into leaf movements. They interact with other pulvinar cells whose pumps, channels, and plasmodesmata regulate ion flow, and with collenchyma cells, whose thickened walls serve as apoplastic reservoirs for ion storage (25).

K^+, Cl^-, and Cell Turgor

Pulvinar extracts contain 0.2M–0.5M K^+ (55, 56, 97, 101). More than 25 years ago, Toriyama (118) demonstrated histochemically that K^+ is released from shrinking cells during seismonastic *Mimosa* leaf movements. Allen (1) confirmed these observations by use of ^{42}K, while we and our colleagues provided evidence for K^+ fluxes in the nonseismonastic species *Albizzia* (101) and *Samanea* (97). Using a flame photometer and an electron microprobe, we documented the redistribution of K^+ (91, 94, 95, 97, 98, 101, 102, 104, 106) and Cl^- (104, 106) within the pulvinar microstructure during endogenously rhythmic (94, 95, 97, 104, 106), light regulated (94, 95, 98, 101, 104), and chemically stimulated (91, 102) movements. Subsequent studies using ^{42}K and flame photometry to analyze K^+ movements in *Trifolium* (107, 108) and *Phaseolus* (55, 56, 66), respectively, support the generalization that K^+ is the major cation involved in osmoregulation during leaf movements. K^+ accounts for 50% of the total osmolality of *P. vulgaris* pulvinar extracts (200 mM in mannitol equivalents) (55). Mg^{2+} and Ca^{2+} make smaller contributions (ca 7% and 10% respectively that of K^+), while Cl^-, NO_3^-, and organic acids make up the remainder. The contribution of fructose and fructose-yielding saccharides is negligible. The methods utilized and data obtained prior to 1978 are reviewed in (89).

As cells swell during the light portion of the light-dark cycle, K^+ changes range from 8% per mg fresh wt and 60% per mg protein in extracts from the extensor half of the *P. vulgaris* pulvinus (56) to 1500% (K^+ content, measured as scintillations/min) in the cells in the middle of the *Samanea* extensor region (25, 104). Fluxes of K^+ in the flexor region are generally smaller (Figure 1; see also 55, 56, 66, 104). The large range of reported K^+ data would appear to depend upon: 1. species; 2. plant age and nutritional status (55); 3. tissue examined; 4. resolution of the detection method; and 5. basis on which K^+ data are expressed (56).

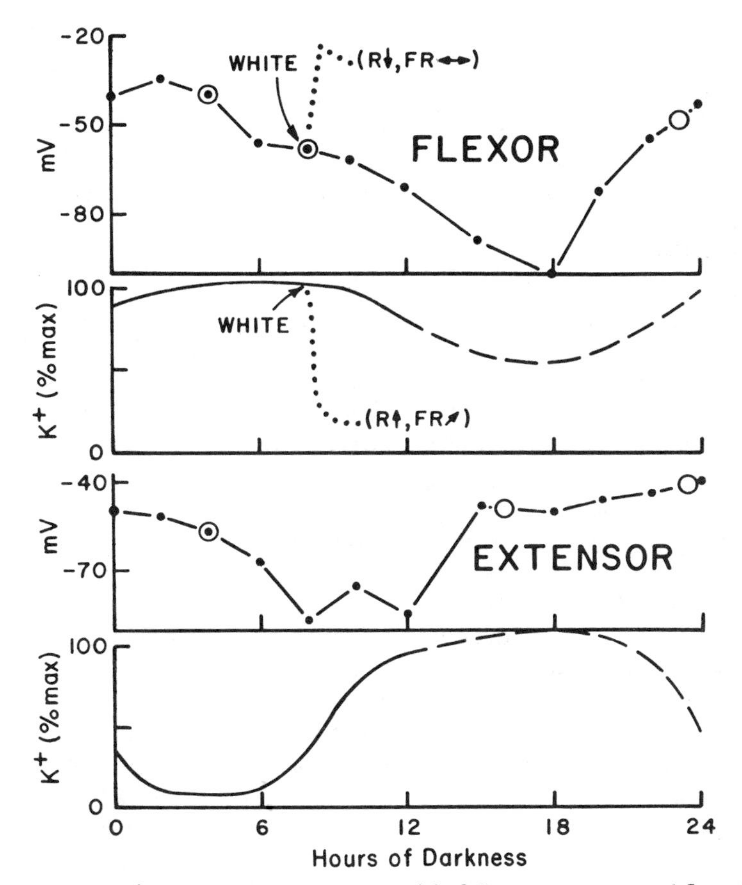

Figure 1 The K^+ content and membrane potential of *Samanea* extensor and flexor cells. Data are from plants darkened for 24 h, except as indicated by dotted lines representing white light followed by R or FR. K^+ values, expressed as % maximum, include both intracellular and extracellular ions. They are based on *Samanea* data in (97, 98, 104), represented by solid lines. *Albizzia* data in (94, 95) contribute toward estimated values, represented by dashed lines. Membrane potential graphs are redrawn from (81b). The large circles indicate times when the cells depolarize in response to external H^+ and sucrose (80). The direction of the small arrows after R and FR indicate the effect of the irradiation.

Cl^- fluxes accompany K^+ fluxes during light-modulated rhythmic movements of *Albizzia* (106), *Samanea* (25, 104), *P. coccineus* (66), and *P. vulgaris* (55), and during seismonastic movements of *Mimosa* (87b). The K^+:Cl^- stoichiometry varies for different species, different plants of the same species, and different regions of the same pulvinus (25, 104). Cl^- fluxes are 75% as high as K^+ fluxes in *Albizzia* (106) but only 40% as high in *P. vulgaris* (55). In *Samanea,* they range from 40% (25) to 80% (104), while in *P. vulgaris,* the concentration of Cl^- increases as the plants age (55). Clearly, anions besides Cl^- must participate in balancing the charge of

K^+. Kiyosawa (55) reported that increases in Cl^-, NO_3^-, and organic acids balance increases in K^+ in swelling cells of *P. vulgaris,* while decreases in these anions balance decreases in K^+ in shrinking cells. Thus both organic and inorganic anions participate in osmoregulation in pulvinar cells, as in stomatal guard cells (reviewed in 82).

Gradmann and colleagues (38, 43, 44, 66), noting the rapid permeation of K^+ through membranes, prefer to interpret turgor pressure changes in terms of less permeable Cl^- and organic anions. Others (46, 82) have challenged this interpretation on the basis that anion and cation movements are electrically coupled and the reflection coefficient for a salt is determined by the less permeant ion. However, if H^+ permeates membranes rapidly in exchange for K^+, Gradmann's analysis would be valid.

These investigators (38, 44, 66) have also proposed that electrogenic Cl^- transport provides the driving force regulating the uptake of part of the K^+ that accumulates in swelling pulvinar cells. The extreme variability of $K^+:Cl^-$ stoichiometry argues against such a mechanism, for it would imply a variable compensatory energy-requiring process.

Evidence for Apoplastic Transport of K^+ and Cl^-

A survey of the extent of the symplast in the *Samanea* pulvinus (71) reveals plasmodesmata in all parts of the cortex examined, although their density (0.7 to 3.9/μm^2 wall area) is not unusually high (84). Thus K^+ and Cl^- might move through the apoplast, symplast, or both, en route from shrinking to swelling cells. Electron microprobe analyses during the 1970s lacked the spatial resolution to discriminate elements in the cell wall from those in the protoplast, but more recent investigations with *Trifolium* (107, 108) and *Samanea* (25) address this question. *Trifolium* leaflets were incubated in ^{42}K for 2 h at different times during 24 h LL. The pulvinus was then severed from the lamina and divided into extensor and flexor halves. Oppositely phased circadian changes in the rate of ^{42}K uptake by each half pulvinus were observed. These results are most readily interpreted in terms of rhythms in the permeability and/or transport properties of the plasmalemma, and thereby imply apoplastic migration.

Newly developed cryo-ultramicrotome techniques (25) permitted preparation of 100 nm freeze-dried sections of *Samanea* pulvini, which were analyzed with a scanning electron microscope (SEM) equipped with an energy-dispersive X-ray analyzer. This type of detector uses a much lower beam current than the wavelength-dispersive X-ray analyzer used for earlier *Albizzia* and *Samanea* elemental analyses; the beam energy could thus be focused on a 20–100 nm diameter region of tissue without volatilization of the elements of interest. It was possible to select wall regions that do not enclose plasmodesmata by avoiding pitfields where plasmodesmata are localized (71). Pitfields are large and readily visible in freeze-dried sections (25) whereas individual plasmodesmata are below the resolution of the methods.

A pulvinus closing 20 min after transfer from white light to darkness has lost K^+ from inside extensor cells to the apoplast and flexor cells (Figure 2a and b). Although K^+ leakage into the apoplast during tissue preparation must be considered, such leakage would not account for the increased ion content in the apoplast of closing leaflets as compared to open controls. The continuous decrease of apoplastic K^+ from extensor to collenchyma to flexor suggests secretion from extensor cells, diffusion through the apoplast, and reabsorption across the plasmalemma of the flexor cells.

However, not all ion distribution patterns can be explained by diffusion along an apoplastic concentration gradient. As shown in Figure 2a and c, K^+ and Cl^- accumulate in the apoplast surrounding extensor cells of open leaflets and flexor cells of closed leaflets. In both situations, scans from the vasculature to peripheral motor cells indicate that extensor-flexor differences in apoplastic K^+ originate in the vascular tissue. It is very unlikely that the latter results are artifactual, since extensor and flexor collenchyma cells with nearly equivalent internal K^+ have very different amounts in the wall (Figure 2c).

A possible structural basis for these data is evident in *Mimosa* (35), where an extensive symplast extends from the middle of the motor organ to the cortex. Transfer cells adjacent to collenchyma cells (35) could move solutes from apoplast to symplast (75). In *Samanea,* vascular collenchyma cells have extraordinarily high levels of Mg^{2+} and phosphorus (R. Satter, unpublished). Since Mg^{2+} is known to activate ATPases, and since Mg^{2+}-dependent ATPases are known to line the ingrowths of some transfer cells (64), a functional link can be envisioned. Furthermore, certain regions of the vascular apoplast stain with the lipid-soluble dye Sudan IV (R. Satter, unpublished), indicating the existence of hydrophobic discontinuities in the apoplast that would prevent free diffusion of solutes. The possible role of vascular transfer cells and symplastic shunt in regulating ion movements is shown in Figure 4, below.

Since plant cell walls have a high density of fixed negative charges and can accumulate 0.3-0.5 M monovalent cation in a medium that lacks divalent cations (29, 82), the apoplast must be considered a reservoir for the storage of ions. The numerous collenchyma cells at the vascular periphery, in apoplastic continuity with neighboring cortical cells, contribute large storage capacity to this reservoir (25, 35). When cells are low in turgor, the apoplast surrounding them is also low in K^+ and Cl^-. What ions have taken their place? We envision their displacement by H^+ ions secreted by pulvinar motor cells, and their movement partly along the apoplast of the collen-

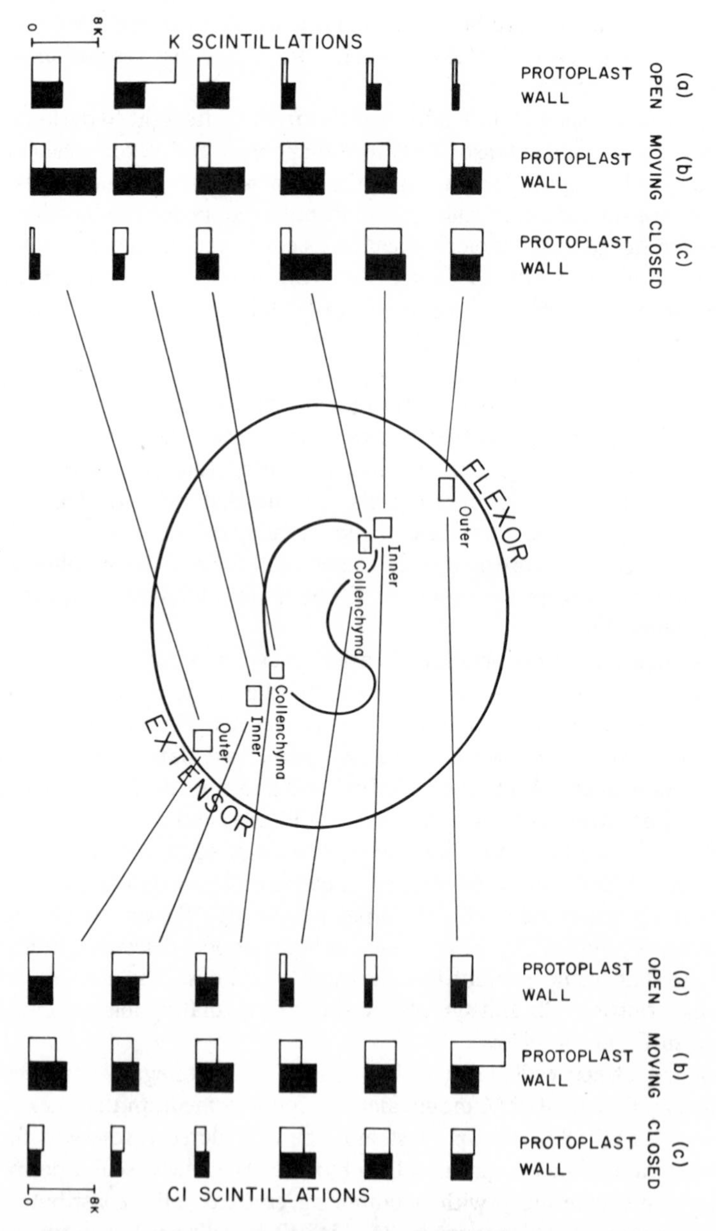

Figure 2 K^+ and Cl^- distributions in cross sections of *Samanea* pulvini from (*a*) open leaflets in white light (angle=120°); (*b*) leaflets in the process of closing after transfer from white light to darkness (angle=60°); (*c*) closed leaflets in the dark (angle=10°). The scale bar corresponds to 8000 scintillations during 150 sec of X-ray detection. From (25).

chyma and partly along a shunt, where transfer cell activity removes them and secretes them on the other side of a suberized barrier (see Figure 4).

Roles of the Tonoplast and the Plasmalemma

Satisfactory methods for separate determination of solutes in vacuole and cytoplasm have not been developed. For example, freeze-dried sections of *Samanea* pulvini do not permit this discrimination, because of structural damage (25). However, osmotically active solutes that regulate leaf movements through turgor change must move into and out of vacuoles. It is thus instructive that the *Albizzia* tonoplast undergoes dramatic changes as motor cells change size (23). The large central vacuole of turgid motor cells breaks up to form many small vacuoles when cells shrink; the small vacuoles coalesce to reform the large central vacuole when cells again swell. Serial sections of extensor cells confirm that the small vacuoles maintain their integrity. Such fragmentation of the central vacuole would provide a mechanism for conserving the surface area of the tonoplast as a cell loses turgor. Furthermore, since cytoplasmic volume remains constant during leaflet movement (23), synchronized changes in the tonoplast and the plasmalemma appear to underlie light-modulated circadian rhythmic leaf movements.

THE MEMBRANE POTENTIAL

Since ions move along electrochemical gradients, measurements of transmembrane potentials in pulvini can contribute toward understanding turgor-based leaf movements. Although circadian rhythms in potentials of pulvinar motor cells have been reported in *Samanea* (80, 81b), *Trifolium* (108), and *P. coccineus* (38, 44), rhythmic waveform, amplitude, and phase with respect to cell turgor differ. These differences might be related in part to the light regimes, since *Samanea* rhythms were monitored during 48 h DD, with plants previously exposed to 16 h L:8 h D cycles, *Trifolium* during 12 h L:12 h D, and *P. coccineus* during 24 h LL, with plants previously exposed to 9 h L:15 h D cycles. As data in Figure 1 and (81b) reveal, both phytochrome and the pigment that controls opening have large effects on the potential of *Samanea* cells.

Methods

Variations in bathing solutions would also account for some reported differences in potential. Total osmolality ranged from 0.2 (108) to ca 200 milliosmoles (38, 44). The medium used for *Samanea* experiments contained 1 mM each of K^+ and Ca^{2+}; since divalent cations bind most effectively to negatively charged wall sites, apoplastic Ca^{2+} might be abnormally high and K^+ abnormally low. The medium used for *P. coccineus* experiments contained 90–99 mM NaCl, on the basis that the water phase of the apoplast in situ has high ionic strength. Measurements of K^+ and Cl^- in the apoplast of *Samanea* confirm this prediction (Figure 2), but reveal that K^+ rather than Na^+ is the dominant cation. Since the addition of 50 mM Na^+ to the solution bathing *Albizzia* leaflets has a large promotive effect on leaflet movement during certain parts of the cycle (92), the bathing solution used for *P. coccineus* measurements might alter rhythms in membrane potential. Thus, these authors may have introduced a new problem in their attempt to solve an old one.

Pulvini are difficult objects for microelectrode impalement, since they are capable of movement and their cells are heavily pigmented, thick-walled, and flexible. Thus, stringent criteria must be used to distinguish cell wall potentials from internal potentials.[4] Comparative studies of different experimental conditions and methods in a single species are clearly required.

Data and Interpretations

Samanea: When *Samanea* plants are darkened for 48 h, extensor cells start to hyperpolarize a few hours before K^+ and Cl^- uptake and attain maximum hyperpolarization just as K^+ and Cl^- accumulation begin (Figure 1). Flexor cells, by contrast, attain maximum hyperpolarization ca 8–10 h later, at a time of minimal K^+ and Cl^-.

These rhythmic patterns are altered considerably by light. The flexor cells of plants exposed to white (81b) or blue (39) light after 8 h of darkness depolarize, whereas similar cells in dark controls continue their rhythmic hyperpolarization (Figure 1). If plants are darkened after 2–3 h of white light, phytochrome photoconversions also alter the potential; R before darkness hyperpolarizes the flexor cells, while subsequent FR reverses the effect of R, restoring the original potential. Thus the membrane potential, like K^+, Cl^-, and leaf movements, is regulated by interactions among a circadian oscillator, phytochrome, and BAP. We interpret light-regulated and rhythmic changes in potential in terms of electrogenic H^+ secretion and back diffusion, as discussed below.

[4] *Samanea* (80, 81b) and *Trifolium* (108) potentials were evaluated by noting electrical characteristics of the preparation in response to current injection. *Samanea* impalements were rejected unless the waveforms contained two distinct components, characteristics of the voltage-charging rates of the electrode and the membrane (80). These are standard criteria used by other electrophysiologists (8, 41). Potentials of *P. coccineus* were rejected unless cells hyperpolarized by 5 mv or more upon reduction of external K^+ from 10 to 1 mM and simultaneous increase of external Na^+ from 90 to 99 mM (38, 44). This should also be a reliable criterion, since internal potentials of higher plant cells are generally more sensitive to K^+ than to Na^+, while the wall potential is insensitive to monovalent cation species but sensitive to total ionic strength (82).

Trifolium: Circadian changes in membrane potential and ^{42}K uptake by extensor and flexor cells of *Trifolium* pulvini were documented during a 12 h L:12 h D cycle. Both extensor and flexor cells start to hyperpolarize several hours before the rate of ^{42}K uptake increases and start to depolarize several hours before it decreases.

P. coccineus: Extensor cells of *P. coccineus* show circadian changes in three electrical parameters during 24 h LL. When turgor is low, the resting potential and CN^--induced depolarizations are large, while K^+-induced voltage changes are small (38). All situations are reversed when cell turgor is high. Gradmann, Mayer, and colleagues therefore conclude that an electrogenic pump in extensor cell membranes is most active when cell turgor is low. Since change in the pH of the bathing solution from 6 to 8 does not alter the potential of the flexor cells, while reduction in external Cl^- from 100 to 20 mM, together with addition of 80 mM NO_3^-, causes a large hyperpolarization (44), they propose that Cl^- rather than H^+ is actively transported through the membranes of *P. coccineus* motor cells. However, lack of response to external pH is not an infallible criterion for lack of H^+ movement across the membrane. H^+ transport is sensitive to external pH in some plants (79) but is insensitive in others (52, 61, 83) where it is apparently under cytoplasmic regulation. Furthermore, reduction in external Cl^- would be expected to depolarize rather than hyperpolarize if there were inward electrogenic Cl^- uniport. The observed hyperpolarization is consistent with $2H^+/Cl^-$ symport, recently proposed for *Chara* (2). Alternatively, hyperpolarization could be a consequence of specific ionic effects of NO_3^-, for large rhythmic changes in the NO_3^- content of both extensor and flexor cells of *P. vulgaris* have recently been reported (55). Thus controls testing the effects of NO_3^- on the potential of *P. coccineus* must be included before drawing conclusions.

Despite differences in data obtained from these three species, an important generalization emerges: Cells on at least one side of the pulvinus hyperpolarize in the low turgor state and depolarize in the high turgor state. To interpret these data, Freudling et al (38) postulate electrogenic Cl^- uptake, followed by electroosmotic H_2O loss coupled to electrically driven Cl^- efflux through wide negatively—or narrow positively—charged pores. Since data supporting electroosmotic water movement are not definitive, and their interpretation controversial (63), we have chosen to omit this mechanism.

MODEL FOR K^+ AND CL^- MOVEMENTS

Having summarized data on leaf movements in several leguminous pulvini, we shall now assemble them into a model. We make no claims of exclusiveness; other models consistent with the data can also be drawn. We include detailed ion transport mechanisms, some demonstrated in other systems but not yet in pulvini, since a useful model should make specific predictions that can be tested experimentally.

Our model is based mainly on data derived from *Albizzia* and *Samanea,* whose ion flux and leaf movements have been investigated in greatest detail, but it should also be applicable to other leguminous pulvini. The model modifies and extends an earlier one (96) proposing that rhythmic behavior results from two competitive processes: active transport, which predominates during half the cycle, and back diffusion, which predominates during the other half. Circadian oscillations, according to this view, depend upon the alternating predominance of membrane-localized pumps and diffusion channels.

Basic Postulates

Postulate 1: Turgor regulation: K^+, Cl^-, and other inorganic and organic anions provide the osmoticum for regulating motor cell turgor. *Supporting data:* The role of K^+ is discussed above. Investigations of leaf movements in *P. vulgaris* (55) and stomatal guard cell movements (82) predict complementary roles for inorganic and organic anions in balancing the charge of K^+.

Postulate 2: Membrane porters: The clock modulates the activity of H^+ pumps, the availability of H^+/sucrose symporters and OH^-/anion antiporters, and the opening of H^+ and K^+ diffusion pathways (Figure 3), while the phytochrome state modulates the opening of Ca^{2+} diffusion pathways, as discussed below. The model postulates rhythmic H^+ pumps and leaks, although either alone would provide a rhythmic driving force for solute uptake. *Supporting data:* Rhythmic pump activity is consistent with the respiratory rhythm in *Samanea* (100), while rhythmic sensitivity of *Samanea* extensor and flexor cell potentials to external H^+ and sucrose (80) provide support for rhythmic H^+/sucrose symporters. In support of rhythmic K^+ diffusion pathways, external K^+ (20) and valinomycin (19) phase shift the rhythm in *P. coccineus* if supplied as leaves are opening, but have little or no effect if supplied as leaves are closing. Both compounds also phase shift a circadian rhythm in nerve impulses in the sea hare *Aplysia* (32).

Several types of evidence suggest that electrogenic H^+ secretion energizes K^+ uptake in pulvinar motor cells (55), as in other plants (60, 83, 112, 114). External sodium acetate inhibits opening (92) and promotes closure (101) in *Albizzia.* This effect could be caused by decrease in transmembrane pH gradients, since weak acids increase cytoplasmic acidity in *Chara* (53). Promotion of opening and inhibition of closure by fusicoccin (6) or auxin (7, 59, 69, 102, 125) in *Cassia* (6), *Phaseolus* (7, 59), *Mimosa* (125), and

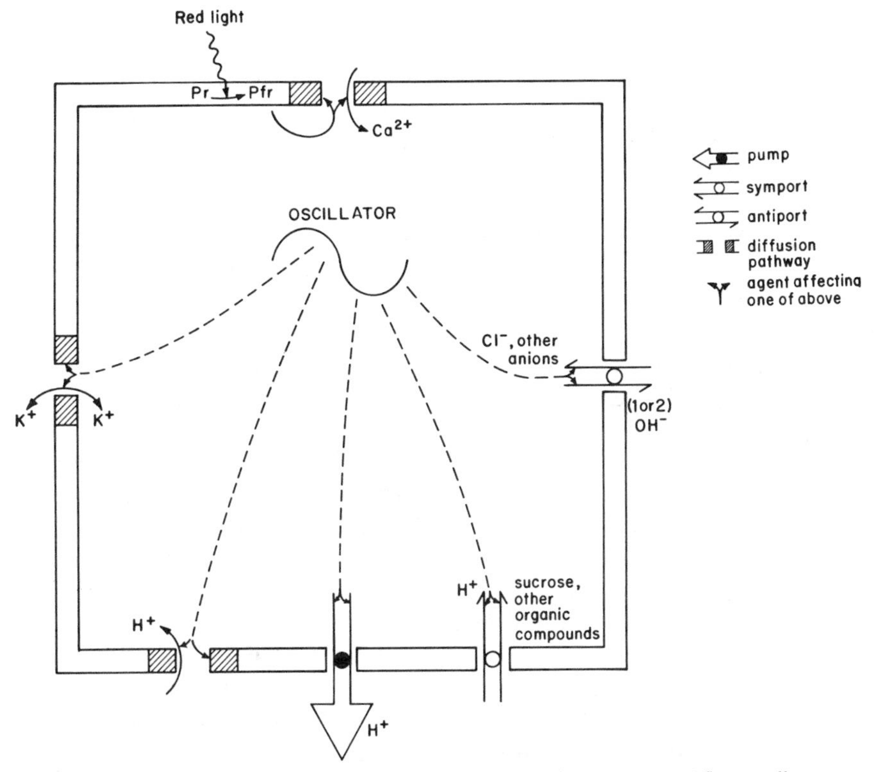

Figure 3 Model for rhythmic and Pfr-modulated porters in extensor and flexor cell membranes. Effects of the clock on proton movements are shown below, other cations left, and anions right. Effects of Pfr on Ca^{2+} movements are shown above.

Albizzia (69, 102) also implicate H^+ in motor cell solute uptake, since both compounds stimulate H^+ secretion from plant cells (65). The effects of auxin on *Albizzia* leaflet movements are correlated with increase in K^+ in extensor cells (102). Vanadate, a potent inhibitor of H^+ secretion from peas (117b) and oats (40, 117b), inhibits rhythmic (105) and Pfr-dependent (93) leaflet closure in *Albizzia.* This result would appear to conflict with the auxin and fusicoccin data discussed above, if these agents all affect hydrogen transport in the same cells. The apparent contradiction could be rationalized if vanadate acts on flexor rather than extensor cells.

The transmembrane proton gradient resulting from electrogenic H^+ secretion appears to energize H^+/sugar co-transport into cells of many higher plants (57). Powering of anion influx by transmembrane pH gradients has also been proposed (49a, 60, 113). Schemes based on OH^-/anion antiport (or H^+/anion symport) (49a, 60) and $2H^+$/anion symport (2, 88) have been suggested. The latter (but not the former) is energetically feasible for *Chara* (88) and supported by electrophysiological data (2).

The model postulates rhythmic pumps and diffusion pathways in cells of the extensor and flexor cortex and in transfer cells in the vascular tissue. Transport structures for other ions (e.g. Ca^{2+}) might also be rhythmic, but available data are insufficient to support such an hypothesis.

Postulate 3: Transport rhythms in extensor and flexor cells: H^+ pump activity in both extensor and flexor cells peaks during K^+ and Cl^- accumulation (in extensor cells during opening and flexor cells during closure). This modifies our previous model (96) which proposed that ion fluxes in both cell types during opening require metabolic energy, while the reverse processes, which occur during closure, are energetically downhill. *Supporting data:* The K^+ content of flexor cells increases ca 15-fold during closure (25), and it is difficult to envision solute accumulation of this magnitude in the absence of metabolic energy input. If the new model is correct, the energy requirements for leaflet movement reflect those of solute accumulation in extensors, but not flexors. Structural differences between extensor and flexor cells are consistent with this interpretation. Extensor cell walls in *Mimosa* (24, 122) and *Albizzia* (N. Campbell, unpublished data) are much thinner than flexor walls, while those in *Mimosa* have a chemical composition that would provide greater elasticity (122). Extensor cells in *Mimosa* (76, 122), *Albizzia* (23, 94, 103), *P. vulgaris* (55, 56), and *Samanea* (97) show the largest changes in ion content and volume during leaf movement, while flexor cells change shape more than size.

The model also postulates solute movements consistent with the greater importance of the extensor cells. Flexor cells hyperpolarize as they lose K^+ and Cl^- during leaflet opening. The electrochemical potential for K^+ would tend to promote net inward K^+ diffusion were it not for active transport processes in extensor and vascular cells, which remove K^+ and Cl^- from the apoplastic continuum. Thus, treatments that reduce the rate of ATP synthesis during leaflet opening, thereby interfering with K^+ and Cl^- accumulation in the extensor cells, necessarily reduce K^+ and Cl^- loss from the flexor cells.

Postulate 4: Phytochrome: Both extensor and flexor cells have phytochrome. Conversion to Pfr decreases K^+ and Cl^- in extensor cells by decreasing H^+ pump activity and/or opening H^+ diffusion pathways while increasing K^+ and Cl^- in flexor cells by the reverse processes. Effects in both cell types are mediated by changes in intracellular free Ca^{2+}. *Supporting data:* Effects of phytochrome conversions on K^+ and Cl^- in *Albizzia* and *Samanea* are well documented (94, 95, 98, 101, 104). Conversion of Pr to Pfr also produces opposite effects in adjacent groups of cells or organs of other plants. For example, R promotes leaf growth in etiolated pea seedlings but inhibits stem growth (74); in mung beans, it promotes the uptake of K^+ (9), Cl^-, and P_i (10) into the hypocotyl tip, but inhibits the uptake of these same ions into subapical segments. When solute uptake and growth are enhanced by Pfr, so is H^+ secretion (10); when these processes are inhibited by Pfr, H^+ secretion is also inhibited. Changes in H^+ flux in mung bean hypocotyls, however, were not detected until 1 h after R.

Samanea flexor cells hyperpolarize 90 sec after R (81b), consistent with Pfr–promoted increase in H^+ secretion. Hyperpolarization precedes massive K^+ entry, suggesting that the increased internal negativity drives K^+ uptake.

Extensor cells lose K^+ after R (98, 101). Activation of H^+/sucrose symporters, permitting H^+ and sucrose to diffuse inward along the electrochemical gradient and depolarize the cells (80), provides a possible mechanism. In support of this postulate, external sucrose is required for differential effects of R and FR on the movements of *Albizzia* (90) and *Samanea* (111) leaflets that have been darkened for 2–3 days. It is also required for the expression of R-FR differences in stem growth (4) and bud expansion (42) of etiolated peas deprived of their main endogenous carbohydrate source.

To understand how the same photoreceptor, activated by quanta of the same wavelength, could produce opposite responses in adjacent tissues, we have considered a recent model (86) for phytochrome action based on Ca^{2+} fluxes. Phytochrome controls Ca^{2+} fluxes in oat coleoptiles (45), *Nitella* (124), and *Mougeotia* (30), leading Roux and colleagues to propose that conversion to Pfr promotes the diffusion of Ca^{2+} from the cellular exterior (45) and mitochondria (26) into the cytoplasm; this would increase free cytoplasmic Ca^{2+}, in turn activating the Ca^{2+}-binding protein calmodulin. Calmodulin activity could be coupled to several transport processes, since intracellular free Ca^{2+} levels have already been shown to regulate membrane phosphorylation, Ca^{2+}-ATPase (27), and the opening of K^+ diffusion pathways (22).

Leaf movements fit this interpretation. Fluxes of Ca^{2+} from the wall to the cellular interior during seismonastic movements of *Mimosa* (119) have been reported. Displacement of Ca^{2+} from the wall could remove mechanical constraints that impede leaf movements (117a), as well as facilitating ion movements by altering membrane structure and function (27). Data suggesting that Ca^{2+} fluxes might phase shift leaf movement rhythms in *P. coccineus* (67, 68) and *Trifolium* (5) are discussed above.

Postulate 5: The apoplast: The wall continuum functions as a reservoir for the storage of K^+, Cl^-, and Ca^{2+}. Protons replace K^+ at negative sites in the wall during certain parts of the rhythmic cycle. *Supporting data:* The apoplast provides the connecting link between cells losing solutes and those absorbing them. The thick walls of the collenchyma cells lying between vascular and motor cells have a high and variable K^+ and Cl^- content (Figure 2) and, by virtue of their large volume, have a considerable adsorptive capacity for these turgor-regulating ions.

Postulate 6: Ion movements through the vasculature: Some K^+ and Cl^- ions diffuse along the concentration gradient through the thick walls of the collenchyma cells. Others follow a shunt pathway in the vascular tissue, as diagrammed in Figure 4. At DD= ca 8 h, outward H^+ pumps in transfer cells at the flexor end of the shunt are activated and K^+ diffusion pathways are open. These cells accumulate high levels of K^+ and Cl^-. Both ions then move through a symplastic route terminating in transfer cells in the extensor region of the vascular tissue. These cells do not actively accumulate K^+ during this part of the cycle, since H^+ pump activity is minimal. Thus the ions diffuse through the symplast from rhythmically active transfer cells in the flexor to oppositely phased transfer cells in the extensor. K^+ and Cl^- then diffuse out of the latter cells into the apoplast. The effect of this mechanism is to deplete ions from one region of the apoplast and to discharge them into another. One half cycle later, the direction of ion flow is reversed. *Supporting data:* This postulate is required because of discontinuities in ion content in the apoplast of the vasculature (Figure 2a and c). Suitably positioned transfer cells equipped with rhythmic pumps and connected to a symplastic pathway could resolve this difficulty. Appropriately located transfer cells and plasmodesmata have been described in *Mimosa* (35). Ion movements could provide the distribution patterns in Figure 2a and c).

K^+ and the Membrane Potential: Comparative Kinetics

Can the model aid in analyzing the relationship between membrane potential and K^+ fluxes depicted in Figure 1? Unfortunately, a rigorous analysis is not possible for several reasons. Measured potentials are summations of potentials across the tonoplast and plasmalemma; this problem is particularly troublesome for pulvinar cells whose central vacuole becomes fragmented during part of the cycle and reforms ca 12 h later (23). In addition, the ionic composition of the apoplast varies with time and space (Figure 2); thus it is not possible to choose a bathing solution that would not alter some part of the apoplastic milieu during each part of the rhythmic cycle. Membrane resistance data would be very useful for understanding permeability changes, but are not available. However, the most serious problem concerns available kinetic K^+ data (Figure 1), which do not discriminate between intracellular and extracellular ions, whereas the equilibrium potential for any ion depends upon the ratio of its intracellular to extracellular concen-

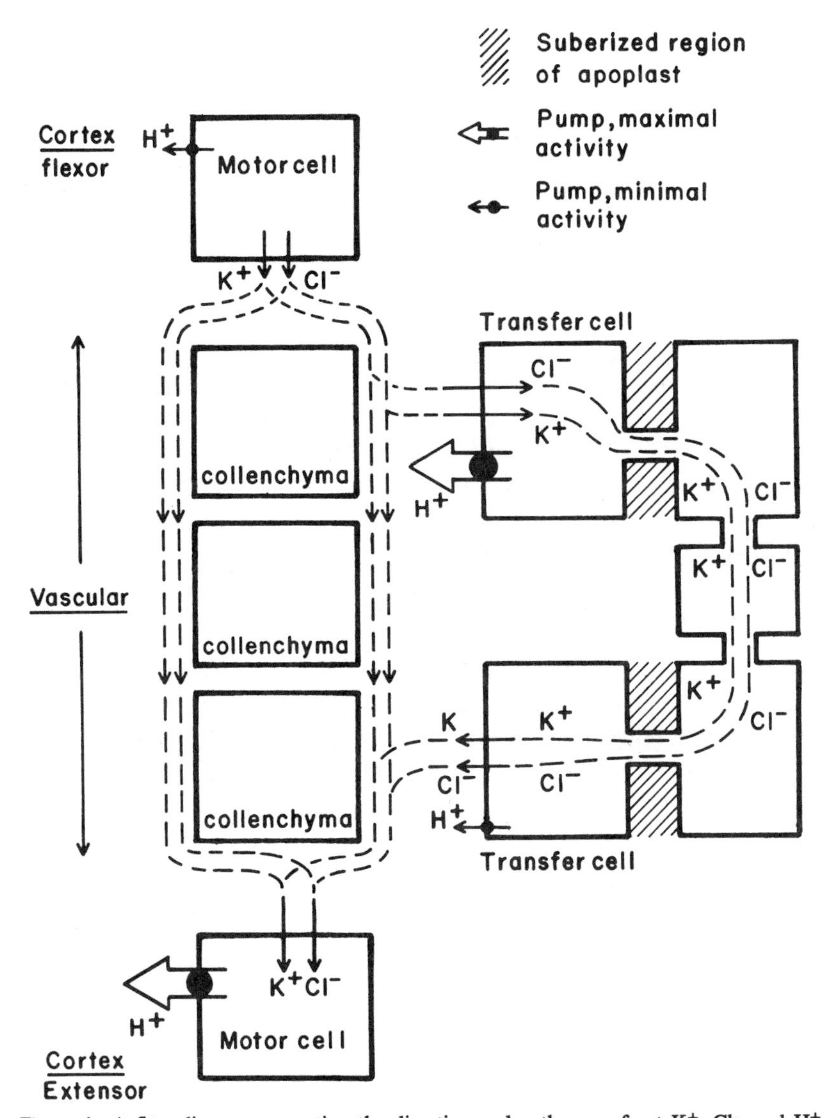

Figure 4 A flow diagram suggesting the direction and pathways of net K^+, Cl^-, and H^+ movements during leaflet opening.

tration. Furthermore, K^+ data are expressed on a mass rather than concentration basis.

Despite these caveats, the model can provide a reasonable explanation for the major anomaly resulting from electrophysiological studies in the three relevant species, i.e. the lack of synchrony between K^+ uptake and hyperpolarization. The following scenario is based on *Samanea* data in Figure 1. It begins with hyperpolarization of the extensor cells from DD=O–8h, which is readily interpreted as due to electrogenic H^+ secretion. The increased internal negativity would be expected to drive K^+ uptake, but the data reveal that K^+ does not start to accumulate until several hours later. If the clock schedules a later opening of K^+ diffusion pathways, the lag is explained. Hyperpolarization is maximum from DD=8–12, as is the rate of K^+ uptake.

Flexor K^+ peaks during the first few hours of darkness, when the potential is at its lowest value. This depolarization would favor outward K^+ diffusion during the next several hours, along the electrochemical gradient. However, rhythmically regulated diffusion pathways may restrict the rate at which this ion redistribution occurs. Once extensor and vascular cells commence massive K^+ absorption, K^+ depletion from the flexor apoplast would steepen diffusion gradients across flexor cell membranes.

The flexor cells hyperpolarize as they lose K^+ and Cl^-. This high potential is a diffusion potential, dependent upon the rapid outward movement of K^+ as compared to Cl^-. Although not a direct consequence of flexor-localized active transport, this hyperpolarization is driven by electrogenic transport in the extensor cells through the shared K^+ reservoir in the apoplastic continuum. K^+ efflux is severely limited in DD [See Figure 1 and (94, 97)], presumably as a result of the accompanying hyperpolarization. If leaflets are illuminated with white light after 8 h of darkness, the flexor cells depolarize, lowering the electrochemical potential for K^+ and promoting its outward diffusion.

CONCLUSIONS

Leaf movements, visual indicators of the redistribution of K^+ and Cl^- between cells on opposite sides of the pulvinus, offer numerous advantages for studying ion transport and turgor regulation. Movements of K^+, Cl^-, Mg^{2+}, and phosphorus are sufficiently massive to be detectable by X-ray analytic techniques that afford subcellular resolution (25). The K^+ and Cl^- fluxes are probably powered by electrogenic H^+ secretion, and their light sensitivity mediated by Ca^{2+} fluxes. These postulates are testable using methods already applied to other plant tissue, e.g. pH sensitive dyes (123) or microelectrodes (80, 117b) to measure H^+ fluxes, Ca^{2+}-binding dyes to measure free Ca^{2+} levels (45, 119), and a vibrating reed electrometer (49b) to monitor the flow of ions through the apoplast linking extensor and flexor. If isolated groups of homogeneous cells or protoplasts derived from them retain their rhythmicity and photosensitivity, they should be particularly useful for examining quantitative relationships between H^+, K^+, and Cl^-

fluxes. Plants grown in Cl^--deficient solutions could help delineate the roles of Cl^- and other anions in turgor regulation.

But can leaf movements also contribute toward understanding the oscillator that acts as driver? Despite two decades of intensive research (77), little is known about the biochemical or biophysical nature of the oscillator, except for the broad generalization that membranes and ion fluxes are involved, directly or indirectly (115).

Although numerous compounds that alter rhythmic phase or frequency have been identified, they have had limited impact on our understanding of the clock mechanism, partly due to their lack of specificity (e.g. D_2O, ethanol) or to inadequate understanding of the mode of action in the system under investigation. It should be possible to test how compounds that alter rhythmic phase and frequency in leguminous pulvini affect K^+, Cl^-, H^+, and Ca^{2+} movements, using methods already applied to pulvini and other plant tissues. Interactions among H^+, Ca^{2+}, and K^+ fluxes would appear capable of producing self-sustained oscillations, for the activities of numerous enzymes are sensitive to pH and intracellular free Ca^{2+}, (27), while membrane configuration and transport properties are sensitive to turgor (126), in turn modulated by K^+ and accompanying anions. Peter Mitchell's (70) prediction that transport can help us understand biochemistry might be extended to physiology as well—perhaps even to the nature of the clock.

ACKNOWLEDGMENTS

We thank the National Science Foundation, whose extended support over many years has made possible our own research and the writing of this manuscript. Recent support from the National Aeronautics and Space Administration has also been helpful. We are grateful to Drs. George W. Bates, Dale Sanders, and Clifford L. Slayman for critical comments on an early draft of the manuscript. We thank Lorraine Klump and Anna Francesconi for competent secretarial assistance.

Literature Cited

1. Allen, R. D. 1969. Mechanism of the seismonastic reaction in *Mimosa pudica*. *Plant Physiol.* 44:1101–7
2. Beilby, M. J., Walker, N. A. 1980. Chloride influx into *Chara:* Electrogenic and probably two-proton coupled. In *Plant Membrane Transport: Current Conceptual Issues,* ed. R. M. Spanswick, W. J. Lucas, J. Dainty, pp. 571–72. Amsterdam: Elsevier
3. Bentrup, F. W. 1979. Reception and transduction of electrical and mechanical stimuli. In *Encyclopedia of Plant Physiology, New Series VII, Physiology of Movements,* ed. W. Haupt, M. E. Feinleib, pp. 42–70. Berlin: Springer
4. Bertsch, W. F., Hillman, W. S. 1961. The photoinhibition of growth in etiolated stem segments. I. Growth caused by sugars in *Pisum. Am. J. Bot.* 48:504–11
5. Bollig, I., Mayer, K., Mayer, W., Engelmann, W. 1978. Effects of cAMP, theophylline, imidazole, and 4-(3,4-dimethoxybenzyl)-2-imidazolidone on the leaf movement rhythm of *Trifolium repens*—a test of the cAMP hypothesis of circadian rhythms. *Planta* 141:255–30

6. Bonnemain, J.-L., Roblin, G., Gaillochet, J., Fleurat-Lessard, P. 1978. Effets de l'acide abscissique et de la fusicoccine sur les réactions motrices des pulvinus du *Cassia fasciculata* Michx. et du *Mimosa pudica* L. *C. R. Soc. Biol. Ser. D* 286:1681–86
7. Brauner, L., Arslan, N. 1951. Experiments on the auxin reactions of the pulvinus of *Phaseolus multiflorus. Rev. Fac. Sci. Univ. Istanbul* 16B:257–300
8. Brennecke, R., Lindemann, B. 1971. A chopped-current clamp for current injection and recording of membrane polarization with single electrodes of changing resistance. *TITJ Life Sci.* 1:53–58
9. Brownlee, C., Kendrick, R. E. 1979. Ion fluxes and phytochrome protons in mung bean hypocotyl segments. I. Fluxes of potassium. *Plant Physiol.* 64:206–10
10. Brownlee, C., Kendrick, R. E. 1979. Ion fluxes and phytochrome protons in mung bean hypocotyl segments. II. Fluxes of chloride, protons and orthophosphate in apical and subhook segments. *Plant Physiol.* 64:211–13
11. Bünning, E. 1959. Allgemeine Gesetze und Phänomene der pflanzlichen Bewegungphysiologie. In *Encyclopedia of Plant Physiology*, XVIII, Pt. 1, ed. W. Ruhland, pp. 8–23. Berlin: Springer
12. Bünning, E. 1959. Die seismonastischen Reaktionen. See Ref. 11, pp. 184–242
13. Bünning, E. 1959. Tagesperiodische Bewegungen. See Ref. 11, pp. 579–656
14. Bünning, E. 1973. *The Physiological Clock.* London: English Univ. Press. 3rd ed.
15. Bünning, E., Baltes, J. 1962. Wirkung von Äthylalkohol auf die physiologische Uhr. *Naturwissenschaften* 49:19
16. Bünning, E., Baltes, J. 1963. Zur Wirkung von schwerem Wasser auf die endogene Tagesrhythmik. *Naturwissenschaften* 50:622
17. Bünning, E., Moser, I. 1966. Response-Kurven bei der circadianen Rhythmik von *Phaseolus. Planta* 69:101–10
18. Bünning, E., Moser, I. 1968. Einfluss des Wassers auf die circadiane Rhythmik von *Phaseolus. Naturwissenschaften* 55:450–51
19. Bünning, E., Moser, I. 1972. Influence of valinomycin on circadian leaf movements of *Phaseolus. Proc. Natl. Acad. Sci. USA* 69:2732–33
20. Bünning, E., Moser, I. 1973. Light-induced phase shifts of circadian leaf movements of *Phaseolus:* comparison with the effects of potassium and of ethyl alcohol. *Proc. Natl. Acad. Sci. USA* 70:3387–89
21. Bünning, E., Tazawa, M. 1957. Über den Temperatureinfluss auf die endogene Tagesrhythmik bei *Phaseolus. Planta* 50:107–21
22. Burgess, G. M., Claret, M., Jenkinson, D. H. 1979. Effects of catecholamines, ATP and ionophore A23187 on potassium and calcium movements in isolated hepatocytes. *Nature* 279:544–46
23. Campbell, N. A., Garber, R. C. 1980. Vacuolar reorganization in the motor cells of *Albizzia* during leaf movement. *Planta* 148:251–55
24. Campbell, N. A., Garber, R. C., Satter, R. L., Stika, K. M., Morrison, G. H. 1980. Potassium movements in the motor organ of the sensitive plant. Submitted for publication
25. Campbell, N. A., Satter, R. L., Garber, R. C. 1981. Apoplastic transport of ions in the motor organ of *Samanea. Proc. Natl. Acad. Sci. USA:* In press
26. Cedel, T. E., Roux, S. J. 1980. Modulation of a mitochondrial function by oat phytochrome *in vitro. Plant Physiol.* 66:704–9
27. Cheung, W. Y. 1980. Calmodulin plays a pivotal role in cellular regulation. *Science* 207:19–27
28. DeMairan, M. 1729. Observation botanique. *Histoire de l'Academie Royale des Sciences Paris,* p. 35
29. DeMarty, M., Ayadi, A., Monnier, A., Morvan, C., Thellier, M. 1977. Electrochemical properties of isolated cell walls of *Lemna minor* L. In *Transmembrane Ionic Exchanges in Plants,* ed. M. Theiller, A. Monnier, M. DeMarty, J. Dainty, pp. 64–73. Rouen: CNRS et Univ. Rouen
30. Dreyer, E. M., Weisenseel, M. H. 1979. Phytochrome-mediated uptake of calcium in *Mougeotia* cells. *Planta* 146:31–39
31. Ehret, C. F., Potter, V. R., Dobra, K. W. 1975. Chronotypic action of theophylline and of pentobarbital as circadian Zeitgebers in the rat. *Science* 188:1212–15
32. Eskin, A. 1979. Circadian system of the *Aplysia* eye: properties of the pacemaker and mechanisms of its entrainment. *Fed. Proc.* 38:2573–79
33. Evans, L. T., Allaway, W. G. 1972. Action spectrum for the opening of *Albizzia julibrissin* pinnules, and the role of phytochrome in the closing movements of pinnules and of stomata of *Vica faba. Aust. J. Biol. Sci.* 25:885–93

34. Feldman, J. F. 1975. Circadian periodicity in *Neurospora:* alteration by inhibitors of cyclic AMP phosphodiesterase. *Science* 190:789–90
35. Fleurat-Lessard, P., Bonnemain, J.-L. 1978. Structural and ultrastructural characteristics of the vascular apparatus of the sensitive plant (*Mimosa pudica* L.). *Protoplasma* 94:127–43
36. Fondéville, J. C., Borthwick, H. A., Hendricks, S. B. 1966. Leaflet movement of *Mimosa pudica* L. indicative of phytochrome action. *Planta* 69:357–64
37. Fondéville, J. C., Schneider, M. J., Borthwick, H. A., Hendricks, S. B. 1967. Photocontrol of *Mimosa pudica* L. leaf movement. *Planta* 75:228–38
38. Freudling, C., Mayer, W., Gradmann, D. 1980. Electrical membrane properties and circadian rhythm in extensor cells of the laminar pulvini of *Phaseolus coccineus* L. *Plant Physiol.* 65:966–68
39. Galston, A. W., Satter, R. L. 1976. Light, clocks and ion flux: an analysis of leaf movement. In *Light and Plant Development,* ed. H. Smith, pp. 159–84. London: Butterworth's
40. Gepstein, S. 1980. The relation between H^+ excretion and senescence of leaves. *Plant Physiol.* 65:62 (Abstr.)
41. Goldsmith, M. H., Fernandez, H. R., Goldsmith, T. H. 1972. Electrical properties of parenchymal cell membranes in the oat coleoptile. *Planta* 102:302–23
42. Goren, R., Galston, A. W. 1966. Control by phytochrome of ^{14}C-sucrose incorporation into buds of etiolated pea seedlings. *Plant Physiol.* 41:1055–64
43. Gradmann, D. 1977. Potassium and turgor pressure in plants. *J. Theor. Biol.* 65:597–99
44. Gradmann, D., Mayer, W. 1977. Membrane potentials and ion permeabilities in flexor cells of the laminar pulvini of *Phaseolus coccineus* L. *Planta* 137: 19–24
45. Hale, C. C. II, Roux, S. J. 1980. Photoreversible calcium fluxes induced by phytochrome in oat coleoptile cells. *Plant Physiol.* 65:658–62
46. Hastings, D. F., Gutknecht, J. 1978. Potassium and turgor pressure in plants. *J. Theor. Biol.* 73:363–66
47. Hillman, W. S., Koukkari, W. L. 1967. Phytochrome effects in the nyctinastic leaf movements of *Albizzia julibrissin* and some other legumes. *Plant Physiol.* 42:1413–18
48. Hoshizaki, T., Hamner, K. C. 1964. Circadian leaf movements: Persistence in bean plants grown in continuous high-intensity light. *Science* 144: 1240–41
49a. Jacoby, B., Rudich, B. 1980. pH-gradient dependent Cl^--flux in ATP depleted root cells. See Ref. 2, pp. 497–98
49b. Jaffe, L. F. 1979. Control of development by ionic currents. In *Membrane Transduction Mechanisms. Soc. Gen. Physiol. Ser. 33,* ed. R. A. Cone, J. E. Dowling, pp. 199–231. New York: Raven
50. Jaffe, M. J., Galston, A. W. 1967. Phytochrome control of rapid nyctinastic movements and membrane permeability in *Albizzia julibrissin. Planta* 77:135–41
51. Johnson, P. N., Inesi, G. 1969. The effect of methylxanthines and local anaesthetics on fragmented sacroplasmic reticulum. *J. Pharmacol. Exp. Ther.* 169:308–14
52. Jones, M. G. K., Novacky, A., Dropkin, V. H. 1975. Transmembrane potentials of parenchyma cells and nematode-induced transfer cells. *Protoplasma* 85:15–37
53. Keifer, D. W. 1980. Alteration of cytoplasmic pH in *Chara,* through membrane transport processes. See Ref. 2, pp. 569–70
54. Keller, S. 1960. Über die Wirkung chemischer Faktoren auf die tagesperiodischen Blattbewegungen von *Phaseolus multiflorus. Z. Bot.* 48:32–57
55. Kiyosawa, K. 1979. Unequal distribution of potassium and anions within the *Phaseolus* pulvinus during circadian leaf movement. *Plant Cell Physiol.* 20:1621–34
56. Kiyosawa, K., Tanaka, H. 1976. Change in potassium distribution in a *Phaseolus* pulvinus during circadian movement of the leaf. *Plant Cell Physiol.* 17:289–98
57. Komor, E., Tanner, W. 1980. Proton-cotransport of sugars in plants. See Ref. 2, pp. 247–60
58. Koukkari, W. L., Hillman, W. S. 1968. Pulvini as the photoreceptors in the phytochrome effect on nyctinasty in *Albizzia julibrissin. Plant Physiol.* 43:698–704
59. Krieger, K. G. 1978. Early time course and specificity of auxin effects on turgor movement of the bean pulvinus. *Planta* 140:107–9
60. Lin, W. 1979. Potassium and phosphate uptake in corn roots. Further evidence for an electrogenic H^+/K^+ exchanger and an OH^-/P_i antiporter. *Plant Physiol.* 63:952–55

61. Lin, W., Hanson, J. B. 1976. Cell potentials, cell resistance and proton fluxes in corn root tissue. Effects of dithioerythritol. *Plant Physiol.* 58:276–82
62. Lörcher, L. 1958. Die Wirkung verschiedener Lichtqualitäten auf die endogene Tagesrhythmik von *Phaseolus. Z. Bot.* 46:209–41
63. MacRobbie, E. A. C. 1971. Phloem translocation. Facts and mechanisms: a comparative survey. *Biol. Rev.* 46: 429–81
64. Maier, K., Maier, U. 1972. Localization of beta-glycerophosphatase and Mg^{++} -activated adenosine triphosphatase in a moss haustorium, and the relation of these enzymes to the cell wall labyrinth. *Protoplasma* 75:91–112
65. Marrè, E., Lado, P., Rasi-Caldogno, F., Colombo, R., De Michelis, M. I. 1974. Evidence for the coupling of proton extrusion to K^+ uptake in pea internode segments treated in fusicoccin or auxin. *Plant Sci. Lett.* 3:365–79
66. Mayer, W. 1977. Kalium-und Chloridverteilung in Laminargelenk von *Phaseolus coccineus* L. während der circadianen Blattbewegung im tagesperiodischen Licht-Dunkelwechsel. *Z. Pflanzenphysiol.* 83:127–35
67. Mayer, W., Gruner, R., Strubel, H. 1975. Periodenverlängerung und Phasenverschiebungen der circadianen Rhythmik von *Phaseolus coccineus* L. durch Theophyllin. *Planta* 125:141–48
68. Mayer, W., Scherer, I. 1975. Phase shifting effect of caffeine in the circadian rhythm of *Phaseolus coccineus* L. *Z. Naturforsch. Teil C* 30:855–56
69. McEvoy, R. C., Koukkari, W. L. 1972. Effects of ethylenediaminetetraacetic acid, auxin and gibberellic acid on phytochrome controlled nyctinasty in *Albizzia julibrissin. Physiol. Plant.* 26:143–47
70. Mitchell, P. 1979. Compartmentation and communication in living systems. Ligand conduction: a general catalytic principle in chemical, osmotic and chemiosmotic reaction systems. *Eur. J. Biochem.* 95:1–20
71. Morse, M. J., Satter, R. L. 1979. Relationships between motor cell ultrastructure and leaf movements in *Samanea saman. Physiol. Plant.* 46:338–46
72. Palmer, J. H., Asprey, G. F. 1958. Studies in the nyctinastic movement of the leaf pinnae of *Samanea saman* (Jacq) Merrill. I. A general description of the effect of light on the nyctinastic rhythm. *Planta* 51:757–69
73. Palmer, J. H., Asprey, G. F. 1958. Studies in the nyctinastic movement of the leaf pinnae of *Samanea saman* (Jacq) Merrill. II. The behaviour of the upper and lower half pulvini. *Planta* 51:770–85
74. Parker, M. W., Hendricks, S. B., Borthwick, H. A., Went, F. W. 1949. Spectral sensitivities for leaf and stem growth of etiolated pea seedlings and their similarity to action spectra for photoperiodism. *Am. J. Bot.* 36:194–204
75. Pate, J. S., Gunning, B. E. S. 1972. Transfer cells. *Ann. Rev. Plant. Physiol.* 23:173–96
76. Pfeffer, W. 1905. *The Physiology of Plants,* Vol. 3. Transl. A. J. Ewart. Oxford: Clarendon
77. Pittendrigh, C. S. 1976. Circadian clocks: What are they? In *The Molecular Basis of Circadian Rhythms,* ed. J. W. Hastings, H.-G. Schweiger, pp. 11–48. Berlin: Abakon Verlagsgesellschaft
78. Pittendrigh, C. S., Minis, D. H. 1964. The entrainment of circadian oscillations by light and their role as photoperiodic clocks. *Am. Nat.* 98: 261–94
79. Poole, R. J. 1974. Ion transport and electrogenic pumps in storage tissue cells. *Can. J. Bot.* 52:1023–28
80. Racusen, R. H., Galston, A. W. 1977. Electrical evidence for rhythmic changes in the cotransport of sucrose and hydrogen ions in *Samanea* pulvini. *Planta* 135:57–62
81a. Racusen, R. H., Galston, A. W. 1980. Phytochrome modifies blue-light–induced electrical changes in corn coleoptiles. *Plant Physiol.* 66:534–35
81b. Racusen, R. H., Satter, R. L. 1975. Rhythmic and phytochrome-regulated changes in transmembrane potential in *Samanea* pulvini. *Nature* 255:408–10
82. Raschke, K. 1979. Movements of stomata, See Ref. 3, pp. 383–441
83. Raven, J. A., Smith, F. A. 1980. The chemiosmotic viewpoint. See Ref. 2, pp. 161–77
84. Robards, A. W. 1976. Plasmodesmata in higher plants. In *Intracellular Communication in Plants: Studies on Plasmodesmata,* ed. B. E. S. Gunning, A. W. Robards, pp. 15–57. Berlin: Springer
85. Robison, G. A., Butcher, R. W., Sutherland, E. W. 1971. *Cyclic AMP.* New York: Academic
86. Roux, S. J., McEntire, K., Slocum, R. D., Cedel, T. E., Hale, C. C. II. 1980. Phytochrome induces photoreversible calcium fluxes in a purified mitochon-

drial fraction from oats. *Proc. Natl. Acad. Sci. USA:* In press

87a. Sachs, J. 1887. *Lectures on the Physiology of Plants.* Transl. H. M. Ward. Oxford: Clarendon

87b. Samejima, M., Sibaoka, T. 1980. Changes in the extracellular ion concentration in the main pulvinus of *Mimosa pudica* during rapid movement and recovery. *Plant Cell Physiol.* 21:467–79

88. Sanders, D. 1980. Control of Cl^- influx in *Chara* by cytoplasmic Cl^- concentration. *J. Membr. Biol.* 52:51–60

89. Satter, R. L. 1979. Leaf movements and tendril curling. See Ref. 3, pp. 442–84

90. Satter, R. L., Applewhite, P. B., Chaudhri, J., Galston, A. W. 1976. P_{fr} phytochrome and sucrose requirement for rhythmic leaflet movement in *Albizzia. Photochem. Photobiol.* 23: 107–12

91. Satter, R. L., Applewhite, P. B., Galston, A. W. 1974. Rhythmic potassium flux in *Albizzia:* effect of aminophylline, cations and inhibitors of respiration and protein synthesis. *Plant Physiol.* 54: 280–85

92. Satter, R. L., Applewhite, P. B., Kreis, D. J. Jr., Galston, A. W. 1973. Rhythmic leaflet movement in *Albizzia julibrissin:* effect of electrolytes and temperature alteration. *Plant Physiol.* 52: 202–7

93. Satter, R. L., Fries, C. 1981. Vanadate inhibition of phytochrome effects on leaflet movement in *Albizzia julibrissin. Plant Sci. Lett.:* In press

94. Satter, R. L., Galston, A. W. 1971. Potassium flux: a common feature of *Albizzia* leaflet movement controlled by phytochrome or endogenous rhythm. *Science* 174:518–20

95. Satter, R. L., Galston, A. W. 1971. Phytochrome-controlled nyctinasty in *Albizzia julibrissin:* III. Interaction between an endogenous rhythm and phytochrome in control of potassium flux and leaflet movement. *Plant Physiol.* 48:740–46

96. Satter, R. L., Galston, A. W. 1973. Leaf movements: rosetta stone of plant behavior? *BioScience* 23:407–16

97. Satter, R. L., Geballe, G. T., Applewhite, P. B., Galston, A. W. 1974. Potassium flux and leaf movement in *Samanea saman.* I. Rhythmic movement. *J. Gen. Physiol.* 64:413–30

98. Satter, R. L., Geballe, G. T., Galston, A. W. 1974. Potassium flux and leaf movement in *Samanea saman.* II. Phytochrome controlled movement. *J. Gen. Physiol.* 64:431–42

99. Satter, R. L., Guggino, S. E., Lonergan, T. A., Galston, A. W. 1981. The effects of blue and far red light on rhythmic leaflet movements in *Samanea* and *Albizzia. Plant Physiol.* In press

100. Satter, R. L., Hatch, A. M., Gill, M. K. 1979. A circadian rhythm in oxygen uptake by *Samanea* pulvini. *Plant Physiol.* 64:379–81

101. Satter, R. L., Marinoff, P., Galston, A. W. 1970. Phytochrome controlled nyctinasty in *Albizzia julibrissin.* II. Potassium flux as a basis for leaflet movement. *Am. J. Bot.* 57:916–26

102. Satter, R. L., Marinoff, P., Galston, A. W. 1972. Phytochrome controlled nyctinasty in *Albizzia julibrissin.* IV. Auxin effects on leaflet movement and K flux. *Plant Physiol.* 50:235–41

103. Satter, R. L., Sabnis, D. D., Galston, A. W. 1970. Phytochrome controlled nyctinasty in *Albizzia julibrissin.* I. Anatomy and fine structure of the pulvinule. *Am. J. Bot.* 57:374–81

104. Satter, R. L., Schrempf, M., Chaudhri, J., Galston, A. W. 1977. Phytochrome and circadian clocks in *Samanea:* rhythmic redistribution of potassium and chloride within the pulvinus during long dark periods. *Plant Physiol.* 59:231–35

105. Saxe, H., Satter, R. L. 1979. Effect of vanadate on rhythmic leaflet movement in *Albizzia julibrissin. Plant Physiol.* 64:905–7

106. Schrempf, M., Satter, R. L., Galston, A. W. 1976. Potassium-linked chloride fluxes during rhythmic leaf movement of *Albizzia julibrissin. Plant Physiol.* 58:190–92

107. Scott, B. I. H., Gulline, H. F. 1975. Membrane changes in a circadian system. *Nature* 254:69–70

108. Scott, B. I. H., Gulline, H. F., Robinson, G. R. 1977. Circadian electrochemical changes in the pulvinules of *Trifolium repens* L. *Aust. J. Plant Physiol.* 4:193–206

109. Sibaoka, T. 1969. Physiology of rapid movements in higher plants. *Ann. Rev. Plant Physiol.* 20:165–84

110. Simon, E., Satter, R. L., Galston, A. W. 1976. Circadian rhythmicity in excised *Samanea* pulvini. I. Sucrose-white light interactions. *Plant Physiol.* 58:417–20

111. Simon, E., Satter, R. L., Galston, A. W. 1976. Circadian rhythmicity in excised *Samanea* pulvini. II. Resetting the clock by phytochrome conversion. *Plant Physiol.* 58:421–25

112. Slayman, C. L. 1974. Proton pumping and generalized energetics of transport:

a review. In *Membrane Transport in Plants,* ed. U. Zimmermann, J. Dainty, pp. 107–19. Berlin: Springer

113. Smith, F. A. 1970. The mechanism of chloride transport in Characean cells. *New Phytol.* 69:903–17
114. Spanswick, R. M. 1972. Evidence for an electrogenic ion pump in *Nitella translucens.* I. The effects of pH, K^+, Na^+, light and temperature on the membrane potential and resistance. *Biochim. Biophys. Acta* 288:73–89
115. Sweeney, B. M. 1976. Evidence that membranes are components of circadian oscillators. See Ref. 77, pp. 267–81
116. Sweet, H. C., Hillman, W. S. 1969. Phytochrome control of nyctinasty in *Samanea* as modified by oxygen, submergence, and chemicals. *Physiol. Plant.* 22:776–86

117a. Tagawa, T., Bonner, J. 1957. Mechanical properties of the *Avena* coleoptile as related to auxin and ionic interactions. *Plant Physiol.* 32:207–12

117b. Taiz, L., Jacobs, M. 1980. Vanadate inhibition of auxin-induced H^+ secretion and elongation in pea epicotyls and oat coleoptiles. *Plant Physiol.* 65:131 (Abstr.)

118. Toriyama, H. 1955. Observational and experimental studies of sensitive plants. VI. The migration of potassium in the primary pulvinus. *Cytologia* 20:367–77
119. Toriyama, H., Jaffe, M. J. 1972. Migration of calcium and its role in the regulation of seismonasty in the motor cell of *Mimosa pudica* L. *Plant Physiol.* 49:72–81
120. Umrath, K. 1959. Mögliche Mechanismen von Krümmungsbewegungen. See Ref. 11, pp. 111–18
121. Watanabe, S., Sibaoka, T. 1973. Site of photo-reception to opening response in *Mimosa* leaflets. *Plant Cell Physiol.* 14:1221–24
122. Weintraub, M. 1952. Leaf movements in *Mimosa pudica* L. *New Phytol.* 50:357–82
123. Weisenseel, M. H., Dorn, A., Jaffe, L. F. 1979. Natural H^+ currents traverse growing roots and root hairs of barley (*Hordeum vulgare* L.). *Plant Physiol.* 64:512–18
124. Weisenseel, M. H., Ruppert, H. K. 1977. Phytochrome and calcium ions are involved in light-induced membrane depolarization in *Nitella. Planta* 137: 225–29
125. Williams, C. N., Raghavan, V. 1966. Effects of light and growth substances on the diurnal movements of the leaflets of *Mimosa pudica. J. Exp. Bot.* 17:742–49
126. Zimmermann, U. 1978. Physics of turgor- and osmoregulation. *Ann. Rev. Plant Physiol.* 29:121–48

Ann. Rev. Plant Physiol. 1981. 32:111–37

CHLOROPLAST PROTEINS: SYNTHESIS, TRANSPORT, AND ASSEMBLY

◆7707

R. John Ellis

Department of Biological Sciences, University of Warwick, Coventry, CV4 7AL, United Kingdom

---

CONTENTS

INTRODUCTION

Life on this planet depends on the photosynthetic ability of chloroplasts, so it is not surprising that a huge literature exists about the mechanism of photosynthesis. But another thread of interest in chloroplasts can be dis-

0066-4294/81/0601-0111$01.00

cerned. Chloroplasts, like mitochondria, contain an entire genetic system in addition to the one located in the nucleus and cytoplasm. The term "genetic system" is used here to embrace the four linked components required to express genetic information: DNA, DNA polymerase, RNA polymerase, and a protein-synthesizing apparatus. It appears that the possession of such extranuclear genetic systems is a fundamental feature of the eukaryotic mode of cell organization. There are no well-authenticated cases of eukaryotic cells which lack such extranuclear genetic systems under natural conditions, although they can be generated in the laboratory (e.g. *Saccharomyces* mutants which lack mitochondrial DNA). The existence of such systems raises questions as to their precise role in cellular differentiation. Thus today it is as important to ask where a chloroplast protein is synthesized and encoded as it is to enquire about its function in photosynthesis.

A second reason can be given for being interested in chloroplast protein synthesis. The major conceptual challenge in biology at the present time is to unravel the molecular basis of differentiation. The leaf is a highly differentiated tissue because of the presence of chloroplasts. Moreover, chloroplasts are easy to isolate and contain large amounts of particular proteins, e.g. the stromal enzyme ribulose bisphosphate carboxylase (RuBP carboxylase) and the thylakoid chlorophyll *a/b* binding protein complex. The hope is that study of the genetic aspects of chloroplasts, as expressed in protein synthesis, will provide insights into both differentiation and into the significance of extranuclear genetic systems in general.

It is now clear that the formation of chloroplasts results from a complex interplay between the nuclear genetic system and that based in the plastid. The fascination of this subject lies in unraveling the molecular events of this interplay. Some chloroplast proteins are synthesized inside the developing organelle, while many others are made in the cytoplasmic compartment. There is thus no prospect that chloroplasts can be cultured like cells. Current evidence indicates that those chloroplast proteins made outside the chloroplast are encoded in the nuclear genome, while those made by chloroplast ribosomes are encoded in the chloroplast genome. It follows that a traffic of nuclear-encoded polypeptides flows across the chloroplast envelope, and recent advances have allowed this process to be studied in cell-free extracts. The incoming polypeptides are assembled into their functional state inside the organelle. This assembly process can also be demonstrated now in cell-free extracts, and involves interactions between nuclear-coded polypeptides and chloroplast-encoded polypeptides.

The purpose of this review is to discuss selected aspects of these synthesis, transport, and assembly processes. Many of the views stated are my own. The emphasis will be on the results of using in vitro methods, including the use of isolated chloroplasts, molecular cloning, and restriction endonuclease and hybridization probe analysis, since most of the recent advances have come from this sort of technology.

The topic of chloroplast protein synthesis has not been considered specifically in this series previously. An excellent introduction to the whole area of plastid biology is the second edition of the book by Kirk & Tilney-Bassett (94). A more specialized book on organelle genetics has been published recently (75). Reviews discussing different aspects of chloroplast development have been published both in this series (8, 10, 14, 20, 91, 95, 109, 111, 133, 134) and elsewhere(76, 88), and as reports of recent conferences (1, 40, 97, 127).

SYNTHESIS

Magnitude of the Problem

Much effort has been concentrated on answering the basic question of where in the cell chloroplast polypeptides are encoded and synthesized. Figure 1 shows a two-dimensional gel analysis of chloroplast polypeptides from *Pisum sativum* chloroplasts. At least 140 spots can be seen when stromal preparations are analyzed, and at least 50 spots and streaks from thylakoid preparations (117). Consideration of the number of enzymes and metabolic pathways known to occur in chloroplasts suggests that the total of about 190 resolvable polypeptides represents just the tip of the plastid protein iceberg.

Two points can be made from such polypeptide analyses:

1. The total molecular weight of chloroplast polypeptides exceeds by at least severalfold the total potential coding capacity of the chloroplast genome, which is about 6×10^6 daltons of polypeptide (59). It can be concluded that not all the chloroplast polypeptides are encoded in the chloroplast genome. This conclusion is in full accord with earlier data derived from genetic studies (75, 94).

2. The majority of the resolved polypeptides are unidentified, especially in the stromal fraction. Some progress has been made recently in identifying the function of several of the thylakoid polypeptides resolved by one-dimensional gel electrophoresis (29, 30, 32, 41, 85, 107, 108, 110, 121, 130, 131). This highlights the main obstacle holding up progress in this field; there is too little collaboration between workers studying chloroplast enzymes at the functional level and those studying chloroplast polypeptides at the physical level. Such collaboration is necessary before we can produce a polypeptide map of identified functions for the chloroplast.

In contrast to the insufficiency of chloroplast DNA to encode all the polypeptides found in this organelle, there is no shortage of chloroplast

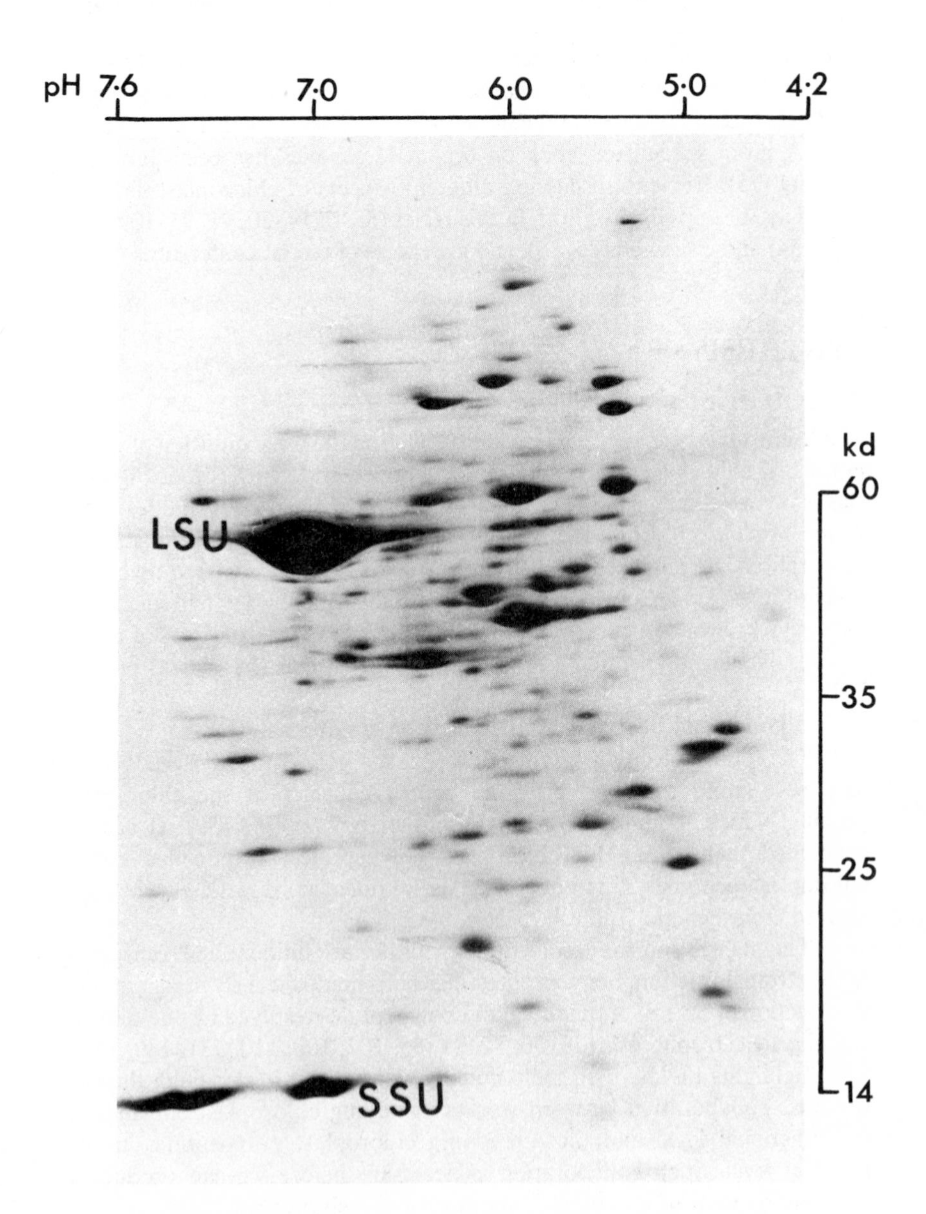
pH 7·6
7·0
6·0
5·0
4·2
kd
60
35
25
14
LSU
SSU

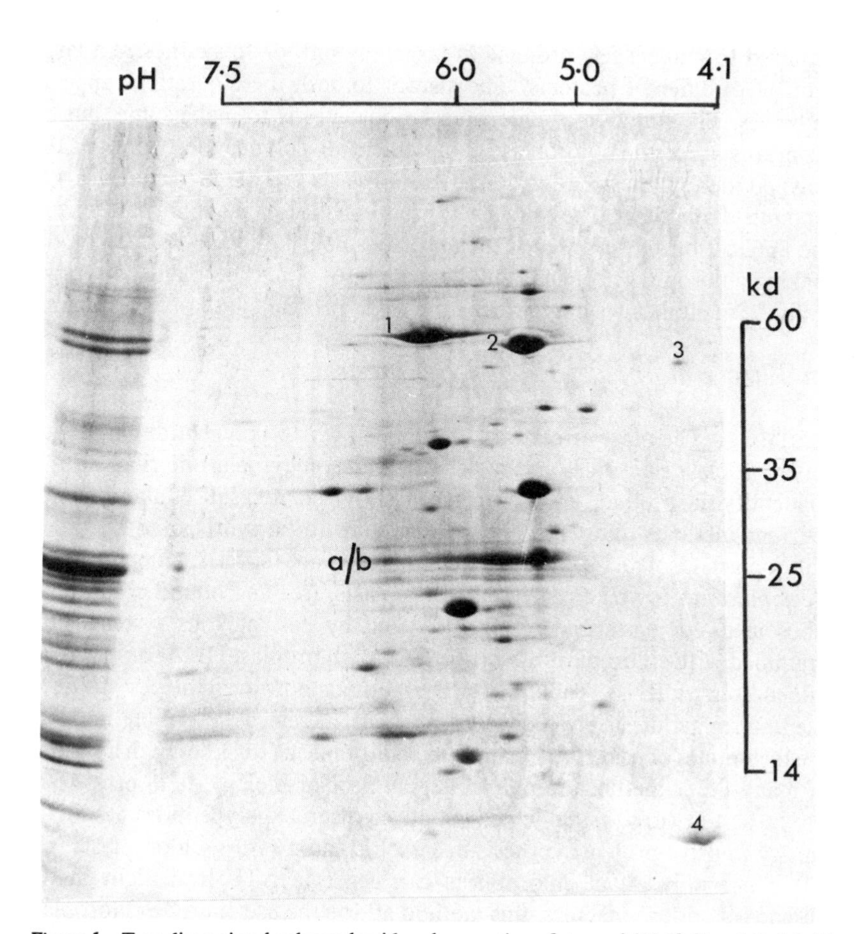

Figure 1 Two-dimensional polyacrylamide gel separation of stromal (A) (*left*) and thylakoid (B) (*above*) fractions from *Pisum sativum* chloroplasts. Chloroplasts were purified by the rapid Percoll method of Mills & Joy (105); fractions were separated by isoelectric focusing in the horizontal dimension and sodium dodecylsulphate electrophoresis in the vertical dimension. Symbols: kd, kilodaltons; LSU and SSU refer to the large and small subunits of RuBP carboxylase respectively; the SSU shows two electrofocusing variants; numbers 1 and 2 refer to the alpha and beta subunits respectively of the chloroplast coupling factor; 3 and 4 refer to two very acidic proteins; a/b marks the chlorophyll *a/b* binding protein which runs as a streak due to its hydrophobicity. Reprinted with permission from Roscoe & Ellis (117).

ribosomes for their synthesis. Chloroplast ribosomes can represent up to 50% of the total ribosomal complement of leaves, depending on the species and the stage of leaf development (54, 59). Is this abundance of ribosomes required to make a few proteins in large amount, or to synthesize a large number of different proteins? The answer to both these questions appears to be yes. The function of chloroplast ribosomes is to make some, but by no means all of the chloroplast complement of polypeptides; some of the polypeptides synthesized inside the chloroplasts are made in much larger amounts than others. The evidence for this conclusion has come partly from the application of heat treatment and ribosomal inhibitors to intact cells making chloroplast proteins, and from analyses of mutant plants, but especially from studies with a variety of in vitro protein-synthesizing systems.

In Vitro Chloroplast Protein Synthesis

SYSTEMS The most direct way to establish which polypeptides are made on chloroplast ribosomes and which are made on cytoplasmic ribosomes is to identify the products of protein synthesis by isolated subcellular systems. Current methods allow discrete polypeptides to be synthesized by intact chloroplasts isolated from several higher plant species, *Euglena* and *Acetabularia,* as well as by lysed chloroplasts, free and bound chloroplast ribosomes, etioplasts and proplastids, and by heterologous systems programmed with chloroplast messenger RNA, chloroplast DNA, or cytoplasmic messenger RNA. Table 1 lists the methods now available for studying the in vitro synthesis of chloroplast proteins. A review describing in detail the techniques of using intact and lysed chloroplasts, together with accounts of many other methods in chloroplast molecular biology, is in press (60).

The first in vitro system which produced discrete polypeptides used light energy to drive protein synthesis in intact isolated *Pisum* chloroplasts (17, 53). Since only intact chloroplasts can generate ATP from light in the absence of added cofactors, this method allows the use of crude chloroplast preparations which can be made very rapidly. With practice, the time taken to extract chloroplasts from their normal environment and place them in an illuminated tube with labeled amino acids can be as short as 3 minutes. Since the envelope around the chloroplasts is intact, the microenvironment around the polysomes is likely to be more normal than in lysed systems. The light-driven system thus provides a baseline against which to judge the results of further fractionating the system. The results of using this rapid method have since been confirmed with spinach chloroplasts purified on silica sol gradients (106).

An essential component of the medium for protein synthesis by both intact and lysed chloroplasts is the presence of potassium ions (53). Re-

Table 1 In vitro chloroplast protein synthetic systems[a]

System	Species	Reference
Intact chloroplasts	*Pisum sativum*	17, 44, 45, 48, 53
	Spinacia oleracea	22, 36, 103, 106, 110, 128, 143
	Hordeum vulgare	53
	Zea mays	53, 78, 79
	Euglena gracilis	16, 112, 137
	Acetabularia	80
Lysed chloroplasts	*Pisum sativum*	53
	Spinacia oleracea	22
Free chloroplast ribosomes	*Pisum sativum*	53
	Hordeum vulgare	3
	Chlamydomonas reinhardii	74
Bound chloroplast ribosomes	*Pisum sativum*	2, 53
	Chlamydomonas reinhardii	104
Etioplasts and proplastids	*Pisum sativum*	126
	Euglena gracilis	43
Chloroplast mRNA plus extracts of *E. coli*, wheat germ, or reticulocytes	*Spinacia oleracea*	23, 47, 84, 128, 141
	Zea mays	11, 99
	Spirodela oligorrhiza	114
	Cucumis sativus	139
Chloroplast DNA plus extracts of *E. coli* and reticulocytes	*Spinacia oleracea*	23
	Zea mays	9, 37
	Chlamydomonas reinhardii	100, 116
Cytoplasmic mRNA plus extracts of wheat germ	*Pisum sativum*	12, 27, 87, 129
	Lemna gibba	135, 136
	Spinacia oleracea	33, 34
	Chlamydomonas reinhardii	42, 123

[a] Updated and reprinted from (63) with permission.

placement of potassium ions by sodium ions prevents both light-dependent protein synthesis in intact chloroplasts and ATP-dependent protein synthesis by free ribosomes. These results suggest that rapid changes in the monovalent cation contents of the stroma occur when chloroplasts are resuspended in different media. The simplest explanation is that resuspension in sodium ion-containing media results in an efflux of stromal potassium ions, perhaps in exchange for sodium ions. This suggestion requires investigation in view of the reported impermeability of isolated mature chloroplasts to monovalent cations [(77); but compare (101)].

PRODUCTS The products of light-driven protein synthesis were first analyzed for *Pisum sativum* chloroplasts incubated with labeled amino acids (17, 48). The analytic method used was electrophoresis on cylindrical polyacrylamide gels which were cut into slices for counting. These studies have since been repeated using high specific activity [^{35}S] methionine, slab gels, and autoradiography, which give greater sensitivity and much higher resolution (63). Polypeptides in all the structural fractions prepared from chloroplasts are labeled, including the chloroplast envelope (92).

In terms of the amount of incorporated methionine, there are two major products and many minor ones. One major product is soluble and has been identified as the large subunit of RuBP carboxylase (17). Subsequent work confirmed that this subunit is a major product of chloroplast protein synthesis in other species, including *Spinacia oleracea* (22, 106), *Hordeum vulgare* (3), and *Euglena gracilis* (137). The abundance of RuBP carboxylase in leaves (55) could explain why chloroplast ribosomes account for such a high proportion of the total leaf complement. As predicted (52), the abundance of this protein makes it ideal for in vitro studies of protein synthesis—I regard it as the globin of the plant biochemist. It is no accident that the first reported in vitro translation of a specific mRNA for a plant enzyme produced the large subunit of RuBP carboxylase (84).

Subsequent work has shown that chloroplasts from *Zea mays* (99), *Spinacia oleracea* (84, 128, 141), and *Euglena gracilis* (118) contain mRNA for the large subunit, but not for the small subunit of RuBP carboxylase. The mRNA is translated by free but not by bound ribosomes (53), and the free polysomes contain only two to five ribosomes (74). All laboratories are in agreement that this mRNA does not possess enough poly(A) to be retained by oligo(dT) cellulose under normal conditions. Non-polyadenylated mRNA for the large subunit has also been isolated from *Chlamydomonas reinhardii* (74, 89) and *Spirodela oligorrhiza* (115).

Direct evidence that the gene for the large subunit is located as a single copy per chloroplast DNA circle has been obtained for *Zea mays* (9, 37) and *Chlamydomonas reinhardii* (73, 100). Detailed mapping and sequencing analyses are in progress on restriction endonuclease fragments of chloroplast DNA cloned in *Escherichia coli* (98). In a tour de force, Bottomley & Whitfeld (23) synthesized large subunits of *Spinacia oleracea* carboxylase from chloroplast DNA incubated in a coupled transcription-translation system obtained from *Escherichia coli.* It is predicted that synthesis of this polypeptide in minicells will soon be achieved; this should permit studies of the properties of native large subunits in the absence of small subunits.

The second major product of in vitro chloroplast protein synthesis is an insoluble polypeptide of molecular weight 32,000, which is firmly bound to the thylakoids. I refer to this product as peak D since it was the fourth labeled peak seen on cylindrical gels in the early studies (48). Time course studies show that peak D is initially made as a precursor of slightly higher molecular weight in *Pisum sativum* (57). The identity of peak D is still unknown. Some of the label in the peak D region of the gel is precipitated by antiserum to the chloroplast coupling factor, suggesting that it may be a membrane component of the ATP-synthase complex (61). However, subsequent work (unpublished) in the author's laboratory shows that peak D is lost early in the purification of this complex from *Pisum* thylakoids by a published method (142). ATP-synthase purified from *Spinacia* thylakoids also shows no polypeptide of molecular weight 32,000 (110).

Although peak D is a major product of chloroplast protein synthesis in terms of incorporation, it does not accumulate as does the carboxylase large subunit. It can be separated from closely running polypeptides by extraction of thylakoids in chloroform/methanol (57). Peak D thus appears to be a minor hydrophobic component of thylakoids which turns over rapidly. Its synthesis can be detected in plastids old enough to have ceased the manufacture of RuBP carboxylase (126).

A polypeptide with similar properties to peak D has been described from *Spirodela oligorrhiza* (49, 114, 115), *Euglena gracilis* (137), *Acetabularia* (80), *Zea mays* (11, 78), *Spinacia oleracea* (47, 106), and *Chlamydomonas reinhardii* (100). In the latter three species, the gene encoding this polypeptide occurs as a single copy per chloroplast DNA circle at the same position relative to the inverted repeat units. Thus a high degree of conservation of the position for this chloroplast gene is suggested from green algae to higher plants.

The major unresolved problem in the field of in vitro chloroplast protein synthesis is to determine the function of the peak D polypeptide and to determine why chloroplasts spend so much time continually making it, only to degrade it. Extensive studies in *Spirodela* have shown that the accumulation of this polypeptide is not essential for photosynthetic carbon dioxide fixation or thylakoid development (140). One possibility is that this protein serves a scavenging function in photosynthesis, for example to prevent activated electrons from occasionally reducing the wrong component. Such a role may result in the protein becoming degraded, and thus it would need to be continually synthesized. Another possibility is that the protein has protease activity and so digests itself. Bennett (13) has suggested that immature chloroplasts require a protease to remove chlorophyll *a/b* binding protein molecules made in the dark when chlorophyll synthesis stops.

What can be said about the minor products of chloroplast protein synthesis? The number of resolvable products has increased with the improvement in techniques. About 80 labeled spots can be seen on a two-dimensional separation of the stromal fraction from *Pisum* chloroplasts incubated with [^{35}S] methionine (61). A similar number of soluble products has been reported for *Spinacia* chloroplasts (36). It is noticeable that most of the labeled spots do not coincide with stained spots, suggesting that they are polypeptides present in lower amounts than the visible set. The identities of these soluble products are unknown for the most part. The published data from in vivo experiments provide no clue as to what they might be. On the other hand, it is largely the enzymes of the Calvin cycle that have been examined in such experiments; enzymes of the many other metabolic pathways found in chloroplasts have not been examined in this regard. Thus a large discrepancy presently exists between the results of in vivo experiments and those of in vitro incorporation experiments as to the function of chloroplast ribosomes.

A number of laboratories in the past few years have embarked on the task of identifying some of the minor products of in vitro chloroplast protein synthesis. Table 2 summarizes the present position. It should be emphasized that early attempts to identify chloroplast coupling factor components and cytochrome *f* as products were unsuccessful (48). This failure can now be attributed to the very small extent of incorporation found in each of the minor products of in vitro chloroplast protein synthesis. Thus the incorporation into cytochrome *f* accounts for only 0.02% of the total incorporation into chloroplast proteins (44). Workers entering this field should be prepared to use large amounts of high specific activity labeled amino acids and to employ efficient methods of concentrating the polypeptides suspected of being made inside the chloroplast.

ARE ALL CHLOROPLAST-SYNTHESIZED POLYPEPTIDES ENCODED IN CHLOROPLAST DNA? The observation that a particular polypeptide is synthesized by isolated chloroplasts does not permit the conclusion that it is encoded in chloroplast DNA. This limitation arises from the fact that protein synthesis is not coupled to RNA synthesis in isolated chloroplasts (17, 22). Despite speculation over many years (24, 25, 90), there is no

Table 2 Identified products of in vitro chloroplast protein synthesis[a]

Polypeptide	Reference
Large subunit of RuBP carboxylase	3, 17, 22, 80, 137
Three or four[b] subunits of the chloroplast coupling factor (CF_1)	53, 78, 103, 110
Elongation factors T and G of chloroplast protein synthesis	36
Cytochrome *f*	44
Cytochrome b_{559}	143
Dicyclohexylcarbodiimide-binding protein	45
Apoprotein of chlorophyll-protein complex I	80, 143

[a] Updated and reprinted with permission from (63).

[b] Nelson et al (110) reported that four of the five subunits were labeled in isolated *Spinacia* chloroplasts, whereas Morgenthaler et al (103) found only three were labeled in the same species. Work in the author's laboratory with *Pisum* chloroplasts also finds only three subunits labeled.

rigorous evidence that RNA crosses the chloroplast envelope in either direction. A claim that *Euglena* chloroplasts export several tRNAs to the cytoplasm (102) has been refuted (124). A critical test of the possibility that chloroplasts import RNA is possible, but has not been reported. This test involves the labeling by in vitro iodination of all the RNA species found in highly purified intact chloroplasts, followed by separation of these RNA species by two-dimensional methods. If all these RNA species are transcribed from chloroplast DNA, then every one of them should hybridize to chloroplast DNA. In the absence of contrary evidence, the current tendency is to assume that all the polypeptides made by isolated chloroplasts are encoded in the chloroplast genome.

An interesting calculation can be made on the assumption that each labeled spot from in vitro chloroplast incubations represents a unique polypeptide. The total molecular weight of detectable polypeptides being synthesized is about 3×10^6, which accounts for about 50% of the total potential coding capacity of the unique sequences in this genome (assuming asymmetric transcription and no overlapping genes). Hybridization studies in *Euglena* show that up to about 50% of the chloroplast DNA is represented as RNA trancripts (28). So we may be within striking distance of being able to account for the information content of chloroplast DNA in terms of resolvable polypeptides. The task for the future is to identify these and to check their encoding in the chloroplast genome.

ARE SORGHUM MESOPHYLL CHLOROPLASTS DIFFERENT? A series of recent papers on protein synthesis by mesophyll chloroplasts isolated from the C4 plant *Sorghum vulgare* deserves special comment (70–72). These papers report the synthesis of both subunits of RuBP carboxylase when *Sorghum* chloroplasts isolated from mesophyll cells (which lack this enzyme) are incubated with total RNA from a C3 plant. The amount of carboxylase made is sufficient to give both a visible precipitin line and detectable enzyme activity. Such finding are remarkable and unique; in vitro protein synthesis normally produces products in only tracer quantities. Certain features of the reported data are difficult to reconcile with our present knowledge of protein synthesis. Three such features are as follows:

1. The incubation medium used for the chloroplasts contains no potassium ions; studies of the effect of added potassium ions reveals only slight stimulation at 10 mM with a decrease of activity at higher concentrations (72). This behavior is completely different from all the other in vitro protein synthetic systems, where potassium or ammonium ions are an absolute requirement (53).

2. The report that carboxylase is made in such massive amounts that it can be detected by its enzymic activity and visualized as a precipitin line

(at least 0.1 μg antigen) is at variance with the observed small extent of incorporation of labeled amino acids (about 10^3 cpm) into the immunoprecipitate.

3. Incubation of the chloroplasts with poly(U) gives polyphenylalanine of a discrete molecular weight (70). This result cannot be reconciled with the polydisperse nature of poly(U) preparations which consist of greatly varying molecular sizes.

In Vivo Systems

USE OF INHIBITORS The application of inhibitors of macromolecular synthesis to intact cells and tissues was in use as a general method for studying the function of the chloroplast genetic system long before the modern preference for in vitro systems became established. The first use of the method (94) became possible once it was realized that there are some similarities between prokaryotic and organellar protein synthesis as regards their inhibition by certain antibiotics such as chloramphenicol and lincomycin. Protein synthesis by the cytoplasmic ribosomes of eukaryotic cells, on the other hand, is inhibited by a different set of antibiotics such as cycloheximide. The hope arose that it might be possible to use such compounds to define the precise contributions of the organellar and extraorganellar genetic systems to chloroplast biogenesis. It tended to be overlooked that their specificities were originally determined on a small number of in vitro systems, but were required to operate in the considerably more complex environment of the intact cell where side effects are possible. Indeed, the first reported use of chloramphenicol on a plant tissue was as an inhibitor of ion uptake (132). The effect of compounds such as chloramphenicol and cycloheximide on processes additional to protein synthesis are now well documented (53). It follows that the inhibitor approach must be used with great caution if incorrect interpretations are to be avoided, and it is not surprising that conflicting conclusions were drawn by different workers in the early literature about the site of synthesis of some chloroplast proteins (24). With the benefit of hindsight it can be said that the use of inhibitors gives suggestive rather than conclusive results, and they have proved to be far more useful with some species than with others. Thus inhibitors must not be used indiscriminately on an untried species; their specificity must be checked in vivo by suitable tests. Criteria for the proper use of inhibitors, together with discussion of the problems involved in interpretation, have been presented (56).

The bulk of published inhibitor experiments suggest that most of the chloroplast proteins so far examined are synthesized on cytoplasmic ribosomes (24, 31, 53, 58, 69, 76, 88). Thus synthesis of the enzymes of the Calvin cycle (except for RuBP carboxylase), ferredoxin, RNA polymerase, aminoacyl-tRNA synthetases, and the majority of the thylakoid and ribosomal polypeptides continues when protein synthesis in the chloroplast is inhibited. Besides RuBP carboxylase, the only other proteins whose accumulation appears to depend on chloroplast ribosomal activity are the elongation factors, the chloroplast coupling factor, nitrite reductase, and several of the chloroplast ribosomal and thylakoid polypeptides, including the cytochromes (31, 88). The excellent agreement between the results of the in vitro and in vivo approaches to the site of synthesis of chloroplast proteins in certain species, e.g. *Pisum sativum* (53), suggests that we can have some confidence in the validity of both approaches when suitable caution is exercized. The major discrepancy between the results of using the two methods is the large number of soluble products of protein synthesis by isolated chloroplasts; the available inhibitor data give little clue as to the possible identities of these products.

EVIDENCE FROM HEAT-TREATED PLANTS A series of papers from Feierabend's laboratory describes the characteristics of plastids present in cereal leaves grown at high temperature (64–68, 86, 113, 119). Growth of *Secale cereale* at 32°C in the light produces chlorotic leaves containing plastids deficient in plastid ribosomes (67). In other respects, the leaves appear to be normal, so this high temperature sensitivity of chloroplast ribosome formation allows the selective elimination of chloroplast protein synthesis. Such leaves present a system for identifying those chloroplast proteins which are synthesized on cytoplasmic ribosomes. It is safe to conclude that any polypeptide present in such tissue cannot have been made on chloroplast ribosomes. It is important to note that the converse conclusion cannot be drawn. The argument that those polypeptides missing from ribosome-deficient plants must be made on chloroplast ribosomes ignores the possibility that some polypeptides may accumulate to detectable levels only if other polypeptides are also present.

The plastid DNA in heat-treated leaves appears normal, as judged by restriction endonuclease analysis (86). This situation contrasts with that in heat-bleached cells of *Euglena,* where chloroplast DNA is undetectable (50). The chlorotic plastids contain RNA polymerase which is enzymically active at 32°C, so the defect in these heat-treated leaves is probably in the assembly of chloroplast ribosomes (26).

It is gratifying to note that the extensive studies of these heat-treated leaves fully confirm the picture gained from the inhibitor and in vitro approaches. Thus plastids isolated from such leaves contain many of the normal components, including chloroplast DNA, RNA polymerase, Calvin cycle and other enzymes (except RuBP carboxylase), the small subunit of RuBP carboxylase, ferredoxin, ferredoxin-NADP reductase, carotenoids, bounding envelope membranes, and many of the thylakoid polypeptides. The large subunit of RuBP carboxylase, several thylakoid polypeptides, the alpha and beta subunits of the chloroplast coupling factor, and the photosynthetic cytochromes are absent. The only discrepancy is that the heat-bleached leaves appear to contain the apoprotein of chlorophyll-protein complex I (65). This protein is made on chloroplast ribosomes according to in vitro studies (Table 2). Further investigation of this point is warranted to establish whether or not the site of synthesis of this protein varies with species.

It can be concluded from these studies that chloroplast protein synthesis is not required for the synthesis, recognition, and transport into an envelope-bound organelle of many chloroplast polypeptides.

EVIDENCE FROM MUTANTS Several reviews are available of research on mutations affecting chloroplast components (75, 76, 83, 93, 94, 96). This work will not be considered in detail here. Two general points are worth making. The first is that most of these mutations in both higher plants and *Chlamydomonas* show Mendelian inheritance in crosses, confirming the major role of nuclear genes in chloroplast biogenesis. The second point is basically a semantic one. A number of higher plant mutants have long been known in which variegated chlorophyll deficiencies show non-Mendelian inheritance (94). Such plants are often termed plastome mutants, because the defective chloroplasts are inherited in a fashion different from nuclear gene mutations and segregate at the somatic level. But in most cases there is no rigorous evidence that the plastome mutation results from an actual change in the chloroplast DNA. An alternative explanation is that the mutation causes the plastids to lose all their ribosomes. This loss could happen either through a nuclear mutation affecting a chloroplast ribosomal protein or a mutation in the chloroplast DNA affecting a different ribosomal protein. In the first case, the nuclear mutation will eventually result in plastids lacking ribosomes, because these are necessary to translate the chloroplast-encoded mRNAs for some of the chloroplast ribosomal proteins. Once a plastid has lost its ribosomes, it cannot regain them even when returned to a normal nuclear background; incomplete loss may permit recovery. Thus some plastome mutants could have an unaltered chloroplast DNA sequence and not be strictly mutants at all.

Recent work on the *iojap* mutation in *Zea mays* supports this conclusion. The recessive nuclear *iojap* gene causes a permanent heritable deficiency in plastid development, which cannot be reversed by transfer to a normal nuclear background. The affected plastids lack ribosomes, but have unaltered chloroplast DNA as judged by restriction endonuclease analysis

(138). Other ribosome-less plastids containing DNA have been reported in so-called plastome mutants of higher plants, e.g. *Hordeum vulgare* (82). Further work is needed to determine whether such plants carry mutations in their chloroplast DNA or should be regarded as genetic equivalents of heat-treated plants.

Does the Site of Synthesis Matter?

The mystery of extranuclear genetic systems is why they exist at all. It seems a high cost to the cell to specify at least 100 extra polypeptides involved in replication, transcription, and translation in order to make a relatively small number of proteins inside the organelle. The answer to this mystery may lie in the origin of these systems. Current arguments favor the endosymbiont hypothesis but are not conclusive (46). Whether organelles originated as endosymbionts or by some autogenous mechanism, it is likely that a massive transfer of genes from organelle to nucleus occurred during evolution. Coupled with this transfer would be a change in the site of synthesis from organelle to cytoplasm.

Implicit in the research reviewed in this section on chloroplast protein synthesis is the belief that the site of synthesis matters. In other words, it is assumed that there are rules which determine whether a given polypeptide is made inside or outside the organelle. It is disappointing that there is no discernible pattern in the current list of chloroplast-synthesized polypeptides (Table 2). If there are rules which dictate that a given polypeptide must be synthesized inside the chloroplast rather than outside it, these are not obviously related to the nature, function, or location of the polypeptides so far identified. What is encouraging, however, is that from the limited evidence available, there is no suggestion that the site of synthesis of a particular polypeptide varies between species. Thus the large subunit of RuBP carboxylase is made in the chloroplast in every species so far examined, from *Euglena* to higher plants. This uniformity may only reflect the paucity of data; it is worrying that there is a difference between two fungal species as to the site of synthesis of a mitochondrial polypeptide (21). This reviewer takes the optimistic view that chloroplast ribosomes have a universal and uniquely defined function; the task for the future is to puzzle out the rules which determine this function.

TRANSPORT

Two polypeptides that are not labeled in isolated chloroplasts are the small subunit of RuBP carboxylase and the chlorophyll *a/b* binding protein. In vivo inhibitor experiments clearly suggest that these are products of cytoplasmic protein synthesis (e.g. 6, 51). When cytoplasmic poly (A)-containing RNA from *Pisum* and *Spinacia* leaves and *Chlamydomonas* cells is translated in a wheat-germ extract, these polypeptides are produced as higher molecular weight precursors. One of the more interesting advances in the field of chloroplast development in the last 2 years has been the demonstration that both these precursors will enter isolated chloroplasts with processing to the mature size, and will then assemble correctly inside the organelle (33–35, 42, 63, 87, 122, 129). Many other unidentified polypeptides made in a wheat-germ extract will also enter isolated chloroplasts (121). This transport has been shown to require ATP (81).

The most striking feature of such studies is that the transport is post-translational, i.e. it occurs after the polypeptide chains have been released from the ribosomes. This type of transport is thus different from that operating for secreted proteins where transport depends on concomitant protein synthesis (19). The observation that polypeptide transport into chloroplasts does not require concomitant protein synthesis explains the failure to find cytoplasmic ribosomes bound to the chloroplast envelope during chloroplast development. A similar post-translational mechanism has subsequently been found to operate for mitochondria (120).

The mechanism of post-translational transport is unknown, but it is presumed to involve a specific interaction between the finished polypeptide precursor and the chloroplast envelope (18, 42, 87). The extra sequence in the precursor is probably involved in this recognition process. The extra sequence has been shown to be entirely at the N-terminus by DNA sequence analysis of cloned cDNA made from mRNA for the carboxylase small subunit from *Pisum sativum* (12). Analysis of the extra sequence in other polypeptides from the same species is needed to determine whether there is a sequence common to all those polypeptides destined to enter the chloroplasts, as predicted by this author (24). The extra sequence in the small subunit precursor from *Chlamydomonas* (123) shows no homology with that for *Pisum sativum* (12), but this is not surprising since the algal precursor will not enter higher plant chloroplasts (33, 34).

A further distinguishing feature of polypeptide transport into chloroplasts is that, at least in the case of the carboxylase small subunit, the processing activity is soluble (42, 129). It is most probable that this protease is located in the stromal compartment, but it cannot yet be ruled out that it is either loosely bound to the inside of the envelope or located between the two envelope membranes. Since heat-treated plants accumulate small subunit inside plastids (68), the processing enzyme itself must be a product of cytoplasmic protein synthesis. This raises the intriguing question as to how the processing enzyme enters the chloroplast. The soluble nature of this enzyme should aid further work on its characterization and transport.

ASSEMBLY

Three multisubunit complexes have been shown to assemble in isolated chloroplasts: the light-harvesting chlorophyll *a/b* protein complex (122), the chloroplast ATP-synthase complex (53, 103, 110), and RuBP carboxylase (7, 33, 34, 129). The latter two complexes contain both cytoplasmically synthesized polypeptides and chloroplast-synthesized polypeptides.

The polypeptides of the light-harvesting complex are highly hydrophobic when isolated from thylakoids. When synthesized as higher molecular weight precursors in wheat-germ extracts, they are soluble (122). One function of the extra sequence may thus be to render the polypeptides soluble to permit their transport through the cytoplasm and stroma. It is not known whether processing precedes assembly, or whether processing intermediates are present at any stage.

The ATP-synthase complex of *Spinacia* chloroplasts contains at least eight nonidentical subunits. Five of these form the extrinsic coupling factor (CF_1) complex, while the remainder are embedded firmly in the thylakoid membrane. Isolated chloroplasts synthesize at least three, and possibly four, of the extrinsic polypeptides (53, 103) as well as two of the membrane subunits (110). The chloroplast-made subunits assemble in isolated chloroplasts into an authentic ATP-synthase complex (110). It is concluded that isolated chloroplasts contain pools of cytoplasmically synthesized subunits, but exchange of newly synthesized subunits with existing subunits in the holoenzyme has not been ruled out.

Recent work in the author's laboratory has resulted in discovery of an abundant stromal protein which binds to newly synthesized large subunits of RuBP carboxylase (7, 63). This large subunit binding protein has a molecular weight in the range 6–7 $\times$ 10^5, and is composed of one type of subunit of molecular weight about 6 $\times$ 10^4. This subunit is termed polypeptide 60. This binding protein can be labeled in vivo with [^{35}S] methionine but not in isolated chloroplasts; it is probably synthesized by cytoplasmic ribosomes and thus is likely to be a nuclear gene product. The binding of the large subunit to this protein also occurs in vivo, so it is not an artefact of isolated chloroplasts (unpublished data). Time course experiments with isolated chloroplasts indicate that newly synthesized large subunits bind to this protein before entering the holoenzyme (7). The aggregate of large subunit with this protein may thus be an intermediate in the assembly of RuBP carboxylase, but other possibilities have not been ruled out (63). The assembly and processing of the small subunit precursor into the carboxylase holenzyme can be demonstrated in the total absence of membranes (129).

Figure 2 summarizes my current model of the events involved in the synthesis, transport, and assembly of RuBP carboxylase. This is the best-understood example of the cooperation between the nuclear and plastid genetic systems.

CONTROL

Much speculation has been made about the nature of the control of protein synthesis with respect to chloroplasts, but only recently have any firm data emerged, primarily due to the use of in vitro protein-synthesizing systems and molecular cloning technology. Information is available for three chloroplast proteins: RuBP carboxylase, peak D, and the light-harvesting chlorophyll *a/b* binding protein complex.

RuBP Carboxylase

Light stimulates the accumulation of this protein, but it is not an essential requirement for its synthesis; the carboxylase is the most abundant soluble protein in etiolated cereal leaves (54). In *Lemna gibba* changes in the in vivo rate of synthesis of the carboxylase small subunit due to light or dark treatments correlate with changes in the content of its mRNA, measured by translation in wheat-germ extracts (135, 136). This type of study does not rule out the possibility that mRNA occurs in an untranslatable form in dark-grown plants. However, work in the author's laboratory, using cloned hybridization probes for the small subunit mRNA in *Pisum sativum,* has shown that the stimulation of small subunit accumulation by light reflects changes in the total content of RNA sequences for this polypeptide (12). The control of synthesis of the small subunit in this species thus operates at the level of either the rate of transcription or the rate of RNA turnover.

Several situations have been described where the relative rates of synthesis of the two subunits of the carboxylase can be altered. Heat-treated *Secale cereale* leaves synthesize small subunit but not large subunit (68). Ribosomal inhibitors uncouple the synthesis of one subunit from that of the other in isolated leaf cells of *Glycine max* (6). Mesophyll cells of the C4 plant *Zea mays* contain the chloroplast gene for the large subunit but not the mRNA; these cells differ from bundle sheath cells in not synthesizing RuBP carboxylase (99). It is not clear whether mesophyll cells of C4 plants contain mRNA for the small subunit.

These reports suggest that there is no tight coupling of the expression of the nuclear and chloroplast genes for the two subunits of RuBP carboxylase. A similar conclusion was reached by Feierabend et al (65) for the synthesis of thylakoid polypeptides in heat-treated plants; the accumulation of cytoplasmically synthesized polypeptides is not prevented by the absence of those made in the chloroplast. In like vein, *Euglena* mutants which lack

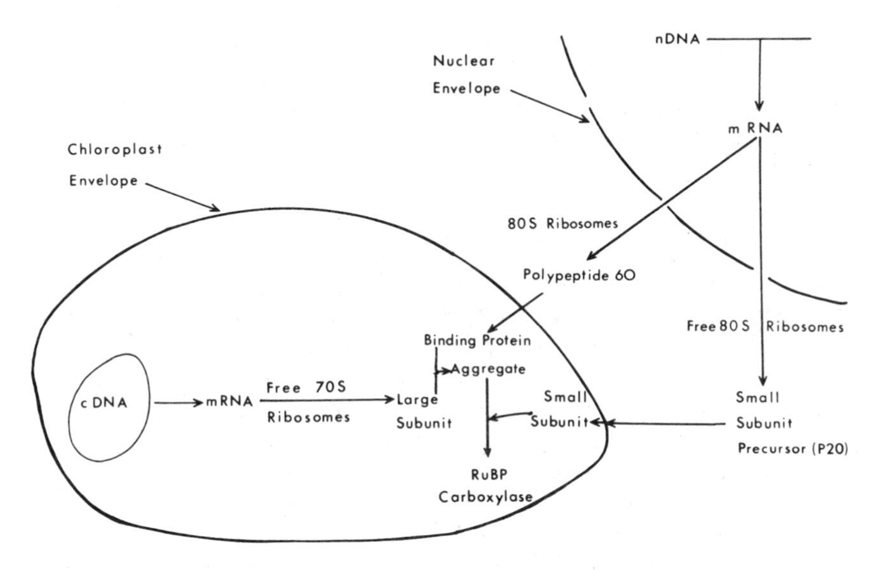

Figure 2 Current model for the cooperation of nuclear and plastid genomes in the synthesis of RuBP carboxylase. All the steps shown have been demonstrated in cell-free extracts, except for the transcription of the small subunit gene. The role of the large subunit binding protein in the assembly process and its encoding in the nucleus are not rigorously established (see text). Reprinted from (62) with permission.

detectable plastid DNA still synthesize many thylakoid polypeptides from nuclear genes (15, 16).

Peak D

Light is an obligatory requirement for the accumulation of peak D in *Pisum sativum* (48), *Spirodela oligorrhiza* (114), and *Zea mays* (11). The increase in the synthesis of peak D is reflected in an increase in the content of both translatable (114) and hybridizable (11) RNA sequences for this polypeptide. The relative rates of synthesis of peak D and carboxylase large subunit change during primary leaf development in *Spinacia oleracea,* and this change is at least partly mirrored in alterations in the translatable mRNA levels (128).

The general picture emerging from these studies on peak D and RuBP carboxylase is that control is mediated through changes in the rate of transcription. Limitations of the current data are the failure to distinguish effects on transcription from effects on RNA turnover (the methods used measure only the steady state concentration of RNA and not its rate of synthesis), and the difficulty in establishing an accurate quantitative relationship between the rate of accumulation of a particular polypeptide and the amount of its mRNA.

Studies on the third chloroplast protein suggest that control can be exerted at the level of protein turnover as well as at the level of transcription.

Chlorophyll a/b Binding Protein Complex

Interest in the two or more polypeptides comprising this complex stems from the fact that it is one of the very few chloroplast components whose accumulation in thylakoids requires continuous illumination. In barley leaves that have received a single flash of red light before being returned to darkness, the mRNA for these polypeptides accumulates over a period of several hours, but the protein cannot be detected (4). Mutants of *Hordeum vulgare* which lack this complex also lack chlorophyll *b* (5). This observation had led to the view that the accumulation of this complex is governed by the synthesis of chlorophyll *b*.

Work in the author's laboratory suggests an alternative explanation. Translationally active mRNA for the chlorophyll *a/b* polypeptides can be detected in polysomes isolated from dark-grown and intermittently illuminated *Pisum sativum* plants (38, 39). Exposure of light-grown plants to darkness results in a decrease in the content of both the chlorophylls and apoprotein of this complex (13). In contrast, the coupling factor subunits and the carboxylase subunits continue to accumulate in darkness, although at a slower rate. In vivo labeling studies show that the light-harvesting chlorophyll *a/b* polypeptides are still being synthesized in darkness, even though the accumulated amount is decreasing (13).

We conclude that the appearance of the mRNAs for the precursors of the chlorophyll *a/b* polypeptides is not strictly dependent on light, although the accumulation of these mRNAs is stimulated by light. The failure of dark-grown or intermittently illuminated plants to accumulate these polypeptides is not due to the absence of mRNAs or to their lack of translation. It is proposed that the light-harvesting complex is subject to rapid turnover in the thylakoid membranes when plants are exposed to darkness (13). The role of light is to provide chlorophylls *a* and *b* which stabilize the apoprotein of the complex.

Do Chloroplasts Export Regulatory Products of Transcription or Translation?

The current picture of chloroplast biogenesis assigns a dominant role to the nucleus. Structural genes for many chloroplast polypeptides, as well as the response of protein synthesis to light (125), are all nuclear features expressed through cytoplasmic protein synthesis and the import of proteins into chloroplasts. Logic demands that we inquire about traffic in the opposite direction.

The operation of a post-translational mechanism for the import of polypeptides implies that the absence of chloroplast ribosomes on the inner side of the chloroplast envelope cannot be used to rule out the possibility that polypeptides are exported from chloroplasts to cytoplasm. Such exported polypeptides could include enzymes that function in the cytoplasm in pathways interacting with chloroplast metabolism, or regulatory molecules which inform the nucleo-cytoplasmic genetic system of the state of affairs in the chloroplast compartment. The only published evidence which suggests that chloroplasts may export proteins comes from the laboratory of Ruth Sager. The replication of nuclear DNA, but not chloroplast DNA, is inhibited in *Chlamydomonas* by antibiotics known to block chloroplast protein synthesis (17a).

The viability of both *Euglena* mutants and heat-treated leaves which lack chloroplast ribosomes makes it unlikely that there is an essential role of chloroplast protein synthesis in heterotrophic growth. Nevertheless, it is possible that exported chloroplast proteins play either some modulatory role or are required for metabolic pathways involved in autotrophic metabolism.

It is to be hoped that further attempts will be made to investigate the possibility that polypeptides are exported from chloroplasts.

A recent paper suggests that chloroplasts export regulatory RNA molecules to the cytoplasm (25). This suggestion was made to explain the failure of plastome mutants of *Hordeum vulgare* to accumulate chloroplast proteins known to be made by cytoplasmic ribosomes. This paper assumes that the steady state concentration of a protein can be taken as a measure of its rate of synthesis. This is an unsafe assumption; for example, the peak D polypeptide is a major product of chloroplast protein synthesis in terms of its rate of synthesis, but a minor component of the thylakoids in terms of the accumulated amount. A further criticism of this work is the assumption it makes that the plastid DNA is altered in the plastome mutant; this may not be the case (see EVIDENCE FROM MUTANTS section). Simpler explanations of the data are that chloroplast proteins are either turned over or subject to feedback control of their synthesis. In neither case is it necessary to postulate export of RNA from the chloroplasts. Rigorous evidence for such a process can come only from in vitro studies of defined RNA species.

THE OVERALL PICTURE

A summary of the present overall picture of chloroplast protein synthesis with respect to chloroplast development can be made in the form of a series of statements:

1. The majority of chloroplast polypeptides are encoded in nuclear genes and synthesized by cytoplasmic ribosomes.
2. These polypeptides are transported into developing chloroplasts by a post-translational mechanism based on the chloroplast envelope.
3. The chloroplast genetic system makes a smaller but essential contribution in the form of about 100 polypeptides.
4. Light is not essential for the synthesis, transport, and assembly of the majority of chloroplast proteins; rather, it has a stimulatory effect at several different levels.
5. Future progress in understanding the mechanism of the control of chloroplast protein synthesis will depend on the development of in vitro transcription systems of high fidelity isolated from both the nucleus and the chloroplast.

ACKNOWLEDGMENTS

I thank Nam-Hai Chua, Jurgen Feierabend, and Jerry Schiff for access to preprints and reprints, and all members of the Warwick Chloroplast Group for stimulating discussion and critical suggestions.

Literature Cited

1. Akoyunoglou, G., Argyroudi-Akoyunoglou, J. H., eds. 1978. *Chloroplast Development,* Amsterdam: Elsevier/ North-Holland. 888 pp.
2. Alscher, R., Patterson, R., Jagendorf, A. T. 1978. Activity of thylakoid-bound ribosomes in pea chloroplasts. *Plant Physiol.* 62:88–93
3. Alscher, R., Smith, M. A., Petersen, L. W., Huffaker, R. C., Criddle, R. S. 1976. *In vitro* synthesis of the large subunit of ribulose bisphosphate carboxylase on 70S ribosomes. *Arch. Biochem. Biophys.* 174:216–25
4. Apel, K. 1979. Phytochrome-induced appearance of mRNA activity for the apoprotein of the light-harvesting chlorophyll *a/b* protein of barley (*Hordeum vulgare*). *Eur. J. Biochem.* 97:183–88
5. Apel, K., Kloppstech, K. 1978. The plastid membranes of barley. Light-induced appearance of mRNA coding for the apoprotein of the light harvesting chlorophyll *a/b* binding protein. *Eur. J. Biochem.* 85:581–88
6. Barraclough, R., Ellis, R. J. 1979. The biosynthesis of ribulose bisphosphate carboxylase. Uncoupling of the synthesis of the large and small subunits in isolated soybean leaf cells. *Eur. J. Biochem.* 94:165–77
7. Barraclough, R., Ellis, R. J. 1980. Protein synthesis in chloroplasts IX. Assembly of newly-synthesized large subunits into ribulose bisphosphate carboxylase in isolated intact pea chloroplasts. *Biochim. Biophys. Acta* 608:19–31
8. Beale, S. I. 1978. δ-Aminolevulinic acid in plants: its biosynthesis, regulation, and role in plastid development. *Ann. Rev. Plant Physiol.* 29:95–120
9. Bedbrook, J. R., Coen, D. M., Beaton, A. R., Bogorad, L., Rich, A. 1979. Location of the single gene for the large subunit of ribulose bisphosphate carboxylase on the maize chloroplast chromosome. *J. Biol. Chem.* 254: 905–10
10. Bedbrook, J. R., Kolodner, R. 1979. The structure of chloroplast DNA. *Ann. Rev. Plant Physiol.* 30:593–620
11. Bedbrook, J. R., Link, G., Coen, D. M., Bogorad, L., Rich, A. 1978. Maize plastid gene expressed during photoregulated development. *Proc. Natl. Acad. Sci. USA* 75:3060–64
12. Bedbrook, J. R., Smith, S. M., Ellis, R. J. 1980. Molecular cloning and sequencing of cDNA encoding the precursor to the small subunit of the chloroplast enzyme ribulose-1,5-bisphosphate carboxylase. *Nature.* 287:692–97

13. Bennett, J. 1981. Biosynthesis of the light-harvesting chlorophyll *a/b* protein. Polypeptide turnover in darkness. *Eur. J. Biochem.* In press
14. Berry, J., Björkman, O. 1980. Photosynthetic response and adaptation to temperature in higher plants. *Ann. Rev. Plant Physiol.* 31:491–543
15. Bingham, S., Schiff, J. A. 1979. Events surrounding the early development of *Euglena* chloroplasts. 15. Origin of plastid thylakoid polypeptides in wild type and mutant cells. *Biochim. Biophys. Acta* 547:512–30
16. Bingham, S., Schiff, J. A. 1979. Events surrounding the early development of *Euglena* chloroplasts. 16. Plastid thylakoid polypeptides during greening. *Biochim. Biophys. Acta* 547:531–43
17. Blair, G. E., Ellis, R. J. 1973. Protein synthesis in chloroplasts I. Light-driven synthesis of the large subunit of Fraction I protein by isolated pea chloroplasts. *Biochim. Biophys. Acta* 319: 223–34
17a. Blamire, J., Flechtner, V. R., Sager, R. 1974. Regulation of nuclear DNA replication by the chloroplast in *Chlamydomonas. Proc. Natl. Acad. Sci. USA* 71:2867–71
18. Blobel, G. 1980. Intracellular protein topogenesis. *Proc. Natl. Acad. Sci. USA* 77:1496–1500
19. Blobel, G., Dobberstein, B. 1975. Transfer of proteins across membranes. *J. Cell Biol.* 67:835–51
20. Boardman, N. K. 1977. Comparative photosynthesis of sun and shade plants. *Ann. Rev. Plant Physiol.* 28:355–77
21. Borst, P., Grivell, L. A. 1978. The mitochondrial genome of yeast. *Cell* 15:705–23
22. Bottomley, W., Spencer, D., Whitfeld, P. R. 1974. Protein synthesis in isolated chloroplasts. Comparison of light-driven and ATP-driven synthesis. *Arch. Biochem. Biophys.* 164:106–17
23. Bottomley, W., Whitfeld, P. R. 1979. Cell-free transcription and translation of total spinach chloroplast DNA. *Eur. J. Biochem.* 93:31–39
24. Boulter, D., Ellis, R. J., Yarwood, A. 1972. Biochemistry of protein synthesis in plants. *Biol. Rev.* 47:113–75
25. Bradbeer, J. W., Atkinson, Y. E., Borner, T., Hagemann, R. 1979. Cytoplasmic synthesis of plastid polypeptides may be controlled by plastid-synthesized RNA. *Nature* 279:816–17
26. Bunger, W., Feierabend, J. 1980. Capacity for RNA synthesis in 70S ribosome-deficient plastids of heat-bleached rye leaves. *Planta* 149:163–69
27. Cashmore, A. R., Broadhurst, M. K., Gray, R. E. 1978. Cell-free synthesis of leaf proteins: identification of an apparent precursor of the small subunit of ribulose-1,5-bisphosphate carboxylase. *Proc. Natl. Acad. Sci. USA* 75:655–59
28. Chelm, B. K., Hallick, R. B. 1976. Changes in the expression of the chloroplast genome of *Euglena gracilis* during chloroplast development. *Biochemistry* 15:593–99
29. Chua, N-H., Bennoun, P. 1975. Thylakoid membrane polypeptides of *Chlamydomonas reinhardtii:* wildtype and mutant strains deficient in photosystem II reaction center. *Proc. Natl. Acad. Sci. USA* 72:2175–79
30. Chua, N-H., Blomberg, F. 1979. Immunochemical studies of membrane polypeptides from spinach and *Chlamydomonas reinhardtii. J. Biol. Chem.* 254:215–23
31. Chua, N-H., Gillham, N. W. 1977. The sites of synthesis of the principal thylakoid membrane polypeptides in *Chlamydomonas reinhardtii. J. Cell Biol.* 74:441–52
32. Chua, N-H., Matlin, K., Bennoun, P. 1975. A chlorophyll-protein complex lacking in photosystem I mutants of *Chlamydomonas reinhardtii. J. Cell Biol.* 67:361–77
33. Chua, N-H., Schmidt, G. W. 1978. Post-translational transport into intact chloroplasts of a precursor to the small subunit of ribulose-1,5-bisphosphate carboxylase. *Proc. Natl. Acad. Sci. USA* 75:6110–14
34. Chua, N-H., Schmidt, G. W. 1978. *In vitro* synthesis, transport and assembly of ribulose bisphosphate carboxylase subunits. See Ref. 127, pp. 325–47
35. Chua, N-H., Schmidt, G. W. 1979. Transport of proteins into mitochondria and chloroplasts. *J. Cell Biol.* 81: 461–83
36. Ciferri, O., Di Pasquale, G., Tiboni, O. 1979. Chloroplast elongation factors are synthesized in the chloroplast. *Eur. J. Biochem.* 102:331–35
37. Coen, D. M., Bedbrook, J. R., Bogorad, L., Rich, A. 1977. Maize chloroplast DNA fragment encoding the large subunit of ribulose bisphosphate carboxylase. *Proc. Natl. Acad. Sci. USA* 74:5487–91
38. Cuming, A. C., Bennett, J. 1981. Biosynthesis of the light-harvesting chlorophyll *a/b* protein: mRNA activity un-

der regulatory light regimes. *Eur. J. Biochem.* In press

39. Cuming, A. C., Bennett, J. 1981. Biosynthesis of the light-harvesting chlorophyll *a/b* protein. Cellular status of mRNA under intermittent illumination. *Eur. J. Biochem.* In press
40. Davies, D. R., Hopwood, D. A., eds. 1980. *The Plant Genome.* Fourth John Innes Symp. Norwich: John Innes Charity. 273 pp.
41. Delepelaire, P., Chua, N-H. 1979. Lithium dodecylsulfate/polyacrylamide gel electrophoresis of thylakoid membranes at 4°C: characterization of two additional chlorophyll *a*-protein complexes. *Proc. Natl. Acad. Sci. USA* 76:111–15
42. Dobberstein, B., Blobel, G., Chua, N-H. 1977. *In vitro* synthesis and processing of a putative precursor for the small subunit of ribulose-1,5-bisphosphate carboxylase of *Chlamydomonas reinhardtii. Proc. Natl. Acad. Sci. USA* 74:1082–85
43. Dockerty, A., Merrett, M. J. 1979. Isolation and enzymic characterization of *Euglena* proplastids. *Plant Physiol.* 63:468–73
44. Doherty, A., Gray, J. C. 1979. Synthesis of cytochrome *f* by isolated pea chloroplasts. *Eur. J. Biochem.* 98:87–92
45. Doherty, A., Gray, J. C. 1980. Synthesis of a dicyclohexylcarbodiimide binding proteolipid by isolated pea chloroplasts. *Eur. J. Biochem.* 108:131–36
46. Doolittle, W. F. 1980. Revolutionary concepts in evolutionary cell biology. *Trends Biochem. Sci.* 5:146–49
47. Driesel, A. J., Speirs, J., Bohnert, H. J. 1980. Spinach chloroplast mRNA for a 32 Kd polypeptide-size and localization on the physical map of the chloroplast DNA. *Biochim. Biophys. Acta.* In press
48. Eaglesham, A. R. J., Ellis, R. J. 1974. Protein synthesis in chloroplasts II. Light-driven synthesis of membrane proteins by isolated pea chloroplasts. *Biochim. Biophys. Acta* 335:396–407
49. Edelman, M., Reisfeld, A. 1980. Synthesis, processing and functional probing of P-32000, the major membrane protein translated within the chloroplast. See Ref. 97, pp. 353–72
50. Edelman, M., Schiff, J. A., Epstein, H. T. 1965. Studies of chloroplast development in *Euglena.* XII. Two types of satellite DNA. *J. Mol. Biol.* 11:769–74
51. Ellis, R. J. 1975. Inhibition of chloroplast protein synthesis by lincomycin and MDMP. *Phytochemistry* 14:89–93
52. Ellis, R. J. 1976. The search for plant messenger RNA. In *Perspectives in Experimental Biology,* ed. N. Sunderland, 2:283–98. Oxford/New York: Pergamon. 523 pp.
53. Ellis, R. J. 1977. Protein synthesis by isolated chloroplasts. *Biochim. Biophys. Acta* 463:185–215
54. Ellis, R. J. 1978. Chloroplast proteins and their synthesis. In *Plant Proteins,* ed. G. Norton, pp. 25–40. London: Butterworths. 352 pp.
55. Ellis, R. J. 1979. The most abundant protein in the world. *Trends Biochem. Sci.* 4:241–44
56. Ellis, R. J. 1981. Inhibitors for studying chloroplast transcription and translation *in vivo.* In *Methods in Chloroplast Molecular Biology,* ed. M. Edelman, R. B. Hallick, N-H. Chua. Amsterdam: Elsevier/North-Holland. In press
57. Ellis, R. J., Barraclough, B. R. 1978. Synthesis and transport of chloroplast proteins inside and outside the cell. See Ref. 1, pp. 185–94
58. Ellis, R. J., Blair, G. E., Hartley, M. R. 1973. The nature and function of chloroplast protein synthesis. *Biochem. Soc. Symp.* 38:137–62
59. Ellis, R. J., Hartley, M. R. 1974. *Biochemistry of Nucleic Acids,* Vol. 6. In *Int. Rev. Sci. Biochem. Ser. 1,* ed. K. Burton, pp. 323–48. Lancaster/London: Med. Tech. Publ. Co. and Butterworths
60. Ellis, R. J., Hartley, M. R. 1980. Isolation of higher plant chloroplasts active in protein and RNA synthesis. See Ref. 56
61. Ellis, R. J., Highfield, P. E., Silverthorne, J. 1977. The synthesis of chloroplast proteins by subcellular systems. *Proc. 4th Int. Congr. Photosynth.,* ed. D. O. Hall, J. Coombs, T. W. Goodwin, pp. 497–506. London: Biochem. Soc. Press. 155 pp.
62. Ellis, R. J., Smith, S. M., Barraclough, R. 1979. Nuclear-plastid interactions in plastid protein synthesis. *Proc. 4th John Innes Symp.,* ed. B. R. Davies, D. A. Hopwood. Norwich: John Innes Charity. 273 pp.
63. Ellis, R. J., Smith, S. M., Barraclough, R. 1980. Synthesis, transport and assembly of chloroplast proteins. See Ref. 97, pp. 321–35
64. Feierabend, J. 1978. Cooperation of cytoplasmic and plastidic protein synthesis in rye leaves. See Ref. 1, pp. 207–13
65. Feierabend, J., Meschede, D., Vogel, K-D. 1980. Comparison of the polypep-

tide compositions of the internal membranes of chloroplasts, etioplasts and ribosome-deficient heat-bleached plastids from rye leaves. *Z. Pflanzenphysiol.* 98:61–78
66. Feierabend, J., Mikus, M. 1977. Occurrence of a high temperature sensitivity of chloroplast ribosome formation in several higher plants. *Plant Physiol.* 59:863–67
67. Feierabend, J., Schrader-Reichhardt, U. 1976. Biochemical differentiation of plastids and other organelles in rye leaves with a high-temperature-induced deficiency of plastid ribosomes. *Planta* 120:133–45
68. Feierabend, J., Wildner, G. 1978. Formation of the small in the absence of the large subunit of RUBP carboxylase in 70S ribosome-deficient rye leaves. *Arch. Biochem. Biophys.* 186:283–91
69. Freyssinet, G. 1978. Determination of the site of synthesis of some *Euglena* cytoplasmic and chloroplast ribosomal proteins. *Exp. Cell Res.* 115:207–19
70. Geetha, V., Gnanam, A. 1980. An *in vitro* protein-synthesizing system with isolated chloroplasts of *Sorghum vulgare. J. Biol. Chem.* 255:492–97
71. Geetha, V., Gnanam, A. 1980. Identification of P700-chlorophyll *a*-protein complex as a product of chloroplast protein synthesis. *FEBS Lett.* 111: 272–76
72. Geetha, V., Mohamed, A. H., Gnanam, A. 1980. Cell-free synthesis of active ribulose-1,5-bisphosphate carboxylase in the mesophyll chloroplasts of *Sorghum vulgare. Biochim. Biophys. Acta* 606:83–94
73. Gelvin, S., Heizmann, P., Howell, S. H. 1977. Identification and cloning of the chloroplast gene coding for the large subunit of ribulose-1,5-bisphosphate carboxylase from *Chlamydomonas reinhardii. Proc. Natl. Acad. Sci. USA* 74:3193–97
74. Gelvin, S., Howell, S. H. 1977. Identification and precipitation of the polyribosomes in *Chlamydomonas reinhardii* involved in the synthesis of the large subunit of ribulose-1,5-bisphosphate carboxylase. *Plant Physiol.* 59:471–77
75. Gillham, N. W. 1978. *Organelle Heredity.* New York: Raven. 602 pp.
76. Gillham, N. W., Boynton, J. E., Chua, N-H. 1978. Genetic control of chloroplast proteins. *Curr. Adv. Bioenerg.* 9:211–60
77. Gimmler, H., Schafer, G., Heber, U. 1974. Low permeability of the chloroplast envelope towards cations. In *Proc. 3rd Int. Congr. Photosynth.*, ed. M. Avron, 2:1381–92. Amsterdam: Elsevier
78. Grebanier, A. E., Coen, D. M., Rich, A., Bogorad, L. 1978. Membrane proteins synthesized but not processed by isolated maize chloroplasts. *J. Cell Biol.* 78:734–46
79. Grebanier, A. E., Steinback, K. E., Bogorad, L. 1979. Comparison of the molecular weights of proteins synthesized by isolated chloroplasts with those which appear during greening in *Zea mays. Plant Physiol.* 63:436–39
80. Green, B. R. 1980. Protein synthesis by isolated *Acetabularia* chloroplasts. *Biochim. Biophys. Acta.* 609:107–20
81. Grossman, A., Bartlett, S., Chua, N-H. 1980. Energy-dependent uptake of cytoplasmically synthesized polypeptides by chloroplasts. *Nature* 285:625–28
82. Hagemann, R., Borner, T. 1978. Plastid ribosome-deficient mutants of higher plants as a tool in studying chloroplast biogenesis. See Ref. 1, pp. 709–20
83. Hallier, U. W., Schmitt, J. M., Heber, U., Chaianova, S. S., Volodorsky, A. D. 1978. Ribulose-1,5-bisphosphate carboxylase-deficient plastome mutants of *Oenothera. Biochim. Biophys. Acta* 504:67–83
84. Hartley, M. R., Wheeler, A. M., Ellis, R. J. 1975. Protein synthesis in chloroplasts V. Translation of messenger RNA for the large subunit of Fraction I protein in a heterologous cell-free system. *J. Mol. Biol.* 91:67–77
85. Henriques, F., Park, R. B. 1976. Identification of chloroplast membrane peptides with subunits of coupling factor and ribulose-1,5-bisphosphate carboxylase. *Arch. Biochem. Biophys.* 176: 472–78
86. Hermann, R. G., Feierabend, J. 1980. The presence of DNA in ribosome-deficient plastids of heat-bleached rye leaves. *Eur. J. Biochem.* 104:603–9
87. Highfield, P. E., Ellis, R. J. 1978. Synthesis and transport of the small subunit of chloroplast ribulose bisphosphate carboxylase. *Nature* 271:420–24
88. Hoober, J. K. 1976. Protein synthesis in chloroplasts. In *Protein Synthesis,* ed. E. H. McConkey, 2:169–248. New York/Basle: Dekker
89. Howell, S. H., Heizmann, P., Gelvin, S., Walker, L. L. 1977. Identification and properties of the messenger RNA activity in *Chlamydomonas reinhardii* coding for the large subunit of riblose-1,5-bisphosphate carboxylase. *Plant Physiol.* 59:464–70

90. Jennings, R. C., Ohad, J. 1972. Biogenesis of chloroplast membranes XI. Evidence for the translation of extrachloroplast RNA on chloroplast ribosomes in a mutant of *Chlamydomonas reinhardii. Arch. Biochem. Biophys.* 153:79–87
91. Jensen, R. G., Bahr, J. T. 1977. Ribulose 1,5-bisphosphate carboxylase-oxygenase. *Ann. Rev. Plant Physiol.* 28:379–400
92. Joy, K. W., Ellis, R. J. 1975. Protein synthesis in chloroplasts IV. Polypeptides of the chloroplast envelope. *Biochim. Biophys. Acta* 378:143–51
93. Kirk, J. T. O. 1972. The genetic control of plastid formation: recent advances and strategies for the future. *Sub-Cell. Biochem.* 1:333–61
94. Kirk, J. T. O., Tilney-Bassett, R. A. E. 1978. *The Plastids: Their Chemistry, Structure, Growth and Inheritance.* Amsterdam/New York/Oxford: Elsevier/North-Holland 960 pp. 2nd ed.
95. Kung, S. D. 1977. Expression of chloroplast genomes in higher plants. *Ann. Rev. Plant Physiol.* 28:401–37
96. Kutzelnigg, H., Stubbe, W. 1974. Investigations on plastome mutants in *Oenothera. Sub-Cell. Biochem.* 3:73–89
97. Leaver, C. J., ed. 1980. *Genome Organization and Expression in Plants.* New York/London: Plenum. 607 pp.
98. Link, G., Bogorad, L. 1980. Sizes, locations, and directions of transcription of two genes on a cloned maize chloroplast DNA sequence. *Proc. Natl. Acad. Sci. USA* 77:1832–36
99. Link, G., Coen, D. M., Bogorad, L. 1978. Differential expression of the gene for the large subunit of ribulose bisphosphate carboxylase in maize leaf cell types. *Cell* 15:725–31
100. Malnoe, P., Rochaix, J-D., Chua, N-H., Spahr, P-F. 1979. Characterization of the gene and messenger RNA of the large subunit of ribulose-1,5-bisphosphate carboxylase in *Chlamydomonas reinhardii. J. Mol. Biol.* 133:417–34
101. Marsho, T. V., Sokolore, P. M., Mackay, A. B. 1980. Regulation of photosynthetic electron transport in intact spinach chloroplasts. I. Influence of exogenous salts on oxaloacetate reduction. *Plant Physiol.* 65:703–6
102. McCrea, J. M., Hershberger, C. L. 1978. Chloroplast DNA codes for tRNA from cytoplasmic polyribosomes. *Nature* 274:717–19
103. Mendiola-Morgenthaler, L. R., Morgenthaler, J-J., Price, C. A. 1976. Synthesis of coupling factor CF_1 protein by isolated spinach chloroplasts. *FEBS Lett.* 62:96–100
104. Michaels, A., Margulies, M. M. 1975. Amino acid incorporation into protein by ribosomes bound to chloroplast thylakoid membranes: formation of discrete products. *Biochim. Biophys. Acta* 390:352–62
105. Mills, W. B., Joy, K. W. 1980. A rapid method for isolation of purified physiologically active chloroplasts and its use to study the intracellular distribution of amino acids in pea leaves. *Planta* 148:75–83
106. Morgenthaler, J. J., Mendiola-Morgenthaler, L. 1976. Synthesis of soluble, thylakoid and envelope membrane proteins by spinach chloroplasts purified from gradients. *Arch. Biochem. Biophys.* 172:51–58
107. Mullet, J. E., Burke, J. J., Arntzen, C. J. 1980. Chlorophyll proteins of photosystem I. *Plant Physiol.* 65:814–22
108. Mullet, J. E., Burke, J. J., Arntzen, C. J. 1980. A developmental study of photosystem I peripheral chlorophyll proteins. *Plant Physiol.* 65:823–27
109. Nasyrov, Y. S. 1978. Genetic control of photosynthesis and improving of crop productivity. *Ann. Rev. Plant Physiol.* 29:215–37
110. Nelson, N., Nelson, H., Schatz, G. 1980. Biosynthesis and assembly of the proton-translocating adenosine triphosphatase complex from chloroplasts. *Proc. Natl. Acad. Sci. USA* 77:1361–64
111. Possingham, J. V. 1980. Plastid replication and development in the life cycle of higher plants. *Ann. Rev. Plant Physiol.* 31:113–29
112. Price, C. A., Reardon, E. M. 1980. Isolation of chloroplasts for protein synthesis from spinach and *Euglena gracilis* by centrifugation in silica sols. See Ref. 56,
113. Rademacher, E., Feierabend, J. 1976. Formation of chloroplast pigments and sterols in rye leaves deficient in plastid ribosomes. *Planta* 129:147–53
114. Reisfeld, A., Gressel, J., Jakob, K. M., Edelman, M. 1978. Characterization of the 32000 dalton membrane protein I. Early synthesis during photoinduced plastid development of *Spirodela. Photochem. Photobiol.* 27:161–65
115. Reisfeld, A., Jakob, K. M., Edelman, M. 1978. Characterization of the 32000 dalton chloroplast membrane protein. II. The molecular weight of chloroplast messenger RNAs translating the precursor to P-32000 and full-size RuBP carboxylase large subunit. See Ref. 1, pp. 669–74
116. Rochaix, J. D., Malnoe, P. 1978. Gene localization on the chloroplast DNA of

Chlamydomonas reinhardii. See Ref. 1, pp. 581–86
117. Roscoe, T. J., Ellis, R. J. 1980. Two dimensional gel electrophoresis of chloroplast proteins. See Ref. 56
118. Sagher, D., Grosfeld, H., Edelman, M. 1976. Large subunit ribulose bisphosphate carboxylase mRNA from *Euglena* chloroplasts. *Proc. Natl. Acad. Sci. USA* 73:722–26
119. Schafers, H. A., Feierabend, J. 1976. Ultrastructural differentiation of plastids and other organelles in rye leaves with a high-temperature-induced deficiency of plastid ribosomes. *Eur. J. Cell Biol.* 14:75–90
120. Schatz, G. 1979. How mitochondria import proteins from the cytoplasm. *FEBS Lett.* 103:203–11
121. Schmid, G. H., Menke, W., Radunz, A., Koenig, F. 1978. Polypeptides of the thylakoid membrane and their functional characterization. *Z. Naturforsch. Teil C* 33:723–30
122. Schmidt, G. W., Bartlett, S., Grossman, A. R., Cashmore, A. R., Chua, N-H. 1980. *In vitro* synthesis, transport and assembly of the constituent polypeptides of the light-harvesting chlorophyll *a/b* protein complex. See Ref. 97, pp. 337–51
123. Schmidt, G. W., Devilliers-Thiery, A., Desruisseaux, H., Blobel, G., Chua, N-H. 1979. NH_2-Terminal amino acid sequences of precursor and mature forms of the ribulose-1,5-bisphosphate carboxylase small subunit from *Chlamydomonas reinhardtii*. *J. Cell Biol.* 83:615–22
124. Schwartzbach, S. D., Barnett, W. E., Hecker, L. I. 1979. Evidence that *Euglena* chloroplasts do not export tRNAs. *Nature* 280:86–87
125. Schwartzbach, S. D., Schiff, J. A. 1979. Events surrounding the early development of *Euglena* chloroplasts 13. Photocontrol of protein synthesis. *Plant Cell Physiol.* 20:827–38
126. Siddell, S. G., Ellis, R. J. 1975. Protein synthesis in chloroplasts. VI. Characteristics and products of protein synthesis *in vitro* in etioplasts and developing chloroplasts from pea leaves. *Biochem. J.* 146:675–85
127. Siegelman, H. W., Hind, G., eds. 1978. *Photosynthetic Carbon Assimilation*. New York/London: Plenum. 445 pp.
128. Silverthorne, J., Ellis, R. J. 1980. Protein synthesis in chloroplasts VIII. Differential synthesis of chloroplast proteins during spinach leaf development. *Biochim. Biophys. Acta* 607: 319–30
129. Smith, S. M., Ellis, R. J. 1979. Processing of small subunit precursor of ribulose bisphosphate carboxylase and its assembly into whole enzyme are stromal events. *Nature* 278:662–64
130. Suss, K.-H. 1976. Identification of chloroplast thylakoid membrane polypeptides: coupling factor of photophosphorylation and cytochrome *f*. *FEBS Lett.* 70:191–96
131. Suss, K-H. 1980. Identification of chloroplast thylakoid membrane polypeptides. *FEBS Lett.* 112:255–59
132. Sutcliffe, J. F. 1960. New evidence for a relationship between ion absorption and protein turnover in plant cells. *Nature* 188:294–97
133. Thomas, H., Stoddart, J. L. 1980. Leaf senescence. *Ann. Rev. Plant Physiol.* 31:83–111
134. Thomson, W. W., Whatley, J. M. 1980. Development of nongreen plastids. *Ann. Rev. Plant Physiol.* 31:375–94
135. Tobin, E. M. 1978. Light regulation of specific mRNA species in *Lemna gibba* L. *Proc. Natl. Acad. Sci. USA* 75:4749–53
136. Tobin, E. M., Suttie, J. L. 1980. Light effects on the synthesis of ribulose-1,5-bisphosphate carboxylase in *Lemna gibba* L. *Plant Physiol.* 65:641–47
137. Vasconcelos, A. C. 1976. Synthesis of proteins by isolated *Euglena gracilis* chloroplasts. *Plant Physiol.* 58:719–21
138. Walbot, V., Coe, E. H. 1979. Nuclear gene *iojap* conditions a programmed change to ribosome-less plastids in *Zea mais*. *Proc. Natl. Acad. Sci. USA* 76:2760–64
139. Walden, R., Leaver, C. J. 1978. Regulation of chloroplast protein synthesis during germination and early development of cucumber (*Cucumis sativus*). See Ref. 1, pp. 251–56
140. Weinbaum, S. A., Gressel, J., Reisfeld, A., Edelman, M. 1979. Characterization of the 32000 dalton chloroplast membrane protein. III. Probing its biological function in *Spirodela*. *Plant Physiol.* 64:828–32
141. Wheeler, A. M., Hartley, M. R. 1975. Major mRNA species from spinach chloroplasts do not contain poly(A). *Nature* 257:66–67
142. Winget, G. D., Kanner, N., Racker, E. 1977. Formation of ATP by the adenosine triphosphatase complex from spinach chloroplasts reconstituted together with bacteriorhodopsin. *Biochim. Biophys. Acta* 460:490–99
143. Zielinski, R. E., Price, C. A. 1980. Synthesis of thylakoid membrane proteins by chloroplasts isolated from spinach. *J. Cell Biol.* 85:435–45

Ann. Rev. Plant Physiol. 1981. 32:139–68

THE CHLOROPLAST ENVELOPE: Structure, Function, and Role in Leaf Metabolism

❖7708

Ulrich Heber

Institute of Botany and Pharmaceutical Biology, University of Würzburg, D-87 Würzburg, Germany

H. W. Heldt

Plant Physiology Institute, University of Göttingen, D-34 Göttingen, Germany

CONTENTS

0066-4294/81/0601-0139$01.00

INTRODUCTION

The last review in this area of interest appeared in this series 7 years ago (38). Since then, many more comprehensive articles dealing with the relationship between chloroplast and extrachloroplast metabolism have appeared elsewhere (13, 20, 35, 39, 41, 50, 55–57, 64, 85, 88, 118, 128, 148–150), not to mention the large number of original research papers which have contributed to our understanding of how the chloroplast envelope separates, links, or controls chloroplast and extrachloroplast metabolism in leaf cells. An excellent and penetrating recent review by Douce & Joyard (19) relieves the present authors of the burden of broad coverage and permits them to concentrate on their own biased views. The following is intended to serve as a brief introduction.

Chloroplasts are limited by an envelope which consists of two membranes. Inside the chloroplasts, closed chlorophyll-containing membranes called thylakoids use light energy to split water, transport ions, and reduce NADP to NADPH. When NADP is unavailable, oxygen acts as the electron acceptor, preventing overreduction of the electron transport chain and facilitating cyclic electron transport (40, 131, 139). The energy of a proton gradient across the thylakoid membranes is believed to be used for the phosphorylation of ADP to ATP. Phosphorylation and reduction take place at the boundary between the chloroplast matrix, or stroma, and the thylakoids. In so-called C_3 plants, such as spinach, and in the bundle sheath of C_4 plants, such as maize, CO_2 is fixed in the chloroplast stroma. The fixation product, 3-phosphoglycerate, is also reduced there, but some of it may escape across the envelope into the cytosol.

In mesophyll cells of C_4 plants, CO_2 fixation occurs in the cytosol (20, 35). However, the cytosolic carbon acceptor molecule phosphoenolpyruvate is synthesized not in the cytosol but in the chloroplasts from pyruvate that originated outside. The carboxylation product oxaloacetate can be transaminated to aspartate both inside and outside the chloroplasts, or is reduced inside the chloroplasts to malate which is further metabolized outside. The main end products of chloroplast photosynthesis are starch, which is stored temporarily inside the chloroplasts, and dihydroxyacetone phosphate, which is exported through the envelope in exchange for phosphate. Sucrose, commonly considered an end product of photosynthesis, is synthesized in the cytosol (120).

If gas exchange is neglected, it is possible to consider chloroplasts of C_3 plants and of the bundle sheath of C_4 plants primarily as phosphate importing and dihydroxyacetone phosphate exporting organelles. In C_3 chloroplasts, export of glycolate is also significant. Bundle sheath chloroplasts of some C_4 species import malate and export pyruvate in addition to exchanging phosphate for dihydroxyacetone phosphate. Mesophyll chloroplasts of C_4 plants are pyruvate, phosphate, and oxaloacetate importing and phosphoenolpyruvate and aspartate or malate exporting organelles. This diversity illustrates specialization and division of labor in different types of chloroplasts and cells.

In the following, we will briefly consider available information on the structure of the chloroplasts envelope before proceeding to a discussion of ways in which the envelope permits and controls transfer and thereby influences and indeed controls leaf cell metabolism.

ISOLATION OF ENVELOPES

Mackender & Leech (94) were the first to report isolation of envelopes from intact chloroplasts by a method employing osmotic chloroplast rupture and differential centrifugation. Their method was subsequently refined (18, 77, 95, 107, 110, 111, 135, 147). Reasonably pure envelope preparations can be obtained by swelling intact chloroplasts purified on a sucrose gradient in a hypotonic buffer and separating an envelope fraction from other material on a discontinuous density gradient. Precautions necessary to avoid contamination particularly by thylakoid membranes have been discussed by Douce & Joyard (19). In contrast to intact chloroplasts which generally have longitudinal diameters ranging from 3 to 10 μm, vesicles formed by isolated envelope membranes are rarely more and usually much less than 0.5 μm in diameter. It is very doubtful whether they are suitable for transport studies (cf 108, 109, 112). The presence of open membrane sheets has also been reported (134).

While the envelope of intact chloroplasts consists of two membranes, isolated envelope vesicles are often bordered by only one membrane (19). Orientation or structure of the isolated membranes appears to be different from that of envelopes in situ as indicated by the observation that phospholipase A attacks only isolated envelopes, not intact chloroplasts (147). Good envelope preparations contain neither significant amounts of chlorophyll nor cytochromes and appear to be essentially free of microsomal, mitochondrial, and thylakoid membranes. The available envelope preparations are mixtures of the inner and outer envelope membranes. All attempts to separate the two membranes (e.g. stripping of the outer membrane in a hypertonic medium, specific labeling of this membrane by antibodies against galactolipids, binding of labeled cationic molecules to the chloroplast surface, partitioning in aqueous two-phase systems) have failed so far (R. Douce, unpublished).

COMPOSITION AND STRUCTURE OF THE ENVELOPE

As in thylakoids, galactosyl glycerols are the predominant lipids in envelope preparations. This has been discussed in detail by Douce & Joyard (19). It appears that these lipids are absent from the other cellular membranes (19, 53). The presence of large amounts of galactolipid in envelope preparations is particularly remarkable as the outer membrane of the envelope is thought to be derived from the endoplasmic reticulum (99). Another major constituent of the envelope is phosphatidylcholine, whereas phosphatidylethanolamine is largely absent from well-purified envelope preparations. The only pigments of good envelope preparations are carotenoids [0.2% by weight of the total envelope lipids (19) and quinones (91)]. Envelope membranes contain surprisingly little protein [protein/polar lipids = 0.5–0.8 (19)] as compared with thylakoids (protein/polar lipids = 2). The protein content of mitochondrial and microsomal membranes is also much higher than that of envelope membranes (98).

In very impressive phase contrast kinematographic studies, Wildman et al (155) have shown the stroma phase of chloroplasts to be in constant motion in situ. Portions of it may even be seen to detach from or to join the chloroplasts which occasionally appear to fuse temporarily. The mobility of the stroma gives testimony to the fluidity of the bordering chloroplast envelope. After staining and cross-sectioning, the chloroplast envelope consists of two dense lines in electronmicrographs which enclose a lighter space. The thickness of the dense lines which correspond to the individual membranes is about 60 Å to 80 Å and that of the lighter space about 100 Å to 200 Å, although wider gaps may also be seen occasionally (32). The gaps become very prominent in vitro on exposure of chloroplasts to hypertonic media (61) in which only the inner membrane of the envelope responds osmotically, decreasing the chloroplast volume. The inner envelope membrane is sometimes extended into a so-called peripheral reticulum which increases the membrane surface. Especially in chloroplasts from C_4 plants, the peripheral reticulum may be highly developed (86).

On freeze-fracturing, which splits membranes in the middle and permits viewing them from "inside," the chloroplast envelope shows the expected four fracture faces which are covered by particles. In spinach chloroplasts (136), the part of the inner envelope membrane facing the stroma was shown to have 70 Å particles with a density of 1820 particles/μm^2, whereas 90 Å particles with a density of 980 particles/μm^2 were found on the outer side of this membrane. Obviously this membrane is asymmetric. The particle densities of both the inner and the outer fracture face of the outer envelope membrane are similar and very low (90 Å, 130 and 150 particles/

μm^2). It is most likely that the particles seen in the freeze-fracture faces represent proteins of the corresponding membranes. Their amount can be estimated if it is assumed that the particles are spherical and the partial specific volume of the membrane proteins is 0.82 ml/g. The molecular weight calculated for the 90 Å particle is close to 280 kD[1] and that for the 70 Å particle 130 kD.

Measurements of the size distribution of isolated spinach chloroplasts by the Coulter Counter technique show that chloroplast size changes as osmotic conditions change (Figure 1). In sorbitol solutions the chloroplast is a near-perfect osmometer above an osmolarity of about 0.1 M. At 0.16 M, sorbitol-isolated chloroplasts are spherical (61) and an average chloroplast occupies a volume of 50 μm^3. At the same osmolarity, the sorbitol-impermeable H_2O space of chloroplasts containing one milligram chlorophyll is 48 μl (61). The number of chloroplasts corresponding to one milligram chlorophyll is then calculated to be $\sim 10^9$. Essentially the same value was obtained earlier from Coulter Counter measurements with pea chloroplasts (102). The total outer surface area of each of the two envelope membranes of 10^9 chloroplasts (1 mg chlorophyll) is then 660 cm^2. From this the amount of protein visible in the envelope is evaluated to be 26 μg in the inner face and 30 μg in the outer face of the inner envelope membrane and 4 μg in the inner face and 5 μg in the outer face of the outer envelope membrane (per milligram of chlorophyll). These data show the outer envelope membrane to have a far lower content of integral protein than the inner envelope membrane. The visible protein in both membranes amounts to 45% of the total protein.

The envelope proteins have been separated by sodium dodecyl sulfate polyacrylamide gel electrophoresis. This method has the disadvantage that proteins are cleaved into subunits. With different envelope preparations a large number of bands ranging from about 14 kD to more than 100 kD were obtained (12, 25, 75, 92, 97, 106, 134). In spinach, major bands have been found at 14, 29, 33, 52, and 125 kD (106). The 52 kD protein was not identical with the large subunit of RuBP carboxylase which would be expected at this position, while the 14 kD protein was electrophoretically indistinguishable from the small subunit of RuBP carboxylase (106). In plastid envelope preparations of *Narcissus* and *Pisum,* bands similar to the 29 kD band of the spinach envelope were found to have molecular weights of 26 and 32 kD respectively (75, 92). As will be discussed, the 29 kD polypeptide band which represents about 20% of the total envelope protein in spinach is involved in phosphate transport across the inner envelope membrane which is the main barrier to transport of hydrophilic

[1]Throughout this chapter, kD will be used to indicate kilodalton.

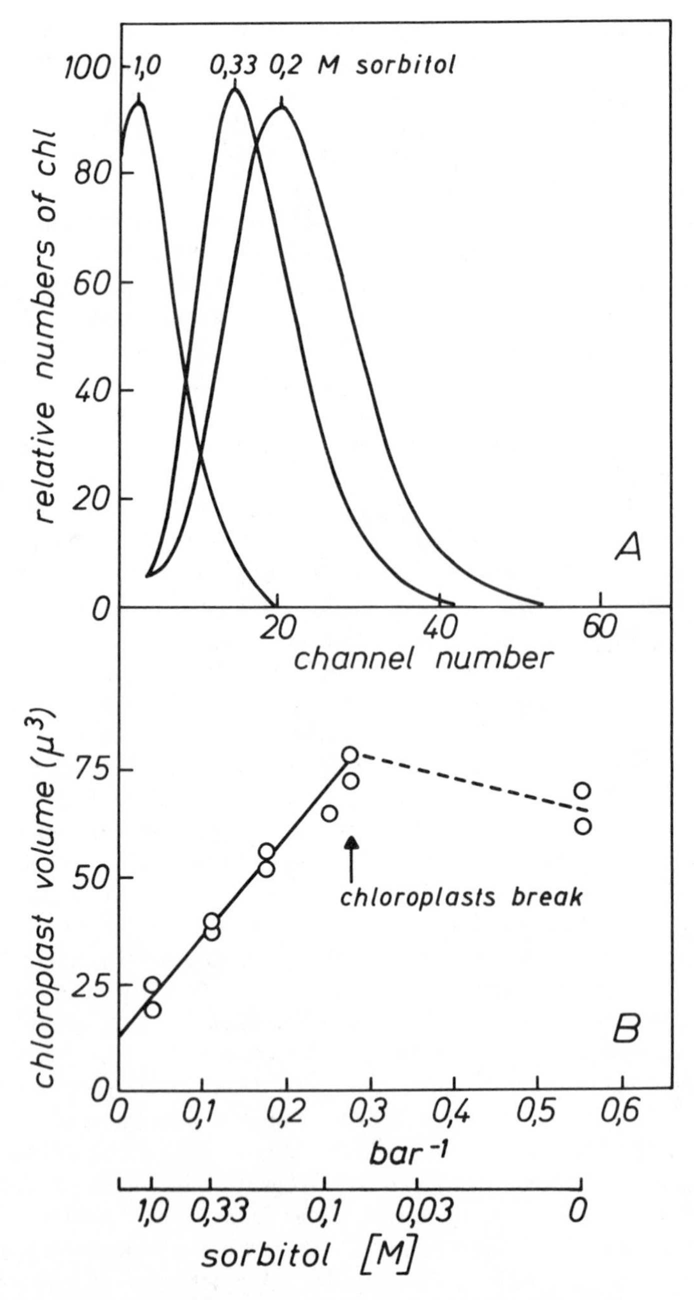

Figure 1. Size distribution of spinach chloroplasts suspended in media of different osmolarity as measured by the Coulter Counter (A) and linear relationship between mean chloroplast volumes and the reciprocal of the medium osmolarity (B). (Data from W. Kaiser, unpublished)

solutes. Other polypeptides occurring in this membrane are also expected to function in transport. The relationship between these polypeptides and the particulate membrane structures visible in the electron microscope needs to be established. The outer envelope membrane, although apparently rich in lipids, is unspecifically permeable to small ions or solute molecules (61). The outer mitochondrial membrane, which is similar in its properties to the outer membrane of the chloroplast envelope, contains a protein producing nonspecific diffusion channels (156).

In the presence of polycationic cytochrome *c,* intact chloroplasts agglutinate (101). Titration experiments revealed considerable binding of cytochrome *c* to the chloroplasts (53 nanomoles per milligram chlorophyll). Addition of Mg^{2+} dissociates the chloroplast-cytochrome *c* complex. Since the envelope contains only about 30 nanomoles of acidic lipid per milligram chlorophyll (sulfolipid and phosphatidylglycerol), negatively charged membrane proteins must be responsible for a considerable part of the binding reaction. It should be noted that 10 to 12 nanomoles cytochrome *c* per milligram chlorophyll would be sufficient to form a monomolecular layer around the outer envelope membrane (diameter of a cytochrome *c* molecule about 30 Å).

ENZYME ACTIVITIES OF THE ENVELOPE

In addition to carriers, a number of enzyme activities are associated with the chloroplast envelope. The function of a Mg^{2+}-dependent ATPase is unknown. In C_3 plants, its activity is usually below 4 μmoles/mg chlorophyll per hour (12, 18, 107), but seasonal increases in activity have been recorded (19). Rathnam & Das (117; see also 5) have reported 80% of the cellular nitrate reductase to be associated with the chloroplast envelope in *Eleusine coracana,* but Douce & Joyard (19) failed to find this enzyme in envelopes of spinach chloroplasts.

The role of the plastid envelope in galactolipid synthesis is well documented (17, 19, 72, 79). Two or even three enzymes responsible for the incorporation of galactose into lipid reside in the chloroplast envelope. Thylakoids contain but do not synthesize galactolipid. One of the galactolipid-synthesizing enzymes transfers the galactosyl moiety from UDP-galactose to diacylglycerol forming monogalactosyldiacylglycerol. Digalactosyldiacylglycerol is synthesized either by galactosyltransfer from one to another molecule of monogalactosyldiacylglycerol (6) or by galactosyltransfer from UDP-galactose to monogalactosyldiacylglycerol (19), or in both modes (19). Diacylglycerol, the substrate from which the galactolipids are formed, is synthesized from glycerol-3-phosphate by two acyltransferases and a phosphatidate phosphatase. One of the transferases and the phosphatase are firmly bound to the chloroplast envelope (19, 78). Also, acyl-CoA serving as substrate for the acylation reaction is synthesized from fatty acid, ATP, and CoASH by a very active long-chain acyl-CoA synthetase which is associated with the envelope (78, 123). Another envelope-bound enzyme catalyzes acyl exchange between phospholipids and galactolipids (52). Envelope preparations were active in the synthesis of plastoquinone (133), γ-tocophenol (132) and carotenoids (14). The envelope has also been suspected of playing a role in desaturation of fatty acids (19).

PERMEABILITY OF THE INNER ENVELOPE MEMBRANE

Gases

Bicarbonate distributes across the chloroplast envelope as predicted from the Henderson/Hasselbalch equation (153). This can be expected only if the permeability of the chloroplast envelope for CO_2 is high and for HCO_3^- low. The permeability of lipid-containing membranes for CO_2 is known to be very high (8). Rapid rates of photosynthetic oxygen production can be sustained under a CO_2 gradient of less than 100 ppm. Since the rate of diffusion is proportional to the gradient from source to sink and this gradient is very small for CO_2, it should not be large for oxygen. As expected, at air levels the concentration of oxygen inside photosynthesizing chloroplasts was comparable to that outside (124, 138). Surprisingly, at low oxygen concentrations significant oxygen gradients were observed which were independent of light intensity (138). Ammonia enters chloroplasts very fast as seen from the immediate response of the intrathylakoid pH when NH_4Cl is added to a suspension of intact chloroplasts (146).

Neutral Compounds

Osmotic changes in the size of small particles lead to changes in light-scattering. The latter are a convenient means of measuring unidirectional penetration. Light-scattering measurements have shown chloroplasts to permit rapid entry of the 3-carbon compound glycerol but not of the 2-carbon amino acid glycine (31). Osmotic swelling in a solution of the C_5 sugar ribose was much slower than in glycerol. The C_6 sugar alcohol sorbitol and the disaccharide sucrose have long been known to provide osmotic support. Uptake studies have indeed shown the envelope of intact chloroplasts to be impermeable to sorbitol or sucrose (61). However, the term "impermeable" must be used in a restricted sense. It only means that the rate of penetration is too low to be measured within the time span suitable for experimentation with fragile organelles such as chloroplasts. In vivo, chloroplasts contain sucrose and raffinose (36, 125), although they do not seem to have the enzymes for their synthesis (7). It must be concluded that the sugars had crossed the chloroplast envelope. The rate of penetration of most neutral amino acids, though sufficient for protein synthesis, is very low (31). Glycine betaine is effective in providing osmotic support to chloroplasts (87).

Ions and Acids

When salts such as KCl, NH_4Cl, KNO_2, or KNO_3 are added stepwise to intact chloroplasts, light-scattering increases stepwise (45). Plots of 1/apparent absorbance against 1/conc (concentration of added salt) yield straight lines indicating that chloroplasts respond osmometrically to salt addition (144). Yet a very slow absorbance decrease following the initial fast increase suggests slow salt penetration. Uptake is limited by the permeability of the chloroplast envelope to K^+ or NH_4^+ (45). When valinomycin is added, which specifically increases the K^+ or NH_4^+ conductivity of biomembranes, fast chloroplast swelling is observed in the presence of anions such as fluoride, chloride, nitrate, nitrite, bromide, and iodide. The rate of swelling follows a Hofmeister power series (49), indicating that the size of the hydrated anions determines the rate of penetration. Both the slow salt uptake in the absence of valinomycin and the fast anion uptake in its presence increase as the gradient across the chloroplast envelope increases suggesting transport by simple diffusion. From the area of the chloroplast envelope ($\sim$ 600 cm^2/mg chlorophyll) and the rate of anion penetration [more than 1000 μeq/mg chlorophyll per hour at a gradient of 100 μeq/ml with chloride and nitrite (45)] approximate permeability constants can be calculated, if it is assumed that membrane potentials are small at high electrolyte concentrations. They are about 5×10^{-8} $msec^{-1}$ or more for chloride and nitrite. For comparison, chloride values between 10^{-8} and 10^{-9} $msec^{-1}$ have been reported for tonoplast membranes (119). Values for the plasmalemma were still lower.

It appears from these data that the chloroplast envelope has an unusually high chloride permeability. The permeability for a number of organic anions and for HCO_3^- is significant, but far lower than that for halogenides and nitrite or nitrate (21, 45). Propionate and acetate anions penetrate faster than the glycolate or glyoxylate anions. Valinomycin did not cause swelling of spinach chloroplasts in the presence of potassium sulfate, phosphate, citrate, or gluconate (45). Obviously, significant net uptake (which must be distinguished from uptake owing to anion exchange) of polyvalent or large hydrophilic anions was not possible. However, chloroplasts of the C_4 plant *Digitaria sanguinalis* appeared to be permeable to the phosphate anion (70).

With salts of weak acids, a small but significant proportion of the anions may be protonated at physiological pH. Penetration of weak acids can be measured osmotically when ammonium salts are added. Ammonium ions react with anions of weak acids to form NH_3 and the protonated anion HA. NH_3 readily permeates biomembranes. If the protonated anion (HA) permeates also, osmotic chloroplast swelling occurs since inside the chloroplasts NH_3 and AH react back to NH_4^+ and A^-. Ammonium salts can also be used to measure anion/OH^- exchange systems. Since NH_3 entering the stroma compartment generates OH^- through $NH_3 + H_2O \rightarrow NH_4^+ + OH^-$, an anion/hydroxyl exchange would result in net anion uptake because the anion follows the NH_3 flux.

Swelling experiments have shown the chloroplast envelope to permit rapid entry of aliphatic acids such as acetic or propionic acid via simple diffusion or, much less likely, by a very active anion/hydroxyl exchange mechanism. Rates of glycolic and glyoxylic acid penetration were low compared with penetration of aliphatic acids (21). Generally the permeability of the chloroplast envelope to protonated anions is much higher than that to anions, but rates of transport via acids may be limited by the low concentration of acids in equilibrium with the corresponding anions.

When the chloroplast envelope is ruptured, penetration of anions or acids across the thylakoid membranes into the intrathylakoid space can be measured by light-scattering in a similar way as with intact chloroplasts. Provided permeabilities and chloroplast concentrations are comparable, rates of valinomycin-induced thylakoid swelling should be expected to be much faster than those of chloroplast swelling because the thylakoid membrane area is larger by a factor of about 25 than the surface area of intact chloroplasts. However, swelling of well-preserved thylakoids in the presence of KCl, KBr, or KI and valinomycin is much slower than that of intact chloroplasts (U. Heber, unpublished), indicating that the permeability coefficients of the envelope for halogenides are much higher than those of the thylakoid membranes. Similar observations were made with other anions. They indicate differences in the anion permeability patterns of the inner envelope membrane and thylakoid membranes.

In general, the cation permeability of the envelope is low (30, 145). Intact isolated chloroplasts photoreduce CO_2 at high rates in a Mg^{2+}-free medium although a number of photosynthetic reactions require Mg^{2+}. Obviously, endogenous Mg^{2+} does not leak out. There is a significant but low permeability for K^+ and Na^+ (80). The permeation of hydrated protons is also slow (30). However, a significant factor in the low rate of passive proton translocation is the very low proton concentration at physiological pH. The few measurements on passive H^+ fluxes from the medium into darkened chloroplasts which are available (30) indicate that the H^+ permeability of the chloroplast envelope is somewhat higher than that of the thylakoids.

SPECIFIC METABOLITE TRANSPORT

The impermeability of the inner envelope membrane to hydrophilic solutes such as phosphate, phosphate esters, dicarboxylates, and glucose is overcome by translocators which catalyze specific transfer of metabolites across the envelope.

Phosphate and Phosphate Ester Transport

The phosphate translocator facilitates the export of fixed carbon in the form of triosephosphate or 3-phosphoglycerate from the chloroplast in exchange for inorganic phosphate. It is the most powerful transport system of the inner envelope membrane. In spinach chloroplasts the V_{max} for this transport was found to be in the range of 300 μmoles per milligram chlorophyll per hour (20°) (22). Apparent k_m values for transport of the main substrates were between 0.1–0.3 mM. Phosphoenolpyruvate, glycerol-1-phosphate, erythrose-4-phosphate, and ribose-5-phosphate are also transported, but K_m values are at least one order of magnitude higher (22). Furthermore, sulfate and sulfite ions were also shown to be transported by this carrier (34). Recent suggestions that the transport of sulfate is due to a special sulfate carrier (100) require further experimental evidence.

Studies of the pH dependence of transport suggested that not only phosphate and triose phosphate but also 3-phosphoglycerate are transported as divalent anions (22), indicating two positive charges at the substrate binding site of the carrier. For each phosphate or phosphate ester anion taken up by spinach chloroplasts, another one is released (22). The rate of unidirectional phosphate transport was found to be three orders of magnitude lower than that of counterexchange (22). Thus, the total amount of phosphate (including phosphate esters) is kept constant in the stroma. Since in contrast to the divalent phosphate and dihydroxyacetone phosphate anions, 3-phosphoglycerate is predominantly a trivalent anion at physiological pH that must equilibrate with the divalent form in a pH-dependent reaction, a proton gradient across the envelope would be expected to influence transport. Indeed, during CO_2 assimilation of isolated chloroplasts mainly triosephosphate is released into the medium, although the stroma level of 3-phosphoglycerate may be ten times higher than that of triosephosphate (93). This apparent restriction of the release of 3-phosphoglycerate is only observed in the light and has been attributed to a pH gradient which is found between the external space and the stroma of illuminated chloroplasts (59, 93). 3-Phosphoglycerate transport was also shown to be inhibited by Mg^{2+} (126). Thus a rise of the stromal Mg^{2+}, as observed with isolated chloroplasts (84, 113), may also contribute to a restriction of 3-phosphoglycerate release from illuminated chloroplasts.

Pyrophosphate and citrate, which bind to the carrier without being transported to a significant extent (22), are suitable competitive inhibitors of phosphate or phosphate ester transport. Other inhibitors are mercurials such as *p*-chloromercuribenzoate (22, 122, 152), which react with sulfhydryl groups, and reagents known to interact with lysyl or arginyl residues [pyridoxal phosphate, trinitrobenzosulfonate, and *p*-diazobenzosulfonate or butandione and phenylglyoxal (25, 26)]. Specific protection by substrates of the phosphate translocator against inhibition indicate that lysine and arginine provide the cationic groups involved in binding the divalent transferable anions (27).

By incubating intact chloroplasts with lysyl reagents and employing radioactive labeling techniques, followed by isolation of membranes and SDS gel electrophoresis of membrane polypeptides, an integral envelope polypeptide possessing an apparent molecular weight of 29 kD was found to be involved in transport. It contains one inhibitor binding site. The molecular activity of transport (turnover number) was estimated to be about 5000 min^{-1} at 20°C (26). The 29 kD polypeptide has recently been solubilized by treatment of the envelope fraction with Triton X-100 and purified to apparent homogeneity (28).

The mechanism of transport has not yet been elucidated. In mitochondria, a polypeptide having a molecular weight of 30 kD is known to be involved in specific ATP/ADP exchange. Recent work has shown exchange to be catalyzed by a dimer which is formed from two 30 kD subunits but contains only one binding site (83). The two subunits are postulated to form a gated pore, where the alternative location of the substrate binding site at each side of the membrane represents a special conformational state. Transport across the membrane is visualized as a conformational change between two states (83). It is unclear whether transport by the chloroplast phosphate translocator functions in a similar manner. Available information on the number of inhibitor binding sites in the inner envelope membrane suggests that each 29 kD polypeptide binds one substrate whereas in mitochondrial ATP/ADP exchange a dimer reacts with one substrate.

So far, the properties of the phosphate translocator have been studied mainly with spinach chloroplasts. Transport with similar specificity was also demonstrated in plastids from etiolated leaves of *Avena sativa* (33). At very early stages of greening, however, phosphate transport of etioplasts was less sensitive to competitive inhibition by dihydroxyacetone phosphate and phosphoglycerate than transport in chloroplasts.

Osmotic swelling of mesophyll chloroplasts from *Digitaria sanguinalis* in the presence of potassium phosphate and valinomycin indicated rapid net uptake of phosphate (70). Furthermore, rapid counterexchange of phosphoenolpyruvate with phosphate was observed. The apparent K_m of phosphoenolpyruvate transport was one order of magnitude lower than in spinach (70). This variation of specificity appears to satisfy the requirement of phosphoenolpyruvate for carboxylation in the cytosol of mesophyll cells of C_4 plants.

Dicarboxylate Transport

A number of dicarboxylates such as L-malate, oxaloacetate, α-ketoglutarate, L-aspartate, and L-glutamate are transported across the envelope (33, 90). This transport is also an exchange, but is less strictly coupled to the simultaneous countertransport of another dicarboxylate ion (90) than substrate exchange by the phosphate translocator. In spinach chloroplasts the V_{max} for the transport of malate was usually about 100 μmoles/mg chl·h (rates extrapolated to 20°) (45). Dicarboxylate transport was initially attributed to a single dicarboxylate translocator of the inner envelope membrane (60). More detailed investigations left the possibility open that this transport might be due to different carriers with overlapping specificity (90). A specific inhibitor of dicarboxylate transport has not yet been reported. Mercurials, e.g. *p*-chloromercuribenzoate, cause some inhibition, but the inhibitory effect of these substances on dicarboxylate transport is less pronounced than on phosphate transport (152).

ATP and ADP Transport

Transport of adenine nucleotides across the chloroplast envelope proceeds by counterexchange and is usually slow (54). It is highly specific for ATP; transport of ADP into the stroma is about one order of magnitude slower. The specificity of transport indicates that the adenylate carrier is not effective in exporting ATP from the chloroplasts, but may rather function in supplying the chloroplasts with ATP, e.g. in the dark. Pyrophosphate is also slowly transported (121). When it is present in the medium, exchange with internal adenylates may deplete stromal adenylate pools and inhibit photosynthesis of isolated chloroplasts (121, 137).

The activity of the adenylate carrier depends on the developmental state of the chloroplasts. Inhibition of CO_2 fixation by pyrophosphate in pea chloroplasts indicated that in young leaves ATP transport is faster than in mature leaves (121). For mesophyll chloroplasts of a C_4 plant (*Digitaria sanguinalis*), a high capacity of the carrier for ATP transport has been reported (40 μmoles/mg chl·h, 20°) (68).

Glucose Transport

Several hexoses and pentoses, such as D-glucose, D-mannose, D-fructose, D-xylose, and D-ribose, but essentially no L-glucose, are transported across the envelope. For D-glucose the approximate K_m was found to be 20 mM and the maximal velocity about 60 μmoles/mg chl·h (20°) (127). Glucose transport appears to be inhibited by a high external glucose concentration (127). Thus glucose can be used as an osmoticum for chloroplasts, although it is specifically transported across the envelope. Specificity, K_m values, and activation energy of transport show striking similarities to the corresponding properties of the glucose carrier in human erythrocytes (89) and rat liver cells (5). In addition to triose phosphate and 3-phosphoglycerate, glucose was a significant product of starch degradation in chloroplasts (58, 105). The glucose carrier appears to function in exporting this glucose from the chloroplasts.

Amino Acid Transport

Early measurements of the osmotic response of pea chloroplasts to various neutral amino acids have led to the proposition that two carriers with apparent K_m values in the order of 100 mM transport glycine and serine across the chloroplast envelope (103, 104). In contrast, with spinach chloroplasts neutral amino acids were found to penetrate the envelope only very slowly (31, 55), and the final concentrations reached in the stroma were similar to the concentration in the medium. Since the rate of amino acid uptake was linearly dependent on concentration (55) and also on the hydrophobicity of the amino acids studied, it was concluded that these amino acids permeate the envelope by simple diffusion (31, 55). McLaren & Barber (96) have reported evidence for a carrier transferring leucine and isoleucine. The dicarboxylate translocator catalyzes rapid transfer of glutamate, aspartate, glutamine, and asparagine (31, 90).

Transport of Pyruvate

In mesophyll chloroplasts from *Digitaria sanguinalis,* pyruvate transport is mediated by a carrier (69). Pyruvate originates in the bundle sheath cells of this C_4 plant and is converted to phosphoenolpyruvate in the mesophyll chloroplasts at rates commensurate with rates of photosynthesis. The apparent K_M of the carrier for pyruvate was found to be 0.6–1 mM and rates of transport were 50 to 100 μmoles/mg chlorophyll per hour. Transport was inhibited by the sulfhydryl mersalyl, by pyruvate analogs such as phenylpyruvate, and by α-cinnamic acid (69). It may be noted that these inhibitors did not affect the uptake of acetate. Pyruvate transport was very slow in spinach chloroplasts (43).

THE pH GRADIENT ACROSS THE CHLOROPLAST ENVELOPE

When intact chloroplasts are illuminated, the stroma pH increases above the pH of the suspending medium (62, 154), mainly due to proton translocation into the intrathylakoid space. However, protons are also excreted into the medium (43, 44, 62). Only after the envelope has been rendered permeable to protons is the alkalinization of the stroma transmitted to the medium (116). Obviously, in illuminated chloroplasts a proton gradient is maintained between stroma and external space, with the stroma being more alkaline than the medium. In contrast, in the dark the stroma pH is below the pH of media usually used for suspending isolated chloroplasts (154). Carbonyl cyanide-*m*-chlorophenylhydrazone (CCCP), a substance known to increase the proton conductivity of biomembranes, abolishes the transenvelope pH gradient of illuminated chloroplasts but does not alter the opposite gradient observed in the dark (62, 154). The latter has therefore been attributed to a Donnan distribution of ions between the proteinaceous stroma and the external medium (30). The pH difference in the dark disappears when sorbitol is replaced as osmoticum by sodium gluconate (A. R. Portis, unpublished) which would be expected to suppress a Donnan potential.

If the Donnan distribution of H^+ is taken into account, the effective ΔpH across the envelope of illuminated chloroplasts is about 0.8, resulting in a stroma pH close to 8. The stroma pH can be lowered and the ΔpH across the envelope decreased or abolished by very low concentrations of CCCP or salts of certain weak acids such as glyoxylic, glycolic, glyceric, formic, acetic, or nitrous acid (21, 24, 39, 41, 50, 116, 154). Prerequisites of the salt effects are that the chloroplast envelope permits penetration of both the undissociated acid and the corresponding anion. In this situation, both species can operate a proton shuttle (Figure 2), which serves to diminish or even abolish gradients in proton activity between chloroplast stroma and medium. The salt effect in diminishing the proton gradient across the envelope is particularly pronounced at lower pH. It is interesting that high concentrations of bicarbonate are also effective in decreasing the ΔpH across the envelope (154). The low permeability of the envelope for the bicarbonate ion (45) appears sufficient to account for this effect. If acidification of the stroma in the presence of bicarbonate were caused by CO_2 influx into the chloroplasts, it should be transient and the original pH should be reestablished after excretion of imported protons. However, the salt-induced decrease in the transenvelope proton gradient is permanent.

The change of the stromal pH between darkness and illumination is an important factor for the regulation of CO_2 fixation. With a stromal pH of 7.0, CO_2 fixation is found to be inhibited, whereas maximal activity is attained at pH 8.0 (154). Apparently the pH changes occurring in the stroma after illumination are sufficient to switch CO_2 fixation from almost zero to maximal activity. Salts of weak acids or very low concentrations of CCCP, in decreasing the light-dependent alkalinization of the stroma, effectively inhibit CO_2 fixation by intact chloroplasts (21, 24, 116, 154). Inhibi-

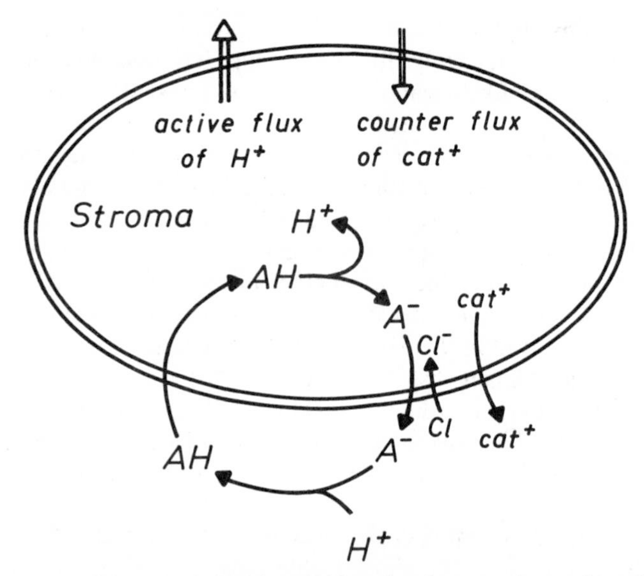

Figure 2. Stroma acidification in illuminated chloroplasts by proton import through a proton shuttle system. Loss of permeable anions (A^-) is facilitated mainly by passive co-transport of cations (cat^+) that have previously been imported in exchange for H^+.

tion is also observed at high concentrations of bicarbonate, although CO_2 is the substrate of photosynthesis (21, 154). From studies of stromal metabolite levels in intact chloroplasts, fructose and sedoheptulose bisphosphatase were identified as pH-sensitive steps of the overall reaction of CO_2 assimilation (21, 24, 116). These enzymes are activated by light, presumably through the ferredoxin-dependent thioredoxin system (9). Experiments with intact chloroplasts indicated that light activation of enzymes is also pH dependent (39, 41). Furthermore, the activity of the activated bisphosphatase is very sensitive to a decrease in pH below the pH optimum (2). It is visualized that the combination of both effects results in rigid control of CO_2 fixation by the stroma pH.

In some cases, inhibition of photosynthesis by salts is not caused by pH effects only. Nitrite which oxidizes ferredoxin during its photoreduction interferes also with redox regulation by the thioredoxins (41), and glyoxylate may inhibit fructose bisphosphatase both through the stroma pH and directly (21).

PROTON TRANSFER BETWEEN THE STROMA AND THE EXTERNAL SPACE

The initial alkalinization of the stroma at the onset of illumination could be a mere consequence of proton transport into the thylakoid space. The resulting proton gradient across the envelope, however, needs to be maintained. Otherwise, proton leakage across the envelope, e.g. mediated by salts of weak acids, would eventually result in a collapse of this gradient. In order to compensate for the passive proton influx, an active extrusion of protons from the stroma to the external space is required. Such extrusion can actually be observed on illumination of intact chloroplasts (43, 44, 62). The net proton efflux from the stroma to the medium is at the onset of illumination compensated in part by a counterflux of K^+ or Na^+ ions (30). Such a counterflux of alkali cations should not be required if the active proton extrusion is fully balanced by passive proton influx. When K^+ is lost from chloroplasts, stroma acidification and inhibition of photosynthesis are observed (80). Photosynthesis can be restored by the addition of K^+ to the medium.

There are two main possibilities to explain proton extrusion by chloroplasts on illumination. H^+ may be exported by an electrogenic pump of the envelope, which would presumably be ATP driven. In the case of a net H^+ efflux, K^+ would flow into the chloroplasts in response to a membrane potential generated by H^+ pumping. An H^+/K^+ antiporter has also been postulated to exist (71). Although an ATPase of unknown function resides in the chloroplast envelope (12, 18, 96, 107), neither evidence for a pump nor for a light-generated membrane potential or the antiporter is presently available. Attempts have failed to explain by ATPase action the report (10) that addition of ATP to darkened chloroplasts causes an alkalinization of the stroma (U. Enser, U. I. Flügge, unpublished). It is feasible, however, that an ATP-driven proton transport across the envelope is inhibited during darkness.

Alternatively, there may be connections, temporary or otherwise, between the thylakoid system and the inner envelope membrane which would permit a limited efflux of protons from the thylakoids to the external space. In this way, passive proton influx from the external space into the stroma can be visualized to be compensated by light-driven proton transport across the thylakoid membrane, followed by a regulated, passive flux of protons from the thylakoid space into the medium. The activity of light-driven proton transport into the thylakoids is almost two orders of magnitude higher than observed rates of active H^+ extrusion from the stroma to the external medium (43, 44, 62). The high capacity of the light-driven proton pump of the thylakoid membrane is also apparent from the fact that salts of weak acids, which lower the stromal pH by permitting passive proton influx across the envelope into the stroma, have no effect on the H^+ gradient across the thylakoid membrane (116).

Net proton extrusion from the stroma appears to be influenced in opposite ways by externally added Mg^{2+} and NH_4^+ ions. Mg^{2+} and other divalent cations have been shown to inhibit CO_2 fixation of intact chloroplasts (3,

65–67), although this effect was not generally observed (113). Since Mg^{2+} penetrates into the stroma only very slowly (30), an effect of this cation on the envelope membrane was likely. It was shown later that added Mg^{2+} causes an acidification of the stroma (16, 71) which is accompanied by an efflux of K^+ (16). External K^+ prevents this effect, which may explain that the Mg^{2+} dependent acidification of the stroma was not generally observed. Similar to salts of weak acids (21, 116), Mg^{2+} was shown to shift the pH optimum of photosynthesis into the alkaline range, i.e. to be more inhibitory at low than at high pH (67, 71). The inhibitory effect of Mg^{2+} on photosynthesis is potentiated when salts of weak acids such as acetate, which increase the passive proton flux into the stroma, are added (71). These findings clearly support the notion that inhibition of photosynthesis in intact chloroplasts by Mg^{2+} is due to acidification of the stroma.

NH_4Cl, a reputed uncoupler of photophosphorylation which is known to decrease the transthylakoid proton gradient, not only did not inhibit but even stimulated CO_2 fixation, which had previously been inhibited by salts of weak acids (116) or Mg^{2+} (U. Enser, unpublished). Apparently, NH_4^+ counteracts the acidification of the stroma by stimulating proton extrusion from the stroma. This effect can be explained in terms of limited connections between the thylakoid and the external space. In the light, NH_3 permeating into the thylakoid space leads to an increase of H^+ transport across the thylakoid membrane and hence to an accumulation of NH_4^+. The resulting NH_4^+ gradient between the thylakoid space and the external medium also stores H^+ ($NH_3 + H^+ \leftrightharpoons NH_4^+$). NH_4^+ discharge from the intrathylakoid space into the medium would increase effectiveness of proton excretion and could explain maintenance of a proton gradient across the chloroplast envelope against increased proton leakage from the medium (U. Enser, unpublished).

GRADIENT COUPLING

Principles of coupling of different gradients to the proton gradient across the chloroplast envelope have been outlined before (37, 39, 50). It has been shown that the proton gradient across the chloroplast envelope results in an opposite anion gradient if the anion is nonpenetrating and its protonation product penetrating. Anion gradients have been measured for bicarbonate and abscisic acid (15, 51, 153). The abscisic acid gradient is particularly interesting as it shows sequestration of a plant hormone in the chloroplasts. It is likely that indole acetate also accumulates in the chloroplasts. The extent of accumulation depends on the magnitude of the proton gradient.

The dicarboxylate carrier catalyzes traffic of malate, aspartate, α-ketoglutarate, and glutamate between the chloroplasts and the cytosol. With isolated chloroplasts a malate/oxaloacetate shuttle reducing or oxidizing external pyridine nucleotide under illumination has been demonstrated (1, 43, 44). Under physiological conditions, the oxaloacetate/malate ratio in equilibrium with the pyridine nucleotides is expected to be between 10^{-3} and 10^{-4}. Such an excess of malate would drastically inhibit oxaloacetate transport by the dicarboxylate carrier. Through the activity of aspartate aminotransferase, which is present on either side of the envelope, the oxaloacetate can react with glutamate to form aspartate and α-ketoglutarate. Thus, the stromal NADP system may be linked to the cytosolic NAD system (37, 43).

Though it is clear that in the cell fluxes occur whose quantitative treatment would require nonequilibrium thermodynamics, redox and energy gradient coupling will be demonstrated for a (in reality nonexisting) equilibrium situation. Equilibrium of the involved aminotransferases and dehydrogenases is represented by:

$$\left[\frac{(\text{Asp})\,(\alpha\text{KG})\,(\text{H}^+)\,(\text{NADH})}{(\text{Glu})\,(\text{Mal})\,(\text{NAD}^+)}\right]_{\text{cytosol}} = \left[\frac{(\text{Asp})\,(\alpha\text{KG})\,(\text{H}^+)\,(\text{NAD(P)H})}{(\text{Glu})\,(\text{Mal})\,(\text{NAD(P)}^+)}\right]_{\text{stroma}} \qquad 1.$$

(Asp, aspartate; αKG, α-ketoglutarate; Glu, glutamate; Mal, malate)

At zero flux across the chloroplast envelope, opposite gradients balance each other:

$$\left[\frac{(\text{Asp})\,(\alpha\text{KG})}{(\text{Glu})\,(\text{Mal})}\right]_{\text{cytosol}} = \left[\frac{(\text{Asp})\cdot(\alpha\text{KG})}{(\text{Glu})\,(\text{Mal})}\right]_{\text{stroma}} \qquad 2.$$

Under these model conditions, a proton gradient across the envelope would stabilize a gradient in the ratio of reduced to oxidized pyridine nucleotides between stroma and cytosol

$$\frac{(\text{H}^+)_{\text{cytosol}}}{(\text{H}^+)_{\text{stroma}}} = \left[\frac{(\text{NAD(P)H})}{(\text{NAD(P)}^+)}\right]_{\text{stroma}} \Big/ \left[\frac{(\text{NADH})}{(\text{NAD}^+)}\right]_{\text{cytosol}} \qquad 3.$$

Deviation from equilibrium of one of the involved enzymes would lead to substrate fluxes that cause the enzymes in the other compartment to react. In this way, electrons not only could flow downhill, but also uphill. Uphill transport would occur if the enzyme in the more acidic compartment is geared to reduction.

Not only the dicarboxylate translocator, but also the phosphate translocator, transfers divalent anions (22, 38, 93). At physiological pH, the substrate phosphoglycerate (PGA) is mainly in the trivalent form which is linked through

$$PGA^{3-} + H^{+} \rightleftharpoons PGA^{2-}$$

with the divalent anion. If only fluxes of phosphoglycerate and dihydroxy-acetone phosphate (DHAP) are considered, the following relation is valid for flux equilibrium (net transfer between chloroplast stroma and cytosol zero):

$$\frac{(H^{+})_{cytosol}}{(H^{+})_{stroma}} = \left[\frac{(DHAP^{2-})}{(PGA^{3-})}\right]_{cytosol} \Big/ \left[\frac{(DHAP^{-2})}{(PGA^{3-})}\right]_{stroma} \quad 4.$$

If the pH of the cytosol is lower than that of the stroma, the ratio of DHAP to PGA must be higher there than in the stroma.

If the equilibrium model has any bearing on the situation in the green cell, important consequences would result. A high PGA/DHAP ratio in the chloroplasts favors CO_2 reduction in photosynthesis even at low phosphorylation potentials (ATP) / (ADP)(P_i), while glycolytic triosephosphate oxidation, which generates ATP and reducing equivalents in the cytosol, is driven by a high cytosolic ratio of DHAP/PGA.

In leaf cells, different shuttles operate simultaneously. As has been shown in detail elsewhere (37, 39, 50), substrate transfer and reaction equilibrium of the shuttles exchanging malate against other dicarboxylates and DHAP against PGA, in addition to the relations presented above, would yield

$$\left[\frac{(DHAP^{2-})}{(PGA^{3-})}\right]_{cytosol} \Big/ \left[\frac{(DHAP^{2-})}{(PGA^{3-})}\right]_{stroma} = \left[\frac{(ATP)}{(ADP)\,(P_i)}\right]_{cytosol} \Big/ \left[\frac{(ATP)}{(ADP)\,(P_i)}\right]_{stroma} \quad 5.$$

The model shows that not only gradients of glycolytic and photosynthetic intermediates can be coupled to the primary proton gradient across the chloroplast envelope (see Equation 4), but also a gradient in the phosphorylation potential. Interestingly, the equation suggests that the stromal phosphorylation potential is lower than that in the cytosol although ATP is generated in the light in the chloroplasts.

Even before gradient coupling in chloroplasts was understood, the secondary gradients which are coupled to a primary proton gradient and are described for equilibrium situations in equations 1 to 5 were observed in leaf cells. There is a high PGA/DHAP ratio in the chloroplast stroma compared with that in the cytosol (48) or, after chloroplast isolation, in the suspending medium (93). The chloroplast NAD and NADP systems were found to be more reduced than the cytoplasmic NAD system (46), and the cytoplasmic phosphorylation potential is higher than that of the chloroplasts even in the light (47, 82, 129). The cytosolic NADP system is kept reduced by a nonphosphorylating glyceraldehydephosphate dehydrogenase (4) and glucose-6-phosphate dehydrogenase and does not appear to be linked to the cytoplasmic NAD system which is much more oxidized (46).

METABOLIC SIGNIFICANCE OF THE PHOSPHATE AND DICABOXYLATE TRANSLOCATORS

The isolated chloroplast is metabolically a very flexible organelle which can under appropriate conditions be induced to emphasize one of a number of possible reactions. Simplified equations of such reactions serve to illustrate the role of the main chloroplast translocators. Of ten reactions listed in Table 1, six require anion exchange by the phosphate translocator (reactions 1 to 6) and two by the dicarboxylate translocator (reactions 8 and 9). Only in two reactions does simple diffusion appear to be sufficient (reactions 7 and 10). In reaction 1, bicarbonate enters the chloroplast as CO_2 leaving OH^- behind (45), and in reactions 2 and 5, PGA^{3-} enters as PGA^{2-} (22), again leaving OH^- outside. Reaction 3 of C_4 mesophyll chloroplasts involves the pyruvate translocator in addition to the phosphate translocator. It is not yet clear whether the phosphate acceptor of C_4 plants accepts PEP^{3-} or rather reacts with PEP^{2-} (70). Transfer of PEP by the phosphate translocator of spinach chloroplasts is very slow (22). In reaction 4, glycerate probably diffuses into the chloroplasts as acid (42), and in reaction 6, glycolate may leave as acid (21). In reaction 10, HNO_2 and NH_3 are probably main transfer species (45).

While Table 1 stresses the importance of the phosphate translocator, it does not reflect its integration into the metabolic network. Its main role is the transport of fixed carbon from the chloroplast stroma to the cytosol. During CO_2 fixation, no more than one-sixth of the newly formed triosephosphate is available for export. Additional export would result in the breakdown of the photosynthetic carbon cycle, as five-sixths of the triosephosphate formed in the light are necessary for the regeneration of ribulose bisphosphate. Thus transport has to be controlled to permit maximal export rates without depleting the CO_2 fixation cycle of its intermediates. Since P_i and triosephosphates are transported by the translocator in both directions and transport occurs via counterexchange, the net influx of P_i and net efflux of triosephosphate will depend on the concentration ratios of both compounds on either side of the inner envelope membrane. The gradient

Table 1 Reactions preferentially performed by isolated chloroplasts under appropriate substrate conditions

1. $3\ HCO_3^- + P_i^{2-} + 2\ H_2O$	→	$DHAP^{2-} + 3\ OH^- + 3\ O_2$
2. $PGA^{3-} + H_2O$	→	$DHAP^{2-} + OH^- + \frac{1}{2}\ O_2$
3. $Pyruvate^- + P_i^{2-}$	→	$PEP^{3-} + H_2O$
4. $Glycerate^- + P_i^{2-}$	→	$DHAP^{2-} + OH^- + \frac{1}{2}\ O_2$
5. $2\ PGA^{3-} + H_2O$	→	(Glucose unit in starch) + $2\ P_i^{2-} + 2\ OH^- + O_2$
6. $2\ DHAP^{2-} + 5\ H_2O + 1\frac{1}{2}\ O_2$	→	$3\ glycolate^- + 2\ P_i^{2-} + 3\ H^+$
7. $2\ HCO_3^- + H_2O$	→	$glycolate^- + OH^- + 1\frac{1}{2}\ O_2$
8. $OAA^{2-} + H_2O$	→	$malate^{2-} + \frac{1}{2}\ O_2$
9. $Ketoglutarate^{2-} + NH_3^+$	→	$glutamate^- + \frac{1}{2}\ O_2$
10. $NO_2^- + 3\ H_2O$	→	$NH_4^+ + 2\ OH^- + 1\frac{1}{2}\ O_2$

situation must be such that accumulation of stromal triosephosphate, which would act back on reduction of PGA, is prevented but sufficient triosephosphate remains available for formation of ribulose bisphosphate.

Efficient control of substrate gradients requires high transport capacity. Indeed, the activity of the phosphate translocator exceeds the rate of net substrate efflux considerably (93). Thus in spinach chloroplasts a carrier activity of about 300 contrasts with a rate of net substrate efflux of 60 at a CO_2 assimilation rate of 180 (rates given in μmoles per milligram chlorophyll per hour at 20°C). The balance between rate of translocation and rate of CO_2 fixation is often disturbed when isolated chloroplasts are kept in an isotonic medium which contains only phosphate and bicarbonate as substrates. With external phosphate concentrations as low as 0.5 or 1 mM, a partial inhibition of CO_2 fixation is frequently observed (23). It is due to substrate depletion of the carbon cycle. Therefore, unless triosephosphate or phosphoglycerate are added to relieve inhibition, phosphate concentrations in the medium must either be kept unphysiologically low or the phosphate translocator must be partially inhibited [e.g. by pyrophosphate or pyridoxalphosphate (23, 151)] for high rates of CO_2 fixation.

Although the mechanism of its action was unknown, pyrophosphate has been used empirically for a long time to improve the performance of isolated chloroplasts (74, 149). In spinach protoplasts, the triosephosphate exported to the cytosol is primarily transformed to sucrose, which is the main product of CO_2 fixation in these cells (29, 120, 143). The phosphate thus released is transported back into stroma in exchange for more triosephosphate. Under optimal conditions of photosynthesis (e.g. high light intensity) the rate of CO_2 fixation may exceed the rate of phosphate liberation in the cytosol that is necessary for triosephosphate formation. In such a metabolic situation, an apparent surplus of the fixed carbon, which is not exported for lack of external phosphate, can be stored within the stroma as starch (58, 63). Sequestration of cytoplasmic phosphate as mannose phosphate after feeding mannose to leaves stimulates photosynthetic starch formation (130). Isolated chloroplasts kept in a medium that contains a very low concentration of phosphate also show a drastically increased rate of starch synthesis (58, 140). Starch formation is stimulated by a high ratio of PGA/P_i (58), which is known to activate ADP glucose pyrophosphorylase, the enzyme controlling starch synthesis (114, 115).

The starch deposited in the chloroplasts is usually degraded during the subsequent night period. An increased stromal phosphate level favors starch breakdown. Glucose-1-phosphate, the product of phosphorolytic starch degradation, is transformed through the oxidative pentosephosphate pathway (141) and also via phosphofructokinase (81) to triosephosphate or further to PGA (58, 105, 142). The phosphate translocator catalyzes export of these phosphate esters into the cytosol. Glucose, the final product of hydrolytic starch degradation (58, 105), is released from the chloroplasts by the glucose translocator.

Another important function of the phosphate translocator is to link intra- and extrachloroplast pyridine nucleotide and adenylate systems through shuttles involving the exchange of dihydroxyacetone phosphate and 3-phosphoglycerate. This has been discussed under "Gradient Coupling." When the intrachloroplast phosphorylation potential is increased in the light by photophosphorylation, export and subsequent oxidation of triosephosphate result in the coupled transfer of reducing equivalents and phosphate energy into the cytosol. When import of ATP exceeds consumption, the cytosolic phosphorylation potential rises above the dark level (47, 129). Since the phosphorylation potentials of cytosol and mitochondria are linked through the mitochondrial adenylate carrier and mitochondrial oxidation is controlled by the phosphorylation potential, a light-induced increase in the mitochondrial energy status over the dark level is expected to inhibit mitochondrial oxidation. Inhibition of dark respiration in the light has indeed often been observed (11, 73).

While the phosphate translocator connects chloroplast and cytosolic NADP systems, transferring electrons uphill through the cytosolic non-phosphorylating glyceraldehydephosphate dehydrogenase (4), it links to the cytosolic NAD system only in a reaction coupled to phosphorylation of ADP (38). In contrast, the dicarboxylate translocator connects chloroplast pyridine nucleotides to the cytosolic NAD system through the aspartate and malate shuttles. In malate-forming C_4 plants, exchange of oxaloacetate for malate by the dicarboxylate translocator of mesophyll chloroplasts appears to be as fast as photosynthesis, since the carbon fixation product oxaloacetate which is synthesized in the cytosol is reduced to malate in the mesophyll chloroplasts. Malate is further metabolized in the cells of the bundle sheath.

CONCLUDING REMARKS

Much of our present knowledge on the chloroplast envelope and its function has been gained during work with spinach. The little information available for chloroplasts of other plant species has revealed interesting differences in transport of adenylates, phosphoenolpyruvate, pyruvate, and phosphate. It appears promising to explore diversity of envelope function in different plastids and in different plant species.

ACKNOWLEDGMENTS

We are very grateful to Professor R. Douce and Drs. U. Enser, I. Flügge, H. Gimmler, K. P. Heise, and M. Stitt for stimulating discussions and helpful comments and to many other colleagues for providing reprints, preprints, and unpublished data to this review.

Literature Cited

1. Anderson, J. W., House, C. M. 1979. Polarographic study of dicarboxylic acid-dependent export of reducing equivalents from illuminated chloroplasts. *Plant Physiol.* 64:1064–69
2. Baier, D., Latzko, E. 1975. Properties and regulation of C1-fructose-1,6-diphosphatase from spinach chloroplasts. *Biochim. Biophys. Acta* 396: 141–48
3. Bamberger, E. S., Avron, M. 1975. Site of action of inhibitors of carbon dioxide assimilation by whole lettuce chloroplasts. *Plant Physiol.* 56:481–85
4. Bamberger, E. S., Ehrlich, B. A., Gibbs, M. 1975. The glyceraldehyde 3-phosphate and glycerate 3-phosphate shuttle and carbon dioxide assimilation in intact spinach chloroplasts. *Plant Physiol.* 55:1023–30
5. Baur, H., Heldt, H. W. 1977. Transport of hexoses across the liver-cell membrane. *Eur J. Biochem.* 74:397–403
6. Besouw, A. van, Wintermans, J. F. G. M. 1978. Galactolipid formation in chloroplast envelopes. I. Evidence for two mechanisms in galactosylation. *Biochim. Biophys. Acta* 529:44–53
7. Bird, J. F., Cornelius, M. J., Dyer, T. A., Keys, A. J., Whittingham, C. P. 1974. Intracellular site of sucrose synthesis in leaves. *Phytochemistry* 13: 59–64
8. Blank, M., Roughton, F. J. W. 1960. The permeability of monolayers to carbon dioxide. *Trans. Farady Soc.* 56: 1832–41
9. Buchanan, B. B., Wolosiuk, R. A., Schürmann, P. 1979. Thioredoxin and enzyme regulation. *Trends Biochem. Sci.* 4:93–96
10. Champigny, M. L., Joyard, J. 1978. Analyse du mechanisme de stimulation, par ATP, de le penetration du carbone inorganique dans les chloroplastes intacts, isolès de l'Epinard. *C. R. Acad. Sci. Paris* 286:1791–94
11. Chevallier, D., Douce, R. 1976. Interactions between mitochondria and chloroplasts in cells. *Plant Physiol.* 57:400–2
12. Cobb, A. H., Wellburn, A. R. 1974. Changes in plastid envelope polypeptides during chloroplast development. *Planta* 121: 273–82
13. Coombs, J. 1976. Interactions between chloroplasts and cytoplasm in C_4 plants. In *The Intact Chloroplast,* ed. J. Barber, pp. 279–313. Amsterdam: Elsevier-North Holland
14. Costes, C., Burghoffer, C., Joyard, J., Block, M., Douce, R. 1979. Occurrence and biosynthesis of violaxanthin in iso-

lated spinach chloroplast envelopes. *FEBS Lett.* 103:17–21
15. Cowan, I. R., Farquhar, G. D., Raven, J. A. 1980. A possible role for abscisic acid in the coupling of biochemical conductance with stomatal conductance in leaves. In manuscript
16. Demmig, B., Gimmler, H. 1979. Effect of divalent cations on cation fluxes across the chloroplast envelope and on photosynthesis of intact chloroplasts. *Z. Naturforsch. Teil C* 34: 233–41
17. Douce, R. 1974. Site of biosynthesis of galactolipids in spinach chloroplasts. *Science* 183:852–53
18. Douce, R., Holtz, R. B., Benson, A. A. 1973. Isolation and properties of the envelope of spinach chloroplasts. *J. Biol. Chem.* 248:7215–22
19. Douce, R., Joyard, J. 1979. Structure and function of the plastid envelope. *Adv. Bot. Res.* 7:1–117
20. Edwards, G. E., Walker, D. A. 1980. *C3, C4—Mechanisms, Cellular and Environmental Regulation of Photosynthesis.* Oxford: Blackwell. In press
21. Enser, U., Heber, U. 1980. Metabolic regulation by pH gradients: Inhibition of photosynthesis by indirect proton transfer across the chloroplast envelope. *Biochim. Biophys. Acta* 592:577–91
22. Fliege, R., Flügge, U. I., Werdan, K., Heldt, H. W. 1978. Specific transport of inorganic phosphate, 3-phosphoglycerate and triosephosphates across the inner membrane of the envelope in spinach chloroplasts. *Biochim. Biophys. Acta* 502:232–47
23. Flügge, U. I., Freisl, M., Heldt, H. W. 1980. Balance between metabolite accumulation and transport in relation to photosynthesis by isolated spinach chloroplasts. *Plant Physiol.* 65:574–77
24. Flügge, U. I., Freisl, M., Heldt, H. W. 1980. The mechanism of the control of carbon fixation by the pH in the chloroplast stroma: Studies with acid mediated proton transfer across the envelope. *Planta* 149:48–51
25. Flügge, U. I., Heldt, H. W. 1977. Specific labelling of a protein involved in phosphate transport of chloroplasts by pyridoxal-5'-phosphate. *FEBS Lett.* 82:29–33
26. Flügge, U. I., Heldt, H. W. 1978. Specific labelling of the active site of the phosphate translocator in spinach chloroplasts by 2,4,6-trinitrobenzene sulfonate. *Biochem. Biophys. Res. Commun.* 84:37–44
27. Flügge, U. I., Heldt, H. W. 1979. Phosphate translocator in chloroplasts: Identification of the functional protein and characterization of its binding site. In *Functions and Molecular Aspects of Biomembrane Transport,* ed. E. Quagliariello et al, pp. 373–82. Amsterdam: Elsevier-North Holland
28. Flügge, U. I., Heldt, H. W. 1980. On the isolation of the phosphate translocator in chloroplasts. *Proc. 1st Eur. Bioenerg. Conf., Urbino, Italy,* pp. 329–30. Bologna: Patron Editore
29. Giersch, Ch., Heber, U., Kaiser, G., Walker, D. A., Robinson, S. 1980. Intracellular metabolite gradients and flow of carbon during photosynthesis of leaf protoplasts. *Arch. Biochem. Biophys.* 205:246–59
30. Gimmler, H., Schäfer, G., Heber, U. 1974. Low permeability of the chloroplast envelope towards cations. *Proc. Int. Congr. Photosynth., 3rd,* 2:1381–92. Amsterdam: Elsevier-North Holland
31. Gimmler, H., Schäfer, G., Kraminer, H., Heber, U. 1974. Amino acid permeability of the chloroplast envelope as measured by light scattering, volumetry and amino acid uptake. *Planta* 120: 47–61
32. Gunning, B. E. S., Steer, M. W. 1975. *Ultrastructure and the Biology of Plant Cells,* pp. 1–312. London: Arnold
33. Hampp, R. 1978. Kinetics of membrane transport during chloroplast development. *Plant Physiol.* 62:735–40
34. Hampp, R., Ziegler, J. 1977. Sulfate and sulfite translocation via phosphate translocator of inner envelope membrane of chloroplasts. *Planta* 137:309–12
35. Hatch, M. D., Osmond, C. B., 1976. Compartmentation and transport in C_4 photosynthesis. In *Encyclopedia of Plant Physiology. Transport in Plants. Intracellular Interactions and Transport Processes,* ed. C. R. Stocking, U. Heber, 3:144–84. Heidelberg: Springer
36. Heber, U. 1959. Beziehung zwischen der Grösse von Chloroplasten und ihrem Gehalt an löslichen Eiweissen und Zuckern im Zusammenhang mit dem Frostresistenzproblem. *Protoplasma* 51:284–98
37. Heber, U. 1974. Energy transfer within leaf cells. See Ref. 30, 2:1335–48.
38. Heber, U. 1974. Metabolite exchange between chloroplasts and cytoplasm. *Ann. Rev. Plant Physiol.* 25:393–421
39. Heber, U. 1980. Metabolic regulation by a transmembrane proton gradient. In *Cell Compartmentation and Metabolic*

Channeling, ed. L. Nover, F. Lynen, K. Mothes, pp. 331–44. Jena: Fischer
40. Heber, U., Egneus, H., Hanck, U., Jensen, M., Köster, S. 1978. Regulation of photosynthetic electron transport and phosphorylation in intact chloroplasts and leaves of *Spinacia oleracea* L. *Planta* 143:41–49
41. Heber, U., Enser, U., Weis, E., Ziem, U., Giersch, Ch. 1979. Regulation of the photosynthetic carbon cycle, phosphorylation and electron transport in illuminated intact chloroplasts. In *Covalent and Non-Covalent Modulation of Protein Function,* ed. D. E. Atkinson, C. F. Fox, pp. 113–39. New York: Academic
42. Heber, U., Kirk, M. R., Gimmler, H., Schäfer, G. 1974. Uptake and reduction of glycerate by isolated chloroplasts. *Planta* 120:31–46
43. Heber, U., Krause, G. H. 1971. Transfer of carbon, phosphate energy, and reducing equivalents across the chloroplast envelope. In *Photosynthesis and Photorespiration,* ed. M. D. Hatch, C. B. Osmond, R. O. Slatyer, pp. 218–25. New York/London: Wiley-Interscience
44. Heber, U., Krause, G. H. 1971. Hydrogen and proton transfer across the chloroplast envelope. *Proc. Int. Congr. Photosynth, 2nd, Stresa,* pp. 1023–33. The Hague: Junk
45. Heber, U., Purczeld, P. 1977. Substrate and product fluxes across the chloroplast envelope during bicarbonate and nitrite reduction. *Proc. Int. Congr. Photosynth., 4th, Reading,* pp. 107–18. London: Biochem. Soc.
46. Heber, U., Santarius, K. A. 1965. Compartmentation and reduction of pyridine nucleotides in relation to photosynthesis. *Biochim. Biophys. Acta* 109:390–408
47. Heber, U. Santarius, K. A. 1970. Direct and indirect transfer of ATP and ADP across the chloroplast envelope. *Z. Naturforsch. Teil B* 25:718–28
48. Heber, U., Santarius, K. A., Hudson, M. A., Hallier, U. W. 1967. Untersuchungen zur intrazellulären Verteilung von Enzymen und Substraten in der Blattzelle. I. Intrazellulärer Transport von Zwischenprodukten der Photosynthese im Photosynthese-Gleichgewicht und im Dunkel-Licht-Dunkel-Wechsel. *Z. Naturforsch. Teil B* 22:1189–99
49. Heber, U., Volger, H., Overbeck, V., Santarius, K. A. 1979. Membrane damage and protection during freezing. *Adv. Chem. Ser.* 180, *Proteins at Low Temperatures,* ed. O. Fennema, pp. 159–89. Washington, DC: Am. Chem. Soc.
50. Heber, U., Walker, D. A. 1979. The chloroplast envelope—barrier or bridge? *Trends Biochem. Sci.* 4:252–56
51. Heilmann, B., Hartung, W., Gimmler, H. 1980. The distribution of abscisic acid between chloroplast and cytoplasm of leaf cells and the permeability of the chloroplast envelope for abscisic acid. *Z. Pflanzenphysiol.* 97:67–78
52. Heinz, E., Bertram, M., Joyard, J., Douce, R. 1978. Demonstration of an acyltransferase activity in chloroplast envelopes. *Z. Pflanzenphysiol.* 87: 325–31
53. Heinz, E., Haas, R. 1980. Use of mesophyll protoplasts to study lipid metabolism in organelles. *Proc. Symp. Recent Adv. Biogen. Funct. Plant Lipids,* ed. P. Beneviste, C. Costes, R. Douce, P. Mazliak, pp. 19–28. Amsterdam: Elsevier-North Holland
54. Heldt, H. W. 1969. Adenine nucleotide translocation in spinach chloroplasts. *FEBS Lett.* 5:11–14
55. Heldt, H. W. 1976. Metabolite transport in intact spinach chloroplasts. See Ref. 13, pp. 215–34
56. Heldt, H. W. 1976. Metabolite carriers of chloroplasts. See Ref. 35, 3:137–43
57. Heldt, H. W. 1976. Transfer of substrates across the chloroplast envelope. In *Horizons in Biochemistry and Biophysics,* ed. E. Quagliariello, F. Palmieri, Th. P. Singer, 2:199–229. Reading Mass: Addison-Wesley
58. Heldt, H. W., Chon, Ch. J., Maronde, D., Herold, A., Stankovic, Z., Walker, D. A., Kraminer, A., Kirk, M. R., Heber, U. 1977. Role of orthophosphate and other factors in the regulation of starch formation in leaves and isolated chloroplasts. *Plant Physiol.* 59:1146–55
59. Heldt, H. W., Flügge, U. I., Fliege, R. 1978. The influence of illumination on the transport of 3-phosphoglycerate across the chloroplast envelope. In *The Proton and Calcium Pumps,* ed. G. F. Azzone, M. Avron, J. C. Metcalfe, E. Quagliariello, N. Siliprandi, pp. 105–14. Amsterdam: Elsevier-North Holland
60. Heldt, H. W., Rapley, L. 1970. Specific transport of inorganic phosphate, 3-phosphoglycerate and dihydroxyacetone phosphate, and of dicarboxylates across the inner membrane of spinach chloroplasts. *FEBS Lett.* 10:143–48
61. Heldt, H. W., Sauer, F. 1971. The inner membrane of the chloroplast envelope as the site of specific metabolite transport. *Biochim. Biophys. Acta* 234:83–91

62. Heldt, H. W., Werdan, K., Milovancev, M., Geller, G. 1973. Alkalization of the chloropast stroma caused by light dependent proton flux into the thylakoid space. *Biochim. Biophys. Acta* 314:224–41
63. Herold, A., Lewis, D. H., Walker, D. A. 1976. Sequestration of cytoplasmic orthophosphate by mannose and its differential effect on photosynthetic starch synthesis in C_3 and C_4 species. *New Phytol.* 76:397–407
64. Herold, A., Walker, D. A. 1979. Transport across chloroplast envelopes. The role of phosphate. In *Membrane Transport in Biology,* ed. G. Giebisch, D. C. Tostenson, J. J. Ussing, 2:411–30. Heidelberg: Springer
65. Huber, S. C. 1978. Regulation of chloroplast photosynthetic activity by exogenous magnesium. *Plant Physiol.* 62:321–25
66. Huber, S. C. 1979. Effect of photosynthetic intermediates on the magnesium inhibition of oxygen evolution by barley chloroplasts. *Plant Physiol.* 63:754–57
67. Huber, S. C. 1979. Effect of pH on chloroplast photosynthesis. Inhibition of O_2 evolution by inorganic phosphate and magnesium. *Biochim. Biophys. Acta* 545:131–40
68. Huber, S. C., Edwards, G. E. 1976. A high-activity ATP translocator in mesophyll chloroplasts of *Digitaria sanguinalis,* a plant having the C-4 dicarboxylic acid pathway of photosynthesis. *Biochim. Biophys. Acta* 440:675–87
69. Huber, S. C., Edwards, G. E. 1977. Transport in C4 mesophyll chloroplasts —characterization of pyruvate carrier. *Biochim. Biophys. Acta* 462:583–602
70. Huber, S. C., Edwards, G. E. 1977. Transport in C4 mesophyll chloroplasts —evidence for an exchange of inorganic phosphate and phosphoenolpyruvate. *Biochim. Biophys. Acta* 462:603–12
71. Huber, S. C., Maury, W. 1980. Effects of magnesium on intact chloroplasts. I. Evidence for activation of (sodium) potassium/proton exchange across the chloroplast envelope. *Plant Physiol.* 65:350–54
72. Hummel, H. C. van, Hulsebos, T. J. M., Wintermans, J. F. G. M. 1975. Biosynthesis of galactosyl diglycerides by nongreen fractions from chloroplasts. *Biochim. Biophys. Acta* 380:219–26
73. Ishii, R., Shibayama, M., Murata, Y. 1979. Effect of light on the CO_2 evolution of C_3 and C_4 plants in relation to the Kok effect. *Jpn. J. Crop Sci.* 48:52–57
74. Jensen, R. G., Bassham, J. A. 1966. Photosynthesis by isolated chloroplasts. *Proc. Natl. Acad. Sci. USA* 56:1095–1101
75. Joy, K. W., Ellis, R. J. 1975. Protein synthesis in chloroplasts. IV. Polypeptides of the chloroplast envelope. *Biochim. Biophys. Acta* 378:143–51
76. Joyard, J., Douce, R. 1975. Mg^{++} dependent ATPase of the envelope of spinach chloroplasts. *FEBS Lett.* 51:335–40
77. Joyard, J., Douce, R. 1976. Prèparation et activitès enzymatiques de l'envelope des chloroplastes d'Epinard. *Physiol. Vèg.* 14:31–48
78. Joyard, J., Douce, R. 1977. Site of synthesis of phosphatidic acid and diacylglycerol in spinach chloroplasts. *Biochim. Biophys. Acta* 486:273–85
79. Joyard, J., Douce, R. 1979. Characterization of phosphatidate phosphohydrolase activity associated with chloroplast envelope membranes. *FEBS Lett.* 102:147–50
80. Kaiser, W. M., Urbach, W., Gimmler, H. 1980. The role of monovalent cations for photosynthesis of isolated intact chloroplasts. *Planta.* 143:170–75
81. Kelly, G. J., Latzko, E. 1977. Chloroplast phosphofructokinase. Proof of phosphofructokinase activity in chloroplasts. *Plant Physiol.* 60:290–94
82. Keys, A. J., Whittingham, C. P. 1969. Nucleotide metabolism in chloroplast and non-chloroplast components of tobacco leaves. *Prog. Photosynth. Res.* 1:352–58
83. Klingenberg, M., Aquila, H., Krämer, R., Babel, W., Feckl, J. 1977. The ADP, ATP translocation and its catalyst. In *Biochemistry of Membrane Transport,* ed. G. Semenza, E. Carafoli. FEBS *Proc. Meet.* 42:567–79. Heidelberg: Springer
84. Krause, G. H. 1977. Light-induced movement of magnesium ions in intact chloroplasts. Spectroscopic determination with eriochrome blue SE. *Biochim. Biophys. Acta* 460:500–10
85. Krause, G. H., Heber, U. 1976. Energetics of intact chloroplasts. See Ref. 13, pp. 171–214
86. Laetsch, W. M. 1974. The C_4 syndrome: A structural analysis. *Ann. Rev. Plant Physiol.* 25:27–52
87. Larkum, A. W. D., Wyn Jones, R. G. 1979. Carbon dioxide fixation by chloroplasts isolated in glycinebetaine. *Planta* 145:393–94
88. Leech, R. M., Murphy, D. J. 1976. The cooperative function of chloroplasts in

the biosynthesis of small molecules. See Ref. 13, pp. 365–401
89. Le Fevre, P. G. 1961. Sugar transport in the red blood cell: Structure activity relationship in substrates and antagonists. *Pharmacol. Rev.* 13:39–70
90. Lehner, K., Heldt, H. W. 1978. Dicarboxylate transport across the inner membrane of the chloroplast envelope. *Biochim. Biophys. Acta* 501:531–44
91. Lichtenthaler, H. A., Prenzel, U., Douce, R., Joyard, J. 1981. Localization of prenylquinones in the envelope of spinach chloroplasts. *Biochim. Biophys. Acta.* In press
92. Liedvogel, B., Sitte, P., Falk, H. 1976. Chromoplasts in the daffodil: fine structure and chemistry. *Cytobiology* 12:155–74
93. Lilley, R., Chon, C. J., Mosbach, A., Heldt, H. W. 1977. The distribution of metabolites between spinach chloroplasts and medium during photosynthesis in vitro. *Biochim. Biophys. Acta* 460:259–72
94. Mackender, R. O., Leech, R. M. 1970. Isolation of chloroplast envelope membranes. *Nature* 228:1347–49
95. Mackender, R. O., Leech, R. M. 1974. The galactolipid, phospholipid, and fatty acid composition of the chloroplast envelope membranes of *Vicia faba* L. *Plant Physiol.* 53:496–502
96. McLaren, J. S., Barber, D. J. 1977. Evidence for carrier mediated transport of L-leucine into isolated pea (*Pisum sativum*) chloroplasts. *Planta* 136:147–51
97. Mendiola-Morgenthaler, L. R., Morgenthaler, J. J. 1974. Proteins of the envelope and thylakoid membranes of spinach chloroplasts. *FEBS Lett.* 49:152–55
98. Moreau, F., Dupont, J., Lance, C. 1974. Phospholipid and fatty acid composition of outer and inner membranes of plant mitochondria. *Biochim. Biophys. Acta* 345:294–304
99. Morrè, D. J., Mollenhauer, H. H. 1976. Interactions among cytoplasm, endomembranes, and the cell surface. See Ref. 35, 3:288–344
100. Mourioux, G., Douce, R. 1979. Transport du sulfate à travers la double membrane limitante, ou enveloppe, des chloroplasts d'epinard. *Biochimie* 61: 1283–92
101. Neuburger, M., Joyard, J., Douce, R. 1977. Strong binding of cytochrome *c* on the envelope of spinach chloroplasts. *Plant Physiol.* 59:1178–81
102. Nobel, P. S. 1968. Light induced chloroplast shrinkage in vivo detectable after rapid isolation of chloroplasts from *pisum sativum. Plant Physiol.* 43: 781–87
103. Nobel, P. S., Cheung, Y. S. 1972. Two amino acid carriers in pea chloroplasts. *Nature* 237:207–8
104. Nobel, P. S., Wang, C. 1970. Amino acid permeability of pea chloroplasts as measured by osmotically determined reflection coefficients. *Biochim. Biophys. Acta* 211:79–87
105. Peavey, D. G., Steup, M., Gibbs, M. 1977. Characterization of starch breakdown in the intact spinach chloroplast. *Plant Physiol.* 60:305–8
106. Pineau, B., Douce, R. 1974. Analyse eletrophoretique des proteines de l'enveloppe des chloroplasts d'Epinard. *FEBS Lett.* 47:255–59
107. Poincelot, R. P. 1973. Isolation and lipid composition of spinach chloroplast envelope membranes. *Arch. Biochem. Biophys.* 159:134–42
108. Poincelot, R. P. 1974. Uptake of bicarbonate ion in darkness by isolated chloroplast envelope membranes and intact chloroplasts of spinach. *Plant Physiol.* 54:520–26
109. Poincelot, R. P. 1975. Transport of metabolites across isolated envelope membranes of spinach chloroplasts. *Plant Physiol.* 55:849–52
110. Poincelot, R. P. 1977. Isolation of envelope membranes from bundle sheath chloroplasts of maize. *Plant Physiol.* 60:767–70
111. Poincelot, R. P. 1980. Isolation of chloroplast envelope membranes. *Methods Enzymol.* 69:121–28
112. Poincelot, R. P., Day, P. R. 1976. Isolation and bicarbonate transport of chloroplast envelope membranes from species of differing net photosynthetic efficiency. *Plant Physiol.* 57:334–38
113. Portis, A. R., Heldt, H. W. 1976. Light-dependent changes of Mg^{2+} concentration in the stroma in relation to the Mg^{2+} dependency of CO_2 fixation in intact chloroplasts. *Biochim. Biophys. Acta* 449:434–46
114. Preiss, J., Ghosh, M. P., Wittkop, J. 1967. Regulation of the biosynthesis of starch in spinach leaf chloroplasts. In *Biochemistry of Chloroplasts,* ed. T. W. Goodwin, 1:131–52. New York: Academic
115. Preiss, J., Levi, C. 1977. Regulation of α1,4-glucan metabolism in photosynthetic systems. *Proc. Int. Congr. Photosynth., 4th, Reading,* pp. 457–68
116. Purczeld, P., Chon, C. J., Portis, A. R., Heldt, H. W., Heber, U. 1978. The

mechanism of the control of carbon fixation by the pH in the chloroplast stroma. Studies with nitrite-mediated proton transfer across the envelope. *Biochim. Biophys. Acta* 501:488–98

117. Rathnam, C. K. M., Das, V. S. R. 1974. Nitrate metabolism in relation to the aspartate type C-4 pathway of photosynthesis in *Eleusine coracana. Can. J. Bot.* 52:2599–2605
118. Raven, J. A. 1976. Division of labour between chloroplast and cytoplasm. See Ref. 13, pp. 403–43
119. Raven, J. A. 1976. Transport in algal cells. In *Encyclopedia of Plant Physiology, New Series. Transport in Plants II. Part A, Cells,* ed. U. Lüttge, M. G. Pitman, pp. 129–88. Heidelberg: Springer
120. Robinson, S. P., Walker, D. A. 1979. The site of sucrose synthesis in isolated leaf protoplasts. *FEBS Lett.* 107:295–99
121. Robinson, S. P. Wiskich, J. T. 1977. Pyrophosphate inhibition of carbon dioxide fixation in isolated pea chloroplasts by uptake in exchange for endogenous adenine nucleotides. *Plant Physiol.* 59:422–27
122. Robinson, S. P., Wiskich, J. T. 1977. Inhibition of CO_2 fixation by adenosine 5'-diphosphate and the role of phosphate transport in isolated pea chloroplasts. *Arch. Biochem. Biophys.* 184:546–54
123. Roughan, P. G., Slack, C. R. 1977. Long-chain acyl-coenzyme A synthetase activity of spinach chloroplasts is concentrated in the envelope. *Biochem. J.* 162:457–59
124. Samisch, Y. B. 1975. Oxygen buildup in photosynthesizing leaves and canopies is small. *Photosynthetica* 9:372–75
125. Santarius, K. A., Milde, H. 1977. Sugar compartmentation in frosthardy and partially dehardened cabbage leaf cells. *Planta* 136:163–66
126. Schäfer, G. 1975. *Permeabilität der Chloroplastenmembranen für physiologisch wichtige Substanzen.* PhD thesis. Univ. Düsseldorf, Germany
127. Schäfer, G., Heber, U., Heldt, H. W. 1977. Glucose transport into spinach chloroplasts. *Plant Physiol.* 60:286–89
128. Schnarrenberger, C., Fock, H. 1976. Interactions among organelles involved in photorespiration. See Ref. 35, 3:185–234
129. Sellami, A. 1976. Evolution des adenosine phosphates et de la charge energetique dans les compartments chloroplastique et nonchloroplastique des feuilles de ble. *Biochim. Biophys. Acta* 423:524–39
130. Sheu-Hwa, Ch., Lewis, D. H., Walker, D. A. 1975. Stimulation of photosynthetic starch formation by sequestration of cytoplasmic orthophosphate. *New Phytol.* 74:383–92
131. Slovacek, R. E., Mills, J. D., Hind, G. 1978. The function of cyclic electron transport in photosynthesis. *FEBS Lett.* 87:73–76
132. Soll, J., Douce, R., Schultz, G. 1980. Site of biosynthesis of α-tocopherol in spinach chloroplasts. *FEBS Lett.* 112:243–46
133. Soll, J., Schultz, G. 1980. Prenyl quinone synthesis in spinach chloroplasts. See Ref. 53
134. Sprey, B., Laetsch, W. M. 1975. Chloroplast envelopes of *Spinacia oleracea* L. I. Polypeptides of chloroplast envelopes and lamellae. *Z. Pflanzenphysiol.* 75:38–52
135. Sprey, B., Laetsch, W. M. 1976. Chloroplast envelopes of *Spinacia oleracea* L. II. Ultrastructure of chloroplast envelopes and lamellae. *Z. Pflanzenphysiol.* 78:146–63
136. Sprey, B., Laetsch, W. M. 1976. Chloroplast envelopes of *Spinacia oleracea* L. III. Freeze-fracturing of envelopes. *Z. Pflanzenphysiol.* 78:360–71
137. Stankovic, Z. S., Walker, D. A. 1977. Photosynthesis by isolated pea chloroplasts. Some effects of adenylates and inorganic pyrophosphate. *Plant Physiol.* 59:428–32
138. Steiger, H. M., Beck, E., Beck, R. 1977. Oxygen concentration in isolated chloroplasts during photosynthesis. *Plant Physiol.* 60:903–6
139. Steiger, H. M., Beck, E. 1980. Photosynthetic formation of H_2O_2 and balance of ATP synthesis to NADP reduction. *Plant Cell Physiol.* Submitted
140. Steup, M., Peavey, D. G., Gibbs, M. 1976. The regulation of starch metabolism by inorganic phosphate. *Biochem. Biophys. Res. Commun.* 72:1554–61
141. Stitt, M., Ap Rees, T. 1980. Carbohydrate breakdown by chloroplasts of *pisum sativum. Biochim. Biophys. Acta* 627:131–43
142. Stitt, M., Heldt, H. W. 1981. Physiological rates of starch breakdown in isolated spinach chloroplasts. *Plant Physiol.* In press
143. Stitt, M., Wirtz, W., Heldt, H. W. 1980. Metabolite levels during induction in the chloroplast and extrachloroplast compartment of spinach chloroplasts. *Biochim. Biophys. Acta* 593:85–102

144. Tedeschi, H., Harris, L. 1958. Some observations on the photometric estimation of mitochondria volume. *Biochim. Biophys. Acta* 28:392–402
145. Telfer, A., Barber, J., Nicolson, J. 1975. Availability of monovalent and the divalent cations within intact chloroplasts for the actions of ionophores nigericin and A23187. *Biochim. Biophys. Acta* 396:301–9
146. Tillberg, J.-E., Giersch, Ch., Heber, U. 1977. CO_2 reduction by intact chloroplasts under a diminished proton gradient. *Biochim. Biophys. Acta* 461:31–47
147. Tuquet, C., De Lubac, M. 1975. Action de le phospholipase A sur l'enveloppe des chloroplastes d'epinard. *C. R. Acad. Sci. Paris* 281:1313–16
148. Walker, D. A. 1974. Chloroplast and cell—The movement of certain key substances, etc., across the chloroplast envelope. In *Biochemistry* (Ser. 1, Plant Biochemistry), 11: pp. 1–49
149. Walker, D. A. 1976. CO_2 fixation by intact chloroplasts: Photosynthetic induction and its relation to transport phenomena and control mechanisms. See Ref. 13, pp. 235–78
150. Walker, D. A. 1976. Plastids and intracellular transport. See Ref. 35, 3:85–136
151. Walker, D. A., Robinson, S. P. 1978. Regulation of photosynthetic carbon assimilation. In *Photosynthetic Carbon Assimilation,* ed. H. W. Siegelman, G. Hind, pp. 43–59. New York/London: Plenum
152. Werdan, K., Heldt, H. W. 1972. The phosphate translocator of spinach chloroplasts. See Ref. 44, pp. 1337–44
153. Werdan, K., Heldt, H. W., Geller, G. 1972. Accumulation of bicarbonate in intact chloroplasts following a pH gradient. *Biochim. Biophys. Acta* 283:430–41
154. Werdan, K., Heldt, H. W., Milovancev, M. 1975. The role of pH in the regulation of carbon fixation in the chloroplast stroma. Studies on CO_2 fixation in the light and in the dark. *Biochim. Biophys. Acta* 396:276–92
155. Wildman, S. G., Hongladarom, T., Honda, S. I. 1966. *Organelles in living plant cells.* 16 mm sound film, Educ. Film Sales and Rentals, Univ. Calif., Berkeley
156. Zalman, L. S., Nikaido, H., Kagawa, Y. 1980. Mitochondrial outer membrane contains a protein producing nonspecific diffusion channels. *J. Biol. Chem.* 255:1771–74

Ann. Rev. Plant Physiol. 1981. 32:169–204

THE ASSIMILATORY NITRATE-REDUCING SYSTEM AND ITS REGULATION

❖7709

Miguel G. Guerrero, José M. Vega, and Manuel Losada

Departamento de Bioquímica, Facultad de Biología y C.S.I.C.,
Universidad de Sevilla, Spain

CONTENTS

INTRODUCTION

The assimilatory reduction of nitrate by plants is a fundamental biological process in which a highly oxidized form of inorganic nitrogen is reduced to ammonia. This ammonia becomes in turn combined with carbon skele-

0066-4294/81/0601-0169$01.00

tons to form the different biological nitrogenous compounds. Not only algae and higher plants but also a variety of bacteria and fungi have the ability to assimilate nitrate. Nevertheless, the assimilation of nitrate by the former organisms deserves special interest from the quantitative point of view and also because it represents a photosynthetic process where the reducing power is provided by water, which acts as the ultimate reductant at the expense of sunlight energy (84).

About 2×10^2 megatons nitrogen per year are assimilated through the process of biological fixation of molecular nitrogen (15). Nevertheless, it can be calculated that the plant kingdom incorporates an amount of nitrogen that exceeds this figure by two orders of magnitude. This means that the operation of the assimilatory nitrate-reducing system can account for the assimilation of as much as 2×10^4 megatons nitrogen on a yearly basis. Elementary analysis of plants yields about 2–10% nitrogen versus 40–50% carbon on a dry weight basis, with a C/N ratio of 5–15 (14). Moreover, the reduction of carbon dioxide to the level of carbohydrate requires four electrons, whereas the reduction of nitrate to ammonia and the conversion of the latter into amino-nitrogen involve ten electrons. Thus, for algae and higher plants, on an average and net basis, the proportion of electrons required for nitrate reduction is as much as about one-fourth of that required for the fixation of carbon dioxide.

It is firmly established (10, 36, 56, 84, 157) that the assimilatory nitrate-reducing system consists of only two metalloproteins, namely nitrate reductase and nitrite reductase, which catalyze the stepwise reduction of nitrate to nitrite and ammonia:

$$NO_3^- \xrightarrow[\text{nitrate reductase}]{2e} NO_2^- \xrightarrow[\text{nitrite reductase}]{6e} NH_4^+$$

Ammonia, the end product of nitrate reduction, is incorporated into organic compounds by way of one or both of the two major routes known to be responsible for the conversion of ammonia-nitrogen into α-amino-nitrogen: the glutamate dehydrogenase pathway and the glutamine synthetase-glutamate synthase pathway (79).

In this article we attempt to summarize and discuss the intensive developments in the field of assimilatory nitrate reduction since the last reviews in this series (9, 54). The scope has been limited primarily to enzymological and regulatory aspects of the process in different types of organisms, with a comparative approach. Several reviews dealing with different aspects of the subject have appeared recently (36, 55, 56, 83–85, 97, 136, 140, 148, 150, 157).

ENZYMES OF THE NITRATE-REDUCING SYSTEM

Reduction of Nitrate to Nitrite

According to the specificity for the electron donor, two main types of assimilatory nitrate reductase can be distinguished: (*a*) ferredoxin-dependent nitrate reductase, which is typically present in cyanobacteria (blue-green algae) and also presumably in chemoergonic and photosynthetic bacteria, and (*b*) pyridine nucleotide-dependent nitrate reductase, which is found in eukaryotic organisms. The differential specificity for the physiological reductant is correlated with sensible differences in their prosthetic groups and physicochemical properties (Table 1). Also, whereas ferredoxin-nitrate reductase of photosynthetic prokaryotes seems to be tightly bound to photosynthetic membranes, the pyridine nucleotide-dependent enzyme of eukaryotes is a soluble enzyme (84, 157).

FERREDOXIN-NITRATE REDUCTASE According to present evidence (18, 47, 89, 107), nitrate reductase from the cyanobacteria *Anabaena cylindrica, Anacystis nidulans,* and *Nostoc muscorum* cannot accept electrons directly from NAD(P)H but is rather dependent on reduced ferredoxin as the physiological electron donor. The reaction catalyzed by cyanobacterial nitrate reductase can thus be written as:

$$NO_3^- + 2\ Fd_{red} + 2\ H^+ \xrightarrow{2e} NO_2^- + 2\ Fd_{ox} + H_2O$$

$$\cdot\ \Delta G'_{o}, \text{ pH } 7 = -38.6 \text{ kcal}\cdot\text{mol}^{-1}$$

Also flavodoxin, a low molecular weight flavoprotein which physiologically substitutes for ferredoxin under conditions of iron starvation, can act as an electron source for the reduction of nitrate to nitrite catalyzed by *Anacystis* nitrate reductase when reduced photosynthetically (18, 88).

Though not yet demonstrated conclusively, there are some indications that nitrate reductases from both chemoergonic bacteria (146, 158) and photosynthetic bacteria (2, 87) can also use reduced ferredoxin as the physiological electron donor. These enzymes are active with reduced viologens, but they cannot use reduced pyridine nucleotides as reductant.

Purification and properties By using affinity chromatography on ferredoxin-Sepharose gel as the main step (90), ferredoxin-nitrate reductase of *Anacystis* has been purified recently to homogeneity and partly characterized. The enzyme is a molybdoprotein having only one polypeptide chain with a molecular weight of 75 kdaltons. From the characteristics of the absorption spectra of homogeneous preparations of the enzyme, the presence of either flavin or cytochrome in its molecule can be excluded (18). Whether this nitrate reductase contains nonheme iron remains to be determined as does the number of molybdenum atoms per enzyme molecule. An essential role of molybdenum in the catalytic activity of the enzyme has, however, been demonstrated on the basis of physiological experiments of competition with tungstate (18).

Ferredoxin-nitrate reductase from *Anacystis* exhibits a surprisingly high pH optimum (about 10.5) with dithionite-reduced methyl viologen as the electron donor (89).

Cyanide and *p*-hydroxymercuribenzoate (*p*HMB) are powerful inhibitors of *Anacystis* nitrate reductase (89). The inhibition of this enzyme by iron-binding agents and the apparent lack of heme in the enzyme molecule suggest the participation of nonheme iron in its catalytic activity (18).

Nitrate reductase from the photosynthetic bacterium *Rhodopseudomonas capsulata* has been purified recently and some of its properties studied (2). However, the low specific activity achieved (1 unit per milligram of protein) poses some questions about the validity of the conclusions. A molecular weight of 185 kdaltons has been reported for a protein which seems to be composed of two subunits of 85 kdaltons and to contain one molybdenum atom and 1.5 heme molecules of the *c*-type.

Nitrate reductases from the aerobic chemoergonic bacteria *Azotobacter chroococcum* and *Acinetobacter calcoaceticus* have been characterized partly and shown to be molybdoproteins of about 100 kdaltons whose activity is inhibited by cyanide and *p*HMB and stimulated by cyanate (43, 44, 158).

NAD(P)H-NITRATE REDUCTASE The nitrate reductase of fungi, green algae, and higher plants catalyzes the reduction of nitrate to nitrite by reduced pyridine nucleotides (36, 54, 57, 84, 157) in accordance with the equation:

$$NO_3^- + NAD(P)H + H^+ \xrightarrow{2e} NO_2^- + NAD(P)^+ + H_2O$$

$$\Delta G_0', \text{pH } 7 = -34 \text{ kcal}\cdot\text{mol}^{-1}$$

According to the specificity shown for NADH or NADPH, three subclasses of pyridine nucleotide-dependent nitrate reductase have been distinguished. EC 1.6.6.1 is specific for NADH and corresponds to the type of enzyme present in the leaves of most higher plants (9, 10, 77) and in some species of green algae such as *Chlorella* (132, 141). Nevertheless, the most prevalent enzyme in other green algae, such as *Ankistrodesmus braunii* (1, 27), and that of yeast (40, 124) can use either of the reduced nucleotides with about the same effectiveness. Accordingly, these nitrate reductases have

Table 1 Properties of highly purified assimilatory nitrate reductases

	Anacystis nidulans	*Chlorella vulgaris*	*Ankistrodesmus braunii*	*Spinacia oleracea*	*Neurospora crassa*	*Rhodotorula glutinis*
Molecular weight (kdaltons)	75	280	460	197	228	230
Subunits	1	3	8	2 + 2 ?	2	2
$s_{20,w}$(S)	5.4	10.0	10.9	8.1	8.0	7.9
Stokes radius (nm)	3.2	6.3	9.8	6.0	7.0	7.0
Electron donor	Fd	NADH	NAD(P)H	NADH	NAD(P)H	NAD(P)H
Specific activity (U/mg)	875[a]	93	61[b]	24	125[b]	148[b]
Turnover number (s^{-1})	1094[a]	434	468[b]	79	475[b]	567[b]
Optimum pH	10.5[a]	7.6	7.5	7.5	7.5	7.5
K_m NO_3^- (μM)	690	84	150[b]	180	200[b]	125[b]
K_m electron donor (μM)	13	nr[c]	13[b]	4.6	62[b]	20[b]
Prosthetic groups (mol/mol)						
FAD	—	2.6	nr*[c]	nr*	nr*	nr*
cyt *b*–557	—	2.5	4	nr*	1.5	1
Mo	nr*	2.0	nr*	nr*	0.9	nr*
E_{m7} cyt *b*–557 (mV)	—	nr	–73; *n* = 1	–60; *n* = 1	nr	nr
Spectral properties						
Absorption maxima	273	279; 413 (ox) 423; 527; 557 (red)	280; 416 (ox) 424; 527; 557 (red)	280; 413 (ox) 424; 527; 557 (red)	280?; 413 (ox) 423; 528; 557 (red)	278; 412 (ox) 423; 527; 557 (red)
A_{280}/A_{Soret}	—	1.7	3.1	4.6	nr	nr
ϵ (Soret) ($mM^{-1} \cdot cm^{-1}$)	—	280	460	nr	nr	124
References	(18, 90)	(37, 132, 133)	(26, 52)	(57, 101)	(36, 109)	(40)

[a] Dithionite-reduced methyl viologen as reductant.
[b] NADPH as reductant.
[c] Definitions: nr, not reported; nr*, present, but no quantitative data available.

been classified as NAD(P)H-dependent (EC 1.6.6.2). Nitrate reductases of the molds *Aspergillus nidulans* and *Neurospora crassa* have been classified as NADPH specific (EC 1.6.6.3) (91, 109). It should, however, be mentioned that the preference for NADPH of mold nitrate reductase is not very exclusive. The simultaneous presence of two different nitrate-reducing enzymes, NADH- and NAD(P)H-dependent, has been demonstrated conclusively for the case of soybean leaves (17). Separation of two different nitrate reductases exhibiting differential specificity for both pyridine nucleotides has also been achieved in young rice seedlings (127). The stereospecificity of nitrate reductases for hydrogen removal from reduced pyridine nucleotides has been determined to be of the A-type regardless the source of enzyme (41).

In addition to catalyzing the normal physiological reaction, i.e. reduction of nitrate by reduced pyridine nucleotides, NAD(P)H-nitrate reductase exhibits two other activities, which can be assayed separately and selectively inhibited, that involve only partial functions of the overall electron transport chain of the enzyme complex. The diaphorase activity represents the function of the first (NAD(P)H-activating) moiety of the complex, its action resulting in the reduction by NAD(P)H of a variety of 1- or 2-electron acceptors. The so-called terminal nitrate reductase constitutes the second (nitrate-activating) moiety of the complex and is expressed as a pyridine nucleotide-independent activity which results in the reduction of nitrate by reduced flavins or viologens. Both moieties participate jointly and sequentially in the transfer of electrons from NAD(P)H to nitrate (84).

Purification and properties Purification to electrophoretic homogeneity of a NAD(P)H-nitrate reductase was first achieved for the enzyme from the green alga *Chlorella vulgaris* (133). Since Solomonson (131) successfully used affinity chromatography on Blue Dextran-Sepharose for the purification of *Chlorella* nitrate reductase, this simple and straightforward procedure has been applied to the enzyme from other sources with outstanding results (26, 40, 41, 77, 101). Affinity chromatography on FAD-Sepharose has also been employed for the purification of fungal nitrate reductase (41, 109).

The pH optima, K_m for substrates and molecular properties of some of the more highly purified pyridine nucleotide nitrate reductases from different sources are summarized in Table 1. The general pattern which can be inferred from the available data is that NAD(P)H-nitrate reductase is an oligomeric enzyme (mol wt from 197 to 460 kdaltons) composed of a variable number of apparently identical subunits.

Inhibitors Chlorate can replace nitrate as a substrate for purified nitrate reductase, either substrate acting as competitive inhibitor of the reduction of the other (101, 157). Assimilatory NAD(P)H-nitrate reductases are generally inhibited by *p*-HMB, cyanide, azide, and cyanate. The diaphorase moiety of the enzyme complex is very sensitive to the action of sulfhydryl-binding reagents, whereas azide and cyanate inhibit, competitively with respect to nitrate, the terminal activity of the complex (157). Cyanide is a potent inhibitor of nitrate reductase activity and forms a stable complex with the reduced enzyme (82). This complex has been proposed to have an important role in modulating in vivo nitrate reductase activity (82, 134, 156)

Nitrite, the reaction product, has been reported to act as an inhibitor of NAD(P)H-nitrate reductase. The inhibition is of the competitive type with respect to nitrate in the enzymes of *Chlorella, Aspergillus,* and *Rhodotorula* (40, 91, 157).

Prosthetic groups From the available data on highly purified NAD(P)H-nitrate reductases from different sources, it appears that FAD, cytochrome *b*-557, and molybdenum are ubiquitous prosthetic groups. Information concerning the absorption spectra of the NAD(P)H-nitrate reductases adequately purified is summarized in Table 1.

Nitrate reductase of the mold *Neurospora* appears to contain one molybdenum atom and about two cytochrome *b*-557 molecules per enzyme molecule (109). Nevertheless, other authors have reported the presence of only one molecule of cytochrome and one molybdenum atom for the *Neurospora* enzyme (63). Also, only one molecule of cytochrome *b* is thought to be present in the enzyme of the yeast *Rhodotorula* (40). Presently available quantitative data about the prosthetic group content of higher plant nitrate reductase are rather scarce. A ratio of two iron to one molybdenum has been reported (101) for the spinach enzyme (mol wt 197 kdaltons), which is thought to be composed of two copies of each of two different types of subunits (57, 101). Only one type of subunit (mol wt about 100 kdaltons) seems to be present in the nitrate reductase of barley leaves, however (77). Regarding the situation in eukaryotic algae, the NADH-dependent enzyme of *Chlorella* (mol wt 280 kdaltons) has been shown to contain about three molecules of FAD and cytochrome *b* and two molybdenum atoms, and to be made up of three equal subunits of about 90 kdaltons (37). In contrast, NAD(P)H-nitrate reductase from *Ankistrodesmus* exhibits a higher molecular weight (460 kdaltons) and appears to be composed of eight similarly sized (58 kdaltons) subunits with an overall content of at least four cytochrome *b* molecules (1, 26).

Experimental evidence obtained mainly from work with fungal systems (36, 54, 101) suggests that molybdenum in nitrate reductase exists as part of a low molecular weight polypeptide, the molybdenum cofactor, which might be common to all molybdoenzymes with the single exception of nitrogenase (65). It has been proposed recently that the active cofactor is composed of molybdenum and a reduced form of a novel pterin (66).

Mechanism of enzyme catalysis The inhibition of nitrate reductase by *p*HMB is generally prevented by reduced pyridine nucleotide (36, 84), suggesting that essential sulfhydryl groups of the protein participate in the binding of NAD(P)H to the enzyme. Recent studies with nitrate reductase from *Neurospora* reinforce this interpretation and suggest also the active participation of sulfhydryl groups in the electron flow from NAD(P)H to the FAD prosthetic group (36).

The participation of FAD in the activity of the diaphorase moiety of the enzyme and of molybdenum in that of the terminal moiety are now well-established facts (36, 57, 84, 157). The molybdenum domain is thought to be the site where nitrate binds and is reduced. No general agreement exists, however, with respect to the oxidation state change of the molybdenum present in nitrate reductase during the reduction of nitrate to nitrite, but the Mo(VI)/Mo(IV) couple may well be the proximal reductant of nitrate (57, 63, 84, 157). Electron spin resonance (ESR) spectra of the oxidized enzymes of *Chlorella* and *Neurospora* exhibit low absorption in the region near $g = 2$, but signals at g values of 1.97 and 1.98, corresponding to an intermediate paramagnetic species of Mo(V), appear upon addition of NAD(P)H (63, 132). The Mo(V) signals disappear upon reoxidation of the enzyme by nitrate. A loss of these signals is also observed upon cyanide addition to the NAD(P)H-reduced enzyme, probably because cyanide blocks the metal in a more highly reduced, nonparamagnetic oxidation state such as Mo(IV) (132).

Active participation of cytochrome *b*-557 in the catalytic activity of the enzyme is sustained by spectrophotometric studies showing that this group is reduced by NAD(P)H and reoxidized by addition of nitrate (84, 157). Oxidized nitrate reductase of *Neurospora* exhibits ESR signals at $g = 2.98$ and 2.27 which have been ascribed to a low-spin ferric form of cytochrome *b*-557. These signals disappear after addition of NADPH. It seems that reduction of the heme and production of the Mo(V) species occur at similar redox potentials (63). The localization of the heme group within the enzymatic electron transport chain remains undefined, although its site of action appears to be placed between FAD and molybdenum. The pathway of electrons from NAD(P)H to nitrate through nitrate reductase from eukaryotes may thus be depicted as:

$$\text{NAD(P)H} \rightarrow (\text{FAD} \rightarrow \text{cyt } b\text{-557} \rightarrow \text{Mo}) \rightarrow NO_3^-$$

Nitrate reduction by reduced pyridine nucleotide catalyzed by *Ankistrodesmus* NAD(P)H-nitrate reductase shows an iso ping pong bi bi steady state kinetic mechanism, with isomerization of the enzymatic form which binds NADH (52). For the nitrate reductase of *Aspergillus,* however, a random order rapid-equilibrium mechanism has been determined (91). On the other hand, random order addition of nitrate or NADH, but ordered (ping pong) mechanism for release of products, has been proposed for the enzymes of squash and spinach (57).

Reduction of Nitrite to Ammonia

Two types of assimilatory nitrite reductase, marked by a well-defined electron donor specificity, have been described: (*a*) ferredoxin-nitrite reductase, characteristic of photosynthetic organisms, and (*b*) NAD(P)H-nitrite reductase found in nonphotosynthetic organisms. Both enzymes catalyze the stoichiometric reduction of nitrite to ammonia, which implies the unusual transfer of six electrons also occurring in the reduction of dinitrogen to ammonia, catalyzed by nitrogenase, and in the reduction of sulfite to sulfide, catalyzed by sulfite reductase.

FERREDOXIN-NITRITE REDUCTASE The photosynthetic nitrite reductase (EC 1.7.7.1) presents a marked specificity for ferredoxin as electron donor (150) and catalyzes the following reaction:

$$NO_2^- + 6\ Fd_{red} + 8\ H^+ \xrightarrow{6e} NH_4^+ + 6\ Fd_{ox} + 2\ H_2O$$

$$\Delta G_0^{'},\ pH\ 7.0 = -103.5\ kcal \cdot mol^{-1}$$

Flavodoxin can substitute for ferredoxin as the immediate electron donor for different nitrite reductases (88, 150, 165). Among the artificial substitutes for ferredoxin, methyl viologen is the most effective one examined. The purified enzyme is inactive with either reduced pyridine nucleotides or reduced flavin nucleotides as electron donors (150).

Ferredoxin-nitrite reductases from several sources can also catalyze the reduction of hydroxylamine to ammonia, but at a lower rate than that for nitrite (54, 165). Sulfite cannot be used as substrate by purified nitrite reductases (59, 155, 165), although there is a marked similarity between this enzyme and sulfite reductase, the latter being able to reduce sulfite, nitrite, and hydroxylamine. The assimilatory ferredoxin-nitrite reductase is usually a soluble enzyme, with the exception of the enzyme from *Anacystis,* which is associated with subcellular particles, a prolonged disruption treatment being needed in order to obtain it in a soluble form (89).

Purification and properties Nitrite reductases from higher plants (60, 150, 155), green algae (165), and red algae (59) have been purified to electrophoretic homogeneity by conventional techniques. In general, these purification procedures provide nitrite reductase preparations with specific activities of 50–110 units per milligram of protein (Table 2). Recently, affinity chromatography on ferredoxin-Sepharose has also been used for purification of the enzyme from spinach leaves (61).

Ferredoxin-nitrite reductase from different organisms appears to be composed of a single polypeptide chain of about 600 amino acid residues, with molecular weight values ranging between 60 and 70 kdaltons. The pH optima, K_m for substrates and molecular properties for purified ferredoxin-nitrite reductase from several organisms, are shown in Table 2.

Inhibitors The inhibition by cyanide appears to be of the competitive type with respect to nitrite for nitrite reductase from different sources (92, 150). Carbon monoxide inhibits nitrite reductase of higher plants (60, 155) and algae (92, 165). Formation or dissociation of the spectrophotometrically detectable CO complex with the spinach enzyme correlates with inhibition, or inhibition reversal, of nitrite reduction (155).

*p*HMB has been reported to inhibit nitrite reductase from higher plants and eukaryotic algae (155, 165), but not the cyanobacterial enzyme (88, 92). Titration of spinach nitrite reductase with *p*HMB and 5,5 dithiobis(2-nitrobenzoic acid) reveals two types of cysteine residues: six freely available to the reagents in the native protein and two accessible only under reducing denaturing conditions (155). *Cucurbita* nitrite reductase is inactivated by treatment with mersalyl (60).

Prosthetic groups Purified ferredoxin-nitrite reductases of higher plants and algae are reddish-brown in color and show similar absorption spectra characteristic of a heme-containing protein (59, 60, 155, 165). These spectra show peaks in the regions 380–390 nm and 570–580 nm (Table 2). The heme prosthetic group of spinach nitrite reductase was identified as "siroheme," previously shown to be a component of sulfite reductase (54, 150). ESR spectroscopy has confirmed the presence of siroheme and provided evidence for the additional presence of an iron-sulfur center in higher plant nitrite reductase (5, 16, 137, 155). Oxidized spinach nitrite reductase shows an ESR spectrum with resonance absorption at g values of 6.72, 5.21, and 2.03, characteristic of high-spin ferric-heme with rhombically distorted tetragonal symmetry. Under strongly reducing conditions (reduced ferredoxin plus CO), ESR signals with g values of 2.04 and 1.94 (reduced iron-sulfur center) appear (78). The ESR spectrum of marrow nitrite reductase exhibits resonance absorption at g values of 6.86, 4.98, and 1.95. Upon addition of dithionite plus methyl viologen, signals at g = 2.04, 1.94, and 1.92 are observed (16). The iron-sulfur center prosthetic group in spinach nitrite reductase has been identified as a tetranuclear cluster 4Fe-4S (78), also found as a typical cluster of sulfite reductases.

Iron appears to be the sole metal component of ferredoxin-nitrite reductase. Analytical and ESR data (78) indicate that the spinach enzyme contains a minimum of five iron atoms. This is compatible with a composition of one siroheme and one tetranuclear, 4Fe-4S, iron-sulfur center per enzyme molecule (78, 150). Values for the midpoint potentials of these prosthetic groups are shown in Table 2.

Mechanism of enzyme catalysis When the siroheme prosthetic group is blocked by CO or cyanide, as determined spectrophotometrically, nitrite reductase is inactive (5, 155). Present evidence (16, 60, 155) supports an interaction between the siroheme prosthetic group and nitrite or these inhibitors. The reaction of nitrite with the oxidized enzyme is relatively slow, and titration of oxidized nitrite reductase with nitrite permits calculations of one mole of nitrite bound per mole of enzyme (78, 155). On the other hand, native nitrite reductase forms a 1:1 complex with ferredoxin (73). Rapid reaction studies show that both the siroheme and the 4Fe-4S center of spinach nitrite reductase are reduced at similar rates by dithionite (78). The relatively slow nature of this reduction may be correlated with the low turnover number of the enzyme in the absence of an electron carrier. The iron-sulfur center of nitrite reductase, prereduced with dithionite, is rapidly reoxidized upon addition of nitrite, a fact that strongly supports a role for the iron-sulfur center in the mechanism of nitrite reduction (78).

Kinetic and steady-state turnover experiments suggest that nitrite is bound to siroheme and reduced to ammonia without the accumulation of free nitrogen compounds of intermediate redox state. The pathway of electron flow from reduced ferredoxin to nitrite, via nitrite reductase, may be represented as follows (150):

$$Fd_{red} \rightarrow [(4Fe\text{-}4S) \rightarrow \text{siroheme}] \rightarrow NO_2^-$$

NAD(P)H-NITRITE REDUCTASE The assimilatory nitrite reductases from nonphotosynthetic organisms present a marked specificity for reduced pyridine nucleotides as electron donors and require FAD for maximal activity. They catalyze the following reaction:

$$NO_2^- + 3\ NAD(P)H + 5\ H^+ \xrightarrow{6e} NH_4^+ + 3\ NAD(P)^+ + 2\ H_2O$$

$$\Delta G_{o'}^{'},\ pH\ 7.0 = -89.5\ kcal{\cdot}mol^{-1}$$

Although this nitrite reductase is generally classified as EC 1.6.6.4, three types of enzyme with different specificity for reduced pyridine nucleotide can be distinguished: NAD(P)H-nitrite reductase, which can use either NADPH or NADH as electron donor and is characteristic of the fungus

Table 2 Properties of highly purified assimilatory nitrite reductases

	Chlorella fusca	*Porphyra yezoensis*	*Spinacia oleracea*	*Cucurbita pepo*	*Neurospora crassa*
Molecular weight (kdaltons)	63	63	61	63	290
Subunits	1	nr[a]	1	nr	2
$s_{29,w}$(S)	nr	nr	4.3	4.9	9.4
Stokes radius (nm)	nr	nr	3.3	nr	7.6
Amino acid residues	600	601	564	565	nr
Electron donor	Fd	Fd	Fd	Fd	NAD(P)H
Specific activity (U/mg)	52	9	108	85	27
Turnover number (s^{-1})	54	10	110	89	130
Optimum pH	7.5	7.5	7.5	7.5	7.5
K_m NO_2^- (μM)	nr	810	360	1	10
K_m electron donor (μM)	20	0.04	70	nr	10
Prosthetic groups (mol/mol)					
FAD	—	—	—	—	nr*[a]
{4Fe–4S}	nr	nr	1	nr*	nr
siroheme	nr*	nr*	1	nr*	nr*
E'_m (mV)					
{4Fe–4S}	nr	nr	–551 (pH = 9.0); *n* = 1?	–570 (pH = 8.1); *n* = 1?	nr
siroheme	nr	nr	–50 (pH = 7.8); *n* = 1	–120 (pH = 8.5); *n* = 1	nr
Spectral properties					
Absorption maxima (nm)	278; 384; 573; 630; 692	278; 385; 580	276; 386; 573; 640; 690	280; 384; 572; 635; 697	280; 390; 580
A_{280}/A_{Soret}	3.3	2.5	1.8	2.0	7.0
ϵ (Soret) ($mM^{-1} \cdot cm^{-1}$)	22.0	28.4	70.1	37.9	46.5
References	(165)	(59)	(78, 137, 150, 155)	(16, 60, 150)	(36, 39, 152)

[a] Definitions: nr, not reported; nr*, present, but no quantitative data available.

Neurospora (36); NADPH-nitrite reductase, with a marked specificity for NADPH as electron donor, characteristic of the yeast *Torulopsis nitratophila* (124); and NADH-nitrite reductase, specific for NADH as electron donor, found in prokaryotic organisms (22, 153). Nitrite reductases from *E. coli* (22) and *Neurospora* (36) can also catalyze the reduction of hydroxylamine to ammonia, but sulfite cannot be used as substrate by these enzymes.

In addition to catalyzing the physiological reaction, i.e. reduction of nitrite to ammonia, the purified NAD(P)H-nitrite reductase of *Neurospora* has been shown to catalyze other electron-transfer reactions. As is the case for NAD(P)H-nitrate reductase (see above), two different activities, namely NAD(P)H-diphorase and terminal nitrite reductase, can also be distinguished in this enzyme (36, 149). Both moieties participate jointly in the sequential transfer of electrons from NAD(P)H to nitrite (36).

Purification and properties NAD(P)H-nitrite reductase from *Neurospora* (39) and NADH-nitrite reductase from *E. coli* (22) have been purified to electrophoretic homogeneity by conventional techniques. Affinity chromatography on Blue Dextran-Sepharose appears to be a decisive step, however, and yields a homogeneous preparation of the *Neurospora* enzyme with a specific activity of 26.9 units per milligram of protein (39). The highest specific activity reported for preparations of nitrite reductase from *E. coli* has been 5.3 units per milligram. Apparently, attempts to obtain preparations with higher specific activity have been unsuccessful because of the instability of this enzyme (22). Fungal nitrite reductase has also been found to be markedly unstable in vitro, probably due to a high sensitivity to oxygen, excellent protection being obtained by the inclusion of dithionite and FAD in all the buffers used during the purification (152). The NADH-nitrite reductase from *Azotobacter* is also rather unstable (153).

NAD(P)H-nitrite reductase from *Neurospora* (mol wt 290 kdaltons) is composed of two similar-sized subunits of 140–150 kdaltons (36). In prokaryotic organisms, nitrite reductase appears to be a smaller molecule. The presence of two subunits of 88 kdaltons each in a native enzyme of 190 kdaltons has been reported for *E. coli* nitrite reductase (22), and a mol wt of 67 kdaltons has been estimated for the NADH-nitrite reductase of *Azotobacter* (153). The pH optima, K_m for substrates and other physicochemical properties of *Neurospora* nitrite reductase, are shown in Table 2.

Inhibitors Assimilatory NAD(P)H-nitrite reductases in general are competitively inhibited by inorganic anions which are similar to nitrite, such as sulfite, arsenite, and cyanide (22, 36). Carbon monoxide inhibits *Neurospora* nitrite reductase parallel to the formation of a complex with the enzyme

(152). *p*-HMB (1 – 50 μM) has been reported to effectively inhibit all assimilatory NAD(P)H-nitrite reductases so far studied (22, 36, 149, 153).

Prosthetic groups Iron has a role in the formation of active nitrite reductase both in *Azotobacter* (43) and in *Neurospora* (152). The absorption spectrum of purified nitrite reductase of *Neurospora* shows wavelength maxima at 280, 390 (Soret) and 580 (α) nm and a shoulder at 450 nm, which are indicative of a hemoprotein containing flavin (39, 152). The flavin component of the enzyme has been identified as FAD, and the heme chromophore as siroheme (36, 149, 152).

No experimental evidence has yet been reported which could indicate the presence of an iron-sulfur center in *Neurospora* nitrite reductase. Nevertheless, in a recent paper, Greenbaum et al (39) suggest that this enzyme might contain nonheme iron. In any case, only iron has been identified so far as a metal component of the enzyme.

The assimilatory NADH-nitrite reductase from *E. coli* appears, however, to be quite different in composition from the fungal enzyme. Coleman et al (22) have reported that no absorbance maximum was detected in the range 380–600 nm in the absorption spectrum of the purified *E. coli* enzyme, but their analytical data show a flavin content of 0.07 mole FMN and 0.4 mole FAD per mole of nitrite reductase.

Mechanism of enzyme catalysis The role of FAD and siroheme as functional prosthetic groups in nitrite reduction by *Neurospora* nitrite reductase is supported by a number of observations (36, 152). The siroheme prosthetic group can be reduced by the physiological electron donor NAD(P)H plus FAD, and absorbance changes occur such that maxima are then found at 556 and 588 nm. In addition, when an excess of nitrite is added to the enzyme reduced by NADPH and FAD, the absorption spectrum shows maxima at 560 and 585 nm as a result of the formation of a complex between reduced enzyme and nitrite (or a reduction product thereof). This complex appears to be the active species of the enzyme during turnover, thus suggesting that siroheme is the site where nitrite is bound (152).

As is the case with ferredoxin-nitrite reductase, present data also suggest that nitrite is reduced to ammonia by NAD(P)H-nitrite reductase while held directly as a ligand of the siroheme prosthetic group, no free nitrogen compound of intermediate redox state having been detected. Although the precise nature of the 6-electron transfer by which this transformation is achieved is as yet unknown, the following minimum linear scheme of electron flow for *Neurospora* nitrite reductase has been proposed (36, 149):

$$\text{NAD(P)H} \rightarrow (\text{FAD} \rightarrow \text{siroheme}) \rightarrow NO_2^-$$

ENZYME LOCALIZATION AND THE PROVISION OF REDUCTANT

The localization of nitrate reductase and nitrite reductase, the nature of the immediate electron donor for the reaction catalyzed by each of these enzymes, and the identity of the specific processes which generate the corresponding reductants will be discussed jointly in this section. An integrated view of this information is valuable for the understanding of the process of nitrate reduction as it takes place in vivo. Because of space limitations, a comprehensive coverage of the different aspects of in vivo nitrate assimilation by different groups of organisms is impossible, but the subject has been reviewed elsewhere (85).

It is widely accepted that in the dark, both photosynthetic and nonphotosynthetic cells that are able to assimilate nitrate can use carbohydrates or other reduced compounds for the reduction of nitrate and nitrite (85). In fungi, both enzymes of the nitrate-reducing pathway depend on reduced pyridine nucleotide and apparently appear free in the cytoplasm (36), where NAD(P)H generated in dissimilatory reactions should be available. Algal cells can assimilate nitrate in the dark provided that they have reserves of carbohydrates or that reduced organic compounds are supplied (42, 140). In dark-grown cells of green algae, NAD(P)H generated from carbohydrate oxidation can directly couple with nitrate reductase (which is NAD(P)H-dependent) and indirectly, by way of ferredoxin-NADP oxidoreductase and ferredoxin, with nitrite reductase (42). A similar situation might also apply for the nonchlorophyllous tissues of higher plants (see below).

Green cells can, in addition, use water as reductant in the light. In any event, the reducing power for the reduction of nitrate to ammonia in every green plant, including its nongreen tissues, comes ultimately from water (84, 85). Nevertheless, although secondary to a certain degree, the question remains debatable how directly the reduced coenzymes generated photosynthetically can couple in vivo with the nitrate-reducing system. The nature of the stimulation by light of nitrate assimilation in photosynthetic cells has been a matter of controversy for a long time (10, 84, 85). Nitrate reduction in green cells has previously been considered as a process far removed from the light reactions of photosynthesis and dependent on organic substrates as the immediate source of reducing power. Now it seems to be well established that the photosynthetic reduction of nitrate is, at least in certain aspects, even more directly coupled to the light reactions of photosynthesis than that of carbon dioxide (84, 85).

This is particularly evident for the blue-green algae, where both nitrate reductase and nitrite reductase are tightly bound to chlorophyll-containing membrane fractions and able to utilize photosynthetically generated reducing power via ferredoxin (88, 89). Unsupplemented lamellar preparations of blue-green algae can carry out the stoichiometric photochemical reduction of nitrate to nitrite and ammonia with water as electron donor (19, 107). Nitrate reduction in these organisms can thus be considered as a genuine photosynthetic process (19, 84).

In eukaryotic algae and green tissues of higher plants, nitrate reductase is a pyridine nucleotide-dependent enzyme whereas nitrite reductase depends on reduced ferredoxin. No direct demonstration seems to be available showing the localization of nitrate reductase and nitrite reductase in eukaryotic algae. Nevertheless, the fact that nitrite decreases chlorophyll fluorescence in green algae more than does nitrate has been taken as evidence of a more direct coupling to the light reactions of photosynthesis of nitrite reduction than of the reduction of nitrate (71, 76). A chloroplastic localization of nitrite reductase, but not of nitrate reductase, in green algae has been inferred from these experiments.

In the leaves of higher plants, nitrite reductase has been localized definitely in the chloroplast (54, 56, 84, 157), and the most recent evidence also confirms this statement (46, 118, 119, 161). Moreover, work from different laboratories has shown that intact chloroplasts isolated from a variety of plants are not only able to photoreduce nitrite to ammonia, but also to convert the latter into amino-nitrogen in light-dependent reactions (4, 79, 100, 113). The intracellular localization of nitrate reductase in the green tissues of higher plants still remains a matter of controversy. Although in some cases nitrate reductase has been shown to be associated, at least in part, with the chloroplast (38, 84, 118, 119), it is frequently referred to as a cytoplasmic enzyme (10, 25, 161). Alternatively, the proposal has been made that nitrate reductase is attached to the chloroplast envelope membranes (54, 118, 123), but other authors have found no evidence for such an association (161). A loose attachment of nitrate reductase to microsomes or microbody-like organelles has also been suggested (54, 81). Reduction of nitrate in the light with concurrent oxygen evolution has been reported for some chloroplast preparations (84, 85, 138). The ability to reduce nitrate by isolated *Wolffia* chloroplasts was reported to be lost after further purification, but it was regained after a soluble component was added back to the purified chloroplast preparation (138). Reconstituted chloroplast systems supplemented with the corresponding cofactors and nitrate-reducing enzymes can carry out the overall reduction of nitrate to ammonia with water as the ultimate reductant (84, 85). Assuming an extraplastidial localization of nitrate reductase, and in order to explain how the photosynthetically generated reducing power comes out of the chloroplast and is transferred to NAD(P)H for nitrate reduction, the participation of a variety of shuttle systems (75) has been proposed.

In the leaves of C_4 plants, nitrate reduction appears to occur predominantly or exclusively (46, 85, 96) in the mesophyll cells. Light-dependent oxygen evolution in the presence of nitrite (but not of nitrate) by isolated crabgrass mesophyll cells has been demonstrated recently (96).

Nitrate reduction in higher plants is not restricted to the leaves, but also takes place in roots, embryos, cotyledons, scutella, aleurone cells, pollen grains, etc (10, 56). Depending on the plant type, the relative contribution of the root to the reduction of nitrate by the plant can be very important (102, 110), and it seems to be especially intensive during the early stages of development (102). The capacity of root tissue to assimilate nitrate appears to be related to its carbohydrate content (93). The characterization and intracellular localization of the enzymes of the nitrate-reducing system in nonchlorophyllous tissue still requires further investigation (54, 56, 85). The properties so far determined for nitrate and nitrite reductase from the roots of higher plants seem to be similar to those of the enzymes in green tissue (10, 62, 102), but the nature of the physiological reductant for the reduction of nitrite in the root tissue remains unclear (10, 54).

REGULATION OF NITRATE REDUCTION

Regulation of nitrate reduction by changes in the activities of the enzymes of the nitrate-reducing pathway is important in the control of the overall process of nitrate assimilation. The level of nitrite reductase in different cells and tissues is usually much higher than that of nitrate reductase and, accordingly, accumulation of nitrite is seldom observed. The reduction of nitrate to nitrite, rather than the further reduction of nitrite to ammonia, appears therefore more likely to be a rate-controlling step in nitrate reduction. These and other considerations have led Beevers & Hageman (9) to conclude that nitrate reductase is the logical point to effect regulation of the input of reduced nitrogen for the organism. In fact, nitrate reductase levels have been shown to fluctuate in response to changes of environmental conditions such as light, temperature, pH, CO_2 and oxygen tensions, water potential, nitrogen source, and other factors (9, 10, 45, 54, 56, 84, 85, 97, 136, 140, 157), changes that usually also influence the capacity of the organisms to assimilate nitrate.

Most of the studies on the regulation of nitrate reduction in different organisms and tissues have therefore been focused on nitrate reductase, especially since evidence is available indicating that significant positive correlations exist between this enzyme and the nitrogen status of some higher plant systems, and that growth, yield, or protein content are sometimes correlated with the enzyme level of seeds or leaves (45, 136).

It should be pointed out that nitrate reductase is only one of several possible controls on nitrate reduction, however. Situations can be envisaged, and in fact do occur, in which, despite the presence of a high enzyme activity level, the potential for reducing nitrate is actually limited by other factors. The provision of substrate to the nitrate-reducing system has an important function in controlling the rate of in vivo nitrate reduction, and hence uptake, storage, and translocation of nitrate can also play outstanding roles in the regulation of the process (10, 45, 85, 102, 148). Feedback inhibition of the enzymes of the nitrate-reducing pathway by the reaction products (157) or by metabolites involved in ammonium assimilation (32, 141) may also have a regulatory role. Besides, the availability of the specific reductants will doubtless have an influence on the rate of in vivo nitrate (and nitrite) reduction. In this context, factors to be taken into account include not only the generation of reductant and its transfer to the site of nitrate reduction, but also the competition of other metabolic processes using the same reducing agent (10, 157). In addition, the participation of intermediate metabolites or final products of different metabolic pathways, especially of those related to carbon metabolism in the control of nitrate reduction, should also be considered (113, 157). Available information about the latter regulatory aspects of the process is unfortunately very scanty, and a clear picture of the controls which are involved in the integration of nitrate reduction and CO_2 fixation in particular, and of nitrogen and carbon metabolism in general, is lacking (85, 157).

For the present purpose, we do not intend to deal comprehensively with all of the different regulatory aspects of nitrate reduction. We will rather restrict ourselves to two main points: (*a*) the control of the amount of active nitrate reductase (and, occasionally, of nitrite reductase), and (*b*) the control of substrate supply to the nitrate-reducing system, especially at the level of nitrate uptake. Emphasis will be laid in each case on the effects of ammonium, the end product of nitrate reduction, which behaves as a very active antagonist of nitrate assimilation in virtually all types of organisms.

Control of the Amount of Active Enzyme

The amount of active nitrate reductase can be considered as a function of the controlled synthesis of active enzyme on the one hand and of further changes in the activity state of preexisting enzyme on the other hand. Modifications of the active enzyme that lead to a loss of its catalytic activity can be reversible (e.g. reversible inactivation) or irreversible (e.g. proteolytic degradation). It should be mentioned that many of the reports regarding fluctuations in the activity level of the nitrate-reducing enzymes do not allow a direct conclusion about the nature of the control mechanism involved in each case, for it has not always been possible to determine whether they were brought about by changes in the rate of enzyme synthesis or by changes in the activity state of preexisting enzyme through inactivation, (re)activation, or degradation.

In striving for clarity, the term nitrate reductase synthesis will be considered here *sensu lato* to imply the ensemble of events involved in the formation of the active enzyme from its basic components, namely amino acids and prosthetic groups. This includes not only synthesis of nitrate reductase-specific mRNA and translation of this mRNA into protein, but also other stages that may be required eventually for expression of full catalytic activity, such as assembly of protein precursors or post-translational modifications of them, as well as incorporation of the corresponding prosthetic groups to the apoprotein.

ENZYME SYNTHESIS In general, the level of nitrate reductase is high in organisms grown on nitrate and low in organisms grown on ammonium, nitrite reductase usually following a similar pattern. Both reductases thus behave as adaptive enzymes.

Although an obligatory requirement for nitrate as the specific inducer of both enzymes has been claimed repeatedly, experimental evidence clearly shows that synthesis of nitrate reductase can take place in the absence of nitrate. De novo synthesis of the enzymes of the nitrate-reducing system has been reported to occur in nitrogen-free medium or in media containing amino acids or nitrogen compounds other than nitrate for different microorganisms such as unicellular algae, both prokaryotic (53) and eukaryotic (27, 51, 97, 120, 139, 154), chemoergonic bacteria (158), and mold fungi (98). In higher plants, nitrate reductase is usually regarded as a substrate-inducible enzyme (10, 54, 56, 136). Nevertheless, in higher plant systems, considerable enzyme levels are sometimes found in the absence of nitrate, as is the case for tobacco cells (99), seedlings of maize (20) and cotton (116), and barley roots (130). In tobacco cells, Heimer & Riklis (49) have demonstrated an RNA synthesis-dependent increase in nitrate reductase activity following removal of organic nitrogen sources from the medium and have stated that the initial development of nitrate reductase activity is independent of the presence of nitrate. Moreover, the stimulatory effect of nitrate on the nitrate reductase level of higher plants seems not to be specific of this compound. De novo synthesis of the enzyme in embryos, seeds, cotyledons, and seedlings of different plant species can also be elicited, in the absence of nitrate, by a variety of unrelated substances, including cytokinins and organic nitrocompounds (54, 56, 74, 136, 157). Therefore, at least in the organisms and systems referred to above, the proposed role of nitrate as an obligatory inducer is difficult to maintain.

The mechanism by which nitrate enhances the level of active nitrate reductase in some organisms is not well defined. Although for mold fungi the term "nitrate induction" of nitrate reductase is frequently used (36), it has been proposed that in *Neurospora* the role played by nitrate is not that of a true inducer. In fact, nitrate appears not to be essential for the transcription of nitrate reductase genes, but its presence results in both enhancement of transcription and protection of the enzyme against inactivation (36, 115). Alternatively, in an example of autogenous regulation (23, 36), *Aspergillus* nitrate reductase has been proposed to act as a repressor of its own synthesis unless nitrate is present. Using tobacco cells, Zielke & Filner (164) provided an elegant demonstration that the increase in nitrate reductase level in the presence of nitrate implied de novo synthesis of the enzyme. These results have been taken as proof of the role of nitrate as an inducer of the enzyme of higher plants (10, 136). Nevertheless, since nitrate reductase synthesis was also shown to take place under "noninducing" conditions in which a decay in activity took place (164), it is not easy to decide whether the effect of nitrate was due to an increased rate of synthesis or to a reduced decay of enzyme activity. It is possible that the situation in higher plants with regard to the positive effect of nitrate on the level of active enzyme resembles the situation in *Neurospora.*

Light also appears to be required for the synthesis of nitrate reductase in photosynthetic tissues (10, 54). It has been demonstrated conclusively, however, that a considerable level of this enzyme is present in *Chlorella* and tobacco cells grown in the dark (42, 99) and in etiolated barley leaves (125) when an adequate carbon source is present. Of course, light-promoted ATP synthesis in photosynthetic tissues can have a positive effect on general protein synthesis, and hence on the production of nitrate reductase. Other possible effects of light might be located in the movement of nitrate to the site of enzyme synthesis (10, 45, 102). It has been suggested that many of the light-induced effects on nitrate metabolism in higher plants, including circadian rhythms of enzyme activity, involve the participation of phytochrome and/or blue light–absorbing pigments (54, 56, 69, 74). Here again it is difficult to determine whether the process affected is enzyme synthesis or if activation-inactivation of preformed enzyme comes into play.

Cycloheximide and other inhibitors of protein synthesis acting at the translation step usually inhibit increases of nitrate and nitrite reductase activity in response to the presence of nitrate or to ammonium removal, both in higher plants (54, 68, 129) and in eukaryotic algae (34, 35, 58, 86, 154), indicating that, in photosynthetic eukaryotes, both enzymes of the nitrate-reducing pathway are synthesized on 80S cytoplasmic ribosomes. In *Chlorella vulgaris,* cycloheximide inhibits the development of nitrate reductase activity following transfer of cells from ammonium to nitrate medium only if the inhibitor is added immediately after the transfer, but not when added once the rise in enzyme activity has commenced (35). These results have been taken as evidence of the existence of several phases in the synthesis of active nitrate reductase, only the first one being sensitive to cycloheximide; the other steps represent assembly of precursors and incorporation of molybdenum to give the active enzyme (156). In the presence of nitrate and absence of molybdenum, *Chlorella* cells synthesize a protein which exhibits diaphorase activity but not nitrate reductase activity. Molybdenum incorporation, in a process which is not sensitive to cycloheximide, results in appearance of active nitrate reductase (154, 156). A protein synthesis-independent development of nitrate reductase activity in response to molybdenum also occurs in higher plant systems (68). These results clearly refute the proposed role of molybdenum as an inducer in nitrate reductase synthesis (9, 54).

The antagonistic effect of ammonium on nitrate metabolism with regard to the synthesis of the enzymes of the nitrate-reducing system is evident in many organisms. The situation appears particularly clear for lower plants where ammonium usually overrides any stimulating effect of nitrate, enzyme activity being totally absent or present only at basal levels both in ammonium- and in ammonium nitrate-containing medium. This is indeed the case for some chemoergonic bacteria (158), unicellular blue-green algae (53), eukaryotic algae (27, 51, 86, 97, 120, 139, 140, 154), and fungi (36). The stage of nitrate reductase synthesis affected by ammonium does not seem to be always the same, however, and significant differences are found in this regard between groups of organisms.

In the case of the mold *Neurospora,* results from recent studies using immunological methods (3) and density labeling (7) indicate that no major stable form or component of nitrate reductase is present in ammonium-grown mycelia. These results strongly suggest that prevention by ammonium of nitrate reductase synthesis affects an early stage of the process, either transcription or translation, thus impeding the appearance of enzyme or enzyme precursor protein(s). In fact, present evidence supports the view that this control is exerted at the transcriptional level in the case of fungi (28, 36, 114, 115). A transcriptional control of nitrate reductase synthesis by ammonium has also been thought to operate in unicellular blue-green algae (53). It is rather interesting that for both groups of organisms, blue-green algae and fungi, not ammonium itself but rather an organic nitrogenous compound whose generation from ammonium requires the participation of glutamine synthetase appears to be the real repressor of nitrate reductase synthesis. Thus, "ammonium repression" is no longer evident in cells without an active glutamine synthetase such as methionine sulfoximine-treated cells of *Anacystis* (53) or certain *Neurospora* mutants

(28, 115). Glutamine has been suggested as the putative repressor of nitrate reductase in *Neurospora* (28, 115). In addition, Premakumar et al (114) have shown that glutamine prevents the synthesis of the short half-life mRNA for nitrate reductase, mRNA which is missing in ammonium-grown *Neurospora.*

The control by ammonium of the synthesis of active nitrate reductase in green algae seems, however, to operate at a post-transcriptional level. From studies with inhibitors, Hipkin & Syrett (58) have concluded that a preformed mRNA for nitrate reductase is present in ammonium-grown cells of different eukaryotic algae. Evidence has also been presented recently (34) that *Chlorella vulgaris* cells grown on ammonium contain enzymatically inactive protein material which binds to antibodies to nitrate reductase, and it has been proposed (34, 35, 156) that in the presence of ammonium, nitrate reductase precursor is constantly being synthesized and degraded. Results from experiments of density labeling with heavy water (64) also indicate the presence of considerable amounts of preformed nitrate reductase precursor in ammonium-grown *Chlorella.* The conclusion has been drawn that control by ammonium of nitrate reductase in *C. vulgaris* is not only post-transcriptional but post-translational as well, i.e. the nitrate reductase-specific mRNA in ammonium-grown cells is not only present but it is also being translated, and that ammonium or a metabolite of it interferes with the assembly of the inactive protein precursor into active enzyme by inhibiting synthesis of an "activator" protein required for this assembly (34, 35, 156).

In certain higher plant systems ammonium and/or amino acids have also been shown to act as negative effectors with regard to the synthesis of active nitrate reductase, but little is known about the underlying mechanisms of these effects. Thus, ammonium or certain amino acids have been reported to prevent the nitrate-promoted increase in nitrate reductase activity in a variety of systems such as roots (33, 102, 103, 117, 130), cell cultures of tobacco, tomato, and soybean (8, 11, 48), and duckweed plants (29, 105). Exceptions to this rather general behavior, however, have been observed for germinating rice seedlings (126) and cell cultures of wheat and rose (8, 95). In plants of the duckweed *Lemna,* ammonium repression of nitrate reductase has been shown conclusively to occur (105), and it is thought that the effect of ammonium is indirect through a product of its assimilation acting at the level of transcription (105). On the other hand, the repressive effect of amino acids on nitrate reductase in tobacco cells has been proposed to take place at the post-transcriptional level (50).

From the information discussed above, it appears that ammonium, or rather a product of ammonium metabolism, plays a crucial role in the regulation of the level of nitrate reductase (and nitrite reductase) in different organisms through negative effects on the synthesis of the active enzyme.

Several types of control mechanisms seem to be involved in this particular aspect of ammonium action. In addition, a not well-defined stimulating effect of nitrate at the level of enzyme synthesis also should be taken into account. For certain organisms, but definitely not for many others, this positive role of nitrate might imply induction of the enzymes of the nitrate-reducing pathway.

ENZYME ACTIVITY In addition to the long-term changes in the amount of enzyme that might be ascribed to controlled synthesis, the occurrence of more or less rapid variations of the nitrate reductase activity level in response to changes in particular environmental factors has likewise to be considered, since it also contributes to determining the actual level of active enzyme. In some cases proteolytic degradation seems to be involved, whereas reversible enzyme inactivation by specific protein factors or metabolites has been shown to be implied in others.

Nitrate reductase decay has been observed after transferring different fungi to ammonium or nitrogen-free medium. The ammonium-induced decay seems to depend upon protein synthesis and is prevented by nitrate (36, 54, 135, 157). Specific nitrate reductase inactivators that exhibit protease activity have been detected in *Neurospora* (135). The occurrence of proteases that preferentially inactivate nitrate reductase has also been shown in higher plant tissues, especially in roots (160, 162). The lower nitrate reductase levels found in older parts of the root appear to be correlated with a higher activity level of inactivating factors (102, 160). The participation of such factors is also thought to be involved in the rapid activity decay of nitrate reductase in leaf tissue in response to transferring leaves to darkness or CO_2-free atmosphere (10, 54).

Other types of proteinaceous nitrate reductase-inactivating factors that inhibit enzyme activity but do not seem to cause a degradation of the nitrate reductase protein have been found to be present in leaf tissue (67, 128), rice seedlings (70), and cell cultures of different plant species (162, 163). Such factors seem to cause nitrate reductase inactivation through binding to the enzyme protein (162), and such action probably is involved in reversible activity changes in response to light-dark transitions (67, 128). In vivo reversible inactivation by ammonium of nitrate reductase in *Lemna* plants has also been explained by a reversible protein-protein interaction (106).

Relevant information about reversible inactivation of nitrate reductase, especially in response to the presence of ammonium, has been obtained from work with algal systems (84, 157) where the so-called "metabolic interconversion," implying the existence of the enzyme in two metabolically interconvertible forms either active or inactive, has been clearly shown to occur. The relative proportion of each form appears to be dependent upon the intracellular concentration of specific metabolites which in turn are affected by changes in environmental conditions. Particularly clear is the effect of ammonium, whose presence in the medium leads to a complete conversion of the active to the inactive form in a middle-term effect (1–2h), in vivo reactivation being achieved in the presence of nitrate upon ammonium removal in a process independent of de novo protein synthesis. Inactivation implies activity loss of the terminal moiety of the NAD(P)H-nitrate reductase complex and seems to involve reduction of a functional group (presumably molybdenum). In vitro reactivation of the inactive form can be achieved upon treatment with oxidizing compounds, nitrate among them. This rather general behavior has been observed for the enzyme of different species of green algae, the blue-green alga *Nostoc,* and the thermophilic red alga *Cyanidium* (27, 51, 82–84, 86, 112, 120–122, 157). In the case of *Cyanidium,* treatments which caused loss of the diaphorase activity or affected the integrity of the enzyme protein have also been reported to reactivate the ammonium-inactivated nitrate reductase (122).

Regarding the mechanism of this ammonium-promoted inactivation of nitrate reductase, different interpretations have been proposed. For *Cyanidium* it has been suggested that in the enzyme complex there is a labile moiety bearing an inhibition site for an organic nitrogen compound resulting from ammonium metabolism (121, 122), treatments which alter the structure of the inhibition site resulting in reactivation of the inactive enzyme (122). For the metabolic interconversion of nitrate reductase in green algae, Losada and coworkers have proposed that the ammonium effect results from its uncoupling action on photosynthetic phosphorylation through the increase in the cellular levels of reducing power and ADP. The active form is regarded as an oxidized state of the enzyme, which is converted into the reduced inactive form, and vice versa, in response to changes in the levels of reduced pyridine nucleotides and in the energy charge of the cell (83, 84, 157). On the other hand, Vennesland and coworkers have proposed that the in vivo inactive form of nitrate reductase represents a cyanide complex of the reduced enzyme, whose generation in response to ammonium results from an indirect effect of this compound on nitrate uptake (112, 134, 156, 157).

Inactivation of *Chlamydomonas* nitrate reductase has been shown to take place in response to transfer of the cells to darkness (143, 151), and activation-inactivation of nitrate reductase following light-dark transitions have also been observed in synchronous cultures of *Chlorella* (144). The dark-inactivated enzyme of *Chlamydomonas* appears to have similar properties to those of the reduced form which results from ammonium treatment in vivo (151).

Evidence of the possible occurrence of nitrate reductase reversible inactivation in higher plants without a direct involvement of protein factors has as yet received only scanty experimental support. Nevertheless, in vitro inactivation by incubation with NADH and reactivation by ferricyanide, properties which might be indicative of the occurrence of metabolic interconversion, have been reported for higher plant nitrate reductase (80, 108, 147, 159). Trinity & Filner (147) have also shown that nitrate reductase extracted from nitrate-grown tobacco cells is partially inactive, activity increasing upon incubation under adequate conditions, and Benzioni & Heimer (12) have reported that temperature affects reversibly nitrate reductase activity in barley leaf strips.

In vitro inactivated nitrate reductase from *Chlorella* and spinach leaves can be rapidly activated by blue light but not by red light. A physiological role for the quality of light in modulating nitrate reductase activity has been suggested, the flavin prosthetic group being the light-absorbing pigment (6).

To date, interconversion between active and inactive forms of the second enzyme of the nitrate-reducing system, i.e. nitrite reductase, has only been reported to occur in vitro, using enzyme preparations of the bacteria *Azotobacter* and *E. coli.* The process is apparently related to a redox change of the enzyme protein (153), but there is no evidence about its occurrence in vivo, and therefore it is not possible to generalize about its significance on metabolic control.

Control of Substrate Supply to Nitrate Reductase

Present evidence strongly supports the participation in nitrate utilization by a variety of cells and tissues of effective mechanisms for nitrate uptake. Entrance of nitrate into different organisms appears to be mediated by specific carriers whose operation, usually dependent on metabolic energy, allows for accumulation of nitrate (31, 85, 104, 140, 148). The significance of the uptake step in the regulation of nitrate reduction is just starting to be widely recognized (84, 85, 140, 148, 157). Once nitrate enters the organism across the barrier of the plasma membrane, and before it is reduced, it can eventually be confined into storage organelles or be transferred to the site where nitrate reduction takes place. The occurrence of these intermediate events, namely storage and translocation, becomes more significant as the complexity of the organism increases (148), and they may also play a relevant role in controlling the access of substrate to the nitrate-reducing system. It is worth noting that, in addition to its role as substrate, nitrate can also have a stabilizing effect on nitrate reductase against inactivation, besides the apparent stimulating effect on nitrate reductase synthesis already stated above. Definitely, a complete picture of the regulation of nitrate reduction will have to include the control of substrate supply to nitrate reductase.

Nitrate uptake appears to play a crucial regulatory role on the control of nitrate reduction under a variety of circumstances in which the level of active nitrate reductase is not the limiting factor of the rate of the process. We will restrict the discussion here mainly to the analysis of the participation of nitrate uptake in the short-term ammonium-driven inhibition of nitrate utilization, a widely recognized fact whose occurrence has been shown for a variety of organisms and systems (85).

Aside from a few recorded exceptions (85), it has been observed repeatedly that uptake of nitrate from solutions containing both ammonium and nitrate only starts after ammonium has been exhausted or its concentration greatly reduced (e.g. 8, 30, 94, 95, 98, 126, 141). Several workers have suggested that such an inhibition might be exerted at the level of nitrate reductase activity (e.g. 29, 98, 141), but the consensus is that, independent of causing inactivation or repression of nitrate reductase in a later stage, the presence of ammonium leads to an immediate inhibition of nitrate uptake (85, 157). This has been particularly well studied in prokaryotic and eukaryotic algae where the addition of low concentrations of ammonium to cells actively consuming nitrate results in a prompt cessation of nitrate uptake (32, 104, 111, 140–142, 145, 148) without a significant change in the cellular level of nitrate reductase activity for the time that nitrate utilization remains inhibited. The inhibition is typically reversible, and the ability of the cells to utilize nitrate is restored immediately after exhaustion of ammonium in the outer medium, both in green algae (140–142, 145, 148) and blue-green algae (32, 104). Similar effects are observed with fungi (98). Confirming previous suggestions, Cresswell & Syrett (24) have clearly demonstrated the nitrate transport step to be the target of the short-term ammonium inhibition of nitrate utilization. Using cells of the diatom *Phaeodactylum tricornutum,* these authors have shown that the presence of ammonium in the medium produces an immediate reversible inhibition of nitrate accumulation within the cells.

The short-term inhibitory effect of ammonium on nitrate uptake by algae appears to respond to ammonium assimilation rather than to ammonium itself. The proposal of an organic product of ammonium assimilation as mediator of ammonium in causing inhibition of nitrate assimilation was raised by Morton (98) and Syrett & Morris (141) on the ground of a lack of effect of ammonium on nitrate uptake by carbohydrate-starved cells unless a carbon source was added. Recent experiments carried out with blue-green algae (32) have shown that in the presence of specific inhibitors of ammonium assimilation (methionine sulfoximine and azaserine), ammonium does not exhibit any inhibitory effect on nitrate uptake by different cyanobacteria. These results suggest an involvement of ammonium incorporation into amino acids in the control of nitrate utilization. The regulation of nitrate uptake by ammonium appears thus to rest on a highly responsive feedback system for the modulation of the rate of nitrate uptake in response to changes in pools of as yet unidentified metabolites involved in ammonium assimilation.

The CO_2 requirement for nitrate utilization by unicellular algae (85, 148) might also be located in the transport of nitrate into the cells. In order to explain the low nitrate uptake activity at low pH values of *Ankistrodesmus* cells in the absence of CO_2 but not in its presence, Ullrich and coworkers (148) have proposed the participation of a carrier-mediated transport of nitrate, the carrier being labile at low pH values but becoming stabilized by CO_2 (or glucose). Preliminary experiments carried out with cells of the blue-green alga *Anacystis,* in which nitrate utilization is strictly dependent upon the presence of CO_2 even at high pH values, also suggest the nitrate transport system to be the locus responsible for the CO_2 requirement. Methionine sulfoximine-treated *Anacystis* cells, which take up and reduce nitrate but excrete most of the resulting ammonium to the medium as their capacity for assimilating it is severely hampered, still exhibit a CO_2 requirement for nitrate uptake, whereas the photosynthetic reduction of nitrate carried out by algal membrane preparations (19) freely proceeds at normal rates in the absence of CO_2 (31). Note that the requirement for CO_2 we are dealing with here is other than the one which can be accounted for by its role as a substrate for the synthesis of the carbon skeletons required for ammonium assimilation.

From the above considerations it can be concluded that nitrate uptake seems to be a very important factor in the regulation of nitrate utilization, determining the rate of the overall process even under conditions in which the cellular level of nitrate reductase activity remains unaffected. In addition to the above-mentioned data for the presence of ammonium, some authors have also concluded that under normal conditions the rate of in vivo nitrate reduction is more a function of the rate of nitrate uptake than of the amount of nitrate reductase present in the cell or tissue (e.g. 21, 148) and not the opposite, as was generally accepted previously. Nevertheless, for those higher plants in which nitrate reduction is totally or predominantly confined to the shoot (110), it appears that nitrate reduction coupled to malate synthesis in the shoot controls the uptake of nitrate by the roots, malate moving down to the root system in which it is oxidized and yields bicarbonate which exchanges for nitrate (13, 72, 110). In other words, even though control of nitrate reduction through availability of substrate by nitrate uptake appears to be a common situation, nitrate reduction can also have a regulatory role in nitrate uptake in certain cases (see 148).

In addition to nitrate uptake, translocation from the uptake site to the main reduction site, e.g. from root to shoot, can also regulate nitrate availability to nitrate reductase and hence the rate of nitrate reduction. Also here the reverse seems to be true, and it appears that nitrate reduction in the shoot regulates not only nitrate uptake but also its translocation from root to shoot (13). Moreover, the existence of compartmentation within cells or tissues may also determine a limited availability of substrate to nitrate reductase. In this context, the existence of two different nitrate pools, namely the storage or nonmetabolic pool (perhaps in the vacuole) and the active or metabolic pool, is a widely accepted concept for the intracellular distribution of nitrate in higher plants (45, 102). Factors that apparently affect permeability of cell membranes, such as light, carbohydrates, hormones, etc, will also affect the rate of in vivo nitrate reduction through substrate supply to the nitrate-reducing system (45, 69, 74, 102).

CONCLUDING REMARKS

A considerable body of information on the enzymology, mechanism, and regulation of nitrate assimilation has accumulated during the last few years. The status of knowledge regarding the molecular properties of nitrate and nitrite reductases of different eukaryotic organisms has already reached the level that will soon allow the unraveling of the fine structure and molecular mechanism of action for both enzymes. Nevertheless, only limited progress has been made in the purification and characterization of the corresponding enzymes in prokaryotes able to assimilate nitrate, and further research is needed in this field.

The intracellular location of nitrate reductase in the green cells of photosynthetic eukaryotes still remains uncertain. New experimental approaches to the problem, such as the use of antibodies to nitrate reductase or other in situ localization procedures, would probably give a clearer and more definite answer than those already obtained from different fractionation techniques following homogenization of the cells or tissues. An accurate and reliable determination of the cell site in which the reduction of nitrate to nitrite takes place will contribute to a better understanding of the nature of the specific process which directly supplies the reducing power for this reaction, and also to an understanding of how the two steps of nitrate reduction are integrated in these cells. A certain structural association without necessarily implying a tight or a permanent binding of nitrate reductase to the chloroplast, the site of nitrite reduction, could better account for the actual effectiveness in the in vivo reduction of nitrate to ammonia without the accumulation of nitrite, a highly toxic compound, in the cytoplasmic compartment.

The demonstration of a direct coupling, through ferredoxin, of nitrate reduction to the photosynthetic noncyclic electron flow in unsupplemented membrane fractions of blue-green algae certainly can be considered of outstanding significance with regard to the elucidation of a direct role of light, without the participation of carbon intermediates, in the provision of reductant for nitrate reduction by photosynthetic organisms. The concept of "nitrate photosynthesis" has thus gained further strong experimental support.

A great deal of progress has also been made in defining different aspects of the regulation of nitrate assimilation, and an understanding is presently being reached of at least the basic underlying mechanisms of some of the processes involved in the control of enzyme synthesis and activity. Other outstanding aspects of control, such as the nature of the mechanisms whereby nitrogen and carbon metabolism are integrated and interact with each other to balance these two elements in the plant, still remain unanswered. Some facets of the well-known—at least at the phenomenological level—antagonistic effects of ammonium on nitrate metabolism have been studied recently in greater detail, and are now known to respond to ammonium assimilation rather than to ammonium itself. The significance of the contribution of nitrate uptake and of the intracellular and intercellular compartmentation of nitrate to the regulation of nitrate assimilation is also sustained by recent experimental evidence. Nitrate uptake is presently considered as a major and early point of control of the process in different organisms. Nevertheless, little is known about the molecular properties of the nitrate uptake system, and this awaits further research.

It thus becomes apparent that despite the many advances made over the past decade in the field of nitrate assimilation, there is still a gamut of fascinating problems which remain to be solved. We hope in the next few years to see a number of very exciting results which will lead to a more complete picture of this process on which ultimately rests the provision of reduced nitrogen for most of the living world.

Acknowledgments

Research cited from authors' laboratory was supported in part by grants from the Centro de Estudios de la Energía (Spain) and the Philips Research Laboratories (The Netherlands). The authors thank Antonia Friend for assistance during the preparation of the manuscript.

Miguel G. Guerrero is grateful for the helpful criticism and discussion which he has enjoyed with Elfriede Pistorius, Candadai S. Ramadoss, and Birgit Vennesland, and thanks also the latter for the hospitality of her laboratory at Berlin, where some of this review was written.

Literature Cited

1. Ahmed, J., Spiller, H. 1976. Purification and some properties of the nitrate reductase from *Ankistrodesmus braunii*. *Plant Cell Physiol.* 17:1–10
2. Alef, K., Klemme, J. H. 1979. Assimilatory nitrate reductase of *Rhodopseudomonas capsulata* AD2: A molybdo-heme-protein. *Z. Naturforsch. Teil C* 34:33–37
3. Amy, N. K., Garrett, R. H. 1979. Immunoelectrophoretic determination of nitrate reductase in *Neurospora crassa*. *Anal. Biochem.* 95:97–107
4. Anderson, J. W., Done, J. 1978. Light-dependent assimilation of nitrite by isolated pea chloroplasts. *Plant Physiol.* 61:692–97
5. Aparicio, P. J., Knaff, D. B., Malkin, R. 1975. The role of an iron-sulfur center and siroheme in spinach nitrite reductase. *Arch. Biochem. Biophys.* 169: 102–7
6. Aparicio, P. J., Roldán, J. M., Calero, F. 1976. Blue light photoreactivation of nitrate reductase from green algae and higher plants. *Biochem. Biophys. Res. Commun.* 70:1071–77
7. Bahns, M., Garrett, R. H. 1980. Demonstration of *de novo* synthesis of *Neurospora crassa* nitrate reductase during induction. *J. Biol. Chem.* 255:690–93
8. Bayley, J. M., King, J., Gamborg, O. L. 1972. The effect of the source of inorganic nitrogen on growth and enzymes of nitrogen assimilation in soybean and wheat cells in suspension cultures. *Planta* 105:15–24
9. Beevers, L., Hageman, R. H. 1969. Nitrate reduction in higher plants. *Ann. Rev. Plant Physiol.* 20:495–522
10. Beevers, L., Hageman, R. H. 1972. The role of light in nitrate metabolism in higher plants. In *Photophysiology*, ed. A. C. Giese, 7:85–113. New York: Academic. 353 pp.
11. Behrend, J., Mateles, R. I. 1975. Nitrogen metabolism in plant cell suspension cultures. I. Effect of amino acids on growth. *Plant Physiol.* 56:584–89
12. Benzioni, A., Heimer, Y. M. 1977. Temperature effect on nitrate reductase activity in vivo. *Plant Sci. Lett.* 9:225–31
13. Benzioni, A., Vaadia, Y., Lips, H. 1971. Nitrate uptake by roots as regulated by nitrate reductase products of the shoot. *Physiol. Plant.* 24:288–90
14. Bowen, H. J. M. 1966. *Trace Elements in Biochemistry*. London: Academic. 241 pp.
15. Burns, R. C., Hardy, R. W. F. 1975. *Nitrogen Fixation in Bacteria and Higher Plants*. Berlin: Springer, 189 pp.
16. Cammack, R., Hucklesby, D. P., Hewitt, E. J. 1978. Electron-paramagnetic-resonance studies of the mechanism of leaf nitrite reductase. *Biochem. J.* 171:519–26
17. Campbell, W. H. 1976. Separation of soybean leaf nitrate reductases by affinity chromatography. *Plant Sci. Lett.* 7:239–47
18. Candau, P. 1979. *Purificación y propiedades de la ferredoxina-nitrato reductasa de la cianobacteria Anacystis nidulans*. PhD thesis. Univ. Sevilla, Sevilla, Spain. 201 pp.
19. Candau, P., Manzano, C., Losada, M. 1976. Bioconversion of light energy into chemical energy through reduction with water of nitrate to ammonia. *Nature* 262:715–17
20. Champigny, M. L. 1963. Sur l'activité et l'induction de la nitrate réductase dans les plantules de maiz. *Physiol. Vég.* 1:139–69
21. Chantarotwong, W., Huffaker, R. C., Miller, B. L., Granstedt, R. C. 1976. In vivo nitrate reduction in relation to nitrate uptake, nitrate content, and in vitro nitrate reductase activity in intact barley seedlings. *Plant Physiol.* 57: 519–22
22. Coleman, K. J., Cornish-Bowden, A., Cole, J. A. 1978. Purification and properties of nitrite reductase from *Escherichia coli* K12. *Biochem. J.* 175:483–93
23. Cove, D. J., Pateman, J. A. 1969. Autoregulation of the synthesis of nitrate reductase in *Aspergillus nidulans*. *J. Bacteriol.* 97:1374–78
24. Cresswell, R. C., Syrett, P. J. 1979. Ammonium inhibition of nitrate uptake by the diatom *Phaeodactylum tricornutum*. *Plant. Sci. Lett.* 14:321–25
25. Dalling, M. J., Tolbert, N. E., Hageman, R. H. 1972. Intracellular location of nitrate reductase and nitrite reductase. I. Spinach and tobacco leaves. *Biochim. Biophys. Acta* 283:505–12
26. De la Rosa, M. A., Diez, J., Vega, J. M., Losada, M. 1980. Purification and properties of assimilatory NAD(P)H-nitrate reductase from *Ankistrodesmus braunii*. *Eur. J. Biochem.* 106:249–56
27. Diez, J., Chaparro, A., Vega, J. M., Relimpio, A. M. 1977. Studies on the regulation of assimilatory nitrate reductase in *Ankistrodesmus braunii*. *Planta* 137:231–34

28. Dunn-Coleman, N. S., Tomset, A. B., Garrett, R. H. 1979. Nitrogen metabolite repression of nitrate reductase in *Neurospora crassa:* Effect of the *gln-1a* locus. *J. Bacteriol.* 139:697–700
29. Ferguson, A. R. 1969. The nitrogen metabolism of *Spirodela oligorrhiza.* II. Control of the enzymes of nitrate assimilation. *Planta* 88:353–63
30. Ferguson, A. R., Bollard, E. G. 1969. Nitrogen metabolism of *Spirodela oligorrhiza.* I. Utilization of ammonium, nitrate and nitrite. *Planta* 88:344–52
31. Flores, E., Guerrero, M. G. 1980. *Properties of nitrate utilization in Anacystis nidulans attributable to a nitrate transport system.* Presented at Congr. Fed. Eur. Soc. Plant Physiol., 2nd, Santiago
32. Flores, E., Guerrero, M. G., Losada, M. 1980. Short-term ammonium inhibition of nitrate utilization by *Anacystis nidulans* and other cyanobacteria. *Arch. Microbiol.* 128:137–44
33. Frith, G. J. T. 1972. Effect of ammonium nutrition on the activity of nitrate reductase in the roots of apple seedlings. *Plant Cell Physiol.* 13: 1085–90
34. Funkhouser, E. A., Ramadoss, C. S. 1980. Synthesis of nitrate reductase in *Chlorella.* II. Evidence for synthesis in ammonia-grown cells. *Plant Physiol.* 65:944–48
35. Funkhouser, E. A., Shen, T. C., Ackermann, R. 1980. Synthesis of nitrate reductase in *Chlorella.* I. Evidence for an inactive protein precursor. *Plant Physiol.* 65:939–43
36. Garrett, R. H., Amy, N. K. 1978. Nitrate assimilation in fungi. *Adv. Microb. Physiol.* 18:1–65
37. Giri, L., Ramadoss, C. S. 1979. Physical studies on assimilatory nitrate reductase from *Chlorella vulgaris. J. Biol. Chem.* 254:1703–12
38. Grant, B. R., Atkins, C. A., Canvin, D. T. 1970. Intracellular location of nitrate reductase and nitrite reductase in spinach and sunflower leaves. *Planta* 94:60–72
39. Greenbaum, P., Prodouz, K. N., Garrett, R. H. 1978. Preparation and some properties of homogeneous *Neurospora crassa* assimilatory NADPH-nitrite reductase. *Biochim. Biophys. Acta* 526: 52–64
40. Guerrero, M. G., Gutierrez, M. 1977. Purification and properties of the NAD(P)H:nitrate reductase of the yeast *Rhodotorula glutinis. Biochim. Biophys. Acta* 482:272–85
41. Guerrero, M. G., Jetschmann, K., Völker, W. 1977. The stereospecificity of nitrate reductase for hydrogen removal from pyridine nucleotides. *Biochim. Biophys. Acta* 482:19–26
42. Guerrero, M. G., Rivas, J., Paneque, A., Losada, M. 1971. Mechanism of nitrate and nitrite reduction in *Chlorella* cells grown in the dark. *Biochem. Biophys. Res. Commun.* 45:82–89
43. Guerrero, M. G., Vega, J. M. 1975. Molybdenum and iron as functional constituents of the enzymes of the nitrate-reducing system of *Azotobacter chroococcum. Arch. Microbiol.* 102: 91–94
44. Guerrero, M. G., Vega, J. M., Leadbetter, E., Losada, M. 1973. Preparation and characterization of a soluble nitrate reductase from *Azotobacter chroococcum. Arch. Microbiol.* 91:287–304
45. Hageman, R. H. 1979. Integration of nitrogen assimilation in relation to yield. See Ref. 55, pp. 591–611
46. Harel, E., Lea, P. J., Miflin, B. J. 1977. The localization of enzymes of nitrogen assimilation in maize leaves and their activities during greening. *Planta* 134:195–200
47. Hattori, A. 1970. Solubilization of nitrate reductase from the blue-green alga *Anabaena cylindrica. Plant Cell Physiol.* 11:975–78
48. Heimer, Y. M., Filner, P. 1970. Regulation of the nitrate assimilation pathway of cultured tobacco cells. II. Properties of a variant cell line. *Biochim. Biophys. Acta* 215:152–65
49. Heimer, Y. M., Riklis, E. 1979. On the mechanism of development of nitrate reductase activity in tobacco cells. *Plant Sci. Lett.* 16:135–38
50. Heimer, Y. M., Riklis, E. 1979. Post-transcriptional control of nitrate reductase of cultured tobacco cells by amino acids. *Plant Physiol.* 64:663–64
51. Herrera, J., Paneque, A., Maldonado, J. M., Barea, J. L., Losada, M. 1972. Regulation by ammonia of nitrate reductase synthesis and activity in *Chlamydomonas reinhardii. Biochem. Biophys. Res. Commun.* 48:996–1003
52. Herrero, A., De la Rosa, M. A., Diez, J., Vega, J. M. 1980. Catalytic properties of *Ankistrodesmus braunii* nitrate reductase. *Plant Sci. Lett.* 17:409–15
53. Herrero, A., Flores, E., Guerrero, M. G. 1981. Regulation of nitrate reductase levels in the cyanobacteria *Anacystis nidulans, Anabaena* sp. strain 7119, and *Nostoc* sp. strain 6719. *J. Bacteriol.* 145: In press

54. Hewitt, E. J. 1975. Assimilatory nitrate-nitrite reduction. *Ann. Rev. Plant Physiol.* 26:73–100
55. Hewitt, E. J., Cutting, C. V., eds. 1979. *Nitrogen Assimilation of Plants.* London: Academic. 708 pp.
56. Hewitt, E. J., Hucklesby, D. P., Notton, B. A. 1976. Nitrate metabolism. In *Plant Biochemistry,* ed. J. Bonner, J. E. Varner, pp. 633–81. New York: Academic. 925 pp.
57. Hewitt, E. J., Notton, B. A. 1980. Nitrate reductase systems in eukaryotic and prokaryotic organisms. In *Molybdenum and Molybdenum-containing Enzymes,* ed. M. Coughlan, pp. 273–325. Oxford: Pergamon. 577 pp.
58. Hipkin, C. R., Syrett, P. J. 1977. Post-transcriptional control of nitrate reductase formation in green algae. *J. Exp. Bot.* 28:1270–77
59. Ho, C. H., Ikawa, T., Nisizawa, K. 1976. Purification and properties of a nitrite reductase from *Porphyra yezoensis* Ueda. *Plant Cell Physiol.* 17:417–30
60. Hucklesby, D. P., James, D. M., Banwell, M. J., Hewitt, E. J. 1976. Properties of nitrite reductase from *Cucurbita pepo. Phytochemistry* 15:599–603
61. Ida, S. 1977. Purification to homogeneity of spinach nitrite reductase by ferredoxin-sepharose affinity chromatography. *J. Biochem.* 82:915–18
62. Ida, S., Mori, E., Morita, Y. 1974. Purification, stabilization and characterization of nitrite reductase from barley roots. *Planta* 121:213–24
63. Jacob, G. S., Orme-Johnson, W. H. 1980. Molecular properties and mechanism of action of nitrate reductase from *Neurospora crassa.* See Ref. 57, pp. 327–44
64. Johnson, C. B. 1979. Activation, synthesis and turnover of nitrate reductase controlled by nitrate and ammonium in *Chlorella vulgaris. Planta* 147:63–68
65. Johnson, J. L. 1980. The molybdenum cofactor common to nitrate reductase, xanthine dehydrogenase and sulfite oxidase. See Ref. 57, pp. 345–83
66. Johnson, J. L., Hainline, B. E., Rajagopalan, K. V. 1980. Characterization of the molybdenum cofactor of sulfite oxidase, xanthine oxidase and nitrate reductase. *J. Biol. Chem.* 255:1783–86
67. Jolly, S. O., Tolbert, N. E. 1978. NADH-nitrate reductase inhibitor from soybean leaves. *Plant Physiol.* 62:197–203
68. Jones, R. W., Abbott, A. J., Hewitt, E. J., Best, G. R., Watson, E. F. 1978. Nitrate reductase activity in Paul's scarlet rose suspension cultures and the differential role of nitrate and molybdenum in induction. *Planta* 141:183–89
69. Jones, R. W., Sheard, R. W. 1979. Light factors in nitrogen assimilation. See Ref. 55, pp. 521–39
70. Kadam, S. S., Gandhi, A. P., Sawhney, S. K., Naik, M. S. 1974. Inhibitor of nitrate reductase in the roots of rice seedings and its effect on the enzyme activity in the presence of NADH. *Biochim. Biophys. Acta* 350:162–70
71. Kessler, E., Zumft, W. G. 1973. Effect of nitrite and nitrate on chlorophyll fluorescence in green algae. *Planta* 111:41–46
72. Kirkby, E. A., Armstrong, M. J. 1980. Nitrate uptake by roots as regulated by nitrate assimilation in the shoot of castor oil plants. *Plant Physiol.* 65:286–90
73. Knaff, D. B., Smith, J. M., Malkin, R. 1978. Complex formation between ferredoxin and nitrite reductase. *FEBS Lett.* 90:195–97
74. Knypl, J. S. 1979. Hormonal control of nitrate assimilation: Do phytohormones and phytochrome control the activity of nitrate reductase? See Ref. 55, pp. 541–56
75. Krause, G. H., Heber, U. 1976. Energetics of intact chloroplasts. In *The Intact Chloroplast,* ed. J. Barber, pp. 171–214. Amsterdam: Elsevier. 476 pp.
76. Kulandaivelu, G., Spiller, H., Böger, P. 1976. Action of nitrite on fluorescence induction in algae. *Plant Sci. Lett.* 7:225–31
77. Kuo, T., Kleinhofs, A., Warner, R. L. 1980. Purification and partial characterization of nitrate reductase from barley leaves. *Plant Sci. Lett.* 17:371–81
78. Lancaster, J. R., Vega, J. M., Kamin, H., Orme-Johnson, N. R., Orme-Johnson, W. H., Krueger, R. J., Siegel, L. M. 1979. Identification of the iron-sulfur center of spinach ferredoxin-nitrite reductase as a tetranuclear center, and preliminary EPR studies of mechanism. *J. Biol. Chem.* 254:1268–72
79. Lea, P. J., Miflin, B. J. 1979. Photosynthetic ammonia assimilation. In *Encyclopedia of Plant Physiology, New Ser.,* ed. M. Gibbs, E. Latzko, 6:445–56. Berlin: Springer. 578 pp.
80. Leong, C. C., Shen, T. C. 1979. Reversible inactivation of the nitrate reductase of rice plants. *Experientia* 35:584–85

81. Lips, H., Avissar, Y. 1972. Plant-leaf microbodies as the intracellular site of nitrate reductase and nitrite reductase. *Eur. J. Biochem.* 29:20–24
82. Lorimer, G. H., Gewitz, H. S., Völker, W., Solomonson, L. P., Vennesland, B. 1974. The presence of bound cyanide in the naturally inactivated form of nitrate reductase of *Chlorella vulgaris. J. Biol. Chem.* 249:6074–79
83. Losada, M. 1976. Metalloenzymes of the nitrate-reducing system. *J. Mol. Catal.* 1:245–64
84. Losada, M., Guerrero, M. G. 1979. The photosynthetic reduction of nitrate and its regulation. In *Photosynthesis in Relation to Model Systems,* ed. J. Barber, pp. 365–408. Amsterdam: Elsevier. 434 pp.
85. Losada, M., Guerrero, M. G., Vega, J. M. 1981. The assimilatory reduction of nitrate. In *Biochemistry and Physiology of Nitrogen and Sulfur Metabolism,* ed. H. Bothe, A. Trebst. Berlin: Springer. In press
86. Losada, M., Paneque, A., Aparicio, P. J., Vega, J. M., Cárdenas, J., Herrera, J. 1970. Inactivation and repression by ammonium of the nitrate reducing system in *Chlorella. Biochem. Biophys. Res. Commun.* 38:1009–15
87. Malofeeva, I. V., Kondratieva, E. N., Rubin, A. B. 1975. Ferredoxin-linked nitrate reductase from the phototrophic bacterium *Ectothiorhodospira shaposhnikovii. FEBS Lett.* 53:188–89
88. Manzano, C. 1977. *La reducción fotosintética del nitrato en el alga verdeazulada Anacystis nidulans.* PhD thesis. Univ. Sevilla, Sevilla, Spain. 164 pp.
89. Manzano, C., Candau, P., Gómez-Moreno, C., Relimpio, A. M., Losada, M. 1976. Ferredoxin-dependent photosynthetic reduction of nitrate and nitrite by particles of *Anacystis nidulans. Mol. Cell. Biochem.* 10:161–69
90. Manzano, C., Candau, P., Guerrero, M. G. 1978. Affinity chromatography of *Anacystis nidulans* ferredoxin-nitrate reductase and NADP reductase on reduced ferredoxin-sepharose. *Anal. Biochem.* 90:408–12
91. McDonald, D. W., Coddington, A. 1974. Properties of the assimilatory nitrate reductase from *Aspergillus nidulans. Eur. J. Biochem.* 46:169–78
92. Méndez, J. M. 1979. *Purificación y caracterización de la ferredoxina-nitrito reductasa de la cianobacteria Nostoc muscorum.* PhD thesis. Univ. Sevilla, Sevilla, Spain. 148 pp.
93. Minotti, P. L., Jackson, W. A. 1970. Nitrate reduction in the roots and shoots of wheat seedlings. *Planta* 95:36–44
94. Minotti, P. L., Williams, D. C., Jackson, W. A. 1969. The influence of ammonium on nitrate reduction in wheat seedlings. *Planta* 86:267–71
95. Mohanty, B., Fletcher, J. S. 1976. Ammonium influence on the growth and nitrate reductase activity of Paul's scarlet rose suspension cultures. *Plant Physiol.* 58:152–55
96. Moore, R., Black, C. C. 1979. Nitrogen assimilation pathways in leaf mesophyll and bundle sheath cells of C_4 photosynthesis plants formulated from comparative studies with *Digitaria sanguinalis* (L.) scop. *Plant Physiol.* 64: 309–13
97. Morris, I. 1974. Nitrogen assimilation and protein synthesis. In *Algal Physiology and Biochemistry,* ed. W. D. P. Steward, pp. 583–609. Oxford: Blackwell, 989 pp.
98. Morton, A. G. 1956. A study of nitrate reduction in mould fungi. *J. Exp. Bot.* 7:97–112
99. Müller, A. J., Grafe, R. 1978. Isolation and characterization of cell lines of *Nicotiana tabacum* lacking nitrate reductase. *Mol. Gen. Genet.* 161: 67–76
100. Neyra, C. A., Hageman, R. H. 1974. Dependence of nitrite reduction on electron transport in chloroplasts. *Plant Physiol.* 54:480–83
101. Notton, B. A., Hewitt, E. J. 1979. Structure and properties of higher plant nitrate reductase, especially *Spinacea oleracea.* See Ref. 55, pp. 227–44
102. Oaks, A. 1979. Nitrate reductase in roots and its regulation. See Ref. 55, pp. 217–26
103. Oaks, A., Aslam, M., Boesel, I. 1977. Ammonium and amino acids as regulators of nitrate reductase in corn roots. *Plant Physiol.* 59:391–94
104. Ohmori, M., Ohmori, Z., Strotmann, H. 1977. Inhibition of nitrate uptake by ammonia in a blue-green alga, *Anabaena cylindrica. Arch. Microbiol.* 114: 225–29
105. Orebamjo, T. O., Stewart, G. R. 1975. Ammonium repression of nitrate reductase formation in *Lemna minor* L. *Planta* 122:27–36
106. Orebamjo, T. O., Stewart, G. R. 1975. Ammonium inactivation of nitrate reductase in *Lemna minor* L. *Planta* 122:37–44

107. Ortega, T., Castillo, F., Cárdenas, J. 1976. Photolysis of water coupled to nitrate reduction by *Nostoc muscorum* subcellular particles. *Biochem. Biophys. Res. Commun.* 71:885–91
108. Palacián, E., De la Rosa, F., Castillo, F., Gómez-Moreno, C. 1974. Nitrate reductase from *Spinacea oleracea.* Reversible inactivation by NAD(P)H and by thiols. *Arch. Biochem. Biophys.* 161:441–47
109. Pan, S. S., Nason, A. 1978. Purification and characterization of homogeneous assimilatory reduced nicotinamide adenine dinucleotide phosphate-nitrate reductase from *Neurospora crassa. Biochim. Biophys. Acta* 523:297–313
110. Pate, J. S. 1980. Transport and partitioning of nitrogenous solutes. *Ann. Rev. Plant Physiol.* 31:313–40
111. Pistorius, E. K., Funkhouser, E. A., Voss, H. 1978. Effect of ammonium and ferricyanide on nitrate utilization by *Chlorella vulgaris. Planta* 141:279–82
112. Pistorius, E. K., Gewitz, H. S., Voss, H., Vennesland, B. 1976. Reversible inactivation of nitrate reductase in *Chlorella vulgaris* in vivo. *Planta* 128: 73–80
113. Plaut, Z., Lendzian, K., Bassham, J. A. 1977. Nitrite reduction in reconstituted and whole spinach chloroplasts during carbon dioxide reduction. *Plant Physiol.* 59:184–88
114. Premakumar, R., Sorger, G. J., Gooden, D. 1978. Stability of messenger RNA for nitrate reductase in *Neurospora crassa. Biochim. Biophys. Acta* 519:275–78
115. Premakumar, R., Sorger, G. J., Gooden, D. 1979. Nitrogen metabolite repression of nitrate reductase in *Neurospora crassa. J. Bacteriol.* 137: 1119–26
116. Radin, J. W. 1974. Distribution and development of nitrate reductase activity in germinating cotton seedlings. *Plant Physiol.* 53:458–63
117. Radin, J. W. 1975. Differential regulation of nitrate reductase induction in roots and shoots of cotton plants. *Plant Physiol.* 55:178–82
118. Rathnam, C. K. M., Das, V. S. R. 1974. Nitrate metabolism in relation to the aspartate-type C-4 pathway of photosynthesis in *Eleusine coracana. Can. J. Bot.* 52:2599–2605
119. Rathnam, C. K. M., Edwards, G. E. 1976: Distribution of nitrate-assimilating enzymes between mesophyll protoplasts and bundle sheath cells in leaves of three groups of C-4 plants. *Plant Physiol.* 57:881–85
120. Rigano, C., Alliotta, G., Violante, U. 1974. Presence of high levels of nitrate reductase activity in *Cyanidium caldarium* grown on glutamate as the sole nitrogen source. *Plant Sci. Lett.* 2:277–81
121. Rigano, C., Rigano, V. D. M., Vona, V., Fuggi, A. 1979. Glutamine synthetase activity, ammonia assimilation and control of nitrate reduction in the unicellular red alga *Cyanidium caldarium. Arch. Microbiol.* 121:117–20
122. Rigano, C., Vona, V., Rigano, V. D. M., Fuggi, A. 1980. Active and inactive nitrate reductase. Effects of mild treatments with denaturing agents of protein. *Biochim. Biophys. Acta* 613:26–33
123. Ritenour, G. L., Joy, K. W., Bunning, J., Hageman, R. H. 1967. Intracellular localization of nitrate reductase, nitrite reductase, and glutamic acid dehydrogenase in green leaf tissue. *Plant Physiol.* 42:233–37
124. Rivas, J., Guerrero, M. G., Paneque, A., Losada, M. 1973. Characterization of the nitrate-reducing system of the yeast *Torulopsis nitratophila. Plant Sci. Lett.* 1:105–13
125. Roth-Bejerano, N., Lips, H. 1973. Induction of nitrate reductase in leaves of barley in the dark. *New Phytol.* 72: 253–57
126. Shen, T. C. 1969. Induction of nitrate reductase and the preferential assimilation of ammonium in germinating rice seedlings. *Plant Physiol.* 44:1650–55
127. Shen, T. C., Funkhouser, E. A., Guerrero, M. G. 1976. NADH- and NAD(P)H-nitrate reductases in rice seedlings. *Plant Physiol.* 58:292–94
128. Sherrard, J. H., Kennedy, J. A., Dalling, M. J. 1979. In vitro stability of nitrate reductase from wheat leaves. *Plant Physiol.* 64:640–45
129. Sluiters-Scholten, C. M. T. H. 1973. Effect of chloramphenicol and cycloheximide on the induction of nitrate reductase and nitrite reductase in bean leaves. *Planta* 113:229–40
130. Smith, F. W., Thompson, J. F. 1971. Regulation of nitrate reductase in excised barley roots. *Plant Physiol.* 48:219–23
131. Solomonson, L. P. 1975. Purification of NADH-nitrate reductase by affinity chromatography. *Plant Physiol.* 56: 853–55
132. Solomonson, L. P. 1979. Structure of *Chlorella* nitrate reductase. See Ref. 55, pp. 199–205

133. Solomonson, L. P., Lorimer, G. H., Hall, R. L., Borchers, R., Bailey, J. L. 1975. Reduced nicotinamide adenine dinucleotide-nitrate reductase of *Chlorella vulgaris. J. Biol. Chem.* 250: 4120–27
134. Solomonson, L. P., Spehar, A. M. 1979. Stimulation of cyanide formation by ADP and its possible role in the regulation of nitrate reductase. *J. Biol. Chem.* 254:2176–79
135. Sorger, G. J., Premakumar, R., Gooden, D. 1978. Demonstration in vitro of two intracellular inactivators of nitrate reductase from *Neurospora. Biochim. Biophys. Acta* 540:33–47
136. Srivastava, H. S. 1980. Regulation of nitrate reductase activity in higher plants. *Phytochemistry* 19:725–33
137. Stoller, M. L., Malkin, R., Knaff, D. B. 1977. Oxidation-reduction properties of photosynthetic nitrite reductase. *FEBS Lett.* 81:271–74
138. Swader, J. A., Stocking, C. R. 1971. Nitrate and nitrite reduction by *Wolffia arrhiza. Plant Physiol.* 47:189–91
139. Syrett, P. J., Hipkin, C. R. 1973. The appearance of nitrate reductase activity in nitrogen-starved cells of *Ankistrodesmus braunii. Planta* 111:57–64
140. Syrett, P. J., Leftley, J. W. 1976. Nitrate and urea assimilation by algae. In *Perspectives in Experimental Biology,* ed. N. Sunderland, 2:221–34. Oxford: Pergamon. 523 pp.
141. Syrett, P. J., Morris, I. 1963. The inhibition of nitrate assimilation by ammonium in *Chlorella. Biochim. Biophys. Acta* 67:566–75
142. Thacker, A., Syrett, P. J. 1972. The assimilation of nitrate and ammonium by *Chlamydomonas reinhardi. New Phytol.* 71:423–33
143. Thacker, A., Syrett, P. J. 1972. Disappearance of nitrate reductase activity from *Chlamydomonas reinhardi. New Phytol.* 71:435–41
144. Tischner, R., Hütermann, A. 1978. Light-mediated activation of nitrate reductase in synchronous *Chlorella. Plant Physiol.* 62:284–86
145. Tischner, R., Lorenzen, H. 1979. Nitrate-uptake and nitrate reduction in synchronous *Chlorella. Planta* 146: 287–92
146. Tortolero, M., Vila, R., Paneque, A. 1975. Ferredoxin-dependent nitrate reductase from *Azotobacter chroococcum. Plant Sci. Lett.* 5:141–45
147. Trinity, P. M., Filner, P. 1979. Activation and inhibition of nitrate reductase extracted from cultured tobacco cells. *Plant Physiol.* 63:133 (Abstr.)
148. Ullrich, W. R. 1981. Uptake and reduction of nitrate in algae. In *Encyclopedia of Plant Physiology, New Ser.,* ed. A. Läuchli, E. Bieleski, Vol. 12. Berlin: Springer. In press
149. Vega, J. M. 1976. A reduced pyridine nucleotides-diaphorase activity associated to the assimilatory nitrite reductase complex from *Neurospora crassa. Arch. Microbiol.* 109:237–42
150. Vega, J. M., Cárdenas, J., Losada, M. 1980. Ferredoxin-nitrite reductase. *Methods Enzymol.* 69:255–70
151. Vega, J. M., Florencio, F. J., De la Rosa, M. A. 1980. *Studies on the regulation of the photosynthetic assimilation of nitrate in green algae.* Presented at Int. Congr. Photosynth., 5th, Kallitea
152. Vega, J. M., Garrett, R. H., Siegel, L. M. 1975. Siroheme: A prosthetic group of the *Neurospora crassa* assimilatory nitrite reductase. *J. Biol. Chem.* 250:7980–89
153. Vega, J. M., Guerrero, M. G., Leadbetter, E., Losada, M. 1973. Reduced nicotinamide-adenine dinucleotide-nitrite reductase from *Azotobacter chroococcum. Biochem. J.* 133:701–8
154. Vega, J. M., Herrera, J., Aparicio, P. J., Paneque, A., Losada, M. 1971. Role of molybdenum in nitrate reduction by *Chlorella. Plant Physiol.* 48:294–99
155. Vega, J. M., Kamin, H. 1977. Spinach nitrite reductase. Purification and properties of a siroheme-containing iron-sulfur enzyme. *J. Biol. Chem.* 252:896–909
156. Vennesland, B. 1980. HCN and the control of nitrate reduction. The regulation of the amount of active nitrate reductase present in *Chlorella* cells. See Ref. 85
157. Vennesland, B., Guerrero, M. G. 1979. Reduction of nitrate and nitrite. See Ref. 79, pp. 425–44
158. Villalobo, A., Roldán, J. M., Rivas, J., Cárdenas, J. 1977. Assimilatory nitrate reductase from *Acinetobacter calcoaceticus. Arch. Microbiol.* 112:127–32
159. Wallace, W. 1975. Effects of a nitrate reductase inactivating enzyme and NAD(P)H on the nitrate reductase from higher plants and *Neurospora. Biochim. Biophys. Acta* 377:239–50
160. Wallace, W. 1978. Comparison of a nitrate reductase-inactivating enzyme from the maize root with a protease from yeast which inactivates tryptophan synthase. *Biochim. Biophys. Acta* 524:418–27

161. Wallsgrove, R. M., Lea, P. J., Miflin, B. J. 1979. Distribution of the enzymes of nitrogen assimilation within the pea leaf cell. *Plant Physiol.* 63:232–36
162. Yamaya, T., Oaks, A., Boesel, I. L. 1980. Characteristics of nitrate reductase-inactivating proteins obtained from corn roots and rice cell cultures. *Plant Physiol.* 65:141–45
163. Yamaya, T., Ohira, K. 1976. Nitrate reductase inactivating factor from rice cells in suspension culture. *Plant Cell Physiol.* 17:633–41
164. Zielke, H. R., Filner, P. 1971. Synthesis and turnover of nitrate reductase induced by nitrate in cultured tobacco cells. *J. Biol. Chem.* 246:1772–79
165. Zumft, W. G. 1972. Ferredoxin:nitrite oxidoreductase from *Chlorella.* Purification and properties. *Biochim. Biophys. Acta* 276:363–75

Ann. Rev. Plant Physiol. 1981. 32:205–36

PHYSICAL AND CHEMICAL BASIS OF CYTOPLASMIC STREAMING

◆7710

Noburô Kamiya

Department of Cell Biology, National Institute for Basic Biology, Okazaki, 444 Japan

CONTENTS

0066-4294/81/0601-0205$01.00

INTRODUCTION

Two decades have elapsed since this author wrote a review on cytoplasmic streaming for Volume 11 of this series (72). As is the case with other types of cell motility, research on cytoplasmic streaming has made great strides during this period—including isolating proteins related to streaming, elucidating its ultrastructural background, and developing new effective methods for studying functional aspects of cytoplasmic streaming.

There are a number of reviews and symposia reports dealing with various aspects of nonmuscular cell motility including cytoplasmic streaming (8, 10, 11, 23, 28, 31, 36, 48, 53, 59, 60, 62, 99, 100, 134, 136, 139, 150, 151, 159). They will provide readers with more comprehensive information on the subject. In this report, I shall limit my consideration to some selected topics in cytoplasmic streaming, focusing on its mechanism at the cellular and molecular levels.

Generally speaking, minute structural shifts in the cytoplasm may be a wide occurrance in living cells, but they will not necessarily develop into significant movement unless they are coordinated. In a variety of cells, cytoplasmic particles are known to make sudden excursions over distances too extensive to be accounted for as Brownian motion (136). Such motions, called "saltatory movements," were described long ago in plant literature as "Glitchbewegung" or "Digressionsbewegung" (cf 71, 73). The movement of the particles is erratic and haphazard, yet it is not totally devoid of directional control as it would be in Brownian motion. According to the degree of orderliness, intracellular streaming manifests various patterns (71, 73).

Cytoplasmic movements exhibited by eukaryotic cells may be classified into two major groups with respect to the proteins involved, i.e. the actin-myosin system and the tubulin-dynein system. Cytoplasmic streaming belongs mostly to the first group. Possible roles for the tubulin-dynein system in cytoplasmic streaming have yet to be investigated. From the phenomenological point of view, it is customary to classify the streaming of cytoplasm at the visual level into two major categories. One is the streaming closely associated with changes in cell form. This type of movement is usually referred to as amoeboid movement and is represented by the amoeba, acellular slime molds, and many other systems. The other is the streaming not dependent upon changes in the external cell shape, as in most plant cells or dermatoplasts.

In the following sections, I shall discuss the two best studied cases as representative models of cytoplasmic streaming. One is shuttle streaming in myxomycete plasmodia of the amoeboid type, and the other is rotational streaming in characean cells of the other type. It is necessary to describe them separately because we still do not know at what organizational level these two major categories of streaming share a common mechanism.

SHUTTLE STREAMING IN THE MYXOMYCETE PLASMODIUM

General

The plasmodium of myxomycetes, especially that of *Physarum polycephalum,* is classic material in which the physiology, biochemistry, biophysics, and ultrastructure of cytoplasmic streaming have been investigated most extensively (20, 35, 36, 71, 99, 133).

The myxomycete plasmodium shows various characteristic features in its cytoplasmic streaming. The rate of flow, as well as the amount of cytoplasm carried along with the streaming, is exceedingly great compared with ordinary cytoplasmic streaming in plant cells (71). Moreover, the direction of streaming alternates according to a rhythmic pattern. There is good evidence to show that the flow of endoplasm is caused passively by a local difference in the intraplasmodial pressure (83). This differential pressure is the motive force responsible for the streaming. It can be measured by the so-called double-chamber method, in which counterpressure just sufficient to keep the endoplasm immobile is applied (70, 71).

The waves representing spontaneous changes in the motive force are sometimes very regular, but the amplitude of the waves often increases and decreases like beat waves. In some other cases, a peculiar wave pattern is repeated over several waves. These waves can be reconstructed closely enough with only a few overlapping sine waves of appropriate periods, amplitudes and phases. This fact is interpreted as showing that physiological rhythms with different periods and amplitudes can simultaneously coexist in a single plasmodium (70, 71).

A variety of physical or chemical agents in the production of the motive force have been investigated (71). Recently, extensive analysis of tactic movements of the slime mold was made by Kobatake and his associates (47, 153–155). Threshold concentrations for the recognition of attractants (glucose, galactose, phosphates, pyrophosphates, ATP, cAMP) and of repellents (such as various inorganic salts, sucrose, fructose) were thus determined (155). It has been suggested that recognition of chemical substances is caused by a change in membrane structure which is transmitted to the motile systems of the plasmodium. Motive force production is closely related to bioelectric potential change (79) as well as anaerobic metabolism (71, 138).

Contractile Properties of the Plasmodial Strand

Since the streaming of the endoplasm is a pressure flow, and the internal hydrostatic pressure of the plasmodium is thought to be produced by contraction of the ectoplasm, the contractile force of the ectoplasm per se should serve as a basis for analyzing cytoplasmic streaming in this organism. Contractility was often assumed to be involved in a variety of movements of nonmuscular cells. Nevertheless, contractile force was not measured directly and precisely in motile systems other than muscles until it was measured in an excised segment of a plasmodial strand (75, 76, 78, 80, 88, 89, 148, 170, 171).

The strand forming the network of the plasmodium is actually a tube or vein with a wall of ectoplasmic gel. The endoplasm flows inside the ectoplasmic gel wall. Though endoplasm and ectoplasm are mutually interconvertible, it is the ectoplasmic gel structure that is mainly responsible for the dynamic activities of the strand (152). The absolute contractile force of the ectoplasm can be measured in this case as a unidirectional force. The maximal contractile force so far measured is 180 g cm^{-2} (78). Dynamic activities of the plasmodial strand can be expressed in terms of spontaneous changes in tension while the length of the strand is kept constant (isometric contraction), or in terms of spontaneous changes in length keeping constant the tension applied to the strand (isotonic contraction).

ACTIVATION CAUSED BY STRETCHING One of the outstanding characteristics of the plasmodial strand is its response to stretching. When a strand segment is stretched, say by 10–20%, the tension and amplitude of the wave increase immediately while the wave period remains constant (76, 80, 176). There is no shift in phase (78, 148, 176). Wohlfarth-Bottermann and his co-workers (2, 101), however, report a phase shift when the strand is stretched as much as 50%. After stretching, the wave train moves downward rather rapidly at first and less rapidly afterwards, showing tension relaxation under isometric conditions. The increase in amplitude of the tension waves of the plasmodial strand by stretching, and its subsequent decrease, are comparable to those shown by the motive force waves of the surface plasmodium inflated with endoplasm by external pressure (74).

ACTIVATION CAUSED BY LOADING Under isotonic conditions, the situation is somewhat different from the above. If the load is increased, the amplitude of the waves increases also. When the load is decreased, the amplitude decreases correspondingly. With constant tension levels, the cyclic contraction waves do not tend to decrease their amplitude. This result is in contrast to isometric contraction waves after stretching. The increase of the amplitude of the isotonic contraction waves under greater tension is not accompanied by an increase in period length. This result indicates that the speed of both contraction and relaxation increases rather than decreases under higher tension (76, 80). In other words, the contracile capacity of a plasmodial strand segment is activated by the tension applied externally. In order to understand this remarkable phenomenon, we have to postulate the presence of some regulatory mechanism by which the plasmodium can "sense" the tension change first, and then control the force output to correspond with the amount of load.

SYNCHRONIZATION OF LOCAL RHYTHMS A strand segment shows no significant rhythmic activities soon after it is excised from the mother plasmodium. It starts rhythmic contraction locally after 10–20 min. Thirty minutes later, small local rhythms become gradually synchronized to form a unified larger rhythm (175).

Takeuchi & Yoneda (141) reported that when individual strand segments of *Physarum* plasmodium having different contraction-relaxation periods were connected by way of a plasmodial mass, the cycles of the two segments became synchronized. To clarify the possible role of the streaming endoplasm as the information carrier for synchronization, Yoshimoto & Kamiya (175) set a single segment of plasmodial strand in a double chamber in such a way that the two halves of the segment were suspended in different compartments of the chamber. The strand penetrating the central septum of the double chamber thus took an inverted U-shape. When the shuttle streaming of the endoplasm occurred freely between the two halves of the strand, they contracted and relaxed in synchrony. But if balancing counterpressure was applied to one of the two compartments to keep the endoplasmic flow in the strand near the septum of the chamber at a standstill, then the contraction-relaxation rhythms of the two halves moved out of phase with each other. When the endoplasm was allowed to stream freely again, the synchrony of their cyclic contractions was reestablished. Thus Yoshimoto & Kamiya (177) concluded that endoplasm must carry some factor(s) which coordinates the period and phase of the contraction-relaxation cycle. It did not control the amplitude of the oscillation. In other words, information necessary for unifying the phases of the local contraction-relaxation cycle is transmitted neither by electric signal nor by direct mechanical tension of the endoplasm. The nature of the factor(s) carried by the endoplasm is still unknown.

Contractile Proteins

PLASMODIUM ACTOMYOSIN Cytoplasmic actomyosin is now known to be present throughout eukaryotic cells (134, 160–162). It is responsible for a variety of cell movements. The presence of an actomyosin (myosin B)-like protein complex in the myxomycete plasmodium was shown by Arial Loewy (108) as early as 1952. This study is a pioneering work on contractile proteins in nonmuscle systems. Since then, the biochemical properties of contractile proteins in these organisms have been studied extensively. Proteins similar to muscle myosin and actin were subsequently extracted from the plasmodium and purified separately. Fortunately, there are comprehensive articles and symposia reports in this area (3, 35, 37, 42, 116); hence we shall describe the matter only briefly here.

Like muscle actomyosin, plasmodium actomyosin shows superprecipitation at low salt concentrations and viscosity drop at high concentrations on addition of Mg^{2+}-ATP (35, 116, 126). ATPase activities of plasmodium actomyosin are basically similar to those of muscle actomyosin. Plasmodium actomyosin having Ca^{2+}-sensitivity has also been isolated (92, 93, 117). Superprecipitation of the protein is observed only in the presence of a micromolar order of free Ca^{2+}; the Mg^{2+}-ATPase is activated two- to sixfold by 1 μM free Ca^{2+}. This Ca^{2+} sensitivity is thought to be caused by the presence of regulatory proteins as in skeletal muscle. A noteworthy characteristic of *Physarum* actomyosin, which is not shared with muscle actomyosin, is that superprecipitation is reversible, i.e. it can be repeated several times on addition of ATP (111).

PLASMODIUM MYOSIN The molecule of plasmodium myosin has a rod-like structure with a globular head on one end just like that of striated myosin (43). Plasmodium myosin can combine with plasmodium F-actin, and muscle F-actin as well, to form actomyosin-like complexes (35, 45). The molecular weight of the heavy chain is 225,000 daltons as determined by SDS gel electrophoresis. Plasmodium myosin has ATPase activity similar to that of myosin from muscle, but in contrast to muscle myosin, plasmodium myosin is soluble at neutral pH and low salt concentrations, including physiological concentration (0.03 M KCl). Hinssen (56), Hinssen & D'Haese (57), Nachmias (114, 115), and D'Haese & Hinssen (26) demonstrated the capacity of *Physarum* myosin to self-assemble into thick, bipolar aggregates or long filaments. In comparison with muscle myosin, plasmodium myosin can form stable thick filaments only under a strictly defined range of conditions with respect to ionic strength, ATP concentration, and pH. Though divalent cations are not absolutely necessary, filament formation is improved by Mg^{2+} concentrations up to 2 mM and by Ca^{2+} up to 0.5 mM. Properties so far known for *Physarum* myosin are listed in a table with the references by Nachmias (116).

PLASMODIUM ACTIN Plasmodium actin was isolated by Hatano & Oosawa (39, 40) from *Physarum* by using its specific binding to muscle myosin; it was purified by salting out with ammonium sulfate. This was the first time actin was isolated from nonmuscle cells. Since then actin has been isolated from many nonmuscle cells. Actin is now known to be a ubiquitous and common protein in eukaryotic cells. The physical and chemical properties of actins from various nonmuscle sources including *Physarum* are all similar to those of muscle actin. The protein is in a monomeric state in a salt-free solution, giving a single sedimentation coefficient of about 3.5 S (4, 40). The molecular weights of actins from muscle and plasmodium are both about 42,000 in the SDS gel electrophoresis system. Analysis of amino acid composition of these two types of actins has indicated some minor disparities (42). The amino acid sequence of *Physarum* actin has recently been determined and shows a difference from mammalian γ-cytoplasmic actin in only 4% of its primary structure (157). The difference in amino acid sequence between *Physarum* actin and rabbit skeletal muscle actin was determined to be 8%. On addition of salts such as KCl, actin monomers polymerize into F-actin with concomitant hydrolysis of ATP. Electron micrographs showed that this polymer takes a form of helical filament identical to those of F-actin from muscle. Plasmodium F-actin also forms a complex with heavy meromyosin (HMM) from muscle to make an arrowhead structure (13, 118, 122).

The actin preparation obtained by Hatano and his collaborators (35, 46) formed an unusual polymer termed "Mg-polymer" with 0.1 to 2.0 mM $MgCl_2$. Although the sedimentation coefficient of this polymer is about the same as that of F-actin, Mg-polymer has a much lower viscosity, less flow birefringence, and appears as a flexible aggregate with an electron microscope. Formation of an Mg-polymer was once believed to be specific for plasmodium actin, but it was shown subsequently that this type of polymer was formed only when the actin preparation contained a cofactor similar to the muscle β-actinin (90). This protein factor was isolated from plasmodium and called "β-actinin-like protein" (110) or plasmodium actinin (41). Recently, Hasegawa et al (33) have further purified this protein factor and found that it is a 1:1 complex of actin and a new protein termed "fragmin" (see later). This protein is shown to have a regulatory function in the formation of F-actin filaments in a Ca^{2+}-sensitive manner. Hence, there is the possibility that the so-called "Mg-polymer" of actin was formed by a trace amount of Ca^{2+}. Detailed properties and polymerizability of plasmodium actin are discussed by Nachmias (116), Hatano et al (37) and Hinssen (55) in a recent book edited by Hatano et al (36).

Tension Production of Reconstituted Actomyosin Threads from Physarum

Physarum actomyosin dissolved in a solution of high ionic strength precipitates in the form of thread if spurted from an injection needle into a solution of low ionic strength. Beck et al (15) showed that the thread consists of a three-dimensional network of filaments, has ATPase activity, and contracts conspicuously on addition of ATP just like muscle actomyosin thread. D'Haese & Hinssen (25) compared thread models made of natural, recombined, and hybridized actomyosin from *Physarum* and rabbit skeletal muscle. Recently, Matsumura et al (112) were able to reconstitute an actomyosin thread from *Physarum* with an orderly longitudinal orientation, and to measure the tension it developed under controlled experimental conditions. To insure orderly longtudinal orientation of actomyosin, which is essential for the thread to generate tension effectively, they developed a special spinning technique applicable to actomyosin.

Thread segments of actomyosin (molar ratio 1 : 1) thus obtained produced little tension below 1 μM ATP, whereas maximum tension (10 g cm^{-2}) was reached at 10 μM ATP. The half-maximum tension was observed at 2–3 μM ATP. Above 20 μM ATP, the thread segment tended to break. Without an ATP-regenerating system, the sensitivity of tension development to ATP concentration was lower by one order.

Full tension at 20 μM ATP decreases as the ATP concentration is decreased stepwise to 1 μM. When the ATP concentration was increased from 1 to 10 μM, the decreased tension rose again to almost the same level as that originally developed. In short, isometric tension generation can be regulated by the micromolar concentration of ATP.

The actomyosin thread from *Physarum* differs from that of skeletal muscle in several ways (24, 112). 1. The *Physarum* actomyosin thread is much more flexible; 2. the concentrations sufficient to produce maximum and half-maximum tension are lower than those reported for muscle actomyosin threads, for which these values are 50 and 8 mM, respectively; 3. tension increase in *Physarum* actomyosin thread is slower than that in muscle actomyosin thread, where the final level of tension is reached within 2 min. These functional differences can probably be ascribed to the difference in properties between *Physarum* and muscle myosin, such as the high solubility of *Physarum* myosin at low ionic strength, a property not found in muscle myosin.

Since their preparation of synthetic actomyosin was highly purified, the thread used by Matsumura et al (112) showed no Ca^{2+} sensitivity in tension generation at micromolar levels. Under their experimental conditions, in which no regulatory proteins were present, there was no sign of oscillation in tension production at constant ATP levels. Whether oscillation in tension production is possible in a reconstituted *Physarum* actomyosin thread in the presence of appropriate regulatory proteins, but without a membrane system, is still an open question. It should be noted that in the demembranated system of *Physarum* plasmodium studied by Kuroda (103, 104), cytoplasmic movement occurred actively, but there was no longer any back and forth movement as is observable in the normal plasmodium having the plasma membrane.

Regulation of Movement

Various regulatory functions can be seen in the force output of the slime mold, as stated in the foregoing pages, such as activation through stretching or loading or phase coordination of tension force production. The cause of rhythmic tension force production may in itself be inseparable from the mechanism regulating interaction between actin and myosin. Though nothing definite is known at present about the regulation mechanism of movement in the slime mold, we should like to consider in the following sections some possible roles of Ca^{2+} and ATP.

THE ROLE OF Ca^{2+} Isolation of Ca^{2+}-sensitive actomyosin complexes from *Physarum* (92, 117) suggests that the actin-myosin interaction is controlled by fluctuation in the concentration of intracellular free Ca^{2+} (174).

The control of motility in *Physarum* by calcium can be demonstrated in various ways. It was shown that in the myxomycete plasmodium there is a calcium storage system analogous to the sarcoplasmic reticulum (91, 94). Calcium-sequestering vacuoles were identified both by histochemical methods and by energy-dispersive X-ray analysis (17, 18, 29, 106, 107). Ca^{2+} is taken up by the vesicles only in the presence of Mg^{2+}-ATP (91, 94). It is possible that there is a shift of calcium between the cytoplasmic and vacuolar compartments during the contraction-relaxation cycle. Teplov et al (149) were successful in detecting oscillations of the free calcium level in the myxomycete plasmodium by injecting murexide in it. Oscillations in the Ca^{2+} level within the period of 1.5–2.0 min were demonstrated by microspectro-fluorometry of injected murexide, although the phase relation between the Ca^{2+} level and motility was unknown. Ridgway & Durham (137) microinjected the calcium-specific photoprotein aequorin and found an oscillation in luminescence related to that in electric potential change. Ca^{2+} regulation is shown also in caffein-derived microplasmodial drops (34, 113), in microinjected strands (152), and in a demembranated system (103).

In a recent attempt to monitor the calcium oscillation in the slime mold and to relate it directly to tension production, Kamiya et al (89) and Yoshimoto et al (178) treated a segment of plasmodial strand with calcium ionophore A 23178. They simultaneously measured tension development and, by means of aequorin luminescence, the calcium efflux into the ambient solution. It was revealed that the amount of calcium coming out of the plasmodial strand pulsates with exactly the same period as that of the tension production, and that the phase of maximal tension production corresponds to the phase of minimal luminescence. With amphotericin B, a channel-forming quasi-ionophore (135), the result was essentially the same. Regular rhythmic changes in both tension and luminescence persisted for hours. In the absence of ionophore, no periodic Ca^{2+} efflux was detected from the strand developing tension rhythmically (79, 109). As is generally the case with the plasma membrane, the surface membrane of the plasmodium also depolarizes when it is stretched. On stretching the strand by 10% or so, the tension level of the strand and amplitude of tension oscillation were immediately increased. But Ca^{2+} efflux was not affected. This result may mean that electric potential difference plays little part, if any, in controlling the efflux of Ca^{2+}. Probably the membrane already has been depolarized in the above conditions. If we interpret the rhythmical efflux of Ca^{2+} in the presence of the ionophores as reflecting corresponding fluctuations of free Ca^{2+} level within the plasmodium, this phase relation is just the opposite of what is expected from the data so far presented. In this connection, it is interesting to note that "fragmin," a new Ca^{2+}-sensitive regulatory factor in the formation of actin filaments, was recently discovered by Hasegawa et al (33). The main function of this protein is to fragment actin polymers into short pieces in the presence of a concentration of free Ca^{2+} higher than 10^{-6}M. Whether or not fragmin plays a part in the regulation by Ca^{2+} of cyclic tension output is unknown.

THE ROLE OF ATP The ATP concentration of plasmodia as a total mass was estimated to be 0.4×10^{-4} M (44). According to the injection experiments in the plasmodial strand performed by Ueda & Götz von Olenhusen (152), the optimal concentration for tension development was found to be around 0.2×10^{-4} M. According to the recent ATP assay by Yoshimoto et al (unpublished), using luminescence of luciferin-luciferase, a considerable part of the ATP in *Physarum* plasmodia is compartmentalized. The free ATP concentration in a carefully prepared homogenate of *Physarum* plasmodium was low, but if the same homogenate was heated in boiling water, the intensity of luminescence was suddenly increased by nearly two orders of magnitude. This result is interpreted as showing that compartmentalized ATP was released by heat. In other words, free ATP available for mechanical work in vivo must be at a much lower level than the total average. The fact that the optimal ATP concentration for tension development by a reconstituted actomyosin thread is as low as 10–20 μM (112) is also in conformity with this notion.

There is some evidence to show that the level of free ATP oscillates with the same period as tension changes. As they had done for Ca^{2+}, Kamiya et al (89) and Yoshimoto et al (178) tried to make the surface membrane of the plasmodial strand leaky for ATP. The combined action of caffeine and arsenate was found to be effective. Simultanously with tension measurement, they measured the amount of ATP diffusing out of the strand by the luminescence of a luciferin-luciferase system. The plasmodial strand could not persist long in the caffeine-arsenate solution, but it could exhibit at least several regular waves of tension production accompanied by simultaneous oscillation in luminescence for 10–20 min before it underwent irreversible damage. In this case, the tension maxima coincided well with the luminescence maxima, and tension minima with the luminescence minima. Under normal conditions, there was no detectable leakage of ATP. Oscillation in light emission caused by ATP released from the strand is provisionally interpreted as reflecting changes in the level of free ATP within the plasmodium. This interpretation is supported by the results obtained recently by T. Sakai (unpublished), who measured ATP levels of the plasmodial strands in contraction and relaxation phases separately with a luciferin-luciferase system; he showed that the ATP level was statistically significantly higher in the phase of maximal tension than in the phase of minimal tension.

The shift in the phases of oscillations in Ca^{2+} and ATP effluxes by just 180° is an interesting problem. Although we do not know the cause of this phase relationship, we are reminded that the calcium-activated ATP pyrophosphohydrolase (APPH) characterized by Kawamura & Nagano (96) exists in the *Physarum* plasmodium. Perhaps if APPH is activated with an increase of Ca^{2+} concentration, the concentration of ATP will be decreased; inversely, if APPH is suppressed with a decrease in Ca^{2+} level, the ATP level will be increased. This is merely speculation on the inverse relation between Ca^{2+} and ATP concentrations.

In analogy to muscle contraction, it is generally believed that micromolar Ca^{2+} stimulates rather than inhibits streaming. The weight of evidence so far known is favorable to this view, but we still cannot rule out the possibility that Ca^{2+} can act as a suppressor of movement by way of controlling the ATP level. As a matter of fact, micromolar Ca^{2+} is inhibitory to streaming in the case of *Nitella,* as we shall see later. Although there are still some ambiguities about the roles of Ca^{2+} and the possibility of ATP limiting the rate of force output, it seems reasonably safe to say that intracellular concentrations of Ca^{2+} and ATP available to actomyosin change in vivo with the same period as tension development.

Structural Basis of Motility

Dynamic activities of the slime mold are closely connected with morphological changes on the part of microfilaments in the cytoplasm. It was shown for the first time by Wohlfarth-Bottermann (168–170) that in the myxomycete plasmodium there are bundles of microfilaments with a diameter of about 7 nm. They are found in the ectoplasm and terminate at the surface of invaginated membrane. There is substantial evidence to show that most of these bundles are composed of plasmodial actomyosin and that they are the morphological entities responsible for the contractility in this organism (13, 14, 122).

CONTRACTION-RELAXATION CYCLE AND ACTIN TRANSFORMATIONS Combining tension monitoring of the plasmodial strand with electron microscopy, various workers have shown that the microfilaments and their assembly undergo remarkable morphological changes in each contraction-relaxation cycle (30, 124, 125, 172). In the shortening phase of the strand under isotonic contractions, the membrane-bound microfilaments with a diameter of 6–7 nm are nearly straight and arranged parallel to one another to form large compact bundles whose adjacent filaments are bridged with cross linkages (124, 125). Among these microfilaments, thicker filaments which are thought to represent myosin bundles are sporadically scattered (14, 124, 125). When the strand approaches the phase of maximal contraction under isotonic conditions, most of the microfilaments become kinky and form networks (124, 125). In the elongation phase, new loose bundles of microfilaments develop from the network. Parallelization of the loose bundles is completed by the time the strand reaches its maximal elongation. They become compact again in the contracting phase, until they lose their parallel order and become entangled to form the network as the maximal contraction is approached. Recent microinjection experiments using phalloidin conducted by Götz von Olenhusen & Wohlfarth-Bottermann (32) also show the involvement of actin transformations in the contraction-relaxation cycle of tension development.

Wohlfarth-Bottermann and his group are of the opinion that actin may undergo G-F transformation in each contraction-relaxation cycle (173). Since endoplasm-ectoplasm conversion is constantly occurring in the migrating plasmodium, especially in its front and rear regions, and the plasmodium contains a considerable amount of G-actin, it is reasonable to suppose that G-F transformation occurs locally. Whether cyclic G-F transformation is a predominant feature occurring with cyclic tension production in a system like the plasmodial strand segment, however, is unknown. Probably plasmodium actinin (41), fragmin (33), or some other regulatory proteins (S. Ogihara and Y. Tonomura, unpublished) play an important role in modifying the state of actin polymers in each cycle of tension development.

FIBRILLOGENESIS In his early observations of the strand hung in the air, Wohlfarth-Bottermann (170) noticed that there are a larger number of fibrils (microfilament bundles) in the ectoplasmic gel tube when the endoplasm flows upward against the gravity than when it flows downward assisted by the gravity. Thus there seemed to be a direct relationship between the amount of motive force needed and the number of fibrils. This notion put forth by Wohlfarth-Bottermann (170) has been supported by further experiments (30, 76). If we apply extra load to the strand, more fibrils emerge. This is a sort of morphologically detectable regulation in response to the load applied.

A favorable material in which to study the de novo formation of the fibrils is a naked drop of endoplasm formed after puncturing the strand (or vein) (61). Soon after the protrusion of the endoplasm, the drop has a low viscosity and no bundles of microfilaments. A considerable amount of actin is thought by some workers to be in a nonfilamentous state at first; but polymerization of actin gradually proceeds to form bundles of F-actin. On the question of whether F-actin is present in an endoplasmic drop from the beginning, opinions are not unanimous (122). At any rate, new bundles of F-actin are developed within 10 min from the "pure" endoplasmic drop to form the ectoplasm. An isolated drop originating from the endoplasm in the mother plasmodium has now become an independent plasmodium having the normal endoplasm-ectoplasm ratio and exhibits shuttle streaming and migration. This fibrillogenesis is an interesting example of regulation on the organizational level of cytoplasm.

BIREFRINGENCE Cyclic changes in filament assembly are demonstrated also by changes in birefringence detectable in the living plasmodium with a highly sensitive polarizing microscope. In a thin fan-like expanse of the spread plasmodium, appearance and disappearance of the birefringent fibers are repeatedly observed in the same loci in the cytoplasm accompanied by back and forth streaming. These birefringent fibrils could be stained with rhodamine-heavy meromyosin (R-HMM). Further, 0.6 KI readily made the birefringent fibrils disappear (127). Thus it was confirmed that these birefringent fibers represent bundles of F-actin. In the contracting phase of the expanse of plasma, in which streaming takes place away from the front (backward streaming), stronger birefringence appears than in the relaxing phase (forward streaming) in which the streaming takes place toward the advancing front. In the latter phase, birefringence fades away or nearly disappears, except that some thicker fibers forming knots or compact regions remain in situ (76). This behavior of the fibrous structure on the optical microscope level agrees well with the observation by electron microscopy. Disappearance or weakening of birefringence is interpreted to reflect deparallelization of the microfilaments or their depolymerization. Localization of the contractile zone in glycerinated specimen (86) also coincides with the distribution and population of the birefringent fibers. Hinssen & D'Haese (58) obtained birefringent synthetic fibrils from *Physarum* actomyosin which resemble in size and fine structure the fibrils in vivo.

Summary

Important points so far known about cytoplasmic streaming in the myxomycete plasmodium may be summarized as follows:

1. The streaming is caused by a local difference in internal pressure of the plasmodium. This pressure difference, or the motive force responsible for the streaming, is measurable by the double-chamber method.
2. The internal pressure is brought about by the active contraction of the ectoplasmic gel.
3. Contractile force of the ectoplasmic gel can be measured as a unidirectional force in a segment of plasmodial strand both under isometric and isotonic conditions.
4. Tension oscillation is augmented conspicuously by stretching or by loading.
5. The structural basis of contraction is the membrane-bound fibrils in the ectoplasm.
6. The fibrils consist mainly of bundles of F-actin and of smaller amounts of myosin and regulatory proteins.
7. Tension development is regulated by Ca^{2+}.
8. Actin filaments undergo cyclic changes in their aggregation pattern and/or G-F transformation in each contraction-relaxation cycle.
9. *Physarum* actin is strikingly similar to rabbit striated muscle actin in its properties, including amino acid composition and sequence.
10. *Physarum* myosin is also similar to rabbit striated muscle myosin in its molecular morphology and ATPase activity, but *Physarum* myosin is more soluble at low salt concentrations than muscle myosin. *Physarum* myosin self-assembles to form bipolar thick filaments which are less stable than those of rabbit skeletal muscle.
11. *Physarum* actomyosin undergoes reversible superprecipitation on addition of Mg-ATP. This characteristic is not shared with muscle actomyosin.

12. The actomyosin thread derived from *Physarum* contracts on addition of Mg^{2+}-ATP. It produces tension as strong as 10 g cm^{-2}, which is comparable to that of a muscle actomyosin thread at the same protein concentration.
13. The aggregation pattern of actin filaments and concentrations of free calcium and free ATP oscillate with the same period as cyclic tension generation. In addition, electric potential difference (79), pH immediately outside the surface membrane Nakamura et al, unpublished), and heat production (12) are also known to change with the same period as shuttle streaming. How these physiological parameters are causally related, and what the origin of oscillation is, still are unknown (16, 172).

ROTATIONAL STREAMING IN CHARACEAN CELLS

General

The other type of streaming which has been studied extensively recently is rotational streaming in giant characean cells. It is in this group of cells that Corti discovered cytoplasmic streaming for the first time in 1774. This type of streaming is quite different from that in the slime mold, in that the cytoplasm flows along the cell wall in the form of a rotating belt, with no beginning or end to the flow. Since the cells of Characeae are large and the rate and configuration of the streaming in them are nearly constant, they lend themselves favorably to physiological and biophysical experiments.

The Site of Motive Force Generation

Each of the two streams going in opposite directions on the opposite sides of the cylindrical cell has the same rate over its entire width and depth to within close proximity of the so-called indifferent lines. There the two opposed streams adjoin and the endoplasm is stationary (73). In other words, the whole bulk of endoplasm moves as a unit, as though it slides along the inner surface of the stationary cortical gel layer. Only the narrow boundary layer between the endoplasm and cortex and the cell sap between the two opposing streams are sheared.

In the endoplasm-filled cell fragment, which can be obtained artificially by gentle centrifugation and subsequent ligation, the endoplasm itself is sheared to show a sigmoid velocity profile similar to that of the cell sap. The highest velocity is found in this case at the region in direct proximity to the stationary cortex. This velocity profile coincides exactly with what is expected when an active shearing force (parallel-shifting force) is produced in opposite directions at the boundary regions on the opposite halves of the cell and the rest of the endoplasm is carried along passively by this peripheral force (81). That the endoplasm in the endoplasm-filled cell, or the cell sap in the normal internodal cell, shows the velocity profile of a sigmoid form is simply a function of the cell's cylindrical geometry. As a matter of fact, when a part of the cylindrical cell is compressed and flattened between two parallel walls to such an extent that the two opposing flows come in contract and fuse, the velocity profile is no longer sigmoid but straight (85). We shall discuss this problem again later.

The simple and unique conclusion derived from these experimental facts is that the motive force driving the endoplasm in the *Nitella* cell is an active shearing force produced only at the interface between the stationary cortex and the endoplasm (71, 81). Naturally this model does not specify the nature of the force or the mechanism of its production, but stresses the importance of interaction between cortex and endoplasm for the production of the motive force. Endoplasm alone is not capable of streaming, as will be described in a later section.

It should be noted, however, that the above model is not the only current view regarding the site of the motive force. N. S. Allen (5, 6, 8) observed the bending waves of the endoplasmic filaments traveling in the direction of streaming and the particle saltation along these filaments. She believes that both play a substantial part in producing the bulk flow, in addition to the force produced at the cortex-endoplasm boundary. Thus she raised an objection to the above model of active shearing. The phenomenon which N. S. Allen described appears in itself fascinating and intriguing. The correctness of a theory on the site of the motive force, however, must be based upon a consistent explanation of the intracellular velocity distributions under defined conditions. If any force participated substantially in driving the endoplasm at loci other than its outermost layer, the flow profiles would be distorted from what we have actually observed.

Subcortical Filaments, Their Identification as Actin, and Their Indispensability for Streaming

Ten years after Kamiya & Kuroda (81) concluded that the motive force is produced in the boundary between the stationary cortex and outer edge of the endoplasm in the form of active shearing, fibrillar structures were discovered at the exact site where the motive force was predicted to be generated. This discovery was made almost simultaneously by light microscopy (66, 68) and electron microscopy (123). The fibrils are attached to the inner surface of files of chloroplasts which are anchored in the cortex. They run parallel to the direction of streaming. Later the subcortical fibrils were visualized also with scanning electron microscopy (9, 98). Each of the fibrils is composed of 50–100 microfilaments with a diameter of 5–6 nm (123, 132). These microfilaments were identified as F-actin by Palevitz et al (130), Williamson (163), Palevitz & Hepler (131), and Palevitz (129) through formation of an arrowhead structure with heavy meromyosin (HMM) or subfragment I of HMM. Further, it was shown by Kersey et al (97) that all these arrowheads point upstream. The subcortical fibrils were also identified as actin by an immunofluorescence technique (37, 167).

An interesting approach to studying functional aspects of the subcortical fibrils is to destroy or dislodge the fibrils locally by such means as microbeam irradiation (67), centrifugation (69), or instantaneous acceleration with a mechanical impact (84). The endoplasmic streaming is either stopped or rendered passive at the site where the filaments disappear. The streaming resumes only after the filaments have regenerated. The new streaming always occurs alongside the newly regenerated filaments and not elsewhere (67–69). All these facts show that subcortical filaments are essential for streaming. Further evidence in support of this view is to be had from differential treatment of an internodal cell with cytochalasin B (121) (see later).

Nitella Myosin and its Localization in the Cell

Voroby' eva & Poglazov (158) showed a slight viscosity drop upon addition of ATP in a myosin B-like protein extracted from *Nitella.* Kato & Tonomura (95) demonstrated the existence of myosin B-like protein in *Nitella* more definitely, and characterized its biochemical properties. At higher ionic strength, ATPase activity of the protein complex was enhanced by EDTA or Ca^{2+} and inhibited by Mg^{2+}. At low ionic strength, superprecipitation occurred with the addition of ATP. Myosin was further purified by Kato and Tonomura from *Nitella* myosin B. The heavy chain of *Nitella* myosin has a molecular weight slightly higher than that of skeletal muscle myosin. Recently myosins were also extracted from the higher plants *Egeria* (*Elodea*) (128) and *Lycopersicon* (156). All these myosins possessed actin-stimulated ATPase activity and the ability to form bipolar aggregates.

An important problem in the mechanism of cytoplasmic streaming is the intracellular localization of myosin and the mode of its action. An effective physiological approach is to treat the streaming endoplasm and the stationary cortex of the living internodal cell of *Nitella* separately with a reagent whose action on actin and myosin is clearly different, and to see how the streaming is affected. A method combining the double-chamber technique with a pair of reciprocal centrifugations made such differential treatment possible (22, 77). When an internodal cell is centrifuged gently, the endoplasm collects at the centrifugal end of the cell while the cortex, including chloroplasts and subortical fibrils in the centripetal half of the cell, remains in situ. Then either the centrifugal half or centripetal half of the cell is treated with an appropriate reagent. Finally, the treated endoplasm is translocated by a second centrifugation in the reverse direction to the other end of the cell to bring it in contact with the untreated cortex; by the same process, untreated endoplasm is brought into contact with treated cortex.

It is known that N-ethylmaleimide (NEM) inhibits F-actin–activated ATPase of myosin, but this reagent has little effect on polymerization or depolymerization of actin. When the cell cortex was treated with NEM and the endoplasm was left untreated, streaming occurred normally. When only endoplasm was treated with the same reagent, no streaming took place (22, 77). If the cell was treated differentially with heat (47.5°C for 2 min) instead of NEM, the result was essentially the same as in the case of NEM (J. C. W. Chen and N. Kamiya, unpublished). This is because myosin is readily denatured by heat while actin is heat-stable. When we used cytochalasin B (CB), whose inhibitory effect on the cytoplasmic streaming in plant cells is well known (19, 21, 164), the situation was exactly the opposite. Cytoplasmic streaming was stopped only when cortex was treated with CB (121); when endoplasm alone was treated with this reagent, streaming continued.

The implication of these results is that the component whose function is readily abolished by NEM and heat must be present in the endoplasm, not in the cortex. It is not incorporated with the subcortical microfilament bundles. This component is presumably myosin. The component whose function is impaired by CB must reside in the cortex, not in the endoplasm. This result shows, as already mentioned, that actin bundles anchored on the cortex are indispensable for streaming.

Motility of Fibrils and Organelles in Isolated Cytoplasmic Droplets

A naked endoplasmic droplet squeezed out of the cell into a solution with an ionic composition similar to the natural cell sap forms a new membrane on its surface and can survive for more than 24 hours, sometimes even for several days.

In the isolated drop of endoplasm containing no ectoplasm, there can be seen what appears as Brownian movement and agitation (saltation) of minute particles; but, as expected, there is no longer any sign of mass streaming of the endoplasm itself. This is a further confirmation of our previous conclusions (71, 81), that rotational cytoplasmic streaming within the cell takes place only when the endoplasm comes in contact with the cortical gel layer and that the bulk of the endoplasm is passively carried along by the force produced at its outer surface.

ROTATION OF CHLOROPLASTS A startling phenomenon in the endoplasmic droplet is the independent and rapid rotation of chloroplasts. This phenomenon was observed by several authors long ago and described in detail by Jarosch for squeezed out cytoplasmic droplets (63–65) and by

Kuroda (102) for isolated endoplasmic droplets containing no cortical gel. The swifter chloroplasts make one rotation in less than one second. When we carefully observe this rotation by means of an appropriate optical system, we find that in the layer immediately outside the rotating body there occurs a streaming in the direction opposite to that of rotation. When rotation is prevented by holding the chloroplast with a microneedle, a counter streamlet becomes manifest along the surface of the chloroplast (102). When the chloroplast is allowed to rotate freely again, the counterstreaming becomes less evident. These observations are further proof for the existence of an active shearing force between sol (endoplasm) and gel (chloroplast) phase. The rotation of chloroplasts is attributable to the attachment to their surface of actin filaments, which are the same as subcortical fibrils.

Using isolated droplets of endoplasm, Hayama & Tazawa (50) investigated the effect of Ca^{2+} and other cations on the rotation of chloroplasts by injecting them iontophoretically. Upon microinjection of Ca^{2+}, chloroplast rotation stopped immediately but recovered with time, suggesting that a Ca^{2+}-sequestering system is present in the cytoplasm. These authors estimated the Ca^{2+} concentrations necessary for halting the rotation to be $> 10^{-4}$M. Sr^{2+} had the same effect as Ca^{2+}. Mn^{2+} and Cd^{2+} also slowed down the rotation, but the effect was gradual and the reversibility was poor. K^{+} and Mg^{2+} had no effect. These results show that Ca^{2+} acts as an inhibitor rather than activator of the movement in *Nitella.* Perfusion experiments (see later) also show the same thing.

MOTILE FIBRILS When we observe a cytoplasmic droplet that has been mechanically squeezed out of a cell of *Nitella* or *Chara* under dark field illumination with high magnification, we find in it a baffling repertory of movements on the part of extremely fine fibrils which may take the form of long filaments or closed circular or polygonal loops. The remarkable behavior of the motile fibrils has been described in detail by Jarosch (63–65) and Kuroda (102).

Important characteristics of these fibrils are their self-motility and their capacity to form polygons. Minute particules (spherosomes) slide (saltate) alongside these fibrils. Their movements are again produced by a mechanism of countershifting relative to the immediately adjacent milieu, as pointed out earlier by Jarosch (63). Kamitsubo (66) confirmed that a freely rotating polygonal loop emerges from the linear stationary fibrils in vivo through "folding off." His observation verifies the view that the motile fibrils and rotating loops or polygons in the droplets are the same in nature as the stationary subcortical fibrils. These motile filaments are identified, through arrowhead formation with HMM (54), as being composed of F-actin. The endoplasmic filaments observed by N. S. Allen (7) in vivo seem to be also of this kind, branching off from the cortical filament bundles.

Demembranated Model Systems

The outstanding merit of using a demembranated cytoplasm is that the chemicals applied presumably diffuse readily into the cytoplasm without being checked by the surface plasma membrane or by the tonoplast.

REMOVAL OF TONOPLAST BY VACUOLAR PERFUSION Taking advantage of the technique of vacuolar perfusion developed by Tazawa (144), experiments were done recently to remove the vacuolar membrane by using media containing ethyleneglycol-bis-(β-aminoethylether)N,N'-tetraacetate (EGTA) (146, 164). This is a sort of cell model, like the demembranated cytoplasmic droplets to be mentioned later. These experiments show that endoplasm remaining in situ shows a full rate of active flow if the perfusing solution fulfils certain requirements. The presence of ATP and Mg^{2+} is indispensable. Williamson (165) showed that generation of the motive force is associated with the subcortical fibrils. On addition of ATP, endoplasmic organelles and particles adhering to the fibrils start moving along the fibrils at speeds up to 50 $\mu m\ s^{-1}$, but are progressively freed from contact, leaving the fibrils denuded. It was shown that streaming requires a millimolar level of Mg^{2+} and free Ca^{2+} at 10^{-7} M or less. Higher concentrations of Ca^{2+} were inhibitory (49, 146). Hayama et al (49) concluded that instantaneous cessation of the streaming upon membrane excitation is caused by a transient increase in the Ca^{2+} concentration in the cytoplasm.

Nitella cytoplasm contains about 0.5 mM ATP (38). Shimmen (140) showed, using tonoplast-free cells, that the ATP concentration for half saturation of streaming velocity was 0.08 mM in *Nitella axillaris* and 0.06 mM in *Chara corallina.* Mg^{2+} is essential at a concentration equal to or higher than that of ATP, whereas Ca^{2+} concentrations in excess of 10^{-7} M are inhibitory to streaming. Inhibition of streaming by depletion of ATP is completely reversible, but that caused by the depletion of free Mg^{2+} is irreversible. This observation suggests that Mg^{2+} may have a role in maintaining some structures(s) necessary for streaming, besides acting as cofactor in the ATPase reaction (140).

DEMEMBRANATED CYTOPLASMIC DROPLETS Taylor et al (142, 143) conducted a series of important observations on amoeboid movement, developing improved techniques of demembranation. In the case of *Nitella* droplets, it is also possible to remove the surface membrane with a tip of a glass needle in a solution with low Ca^{2+} concentration [80 mM KNO_3, 2 mM NaCl, 1 mM $Mg(NO_3)_2$, 1 mM $Ca(NO_3)_2$, 30 mM EGTA, 1 mM ATP, 2 mM dithiothreithol (DTT), 160 mM sorbitol, 3% Ficoll, 5 mM

Pipes buffer (pH 7.0)] (105). As the cytoplasm gradually dispersed into the surrounding solution, the earlier clear demarcation of the droplet disappeared, but chloroplasts continued to rotate almost as actively as they did before the surface membrane was removed. When a chloroplast came out spontaneously from the cytoplasm-dense region into the cytoplasm-sparse area beyond the original boundary, the rotation became slow and sporadic.

The rotation of chloroplasts that emerged from the core area stopped completely on addition of 1 mM N-ethylmaleimide (NEM) to the above solution. Chloroplast rotation did not recover even though free NEM was removed with an excess amount of DTT. This was probably because NEM inhibited F-actin-activated ATPase of myosin, forming a covalent bond with SH-1 of myosin. Kuroda & Kamiya (105), however, observed that the rotation of chloroplasts resumed, even though the rate of rotation was slow and somewhat sporadic, if rabbit-heavy meromyosin (HMM) was added to the solution in the presence of Mg^{2+}-ATP. It took 1 min or longer for a chloroplast to complete one revolution. The rotation of chloroplasts lasted only 10 min or so, but this fact implies that chloroplast rotation in vitro, and hence cytoplasmic streaming in vivo, results from the interaction between F-actin attached to the chloroplast and myosin in the presence of Mg^{2+}-ATP. It also shows that *Nitella* myosin can be functionally replaced, to some extent at least, with muscle HMM (105).

ACTIVE MOVEMENT IN VITRO OF CYTOPLASMIC FIBRILS Another interesting observation on the movement of chloroplasts and cytoplasmic fibrils were made recently by Higashi-Fujime (54). In the demembranated system in an activating medium (1.5 mM ATP, 2 mM $MgSO_4$, 0.2 M sucrose, 4 mM EGTA, 0.1 mM $CaCl_2$, 60 mM KCl, 10 mM imidazole buffer pH 7.0), chloroplast chains linked together by cytoplasmic fibrils moved around at the rate of about 10 μm/sec for 5–10 min. Dark field optics showed that free fibrils curved and turned but never reversed. Traveling fibrils occasionally were converted into a rotating ring. All of the rotating rings, moving fibrils, and fibrils connecting chloroplasts in chains were shown to be composed of bundles of F-actin. Each bundle of F-actin had the same polarity as revealed by decoration with rabbit muscle HMM. Addition of HMM from rabbit skeletal muscle did not accelerate their swimming velocity (S. Higashi-Fujime, personal communications).

Measurement of the Motive Force

The rate of flow, which is often used as the criterion for the activity of cytoplasmic streaming, is a function of two variables: motive force responsible for the streaming, and the viscosity of the cytoplasm. Therefore, if the rate of flow is changed under a certain experimental condition, we do not know to what extent it is due to the change in the motive force and to what extent to the change in viscosity. Since there is no method for measuring exactly the viscosity of cytoplasm in vivo, it is extremely important for understanding the dynamics of cytoplasmic streaming to measure the motive force. So far three methods have been developed for measuring the motive force responsible for rotational streaming in *Nitella*.

CENTRIFUGATION METHOD (82) When an internodal cell is centrifuged along its longitudinal axis, the streaming toward the centrifugal end is accelerated while the streaming toward the centripetal end is retarded. Through centrifuge microscope observation it is possible to determine the centrifugal acceleration at which the streaming endoplasm toward the centripetal end is brought just to a standstill. Measuring this balance-acceleration and the thickness of the layer of the streaming endoplasm, and knowing the difference in density between the endoplasm and cell sap, we can calculate the motive force, i. e. the sliding force at the endoplasm-cortex boundary per unit area. It is found usually to be between 1.0–2.0 dyn cm^{-2} (82). This is the method which made possible the measurement of the motive force of rotational cytoplasmic streaming for the first time.

PERFUSION METHOD (145) When both ends of an adult *Nitella* internode are cut off in an isotonic balanced solution and a certain pressure difference is established between the two openings of the cell, the vacuole can be perfused with that solution. This flow of fluid exerts a shear force upon the vacuolar membrane (tonoplast) and accelerates the endoplasmic streaming in the direction of perfusion on one side of the cell, but retards the streaming on the other side. The greater the speed of perfusion, the slower is the rate of streaming against it. At a certain perfusion rate, the endoplasmic streaming against it is brought to a standstill. The shearing force produced at the tonoplast under this state must be nearly equal to the motive force produced at the endoplasm-ectoplasm interface, because the thickness of the endoplasmic layer of the adult *Nitella* internode is small as compared with the radius of the vacuole. The motive force thus measured by Tazawa was found mostly in the range of 1.4–2.0 dyn cm^{-2}. This range is in excellent agreement with the values obtained by the centrifugation method in the same material (82). By using a theoretical model and assumptions regarding the form of stress/rate of strain curve (85), Donaldson (27) determined possible velocity distributions for the streaming cytoplasm in the normal cell and in the cell whose vacuole is perfused. Based on a mathematical analysis, he estimated the motive force to be 3.6 dyn cm^{-2} and the thickness of the motive force field (high shear zone) between the endoplasm and cortical gel layer to be 0.1 μm.

With the perfusion method, it was shown that the motive force is almost independent of temperature within the range of 10°–30°C, while flow velocity increases conspicuously with the rise in temperature (145). This result shows that the change in the rate of flow as the temperature varies, is, at least in the case of characean cells, mainly the result of a change in viscosity of the fluid in the high shear zone at the endoplasm-ectoplasm boundary and not to the change in the motive force.

Increase in viscosity of the cytoplasm, however, is not always the cause of the slowing down or cessation of flow. For instance, the slackening of the streaming through application of an SH reagent, *p*-chloromercuribenzoate (1), is the result of the disappearance of the motive force (145). It has long been known that when an action potential is evoked in characean cells the streaming halts transiently. The inhibiting effect of the action potential upon streaming has been demonstrated to be due to a momentary interruption of the motive force and not to a increase in the viscosity of the cytoplasm (147). As has been stated before, this change is triggered by a transient increase in Ca^{2+} at the site of motive force generation (49).

METHOD OF LATERAL COMPRESSION (87) There is one further approach to measuring the motive force. When a part of the large cylindrical cell of *Nitella* showing vigorous cytoplasmic streaming is compressed and flattened between a pair of parallel flat walls, the two opposed streamings on the opposite sides of the cell come in contact and fuse. The velocity and the velocity gradient of the fused region of the endoplasm are changed as the width of the fused region is modified (87). From these measurements it is possible to calculate not only the motive force responsible for streaming, but also the force resisting the sliding at the endoplasm-ectoplasm boundary, by the simultaneous flow equation: $-F=Rv+\eta_a \frac{dv}{dy}$, where F represents the motive force, R the sliding resistance per unit velocity, v the marginal velocity, dv/dy the shear rate of the endoplasm, and η_a the apparent viscosity of the endoplasm. For an arbitrary dv/dy, η_a can be obtained experimentally by the data of the isolated endoplasm (85, 87). We can measure directly v and dv/dy for each width of the stream. Since both the motive force F and the resisting force R are unknown quantities, we calculate them using two simultaneous flow equations under two different widths of the stream. Again the motive force is found to be 1.7 dyn cm^{-2}, in good agreement with the data previously reported. Further, it is known that the major part of the motive force is used in overcoming the resistance to sliding at the endoplasm-ectoplasm boundary. Since the viscosity of the cell sap is low (1.4–2.0 cp) (unpublished data of N. Kamiya & K. Kuroda), less than 1% of the motive force is used for bringing about its shearing in the normal cell.

The resistance per unit sliding velocity is known to be of the order of 230 dyn s cm^{-3} (87).

Recently, Hayashi (51) investigated a theoretical model of cytoplasmic streaming in *Nitella* and *Chara.* On the basis of experimental results obtained by Kamiya & Kuroda (81, 82, 85, 87), he derived general rheological equations for the non-Newtonian fluid of cytoplasm, constructed a fluid-dynamic model of the streaming, derived a general expression for the velocity distribution in the cell, and showed how the rheological properties of the cytoplasm are estimated in vivo. He (52) also constructed a theoretical model of cytoplasmic streaming in a boundary layer where the motive force is supposed to exert directly to cytoplasm, and estimated several important parameters such as the depth of the boundary region, rheological constants as well as the maximum velocity in the boundary layer, and the magnitude of motive force generated by a unit length of a single subcortical fibril. The depth of the layer was 1.3–2.0 μm, while the viscosity was a small fraction of that estimated in the bulk layer, and the maximum velocity was six or seven times that of the bulk layer. The magnitude of the motive force was ca 0.8 dyn, which can be converted to a sliding force 1.2 dyn/cm^2, showing a good coincidence with those obtained by Kamiya & Kuroda (82) and Tazawa (145).

Molecular Mechanism of Rotational Streaming

It is now established that the subcortical filaments are composed of bundles of actin filaments all with the same polarity over their entire length (97). The streaming occurs steadily in the form of an endless belt in only one direction alongside these filaments. There is no chance for the endoplasm to move backward; the track is strictly "one-way." The sliding force (motive force) responsible for the streaming was measured to be 1–2 dyn cm^{-2} (82, 87, 145) or 3.6 dyn cm^{-2} (27).

The ATP content in *Nitella* cytoplasm, which is around 0.5 mM (38), is already saturated for streaming. *Nitella* myosin has been isolated and characterized (95), and it has been suggested that it resides in the endoplasm (22). It would be of great interest to know what kind of aggregation pattern the *Nitella* myosin has in vivo, and what kind of interaction occurs with the subcortical fibrils to produce the continuously active shear force responsible for this endless rotation. With the electron microscope, N. F. Allen (7) has shown putative myosin molecules with bifurcating ends in the endoplasm of *Nitella.* Bradley (19) and Williamson (164) suggested that the putative *Nitella* myosin may link with endoplasmic organelles. Williamson's observation of the subcortical fibrils in the perfused cell (165) shows cogently that the streaming is brought about through interaction between cortex-attached F-actin bundles and myosin.

Nagai & Hayama (119, 120) confirmed the existence of many balloon-shaped, membrane-bound endoplasmic organelles linked to the bundles of microfilaments under ATP-deficient conditions. These organelles, whose shape and size are diverse, have one or more protuberances in the form of rods or horns (119, 120). Electron-dense bodies 20–30 nm in diameter are located on the surface of these protuberances in an ordered array with a spacing of 100–110 nm. It is these electron-dense bodies which link the endoplasmic organelles to F-actin bundles. The dense bodies on the organelles detach from the F-actin bundles on application of ATP. Filaments 4 nm or less in diameter, and different from F-actin in their ability to react with muscle HMM, are attached to the surface of these protuberances. They are probably the same as the 4 nm filaments observed and presumed to be myosin filaments by Williamson (166) and N. S. Allen (7). Nagai & Hayama suggested that the 20 to 30 nm globular bodies attached to the protuberances of the organelles may act as functional units when the organelles move along the microfilaments, and that they are composed, at least in part, of the functional head of myosin or myosin aggregates, possibly with some other unknown protein(s).

There are several contemporary views regarding the mechanism of rotational streaming other than the above, views which are instructive and suggestive in some respects. But taking into consideration the facts so far known, the most plausible and resonable picture of the streaming in *Nitella* is as follows. The rotational streaming is caused by the unidirectional sliding force of the endoplasmic organelles loaded with myosin along the stationary subcortical fibrils composed of F-actin having the same polarity. The motive force, which is the active shearing force, is produced through the interaction of the myosin with the F-actin on the cortex. Endoplasmic organelles, with which *Nitella* myosin is associated, presumably effectively help drag the rest of the endoplasm so that the whole endoplasmic layer slides alongside the cortex of the cell as a mass. Important problems left to be solved are the molecular mechanism of the active shearing, and its control.

CONCLUDING REMARKS

In the foregoing pages, we reviewed molecular and dynamic aspects of cytoplasmic streaming and related phenomena in the two representative materials, myxomycete plasmodium and characean cells. These two types of streaming both utilize an actomyosin-ATP system for their force output, but their modes of movement are quite different.

The streaming in *Physarum* plasmodium is a pressure flow, caused by contractions of actomyosin filament bundles, while streaming in *Nitella* is brought about by the shearing force produced at the interface between subcortical actin bundles and putative myosin attached to the endoplasmic organelles. In spite of extensive biochemical studies on contractile and regulatory proteins, and of morphological studies on contractile structures, there has been so far no direct evidence to show that the contraction-relaxation cycle of the slime mold operates on the basis of a sliding filament mechanism. Nor is there any convincing evidence that the changes in aggregation patterns on the part of the microfilaments produce the tension force. There is a possibility that entanglement and network formation are the visual manifestations of the events accompanying the sliding of antiparallel microfilaments against each other, the sliding being effected by myosin dimers or oligomers in the presence of Mg^{2+}-ATP and regulatory proteins.

Another important feature of the streaming in the acellular slime mold is its rhythmicity. Wohlfarth-Bottermann discussed four possibilities as a source of oscillation (172). As described before, various physical and chemical parameters of the streaming change with the same period in association with the periodic force production. The origin of oscillation is, however, still unknown.

In rotational streaming in *Nitella,* the site of the motive force and its mode of action appear to be quite well known now. Probably the sliding mechanism of the cytoplasmic streaming in *Nitella* is similar in principle to the mechanism of less organized types of streaming in plant cells and saltatory particle movement. But it is not known yet how far the active shear mechanism is applicable to other systems like the myxomycete plasmodium or amoeba.

The basic mechanism of actomyosin-based movement may be common in many different kinds of cells, from muscle to amoeba to plant cells. Our knowledge is, however, still too meager to obtain a unified concept of a variety of cytoplasmic streaming patterns in terms of their molecular organization.

Literature Cited

1. Abe, S. 1964. The effect of *p*-chloromercuribenzoate on rotational protoplasmic streaming in plant cells. *Protoplasma* 58:483–92
2. Achenbach, U., Wohlfarth-Bottermann, K. E. 1980. Oscillating contractions in protoplasmic strands of *Physarum. J. Exp. Biol.* 85:21–31
3. Adelman, M. R., Taylor, E. W. 1969. Isolation of an actomyosin-like protein complex from slime mold plasmodium and the separation of the complex into actin- and myosin-like fractions. *Biochemistry* 8:4964–75
4. Adelman, M. R., Taylor, E. W. 1969. Further purification and characterization of slime mold myosin and slime mold actin. *Biochemistry* 8:4976–88
5. Allen, N. S. 1974. Endoplasmic filaments generate the motive force for rotational streaming in *Nitella. J. Cell Biol.* 63:270–87
6. Allen, N. S. 1976. Undulating filaments in *Nitella* endoplasm and motive force generation. See Ref. 31, pp. 613–21
7. Allen, N. S. 1980. Cytoplasmic streaming and transport in the characean alga *Nitella. Can. J. Bot.* 58:786–96
8. Allen, N. S., Allen, R. D. 1978. Cytoplasmic streaming in green plants. *Ann. Rev. Biophys. Bioeng.* 7:497–526

9. Allen, R. D. 1977. Concluding remarks. Symp. Mol. Basis Motility. In *International Cell Biology 1976–1977,* ed. B. R. Brinkley, K. R. Porter, pp. 403–6. New York: Rockefeller Univ. Press. 694 pp.
10. Allen, R. D., Allen, N. S. 1978. Cytoplasmic streaming in amoeboid movement. *Ann. Rev. Biophys. Bioeng.* 7:469–95
11. Allen, R. D., Kamiya, N., eds. 1964. *Primitive Motile Systems in Cell Biology.* New York: Academic. 642 pp.
12. Allen, R. D., Pitts, W. R., Speir, D., Brault, J. 1963. Shuttle streaming: Synchronization with heat production in slime mold. *Science* 142:1485–87
13. Alléra, A., Beck, R., Wohlfarth-Bottermann, K. E. 1971. Weitreichende, fibrilläre Protoplasmadifferenzierungen und ihre Bedeutung für die Protoplasmaströmung. VIII. Identifizierung der Plasmafilamente von *Physarum polyceplalum* als F-Actin durch Anlagerung von heavy meromyosin in situ. *Cytobiologie* 4:437–49
14. Alléra, A., Wohlfarth-Bottermann, K. E. 1972. Weitreichende fibrilläre Protoplasmadifferenzierungen und ihre Bedeutung für Protoplasmaströmung. IX. Aggregationszustände des Myosins und Bedingungen zur Entstehung von Myosinfilamenten in den Plasmodien von *Physarum polycephalum. Cytobiologie* 6:261–86
15. Beck, R., Hinssen, H., Komnick, H. 1970. Weitreichende, fibrilläre Protoplasmadifferenzierungen und ihre Bedeutung für die Protoplasmaströmung. V. Kontraktion, ATPase-Aktivität und Feinstructur isolieter Actomyosin-Fäden von *Physarum polycephalum. Cytobiologie* 2:259–74
16. Berridge, M. J., Rapp, P. E. 1979. A comparative survey of the function, mechanism and control of cellular oscillators. *J. Exp. Biol.* 81:217–79
17. Braatz, R. 1975. Differential histochemical localization of calcium and its relation to shuttle streaming in *Physarum. Cytobiologie* 12:74–78
18. Braatz, R., Komnick, H. 1970. Histochemischer Nachweis eines Calciumpumpenden Systems in Plasmodien von Schleimpilzen. *Cytobiologie* 2:457–63
19. Bradley, M. O. 1973. Microfilaments and cytoplasmic streaming: Inhibition of streaming with cytochalasin. *J. Cell Sci.* 12:327–43
20. Britz, S. J. 1979. Cytoplasmic streaming in *Physarum.* See Ref. 48, pp. 127–49
21. Chen, J. C. W. 1973. Observations of protoplasmic behavior and motile protoplasmic fibrils in cytochalasin B treated *Nitella* rhizoid. *Protoplasma* 77:427–35
22. Chen, J. C. W., Kamiya, N. 1975. Localization of myosin in the internodal cell of *Nitella* as suggested by differential treatment with N-ethylmaleimide. *Cell Struct. Funct.* 1:1–9
23. Clarke, M., Spudich, J. A. 1977. Nonmuscle contractile proteins: The role of actin and myosin in cell motility and shape determination. *Ann. Rev. Biochem.* 46:797–822
24. Crooks, R., Cooke, R. 1977. Tension generation by threads of contractile proteins. *J. Gen. Physiol.* 69:37–55
25. D'Haese, J., Hinssen, H. 1978. Kontraktionseigenschaften von isolierten Schleimpilzactomyosin. 1. Vergleichende Untersuchungen an Fadenmodellen aus natürlichen, rekombinierten und hybridisierten Actomyosinen von Schleimpilz und Muskel. *Protoplasma* 95:273–95
26. D'Haese, J., Hinssen, H. 1979. Aggregation properties of nonmuscle myosin. See Ref. 36, pp. 105–18
27. Donaldson, I. G. 1972. Cyclic longitudinal fibrillar motion as a basis for steady rotational protoplasmic streaming. *J. Theor. Biol.* 37:75–91
28. Dryl, S., Zurzycki, J., eds. 1972. *Motile Systems of Cells (Acta Protozool.* 11). Warszawa: Nencki Inst. 418 pp.
29. Ettienne, E. 1972. Subcellular localization of calcium repositories in plasmodia of the acellular slime mold *Physarum polycephalum. J. Cell Biol.* 54:179–84
30. Fleischer, M., Wohlfarth-Bottermann, K. E. 1975. Correlation between tension force generation, fibrillogenesis and ultrastructure of cytoplasmic actomyosin during isometric and isotonic contractions of protoplasmic strands. *Cytobiologie* 10:339–65
31. Goldman, R., Pollard, T., Rosenbaum, J. 1976. *Cell Motility.* New York: Cold Spring Harbor Lab. 1373 pp.
32. Götz von Olenhusen, K., Wohlfarth-Bottermann, K. E. 1979. Evidence of actin transformation during the contraction-relaxation cycle of cytoplasmic actomyosin: Cycle blockade by phalloidin injection. *Cell Tissue Res.* 196:455–70
33. Hasegawa, T., Takahashi, S., Hayashi, H., Hatano, S. 1980. Fragmin: A calcium ion sensitive regulatory factor on the formation of actin filaments. *Biochemistry* 19:2677–83

34. Hatano, S. 1970. Specific effect of Ca^{2+} on movement of plasmodial fragment obtained by caffeine treatment. *Exp. Cell Res.* 61:199–203
35. Hatano, S. 1973. Contractile proteins from the myxomycete plasmodium. *Adv. Biophys.* 5:143–76
36. Hatano, S., Ishikawa, H., Sato, H., eds. 1979. *Cell Motility: Molecules and Organization.* Tokyo: Univ. Tokyo Press. 696 pp.
37. Hatano, S., Matsumura, F., Hasegawa, T., Takahashi, S., Sato, H., Ishikawa, H. 1979. Assembly and disassembly of F-actin filaments in *Physarum* plasmodium and *Physarum* actinin. See Ref. 36, pp. 87–104
38. Hatano, S., Nakajima, H. 1963. ATP content and ATP-dephosphorylating activity of *Nitella. Ann. Rep. Sci. Works Fac. Sci. Osaka Univ.* 11:71–76
39. Hatano, S., Oosawa, F. 1966. Extraction of an actin-like protein from the plasmodium of a myxomycete and its interaction with myosin A from rabbit striated muscle. *J. Cell Physiol.* 68:197–202
40. Hatano, S., Oosawa, F. 1966. Isolation and characterization of plasmodium actin. *Biochim. Biophys. Acta* 127:489–98
41. Hatano, S., Owaribe, K. 1976. Actin and actinin from myxomycete plasmodia. See Ref. 31, pp. 499–511
42. Hatano, S., Owaribe, K., Matsumura, F., Hasegawa, T., Takahashi, S. 1980. Characterization of actin, actinin, and myosin isolated from *Physarum. Can. J. Bot.* 58:750–59
43. Hatano, S., Takahashi, K. 1971. Structure of myosin A from the myxomycete plasmodium and its aggregation at low salt concentrations. *J. Mechanochem. Cell Motil.* 1:7–14
44. Hatano, S., Takeuchi, I. 1960. ATP content in myxomycete plasmodium and its levels in relation to some external conditions. *Protoplasma* 52:169–83
45. Hatano, S., Tazawa, M. 1968. Isolation, purification and characterization of myosin B from myxomycete plasmodium. *Biochim. Biophys. Acta* 154:507–19
46. Hatano, S., Totsuka, T., Oosawa, F. 1967. Polymerization of plasmodium actin. *Biochim. Biophys. Acta* 140:109–22
47. Hato, B., Ueda, T., Kurihara, K., Kobatake, Y. 1976. Phototaxis in true slime mold *Physarum polycephalum. Cell Struct. Funct.* 1:269–78
48. Haupt, W., Feinleib, M. E. 1979. *Physiology of Movements. Encyclopedia of Plant Physiology, New Ser. 7.* Berlin/Heidelberg/New York: Springer. 731 pp.
49. Hayama, t., Shimmen, T., Tazawa, M. 1979. Participation of CA^{2+} in cessation of cytoplasmic streaming induced by membrane excitation in Characeae internodal cells. *Protoplasma* 99: 305–21
50. Hayama, T., Tazawa, M. 1980. Ca^{2+} reversibly inhibits active rotation of chloroplasts in isolated cytoplasmic droplets of *Chara. Protoplasma* 102:1–9
51. Hayashi, Y. 1980. Fluid-dynamical study of protoplasmic streaming in a plant cell. *J. Theor. Biol.* 85:451–67
52. Hayashi, Y. 1980. Theoretical study of motive force of protoplasmic streaming in a plant cell. *J. Theor. Biol.* 85: 469–80
53. Hepler, P. K., Palevitz, B. A. 1974. Microtubules and microfilaments. *Ann. Rev. Plant Physiol.* 25:309–62
54. Higashi-Fujime, S. 1980. Active movement in vitro of bundles of microfilaments isolated from *Nitella* cell. *J. Cell Biol.* 87:569–78
55. Hinssen, H. 1970. Synthetische Myosinfilamente von Schleimpilz-Plasmodien. *Cytobiologie* 2:326–31
56. Hinssen, H. 1979. Studies on the polymer state of actin in *Physarum polycephalum.* See Ref. 36, pp. 59–85
57. Hinsen, H., D'Haese, J. 1974. Filament formation by slime mould myosin isolated at low ionic strength. *J. Cell Sci.* 15:113–29
58. Hinsen, H., D'Haese, J. 1976. Synthetic fibrils from *Physarum* actomyosin—Self assembly, organization and contraction. *Cytobiologie* 13:132–57
59. Hitchcock, S. E. 1977. Regulation of motility in nonmuscle cells. *J. Cell Biol.* 74:1–15
60. Inoúe, S., Stephens, R. E. 1975. *Molecules and Cell Movement.* New York: Raven. 450 pp.
61. Isenberg, G., Wohlfarth-Bottermann, K. E. 1976. Transformation of cytoplasmic actin. Importance for the organization of the contractile gel reticulum and the contraction-relaxation cycle of cytoplasmic actomyosin. *Cell Tissue Res.* 173:495–528
62. Jahn, T. L., Bovee, E. C. 1969. Protoplasmic movements within cells. *Physiol. Rev.* 49:793–862
63. Jarosch, R. 1956. Plasmaströmung und Chloroplastenrotation bei Characeen. *Phyton* 6:87–107
64. Jarosch, R. 1958. Die Protoplasmafi-

brillen der Characeen. *Protoplasma* 50:93–108
65. Jarosch, R. 1976. Dynamisches Verhalten der Actinfibrillen von *Nitella* auf Grund schneller Filament-Rotation. *Biochem. Physiol. Pflanz.* 170:111–31
66. Kamitsubo, E. 1966. Motile protoplasmic fibrils in cells of characeae. II Linear fibrillar structure and its bearing on protoplasmic streaming. *Proc. Jpn. Acad.* 42:640–43
67. Kamitsubo, E. 1972. A 'window technique' for detailed observation of characean cytoplasmic streaming. *Exp. Cell Res.* 74:613–16
68. Kamitsubo, E. 1972. Motile protoplasmic fibrils in cells of the Characeae. *Protoplasma* 74:53–70
69. Kamitsubo, E. 1980. Cytoplasmic streaming in characean cells: role of subcortical fibrils. *Can. J. Bot.* 58:760–65
70. Kamiya, N. 1953. The motive force responsible for protoplasmic streaming in the myxomycete plasmodium. *Ann. Rep. Sci. Works Fac. Sci. Osaka Univ.* 1:53–83
71. Kamiya, N. 1959. Protoplasmic streaming. *Protoplasmatologia* 8, 3 a. Vienna: Springer. 199 pp.
72. Kamiya, N. 1960. Physics and chemistry of protoplasmic streaming. *Ann. Rev. Plant Physiol.* 11:323–40
73. Kamiya, N. 1962. Protoplasmic streaming. *Handb. Pflanzenphysiol.* 17(2): 979–1035
74. Kamiya, N. 1968. The mechanism of cytoplasmic movement in a myxomycete plasmodium. In *Aspects of Cell Motility.* Symp. Soc. Exp. Biol. 22, pp. 199–214. Cambridge: Cambridge Univ. Press
75. Kamiya, N. 1970. Contractile properties of the plasmodial strand. *Proc. Jpn. Acad.* 46:1026–31
76. Kamiya, N. 1973. Contractile characteristics of the myxomycete plasmodium. *Proc. 4th Int. Biophys. Congr. (Moscow) Symp. III Biophysics of Motility,* pp. 447–65
77. Kamiya, N. 1977. Introductory remarks. Symp. Mol. Basis Motility. See Ref. 9, pp. 361–66
78. Kamiya, N. 1979. Dynamic aspects of movement in the myxomycete plasmodium. See Ref. 36, pp. 399–414
79. Kamiya, N., Abe, S. 1950. Bioelectric phenomena in the myxomycete plasmodium and their relation to protoplasmic flow. *Colloid Sci.* 5:149–63
80. Kamiya, N., Allen, R. D., Zeh, R. 1972. Contractile properties of the slime mold strand. *Acta Protozool.* 11:113–24
81. Kamiya, N., Kuroda, K. 1956. Velocity distribution of the protoplasmic streaming in *Nitella* cells. *Bot. Mag.* 69:544–54
82. Kamiya, N., Kuroda, K. 1958. Measurement of the motive force of the protoplasmic rotation in *Nitella. Protoplasma* 50:144–48
83. Kamiya, N., Kuroda, K. 1958. Studies on the velocity distribution of the protoplasmic streaming in the myxomycete plasmodium. *Protoplasma* 49:1–4
84. Kamiya, N., Kuroda, K. 1964. Mechanical impact as a means of attacking structural organization in living cells. *Ann. Rep. Sci. Works Fac. Sci. Osaka Univ.* 12:83–97
85. Kamiya, N., Kuroda, K. 1965. Rotational protoplasmic streaming in *Nitella* and some physical properties of the endoplasm. *Proc. 4th Int. Congr. Rheology Part 4, Symp. Biorheol.,* pp. 157–71. New York: Wiley
86. Kamiya, N., Kuroda, K. 1965. Movement of the myxomycete plasmodium. I. A study of glycerinated models. *Proc. Jpn. Acad.* 41:837–41
87. Kamiya, N., Kuroda, K. 1973. Dynamics of cytoplasmic streaming in a plant cell. *Biorheology* 10:179–87
88. Kamiya, N., Yoshimoto, Y. 1972. Dynamic characteristics of the cytoplasm —A study on the plasmodial strand of a myxomycete. In *Aspects of Cellular and Molecular Physiology,* ed. K. Hamaguchi, pp. 167–89. Tokyo: Univ. Tokyo Press, 257 pp.
89. Kamiya, N., Yoshimoto, Y., Matsumura, F. 1981. Physiological aspects of actomyosin-based cell motility. In *International Cell Biology 1980–1981,* ed. H. G. Schweiger, pp. 346–58. Heidelberg/New York: Springer
90. Kamiya, R., Maruyama, K., Kawamura, M., Kikuchi, M. 1972. Mg-polymer of actin formed under the influence of β-actinin. *Biochim. Biophys. Acta* 256:120–31
91. Kato, T. 1979. Ca^{2+} uptake of *Physarum* microsomal vesicles. See Ref. 36, pp. 211–23
92. Kato, T., Tonomura, Y. 1975. Ca^{2+}-sensitivity of actomyosin ATPase purified from *Physarum polycephalum. J. Biochem.* 77:1127–34
93. Kato, T., Tonomura, Y. 1975. *Physarum* tropomyosin-troponin complex. Isolation and properties. *J. Biochem.* 78:583–88

94. Kato, T., Tonomura, Y. 1977. Uptake of calcium ion into microsomes isolated from *Physarum polycephalum*. *J. Biochem.* 81:207–13
95. Kato, T., Tonomura, Y. 1977. Identification of myosin in *Nitella flexilis*. *J. Biochem.* 82:777–82
96. Kawamura, M., Nagano, K. 1975. A calcium ion-dependent ATP pyrophosphohydrolase in *Physarum polycephalum*. *Biochim. Biophys. Acta* 397:207–19
97. Kersey, Y. M., Hepler, P. K., Palevitz, B. A., Wessells, N. K. 1976. Polarity of actin filaments in Characean algae. *Proc. Natl. Acad. Sci. USA* 73:165–67
98. Kersey, Y. M., Wessells, N. K. 1976. Localization of actin filaments in internodal cells of characean algae. *J. Cell Biol.* 68:264–75
99. Komnick, H., Stockem, W., Wohlfarth-Bottermann, K. E. 1973. Cell Motility: Mechanism in protoplasmic streaming and ameboid movement. *Int. Rev. Cytol.* 34:169–249
100. Korn, E. D. 1978. Biochemistry of actomyosin-dependent cell motility (a review). *Proc. Natl. Acad. Sci. USA* 75:588–99
101. Krüger, J., Wohlfarth-Bottermann, K. E. 1978. Oscillating contractions in protoplasmic strands of *Physarum:* stretch-induced phase shifts and their synchronization. *J. Interdiscip. Cycle Res.* 9:61–71
102. Kuroda, K. 1964. Behavior of naked cytoplasmic drops isolated from plant cells. See Ref. 11, pp. 31–41
103. Kuroda, K. 1979. Movement of cytoplasm in a membrane-free system. See Ref. 36, pp. 347–61
104. Kuroda, K. 1979. Movement of demembranated slime mould cytoplasm. *Cell Biol. Int. Rep.* 3:135–40
105. Kuroda, K., Kamiya, N. 1975. Active movement of *Nitella* chloroplasts in vitro. *Proc. Jpn. Acad.* 51:774–77
106. Kuroda, R., Kuroda, H. 1980. Calcium accumulation in vacuoles of *Physarum polycephalum* following starvation. *J. Cell Sci.* 44:75–85
107. Kuroda, R., Kuroda, H. 1981. Relation of cytoplasmic calcium to contractility in *Physarum polycephalum*. *J. Cell Sci.* In press
108. Loewy, A. G. 1952. An actomyosin-like substance from the plasmodium of a myxomycete. *J. Cell. Comp. Physiol.* 40:127–56
109. Ludlow, C. T., Durham, A. C. H. 1977. Calcium ion fluxes across the external surface of *Physarum polycephalum*. *Protoplasma* 91:107–13
110. Maruyama, K., Kamiya, R., Kimura, S., Hatano, S. 1975. β-Actinin-like protein from plasmodium. *J. Biochem.* 79:709–15
111. Matsumura, F., Hatano, S. 1978. Reversible superprecipitation and bundle formation of plasmodium actomyosin. *Biochim. Biophys. Acta* 533:511–23
112. Matsumura, F., Yoshimoto, Y., Kamiya, N. 1980. Tension generation by actomyosin thread from a non-muscle system. *Nature* 285:169–71
113. Matthews, L. M. 1977. Ca^{++} regulation in caffeine-derived microplasmodia of *Physarum polycephalum*. *J. Cell Biol.* 72:502–5
114. Nachmias, V. T. 1972. Filament formation by purified *Physarum* myosin. *Proc. Natl. Acad. Sci. USA* 69:2011–14
115. Nachmias, V. T. 1974. Properties of *Physarum* myosin purified by a potassium iodide procedure. *J. Cell Biol.* 62:54–65
116. Nachmias, V. T. 1979. The contractile proteins of *Physarum polycephalum* and actin polymerization in plasmodial extracts. See. Ref. 36, pp. 33–57
117. Nachmias, V. T., Ash, A. 1976. A calcium-sensitive preparation from *Physarum polycephalum*. *Biochemistry* 15: 4273–78
118. Nachmias, V. T., Huxley, H. E., Kessler, D. 1970. Electron microscope observations on actomyosin and actin preparations from *Physarum polycephalum,* and on their interaction with heavy meromyosin subfragment from muscle myosin. *J. Mol. Biol.* 50:83–90
119. Nagai, R., Hayama, T. 1979. Ultrastructure of the endoplasmic factor responsible for cytoplasmic streaming in *Chara* internodal cells. *J. Cell Sci.* 36:121–36
120. Nagai, R., Hayama, T. 1979. Ultrastructural aspects of cytoplasmic streaming in characean cells. See Ref. 36, pp. 321–37
121. Nagai, R., Kamiya, N. 1977. Differential treatment of *Chara* cells with cytochalasin B with special reference to its effect on cytoplasmic streaming. *Exp. Cell Res.* 108:231–37
122. Nagai, R., Kato, T. 1975. Cytoplasmic filaments and their assembly into bundles in *Physarum plasmodium*. *Protoplasma* 86:141–58
123. Nagai, R., Rebhun, L. I. 1966. Cytoplasmic microfilaments in streaming *Nitella* cells. *J. Ultrastruct. Res.* 14:571–58

124. Nagai, R., Yoshimoto, Y., Kamiya, N. 1975. Changes in fibrillar structures in the plasmodial strand in relation to the phase of contraction-relaxation cycle. *Proc. Jpn. Acad.* 51:38–43
125. Nagai, R., Yoshimoto, Y., Kamiya, N. 1978. Cyclic production of tension force in the plasmodial strand of *Physarum polycephalum* and its relation to microfilament morphology. *J. Cell Sci.* 33:121–36
126. Nakajima, H. 1960. Some properties of a contractile protein in a myxomycete plasmodium. *Protoplasma* 52:413–36
127. Ogihara, S., Kuroda, K. 1979. Identification of birefringent structure which appears and disappears in accordance with the shuttle streaming in *Physarum* plasmodia. *Protoplasma* 100:167–77
128. Ohsuka, K., Inoue, A. 1979. Identification of myosin in a flowering plant, *Egeria densa. J. Biochem.* 85:375–78
129. Palevitz, B. A. 1977. Actin cables and cytoplasmic streaming in green plants. See Ref. 31, pp. 601–11
130. Palevitz, B. A., Ash, J. F., Hepler, P. K. 1974. Actin in the green algae, *Nitella. Proc. Natl. Acad. Sci. USA* 71:363–66
131. Palevitz, B. A., Hepler, P. K. 1975. Identification of actin in situ at the ectoplasm-endoplasm interface of *Nitella.* Microfilament-chloroplast association. *J. Cell Biol.* 65:29–38
132. Pickett-Heaps, J. D. 1967. Ultrastructure and differentiation in *Chara sp.* I. Vegetative cells. *Aust. J. Biol. Sci.* 20:539–51
133. Poff, K. L., Whitaker, B. D. 1979. Movement of slime molds. See Ref. 48, pp. 356–82
134. Pollard, T. D., Weihing, R. R. 1974. Actin and myosin and cell movement. *CRC Biochem.* 2:1–65
135. Pressman, B. C. 1976. Biological applications of ionophores. *Ann. Rev. Biochem.* 45:501–30
136. Rebhun, L. I. 1972. Polarized intracellular particle transport: Saltatory movements and cytoplasmic streaming. *Int. Rev. Cytol.* 32:93–137
137. Ridgway, E. B., Durham, A. C. H. 1976. Oscillations of calcium ion concentrations in *Physarum polycephalum. J. Cell Biol.* 69:223–26
138. Sachsenmaier, W., Blessing, J., Brauser, B., Hansen, K. 1973. Protoplasmic streaming in *Physarum polycephalum.* Observation of spontaneous and induced changes of the oscillatory pattern by photometric and fluorometric techniques. *Protoplasma* 77:381–96
139. Seitz, K. 1979. Cytoplasmic streaming and cyclosis of chloroplasts. See Ref. 48, pp. 150–69
140. Shimmen, T. 1978. Dependency of cytoplasmic streaming on intracellular ATP and Mg^{2+} concentrations. *Cell. Struct. Funct.* 3:113–21
141. Takeuchi, Y., Yoneda, M. 1977. Synchrony in the rhythm of the contraction-relaxation cycle in two plasmodial strands of *Physarum polycephalum. J. Cell Sci.* 26:151–60
142. Taylor, D. L., Condeelis, J. S. 1979. Cytoplasmic structure and contractility in amoeboid cells. *Int. Rev. Cytol.* 56:57–144
143. Taylor, D. L., Condeelis, J. S., Moore, P. L., Allen, R. D. 1973. Contractile basis of amoeboid movement. I. The chemical control of motility in isolated cytoplasm. *J. Cell Biol.* 59:378–94
144. Tazawa, M. 1964. Studies on *Nitella* having artificial cell sap. I. Replacement of the cell sap with artificial solutions. *Plant Cell Physiol.* 5:33–43
145. Tazawa, M. 1968. Motive force of the cytoplasmic streaming in *Nitella. Protoplasma* 65:207–22
146. Tazawa, M., Kikuyama, M., Shimmen, T. 1976. Electric characteristics and cytoplasmic streaming of Characeae cells lacking tonoplast. *Cell Struct. Funct.* 1:165–76
147. Tazawa, M., Kishimoto, U. 1968. Cessation of cytoplasmic streaming of *Chara* internodes during action potential. *Plant Cell Physiol.* 9:361–68
148. Teplov, V. A., Budnitsky, A. A. 1973. Device for registration of mechanical activity of plasmodium strands (Russian). In *Biophysics of Living Cell,* Chief ed. G. M. Frank, ed. V. N. Karnaukhov, 4:165–70. Pushchino. 207 pp.
149. Teplov, V. A., Matveeva, N. B., Zinchenko, V. P. 1973. Free calcium level oscillation in myxomycete plasmodium during protoplasm shuttle-moving (Russian). See Ref. 148, pp. 110–15
150. Thornburg, W. 1967. Mechanisms of biological motility. *Theor. Exp. Biophys.* 1:77–127
151. Tonomura, Y., Oosawa, F. 1972. Molecular mechanism of contraction. *Ann. Rev. Biophys. Bioeng.* 1:159–90
152. Ueda, T., Götz von Olenhusen, K. 1978. Replacement of endoplasm with artificial media in plasmodial strands of *Physarum polycephalum. Exp. Cell Res.* 116:55–62
153. Ueda, T., Kobatake, Y. 1977. Changes in membrane potential, zeta potential and chemotaxis of *Physarum polyceph-*

alum in response to n-alcohols, n-aldehydes and n-fatty acids. *Cytobiologie* 16:16–26
154. Ueda, T., Muratsugu, M., Kurihara, K., Kobatake, Y. 1976. Chemotaxis in *Physarum polycephalum.* Effects of chemicals on isometric tension of the plasmodial strand in relation to chemotactic movement. *Exp. Cell Res.* 100:337–44
155. Ueda, T., Terayama, K., Kurihara, K., Kobatake, Y. 1975. Threshold phenomena in chemoreception and taxis in slime mold *Physarum polycephalum. J. Gen. Physiol.* 65:223–34
156. Vahey, M., Scordilis, S. P. 1980. Contractile proteins from the tomato. *Can. J. Bot.* 58:797–801
157. Vandekerckhove, J., Weber, K. 1978. The amino acid sequence of *Physarum* actin. *Nature* 276:720–21
158. Vorobyeva, I. A., Poglazov, B. F. 1963. Isolation of contractile protein from alga *Nitella flexilis. Biofizika* 8:427–29
159. Wagner, G. 1979. Actomyosin as a basic mechanism of movement in animals and plants. See Ref. 48, pp. 114–26
160. Weihing, R. R. 1976. Occurrence of microfilaments in non-muscle cells and tissues. In *Biological Data Book,* ed. P. L. Altman, D. D. Katz, pp. 340–46. Bethesda, Md: Fed. Am. Soc. Exp. Biol.
161. Weihing, R. R. 1976. Physical and chemical properties of microfilaments in non-muscle cells and tissues. See Ref. 160, pp. 346–52
162. Weihing, R. R. 1976. Biochemistry of microfilaments in cells. See Ref. 160. pp. 352–56
163. Williamson, R. E. 1972. A light-microscope study of the action of cytochalasin B on the cells and isolated cytoplasm of the Characeae. *J. Cell. Sci.* 10:811–19
164. Williamson, R. E. 1974. Actin in the alga, *Chara corallina. Nature* 248: 801–2
165. Williamson, R. E. 1975. Cytoplasmic streaming in *Chara*: A cell model activated by ATP and inhibited by cytochalasin B. *J. Cell Sci.* 17:655–68
166. Williamson, R. E. 1980. Actin in motile and other processes in plant cells. *Can. J. Bot.* 58:766–72
167. Williamson, R. E., Toh, B. H. 1979. Motile models of plant cells and the immunofluorescent localization of actin in a motile *Chara* cell model. See Ref. 36, pp. 339–46
168. Wohlfarth-Bottermann, K. E. 1962. Weitreichende, fibrilläre Protoplasmadifferenzierungen und ihre Bedeutung für die Protoplasmaströmung. I. Elektronenmikroskopischer Nachweis und Feinstructur. *Protoplasma* 54: 514–39
169. Wohlfarth-Bottermann, K. E. 1963. Weitreichende, fibrilläre Protoplasmadifferenzierungen und ihre Bedeutung für die Protoplasmaströmung. II. Lichtmikroskopische Darstellung. *Protoplasma* 57:747–61
170. Wohlfarth-Bottermann, K. E. 1964. Differentiations of the ground cytoplasm and their significance for the generation of the motive force of amoeboid movement. See Ref. 11, pp. 79–109
171. Wohlfarth-Bottermann, K. E. 1975. Tensiometric demonstration of endogenous, oscillating contractions in plasmodia of *Physarum polycephalum. Pflanzenphysiol.* 76:14–27
172. Wohlfarth-Bottermann, K. E. 1979. Oscillatory contraction activity in *Physarum. J. Exp. Biol.* 81:15–32
173. Wohlfarth-Bottermann, K. E., Fleischer, M. 1976. Cycling aggregation patterns of cytoplasmic F-actin coordinated with oscillating tension force generation. *Cell Tissue Res.* 165:327–44
174. Wohlfarth-Bottermann, K. E., Götz von Olenhusen, K. 1977. Oscillating contractions in protoplasmic strands of *Physarum:* Effects of external Ca^{++}-depletion and Ca^{++}-antagonistic drugs on intrinsic contraction automaticity. *Cell Biol. Int. Rep.* 1:239–47
175. Yoshimoto, Y., Kamiya, N. 1978. Studies on contraction rhythm of the plasmodial strand. I. Synchronization of local rhythms. *Protoplasma* 95:89–99
176. Yoshimoto, Y., Kamiya, N. 1978. Studies on contraction rhythm of the plasmodial strand. II. Effect of externally applied forces. *Protoplasma* 95:101–9
177. Yoshimoto, Y., Kamiya, N. 1978. Studies on contraction rhythm of the plasmodial strand. III. Role of endoplasmic streaming in synchronization of local rhythms. *Protoplasma* 95:111–21
178. Yoshimoto, Y., Matsumura, F., Kamiya, N. 1980. *ATP and calcium rhythm in Physarum plasmodium.* Presented at Ann. Meet. Jpn. Soc. Cell Biol., Tokyo

Ann. Rev. Plant Physiol. 1981. 32:237–66

PLANT PROTOPLASTS AS PHYSIOLOGICAL TOOLS

❖7711

Esra Galun

Department of Plant Genetics, The Weizmann Institute of Science, Rehovot, Israel

---

CONTENTS

0066-4294/81/0601-0237$01.00

INTRODUCTION

The term protoplast is commonly used to describe a cell devoid of its cell wall. Plant protoplasts are obtained either from intact tissues such as root tips, coleoptiles, epidermis, and leaf mesophyll, or from calli and suspension cultures, and are isolated by either mechanical or enzymatic means (150). Irrespective of source, method of isolation, and future use, the cells and subsequently the protoplasts are maintained in concentrated solutions, usually of sugar, to which the plasmalemma is at least partially impermeable and which cause partial dehydration of the protoplasts. It should be noted that osmotic stress, removal of cell wall, and new solute environment are not the only hardships encountered during isolation; the protoplasts are also removed from their original cell environment, the plasmodesmata are disconnected, and important proteins situated between the plasma membrane and the wall may be removed with the cell wall. Moreover, the cell may be severely damaged by enzyme contaminants in the preparations employed for wall degradation because these enzymes are rarely pure pectinases and cellulases. The question of whether or not protoplasts can be regarded as plant cells without a wall has therefore been raised repeatedly (see 17). This question has great relevance when protoplasts are to be used as physiological tools, but it is less meaningful for studies in which protoplasts are cultured and used for plant regeneration, somatic hybridization, uptake of informative molecules, and virus research. These latter uses of protoplasts are outside the scope of our present discussion and have been thoroughly documented in recent years (7, 33, 46, 52, 131, 144, 145, 149, 150). With due caution, as will be indicated below, protoplasts comprise an attractive tool to study several aspects of plant cell biochemistry and physiology.

This review will first evaluate structural and physiological aspects of the transition from cell to protoplast, pointing toward the question of whether protoplasts are useful to study the process of dedifferentiation in plant cells. The existing methodologies to isolate cell components via protoplasts will also be reviewed briefly, focusing on those components for which protoplasts seem to serve as a better source than cells. I shall then look at the metabolism of protoplasts derived from highly differentiated cells, e.g. photosynthetic tissue and stomatal complexes attempting to evaluate the usefulness of protoplasts to better understand the in situ metabolism of these cells.

The exposed plasmalemma prompted many investigators to use protoplasts for studies on the interactions of this membrane with ions and lectins as well as on the onset of cell wall formation. In surveying these studies we shall look at correlations between cell wall formation, cytokinesis, and karyokinesis. I shall finally refer to a number of investigations in which protoplasts were used to study the mode of action of auxins and pesticides, trying to evaluate the advantages and pitfalls related to the use of protoplasts rather than whole cells.

Protoplast studies were thoroughly reviewed in this series in 1972 (29); therefore the present review is based almost exclusively on literature of the recent decade.

TRANSITION FROM CELL TO PROTOPLAST

Leaf mesophyll served as the cell source for the first "protoplast system" in which in vitro cultured protoplasts resulted in functional plants (137). The majority of the subsequently reported "protoplast systems" (150) were derived from either mesophyll cells or other fully differentiated tissues, while in only relatively few such systems do in vitro cultured cells serve as a source for protoplasts. Obviously, this abrupt transition from fully differentiated to a dedifferentiated state poses an outstandingly interesting phenomenon. This review will therefore look at the available information concerning the ultrastructure and biochemistry of this phenomenon, focusing attention on processes which may be related to changes in the expression of genetic information.

Ultrastructure

A considerable number of authors investigated the ultrastructure of isolated protoplasts and their transition into dividing cells [(41, 42, 53, 56, 64, 80, 138, 142, 154); see also previous review (29)]. In addition to the complete removal of cell walls, conspicuous changes in the chloroplasts were reported repeatedly. Pseudocrystallinic arrays were observed as well as spherical, multimembrane inclusions and other malformations. At least some of these changes were attributed to the osmoticum used during isolation rather than to the effect of maceration enzymes (54, 143).

Unfortunately, the above mentioned studies paid little attention to structural alterations which may be related to the abrupt change in genetic expression, e.g. disappearance of polysomes or reduction in ribosome density in either the cytoplasm or the organelles. There are only occasional notes that the number of ribosomes is drastically reduced at or soon after protoplast isolation (e.g. 53). With one notable exception (43), it is generally rather clear from the micrographs that upon further culture, the cytoplasm which is relatively clear at isolation is later filled with polysomes and other cytoplasmic constituents (54). Are we faced with a situation of erasing existing transcription and translation of genetic products in preparation for a completely different cell function? It should be noted that such a possibility was suggested in respect to other cases of abrupt changes in function, e.g. during microsporogenesis (66) and in androgenesis (136). Investigations specifically aimed at this question may furnish an appropriate answer. Nevertheless, such a possibility is supported by studies on changes in RNase and polysome profiles which will be mentioned below.

Hydrolases

Early information on protoplast hydrolases resulted from studies which were not aimed to study those degradative processes in protoplasts which are involved with the change in gene expression. Thus, Pilet et al (104) studied RNase levels in onion root tips and found no significant differences between mechanically and enzymatically isolated protoplasts. The RNase levels in protoplasts were also similar to those of equivalent intact root tissue, increasing basipetally. Galston and his collaborators studied nucleases of oat protoplasts in reference to an attempt to extend protoplast vitality and eventually enable the latter's division in vitro (2, 44, 47, 74, 75). They found that the nucleases were substantially higher in protoplasts than in the tissue from which they were derived and that RNase levels increased during incubation in culture medium. The level of these nucleases could be reduced, but not eliminated, by several antisenescence compounds, especially by polyamines. Because the oat protoplasts used in the above studies did not divide even after the appropriate antisenescence treatment (although some cell division was recently observed; A. W. Galston, personal communication), the correlation between increase of nucleases and reduction in vitality was not established unequivocally. Moreover, studies which will reveal which types of cellular RNAs and DNAs are actually degraded in situ are desirable.

Farkas and his collaborators studied hydrolases in tobacco protoplasts, a system which readily results in cell division and colony formation. Lazar et al (83) found a 12- to 15-fold increase in RNase during protoplast incubation (which did not prevent subsequent division). Interestingly, the RNases of the source tissue and the protoplasts, respectively, were qualitatively different. In conformity with studies on onion protoplasts (104), RNases in tobacco protoplasts also seem to be induced by exposure to high osmoticum rather than by the maceration enzymes (105). The level of RNase activity in tobacco protoplasts could be reduced by cycloheximide and by kinetin, and the authors concluded that the increase of RNase is based on stress-induced synthesis of new protein (105, 106). Direct evidence on the destruction of DNA in cultured protoplasts was reported only in callus-derived protoplasts (108), but such a degradation could be the norm during early cultivation of protoplasts. Are hydrolases compartmented in lysosomes and/or in the vacuole and released into the protoplast cytosol upon isolation? Homogenization of whole protoplasts will not reveal such a mechanism. The approach of Wagner & Siegelman (153) and Butcher et al (25), who developed techniques to clearly separate vacuoles from protoplast and hence to obtain protoplast "sap" and "cytosol," may be useful to answer this question. These authors found, by analyzing flower petal protoplasts, that the cytosol contained several hydrolases such as esterase, carboxypeptidase, β-galactosidase, α-glycosidase, and β-glucoronidase which could not be detected in the vacuole, while acid phosphatase, RNase, and DNase were detected in both the vacuole sap and in the cytosol. The publication of Butcher et al (25) on lysosomal and vacuolar compartmented hydrolases and their possible release during developmental changes is recommended as a good introduction to this question.

RNA, Protein, and DNA Synthesis

Soon after Nagata & Takebe (95) showed that isolated tobacco leaf mesophyll protoplasts have the capability to divide in vitro, Sakai & Takebe (121) reported that these protoplasts incorporated precursors into acid-insoluble fractions of RNA and protein. Several subsequent studies were aimed at investigating further the capacity of isolated protoplasts to synthesize RNA and protein. Thus Wasilewska & Kleczkowski (155) used protoplasts from maize seedling shoot tips to follow RNA synthesis, and reported that gibberellic acid promoted this synthesis and increased the poly $(A)^+$ RNA fraction. As their protoplasts did not show cell division, the incorporation of uridine was terminated after 2 to 3 hours. Using cotyledon and leaf cucumber protoplasts, Coutts et al (31) reported that uridine was incorporated into ribosomal RNA and that this incorporation was rather low at first but increased gradually up to 1 or 2 days. Such an increase of precursors' incorporation was also indicated for tobacco mesophyll protoplasts (113). Further examination of this and other protoplast systems indicated that the density of the protoplast suspension during culture strongly affects RNA and protein precursor incorporation (117), that lectins improve leucine, uridine, and thymidine incorporation (28), and that although the incorporation of these precursors increases steadily during early culture, it is strongly reduced by osmotic stress (106). An osmotic-stress inhibition of leucine uptake was reported in another system, *Convolvulus* protoplasts (119), but it should be noted that the latter protoplasts are unable to divide.

I mentioned above that chloroplasts are strongly damaged upon protoplast isolation. The biosynthetic capacity of chloroplasts in cultured protoplasts was therefore investigated. Fleck et al (40) reported that the biosynthesis of the large subunit of Fraction I protein, which is chloroplast coded, stopped almost completely after protoplast isolation. Different results were obtained in a similar system by Hirai & Wildman (67), indicating that rRNA and mRNA are indeed synthesized in chloroplasts of cultured protoplasts and that ^{35}S-methionine incorporation into whole protoplasts was reduced to 70 or 30% of control when the protoplasts were exposed to chloramphenicol or cycloheximide, respectively. Moreover, ^{35}S-methionine was found in both the large and the small subunit of Fraction I protein. Evidence for the labeling of the large subunit of Fraction I, following exposure of protoplasts to protein precursors, was also reported by Zelcer (163); but it was observed that when protoplasts were cultured for 0, 20, or 40 hr, the labeling of the large subunit was gradually reduced.

In an attempt to compare RNA synthesis in isolated protoplasts to that of intact cells, Kulikowski & Mascarenhas (79) used *Centaurea* cell cultures and protoplasts. They found that protoplasts differed from cells in several respects as to poly $(A)^+$ RNA profiles, levels of rRNA synthesis, and rRNA processing, but protoplasts were similar to cells in the overall rate of RNA synthesis. Unfortunately, in none of the above mentioned studies was the rate of protein and DNA precursor incorporation accompanied by information on the cell division cycle. Actually, in most cases it is not clear whether the experimental procedure brought about any cell division. Obviously, if we are faced with a system which is actually losing its vitality during the experimental procedure, the results are of limited usefulness.

Cereals are some of our most important crops. In spite of efforts by many investigators—with one recent exception (151)—none of the cereals furnished a "protoplast system" (144). The studies of Galston and his collaborators (2, 44, 47, 48, 76, 77, 135) with oat protoplasts should therefore represent mainly efforts to improve oat protoplast vitality by following their capacity to synthesize macromolecules under different culture conditions, rather than to clarify the patterns of protein, RNA, and DNA synthesis during transition from a differentiated cell into a dividing protoplast. Nevertheless, some of the findings with oat mesophyll protoplasts should be noted. Polyamines, and to some extent kinetin, were found to increase the rate and extent of precursor incorporation into protein, RNA, and DNA. Pretreatment of leaves with cycloheximide (up to 10 $\mu g/ml$) improved protoplast yield, and low cycloheximide concentrations (0.5–1.0 $\mu g/ml$) improved leucine and uridine incorporation into the TCA precipitable fraction. There is yet no information on the effect of polyamines on fully viable protoplasts, such as tobacco mesophyll protoplasts, which are capable of readily entering into sustained cell division.

Zelcer & Galun (163, 165, 166) investigated the kinetics of protein, RNA, and DNA precursor incorporation into tobacco mesophyll protoplasts during the first few days of in vitro culture. Culture methodologies were amended to result in 70% cell division after 70 and 35 hours when incubated at 27 and 32°C, respectively. Increase in both ^{14}C-leucine and ^{3}H-uridine incorporation began at commencement of culture—or soon afterwards—while DNase sensitive/RNase resistant ^{14}C-thymidine incorporation started after a lag of several hours. Moreover, the first peak of thymidine incorporation preceded the rise in percent of divided cells by a few hours. A similar pattern was reported by Gigot et al (54) with respect to the incorporation of protein and RNA precursors. These authors also reported that at 48 hr most of the RNA synthesis was ribosomal. The only direct evidence of DNA increase, as well as degradation, in cultured protoplasts was provided by Radojevic & Kovoor (108), who used *Corylus avellana* protoplasts from callus cultures.

Ruzicska et al (120) followed the changes in polysomal profiles in protoplasts during culture and compared them to polysomal profiles of comparable intact tissue. They reported that right after isolation there is a drastic reduction of polysomes relative to intact cells, and that with the onset of cell division this profile changes back to normal high-polysomal profiles. In spite of the existing RNAses in isolated protoplasts, these could not account solely for all the disappearance of polysomes since exposure of leaves to high osmoticum induces similar RNase levels in their cells without affecting the polysomal profiles. These authors concluded: "This process (dedifferentiation) is certainly associated with a change in the RNA pattern and/or amount. The marked accumulation of ribosomal subunits in protoplasts might be associated with a decreased rate of protein initiation due to a lower amount of mRNA available." This statement is fully accepted as a working hypothesis by this reviewer.

ISOLATION OF CELL COMPONENTS

The danger of erroneous interpretation of protoplast metabolism as truly representing intact cell processes is hardly encountered when the aim is merely the isolation of cell components. The rationale of isolating such components as plasmalemma, nuclei, and chloroplasts from protoplasts, rather than from cells or intact tissues, is in most cases to obtain them by convenient methods which will cause the least destruction. Therefore, if the functionality of these components is at stake, the evaluation of the rate and quality of damage (e.g. in nuclei) as well as their unquestionable identification and purity (e.g. of plasmalemmae) are important considerations. Most of the problems involved in the isolation of cell components from protoplasts are specific to plant cells because of the latter's rigid cell wall and their specific organelles, e.g. chloroplasts and vacuoles. These last are very rare in animal cells but common to all but a few plant cell types.

Plasmalemma

Hall & Taylor (63) have recently reviewed the methods for plasmalemma isolation from higher plants. They noted that this isolation is hampered by two major factors: that the high shear force used for cell homogenization will damage the plasmalemma and that good markers are needed to further identify the plasmalemma fraction. The high shear force can be eliminated by the use of protoplasts. Hence, protoplasts obtained from either leaf mesophyll (143) or from cell suspensions (11) were isolated by enzyme treatment. Plant protoplasts bind concanavalin A (con A) (57); therefore this lectin can be used as a plasmalemma marker (159). Hence, binding of ^{14}C-acetyl-con A before homogenization and subsequent separation on discontinuous (10 to 38%) or continuous (5 to 50%) Renographin gradients resulted in labeled fractions which peaked at about 1.14 $g \cdot cm^{-3}$ apparent density and had enzyme activities attributed to the plasmalemma (11). A similar labeling technique by which protoplasts were preincubated in diazotized ^{35}S-sulphanilic acid was suggested by Galbraith & Northcote (45). This method was modified recently (103) and used with maize protoplasts. Taylor & Hall (143) questioned the validity of some plasma membrane markers such as phosphotungstic acid/chromic acid and silicotungstic acid/chromic acid, since these markers may be not specific for plasmalemma, and they further indicated that lanthanum ($LaCl_3$) staining, which is rather specific to the plasma membrane, was distributed in all fractions of a sucrose gradient which, according to other markers, was supposed to isolate this membrane. They therefore suggested that further separation improvements are still required before pure plasmalemma fractions can be isolated.

Chloroplasts

The isolation of fully functional chloroplasts from leaves of most angiosperms by mechanical homogenization has usually met with difficulties. Chloroplasts were first isolated from protoplasts by Wagner & Siegelman (153) as a "by-product" of their method to isolate vacuoles. The protoplasts were ruptured osmotically by suspending them into a phosphate buffer containing Mg^{2+} and dithiothreitol. Chloroplasts were then separated on discontinuous sucrose gradients and were found to be photochemically active. Isolation of cereal chloroplasts was especially problematic; Rathnam & Edwards (109) therefore utilized protoplasts as a chloroplast source. Basically the chloroplasts were obtained by suspending the protoplasts in an appropriately buffered osmoticum (containing bovine serum albumin and dithiothreitol) and passed through a 20 μm nylon net. The chloroplasts were then collected by centrifugation without further fractionation. Several tests indicated that this procedure can produce in cereal (*Agropyron, Hordeum, Panicum,* and *Triticum*) chloroplasts with a reasonably high CO_2 fixation capacity, i.e. over 100 μmoles $CO_2 \cdot mg\ chl^{-1} \cdot hr^{-1}$, and in *Nicotiana* chloroplasts with 86 μmole $CO_2 \cdot mg\ chl^{-1} \cdot hr^{-1}$. Moreover, the protoplasts could be stored in the cold for 20 hr before chloroplast extraction without causing an appreciable loss of chloroplast activity. The isolation of functional chloroplasts from certain plants was found to be increased by specific conditions such as the addition of chelating agents (37).

A similar method was independently developed by Nishimura et al (100) for spinach leaf protoplasts, using a syringe needle to rupture the protoplasts and subsequently purifying the chloroplasts by a sucrose gradient. Various enzyme assays (e.g. ribulose-1,5-biphosphate carboxylase, NAD-triose P-dehydrogenase) were used to verify the purity of the chloroplast preparation. The sucrose gradient step presumably caused some damage to the function of the chloroplasts but resulted in a highly purified preparation. Protoplasts were also used by Nishimura & Beevers (98) to obtain intact plastids from castor bean endosperm. The protocol was basically as reported by Nishimura et al (100) for spinach leaves, but castor endosperm protoplasts were found to be more fragile than spinach leaf protoplasts. A rapid separation of chloroplasts from the rest of the cytoplasm was devised by Robinson & Walker (115). They also used the nylon screen technique to rupture the protoplasts, but placed the screen directly above a centrifuge tube which contained a silicon oil layer over a 0.4 M sucrose cushion. Thus the purified chloroplasts could be pelleted in one quick procedure. Chlorophyll and enzyme markers indicated that the separation into pellet/supernatant by this technique was 96/4 and 86/14 for chlorophyll and RUBP carboxylase, respectively (chloroplast markers), and 15/85 and 13/87 for cytochrome *c* oxidase and fumarase, respectively (mitochondrial markers). Some functional features of chloroplasts isolated from specific protoplasts will be discussed below.

Mitochondria

High yields of intact mitochondria normally can be obtained from plant tissues and cultured cells by conventional methods; short, high-speed shearing (with sharp knives) is commonly employed with leaf tissue, while disruption in a pressure cell (French press, ca 3000 psi) will give adequate results with calli and cell suspensions. Nevertheless, some of the methods mentioned above for chloroplast isolation from protoplasts may be also useful for mitochondria (98, 100), and the two types of organelles can be separated from the same sample of plant materials. When 0.05–0.1% bovine serum albumin is included and a sucrose gradient is employed for fractionation of the disrupted protoplasts, mitochondria will band quite sharply at a density of 1.18 $g \cdot cm^{-3}$, and they can be identified by their fumarase activity, while chloroplasts will peak at an appreciably higher sucrose density (1.22 $g \cdot cm^{-3}$).

Vacuoles

Vacuoles are very fragile cell components; therefore no large-scale method of isolation from plant tissue was available before the protoplast "era." Wagner & Siegelman (153) found that when protoplasts from various plant tissues are ruptured gently by transfer into a phosphate buffer, they could then be separated by a low speed centrifugation and be further purified by layering over 5% Ficol containing 0.55 M sorbitol and 1 mM tris-MES buffer. The homogeneity of the vacuole preparation was nicely documented by the use of anthocyanin-containing tissues, e.g. *Tulipa* petals. Butcher et al (25) modified the above procedure to analyze the enzymatic content of vacuoles and reported that in mature cells the vacuoles contained some RNases, DNases, and acid phosphatase, but most of the activity of these enzymes, as well as probably all the activity of β-galactosidase, β-glycosidase, or protease, resided in the "cytoplasmic fraction," where these hydrolases may be compartmented in small lysosomal-like organelles. Mettler & Leonard (88) used tobacco cell suspension protoplasts for the isolation of vacuoles and tonoplast membranes, and discussed previous methods for the isolation of these constituents (e.g. 84). Their procedures differed from previous techniques in two major points: the protoplasts were gently ruptured in a solution containing 0.3 M sorbitol and 2 mM ethylenediamine tetraacetate (EDTA) and the vacuoles were purified on a Ficol step gradient. They concluded that in tobacco suspension cultures the vast majority of acid phospatase is located in the vacuole. In a further study (5), the tonoplasts were further purified on a sucrose gradient and were found to have an apparent density of 1.12 $g \cdot cm^{-3}$ and to completely lack ATPase activity. It appears to this reviewer that the location of degrading enzymes is still an open question and further studies on this subject are required. Recently Admon & Jacoby (1) used the cytoplasmic vital fluorescent dye fluorescencein diacetate and demonstrated that cytoplasmic impurities overlay the tonoplasts of vacuoles, because the tonoplast itself was not stained by this dye. This method should be instrumental in improving vacuole purification.

Nuclei

Early experimental procedures for the isolation of nuclei from protoplasts were undertaken with the intention of using these nuclei for genetic manipulations, i.e. transferring them into alien protoplasts. Thus, Blaschek et al (8) isolated nuclei from leaf protoplasts of *Petunia hybrida, Nicotiana glauca,* and *N. langsdorffii.* The main features of their technique were lysing the protoplast with Triton X 100 and further fractionation of the crude nuclear preparation on a sucrose gradient. Nuclei (*N. glauca*) isolated by this method were able to transcribe RNA, as indicated by UTP incorporation for 30 min. The addition of actinomycin D during the assay, as well as RNase pretreatment, completely inhibited the transcription. Triton X lysis was also used by Ohyama et al (101, 102) to isolate nuclei from protoplasts derived from soybean suspension cultures. The purification technique of the latter authors differed in some respects from that of Blaschek et al (8) in their use of lower Triton X concentration, Teflon-homogenization of the lysed preparation, and sorbitol rather than sucrose during isolation and purification. The nuclear fraction thus obtained did incorporate protein—and to a lesser extent RNA precursors—for at least 2 hr, but no ^{3}H-dGTP incorporation into DNA could be detected. Gigot et al (55) isolated tobacco leaf nuclei as a step in the isolation of chromatin. The concentration of Triton X was substantially higher (5%), and nylon screen filtration (25 and 10 μm) rather than sugar gradient centrifugation was employed to obtain the nuclear fraction. This procedure resulted in a yield of approximately 70 to 80%, but no data were available on the biosynthetic capacity of these nuclei. Zuily-Fodil et al (169) recently analyzed various methods for isolation of nuclei using protoplasts from *Parthenocissus* crown gall callus. The optimal concentrations of Triton X, polyvinylpyrrolidone (PVP), and glycerol were 0.005, 6, and 20%, respectively. The application of the protectants PVP and glycerol was reported to be very beneficial and resulted in preparation with RNA polymerase activity and a 80 to 90% yield.

The use of a detergent in all the above mentioned procedures may cause substantial damage to the nuclei; in order to avoid such damage, Tallman & Reeck (140) recently devised an isolation procedure without detergents. The nuclei were obtained by suspending the protoplasts in hypertonic buffer. This relatively mild treatment resulted in nuclei which retained their double membrane and resembled nuclei of intact protoplasts as judged from transmission electron micrographs.

Chromosomes

The same gentle lysis procedures used for the isolation of vacuoles and nuclei could be useful for metaphase chromosome isolation from plant protoplasts, provided a large fraction of the protoplasts is in the appropriate cell cycle phase. A recently reported technique is based on these two considerations (85). Frequently transferred suspension cultures of tobacco and tomato were partially synchronized by a transient transfer to 2 μg·ml^{-1} fluorodeoxiuridine (in the presence of 1 μg·ml^{-1} uridine). The cultures were subsequently transferred to 2 μg·ml^{-1} thymidine and finally maintained for several hours in 0.005% colchicine. This resulted in an increase of the mitotic index from 3 to 15–35%. A similar procedure enabled the isolation of meiocytes of lily and daylily chromosomes. Although the authors provided evidence that the isolated fraction constitutes chromosomes, the vitality of the chromosomes, i.e. their ability to replicate DNA and transcribe RNA and hence being of use for genetic manipulations, has still to be verified.

METABOLISM OF PROTOPLASTS FROM SPECIFIC TISSUES

The usefulness of protoplast techniques for physiological investigations is especially evident in studies with cells which have specific functions and are surrounded by other cells which may interfere with these studies. I shall therefore shortly review the use of protoplasts in physiological studies of photosynthetically active leaf cells, of stomatal guard cells, and of cells from legume root nodules. I intend to focus on the utilization of protoplasts rather than providing full physiological accounts of detailed studies. The reader is thus referred to the cited literature for further information.

Leaf Protoplasts

Sunflower leaves were the source of protoplasts in an early attempt to follow carbon dioxide fixation in protoplasts (156). Yields were rather low, ca 6 μmole $CO_2 \cdot$lmg $chl^{-1} \cdot hr^{-1}$, i.e. less than 10% of intact cells. Since these protoplasts were unable to divide in culture, it is not clear whether the low yield was due to protoplast damage during isolation or to other deficiencies in this system. Much higher yields of CO_2 fixation were obtained with spinach leaf protoplasts (97, 99), although as with sunflower protoplasts, they were also unable to divide in culture. The studies with spinach protoplasts showed that this system has several features: light-dependent O_2 evolution, a rather high apparent "K_m (CO_2)" comparable to the values found in intact leaves, and a strong dependence of the quantity and quality of assimilates on the $CO_2/O_2/N_2$ ratio in the atmosphere above the reaction mixture. It should be noted that spinach is a favorite plant for photosynthesis studies, and functional chloroplasts can be obtained directly from its leaves by mechanical means.

In preparation of a research system for in vitro studies of C_4 photosynthesis (e.g. in plants having a "Kranz" leaf anatomy such as *Pancium* spp. and *Sorghum bicolor*) and C_3 photosynthesis [e.g. in plants having a "not-Kranz" anatomy, such as barley, wheat, and spinach; see Hatch & Osmond (65) for details and nomenclature], Kanai & Edwards (71–73) developed techniques for the isolation of photosynthetically active mesophyll protoplasts of C_3 plants and for the separation of mesophyll protoplasts from bundle sheath strands of C_4 plants belonging to the Gramineae. In further studies (e.g. 69, 70, 78) reviewed by Edwards & Huber (34, 35), Edwards and his collaborators revealed several physiological features of these protoplasts and cells concerning compartmentation of enzymatic activities and metabolic regulations. Thus it was shown that the carboxylation phase of the C_4 pathway is located in mesophyll protoplasts while the decarboxylative phase of this pathway as well as the carboxylative phase of the Calvin-Benson pathway is located in the bundle sheath cells. Ribulose-1,5-bisphosphate carboxylase was found only in bundle sheath cells and not in mesophyll protoplasts. By further isolation of chloroplasts from leaf protoplasts, intracellular compartmentation was also studied. These studies indicated that PEP (phosphoenolpyruvate) carboxykinase is restricted to the cytosol while NADP-triose P-dehydrogenase is located in the chloroplasts. Furthermore, by adjusting the reaction mixture to include the appropriate precursors and metabolic inhibition, cyclic, pseudocyclic, and noncyclic photophosphorylation of C_4 mesophyll protoplasts was measured. Slight differences in density allowed the separation of bundle sheath protoplasts from mesophyll protoplasts of C_4 plants, and assays in the presence of various metabolites indicated subtle differences in enzyme activities and transport mechanisms between mesophyll and bundle sheath derived chloroplasts (36). Based on these and other studies, a scheme of enzyme interactions and translocator mechanisms across the chloroplast membrane for photosynthesis in the C_4 mesophyll cell was derived. With respect to the carboxylation and decarboxylation phases, three groups of C_4 plants can be recognized: NADP-malic enzyme, NAD-malic enzyme, and PEP carboxykinase enzyme—species represented by *Digitaria sanguinalis, Erichloa borumensis,* and *Panicum miliaceum,* respectively. Studies with mesophyll protoplasts and bundle sheath cells of various C_4 plant species could furnish additional information on these enzymatic differences. The C_4 plants were represented only by monocots since functional mesophyll protoplasts could not be derived from C_4 dicots.

One of the deficiencies of earlier studies with isolated protoplasts and chloroplasts was probably inappropriate assay conditions. In a study with isolated wheat protoplasts and their nonparticulate and chloroplast components, Edwards et al (37, 38) adjusted the assay condition to obtain maximum photosynthetic yields. The presence of a chelating agent (EDTA) and a rather narrow range of P_i (orthophosphate) was required by chloroplasts. In intact tissue but not in the in vitro system, the recycling of P_i probably provides the chloroplasts with adequate P_i levels.

While protoplasts could be assayed in the presence of an appreciable PP_i (pyrophosphate) concentration, it inhibited photosynthesis of chloroplasts from young (6 days) wheat seedlings and to a lesser extent also of chloroplasts from older seedlings. This inhibition was reversible by ADP or ATP. These and other findings indicated that wheat chloroplasts are much more sensitive to phosphate levels than the traditional tools for studies of photosynthesis: spinach and pea chloroplasts. The methodological lesson from these studies is that assays should be carefully adapted to each system since decompartmentation may not have the same effect in different species.

The endogenous compartmentation of a CAM (Crassulacean acid metabolism) plant, *Sedum praealtum*, was also studied by the use of isolated protoplasts, and key enzymes involved in carbon assimilation of this CAM plant were assigned to organelles and nonparticulate cell components (134). Chloroplasts isolated from *S. praealtum* had a relatively low photosynthetic yield, 20–30 μmole $CO_2 \cdot mg\ chl^{-1} \cdot hr^{-1}$, but this level could be increased by the addition of Calvin cycle intermediates, and the lag in CO_2 fixation could be shortened by the addition of dihydroxyacetone phosphate or ribose-5-phosphate (133). Interactions between chloroplasts and nonparticulate cytoplasm was also studied in spinach, a traditional model for C_3 photosynthesis (110). The chloroplasts in this study were derived from protoplasts which had a rather high CO_2 fixation yield.

A very rapid procedure to isolate chloroplasts from protoplasts (114) was instrumental in following the location of labeled sucrose and earlier metabolites after short $NaH^{14}CO_3$ labeling of protoplasts. Studies with wheat and spinach protoplasts in which the above procedure was implemented (116) indicated that soon after the exposure of protoplasts to light, the label appeared in the chloroplasts, but it leveled off after a few minutes and accumulated in the cytoplasm. This suggests a rapid transfer of sugar phosphates out of the chloroplasts. The existence of such a mechanism, i.e. terminal steps of sucrose formation in the cytoplasm rather than in the chloroplasts, is supported by a recent investigation with other plants (146).

Leaf protoplasts were recently used to analyze the enhanced photorespiration of the tobacco *su* mutation. The protoplasts were derived from either wild type tobacco or *Su/su* plants. When protoplasts of the latter plant type were exposed to light in the absence of bicarbonate, they had a severalfold higher oxygen consumption than wild type protoplasts, as indicated by ^{18}O uptake. This high level of oxygen consumption seemed to be closely related to photosystem I (124). The delayed light emission and fluorescence induction characteristic of photosystem II were employed to look into differences in the activity of the two photosystems between chloroplasts from mesophyll protoplasts and bundle sheath cells of maize (59). By varying the assay system, results were obtained which support the notion that bundle sheath cells of maize as well as other malic acid C_4 plants have an inactive photosystem II.

The various studies discussed above obviously cannot be summarized presently in a coherent manner, but they clearly demonstrate the manifold utilizations of protoplasts and organelles derived from protoplasts in recent photosynthesis investigations.

Stomatal Guard Cell Protoplasts

In order to furnish guard cell protoplasts (GCP) for future physiological studies, Zeiger & Hepler (161) devised a microchamber in which the enzymatic digestion of the cell walls could be monitored. Gentle enzymatic procedures were developed to obtain tobacco and onion GCP which retained their vitality for several days. Using such onion GCP, these authors (162) showed that the blue light effect—causing stomatal opening—can be followed at the cellular (GCP) level: GCP but not leaf epidermal protoplasts were stimulated to swell by blue light. A blue light photoreceptor attached to GCP plasmalemma was thus suggested. Guard cell swelling and contraction (i.e. opening and closing the stomata) which is not directly light induced was also studied with GCP. Abscisic acid (ABA) and fusicoccin (FC) cause contraction and swelling, respectively, of onion and broadbean (*Vicia faba*) stomata. The same effect, but with faster kinetics, was found in the respective GCP (128). Using this system, Schnabl (125) could further analyze the physiology of guard cells. A clear difference, with respect to chloride ion interference with ABA, FC, and K^+ ion function, was detected between GCP from starch-containing (broadbean) and starch-deficient (onion) guard cells. Only the latter were chlorine sensitive.

Recent extensive studies by Schnabl and her collaborators (126, 127, 129, 130, 168) were performed with the starch-containing and starch-lacking GCPs of broadbeans and onion, respectively. They devised methodologies to conduct experiments with single GCPs and studied the ultrastructure of the guard cell plasmalemma. A hexagonal array of particles was revealed in the plasmalemma which might be involved in the transport of metabolites across the membrane. Of special interest is the finding that GCP of neither broadbeans nor onion incorporate $^{14}CO_2$ into phosphorylated metabolites although in broadbeans the starch was ultimately labeled. Onion GCP did not incorporate the label unless PEP was added. Thus there is some similarity between the mesophyll and guard cell duality and the bundle sheath and mesophyll duality in C_4 plants. The above studies led to an overall scheme of the swelling of starch-containing and starch-lacking guard cells. Basically it was assumed that in starch-containing guard cells the cycle malate-PEP-starch-PEP is the driving force of volume changes and the change is induced by an influx of K^+, while malate contributes H^+ for the exchange of K^+. For the starch-lacking guard cells it was suggested that malate is imported from mesophyll or epidermal cells. Its role is probably indirect, providing H^+ to stimulate the proton pump and Cl^- to trap the H^+ released outside of the guard cell. This scheme did not incorporate the blue light effect which stimulates guard cell swelling; hence, for a comprehensive appreciation of stomatal mechanisms, more studies are required.

Legume Root Nodule Protoplasts

Legume root nodules, which are the result of a very intimate legume-rhizobia symbiosis, comprise the site of N_2 fixation. Thus much attention has been focused on these tissues by plant physiologists. Early work with legume root nodule protoplasts (32) did not demonstrate in vitro nitrogenase activity. A modified isolation procedure was employed by Broughton et al (16) to obtain *Vigna anguiculata* nodule protoplasts. By measuring acetylene reduction, these authors report that such protoplasts, freed from external rhizobia, start to have nitrogenase activity after one day in culture, reaching a maximum after 9 days. The viability of these protoplasts was not evaluated. The possibility that this nitrogenase activity was due to rhizobia temporarily entrapped in nonviable protoplasts can therefore not be ruled out. Another report (122) also dealt with nitrogenase activity in isolated root nodule protoplasts (soybean). Anaerobic conditions were employed for protoplast isolation to ensure nitrogenase activity, and the protoplasts were continuously washed to free them from rhizobia set free by the bursting of some of the cultured protoplasts. It was thus suggested that most of the detected nitrogenase activity resulted from the protoplasts rather than from the freed rhizobia. An attempt to fuse such nodule protoplasts with regular soybean suspension-cultured protoplasts revealed fusion products, but these could not be followed further. Obviously, future work along this line would be profitable only with legume protoplasts which can undergo cell division and regeneration in culture, e.g. from *Medicago sativa.*

PLASMALEMMA, CELL WALL FORMATION, AND CELL DIVISION

While protoplasts have certain advantages, they are not obligatory intermediates for the isolation and study of many cell components. Studies on plasmalemma character, however, could be performed only after methods to isolate protoplasts became available. Hence valuable information on the plasmalemma was accumulated during the first decade which followed the mass isolation of protoplasts. These studies were reviewed comprehensively by Cocking (29), and I shall consider only studies of the last decade. Studies on wall formation and its relation to cell division could be initiated only after Nagata & Takebe (95) showed that protoplasts have the capability of wall formation and cell division. These investigations have been conducted in recent years and will be discussed below.

Plasmalemma

From structural studies by either freeze-etching (e.g. 61, 128) or by transmission electron microscopy (e.g. 24), the plasmalemma emerges as a typical biological bilayer membrane containing randomly distributed particles of about 10 nm in diameter having the ability to "stretch" locally (e.g. during endocytosis) and to change its surface area (e.g. resulting from changes in osmotic pressure).

MEMBRANE POTENTIAL The membrane potential of protoplasts was estimated by several methods. Microelectrode techniques used with a number of leaf and cell suspension protoplasts (107) indicated inside positive potentials of +10 to +15 mV. Using a lipophilic cation whose concentration inside and outside, after equilibrium, is evaluated by the Nernst equation, Rubinstein (118) estimated the membrane potential of oat protoplasts to be negative (60mV) relative to the outside solution. Recently the microelectode method to evaluate membrane potential was employed (14) with several technical improvements, and the results of Rubinstein (118) were confirmed. Grout et al (61) Nagata & Melchers (94) estimated membrane potential by microelectrophoresis. In this method the values are expressed as zeta potentials (ζ) and based on the equation:

$$u_E = \zeta D\, f(ka)/6\pi\eta$$

where u_E is mobility ($m \cdot sec^{-1} \cdot Vcm^{-1}$); D is dialectric constant; η is viscosity (centipoise) and $f(ka)$ is Henry's constant. Mobility measurements indicated zeta values of –10 to –35 mV, i.e. negative relative to their external medium. This potential could be changed to less negative by various additions to the medium such as $CaCl_2$ and positively charged polymers (e.g. poly-L-lysine), but not with α-neuraminidase which is known to change the charge of mammalian cells. Pronase treatment, which probably removes proteins from the membrane, increased the negative surface charge. The negative surface charge is probably the cause of repulsion between protoplasts which is reversed by polyethyleneglycol during induced aggregation (and subsequent fusion) of protoplasts. Measurements of changes in electrophoretic mobility were also employed recently to obtain information on plasmalemma composition. Thus the application of sodium dodecyl sulphate caused mobility changes in beet leaf protoplasts but not of tobacco leaf protoplasts, indicating that only the former had surface lipids (39).

LECTINS The notion that plant cells should have recognition capacities functionally similar to those shown to exist in animal cells is based on several obvious considerations. Carbohydrates or complexes containing carbohydrates are reasonable candidates as recognition determinants. Hence a search was initiated for recognition sites by several laboratories, and lectins were used to identify them. There is a conceptual difficulty in locating recognition sites on the plasmalemma, since the latter is normally covered by a cell wall. Larkin (82) furnished several arguments for considering the plasmalemma a reasonable candidate for recognition site carrier.

Agglutination of plant protoplasts by concanavalin A (con A) was first used by Gimelius et al (57) as a means to monitor enzyme isolation of carrot protoplasts. Using peroxidase staining after con A binding, combined with electron microscopy, these authors (58) studied the binding sites of con A on the plasmalemma and explained several characteristics of the membrane, e. g. its fluidity. Similar results were obtained by Williamson et al (159) using soybean protoplasts and haemocyanin as the con A probe. Further studies, with improved methods, suggested (158) that con A binds to the plasmalemma itself rather than to nascent wall sugars. This latter possibility was raised by Burgess & Linstead (19), who found that protoplasts which quickly regenerate cell walls (e.g. tobacco) bind to con A better than protoplasts which are very slow in wall formation (e.g. vine). In a further study (20), con A attachment to tobacco protoplasts was investigated by scanning electron microscopy of gold-labeled con A, and, with some reservations, about 10^8 binding site per plasmalemma were estimated, i. e. neglecting the previously mentioned nascent wall binding and suggesting that the plasmalemma has "fluid mosaic" (132) characteristics.

Con A binds to several sugar moities; it has therefore obvious limitations as a discriminating tool for mapping sugars on membrane surfaces or looking into possible differences between plasmalemmata. Larkin (81) employed several additional lectins to agglutinate protoplasts from 11 species (monocots and dicots) and still found that the lectins failed to differentiate between these species. Similar results were obtained by Chin & Scott (28). We are thus still left with two possibilities: either the binding sites of angiosperm plasmalemmata are rather similar if not identical, or the present lectin agglutination methods fail to disclose existing differences. Indications on similarity emerge from studies of Larkin [(82); see also review in (81)], who studied plant plasmalemma lectins ("β-lectins") and used antigens with defined externally exposed sugar determinants (Yariv antibodies) to probe the binding capacities of those lectins. Again, no differences could be found among a wide range of protoplasts. Nevertheless, the mere existence of such β-lectins is of significant interest in the recognition process in plants (82).

APPLICATION IN PHYSIOLOGICAL STUDIES Protoplasts are favorable tools to study transport physiology as well as other physiological entities of the plasmalemma, provided there is reasonable assurance that this membrane was not altered with respect to the investigated phenomenon during protoplast isolation. This requirement was probably satisfied for sugar and ion transport across the plasmalemma of peas and *Nicotiana glutinosa*, respectively (62, 86, 87). The mechanical integrity of the plasmalemma and changes in its fluidity were utilized to estimate cell damage. Thus by following the bursting of radish root protoplasts, Cailloux et al (26) found a positive correlation between bursting and ozone concentrations, in the range of 0.2 to 2.0 ppm, and the protecting action of sugars against freeze-thaw rupture of the plasmalemma was reported (157). Membrane fluidity changes evaluated from microviscosity can be estimated rather swiftly, provided appropriate instrumentation is available. This approach was utilized in two independent studies on horticultural and agronomic problems. In the first study (9, 10) it was observed that the microviscosity of plasmalemmae gradually increased with the age of the rose petals from which the protoplasts were obtained. Using a similar methodology, Vigh et al (152) found that those wheat varieties which showed a gradual increase in membrane fluidity during the hardening program constituted the more winter-hardy ones. Microviscosity was also increased by raising the temperature and Ca^{2+} concentration or lowering the pH. Results of several tests suggested that the increase in viscosity resulted from an increase of the sterol to phospholipid ratio.

Cell Wall Formation

STRUCTURE AND BIOCHEMISTRY The time course of initial cell wall formation was unclear until recent years. Conflicting reports set the initiation of cellulose fiber formation at various times from 10 min after isolation (160) to 72 hr (18) or even later. When the protoplasts used for studying wall formation were unable to divide in culture [e.g. *Convolvulus arvensis*, see (68)], the assumption had to be made that budding indicates preparation for cell division and wall formation. Since cytokinesis seems correlated with wall formation—as will be indicated below—studies on wall formation in these systems should be more meaningful. Moreover, early studies were probably technically deficient; thus the same authors had to correct their earlier reported times of cell wall initiation (e.g. 18 vs 22). After these deficiencies were eliminated, Burgess et al (22, 23) could study the time course of wall formation in several species. In tobacco, fibers appeared after 10 hr of culture, as observed by scanning electron microscopy. These fibers could be removed by cellulase and were most probably cellulosic. When the protoplasts were freed from the enzyme and recultured, fibers reappeared without any lag. A few hours after initial fiber formation, a mat of fibers covered the protoplasts' surface, while transfer to unfavorable growth media inhibited fiber formation. It should be noted that the exact timing of wall formation is of little relevance since culture conditions which facilitate cell division [e.g. from 4 to 2 days after start of culture initiation—see Zelcer & Galun (165, 166)] should also facilitate cellulose production. The cellulosic character of the fibers was indicated already by early electron-microscopic observations (96) on tobacco mesophyll protoplasts. A chemical characterization of cellulose formation in protoplasts from carrot cell suspensions was furnished by Asamizu et al (4). These authors found that cellulose was synthesized within 24 hr of culture and that initially short polymers were mostly formed. Cellulose with normal chain length was formed at a later stage but always before the first cell division. Biochemical details on wall formation were subsequently studied in the same system (3), indicating that during the early period of protoplast cultivation, cell wall components differ from those of intact cells. The arabinose in the noncellulosic fraction was lower, while mannose, xylose, and glucose were higher in the cultured protoplasts. The latter also released abundant polysaccharides into the medium, as reported earlier (64). Actually, most of the polysaccharides, labeled by a pulse of ^{14}C-glucose and ^{3}H-myo inositol, were released into the medium during the early days of culture. Takeuchi & Komamine (139) reported similar observations, indicating differences in wall composition between intact cells and cultured protoplasts derived from suspension cultures. Since no comparable studies were performed with leaf protoplasts, it is not clear whether findings with cultured protoplasts derived from suspension cultures represent the norm, or whether the two systems differ substantially.

Protoplasts of *Vinca rosea* cell suspension cultures regenerate cell walls within 24 hr and were thus used to study the hydroxylation of proline and glycoprotein synthesis in the wall. This study indicated that following ^{14}C-proline incorporation, ^{14}C-hydroxyproline became a wall component and that arabinose was incorporated into the hydroxyproline-containing macromolecules, forming glycoproteins (141).

WALL FORMATION AS PREREQUISITE FOR CELL DIVISION The question of whether or not wall formation is a prerequisite for cell division in cultured protoplasts has prompted several investigations. Meyer & Abel (89, 90) found that when tobacco protoplasts were cultured in a medium which enabled the formation of "pseudo walls," i.e. an incomplete deposition of wall carbohydrates, no cell division followed. These findings indicated a correlative, but not a causal relationship, between wall formation and division. In the same manner, Schilde-Rentschler (123) found that the addition of cellulase to cultured tobacco protoplasts inhibited wall formation as well as division. Inhibiting the added cellulase with cellobiose caused initiation of cell division. Cytokinesis and nuclear division were similarly affected. The inhibitory effect of coumarin on microfibril formation on tobacco protoplasts (21) was used by Zelcer & Galun (166) to transiently arrest cell division. Since cell division occurred soon after coumarin removal, it can be assumed that in higher plants this inhibitor affects cellulose specifically rather than having a more general effect as found in *Acetobacter xylinum* (30). In such studies the use of a cell wall inhibitor which is both specific and reversible is a prerequisite for meaningful results. Hence Meyer & Herth (92) screened several inhibitors and found that low concentrations of 2,6-dichlorobenzonitrile reversibly inhibited wall formation of cultured tobacco protoplasts. Structural analysis indicated clearly that cytokinesis, but not nuclear division, was inhibited by those concentrations which prevented wall formation. The sum of the available information thus indicates a clear correlation between cytokinesis and wall formation, but the biochemical and morphogenetic basis of this correlation is still an open question.

GROWTH REGULATORS AND PESTICIDES

Growth Regulators: Metabolism and Cell Division

It is common knowledge among those engaged in protoplast research that when auxins and cytokinins are omitted from the medium no cell division will occur and the protoplasts derived from cells which do not require phytohormones are probably self-sufficient with respect to these growth regulators. Hence *Citrus* protoplasts derived from nucellar callus are not only independent of an outside supply of auxin (50, 147, 148), but may even be killed by relatively low auxin concentrations (A. Vardi, personal communication). In spite of this general knowledge that hormones are essential for division, little information on the mode of auxin and cytokinin action in protoplast division is available. Zelcer & Galun (165, 166) reported that incorporation of leucine and uridine into protein and RNA, respectively, was not inhibited in the absence of hormones up to several hours after commencement of tobacco protoplasts culture, but lack of hormones after the tenth hour drastically inhibited incorporation. Moreover, auxin (NAA) and to a lesser extent cytokinin (BAP) were found to be essential during the early hours of culture; when added after 10 hr, about 50% reduction of cell division was recorded while a 95% reduction occurred when either auxin or cytokinin or both were witheld up to the fifteenth hour of culture. Meyer & Cooke (91) also used tobacco protoplasts to study the involvement of auxins and cytokinins in cell division as well as in wall formation. In conformation with the results mentioned above, they found that addition of both 2,4-D and BA 3 hr after beginning the culture caused no change in cell division rate, while delaying their addition for 24 hr drastically reduced the protoplasts' ability to divide. Division was almost normal when the protoplasts were cultured in 2,4-D for 24 hr before addition of BA, but a further delay of BA addition reduced cell division substantially. This inhibitory effect was probably caused by the release of a toxic substance by protoplasts cultured in the absence of hormones, since washing the protoplasts after growth without hormones, followed by reculture in medium containing hormones, vastly improved further division. Moreover, when protoplasts were cultured in the absence of 2,4-D but in the presence of an inhibitor of wall formation and then transferred into normal hormone-containing medium, division occurred even after the cells had been maintained for 96 hr in hormone-deficient medium. These authors therefore suggested that during hormone starvation an inhibitor, related to incomplete wall formation, accumulated in the medium.

Use of Protoplasts in Pesticide Research

Although protoplasts have several attractive features as tools for pesticide research—especially for pesticides which may affect photosynthesis, the plasmalemma integrity, or the synthesis of specific macromolecules—there is very little information available on this subject. Mumma & Hamilton (93), who recently reviewed the use of plant tissue and cell cultures in xenobiotic metabolism, made no specific references to protoplasts.

Boulware & Camper (12) used tomato protoplasts to analyze the effect of several herbicides on the cellular integrity and found that Paraquat [(1-1'-dimethyl(-4-4'bipyridinium ion)] initially caused segregation of the cytoplasm into isolated areas near the plasmalemma and ultimately caused rupture of this membrane. Several other herbicides caused no visually characteristic damage (e.g. Perforam, Trifluralin). Using radioactive labeled herbicides, these authors showed (13) that protoplasts could be used to locate the herbicide in cell components of protoplasts exposed to these compounds and that certain herbicides (e.g. Fluorodifen = *p*-nitrophenyl-*α*, *α*, *α*-trifluoro-2-nitro-*p*-tolyl ether) were relatively abundant in the chloroplast/nuclear fraction, while other herbicides were apparently uniformly distributed among the cell components. Having indications that the "pinching" growth regulator Dikegulac (2,3:4,6-di-*o*-isopropylidine-2-keto-1-

... integrity of the plasmalemma (see 60), Zilkah & ... iorescein diacetate staining of tobacco protoplasts, in ... posure to Dikegulac, to further study this growth ... ved that Dikegulac concentrations which completely ... i.e. 10^{-5}M) caused no dye leakage. Only a tenfold ... aused dye leakage and bursting of the plasmalemma, ... n of the protoplasts. These findings are compatible ... hole plants in which low Dikegulate concentrations ... growth retardation but high concentration resulted in "pinching" of apical meristems.

CONCLUDING REMARKS

Evidence emerging from numerous studies indicates that protoplasts are not merely cells devoid of walls. The process of separation from the original tissue, together with their maintenance as isolated cells, caused substantial changes, most of which have not yet been thoroughly studied. Some of these changes still constitute a complete mystery. The latter kind of changes is exemplified by the "density requirement": protoplasts will degenerate rather than divide if cultured below a given density. The reason for this requirement is still not known, but it is characteristic of protoplasts derived from leaves as well as from in vitro cultured cells. It can be satisfied partially by adjustments in the culture medium and completely by culture over a layer of protoplasts or cells in which cell division is arrested by X-irradiation (27, 51, 111, 112). The protoplasts may then be used in investigations where very low densities are unavoidable (e.g. 5, 6, 49, 164).

The wealth of new information which has accumulated during the last 10 years through the use of plant protoplasts clearly indicates that this approach may open the way to further important and meaningful results in plant biology, provided the specific properties of the protoplasts are recognized and accounted for during planning and execution of further research.

Acknowledgments

I am grateful to Dr. G. W. Schaeffer, Chief, Cell Culture and Nitrogen Fixation Laboratory, BARC-West, U. S. Department of Agriculture, for his generous hospitality during the writing of this review, and to Miss Jane A. Wooster for typing the manuscript.

Literature Cited

1. Admon, A., Jacoby, B. 1980. Assessment of cytoplasmic contamination in isolated vacuole preparations. *Plant Physiol.* 65:85–87
2. Altman, A., Kaur-Sawhney, R., Galston, A. W. 1977. Stabilization of oat leaf protoplasts through polyamine-mediated inhibition of senescence. *Plant Physiol.* 60:570–74
3. Asamizu, T., Nishi, A. 1980. Regenerated cell wall of carrot protoplasts isolated from suspension-cultured cells. *Physiol. Plant.* 48:207–12
4. Asamizu, T., Tanaka, K., Takebe, I., Nishi, A. 1977. Change in molecular size of cellulose during regeneration of cell wall on carrot protoplasts. *Physiol. Plant.* 40:215–18
5. Aviv, D., Fluhr, R., Edelman, M., Galun, E. 1980. Progeny analysis of the interspecific somatic hybrids: *Nicotiana tabacum* (CMS) + *Nicotiana sylvestris* with respect to nuclear and chloroplast markers. *Theor. Appl. Genet.* 56:145–52
6. Aviv, D., Galun, E. 1980. Restoration of fertility in cytoplasmic male sterile (CMS) *Nicotiana sylvestris* by fusion with x-irradiated *N. tabacum* protoplasts. *Theor. Appl. Genet.* 58:185–90
7. Bhojwani, S. S., Evans, P. K., Cocking, E. C. 1977. Protoplast technology in relation to crop plants: progress and problems. *Euphytica* 26:343–60
8. Blaschek, W., Hess, D., Hoffmann, F. 1974. Transcription in nuclei prepared from isolated protoplasts of *Nicotiana* and *Petunia. Z. Pflanzenphysiol.* 72:262–71
9. Borochov, A., Halevy, A. H., Borochov, H., Shinitzki, M. 1978. Microviscosity of plasmalemma in rose petals as affected by age and environmental factors. *Plant Physiol.* 61:812–15
10. Borochov, A., Halevy, A. H., Shinitzki, M. 1976. Increase in microviscosity with aging in protoplast plasmalemma of rose petals. *Nature* 263:158–59
11. Boss, W. F., Ruesink, A. W. 1979. Isolation and characterization of concanavalin A-labeled plasma membranes of carrot protoplasts. *Plant Physiol.* 64:1005–11
12. Boulware, M. A., Camper, N. D. 1972. Effects of selected herbicides on plant protoplasts. *Physiol. Plant.* 26:313–17
13. Boulware, M. A., Camper, N. D. 1973. Sorption of some ^{14}C-herbicides by isolated plant cells and protoplasts. *Weed Sci.* 21:145–49
14. Briskin, D. P., Leonard, R. T. 1979. Ion transport in isolated protoplasts from tobacco suspension cells. III. Membrane potential. *Plant Physiol.* 64:959–62
15. Briskin, D. P., Leonard, R. T. 1980. Isolation of tonoplast vesicles from tobacco protoplasts. *Plant Physiol.* In press
16. Broughton, W. J., Wooi, K. C., Hoh, C. H. 1976. Acetylene reduction by legume root protoplasts. *Nature* 262:208–9
17. Burgess, J. 1978. Plant cells without walls. *Nature* 275:588–89
18. Burgess, J., Fleming, E. N. 1974. Ultrastructural observations of cell wall regeneration around isolated tobacco protoplasts. *J. Cell Sci.* 14:439–49
19. Burgess, J., Linstead, P. J. 1976. Ultrastructural studies of the binding of concanavalin A to the plasmalemma of higher plant protoplasts. *Planta* 130:73–79
20. Burgess, J., Linstead, P. J. 1977. Membrane mobility and concanavalin A binding system of the plasmalemma of higher plants. *Planta* 136:253–59
21. Burgess, J., Linstead, P. J. 1977. Coumarin inhibition of microfibril formation at the surface of cultured protoplasts. *Planta* 133:267–73
22. Burgess, J., Linstead, P. J., Bonsall, V. E. 1978. Observations on the time course of wall development at the surface of isolated protoplasts. *Planta* 139:85–91
23. Burgess, J., Linstead, P. J., Fisher, V. E. L. 1977. Studies on higher plant protoplasts by scanning electron microscopy. *Micron* 8:113–22
24. Burgess, J., Watts, J. W., Fleming, E. N., King, J. M. 1973. Plasmalemma fine structure in isolated tobacco mesophyll protoplasts. *Planta* 110:291–301
25. Butcher, H. C., Wagner, G. J., Siegelman, H. W. 1977. Localization of acid hydrolases in protoplasts. Examination of the proposed lysosomal function of the mature vacuole. *Plant Physiol.* 59:1098–1103
26. Cailloux, M., Phan, C. T., Chung, Y. S. 1978. Damage by ozone to the mechanical integrity of the protoplast plasmalemma. *Experientia* 34:730–31
27. Cella, R., Galun, E. 1980. Utilization of irradiated carrot cell suspensions as feeder layer for cultured *Nicotiana* cells and protoplasts. *Plant Sci. Lett.* 19:243–52
28. Chin, J. C., Scott, K. J. 1979. Effect of phytolectins on isolated protoplasts from plants. *Ann. Bot.* 43:33–44

29. Cocking, E. C. 1972. Plant cell protoplasts—isolation and development. *Ann. Rev. Plant Physiol.* 23:29–50
30. Colvin, J. R., Witter, D. E. 1980. On the inhibition of cellulose biosynthesis by coumarin. *Plant Sci. Lett.* 18:33–38
31. Coutts, R. H. A., Barnett, A., Wood, K. R. 1975. Ribosomal RNA metabolism in cucumber leaf mesophyll protoplasts. *Nucleic Acid Res.* 2:1111–12
32. Davey, M. R., Cocking, E. C., Bush, E. 1973. Isolation of legume root nodule protoplasts. *Nature* 244:460–61
33. Day, P. R. 1977. Plant genetics: increasing crop yield. *Science* 197:1334–39
34. Edwards, G. E., Huber, S. C. 1978. Usefulness of isolated cells and protoplasts for photosynthetic studies. *Proc. 4th Int. Congr. Photosynth.*, ed. D. O. Hall, J. Coobs, T. W. Goodwin, pp. 95–106. London: Biochem. Soc.
35. Edwards, G. E., Huber, S. C. 1979. C_4 metabolism in isolated cells and protoplasts. In *Encyclopedia of Plant Physiology, New Ser., Vol. 6: Photosynthesis II,* ed. M. Gibbs, E. Latzko, pp. 102–12. Berlin: Springer
36. Edwards, G. E., Lilley, R. McC., Craig, S., Hatch, M. D. 1979. Isolation of intact and functional chloroplasts from mesophyll and bundle sheath protoplasts of the C_4 plant *Panicum miliaceum. Plant Physiol.* 63:871–77
37. Edwards, G. E., Robinson, S. P., Tyler, N. J. C., Walker, D. A. 1978. A requirement for chelation in obtaining functional chloroplasts of sunflower and wheat. *Arch. Biochem. Biophys.* 190: 421–33
38. Edwards, G. E., Robinson, S. P., Tyler, N. J. C., Walker, D. A. 1978. Photosynthesis by isolated protoplasts, protoplast extracts and chloroplasts of wheat. Influence of orthophosphate, pyrophosphate and adenylates. *Plant Physiol.* 62:313–19
39. Fisher, D. J. 1979. Studies of plant membrane components using protoplasts. *Plant Sci. Lett.* 15:127–33
40. Fleck, J., Durr, A., Lett, M. C., Hirth, L. 1979. Changes in protein synthesis during the initial stage of life of tobacco protoplasts. *Planta* 145:279–85
41. Fowke, L. C., Bech-Hansen, C. W., Constabel, F., Gamborg, O. L. 1974. A comparative study of the ultrastructure of cultured cells and protoplasts of soybean during cell division. *Protoplasma* 81:189–203
42. Fowke, L. C., Bech-Hansen, C. W., Gamborg, O. L. 1974. Electron microscopic observations of cell regeneration from cultured protoplasts of *Ammi visnaga. Protoplasma* 79:235–48
43. Fowke, L. C., Bech-Hansen, C. W., Gamborg, O. L., Shyluk, J. P. 1973. Electron microscopic observations of cultured cells and protoplasts of *Ammi visnaga. Am. J. Bot.* 60:304–12
44. Fuchs, Y., Galston, A. W. 1976. Macromolecular synthesis in oat leaf protoplasts. *Plant Cell Physiol.* 17:475–82
45. Galbraith, D. W., Northcote, D. H. 1977. The isolation of plasma membranes from protoplasts of soybean suspension cultures. *J. Cell Sci.* 24:295–310
46. Galston, A. W. 1978. The use of protoplasts in plant propagation and improvement. In *Propagation of Higher Plants Through Tissue Culture,* ed. K. W. Hughes, R. Henke, M. Constantin, pp. 200–12. Washington DC: Tech. Inf. Ctr. US Dep. Energy
47. Galston, A. W., Altman, A., Kaur-Sawhney, R. 1978. Polyamines, ribonuclease and the improvement of oat leaf protoplasts. *Plant Sci. Lett.* 11:69–79
48. Galston, A. W., Kaur-Sawhney, R., Altman, A., Flores, H. 1979. Polyamines, macromolecular syntheses and the problem of cereal protoplast regeneration. Advances in protoplast research. *Proc. 5th Int. Protoplast Symp., Szeged, Hung.,* pp. 485–97. Budapest: Hung. Acad. Sci.
49. Galun, E., Aviv, D. 1979. Plant cell genetics in *Nicotiana* and its implication to crop plants. *Monogr. Genet. Agric.* 6:153–75
50. Galun, E., Aviv, D., Raveh, D., Vardi, A., Zelcer, A. 1977. Protoplasts in studies on cell genetics and morphogenesis. In *Plant Tissue Culture and its Biotechnological Application,* ed. W. Barz, E. Reihard, H. M. Zenk, pp. 302–12. Berlin: Springer
51. Galun, E., Raveh, D. 1975. *In vitro* culture of tobacco protoplasts: survival of haploid and diploid protoplasts exposed to x-ray radiation at different times after isolation. *Radiat. Bot.* 15:79–82
52. Gamborg, O. L., Kartha, K. K., Ohyama, K., Fowke, L. 1978. Protoplasts and tissue culture methods in crop plant improvement. *Proc. Symp. Plant Tissue Cult.,* Peking, pp. 265–78. Peking: Sci. Press
53. Gigot, C., Kopp, M., Schmitt, C., Milne, R. G. 1975. Subcellular changes during isolation and culture of tobacco mesophyll protoplasts. *Protoplasma* 84:31–41

54. Gigot, C., Philipps, G., Hirth, L. 1976. Evenements biochimiques accompagnant les étapes chapes initiales de la dedifferenciation des protoplastes de tabacs en culture. In *Memoires Originaux* ed. G. Morel, pp. 186–911. Paris: Masson
55. Gigot, C., Philipps, G., Nicolaieff, A., Hirth, L. 1976. Some properties of tobacco protoplast chromatin. *Nucleic Acids Res.* 3:2315–29
56. Gigot, C., Schmitt, C., Hirth, L. 1972. Modifications ultrastruturales observees au cours de la preparation de protoplastes a partir de cultures de tissue de tabac. *J. Ultrastruct. Res.* 41:418–32
57. Gimelius, K., Wallin, A., Eriksson, T. 1974. Agglutinating effects of concanavalin A on isolated protoplasts of *Daucus carota. Physiol. Plant.* 31:225–30
58. Gimelius, K., Wallin, A., Eriksson, T. 1978. Ultrastructural visualization of sites binding concanavalin A on the cell membrane of *Daucus carota. Protoplasma* 97:291–300
59. Gregory, R. P. F., Droppa, M., Horvath, G., Evans, E. H. 1979. A comparison based on delayed light emission and fluorescence induction of intact chloroplasts isolated from mesophyll protoplasts and bundle sheath cells of maize. *Biochem. J.* 180:253–55
60. Gressel, J. 1980. Uses and drawbacks of cell cultures in pesticide research. In *Plant Cell Cultures: Results and Perspectives,* ed. F. Sala, B. Parisi, R. Cella, O. Ciferri, pp. 379–88 Amsterdam: Elsevier/North-Holland Biomed.
61. Grout, B. W. W., Willison, J. H. M., Cocking, E. C. 1972. Interactions at the surface of plant cell protoplasts: an electrophoretic and freeze-etch study. *J. Bioenerg.* 4:311–28
62. Guy, M., Reinhold, L., Laties, G. G. 1978. Membrane transport of sugars and amino acids in isolated protoplasts. *Plant Physiol.* 61:593–96
63. Hall, J. L., Taylor, A. R. D. 1979. Isolation of plasma membrane from higher plant cells. In *Plant Organelles-Methodological Surveys* (B) *Biochemistry, Vol. 9,* ed. E. Reid, pp. 103–11. Chichester, UK: Horwood
64. Hanke, D. E., Northcote, D. H. 1974. Cell wall formation by soybean callus protoplasts. *J. Cell Sci.* 14:29–50
65. Hatch, M. D., Osmond, C. B. 1976. Compartmentation and transport in C_4 photosynthesis. In *Transport in Plants, Encyclopedia of Plant Physiology, New Series, Vol. 3,* ed. C. R. Stocking, U. Heber, pp. 144–84. New York: Springer
66. Heslop-Harrison, J. 1972. Sexuality in angiosperms. In *Plant Physiology-A Treatise,* ed. F. C. Steward, 6C:133–289. New York: Academic
67. Hirai, A., Wildman, S. G. 1977. Kinetic analysis of fraction I protein biosynthesis in young protoplasts of tobacco leaves. *Biochim. Biophys. Acta* 479: 39–52
68. Horine, R. K., Ruesink, A. W. 1972. Cell wall regeneration around protoplasts isolated from *Convolvulus* tissue culture. *Plant Physiol.* 50:438–45
69. Huber, S. C., Edwards, G. E. 1975. C_4 photosynthesis: light dependent CO_2 fixation by mesophyll cells, protoplast and protoplast extracts of *Digitaria sanguinalis. Plant Physiol.* 55:834–44
70. Huber, S. C., Edwards, G. E. 1975. An evaluation of some parameters required for the enzymatic isolation of cells and protoplasts with CO_2 fixation capacity from C_3 and C_4 grasses. *Physiol. Plant.* 35:203–9
71. Kanai, R., Edwards, G. E. 1973. Enzymatic separation of mesophyll protoplasts and bundle sheath cells from leaves of C_4 plants. *Naturwissenschaften* 60:157–8
72. Kanai, R., Edwards, G. E. 1973. Separation of mesophyll protoplasts and bundle sheath cells from maize leaves for photosynthetic studies. *Plant Physiol.* 51:1133–37
73. Kanai, R., Edwards, G. E. 1973. Purification of enzymatically isolated mesophyll protoplasts from C_3, C_4 and CAM plants using an aqueous dextran-polyethylene glycol two phase system. *Plant Physiol.* 52:482–90
74. Kaur-Sawhney, R., Adams, W. R. Jr., Tsang, J., Galston, A. W. 1977. Leaf pretreatment with senescence retardants as a basis for oat protoplast improvement. *Plant Cell Physiol.* 18: 1309–17
75. Kaur-Sawhney, R., Altman, A., Galston, A. W. 1978. Dual mechanisms in polyamine-mediated control of ribonuclease activity in oat leaf protoplasts. *Plant Physiol.* 63:158–60
76. Kaur-Sawhney, R., Flores, H. H., Galston, A. W. 1980. Polyamine-induced DNA synthesis and mitosis in oat leaf protoplasts. *Plant Physiol.* 65:368–71
77. Kaur-Sawhney, R., Rancillac, M., Staskawicz, B. , Adams, W. R. Jr., Galston, A. W. 1976. Effect of cycloheximide and kinetin on yield, integrity and meta-

bolic activity of oat leaf protoplasts. *Plant Sci. Lett.* 7:57–67

78. Ku, S. B., Gutierrez, M., Kanai, R., Edwards, G. E. 1974. Photosynthesis in mesophyll protoplasts and bundle sheath cells of C_4 plants. Chlorophyll and Hill reaction studies. *Z. Pflanzenphysiol.* 72:320–27
79. Kulikowski, R. R., Mascarenhas, J. P. 1978. RNA synthesis in whole cells and protoplasts of *Centaurea*—a comparison. *Plant Physiol.* 61:575–80
80. Lai, K-L., Liu, L. F. 1978. Studies on the rice protoplasts—ultrastructural changes during enzymatic isolation. *J. Agric. Assoc. China New Ser. (Taiwan)* 102:11–23
81. Larkin, P. J. 1978. Plant protoplast agglutination by artificial carbohydrate antigens. *J. Cell Sci.* 30:283–92
82. Larkin, P. J. 1980. Plant protoplast agglutination and immobilization. *Proc. Symp. Phytochem. Cell Recognition Cell Surf. Interactions, Pullman, Wash.* In press
83. Lazar, G., Borbely, G., Udvardy, J., Premecz, G., Farkas, G. L. 1973. Osmotic shock triggers an increase in ribonuclease level in protoplasts isolated from tobacco leaves. *Plant Sci. Lett.* 1:53–57
84. Lorz, H., Harms, C. T., Potrykus, I. 1976. Isolation of "vacuolplasts" from protoplasts of higher plants. *Biochem. Physiol. Pflanz.* 169:617–20
85. Malmberg, R. L., Griesbach, R. J. 1980. The isolation of mitotic and meiotic chromosomes from plant protoplasts. *Plant Sci. Lett.* 17:141–47
86. Mettler, I. J., Leonard, R. T. 1979. Ion transport in isolated protoplasts from tobacco suspension cells. I. General characteristics. *Plant Physiol.* 63: 183–90
87. Mettler, I. J., Leonard, R. T. 1979. Ion transport in isolated protoplasts from tobacco suspension cells. II. Selectivity kinetics. *Plant Physiol.* 63:191–94
88. Mettler, I. J., Leonard, R. T. 1979. Isolation and partial characterization of vacuoles from tobacco protoplasts. *Plant Physiol.* 64:1114–20
89. Meyer, Y., Abel, W. O. 1975. Importance of the wall for cell division and in the activity of the cytoplasm in cultured tobacco protoplasts. *Planta* 123:33–40
90. Meyer, Y., Abel, W. O. 1975. Budding and cleavage division of tobacco mesophyll protoplasts in relation to pseudo wall and wall formation. *Planta* 125:1–13
91. Meyer, Y., Cooke, R. 1979. Time course of hormonal control of the first mitosis in tobacco mesophyll protoplasts cultivated *in vitro. Planta* 147:181–85
92. Meyer, Y., Herth, W. 1978. Chemical inhibition of cell wall formation and cytokinesis, but not of nuclear division in protoplasts of *Nicotiana tabacum* cultivated *in vitro. Planta* 142:253–62
93. Mumma, R. O., Hamilton, R. H. 1979. Xenobiotic metabolism in higher plants: *in vitro* tissue and cell culture techniques. In *Xenobiotic Metabolism: In Vitro Methods ACS Sym. Ser. 97,* ed. G. D. Paulson, S. Frear, E. P. Marks, pp. 35–76. Am. Chem. Soc.
94. Nagata, T., Melchers, G. 1978. Surface charges of protoplasts and their significance in cell-cell interaction. *Planta* 142:235–8
95. Nagata, T., Takebe, I. 1970. Cell wall regeneration and cell division in isolated tobacco mesophyll protoplasts. *Planta* 92:301–8
96. Nagata, T., Yamaki, T. 1973. Electron microscopy of isolated tobacco mesophyll protoplasts cultrued *in vitro. Z. Pflanzenphysiol.* 70:452–59
97. Nishimura, M., Akazawa, T. 1975. Photosynthetic activity of spinach leaf protoplasts. *Plant Physiol.* 55:712–16
98. Nishimura, M., Beevers, H. 1978. Isolation of intact plastids from protoplasts from castor bean endosperm. *Plant Physiol.* 62:40–43
99. Nishimura, M., Graham, D., Akazawa, T. 1975. Effect of oxygen on photosynthesis by spinach leaf protoplasts. *Plant Physiol.* 56:718–22
100. Nishimura, M., Graham, D., Akazawa, T. 1976. Isolation of intact chloroplasts and other cell organelles from leaf protoplasts. *Plant Physiol.* 58:309–14
101. Ohyama, K., Pelcher, L. E., Horn, D. 1977. DNA binding and uptake by nuclei isolated from plant protoplasts. *Plant Physiol.* 60:98–101
102. Ohyama, K., Pelcher, L. E., Horn, D. 1977. A rapid, simple method for nuclei isolation from plant protoplasts. *Plant Physiol.* 60:179–81
103. Perlin, D. S., Spanswick, R. M. 1980. Labeling and isolation of plasma membranes from corn leaf protoplasts. *Plant Physiol.* 65:1053–57
104. Pilet, P. E., Prat, R., Roland, J. C. 1972. Morphology, RNAse and transaminase of root protoplasts. *Plant Cell Physiol.* 13:297–309
105. Premecz, G., Olah, T., Gulyas, A., Nyitrani, A., Palfi, G., Farkas, G. L. 1977.

Is the increase in ribonuclease level in isolated tobacco protoplasts due to osmotic stress? *Plant Sci. Lett.* 9:195–200
106. Premecz, G., Ruzicska, P., Olah, T., Farkas, G. L. 1978. Effect of "osmotic stress" on protein and nucleic acid synthesis in isolated tobacco protoplasts. *Planta* 141:33–36
107. Racusen, R. H., Kinnersley, A. M., Galston, A. W. 1977. Osmotically induced changed charges in electrical properties of plant protoplast membranes. *Science* 198:405–7
108. Radojevic, Lj., Kovoor, A. 1978. Characterization and estimation of newly synthesized DNA in higher plant protoplasts during the initial period of culture. *J. Exp. Bot.* 29:963–68
109. Rathnam, C. K. M., Edwards, G. E. 1976. Protoplasts as a tool for isolating functional chloroplasts from leaves. *Plant Cell Physiol.* 17:177–86
110. Rathnam, C. K. M., Zilinskas, B. A. 1977. Reversal of 3-(3,4,-dichlorophenyl)-1,1-dimethylurea inhibition of carbon dioxide fixation in spinach chloroplasts and protoplasts by dicarboxylic acids. *Plant Physiol.* 60:51–53
111. Raveh, D., Galun, E. 1975. Rapid regeneration of plants from tobacco protoplasts plated at low densities. *Z. Pflanzenphysiol.* 76:76–79
112. Raveh, D., Huberman, E., Galun, E. 1973. *In vitro* culture of tobacco protoplasts: Use of feeder techniques to support division of cells plated at low densities. *In Vitro* 9:216–22
113. Robinson, D. J., Mayo, M. A. 1975. Changing rates of uptake of [^{3}H] leucine and other compounds during culture of tobacco mesophyll protoplasts. *Plant Sci. Lett.* 8:197–204
114. Robinson, S. P., Walker, D. A. 1979. The site of sucrose synthesis in isolated leaf protoplasts. *FEBS Lett.* 107:295–99
115. Robinson, S. P., Walker, D. A. 1979. Rapid separation of the chloroplast and cytoplasmic fractions from intact leaf protoplasts. *Arch. Biochem. Biophys.* 196:319–23
116. Robinson, S. P., Walker, D. A. 1980. Distribution of metabolites between chloroplast and cytoplasm during the induction phase of photosynthesis in leave protoplasts. *Plant Physiol.* 65:902–5
117. Rubin, G., Zaitlin, M. 1976. Cell concentration as a factor in precursor incorporation by tobacco leaf protoplasts or separated cells. *Planta* 131:87–89
118. Rubinstein, B. 1978. Use of lipophilic cations to measure the membrane potential of oat leaf protoplasts. *Plant Physiol.* 62:927–29
119. Ruesink, A. W. 1978. Leucine uptake and incorporation by *Convolvulus* tissue culture cells and protoplasts under severe osmotic stress. *Physiol. Plant.* 44:48–56
120. Ruzicska, P., Mettrie, R., Dorokhov, Y. L., Premecz, G., Olah, T., Farkas, G. L. 1979. Polyribosomes in protoplasts isolated from tobacco leaves. *Planta* 145:199–203
121. Sakai, F., Takebe, I. 1970. RNA and protein synthesis in protoplasts isolated from tobacco leaves. *Biochim. Biophys. Acta* 224:531–40
122. Schetter, C., Hess, D. 1977. Nitrogenase activity in protoplasts isolated from root nodules of *Glycine max. Biochem. Physiol. Pflanz.* 171:63–67
123. Schilde-Rentschler, L. 1977. Role of the cell-wall in the ability of tobacco protoplasts to form callus. *Planta* 135:177–81
124. Schmid, G. H., Thibault, P. 1979. Characterization of a light induced oxygen-uptake in tobacco protoplasts. *Z. Naturforsch. Teil C* 34:570–75
125. Schnabl, H. 1978. The effect of Cl^- upon the sensitivity of starch-containing and starch-deficient stomata and guard cell protoplasts towards potassium ions, fusicoccin and abscisic acid. *Planta* 144:95–100
126. Schnabl, H. 1980. Anion metabolism as correlated with volume changes of guard cell protoplasts. *Z. Naturforsch. Teil C* 34: In press
127. Schnabl, H. 1980. CO_2 and malate metabolism in starch-containing and starch-lacking guard-cell protoplasts. *Planta* 149:52–58
128. Schnabl, H., Bornman, C. H., Ziegler, H. 1978. Studies on isolated starch-containing (*Vicia faba*) and starch-deficient (*Allium cepa*) guard cell protoplasts. *Planta* 143:33–39
129. Schnabl, H., Scheurich, P., Zimmermann, U. 1980. Mechanical stabilization of guard cell protoplasts of *Vicia faba. Planta* 149:280–82
130. Schnabl, H., Vienken, J., Zimmermann, U. 1980. Regular arrays of intermembraneous particles in the plasmalemma of guard cell and mesophyll cell protoplasts of *Vicia faba. Planta* 148:231–37
131. Scowcroft, W. R. 1977. Stomatic cell genetics and plant improvement. *Adv. Agron.* 29:39–81

132. Singer, S. J., Nicolson, G. L. 1972. The fluid mosaic model of the structure of cell membranes. *Science* 175:720–31
133. Spalding, M.H., Edwards, G. E. 1980. Photosynthesis in isolated chloroplasts of the Crassulacean acid metabolism plant *Sedum praealtum*. *Plant Physiol.* 65:1044–48
134. Spalding, M. H., Schmitt, M. R., Ku, S. B., Edwards, G. E. 1979. Intracellular localization of some key enzymes of Crassulacean acid methabolism in *Sedum praealtum*. *Plant Physiol.* 63:738–43
135. Staskawicz, B., Kaur-Sawhney, R., Slaybaugh, R., Adams, W. Jr., Galston, A. W. 1978. The cytokinin-like action of methyl-2-benzimidazole-carbamate on oat leaves and protoplasts. *Pestic. Biochem. Physiol.* 8:106–10
136. Sunderland, N., Dunwell, J. M. 1974. Pathways in pollen embryogenesis. In *Tissue Culture and Plant Science*, ed. H. Street, pp. 141–67. London/New York: Academic
137. Takebe, I., Labib, G., Melchers, G. 1971. Regeneration of whole plants from isolated mesophyll protoplasts of tobacco. *Naturwissenschaften* 58: 318–20
138. Takebe, I., Otsuki, Y., Honda, Y., Nishio, T., Matsui, C. 1973. Fine structure of isolated mesophyll protoplasts of tobacco. *Planta* 113:21–27
139. Takeuchi, Y., Komamine, A. 1978. Composition of the cell wall formed by protoplasts isolated from cell suspension culture of *Vinca rosea*. *Planta* 140:227–32
140. Tallman, G., Reeck, G. R. 1980. Isolation of nuclei from plant protoplasts without the use of a detergent. *Plant Sci. Lett.* 18:271–75
141. Tanaka, M., Uchida, T. 1979. Heterogeneity of hydroxyproline-containing glycoproteins in protoplasts from a *Vinca rosea* suspension culture. *Plant Cell Physiol.* 20:1295–1306
142. Taylor, A. R. D., Hall, J. L. 1978. Fine structure and cytochemical properties of tobacco leaf protoplasts and comparison with the source tissue. *Protoplasma* 96:113–26
143. Taylor, A. R. D., Hall, J. L. 1979. An ultrastructural comparison of lanthanum and silicotungstic acid/chronic acid or plasma membrane stains of isolated protoplasts. *Plant Sci. Lett.* 14:139–44
144. Thomas, E., King, P. J., Potrykus, I. 1979. Improvement of crop plants via single cells *in vitro*—an assessment. *Z. Pflanzenzuecht.* 82:1–30
145. Uchimiya, H. 1979. Progress in plant protoplast technology—isolation, culture and genetic manipulation. *Rep. Inst. Agric. Res. Tohoku Univ.* 30:29–53
146. Usuda, H., Edwards, G. E. 1980. Localization of glycerate kinase and some enzymes for sucrose synthesis in C_3 and C_4 plants. *Plant Physiol.* 65:1017–22
147. Vardi, A., Raveh, D. 1976. Cross feeder experiments between tobacco and orange protoplasts. *Z. Pflanzenphysiol.* 78:350–59
148. Vardi, A., Spiegel-Roy, P., Galun, E. 1975. *Citrus* cell culture; isolation of protoplasts, plating densities, effect of mutagens and regeneration of embryos. *Plant Sci. Lett.* 4:231–36
149. Vasil, I. K. 1976. The progress, problems and prospects of plant protoplast research. *Adv. Agron.* 28:119–60
150. Vasil, I. K., Ahuja, M. R., Vasil, V. 1979. Plant tissue cultures in genetics and plant breeding. *Adv. Genet.* 20:127–215
151. Vasil, V., Vasil, I. K. 1980. Isolation and culture of cereal protoplasts. 2: Embryogenesis and plantlet formation from protoplasts of *Pennisetum americanum*. *Theor. Appl. Genet.* 56: 97–99
152. Vigh, L., Horvath, I., Horvath, L. I., Dudits, D., Farkas, T. 1979. Protoplast plasmalemma fluidity of hardened wheats correlates with frost resistance. *FEBS Lett.* 107:291–94
153. Wagner, G. J., Siegelman, H. W. 1975. Large scale isolation of intact vacuoles and isolation of chloroplasts of mature plant tissues. *Science* 190:1298–99
154. Wallin, A., Eriksson, T. 1973. Protoplast cultures from cell suspensions of *Daucus carota*. *Physiol. Plant.* 28:33–39
155. Wasilewska, L. D., Kleczkowski, K. 1974. Phytohormone induced changes in the nuclear RNA population of plant protoplasts. *FEBS Lett.* 44:164–68
156. Wegman, K., Muhlbach, H. P. 1973. Photosynthetic CO_2 incorporation by isolated leaf cell protoplasts. *Biochim. Biophys. Acta* 314:79–82
157. Wiest, S. C., Steponkus, P. L. 1978. Freeze thaw injury to isolated spinach protoplasts and its simulation at above freezing temperatures. *Plant Physiol.* 62:699–705
158. Williamson, F. A. 1979. Concanavalin A binding sites on the plasma membrane of leek stem protoplasts. *Planta* 144:209–15

159. Williamson, F. A., Fowke, L. C., Constabel, F. C., Gamborg, O. L. 1976. Labeling of concanavalin A sites on the plasma membrane of soybean protoplasts. *Protoplasma* 89:305–16
160. Williamson, F. A., Fowke, L. C., Weber, G., Constabel, F. C., Gamborg, O. L. 1977. Microfibril deposition on cultured protoplasts of *Vicia hajastama*. *Protoplasma* 91:213–19
161. Zeiger, E., Hepler, P. K. 1976. Production of guard cell protoplasts from onion and tobacco. *Plant Physiol.* 58: 492–98
162. Zeiger, E., Hepler, P. K. 1977. Light and stomatal function: blue light stimulates swelling of guard cell protoplast. *Science* 196:887–89
163. Zelcer, A. 1978. *Metabolic changes related to the induction of mitosis in isolated plant protoplasts.* PhD thesis. Weizmann Inst. Sci., Rehovot, Israel. Hebrew, English summary, 89 + v pp.
164. Zelcer, A., Aviv, D., Galun, E. 1978. Interspecific transfer of cytoplasmic male sterility by fusion between protoplasts of normal *Nicotiana sylvestris* and x-ray irradiated protoplasts of male-sterile *N. tabacum*. *Z. Pflanzenphysiol.* 90:397–407
165. Zelcer, A., Galun, E. 1976. Culture of newly isolated tobacco protoplasts: precursor incorporation into protein, RNA and DNA. *Plant Sci. Lett.* 7:331–36
166. Zelcer, A., Galun, E. 1980. Culture of newly isolated tobacco protoplasts: cell division and precursor incorporation following a transient exposure to coumarin. *Plant Sci. Lett.* 18:185–90
167. Zilkah, S., Gressel, J. 1980. Multistage disruption of protoplasts by dikegulac. *Planta* 147:274–76
168. Zimmermann, U., Groves, M., Schnabl, H., Pilwat, G. 1980. Development of a new Coulter counter system: measurement of the volume, internal conductivity and dielectric breakdown voltage of a single guard cell protoplast of *Vicia faba*. *J. Membr. Biol.* 52:37–50
169. Zuily-Fodil, Y., Passaquet, C., Esnault, R. 1978. High yield isolation of nuclei from plant protoplasts. *Physiol. Plant.* 43:201–4

Ann. Rev. Plant Physiol. 1981. 32:267–89

ELECTROGENIC ION PUMPS

♦7712

Roger M. Spanswick

Section of Plant Biology, Division of Biological Sciences, Cornell University, Ithaca, New York 14853

CONTENTS

INTRODUCTION

Despite pioneering work by Blinks (6), Osterhout (75), and Umrath (122), the study of the electrophysiology of plant cell membranes initially lagged far behind similar work on animal systems. Thus it is not surprising that the first review in this series to deal with the relationship between ion transport and electrophysiology (18) was heavily influenced by the prevailing view of ion transport in animal cells. According to this view, the electrical properties of cell membranes could be accounted for by the passive diffusion of ions down gradients established by ion pumps that were assumed to be neutral. (Neutral pumps transport zero net charge across the membrane and therefore have no direct effect on the membrane potential.)

0066-4294/81/0601-0267$01.00

Electrogenic pumps, which transport a net charge across the membrane, were first well characterized in animal epithelia such as frog skin (123), but the ion pumps in most animal cells appeared to be neutral or only weakly electrogenic (53, 121). It quickly became evident that it was not possible to account for the electrical properties of plant cell membranes in terms of neutral ion pumps and passive diffusion (38, 54, 95, 115) except under special conditions such as Ca^{2+}-free media (47) or very high concentrations of monovalent cations (54).

The recognition of the existence of electrogenic pumps (21, 31, 56, 86, 96, 108) has made it possible to explain the basic electrical properties of plant cell membranes. That the pump in nonmarine systems probably transports protons has created widespread interest both because it makes it possible to apply a chemiosmotic scheme to transport across plant cell membranes (70, 98), which can account for the transport of organic molecules in terms of cotransport systems (2, 23, 74, 103), and because the proton pump may play an important role in auxin action (14, 15).

In this review I shall attempt to evaluate alternative theoretical treatments of electrogenic pumps and show how information about the electrical properties of membranes may be used to elucidate the properties of the pump. I shall also examine the characteristics of electrogenic pumps in particular systems and their relationship to metabolism. Finally, I shall consider briefly the relationship between proton pumps and a chemiosmotic scheme for transport across the plasmalemma.

DEMONSTRATION OF THE EXISTENCE OF ELECTROGENIC PUMPS

In principle, an electrogenic pump may be detected by halting the pump and looking for a change in the membrane potential. In practice, the situation is complicated by the fact that any treatment used to inhibit the pump may also affect the diffusion potential which is always present across the membrane. To deal with this problem it is usually necessary to estimate the diffusion potential E from an equation such as the Goldman equation:

$$E = \frac{RT}{F} \ln \frac{P_K [K_o^+] + P_{Na} [Na_o^+] + P_{Cl} [Cl_i^-]}{P_K [K_i^+] + P_{Na} [Na_i^+] + P_{Cl} [Cl_o^-]} \qquad 1.$$

where P_K etc are permeability coefficients, $[K_o^+]$ etc are external, and $[K_i^+]$ etc are internal concentrations, R is the gas constant, T is the absolute temperature, and F is the Faraday. This in turn requires an estimate of the permeability coefficient for each ion that may be important in controlling the diffusion potential; the internal ion concentrations may be assumed to remain constant during brief treatments. Measurements of the passive ion fluxes and the membrane potential must be made to calculate the permeability coefficients from the constant field equations that are consistent with Equation 1 (18). It is assumed that the fluxes in the opposite direction to those containing an active component are entirely passive (18, 36). This approach was used by Higinbotham et al (39) to prove the existence of a CN^--sensitive electrogenic pump in the membranes of oat coleoptiles and pea hypocotyls, the existence of which was suspected from the effect of DNP on the membrane potential observed during the first microelectrode measurements on higher plants (21). The main drawback of this approach is the large number of measurements required, particularly when effluxes must be estimated by compartmental analysis.

The situation is much clearer for systems in which the electrogenic pump hyperpolarizes the membrane potential beyond the negative limit of the diffusion potential. To understand this it is necessary to recognize that even in the absence of any restrictions on the values of the permeability coefficients, the value of the diffusion potential given by Equation 1 can only vary within limits set by the most extreme values of the Nernst potentials for the individual ions. In most systems the K^+ ion has the largest permeability coefficient and the most negative Nernst potential. In the limiting case, when the terms for all other ions are negligible compared with those for K^+, Equation 1 reduces to the Nernst equation for K^+:

$$E_K = \frac{RT}{F} \ln \frac{[K_o^+]}{[K_i^+]} \qquad 2.$$

Thus, if the internal concentration is known, the negative limit of the diffusion potential can be calculated. Hyperpolarization of the potential beyond this limit may be postulated to be due to the presence of an electrogenic pump.

A distinct hyperpolarization of the membrane potential was first demonstrated by Slayman (95) in *Neurospora crassa.* At an external K^+ concentration of 1 mM, the Nernst potential for K^+ was about –133 mV while the measured potential was –200 mV in the presence of Ca^{2+}. Furthermore, the membrane potential was very sensitive to respiratory inhibitors (96). Slayman (96) suggested that uphill extrusion of H^+ could be the electrogenic process.

In most plant cells the membrane potential is not markedly more negative than the Nernst potential for K^+. The leaf cells of many aquatic plants appear to be an exception. For example, the leaf cells of *Elodea canadensis* have a membrane potential of –257 mV in the presence of 0.1 mM K^+ (109), a potential that is clearly outside the range of possible diffusion potentials.

The same is true for *Lemna* (60, 74, 128) with a membrane potential of about –200 mV. Some terrestrial plant tissues also have cells with membrane potentials more negative than E_K (1, 37, 80).

In the case of the Characeae, for which the cytoplasmic ion concentrations can be determined directly (116), the situation was originally less clear because the membrane potential lay within the range of possible diffusion potentials (47, 115). However, the response of the membrane potential to added bicarbonate, later shown to be primarily a pH effect (106), led Hope (46) to postulate the existence of an electrogenic bicarbonate influx. The small response of the membrane potential to changes in external cation concentrations also led to the suggestion of a possible electrogenic pump (115), and Kitasato (56) put forward a more concrete hypothesis when he demonstrated strong pH dependence of the membrane potential in *Nitella clavata* which he suggested was the result of high passive permeability of the membrane to H^+. Since the addition of terms for H^+ to the Goldman equation (Equation 1) with a high value for P_H, the permeability coefficient for H^+, would predict a membrane potential much more positive than that observed, Kitasato (56) also postulated that the large passive influx of H^+ was balanced by an electrogenic active efflux of H^+ which polarized the membrane potential to its observed value.

The existence of an electrogenic pump in *Nitella* was confirmed by raising the external K^+ concentration to 0.5 mM, at which concentration the membrane potential in *Nitella translucens* is more negative than E_K (108). Under these conditions the electrogenic component of the membrane potential is light-stimulated. However, the value to which the potential falls under conditions in which the electrogenic pump is inhibited is the value predicted by the Goldman equation without terms for H^+, i.e. a value slightly more positive than E_K (111). This is also true for Kitasato's own experiments with DNP (56) and has been confirmed for *Nitella axilliformis* (88). Thus, although Kitasato's hypothesis of an electrogenic pump is valid, the assumption of a large passive H^+ influx to explain the effect of external pH on the membrane potential may not be valid; an alternative explanation (108) will be considered below.

The existence of electrogenic H^+ pumps in the membranes of nonmarine plants is now widely accepted (82). In marine plants, however, the predominant electrogenic process appears to be anion transport. In *Acetabularia* the membrane potential in artificial sea water (–170 mV) is more negative than E_K [–93mV(86)] and is sensitive to temperature and inhibitors (31, 86). A strong depolarization is also produced by reduction of the external Cl^- concentration (86), indicating that Cl^- is the main ion transported electrogenically. With *Limonium,* which possesses salt glands on the leaf surface, Hill (40) was able to perform experiments using the short-circuit technique originally developed for frog skin (123). With an identical solution on either side of the leaf and the potential difference across the leaf clamped at zero, there was a net (active) flux of several monovalent anions and cations. In the presence of NaCl, the short-circuit current was equal to the difference between the net chloride and sodium fluxes, the Cl^- flux being greater than that for Na^+. Further analysis of the system (41) indicates that the primary active transport system is a Cl^- efflux across the gland cell membranes to which the Na^+ flux is electrically coupled. An interesting feature of this system is that Cl^- pumping can be induced by addition of NaCl to plants grown under low salt conditions (92). An increase in Cl^--stimulated ATPase activity accompanies the increase in salt secretion (42).

Bowling (8, 9) has demonstrated a correlation between the membrane potential and the rate of phosphate uptake in white clover and sunflower roots. He suggests that this is evidence for an electrogenic phosphate pump. However, the depolarization that occurs when phosphate is removed from the external medium takes 2 h to complete. The depolarization resulting from removal of substrate from a primary electrogenic phosphate pump would be immediate, so it is possible that the effect on the potential may be a secondary effect resulting from the operation of control mechanisms.

THE ELECTRICAL CHARACTERISTICS OF ELECTROGENIC PUMPS

Attempts to describe the effect of electrogenic pumps on the membrane potential appear to fall into two categories. In one the Goldman equation (Equation 1) is modified to take into account the current generated by the electrogenic pump. Thus, instead of the sum of the passive ion fluxes being set equal to zero in the steady state, it is set equal to the flux through the pump (10, 71, 72). The usefulness of the resulting equations is limited by the fact that in a case involving the electrogenic transport of a single ion, the equation is not explicit for the membrane potential and requires a knowledge of the flux through the pump. In the case of an electrogenic pump with linked fluxes having a fixed stoichiometry (72), it is sufficient to know the ratio of the fluxes.

An alternative approach, perhaps less rigorous but much easier to interpret in terms of membrane potentials and conductances, makes use of an electrical analog (equivalent circuit) for the membrane. The simplest version, shown in Figure 1, consists of two branches connected in parallel (25, 96, 108). The left-hand branch represents the passive diffusion channels in the membrane lumped together so that the total electrical conductance for the channel may be represented by g_D. The EMF of the passive channels,

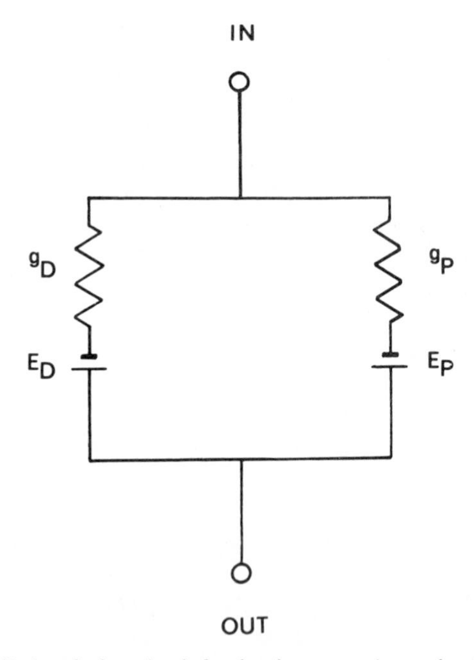

Figure 1 A simplified equivalent circuit for the plasma membrane showing the electrogenic pump (*P*) in parallel with the channels for passive diffusion. The pump has been assigned a conductance g_p, and an EMF, E_p. The conductances of the passive channels are lumped together as a single conductance g_p, which is in series with the diffusion potential E_D.

E_D, is equivalent to the diffusion potential (Equation 1). The right-hand branch of the circuit represents the electrogenic pump which is also assigned a conductance, g_P, and EMF, E_P. The application of this approach will be illustrated using studies on characean internodal cells, for which most information is available and for which it is most useful, before other systems are examined.

Characean Cells

MEMBRANE POTENTIAL In the presence of Ca^{2+}, the membrane potential of characean cells is much less sensitive to changes in external ion concentrations than would be expected if it were a diffusion potential described by Equation 1 (47, 56, 115). Another major problem with the electrophysiology of characean cells is that the measured electrical conductance is much greater than the sum of the conductances calculated from the passive ion fluxes (64, 107, 127). In principle, Kitasato (56) solved both problems when he demonstrated the pH-dependence of the membrane potential and suggested the existence of a large passive H^+ influx and an electrogenic H^+ pump. Assuming further that the electrogenic pump acts as a current source, he described the membrane potential by the equation:

$$E = E_D - \frac{(I_j \pm)_{\text{active}}}{g_m} \qquad 3.$$

where $(I_J \pm)_{\text{active}}$ is the current carried by actively transported ions and g_m is the chord conductance of the membrane. If, in addition, most of the passive current is carried by H^+, this equation can be approximated by:

$$E = E_H - \frac{F\phi_H}{g_m} \qquad 4.$$

where E_H is the Nernst potential for H^+ and $\o_H$ is the active flux of H^+. Under these circumstances it could also be assumed that $g_m \simeq g_H$, the conductance due to the passive movement of H^+.

One problem with this model, that the membrane potential tends toward E_K when the pump is inhibited (56, 108, 111), has already been mentioned. Other problems are that the membrane conductance increases at high external pH, whereas a decrease would be expected as a result of the reduction in the passive H^+ influx, and the membrane potential does not respond to changes in external K^+ concentration at high pH when $P_H H_o$ would be small compared to $P_K K_o$ in Equation 1. On the basis of these inconsistencies and also the large decreases in membrane conductance that accompanied pump inhibition (108), an alternative model for the electrogenic pump was presented (108) in which the pump was postulated to have the property of electrical conductance, i.e. the flux through the pump would vary with the membrane potential. If the conductance of the pump were sufficiently large, it would be possible to account for the apparent discrepancy between the measured membrane conductance and that calculated from the passive fluxes of the major ions (64, 127).

There remains the problem of accounting for the dependence of the membrane potential on external pH. However, Equations 3 and 4 would no longer be appropriate and the equation giving the membrane potential for the equivalent circuit in Figure 1 is:

$$E = \frac{E_P g_P + E_D g_D}{g_P + g_D} \qquad 5.$$

If $g_P > g_D$, as postulated, the membrane potential will tend toward E_P, the EMF of the electrogenic pump.

An expression for E_P may be obtained from Rapoport's theory (83) for the electrogenic pump (108) or by simply considering the equilibrium thermodynamic situation (125). For an H^+ pump:

$$E_P = \frac{\Delta\bar{\mu}_P}{\nu_H F} - RT \ln \frac{H_i^+}{H_o^+} \qquad 6.$$

where $\Delta\bar{\mu}_P$ is the free energy of the driving reaction, ν_H is a stoichiometric coefficient, and H_i^+ and H_o^+ are the internal and external hydrogen ion concentrations respectively. Thus E_p depends on external pH according to the second term on the right-hand side of Equation 6, and this may in turn account for the dependence of the membrane potential on external pH. It seems possible, therefore, that an alternative explanation for the electrical behavior of characean plasma membranes is possible in terms of an electrogenic proton pump having the property of electrical conductance (108). An important corollary of this hypothesis is that it is no longer necessary to postulate the existence of a large, energy-wasting passive influx of H^+. This type of electrogenic pump would therefore be compatible with a chemiosmotic transport scheme (70).

Equation 6 also suggests a dependence of E_P on cytoplasmic pH. Although Fujii et al (26) have demonstrated a dependence of about 25 mV/pH unit on internal pH in the pH range 6 to 9 in perfused cells of *Chara corallina,* the situation in *Nitella translucens* appears to be more complicated because acidification of the cytoplasm has no effect when the membrane potential is already hyperpolarized (113), but it can cause the membrane potential to repolarize if the light-induced hyperpolarization has decayed spontaneously (113) or as a result of exposure to Cl^--free solutions (112). This may indicate that internal pH also exerts kinetic control over the pump. Such kinetic control may also be necessary to account for the depolarization of the membrane potential at high external pH (56, 108) since the observed changes in cytoplasmic pH (125; R. M. Spanswick, unpublished information) are not sufficient to account for the depolarization in terms of the effect on E_P predicted by Equation 6.

MEMBRANE CONDUCTANCE The idea that electrogenic pumps may contribute a major fraction of the membrane conductance is still not generally accepted, although theoretical treatments show clearly that an electrogenic pump will contribute to the conductance (25, 83) and a correlation between conductance and pump activity is well established in animal epithelia (11, 45). It is more common to treat the electrogenic pump as a current source (56, 102), i.e. as though it has infinite impedance, in which case the flux through the pump would be independent of the membrane potential.

The strong correlation between inhibition of the electrogenic pump and a decrease in membrane conductance in *Nitella* (87–89, 108, 111) seemed consistent with the idea that the pump contributes significantly to the membrane conductance. The utility of this approach may be illustrated by the ability to fit the curvilinear relationship between the membrane potential and membrane conductance observed during the course of inhibition of the electrogenic pump by CO_2 in *Nitella translucens* using Equation 5 and assuming that the observed increase in membrane conductance was attributable entirely to a change in g_P (112). In fitting the curve, the value assumed for E_D (-120 mV) was 4 mV more positive than E_K and the value for E_P was -180 mV. The value used for the conductance of the passive channels (8 μS cm^{-2}) was of similar magnitude to that calculated from the passive ion fluxes [5 μS cm^{-2} (107)]. Thus the conducting pump model appears to be able to account for the change in membrane potential, in this case, on the assumption that only the pump conductance is affected. Since CO_2 does not affect the ATP level in *Nitella* (113) and therefore probably has no effect on E_P (Equation 6), the effect of CO_2 may be explained by a control mechanism that switches the pump sites from a conducting to a nonconducting state. Thus, as in animal epithelia (11), pump conductance appears to provide a measure of pump activity, though in this case the difficulty of measuring H^+ fluxes makes it impossible to verify this hypothesis directly.

In the example given above, inhibition of the electrogenic pump resulted in an apparent reduction of the pump conductance g_P from 44 to 3.8 μS cm^{-2}. Thus the pump conductance in the activated state is greater than that of the passive channels (8 μS cm^{-2}) by a factor of 5.5. It should be noted, however, that results at variance with this conclusion have also been obtained. Kishimoto et al (55) have analyzed the conducting pump model (Figure 1) using the inhibitors DNP (0.2 mM) and triphenyltin chloride (2 μM). They assumed that the inhibitors affected only the pump and that the steady values of the membrane potential and conductance were equal to those of the passive channels alone (i.e. g_P is assumed to become zero under inhibited conditions). They deduced that g_P was slightly larger than g_D under normal conditions. However, one may question whether g_P did fall to zero in their experiments since the steady value of the conductance in the presence of DNP (85 μS cm^{-2}) or triphenyltin chloride (60 μS cm^{-2}) was much greater than the conductance calculated from the passive ion fluxes in the same species [about 3.5 μS cm^{-2} (51)], and larger than the value observed with 0.1 mM DNP [17 μS cm^{-2} (51)]. This result suggests that g_D may have been overestimated in these experiments and g_P consequently underestimated relative to the total membrane conductance which is, however, unusually high.

Tazawa & Shimmen (120) present experiments in which the electrogenic pump in *Chara australis* was inhibited by perfusion of tonoplast-free cells with medium containing no ATP. Although the membrane potential declined instantly, the membrane conductance declined much more slowly. From this result they concluded that the pump conductance was not directly coupled with the electrogenic pump activity. However, such a direct coupling is not a necessary consequence of the conducting pump model. Decreasing the ATP concentration will affect the membrane potential via an effect on E_p. A change in the ratio g_P:g_D will also affect the membrane potential, but there is no theoretical basis for predicting a dependence of g_P on the ATP concentration (83, 108, 112). However, the relationship between the membrane conductance and the ATP level in *Nitella translucens* during the course of inhibition by CCCP has led to the suggestion that the pump may have an allosteric site for ATP (112) that activates the pump in a manner analogous to the activation that is observed with several transport ATPases (19, 28, 124). Tazawa & Shimmen (120) also demonstrated that the response of the membrane potential to a change in external pH is reduced a short time after the perfusion, when the membrane conductance is still high, to the same extent as later when the conductance is low. They interpret this observation to mean also that the pump conductance is small compared with g_D. However, the membrane potential in the absence of ATP is initially about -80 mV, which is in the region of the action potential threshold. It is quite possible that g_D is consequently enhanced, a possibility that should be explored by the measurement of K^+ and Cl^- fluxes. Thus it is possible that the decline in conductance following perfusion reflects a decrease in g_D which may be correlated with the decline in excitability that occurs with a similar time course; g_P may already be at a low value.

Neurospora crassa

Although Slayman (96) initially considered the electrical properties of *Neurospora* in terms of an equivalent circuit having a pump with conductance, subsequent analysis of current-voltage curves before and after inhibition by CN^- (32, 101) has shown that in the region of the resting potential the conductance of the pump accounts for only 5 to 10% of the total membrane conductance. Thus, to a close approximation, the electrogenic pump in *Neurospora* behaves as a current source. Gradmann et al (32) point out that the difference between *Neurospora* and the giant algal cells in this respect arises because the resting potential in *Neurospora* (-175 mV) is rather far from the reversal potential of the pump (-390 mV), which is equivalent to E_p. The slope of the current-voltage curve for the pump (the conductance) tends to be greater near the reversal potential and in the Characeae the resting potential appears to lie very close to the reversal potential (112). The difference in the apparent pump conductance in *Neurospora* and the Characeae may stem from a difference in stoichiometry. The reversal potential in *Neurospora* is so large that it is only compatible with a stoichiometry of 1 H^+/ATP (32) whereas in the Characeae it would be possible to have a stoichiometry of 2 H^+/ATP (112). The behavior of the membrane potential in wild type and mutant strains under a variety of conditions (99, 100) appears to be consistent with pump and leakage pathways which are both subject to control mechanisms. Nevertheless, it would be useful to have an assessment of the leakage pathways that is independent of conductance measurements.

Marine Algae

The electrogenic ion pump in *Acetabularia* (31, 86) has been analyzed in detail by Gradmann (29). The electrical responses to applied current pulses are rather complex under normal conditions because the initial voltage response is followed by a second voltage change in the same direction but with a much longer time constant. The slow response could be blocked by low temperatures and was therefore attributed to the electrogenic pump. The pump channel appears to consist of an EMF in series with two conductances, g_{P1} and g_{P2}. In parallel with the conductance g_{P2} there is a capacitance, C_P, which accounts for the second (slow) response of the membrane potential to a current pulse at normal temperatures. The value of C_P is much larger than that of a normal membrane capacitance and it is increased by light. Gradmann (29) compared the energy stored by this capacitance with the energy available from ATP hydrolysis and showed that they were approximately equal. On this basis, he suggested that the element *P*1 of the analog circuit is the pump itself while *P*2 is associated with the phosphorylating reactions. The slow action potentials observed in this species also seem to involve changes in *P*2 rather than in the passive transport channels (30).

Although not interpreted specifically in terms of electrogenic ion pumps, some of the first evidence for electrogenesis was obtained by Blinks (5, 6), using cells of *Halicystis* with their vacuoles perfused with solutions identical to those outside the cell. Under these conditions the cells could maintain a potential which was sensitive to anaerobiosis but could be restored by light in the presence of nitrogen. More recent work (33, 34) has shown that the membrane potential is generated by an electrogenic Cl^- pump and has confirmed the earlier observation of Blinks et al (7) that inhibition of the pump increases the membrane resistance (34). In this case it was possible to measure the Cl^- influx into voltage-clamped cells and demonstrate that the Cl^- influx is in fact voltage sensitive, though in the opposite direction to that predicted by theories for the electrogenic pump (25, 83). It is this phenomenon that gives rise to a negative conductance region in the current-voltage relationship (34).

Higher Plants

Investigation of the electrical characteristics of the membranes of higher plants is complicated by several factors in addition to the difficulties posed by the small size of the cells. One, the difficulty of inserting a microelectrode into the thin layer of cytoplasm, generally precludes separating the electrical characteristics of the plasmalemma and tonoplast. Even more serious is the difficulty encountered in making reliable measurement of membrane conductance. Insertion of separate electrodes for current injection and potential measurement into higher plant cells is very difficult, and single electrode methods appear to be unreliable because of uncontrollable changes in electrode resistance during the course of experiments (22). Even when a reliable technique has been devised, problems will remain as a result of the plasmodesmatal connections between cells. In the case of *Elodea canadensis* (109) it was estimated that about 80% of the current injected into a cell passed into neighboring cells via the plasmodesmata. As a result, the resistance of the plasmalemma will be underestimated and, furthermore, the contribution of the tonoplast will be overestimated since all the current passes across this membrane if the electrodes are located in the vacuole. The use of isolated protoplasts might avoid some of these problems.

In view of the technical problems involved, it is not surprising that no detailed analyses of the electrical characteristics of higher plant cell membranes have been performed. However, resistance measurements have been made on the long rhizoid cells of the aquatic liverwort *Riccia fluitans* (24). The response of these cells to light is a hyperpolarization similar to that observed in the higher aquatic plants *Elodea canadensis* (110) and *Vallisneria* (4). Unlike *Nitella,* however, the membrane conductance is higher in the dark than in the light. Although this difference may at first sight appear to be at variance with a model in which the pump conductance is high compared to that of the passive channels, examination of the current-voltage relationships in the light and dark revealed that the high conductance in the dark is a consequence of the nonlinearity of the current-voltage curve; the membrane conductance at the dark resting potential is indeed high but, at a potential in the dark equal to that in the light, the conductance in the dark is lower. The high value of the conductance at the resting potential in the dark probably results from an increase in the passive ion conductances at this relatively positive potential. Taking this increase into account, Felle & Bentrup (24) were also able to interpret their results using the circuit shown in Figure 1.

Because of the problems of making resistance measurements, most work on higher plants has been restricted to potential measurements (37, 81). As a consequence, little is known about the conductance characteristics of electrogenic pumps in higher plants, though the insensitivity of the membrane potential to external pH in most species might be taken as indirect evidence for the pump acting as a current source if indeed H^+ is the ion pumped electrogenically. This is because the external pH affects E_p (Equation 6) and would therefore be expected to affect the membrane potential only if g_p is significant compared to g_D. Beet tissues (80) and *Lemna* (59) are the only species for which substantial pH dependence of the membrane potential has been reported. However, recent work on beet (69) suggests that the electrogenic pump in this tissue may act as a constant current source. The pump was inhibited by 50 μM CN^- but the membrane conductance was unchanged. Although a single electrode technique was used in this work, the absence of a consistent change in conductance can be taken as a reliable indication of a constant membrane conductance even if the absolute magnitude of the conductance reported may not be completely accurate. The current-voltage relationship was linear in inhibited cells and thus the constancy of the membrane conductance is unlikely to be caused by a potential-dependent increase in the conductance of the passive channels compensating for a decrease in the pump conductance during inhibition. As the authors point out, however, it is still possible that changes in the plasmalemma resistance could be obscured by a high tonoplast resistance and, as mentioned above, the resistance of the tonoplast relative to that of the plasmalemma is exaggerated by leakage of current to neighboring cells through the plasmodesmata (109).

Cheeseman & Hanson (12) have developed an analysis in which they express the membrane potential as the sum of the diffusion potential plus the hyperpolarization produced by the pump:

$$E = E_D + E_{HYP} \qquad 7.$$

From Equation 1 it will be seen that $\exp(FE/RT)$ is a linear function of K_o^+, and from Equation 7:

$$\exp(FE/RT) = \exp(FE_D/RT)\exp(FE_{HYP}/RT) \qquad 8.$$

For corn roots grown in the absence of Na, and assuming $Cl_o^- << K_i^+$, it was possible to write:

$$\exp(FE_D/RT) = \frac{K_o^+ + (P_{Cl}/P_K)\,Cl_i^-}{K_i^+} \tag{9}$$

By assuming also that the plot of $\exp(FE/RT)$ v. K_o^+ deviated from the linearity exhibited during inhibition by FCCP due to the term $\exp(FE_{HYP}/RT)$ and that the deviation could be described as a saturating function of K_o^+, it was possible to obtain an equation for $\exp(FE_{HYP}/RT)$ which could be used to obtain values for E_{HYP}. These values in turn could be used to fit the curve of E v. K_o^+ with reasonable accuracy.

In later work (13), the sensitivity of the hyperpolarization to anaerobiosis in the presence and absence of DCCD and as a function of K_o^+ has been studied. DCCD appears to inhibit the hyperpolarization at low K_o^+ (1 mM) but not at higher K_o^+.

While this approach may be useful in describing the hyperpolarization produced by an electrogenic pump, this parameter is not in itself of fundamental importance. In the simplest case, that of a constant current pump, the hyperpolarization will be a function of both the current through the pump and the conductance of the passive channels. For a conducting pump the hyperpolarization is given by:

$$E_{HYP} = E - E_D \tag{10}$$

Substituting for E from Equation 5 gives:

$$E_{HYP} = \frac{g_P\,(E_P - E_D)}{g_P + g_D} \tag{11}$$

The hyperpolarization will therefore be dependent on the EMFs and conductances of both the active and passive channels. Thus, the hyperpolarization alone does not provide information about the underlying processes unless special assumptions are made.

IDENTIFICATION OF THE TRANSPORTED ION

In principle it is possible to identify the ion transported by an electrogenic pump by removing it from the side of the membrane from which it is transported, assuming that any effect on the diffusion potential can be taken into account. In the case of inwardly directed pumps, such as the Cl^- pump in *Acetabularia* (31, 86), this is a straightforward matter. However, for an ion pumped from the cytoplasm it is usually impossible to manipulate the concentration in this way, and it is necessary to look for correlations between the efflux and the electrogenic component of the membrane potential under conditions in which the pump is activated or inhibited. In the particular case of an electrogenic H^+ efflux, the situation is made even more difficult because it is only possible to detect net H^+ fluxes.

From measurement of cytoplasmic pH (57, 114, 125) it is clear that H^+ is pumped out of the cell but, although Kitasato (56) postulated in 1968 that H^+ was the ion pumped out of the cell, strong positive evidence for this hypothesis has been difficult to obtain. Lack of inhibition of the pump by inhibition or elimination of the Na^+, Cl^-, and HCO_3^- fluxes was consistent with this hypothesis (111). Furthermore, an estimate of the current passing through the pump, obtained by passing sufficient current across the membrane to depolarize the potential to E_D (at which point the net current through the passive channels should be zero), gave a value (approximately 22 pmol cm^{-2} s^{-1}) which was an order of magnitude larger than the passive fluxes of the major ions (111). Regions of acidification were also observed on the cell surface (117) but, particularly in the presence of HCO_3^-, these appear to represent smaller fluxes than the intervening alkaline regions which most probably result from an efflux of OH^- (62). Nevertheless, Spear et al (117) were able to estimate that the initial H^+ efflux in the light in these regions was in the range of 5 to 20 pmol cm^{-2} s^{-1}. More recently, it has been demonstrated that the efflux of H^+ from perfused cells of *Chara corallina* is stimulated by addition of ATP to the perfusion medium (63). Shimmen & Tazawa (94) have demonstrated, by back titration of a small volume of solution surrounding perfused cells, that the ATP-stimulated H^+ efflux in *Chara australis* is 2 to 6 pmol cm^{-2} s^{-1}.

Accurate estimation of the proton flux through the electrogenic pump is subject to considerable uncertainty as a consequence of the possible contribution of several processes to the external pH changes upon which estimates are based. These include changes caused by CO_2 fluxes (73), OH^- efflux, HCO_3^- uptake (61, 117), passive H^+ fluxes and H^+ movements via symports and antiports. In addition, to observe a net efflux of protons it seems probable that it would be necessary for the cell to have a net source of protons. In fungi (97), there does appear to be a correlation between membrane hyperpolarization and H^+ efflux. However, the most extensive measurements have been made in higher plants where net acidification of the medium surrounding roots is readily observed in the presence of a permeable cation such as K^+. In this case the source of protons is organic acid synthesis in the cytoplasm (105). Study of H^+ excretion from higher plant tissues has received considerable attention in recent years because of the demonstrated importance of this process in auxin action (14). Although it may have a different mode of action from auxin, the stimulation of H^+ excretion by the toxin fusicoccin (66) has been most useful. Of particular importance is the fact that these substances bring about a hyperpolarization of the membrane potential (16, 17, 20, 67, 79), providing further evidence that it is the H^+ pump that is electrogenic.

K^+ uptake is often observed to be correlated with H^+ excretion, and it has been postulated that K^+ may be transported via the same system that is responsible for H^+ transport (81). However, the work of Bellando et al (3), which demonstrated that K^+ uptake is inhibited and H^+ excretion stimulated in the presence of lipophilic cations, suggests that the coupling between the H^+ and K^+ fluxes may be electrical and not due to transport via a common system.

Some of the complications encountered when using intact tissues may be avoided by using plasma membrane vesicles. For use in transport studies the vesicles must be both nonleaky and oriented such that the catalytic site for ATP is on the exterior of vesicle (i.e. inside-out relative to the membrane from which the vesicle was derived). Membrane preparations obtained by the method of Hodges & Leonard (44) contain vesicles that are mostly leaky to protons as determined by the rapid decay of the fluorescence quench of 9-aminoacridine resulting from an imposed pH gradient (78). However, a population of nonleaky vesicles may be obtained by the method of Steck (118) using a discontinuous ficoll gradient (77). A similar result, as judged by nigericin-stimulated ATPase activity of the vesicles, may be obtained using a dextran gradient (119). Addition of ATP to vesicles of this type results in the accumulation of the permeable anion SCN^- in vesicles from *Neurospora* (91) and tobacco callus (119 & H. Sze, personal communication), indicating the generation of an interior-positive membrane potential that would be consistent with the operation of an electrogenic proton pump. ATP-stimulated proton transport has now been detected in plasma membrane vesicles from *Neurospora* (91) and also from both corn roots and tobacco suspension cells (F. M. DuPont, A. B. Bennett, and R. M. Spanswick, unpublished).

ENERGY SOURCES AND THE CONTROL OF PUMP ACTIVITY

Investigations of the light-stimulation of the influxes of K^+ and Cl^- in *Nitella translucens* and *Hydrodictyon africanum* and the differential response of these fluxes to uncoupling agents and inhibitors of photosynthetic electron flow (64, 65, 85) led to an hypothesis in which it was suggested that the K^+ influx was ATP-dependent and the Cl^- influx was dependent on photosynthetic electron flow. The stimulation of the membrane potential by light (111) and its sensitivity to DNP (56) and a variety of inhibitors other than DCMU (87, 110, 111) pointed to the conclusion that the electrogenic pump was dependent on cyclic photophosphorylation. The conclusion that the pump is ATP-dependent is probably correct since inhibitors that reduce the cellular ATP level in *Nitella translucens* (R. M. Spanswick and A. G. Miller, unpublished) and *Chara corallina* (52) also inhibit the electrogenic pump. Dependence of the hyperpolarization in perfused cells on ATP supports this conclusion (93). However, Penth & Weigl (76) showed that light had little effect on the ATP level in *Chara foetida* and the same is true for *Nitella translucens* (113) in which the electrogenic pump is light-stimulated and *Chara corallina* in which effects of light are small. Similarly, 1 mM CO_2/HCO_3^- at pH 6 inhibits the electrogenic pump in *Nitella translucens* but has no effect on the ATP level (113).

The results of these experiments may be used to emphasize the importance of verifying directly the implied effects of light and inhibitor treatments. They also demonstrate the existence and importance of control systems which affect transport rates by mechanisms independent of the chemical potential of the reactions driving the pumps.

The existence of control mechanisms has also been investigated in *Nitella mucronata* (35, 68) by measuring the response of the membrane potential to sinusoidally modulated light. The results have been interpreted in terms of a model with three parallel reaction pathways between the chloroplasts and the electrogenic pump in the membrane. However, the actual reactions in the pathway have yet to be identified.

The energy source in *Neurospora crassa* was investigated by examining the time-course of the decay of the membrane potential, electron flow, and ATP levels during inhibition by CN^- (102). Electron flow was inhibited more rapidly than the decay of the membrane potential, but the ATP level decayed with a time constant nearly identical to that for the decline of the membrane potential. Despite the apparent simplicity of the relationship between the membrane potential and the ATP level, there is considerable evidence for the existence of control mechanisms in *Neurospora* (99, 100). These are particularly evident under conditions of restricted energy supply when the membrane potential and the cellular ATP level are maintained but the growth rate is reduced, indicating regulation of many energy-consuming processes including the electrogenic pump which has been estimated to be responsible for about 20% of the normal ATP turnover (32). An interesting example is the *poky f* strain which is depleted in cytochromes *a* and *b* but which contains a high level of the alternate oxidase (126). Addition of cyanide results in oscillations of the membrane potential which stabilize at a level only slightly more positive than the control level. The ATP level is reduced by only 26% compared to 78% in the wild type. Examination of the current-voltage curves with and without CN^- in the wild type shows that the pump has a reversal potential of –390 mV. Comparison of the current-voltage curves in the *poky f* strain in the presence of CN^- and CN^- + SHAM gives a reversal potential of –196 mV. Warncke & Slayman (126) interpret this result as indicating a change in the stoichiometry of the pump from 1 H^+/ATP to 2 H^+/ATP under conditions of energy restriction. They also point out that in terms of electrical circuits this represents a change from maximal power transfer (load voltage equals half source voltage) to maximal energy transfer (load voltage equals source voltage) and hence maximum efficiency.

The effects of inhibitors on the membrane potential in higher plants are generally consistent with ATP being the energy source for the electrogenic pump (36, 69, 82, 110), but the effect of the treatment on the ATP level has only been investigated in one case (69). Current evidence in favor of involvement of an ATPase with H^+ transport should not be allowed to overshadow the possibility that electron transport may also occur in plasma membranes (48).

It is clear that study of the regulation of the activity of electrogenic pumps is still in its early stages. For systems in which the pump acts as a current source the current is proportional to the deviation of the membrane potential from the diffusion potential. In the case of a conducting pump, the deviation of the membrane potential from the diffusion potential will be a function of both E_P and the relative values of g_P and g_D. Thus thermodynamic factors will affect the value of E_p and kinetic controls may be thought of as affecting g_P. In the case of the effects of darkness or CO_2 on *Nitella,* it was only necessary to postulate an effect on g_P. Also, it appears that the cytoplasmic pH and ATP level may affect g_P in addition to their more direct effect on E_p (112).

CONCLUSIONS

Electrogenic ion pumps are clearly of much greater importance in plants than in animals. For plants living in nonsaline environments, the electrogenic proton pump serves two vital roles. First, it is of major importance in the regulation of cytoplasmic pH (105), and second, it appears to be the primary active transport system in a chemiosmotic scheme in which the electrochemical potential difference for protons may be responsible for driving the fluxes not only of major ions such as Na^+ (49, 50, 84) and Cl^- (43, 90, 104) via symports or antiports, but also the transport of sugars and amino acids via symports (23, 74, 103). The importance of proton cotransport systems extends to phloem loading (2, 27) and the accumulation of organic molecules by developing embryos (58). Thus the electrogenic proton pump may well be the central component of a scheme which makes it possible to provide a more unified view of solute transport in plants than has previously been possible. Nevertheless, the chemiosmotic viewpoint should not be accepted uncritically; it should be recognized that the existence of each component must be individually tested and that the presence of multiple kinetic controls governing each process will make this a lengthy and arduous task.

ACKNOWLEDGMENTS

I wish to thank Professor Tazawa and Drs. Lucas, Shimmen, and Sze for information on their research prior to publication.

Literature Cited

1. Anderson, W. P., Robertson, R. N., Wright, B. J. 1977. Membrane potentials in carrot root cells. *Aust. J. Plant Physiol.* 4:241–52
2. Baker, D. A. 1978. Proton co-transport of organic solutes by plants. *New Phytol.* 81:485–97
3. Bellando, M., Trotta, A., Bonetti, A., Colombo, R., Lado, P., Marrè, E. 1979. Dissociation of H^+ extrusion from K^+ uptake by means of lipophilic cations. *Plant Cell Environ.* 2:39–47
4. Bentrup, F. W., Gratz, H. J., Unbehauen, H. 1973. The membrane potential of *Vallisneria* leaf cells: evidence for light-dependent proton permeability changes. In *Ion Transport in Plants,* ed. W. P. Anderson, pp. 171–82. London: Academic. 630 pp.
5. Blinks, L. R. 1935. Protoplasmic potentials in *Halicystis.* IV. Vacuolar perfusion with artificial sap and sea water. *J. Gen. Physiol.* 18:409–20
6. Blinks, L. R. 1949. The source of the bioelectric potentials in large plant cells. *Proc. Natl. Acad. Sci. USA* 35:566–75
7. Blinks, L. R., Darsie, M. L., Skow, R. K. 1938. Bioelectric potentials in *Halicystis.* VII. The effects of low oxygen tension. *J. Gen. Physiol.* 22: 255–79
8. Bowling, D. J. F. 1980. An electrogenic phosphate pump in sunflower roots. In *Plant Membrane Transport: Current Conceptual Issues,* ed. R. M. Spanswick, W. J. Lucas, J. Dainty, pp. 405–6. Amsterdam: Elsevier/North-Holland Biomed. Press. 670 pp.
9. Bowling, D. J. F., Dunlop, J. 1978. Uptake of phosphate by white clover. I. Evidence for an electrogenic phosphate pump. *J. Exp. Bot.* 29:1139–46
10. Briggs, G. E. 1962. Membrane potential differences in *Chara australis. Proc. R. Soc. London Ser. B* 156: 573–77
11. Caplan, S. R., Essig, A. 1977. A thermodynamic treatment of active sodium transport. *Curr. Top. Membr. Transp.* 9:145–76
12. Cheeseman, J. M., Hanson, J. B. 1979. Mathematical analysis of the dependence of cell potential on external potassium in corn roots. *Plant Physiol.* 63:1–4
13. Cheeseman, J. M., LaFayette, P. R., Gronewald, J. W., Hanson, J. B. 1980. Effect of ATPase inhibitors on cell potential and K^+ influx in corn roots. *Plant Physiol.* 65:1139–45
14. Cleland, R. 1971. Cell wall extension. *Ann. Rev. Plant Physiol.* 22:197–222
15. Cleland, R. E., Lomax, T. 1977. Hormonal control of H^+excretion from oat cells. In *Regulation of Cell Membrane Activities in Plants,* ed. E. Marrè, O. Ciferri, pp. 161–71. Amsterdam: North-Holland. 332 pp.
16. Cleland, R. E., Prins, H. B. A., Harper, J. R., Higinbotham, N. 1977. Rapid hormone-induced hyperpolarization of the oat coleoptile transmembrane potential. *Plant Physiol.* 59:395–97
17. Cocucci, M., Marrè, E., Ballarin-Denti, A.,Scacchi, A. 1976. Characteristics of fusicoccin-induced changes of transmembrane potential and ion uptake in maize root segments. *Plant Sci. Lett.* 6:143–56
18. Dainty, J. 1962. Ion transport and electrical potentials in plant cells. *Ann. Rev. Plant Physiol.* 13:379–402
19. DuPont, Y. 1977. Kinetics and regulation of sarcoplasmic reticulum ATPase. *Eur. J. Biochem.* 72:185–90
20. Etherton, B. 1970. Effect of indole-3-acetic acid on membrane potentials of oat coleoptile cells. *Plant Physiol.* 45:527–28
21. Etherton, B., Higinbotham, N. 1960. Transmembrane potential measurements of cells of higher plants. *Science* 131:409–10
22. Etherton, B., Keifer, D. W., Spanswick, R. M. 1977. A comparison of three methods for the measurement of electrical resistances of plant cell membranes. *Plant Physiol.* 60:684–88
23. Etherton, B., Rubinstein, B. 1978. Evidence for amino acid-H^+ co-transport in oat coleoptiles. *Plant Physiol.* 61:933–37

24. Felle, H., Bentrup, F. W. 1976. Effect of light upon membrane potential, conductance, and ion fluxes in *Riccia fluitans*. *J. Membr. Biol.* 27:153–70
25. Finkelstein, A. 1964. Carrier model for active transport of ions across a mosaic membrane. *Biophys. J.* 4:421–40
26. Fujii, S., Shimmen, T., Tazawa, M. 1979. Effect of intracellular pH on the light-induced potential change and electrogenic activity in tonoplast-free cells of *Chara australis*. *Plant Cell Physiol.* 20:1315–28
27. Giaquinta, R. 1980. Sucrose/proton cotransport during phloem loading and its possible control by internal sucrose concentration. See Ref. 8, pp. 273–82
28. Glynn, I. M., Karlish, S. D. J. 1976. ATP hydrolysis associated with an uncoupled sodium flux through the sodium pump: evidence for allosteric effects of intracellular ATP and extracellular sodium. *J. Physiol.* 256:465–96
29. Gradmann, D. 1975. Analog circuit of the *Acetabularia* membrane. *J. Membr. Biol.* 25:183–208
30. Gradmann, D. 1976. "Metabolic" action potentials in *Acetabularia*. *J. Membr. Biol.* 29:23–45
31. Gradmann, D., Bentrup, F. W. 1970. Light-induced membrane potential changes and rectification in *Acetabularia*. *Naturwissenschaften* 57: 46–47
32. Gradmann, D., Hansen, U.-P., Long, W. S., Slayman, C. L., Warncke, J. 1978. Current-voltage relationships for the plasma membrane and its principal electrogenic pump in *Neurospora crassa:* I: Steady-state conditions. *J. Membr. Biol.* 39:333–67
33. Graves, J. S., Gutknecht, J. 1977. Chloride transport and the membrane potential in the marine alga, *Halicystis parvula*. *J. Membr. Biol.* 36:65–81
34. Graves, J. S., Gutknecht, J. 1977. Current-voltage relationships and voltage sensitivity of the Cl^- pump in *Halicystis parvula*. *J. Membr. Biol.* 36:83–95
35. Hansen, U.-P. 1978. Do light-induced changes in the membrane potential of *Nitella* reflect the feed-back regulation of a cytoplasmic parameter? *J. Membr. Biol.* 41:197–224
36. Higinbotham, N. 1973. Electropotentials of plant cells. *Ann. Rev. Plant Physiol.* 24:25–46
37. Higinbotham, N., Anderson, W. P. 1974. Electrogenic pumps in higher plant cells. *Can. J. Bot.* 52:1011–21
38. Higinbotham, N., Etherton, B., Foster, R. J. 1964. Effect of external K, NH_4, Na, Ca, Mg, and H ions on the cell transmembrane electropotential of *Avena* coleoptile. *Plant Physiol.* 39:196–203
39. Higinbotham, N., Graves, J. S., Davis, R. F. 1970. Evidence for an electrogenic ion transport pump in cells of higher plants. *J. Membr. Biol.* 3:210–22
40. Hill, A. E. 1967. Ion and water transport in *Limonium*. I. Active transport by the leaf gland cells. *Biochim. Biophys. Acta* 135:454–60
41. Hill, A. E., Hill, B. S. 1973. The electrogenic chloride pump of the *Limonium* salt gland. *J. Membr. Biol.* 12:129–44
42. Hill, B. S., Hill, A. E. 1973. ATP-driven chloride pumping and ATPase activity in the *Limonium* salt gland. *J. Membr. Biol.* 12:145–58
43. Hodges, T. K. 1973. Ion absorption by plant roots. *Adv. Agron.* 25:163–207
44. Hodges, T. K., Leonard, R. T. 1974. Purification of a plasma membrane-bound adenosine triphosphatase from plant roots. *Methods Enzymol.* 32:392–406
45. Hong, C. D., Essig, A. 1976. Effects of 2-deoxy-D-glucose, amiloride, vasopressin, and ouabain on active conductance and E_{Na} in the toad bladder. *J. Membr. Biol.* 28:121–42
46. Hope, A. B. 1965. Ionic relations of cells of *Chara australis*. X. Effects of bicarbonate ions on the electrical properties. *Aust. J. Biol. Sci.* 18:789–801
47. Hope, A. B., Walker, N. A. 1961. Ionic relations of *Chara australis* R. Br. IV. Membrane potential differences and resistances. *Aust. J. Biol. Sci.* 14:26–44
48. Ivankina, N. G., Novak, V. A. 1980. H^+-transport across plasmalemma. H^+-ATPase or redox-chain? See Ref. 8, pp. 503–4
49. Jeschke, W. D. 1980. Roots: cation selectivity and compartmentation, involvement of protons. See Ref. 8, pp. 17–28
50. Jeschke, W. D. 1980. Involvement of proton fluxes in the K^+-Na^+ selectivity at the plasmalemma; K^+-dependent net extrusion of sodium in barley roots and the effect of anions and pH on sodium fluxes. *Z. Pflanzenphysiol.* 98:155–75
51. Keifer, D. W., Spanswick, R. M. 1978. The activity of the electrogenic pump in *Chara corallina* as inferred from measurements of the membrane potential, conductance and potassium permeability. *Plant Physiol.* 62:653–61
52. Keifer, D. W., Spanswick, R. M. 1979. Correlation of ATP levels in *Chara corallina* with the activity of the elec-

trogenic pump. *Plant Physiol.* 64: 165–68
53. Kerkut, G. A., York, B. 1971. *The Electrogenic Sodium Pump.* Bristol: Scientechnica. 182 pp.
54. Kishimoto, U. 1959. Electrical characteristics of *Chara corallina. Ann. Rep. Sci. Works Fac. Sci. Osaka Univ.* 7:115–46
55. Kishimoto, U., Kami-ike, N., Takeuchi, Y. 1980. The role of the electrogenic pump in *Chara corallina. J. Membr. Biol.* 55:149–56
56. Kitasato, H. 1968. The influence of H^+ on the membrane potential and ion fluxes of *Nitella. J. Gen. Physiol.* 52:60–87
57. Kurkdjian, A., Guern, J. 1978. Intracellular pH in higher plant cells. I. Improvements in the use of the 5,5-dimethyloxazolidine-2[^{14}C], 4-dione distribution technique. *Plant Sci. Lett.* 11: 337–44
58. Lichtner, F. T., Spanswick, R. M. 1980. Sucrose/proton cotransport in developing soybean cotyledons. See Ref. 8, pp. 545–46
59. Löppert, H. 1979. Evidence for electrogenic proton extrusion by subepidermal cells of *Lemna paucicostata* 6746. *Planta* 144:311–15
60. Löppert, H., Kronberger, W., Kandeler, R. 1978. Phytochrome-mediated changes in membrane potential of subepidermal cells of *Lemna paucicostata* 6746. *Planta* 138:133–36
61. Lucas, W. J. 1975. Photosynthetic fixation of ^{14}carbon by internodal cells of *Chara corallina. J. Exp. Bot.* 26:331–46
62. Lucas, W. J. 1979. Alkaline band formation in *Chara corallina.* Due to OH^- efflux or H^+ influx? *Plant Physiol.* 63:248–54
63. Lucas, W. J., Shimmen, T. 1981. Intracellular perfusion and cell centrifugation studies on plasmalemma transport processes in *Chara corallina. J. Membr. Biol.* In press
64. MacRobbie, E. A. C. 1962. Ionic relations of *Nitella translucens. J. Gen. Physiol.* 45:861–78
65. MacRobbie, E. A. C. 1965. The nature of the coupling between light energy and active ion transport in *Nitella translucens. Biochim. Biophys. Acta* 94: 64–73
66. Marrè, E. 1979. Fusicoccin: a tool in plant physiology. *Ann. Rev. Plant Physiol.* 30:273–88
67. Marrè, E., Lado, P., Ferroni, A., Ballarin-Denti, A. 1974. Transmembrane potential increase induced by auxin, benzyladenine and fusicoccin. Correlation with proton extrusion and cell enlargement. *Plant Sci. Lett.* 2:257–65
68. Martens, J., Hansen, U.-P., Warncke, J. 1979. Further evidence for the parallel pathway model of the metabolic control of the electrogenic pump in *Nitella* as obtained from the high frequency slope of the action of light. *J. Membr. Biol.* 48:115–39
69. Mercier, A. J., Poole, R. J. 1980. Electrogenic pump activity in red beet: its relation to ATP levels and to cation influx. *J. Membr. Biol.* 55:165–74
70. Mitchell, P. 1967. Translocations through natural membranes. *Adv. Enzymol.* 29:33–87
71. Moreton, R. B. 1969. An investigation of the electrogenic Na pump in snail neurones, using the constant field theory. *J. Exp. Biol.* 51:181–201
72. Mullins, L. J., Noda, K. 1963. The influence of sodium-free solutions on the membrane potential of frog muscle fibers. *J. Gen. Physiol.* 47:117–32
73. Neumann, J., Levine, R. P. 1971. Reversible pH changes in cells of *Chlamydomonas reinhardii* resulting from CO_2 fixation in the light and its evolution in the dark. *Plant Physiol.* 47:700–4
74. Novacky, A., Ullrich-Eberius, C. I., Lüttge, U. 1978. Membrane potential changes during transport of hexoses in *Lemna gibba* Gl. *Planta* 138:263–70
75. Osterhout, W. J. V. 1931. Physiological studies of single plant cells. *Biol. Rev.* 6:369–411
76. Penth, B., Weigl, J. 1971. Anionen-Influx, ATP-Spiegel und CO_2-Fixierung in *Limnophila gratioloides* und *Chara foetida. Planta* 96:212–23
77. Perlin, D. S. 1980. *Isolation and characterization of plasma membranes from higher plants.* PhD thesis. Cornell Univ., Ithaca, NY. 193 pp.
78. Perlin, D. S., Spanswick, R. M. 1980. Proton transport in plasma membrane vesicles? See Ref 8, pp. 529–30
79. Pitman, M. G., Anderson, W. P., Schaefer, N. 1977. H^+ ion transport in plant roots. See Ref. 15, pp. 147–60
80. Poole, R. J. 1966. The influence of the intracellular potential on potassium uptake by beetroot tissue. *J. Gen. Physiol.* 49:551–63
81. Poole, R. J. 1974. Ion transport and electrogenic pumps in storage tissue cells. *Can. J. Bot.* 52:1023–28
82. Poole, R. J. 1978. Energy coupling for membrane transport. *Ann. Rev. Plant Physiol.* 29:437–60

83. Rapoport, S. I. 1970. The sodium-potassium exchange pump: relation of metabolism to electrical properties of the cell. I. Theory. *Biophys. J.* 10: 246–59
84. Ratner, A., Jacoby, B. 1976. Effect of K^+, its counter anion, and pH on sodium efflux from barley root tips. *J. Exp. Bot.* 27:843–52
85. Raven, J. A. 1967. Light stimulation of active transport in *Hydrodictyon africanum. J. Gen. Physiol.* 50:1627–40
86. Saddler, H. D. W. 1970. The membrane potential of *Acetabularia mediterranea. J. Gen. Physiol.* 55:802–21
87. Saito, K., Senda, M. 1973. The light-dependent effect of external pH on the membrane potential of *Nitella. Plant Cell Physiol.* 14:147–56
88. Saito, K., Senda, M. 1973. The effect of external pH on the membrane potential of *Nitella* and its linkage to metabolism. *Plant Cell Physiol.* 14:1045–52
89. Saito, K., Senda, M. 1974. The electrogenic ion pump revealed by the external pH effect on the membrane potential of *Nitella.* Influences of external ions and electric current on the pH effect. *Plant Cell Physiol.* 15:1007–16
90. Sanders, D. 1980. The mechanism of Cl^- transport at the plasma membrane of *Chara corallina* I. Cotransport with H^+. *J. Membr. Biol.* 53:129–41
91. Scarborough, G. A. 1980. Proton translocation catalysed by the electrogenic ATPase in the plasma membrane of *Neurospora. Biochemistry* 19:2925–31
92. Schacher-Hill, B., Hill, A. E. 1970. Ion and water transport in *Limonium.* VI. The induction of chloride pumping. *Biochim. Biophys. Acta* 211:313–17
93. Shimmen, T., Tazawa, M. 1977. Control of membrane potential and excitability of *Chara* cells with ATP and Mg^{2+}. *J. Membr. Biol.* 37:167–92
94. Shimmen, T., Tazawa, M. 1981. ATP-dependent H^+-extrusion in *Chara australis. Plant Cell Physiol.* In press
95. Slayman, C. L. 1965. Electrical properties of *Neurospora crassa.* Effects of external cations on the intracellular potential. *J. Gen. Physiol.* 49:69–92
96. Slayman, C. L. 1965. Electrical properties of *Neurospora crassa.* Respiration and the intracellular potential. *J. Gen. Physiol.* 49:93–116
97. Slayman, C. L. 1970. Movement of ions and electrogenesis in microorganisms. *Am. Zool.* 10:377–92
98. Slayman, C. L. 1974. Proton pumping and generalized energetics of transport: a review. In *Membrane Transport in Plants,* ed. U. Zimmermann, J. Dainty, pp. 107–19
99. Slayman, C. L. 1977. Energetics and control of transport in *Neurospora.* In *Water Relations in Membrane Transport in Plants and Animals*, ed. A. M. Jungreis, T. K. Hodges, A. Kleinzeller, S. G. Schultz, pp. 69–86. New York: Academic. 393 pp.
100. Slayman, C. L. 1980. Transport control phenomena in *Neurospora.* See Ref. 8, pp. 179–90
101. Slayman, C. L., Gradmann, D. 1975. Electrogenic proton transport in the plasma membrane of *Neurospora. Biophys. J.* 15:968–71
102. Slayman, C. L., Long, W. S., Lu, C. Y.-H. 1973. The relationship between ATP and an electrogenic pump in the plasma membrane of *Neurospora crassa. J. Membr. Biol.* 14:305–38
103. Slayman, C. L., Slayman, C. W. 1974. Depolarization of the plasma membrane of *Neurospora* during active transport of glucose: evidence for a proton-dependent cotransport system. *Proc. Natl. Acad. Sci. USA* 71:1935–39
104. Smith, F. A. 1970. The mechanism of chloride transport in characean cells. *New Phytol.* 69:903–17
105. Smith, F. A., Raven, J. A. 1979. Intracellular pH and its regulation. *Ann. Rev. Plant Physiol.* 30:289–311
106. Spanswick, R. M. 1970. The effects of biocarbonate ions and external pH on the membrane potential and resistance of *Nitella translucens. J. Membr. Biol.* 2:59–70
107. Spanswick, R. M. 1970. Electrophysiological techniques and the magnitudes of the membrane potentials and resistances of *Nitella translucens. J. Exp. Bot.* 21:617–27
108. Spanswick, R. M. 1972. Evidence for an electrogenic ion pump in *Nitella translucens.* I. The effects of pH, K^+, Na^+, light and temperature on the membrane potential and resistance. *Biochim. Biophys. Acta* 288:73–89
109. Spanswick, R. M. 1972. Electrical coupling between the cells of higher plants: a direct demonstration of intercellular transport. *Planta* 102:215–27
110. Spanswick, R. M. 1973. Electrogenesis in photosynthetic tissues. See Ref. 4, pp. 113–28
111. Spanswick, R. M. 1974. Evidence for an electrogenic ion pump in *Nitella translucens.* II. Control of the light-stimulated component of the membrane potential. *Biochim. Biophys. Acta* 332:387–98

112. Spanswick, R. M. 1980. Biophysical control of electrogenicity in the Characeae. See Ref. 8, pp. 305–16
113. Spanswick, R. M., Miller, A. G. 1977. The effect of CO_2 on the Cl^- influx and electrogenic pump in *Nitella translucens.* In *Transmembrane Ionic Exchanges in Plants,* ed. M. Thellier, A. Monnier, M. Demarty, J. Dainty, pp. 239–45. Paris/Rouen: C.N.R.S. 607 pp.
114. Spanswick, R. M., Miller, A. G. 1977. Measurement of the cytoplasmic pH in *Nitella translucens. Plant Physiol.* 59:664–66
115. Spanswick, R. M., Stolarek, J., Williams, E. J. 1967. The membrane potential of *Nitella translucens. J. Exp. Bot.* 17:1–16
116. Spanswick, R. M., Williams, E. J. 1964. Electrical potentials and Na, K, and Cl concentrations in the vacuole and cytoplasm of *Nitella translucens. J. Exp. Bot.* 15:422–27
117. Spear, D. G., Barr, J. K., Barr, C. E. 1969. Localization of hydrogen ion fluxes in *Nitella. J. Gen. Physiol.* 54:397–414
118. Steck, T. L. 1974. Preparation of impermeable inside-out and right-side-out vesicles from erythrocyte membranes. *Methods Membr. Biol.* 2:245–81
119. Sze, H. 1980. Nigericin-stimulated ATPase activity in microsomal vesicles of tobacco callus. *Proc. Natl. Acad. Sci. USA* 77:5904–8
120. Tazawa, M., Shimmen, T. 1980. Action potential in Characeae; some characteristics revealed by internal perfusion studies. See Ref. 8, pp. 349–62
121. Thomas, R. C. 1972. Electrogenic sodium pump in nerve and muscle cells. *Physiol. Rev.* 52:563–94
122. Umrath, K. 1934. Der Einfluss der Temperatur auf das elektrische Potential, den Aktionsstrom und die Protoplasmastromung bei *Nitella mucronata. Protoplasma* 21:329–34
123. Ussing, H. H., Zerahn, K. 1951. Active transport of sodium as the source of electric current in the short-circuited isolated frog skin. *Acta Physiol. Scand.* 23:110–27
124. Verjovski-Almeida, S., Inesi, G. 1979. Fast-kinetic evidence for an activating effect of ATP on the Ca^{2+} transport of sarcoplasmic reticulum ATPase. *J. Biol. Chem.* 254:18–21
125. Walker, N. A., Smith, F. A. 1975. Intracellular pH in *Chara corallina* measured by DMO distribution. *Plant Sci. Lett.* 4:125–32
126. Warncke, J., Slayman, C. L. 1980. Metabolic regulation of stoichiometry in a proton pump. *Biochim. Biophys. Acta* 591:224–33
127. Williams, E. J., Johnston, R. J., Dainty, J. 1964. The electrical resistance and capacitance of the membranes of *Nitella translucens. J. Exp. Bot.* 15:1–14
128. Young, M., Sims, A. P. 1973. The potassium relations of *Lemna minor* L. II. The mechanism of potassium uptake. *J. Exp. Bot.* 24:317–27

Ann. Rev. Plant Physiol. 1981. 32:291–311

CONTROL OF MORPHOGENESIS IN IN VITRO CULTURES

◆7713

Kiem M. Tran Thanh Van

Laboratoire du Phytotron, CNRS, F-91190, Gif sur Yvette, France

---

CONTENTS

0066-4294/81/0601-0291$01.00

INTRODUCTION

In order to include the de novo genesis of structures and functions already known or yet unknown, as well as structures without apparent function such as callus or unicellular hair formation, I will use the term "morphogenesis." The term "regeneration" is defined as the structuring of any part which has been removed or physiologically isolated from the organism.

This review does not attempt to cover all aspects of morphogenesis in in vitro growing systems. There are comprehensive reviews, books, and proceedings (12, 15, 26, 30, 33, 36, 51) treating morphogenesis on different in vitro-cultured systems—organ, tissue, cell, protoplast, sporocyte, etc—under various aspects such as morphological, histological, biochemical, and molecular analysis, as well as the technology of in vitro culture itself.

The following specific aspects have already been reviewed: different factors (inherent and exogenously applied) quoted as controlling morphogenesis in vitro (37, 41), the genetic determinant of the morphogenetic expressions being revealed in vitro (according to our up-to-date status of knowledge) (46), as well as an original method to force the cellular regions "incapable of flower formation" (i.e. senescing meristems, perpetual vegetative meristems of perennial plants) to flower by simply suppressing an inhibition (39, 40). Trusting that the general body of knowledge relevant to morphogenesis and in vitro culture technique is well known, in this review I will examine the conceptual and technological reasons which could have limited our control of morphogenesis. In the last part, I will develop some aspects which I think are in need of more intensive and more concerted investigation.

ENTIRE PLANT AND EXPLANTS

A seed has left the parent plant with the message to reconstitute an entire plant with similar shape, structure, and function. How this message is kept in the embryo and how it is realized is not yet known in detail. All we know is that a complex adult multicellular organism has emerged from a relatively simply organized zygote through a sequence of mitoses, the very first division being an unequal one. We are now facing an orderly patterned assembly of cells, tissues, and organs harmoniously structured with specialized functions. These have the following property, so particular to living organisms: that is that time and space are anchored in each of their cells through a network of interrelation which links together every elementary particle, every atom, molecule, and macromolecule constituting the complex edifice. Changes occurring at any part of this ensemble result in a different arrangement of components and the appearance of new ones in this interrelation network.

With this moving and modulating spatial and temporal network in mind, we are unable to excise some parts from a mother plant and put them in the culture medium without taking into account the new interrelation network now found in the explant and in the remaining mother plant. The "inherent cellular state" which the explant has brought along into the culture medium after its excision, as well as "the past of the mother plant," are as important components as the nutritional, phytohormonal, or environmental factors being applied to the explant. The "inherent cellular state" is clearly demonstrated in the tobacco (9b) and *Torenia* (35) stem fragments or thin cell layers (39a, 44) excised along the stem axis and along the floral branches. We can modulate the "inherent cellular state" by pretreating the mother plant and modifying its physiological, biochemical, and genetic traits.

PRESENT STATE OF IN VITRO MORPHOGENESIS

There are two categories of plants classified according to their ability to differentiate in vitro new shapes, structures, and/or functions: the easily regenerating species and the reluctant ones. Where lies the difference between them? Before trying to answer this question, let us outline the present state of in vitro morphogenesis.

Auxin-Cytokinin Ratio and Other Factors

Since the discovery of phytohormones and the hypothesis of regulation by an auxin/cytokinin ratio (31), significant progress has been made leading to important applications in agronomy and industry and to the promising technology of haploidization and of protoplast fusion leading to genetic engineering. We have been so fascinated by the apparent totipotentiality of plant cells and by the ease in regenerating organs in certain species through changing the growth hormone ratio, that we have almost been convinced since 1957 that this ratio is the basis of control of morphogenesis. In fact, when confronted by some reluctant cases, we can somehow overcome the difficulties (*a*) by changing the type of growth hormones and growth inhibitors, e. g. by using Pichloram or TIBA (triiodobenzoic acid) instead of 2,4-D (21, 28), by using GA and BA (benzyladenine) instead of NAA and BA (8), or by using auxin conjugates instead of free auxin (18); (*b*) by removing 2,4-D or agar from the medium (43). However, changes in the macro-micronutrient ratio (43) and the addition of various substances such as charcoal or organic compounds (vitamins, amino acids, polyamines, phenolamines, polypeptides, steroids, or diverse plant extracts) can affect morphogenesis. Some substances such as tyrosine, polyamines (1, 14), or phenolamines (23) interfere in shoot formation and cell division. Others like carbohydrates, sucrose, or glucose have been shown to control tracheid formation specifically or de novo flower formation in thin cell layers of tobacco (39b, 44). To this list can be added a great number of combinations and variations of light (quality and quantity), pH (9c), water potential (39b), temperature, gaseous atmosphere (40), container shape (40), etc. Yet neither the outburst of miscellaneous factors nor the unique hypothesis of auxin-cytokinin ratio can bring one closer to a basic understanding of the whole process of morphogenetic differentiation as long as the target cells are scattered among a heterogeneous mass of cells. From these considerations emerges the need for having a common cell origin for all morphogenetic patterns in order to localize more closely the target cells and the role of morphogenetic signals.

Totipotentiality, Unipotentiality, or Nullipotentiality; Multiprogrammable Potentiality of Thin Cell Layers

Whether all possible morphogenetic patterns have been obtained in vitro from the same experimental system is still a question open to discussion. The most frequently obtained patterns are callus, tracheid, or nodule with subsequent formation of buds and/or roots. Flowers or embryos were not obtained in these systems. The only morphogenetic pattern one can obtain from a microspore or macrospore isolated from the mother plant is an embryo (via callus formation or not). Although embryos can be considered as totipotent, micro or macrospores seem to be rather unipotent. For a few experimental systems in which de novo floral buds can be obtained, there are only two alternatives: floral buds or vegetative buds; neither embryos nor roots are obtained. In some species guard cells can be considered as nullipotent. Thus, in general, not all morphogenetic patterns (embryo, root, bud, flower) can be initiated from the same material, that is to say from the same cell layer or from the same cell. The totipotentiality may perhaps exist, but up to now there are only one or two organogenetic potentials being revealed.

The purpose of using a cell layer system is to force the same cell layer to initiate all patterns of morphogenesis, but only one pattern at a time (pure morphogenetic program). Experiments have shown that a monolayer of *Begonia* leaf epidermis can be "programmed" to form de novo unicellular hairs (a new morphogenetic pattern if we consider the origin of the tissue), roots, or buds, but not flowers or embryos (39b). Thin cell layers of tobacco form de novo buds, roots, or flowers, as well as calluses without any subsequent organogenesis, from the same cell layer if one varies quantitative combinations of glucose, indolylbutyric acid (IBA), and kinetin (39a, 41, 44).

The same morphogenetic programs were obtained in *Cichorium* and in *Torenia.* These examples illustrate the possibility of multiprogrammable morphogenetic potentiality and therefore real control of all patterns of morphogenetic differentiation. It has been shown that up to a certain stage, thin cell layers can be converted from one program to another, leading to the concepts of determination, reversibility, and transdetermination. The latter has already been proposed for differentiation in *Drosophila* research (17). Furthermore, in one of the reluctant species such as *Psophocarpus* (Leguminosae), thin cell layers have been found to be more morphologically competent than the corresponding internode. There can be induced to form directly from the epidermal layer, within 10 days, all types of organs: buds and roots as well as friable or compact calluses (50).

Excision, Wound, and Healing

As long as tissues or cells are maintained in the network of interrelationships developed within an entire plant during the ontogenetic differentiation, those tissues or cells behave as matured cells, expressing the morphogenetic program they are confined to express. In certain cases, isolation of these tissues or cells is sufficient to trigger de novo bud, root, or callus formation without any exogenous factors. It is therefore important to consider the multiple consequences of the routine act of excision. What is triggered by the excision itself and which process is ultimately morphogenetic? It is difficult to conduct experiments in order to answer both questions since the morphogenetic differentiation in in vitro systems always occurs after a compulsory step of excision. The only known case of de novo bud formation without excision was shown on intact and attached *Begonia* leaves consequent to 6(γ,γ-dimethylallylamino-)purine (DMAAP) treatment (9a). It was shown (18a) on dormant tissue of a potato tuber that slicing initiates important metabolic and genetic changes leading to mitotic activities. Several authors noted numerous modifications prior to cell division: organelle and nuclear displacement, increase in nuclear size, in extent of endoplasmic reticulum, in polysome formation, and in closing of sieve plates by callose plugs. In addition, other events were observed very early after excision: increase in membrane permeability, changes in ionic partition, synthesis of the DNA satellite (1a), destruction or synthesis of growth hormones consequent to differential gene activation, increase in peroxidase activity, active synthesis of proteins and phenolic compounds, and an important increase in respiration and in ethylene production. These complex events occur at the initial time of culture and therefore strongly change the "inherent cellular state."

However, the excision alone is not sufficient to trigger cell division in tobacco pith tissue (31) and in thin cell layers of tobacco (44). For the reluctant species, it is particularly important to study carefully the metabolic and genetic changes occurring after excision: DNA amplification, changes in peroxidase activity, in isozymic pattern (14a), and in metabolism of phenolic compounds. Some of these compounds, considered as auxin protectors, may counteract IAA destruction (19) as well as act on the polarization of cell membranes and on the inhibition of ion uptake. The precise understanding of the effect of a wound and the determination of endogenous growth hormone content, of enzymic modifications and DNA modification, would help in determining whether exogenous growth hormones are to be supplied and, if so, when and at what ratio. In fact, there are several cases in which buds were initiated with a higher auxin than kinetin level, whereas roots were formed with a reverse ratio (52). Furthermore, one of the difficulties encountered in woody plants, orchid plants, and in some species of legumes is the deposition of polyphenols and of oxidation products such as melanin. There can also be an increase in lignin, suberin, cutin, and callose around the cut surface which may greatly modify both the composition of the culture medium and the uptake of metabolites.

Changes Induced in In Vitro Culture

GENETIC AND EPIGENETIC VARIABILITIES In addition to the cases in which both morphogenetic potential and the absence of morphogenesis can be maintained during prolonged periods, there are several instances where a loss of morphogenetic capacity was gradually observed in callus cultures. Genetic and epigenetic changes were recorded during cell culture. Important genetic changes occurred during excision and during in vitro culture (10): chromosomal endoreduplication, nuclear fragmentation resulting in multinucleate cells, and normal or abnormal mitosis. These changes resulted in heterogeneous cell populations with ploidy levels ranging from diploid, to aneuploid, and/or polyploid. It was suggested that diploid cells can be selected from frequent subcultures of callus suspension. From the observation that only diploid plants were regenerated from callus of mixed ploidy level, a hypothesis has been postulated that only diploid cells were morphologically competent. This may not be a general process since it was shown that haploid and hypohaploid (resulting from embryogenesis obtained from pollen-grain formed in vitro on haploid cell layers) thin epidermal and subepidermal cell layers are highly morphogeneticaly competent. Under dark conditions, 45% of the diploid cell layers formed vegetative and floral buds, compared to 80% of the haploid cell layers (49).

However, changes in morphogenetic ability do not always result from genetic changes. The following factors are relevant to epigenetic changes: alteration in phytohormone level and in polypeptides (6) exhibited by habituated versus nonhabituated cell lines; phenotypic variations maintained stable during several generations through maternal transmission (10a). On the other hand, cytoplasm-genome interaction was observed in *Nicotiana tabaccum* when the frequency in chromosomal pairing at ana-metaphase I on pollen mother cells of flowers formed in vivo on the haploid mother plants (N=X+X'=24) and in vitro on haploid thin cell layers was compared. A striking observation was that embryo formation could be obtained as a result of this difference in chromosomal behavior when microspores obtained in vitro were brought into culture (45). These embryos and subsequent plants were both found to be aneuploid and hypohaploid with chromosome number ranging from 2 to 24 (39b). Some hypohaploid tobacco plants have lost their flowering capacity and remain vegetative.

On haploid plants, abnormal meiosis occurring in vivo did not give rise to viable pollen grains. The hypohaploidization now becoming possible with the in vitro meiosis was extended to other species of *Nicotiana* (2N=2X=20). It was shown that in vitro anthers are more embryogenetic than in vivo anthers (47). This finding would permit production of haploid plants from less favorable species. Since in vitro flower neoformation is genotypically dependent, two alternatives remain: hybridization with a favorable genotype or in vitro development of floral primordia on suitable culture medium. The latter process is aimed at allowing meiotic division to occur under different cytoplasmic conditioning (different interrelation network). Regarding the first process, it was shown that sexual hybridization (18b) or somatic hybridization can restore some morphogenetic potential as a result of genetic complementation and induced regulatory changes.

In conclusion, in vitro culture offering different interrelation networks has opened up new genetic and epigenetic traits. This is an important area of fundamental and applied research which deserves to be developed on systems with appropriate nuclear and cytoplasmic markers [e.g. cytoplasmic male sterility (45)].

PHOTOSYNTHESIS, RESPIRATION, AND PHOTORESPIRATION IN IN VITRO SYSTEMS In vitro systems differ essentially from in vivo systems in the need both for light and for exogenous carbohydrates exhibited by green organs or tissues. It has been possible to select an autotrophic cell line from a mixotrophic culture, using an appropriate combination of NAA, CO_2 enriched air, and high illumination (55). This success is very important, although in some cases, systems in which the photosynthetic function is not operating (without the intervention of exogenous inhibitors) would be useful in determining the part played in the morphogenetic process by chloroplasts in terms of photoreception (photosynthesis and energetic process) and in terms of intergenomic chloroplast-nucleus cooperation.

In order to form flowers, tobacco thin cell layers have an obligate requirement for light when either glucose or sucrose is supplied separately. However, a light requirement can be partly suppressed when an equivalent osmolarity of carbohydrates is supplied using a combination of sucrose and glucose. This would indicate a differential transport of these compounds in the light and in the dark. Glucose and light can be supplied sequentially from day 4 to day 12 of culture (43). CO_2 measurement revealed that the photosynthesis rate is low, although the presence of chloroplasts seems to be an important factor, as suggested by the following results. In tobacco cell layers, a very low percentage of explants forming flowers (8%) and a very low number of flowers (one) per explant were observed when using a periclinal tobacco chlorophyll mutant lacking functional chloroplasts in the subepidermal layer. In addition, flower and bud programs are closely controlled by the relative CO_2/O_2 concentration (42). On tobacco root fragments initiated to form de novo buds, chloroplast formation preceded organogenesis.

On cotyledons and hypocotyls of the Douglas fir, direct root and/or bud formation has been shown to be sharply controlled by intensity at the energetic light irradiances of 0.2 W/m^2, 2 W/m^2, and 20 W/m^2. These experiments were carried out in a control chamber where the temperature was maintained at 20°C inside the container. Different wavelengths from 400 nm to 750 nm were used. Direct root formation was initiated in the dark and direct bud formation under high light intensity (light of 560 nm and 660nm). The control of root and bud formation was obtained on a culture medium comprising two alternate steps: the first including growth hormones, the second being deprived of them (48). All these results call for more investigation relevant to nuclear-organelle (chloroplasts, mitochondria, etc) interrelationships, photosynthesis, respiration, photorespiration, and participation of other photoreceptors. Furthermore, as all morphogenetic process may require energy input, energy charge modification needs to be studied.

ORGAN AND TISSUE-SPECIFIC DETERMINANTS Other changes introduced by in vitro culture techniques are related to regulation of gene expression. In tobacco protoplasts, for example, the biosynthesis of Fraction I protein stopped almost completely (13).

One of the most exciting questions is whether plant cells produce any protein specific to a given step of differentiation. Immunodiffusion and immunoelectrophoretic methods have shown that callus cells derived from different organs exhibit a wider range of determinants than the organ of origin. These comprise specific as well as common determinants (29). Furthermore, characteristic developmental phase-specific protein in tobacco (2), synthesis of storage protein during somatic and androgenetic embryogenesis (34), appearance of specific proteins for vascular differentiation (20), and quantitative variation of lectins during differentiation (24) have also been reported. These markers, if precisely located in the cell or tissue could be of great value in optimizing the environmental factors and the composition of the culture medium. Thus arbitrary conditions not necessarily appropriate for a given differentiation pattern could be avoided. In addition, other changes such as appearance of various secondary products could in some cases be correlated either with a certain type of cell division or with a certain stage of morphogenetic differentiation.

ATTEMPTS IN CONTROLLING MORPHOGENESIS

Two-level controls of morphogenesis can be conducted at the level of the mother plant to try to modulate the inoculum's "inherent cellular state" and at the level of the inoculum itself.

At the Level of the Mother Plant

It is well known that the physiological stage of the mother plant, its nutritional and environmental conditioning, and its spatial gradient are important parameters. Numerous factors were reported to be correlated with the gradient [proline, peroxidase (38)] or with the physiological stage [aromatic amides such as caffeyl-putrescine or caffeyl-spermidine (23), water potential, nicotine]. Thus a well-controlled environment of the mother plant could provide more reproducible results even though some endogenous rhythm may be irreducible. Another question to be debated is whether primary explants (taken from a well-conditioned mother plant) are more homogeneous than cells derived from serial subculture. The choice of primary explant or cells or callus subculture depends upon the problem studied. However, when the choice is possible, it is my belief that entire plants with their integrated interrelation network would be a less heterogeneous source of material, provided they are grown under well-defined environmental conditions.

For woody species, it is possible to regenerate some types of organs only when culturing embryos or young inflorescences. In general, the inoculum must comprise actively dividing cells or must belong to a juvenile stage. The oldest stage in *Abies* giving rise to some weak organogenesis was recorded as 15–20 years old. We have shown that stems taken from an 80-year-old Douglas fir can be induced to form growing shoots. There are known cases in which mature plant parts have regenerated new shoots only after successive grafting onto a juvenile plant. We have neither been able as yet to define at the biochemical and molecular level the differences between juvenile and adult stages, nor have we found what is transmitted through a graft union (7). This process is an interesting step in trying to overcome the poor regenerating ability of woody species.

When plant sources are taken from their original habitat, the following traits need to be considered: (*a*) the influence of air pollutants on stomatal aperture and therefore on the photosynthesis rate; (*b*) any host-parasite interrelation; (*c*) the different types of symbiosis. This would suggest that in order to deal with better defined material, it would be useful to create some intermediate steps either by grafting the plant part to a well-conditioned stock plant or by growing cuttings in in vitro conditions before the in vitro culture itself. Another advantage of this method is the adaptation of the enzymatic systems to the synthetic components brought into the culture medium. In addition, the preparation of the mother plant can be conducted at the genome level by selecting appropriate genotypes, by sexual or somatic hybridization, by haploidization, or by hypohaploidization followed or not by doubling the chromosome number.

At the Explant Level

The various factors which have been shown to influence morphogenetic differentiation on organs, tissues, cells, and protoplasts in culture are surprisingly similar: growth hormone concentration, temperature, light conditions, or other various substances. This means that the poor specificity is a handicap to conducting further research in determining the mechanism of control of morphogenesis. In order to closely define the problem, it would be necessary to create a mutant of morphogenesis or to compare within the same genus two genotypes, one favorable and one reluctant, using the most simplified system and the most simplified medium. Thin epidermal and subepidermal layers of *Nicotiana* (*tabacum* and RMB_7, a reluctant hybrid) were shown to give relatively specific responses.

Five morphogenetic patterns: flowers (100% of explants forming 20 to 50 flowers each), buds, roots (39a, 44), embryo-like structures (40), calluses, or complete absence of morphogenesis [the explant remained unchanged—absence of cell division—in spite of an appropriate growth hormone supply (43)] were selectively programmed from the same layer of cells, the subepidermal layers containing chloroplasts. This selection can be obtained by manipulating these factors: 1. for a precise genotype, the inherent cellular state characterized by a certain pattern of peroxidase isozyme is operating; 2. for a given inherent cellular state, it is now the relative proportion of three factors (glucose, indolylbutyric acid, kinetin) added in combination to agar media which decides the morphogenetic pattern; 3. of all cytokinins tested [zeatin, benzyladenine, 6(γ,γ-dimethylallylamino-purine, kinetin)], only kinetin was proved effective in flower formation; 4. for an appropriate combination of glucose, IBA, and kinetin for flower formation, the removal of agar from the culture medium deviates development to bud formation; 5. the introduction of glass beads of a certain diameter into a liquid medium reestablishes flower formation; 6. this effect is partly correlated with the intracellular and medium pH (9c). Even though pH is a common factor, the selectivity of the morphogenetic responses obtained here (all-or-none effect) suggested an important impact of the membrane on the morphogenetic control.

Other factors have been shown to interfere with this control. For a given pH of the medium, changes in ionic partition (microelement, macroelement, micro-macro element ratio), the presence or the absence of chloroplasts, and different CO_2 and O_2 ratios in the atmosphere surrounding the explant (42) all provoke an all-or-none effect on morphogenesis. Furthermore, the following observations led to the consideration that during the course of differentiation a certain interrelation network is being established within the thin cell layers or between them, the medium, and the atmosphere: (*a*) the possibility of a sequential supply of glucose or light, and the absence of flower formation as a consequence of a sequential supply of 5-bromodeoxyuridine, suggested the existence of a very critical phase for a given pattern of morphogenesis to be decided, the determination phase; (*b*) the shape of the container and changes in the medium-explant volume ratio interfere strongly with this control. In order to determine whether this latter effect could be correlated with the differences in the atmosphere surrounding the explant, containers of different shapes (test tubes and petri dishes) were put in a microchamber specially designed for this purpose. In these conditions, the different morphogenetic responses were correlated with both the absence of gradients in the culture medium and the accumulation in the atmosphere of CO_2 and/or ethylene (40) when petri dishes were used. Consequently, experiments with a continuous flow of CO_2-free air should be conducted.

In conclusion, thin cell layers can be considered as a system on which one can selectively control virtually all morphogenetic programs (a multiprogrammable system). The speed of obtaining different organs (8 to 12 days, depending on the genotype), the uniformity of the morphogenetic responses (pure program), and the possibility of determining a specific morphogenetic response to one definite factor are important traits. The small size of thin cell layers (1 mm x 5 mm) could be considered disadvantageous for biochemical analysis. In fact, if we compare the number of flowers formed on each tiny explant (20–50) to the number of flowers formed in vivo on one mature plant (30–60), we would find that the proportion of cells implicated in one step of differentiation is much higher in thin cell layers than that found in the voluminous explant (see Figure 1). Besides, somatic embryogenesis is another system on which biochemical studies can also be carried out, but the morphogenetic pattern of the latter is limited to embryo formation.

Before we know the mechanism of morphogenetic differentiation, we can only consider the control of morphogenesis but not its regulation. The main limitation is therefore the lack of suitable experimental systems and not so much the fine analytical techniques.

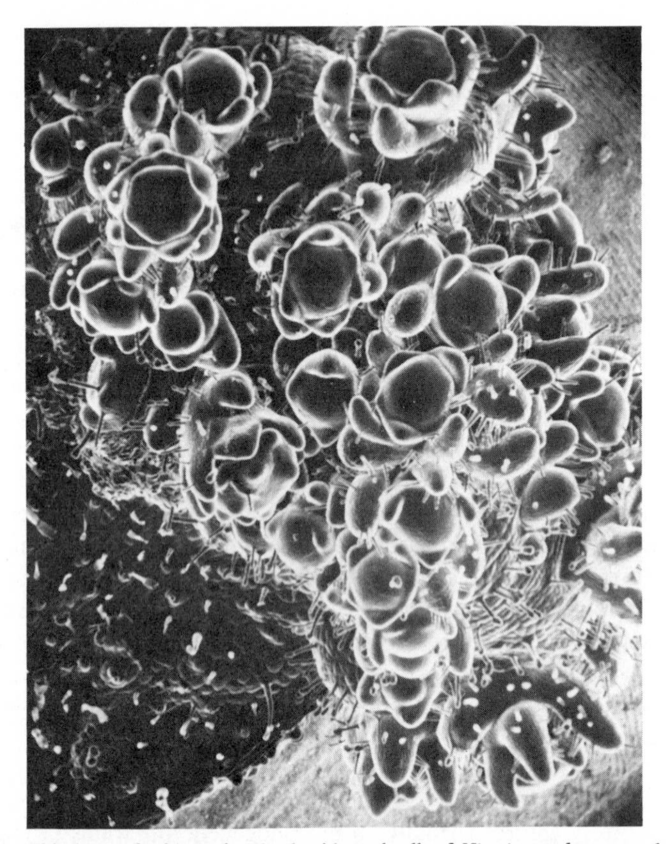

Figure 1 Thin layer of epidermal and subepidermal cells of *Nicotiana tabacum* at day 10 of in vitro culture showing numerous (around 30) floral primordia formed directly (without intermediate callus) on the epidermal surface of an explant of 1 mm X 5 mm comprising 3 to 6 layers of cells. Electron scanning view. X:70.

SUGGESTED PROSPECTS FOR FUTURE RESEARCH

General Considerations

From the outset, there are some general remarks which may have far-reaching consequences. The history of any type of research is full of exciting periods followed by depressed periods or the reverse. This is true especially in experimental morphogenesis, the history of which stretches back to the last century.

INHIBITION OR STIMULATION? For one of the most striking types of morphogenesis—the flowering of entire plants—we are still unable to propose a rational explanation for all of the confusing data accumulated in the literature. It is my belief that one of the main causes is the "one directional" way of tackling the phenomena. The results I have obtained brought me to the concept of inhibition of flowering (39, 40) instead of its stimulation by a hypothetical florigen. More specifically, if a plant cannot flower or cannot express one of the desired morphological patterns, it may be thought that this absence may result from an inhibition. After 40 years, as attempts to extract the flowering hormone have failed, plant physiologists have turned only recently to antiflorigen grafts (22).

MODULATION OF THE INHIBITORY NETWORK: CONCEPT OF THIN CELL LAYER This concept of network of organ, tissue, and cell interrelation, the result of which is an "inhibitory process," received experimental confirmation when extended to other species or meristems reluctant to flower (39, 40). This concept of an inhibitory network brought me to the concept of a thin cell layer whereby I have tried to separate gradually the organ tissue and cell interrelation network and have attempted to reveal its new morphogenetic potentials.

MULTICELLULAR OR UNICELLULAR? When working on morphogenesis in plants or animals, we are facing a multicellular system which exhibits forms, structures, and functions through different temporal and spatial cell cycles. This holds true even when the multicellular system is temporarily reduced to isolated cells or to protoplasts that soon form clumps of cells. Since the discovery of gene regulation in bacteria, although we are tempted to rediscover this general mechanism of control in multicellular systems, the multicellular organization with its inherent gradient and therefore the lack of perfect synchronization would bring us to consider strongly the spatial dimension fusioned together with the temporal dimension. The observation that the DNA in higher organisms is contained within the nuclear membrane may raise different problems from those found in bacteria.

SPECIFIC MARKERS OF MORPHOGENESIS In most cases, as differentiation of a given morphogenetic pattern may result from a gradient established first within one original cell, biochemical and/or molecular changes that we have almost arbitrarily decided to analyze may not be directly related to the precise morphogenetic process we are considering. These changes may rather be relevant to some other parts of this same inoculum which has evolved differently from the "visible morphogenetic developing zone," or they may result from both zones. This is the problem of specific markers. The lack of synchronization and of homogeneity, which is an inherent consequence of the gradient, makes the interpretation of biochemical and molecular data difficult. These considerations brought us to the necessity of associating biochemical, biophysical, and molecular analysis with morphological, anatomical, histological, histocytochemical, and infrastructural studies.

In concluding these general considerations, it seems that an attempt should be made to visualize all the facets of the same problem and to analyze the organogenetic zones as well as the nonorganogenetic ones. It is useful to concentrate our efforts not only on the genotypes capable of exhibiting morphogenesis in vitro, but also on the recalcitrant ones.

Some Problems in Need of More Intensive Investigation

Before conducting a systematic analysis at the biochemical and molecular level of regulation of cell division, growth, and differentiation on more or less suitable experimental systems, some basic questions still remain to be answered.

CHANGE IN CELL SHAPE The shape of a cell is as important as its function. Although shapes, structures, and functions interact in living systems, some basic properties thought to characterize them, such as self-regulation, memory, or self-duplication, are known in nonliving systems. It is not untrue to think that the shape and the structure can in some cases create function. One should pay more attention to the common phenomenon of change in cell shape, for example, rounded cells becoming cylindrical or kidney shaped. Besides the guard-cell initiation, unicellular hairs experimentally induced from epidermal cells (39b) would offer an interesting system to study the rearrangement and synthesis of cell wall and cytoskeletal components. Furthermore, one of the ways by which one can detect organized patterns being formed either from a callus or from an organ fragment is the emergence of tunica structure. This organization would imply a certain quality of cell wall (sufficient strength to maintain the structures). Besides, a certain preferential orientation of cell division and therefore a certain quality of cell contact are important steps in morphogenetic differentiation. Compact calluses or friable calluses are often associated with the presence or the absence of morphogenetic competence and in certain cases with the retention in the cell or the release into the medium of specific surface proteins such as lectins (24).

EXTERNAL AND INTERNAL GRADIENT, TRANSMEMBRANE CONTROL EVENTS, AND SURFACE SIGNALS How are various outside signals perceived by the cell interior, and how can the cell interior modulate the cell surface? In plant material many different systems are available and are to a certain extent different from animal systems. Plant growth hormones acting on membrane permeability may provide an interesting tool, as well as plant lectins which have been shown to stimulate lymphocyte division (53). It would be interesting to assess the role of lectins (if any) in plant cell and the role of plant growth hormones (if any) on the growth and on the development of animal systems. In fact, innumerable plant products act not only on the metabolism of animal cells but also on their division: for example, alkaloids of the ellepticine group inhibiting proliferation of leukemia cells. The impact of the intracellular pH on cell division (22a), that of the culture medium pH, and the influence of ionic partition on the control of morphogenesis on thin cell layers (9c, 43) suggest that it is important to study the transmembrane control.

CELL-CELL INTERRELATION, CELL RECOGNITION

Plasmodesmata The presence and the number of plasmodesmata in relation to the degree of morphogenetic differentiation must be determined. In somatic embryogenesis, it was shown that the group of daughter cells within the mother cell are joined by numerous plasmodesmata and that the spherical cell clusters are delimited by an outer wall. Communication as well as isolation seem to be the selective conditions of embryo formation (54).

Positional information The whole process of differentiation may depend on how differently two adjacent cells respond to the same signal. If the signal appears the same for the investigator, it may nevertheless be seen differently by each of a pair of cells, depending upon their reciprocal relation. This is illustrated in *Mimosa pudica,* where turgor pressure variation is different between the upper part and the lower part of the pulvini. A high frequency of plasmodesmata is observed in the pulvini cells. Furthermore, the classical "positional information" described in animal embryogenesis and the loss of contact inhibition in transformed cells are both very interesting problems. All of these lead to the fundamental question concerning the nature of the cell-cell interrelation in regulating cell division. We have shown that not all epidermal cells of a monolayer of *Torenia* underwent mitosis when cultured in vitro (9, 40). An increased number of cell division centers was observed from the apical to the basal pole of the explant (41).

CELL DIVISION (MITOSIS AND MEIOSIS) What has caused a cell to become committed to mitosis or meiosis? Is cell division a process of distributing the same genome into two daughter cells, or has cell division opened up some properties not existing in the mother plant? Hormonal control of cell division is well demonstrated in tobacco pith cells requiring both auxin and cytokinin for division (31). The study of its mechanism is being developed at the biochemical and molecular level. Apart from stressing the fundamental problems of commitment to cell division, of control of cell cycle, and the need to study other cytoplasmic and nuclear events which are essential, I only mention here some aspects which are related to morphogenesis and which are studied less frequently.

Nuclear migration and unequal division Nuclear migration may be a determinant step in cell differentiation since unequal divisions which are almost always associated with morphogenetic changes are preceded by nuclear migration to a defined position in the cytoplasm. The future pattern of differentiation may have been programmed in the cell before the very first division occurs. What causes, decides, and allows this journey? The concomitant cytoskeletal modifications, the rate of cytoplasmic streaming, and therefore the implication of mobility proteins would be interesting factors to investigate. The unequal division seems to correspond to the very first choice of the mother cell in preparing the next step for the two daughter nuclei to be adequately positioned at different cytoplasmic regions of the mother cell. This common phenomenon suggests that a detailed study of organelle repartition in the cytoplasm should be conducted as well as the involvement of cytoplasmic determinants in cell differentiation. A shift of the spindle or a change of the cell plate position may induce changes in the nucleo-cytoplasmic interrelation network. This is observed in tobacco pollen grains cultured in vitro where a shift from unequal to equal division corresponds to embryo formation.

Some problems related to cell division (mitosis and meiosis) and morphogenetic differentiation which could be studied using neoplastic cells Tumor cells induced by Ti plasmid DNA, by RNA virus, by chromosomal imbalance, or by habituation all develop a persisting capacity for cell division as shown on hormone-free synthetic media. This property is absent in normal cells. Thus the regulation of the cell cycle has changed in the neoplastic cells concomitant with the acquisition of the persistent ability to synthesize the essential growth regulator substances, the auxins, and the cell division promoting factors (3).

In tobacco crown gall induced by Ti plasmid DNA, several important aspects relevant to morphogenesis are of interest. Tumor tissues induced by the B_6 strain of *Agrobacterium tumefaciens* consist of highly disorganized masses of actively dividing cells and chromosomal abnormalities in some species. Other interesting traits of these tumors are the following: 1. the capacity to form tumor shoots and the incapacity to form roots over long periods of time when induced by nopaline Ti plasmid, T_{37} strain (27); 2. the necessity to be grafted successively onto a growing plant to perform any organized growth, thus indicating the importance of a gradient; 3. the recovery from a tumorous state observed on the plants issued from mature seeds of these grafted plants deriving from teratoma tissues of single cell origin (5). This last aspect would shed light on the way the meiotic division itself would interfere with the foreign plasmid. In fact, it is an observation generally admitted—but not yet understood—that meiosis brings about recovery in many cases of infection. With the now recognized possibility to realize in vitro meiosis in thin cell layers (39a) and the demonstration of a higher frequency of chromosomal pairing in in vitro meiosis compared to in vivo meiosis (45), it would be interesting to attempt to induce de novo flower formation from tumorous tissues excised from these grafted plants and to compare the progeny thus obtained in vitro to those obtained in vivo.

Furthermore, habituated tissues (16) with tumor-like properties provide a useful tool to evaluate the epigenetic control mechanism which could be superimposed on the genomic changes. The possibility of inducing habituation by temperature treatment suggests that plant cells possess an inherent potential to form tumors (25), and that this tendency normally would be inhibited by some regulation process (4). This suggestion is to be compared to the hypothesis which stipulates that all cellular territories possess the potential ability to flower and that this potential is inhibited during the ontogenetic regulatory processes (39, 40).

One other interesting type of tumor formation is shown by some hybrids within *Nicotiana, Brassica, Bryophyllum,* and *Lycopersicon* genera (32). Hybrids (2N=21) resulting from a cross between *Nicotiana glauca* (2N=24) and *N. langsdorfii* (2N=18) develop a normal phenotype during the vegetative phase although the potential for tumorigenesis is present at any stage. It can be expressed after a wound or a treatment by stress agents. One of the most striking observations is the interrelation between flowering or maturity stages and the spontaneous proliferation of tumors on all parts of the plant. This property is also exhibited on parasexual hybrids after fusion of mesophyll protoplasts. Whether this profusion of tumors is correlated with an arrest of active growth, with a particular growth hormone or amino acid content of tumor tissues, or with possible DNA amplification during flower initiation is not known yet.

Results obtained on thin cell layers of tobacco have shown that thymidine analog 5-bromodeoxyuridine (5BrdU) specifically inhibited flower formation in vitro when applied during 24 hr at the 8-day stage. It has been shown that incorporation of 5BrdU into DNA inhibits pith tissue proliferation of *Nicotiana glauca* (11). Besides the very interesting molecular aspect of the process of tumorigenesis, the triggering of cell division—either anarchic or capable of a certain specific morphogenetic pattern—on arrested cell cycles is interesting in itself. The modulation of this triggering by the physiological stages of growth and development, the existence of different types of neoplastic induction, and the recovery of normal phenotype after meiotic events would allow a study of the regulatory mechanisms. This recovery occurs with or without the intervention of a foreign plasmid. The in vitro liposome-mediated uptake of Ti plasmid DNA by protoplasts is a very promising technique for studying the development of gene vector systems which could be used as valuable markers of cell differentiation, not to mention its importance for genetic engineering.

Absence of cell division There are some other aspects of cell division which could be investigated using very specific conditions in which cell division does not occur in spite of a supply of phytohormone, nutrient, and organic components. On thin cell layers of *Nicotiana tabacum* normally capable of forming all known patterns of morphogenesis ("multiprogramable potentialities"), no cell division was observed under specific combination of IBA kinetin ratio and pH value of the culture medium (9c).

Another example of absence of cell division was found in thin cell layers or fragment excised from a *Nicotiana* hybrid (RMB_7 obtained by H. Smith from the seventh backcross on *N. rustica* by *N. tabacum*) when cultured in the presence of specific IBA/kinetin ratio. This hybrid is an opposite example to tumor hybrids where the chromosomal imbalances induce on the contrary, the ability to proliferate indefinitely without any supply of growth substances. This tobacco hybrid is particularly interesting because of its incapacity to form de novo buds or flowers in a wide range of experimental conditions tested. It would add significantly to the study of the control of morphogenesis when this kind of finding is extended to other species such as woody plants, legumes, and grass. The absence of cell division and of subsequent morphogenesis in these two systems makes them useful controls for biochemical and molecular analysis when we need to distinguish the early events related to the excision from the ones relevant to cell division or morphogenetic processes or both.

CONCLUDING REMARKS

Given the great complexity of developmental processes, it would benefit all of us to unify our efforts in a research program as integrated as the multicellular system we are studying. Perhaps, as for the amoeboid slime mold, we may need or we are awaiting a "cell aggregation signal." We learn from it that for cell cohesion—a compulsory stage to achieve a cell cycle and fructification—almost each cell in the group has to secrete a part of the signal and transmit to its neighbors the signal perceived. Cohesion and complementarity (but not unidirectionality) would therefore eliminate fragmentary or dispersed results and would shed more light (not just instantaneous flashes) on the analysis of differentiation when for some reason two cells are linked together.

ACKNOWLEDGMENTS

I am grateful for the enthusiasm of my professor, P. Chouard, and for the contributions of all my co-workers, past and present, to my research. I also wish to thank those colleagues who sent reprints, which have been of great help in this review. Because of limited space, some works now considered classics have not been included in the literature references. My special thanks to Emmanuelle Thanh Tam and to Binh Minh for their devoted efforts in processing the manuscript.

Literature Cited

1. Bagni, N., Fracassini, D. S. 1973. The role of polyamines as growth factors in higher plants and their mechanism of action. In *Plant Growth Substances,* pp. 1205–17. Tokyo: Hirokawa

1a. Buiatti, M. 1977. DNA amplification and tissue culture. See Ref. 29a, pp. 358–74, 442–64

2. Boutenko, R. G., Volodarsky, A. D. 1968. Analyse immunochimique de la differenciation cellulaire dans les cultures de tissus de Tabac. *Physiol. Veg.* 6:299–309

3. Braun, A. C. 1953. Bacterial and host factors concerned in determining tumor morphology in crown gall. *Bot. Gaz.* 114:363–71

4. Braun, A. C. 1969. Abnormal growth in plants. In *Plant Physiology, A Treatise,* ed. F. C. Steward, 5b:379–420. New York: Academic

5. Braun, A. C., Wood, H. N. 1976. Suppression of the neoplastic state with the acquisition of specialized functions in cells, tissues, and organs of crown gall teratomas of tobacco. *Proc. Natl. Acad. Sci. USA* 73:496–500

6. Carlson, P. S. 1979. Peptides of normal and variant cells of Tobacco. *Dev. Genet.* 1:3–12

7. Champagnat, P. 1978. Exposé: La greffe chex les végétaux. *C.R. Acad. Sci.* 9 Oct.

8. Chang, W. C., Hsing, Y. I. 1980. In vitro flowering of embryoids derived from mature root callus of ginseng (*Panax Ginseng*). *Nature* 284:341–42

9. Chlyah, H. 1978. Intercellular correlation: relation between DNA synthesis and cell division in early stages of in vitro bud formation. *Plant Physiol.* 62:482–85

9a. Chlyah-Arnasson, A., Tran Thanh Van, K. 1968. Budding capacity of undetached *Begonia rex* leaves. *Nature* 218:493

9b. Chouard, P., Aghion, D. 1961. Modalité de la formation de bourgeons floraux sur des cultures de segments de tige de Tabac. *C. R. Acad Sci. Paris* 252:3864–66

9c. Cousson, A., Tran Thanh Van, K. 1980. In vitro control of de novo flower differentiation from tobacco thin cell layer on a liquid medium. *Physiol. Plant.* 51:77–84

10. D'Amato, F. 1978. Chromosome number variation in cultured cells and regenerated plants. See Ref. 36, pp. 287–95

10a. Demarly, Y. 1976. La notion de programme génétique chez les végétaux supérieurs. *Ann. Amélior. Plant.* 26: 117–38

11. Durante, M., Geri, C., Nuti-Ronchi, V., Martini, G., Guillé, E., Grisvard, J., Giorgi, L., Parenti, R., Buiatti, M. 1977. Inhibition of *Nicotiana glauca* pith tissue proliferation through incorporation of 5-BrdU into DNA. *Cell Differ.* 6:53–63

12. Fiechter, A., ed. 1980. *Plant Tissue Culture.* Berlin/Heidelberg/New York: Springer. 193 pp.

13. Fleck, J., Durr, A., Lett, M. C., Hirth, L. 1979. Changes in protein synthesis during the initial stage of life of tobacco protoplasts. *Planta* 145:279–85

14. Galston, A. W., Altman, A., Kaur-Sawhney, R. 1978. Polyamines, ribonuclease and the improvement of oat leaf protoplasts. *Plant Sci. Lett.* 11:67–77

14a. Gaspard, T., Penel, C., Tran Thanh Van, K., Greppin, H. 1979. Des isoperoxydases comme marqueurs de la differenciation cellulaire et du développement chex les végataux. In *Xème Rencontre de Méribel sur la Différenciation Cellulaire,* ed. J. Tavlitzki, K. Tran Thanh Van, pp. 175–93. CNRS. 246 pp.

15. Gautheret, R. J. 1942. *Manuel Tech-*

nique de Culture des Tissus Végétaux. Paris: Masson. 863 pp.
16. Gautheret, R. J. 1946. Comparaison entre l'action de l'acide indoleacétique et celle du *Phytomonas tumefaciens* sur la croissance des tissus végétaux. *C. R. Seances Soc. Biol. Paris* 140:169–71
17. Hadorn, E. 1978. Transdetermination. In *The Genetics and Biology of Drosophila,* ed. M. Ashburner, T. R. F. Wright, 2c:556–617. London/New York: Academic
18. Hangarter, R. P., Peterson, R. D., Good, N. E. 1980. Biological activities of indoleacetylamino acids and their use as auxins in tissue culture. *Plant Physiol.* 65:761–67
18a. Kahl, G. 1973. Genetic and metabolic regulation in differentiating plant storage tissue cells. *Bot. Rev.* 39:274–99
18b. Kamate, K., Cousson, A., Trinh, T. H., Tran Thanh Van, K. 1980. Influence des facteurs génétiques et physiologiques chez le *Nicotiana* sur la néoformation *in vitro* des fleurs à partir d'assises cellulaires épidermiques et sous épidermiques. *Can. J. Bot.* In press
19. Kevers, C., Coumans, M., De Greef, W., Hoffinger, M., Gaspar, T. 1981. Habituation in sugarbeet callus: auxin content, auxin protectors, peroxidase pattern and inhibitors. *Physiol. Plant.* In press
20. Khavkin, E. E., Markov, E. Yu., Misharin, S. I. 1980. Evidence for proteins specific for vascular elements in intact and cultured tissues and cells of maize. *Planta* 148:116–23
21. Ku, M. K., Cheng, W. C., Kuo, L. C., Kuan, Y. L., An, H. P., Huang, C. H. 1978. Induction factors and morphocytological characteristics of pollen-derived plants of maize (*Zea mays*). *Proc. Symp. Plant Tissue Cult.,* pp. 35–44. Peking: Science Press
22. Lang, A., Chailakhyan, M. Kh., Frolova, I. A. 1977. Promotion and inhibition of flower formation in a day-neutral plant in grafts with a short-day and a long-day plant. *Proc. Natl. Acad. Sci. USA* 74:2412–16
22a. Leguay, J. J. 1979. Regulation de la division cellulaire par les auxines: Recherche du mode d'action du 2,4-D sur la division de cellules d'Erable (*Acer pseudoplatanus* L.) cultivés en suspension. Thesis Doct. ès Sciences. Paris. 157 pp.
23. Martin-Tanguy, J., Cabanne, F., Perdrizet, E., Martin, C. 1978. The distribution of hydroxycinnamic amides in flowering plants. *Phytochemistry* 17: 1927–28
24. Meimeth, T., Tran Thanh Van, K., Marcotte, J. L., Trinh, T. H., Clarke, A. 1980. Nature and distribution of lectins in tissues, derived callus and de novo roots of *Psophocarpus tetragonolobus* (Winged bean). In press
25. Meins, F. 1974. Mechanisms underlying the persistence of tumour autonomy in crown-gall disease. In *Tissue Culture and Plant Science,* ed. H. E. Street, pp. 233–64. London/New York: Academic
26. Murashige, T. 1974. Plant propagation through tissue cultures. *Ann. Rev. Plant Physiol.* 25:135–66
27. Petit, A., Delhaye, S., Tempé, J., Morel, G. 1970. Recherches sur les guanidines des tissus de crown gall. Mise en évidence d'une relation biochimique spécifique entre les souches d'*Agrobacterium tumefaciens* et les tumeurs qu'elles induisent. *Physiol. Veg.* 8:205–13
28. Phillips, G. C., Collins, G. B. 1980. Somatic embryogenesis from cell suspension cultures of red clover, *Trifolium pratense* cv Arlington. *Crop. Sci.* 20(3):323–27
29. Raff, J. W., Hutchinson, J. F., Knox, R. B., Clarke, A. 1979. Cell recognition: Antigenic determinants of plant organs and their cultured callus cells. *Differentiation* 12:179–86
29a. Reinert, J., Bajaj, Y. P. S. 1977. *Applied and Fundamental Aspects of Plant Cell Tissue and Organ Culture.* Berlin: Springer. 803 pp.
30. Reinert, J., Bajaj, Y. P. S., Zbell, B. 1977. Aspects of organization—Organogenesis, embryogenesis, cytodifferentiation. See Ref. 33, pp. 389–427
31. Skoog, F., Miller, C. O. 1957. Chemical regulation of growth and organ formation in plant tissue cultured in vitro. *Symp. Soc. Exp. Biol.* 11:118–31
32. Smith, H. H. 1972. Plant genetic tumors. *Prog. Exp. Tumor Res.* 15:138–64
33. Street, H. E., ed. 1977. *Plant Tissue and Cell Culture.* Berkeley/Los Angeles: Univ. Calif. Press. 614 pp.
34. Sussex, I. M., Dale, R. M. K., Crouch, M. L. 1980. Developmental regulation of storage protein synthesis in seeds. In *Genome Organization and Expression in Plants,* ed. C. J. Leaver, pp. 238–89. New York: Plenum
35. Tanimoto, S., Harada, H. 1979. Influences of environmental and physiological conditions on floral bud formation of *Torenia* stem segments cultured in vitro. *Z. Pflanzenphysiol.* 95:33–41

36. Thorpe, T. A., ed. 1978. *Frontiers of Plant Tissue Culture 1978.* Calgary: Int. Assoc. Plant Tissue Cult. 556 pp.
37. Thorpe, T. A. 1978. Regulation of organogenesis in vitro. In *Propagation of Higher Plants Through Tissue Culture,* ed. K. W. Hughes, R. Henke, M. Constantin, pp. 87–101. Springfield: Natl. Tech. Inf. Serv. 305 pp.
38. Thorpe, T. A., Tran Thanh Van, K., Gaspar, T. 1978. Isoperoxydases in epidermal layers of Tobacco and changes during organ formation in vitro. *Physiol. Plant.* 44:388–94
39. Tran Thanh Van, K. 1965. La vernalisation du "Geum urbanum" L. Etude experimentale de la mise à fleur chez une plante vivace en rosette exigeant le froid vernalisant pour fleurir. *Ann. Sci. Nat. Bot. Biol. Veg.* 6:373–594
39a. Tran Thanh Van, K. 1973. Direct flower neoformation from superficial tissues of small explant of *Nicotiana tabacum* L. *Planta* 115:87–92
39b. Tran Thanh Van, K. 1977. Regulation of morphogenesis. In *Plant Tissue Culture and Its Bio-technological Application,* ed. W. Barz, E. Reinhard, M. H. Zenk, pp. 367–85. Berlin: Springer. 419 pp.
40. Tran Thanh Van, K. 1980. Control of morphogenesis or what shapes a group of cells? See Ref. 12, pp. 151–71
41. Tran Thanh Van, K. 1980. Thin cell layers: control of morphogenesis by inherent factors and exogenously applied factors. *Int. Rev. Cytol.* Suppl. 11A:175–94
42. Tran Thanh Van, K., Cornic, G., Louason, G., Marcotte, J. L., Kamaté, K., Cousson, A. Unpublished results
43. Tran Thanh Van, K., Cousson, A. 1980. Microenvironment-genome interactions on *de novo* morphogenetic differentiation on thin cell layers of *Nicotiana. NSF-CNRS Colloque Proceedings.* New York: Praeger. In press
44. Tran Thanh Van, K., Nguyen, T. D., Chlyah, A. 1974. Regulation of organogenesis in small explants of superficial tissues of *Nicotiana tabaccum* L. *Planta* 119:149–59
45. Tran Thanh Van, K., Trinh, T. H. 1978. Plant propagation: non-identical and identical copies. See Ref. 37, pp. 134–58
46. Tran Thanh Van, K., Trinh, T. H. 1978. Morphogenesis in thin cell layers: concept methodology and results. See Ref. 36, pp. 37–48
47. Tran Thanh Van, K., Trinh, T. H. 1980. Capacité embryogénétique des anthères des fleurs néoformeés à partir des couches cellulaires minces et celle des anthères des fleurs prélevées sur la plante-mère chez le *Nicotiana tabaccum* L. et *Nicotiana plumbaginifolia* Viv. *Z. Pflanzenphysiol.* 100:379–88
48. Tran Thanh Van, K., Yilmaz, D. Unpublished results
48a. Trinh, T. H., Gaspar, T., Tran Thanh Van, K., Marcotte, J. L. 1980. Genotype, ploidy and physiological state in relation to isoperoxidases in *Nicotiana. Physiol. Plant.* In press
49. Trinh, T. H., Tran Thanh Van, K. 1981. Formation in vitro de fleurs à partir de couches cellulaires minces épidermiques et sous-épidermiques diploides et haploides chez le *Nicotiana tabaccum* L. et chez le *Nicotiana plumbaginifolia* Viv. *Z. Pflanzenphysiol.* 101:1–8
50. Trinh, T. H., Lie-Schricke, H., Tran Thanh Van, K. 1980. Direct in vitro bud formation from fragments and thin cell layers of different organs of the Winged-beans (*Psophocarpus tetragonolobus. Z. Pflanzenphysiol.* In press
51. Vasil, I. K. 1980. Perspectives in plant cell and tissue culture. *Int. Rev. Cytol.* Suppl 11A. 253 pp.
52. Walker, K. A., Yu, P. C., Sato, S. J., Jaworski, E. G. 1978. The hormonal control of organ formation in callus of *Medicago sativa* L cultured in vitro. *Am. J. Bot.* 65:650–59
53. Wang, J. L., McClain, D. A., Edelman, G. M. 1975. Modulation of lymphocyte mitogenesis. *Proc. Natl. Acad. Sci. USA* 72:1917–21
54. Wetherell, D. F. 1978. In vitro embryoid formation in cells derived from somatic plant tissues. See Ref. 37, pp. 102–24
55. Yamada, Y., Sato, F., Hagimori, M. 1978. Photoautotrophism in green cultured cells. See Ref. 36, pp. 453–62

Ann. Rev. Plant Physiol. 1981. 32:313–25

VIROIDS: ABNORMAL PRODUCTS OF PLANT METABOLISM

❖7714

T. O. Diener[1]

Plant Virology Laboratory, Plant Protection Institute, Science and Education Administration, U.S. Department of Agriculture, Beltsville Agricultural Research Center, Beltsville, Maryland 20705

CONTENTS

INTRODUCTION

Viroids are distinct low molecular weight nucleic acids (1.1 to 1.3 $\times$ 10^5) that can be isolated from certain plants afflicted with specific diseases. They are not detectable in healthy individuals of the same species but, when introduced into such individuals, they are replicated autonomously in spite of their small size and cause the appearance of the characteristic disease syndrome. In certain other species, some viroids are replicated but cause no obvious symptoms. Unlike viral nucleic acids, viroids are not encapsidated; that is, no virus-like nucleoprotein particles occur in infected tissue. Viroids constitute a novel class of subviral pathogens; they are the smallest known agents of infectious disease. So far viroids are definitely known to exist only in higher plants and to consist of RNA with a unique, previously unknown structure.

The first viroid was discovered in attempts to purify and characterize the causative agent of potato spindle tuber, a disease that for many years had been assumed to be of viral etiology. In 1967 we reported that the transmissible agent of this disease is a free RNA and that no viral nucleoprotein particles (virions) are detectable in infected tissue (16). By 1971, sedimentation and gel electrophoretic analyses had shown conclusively that the infectious RNA is far smaller than the smallest genomes of autonomously replicated viruses (9). No evidence for the involvement of helper viruses in the replication of the RNA could be obtained (9); thus, despite its small size, the RNA appeared to be replicated autonomously in susceptible cells. Because of the basic differences between the potato spindle tuber disease agent and conventional viruses, the term *viroid* was introduced to denote it and agents with similar properties (9).

Since that time, viroids and viroid diseases have been given steadily increasing attention, with the result that almost a dozen diseases of higher plants are now known or believed to be caused by viroids and our knowledge of the physical-chemical properties of viroids has increased dramatically. The subject has been reviewed extensively in recent years (12, 13, 44), and it appears unproductive to attempt here another overall review.

From a plant physiological standpoint, viroids may be regarded as abnormal products of plant metabolism. Accordingly, in this review, emphasis is placed on viroid synthesis and interaction of viroids with host cells, in short, on viroid function as opposed to viroid structure. Unfortunately, in contrast to the extensive knowledge that has been achieved regarding physical-chemical properties of viroids, knowledge of viroid-host plant interactions, that is, of the dynamic aspects of the subject, has accrued at a far slower rate. Thus, the present review represents not so much a summary of knowledge achieved in this area, but rather an attempt to analyze critically information presented in the relatively few, but often contradictory, reports that have been published and, based on this analysis, to try to arrive at a more or less cohesive, but necessarily speculative, model of viroid function.

SOME BIOLOGICAL PROPERTIES

Detailed discussion of the biological properties of viroids and viroid diseases is beyond the scope of this review; only those properties helpful to a discussion of viroid function are considered. A more detailed description of these properties is available (13).

The following plant diseases are now known to be a result of viroid infection: potato spindle tuber (9), citrus exocortis (40a, 49), chrysanthemum stunt (15), chrysanthemum chlorotic mottle (40), cucumber pale fruit (42, 53), hop stunt (43), and tomato planta macho (J. Galindo, D. R. Smith, and T. O. Diener, unpublished) diseases. With the following diseases, viroid etiology has not yet been demonstrated rigorously, but is likely in view of the association of viroid-like RNAs with the diseases: coconut cadang-cadang (38) and avocado sunblotch (31, 37). Furthermore, a viroid has been isolated from apparently healthy *Columnea erythrophae* plants that causes symptoms similar to, it not identical with, those of the potato spindle tuber viroid (PSTV) in tomato or potato, yet is distinct from PSTV in its primary structure (35).

All known viroids infect their hosts in a persistent manner; that is, no recovery occurs and viroids can be isolated from infected plants as long as the plants live. Furthermore, with PSTV at least, vertical transmission through both seed and pollen of infected plants has been documented (19, 50). All known viroids are transmissible by mechanical means, either readily or with some difficulty. With PSTV, mechanical transmission by contact with farm implements is mainly responsible for the spread of the disease in nature; with other viroids, contaminated budding knives and other tools have been implicated (13). No insect vectors have been identified with any of the known viroids.

In their symptomatology, viroid diseases do not differ significantly from virus diseases. Although stunting of plants is a predominant symptom of most viroid diseases, stunting is also known as a consequence of infection of plants by various conventional viruses. This also holds true for other types of symptoms observed with viroid diseases. Conversely, most of the types of symptoms observed with virus diseases also occur with viroids. These include stunting, epinasty, veinal discolorations, leaf distortions, vein clearing, localized chlorotic or necrotic spots, mottling of leaves, necrosis of leaves, and death of whole plants.

Another property viroids share with viruses is the phenomenon of cross-protection in which infection of plants with a mild strain of viroid protects the plants from the effects of superinfection with a severe strain of the same viroid. As with viruses, cross-protection is believed to occur only if the mild and superinfecting pathogens are related (33).

MOLECULAR STRUCTURE

For a detailed description of viroid structure, the reader is referred to a recent extensive review (24). Here it must suffice to state that viroids are single-stranded, covalently closed circular as well as linear RNA molecules

(30) that occur, because of extensive regions of intramolecular complementarity, in the form of collapsed circles and hairpin structures with the appearance in the electron microscope of double-stranded molecules (42, 51). The complete primary sequence of PSTV has been determined (23) and, on the basis of this sequence as well as of that of quantitative thermodynamic and kinetic studies of its thermal denaturation (28, 29), a model for the secondary structure of PSTV has been proposed (23). According to this model, viroids exist in their native configuration as extended rodlike structures characterized by a series of double-helical sections and internal loops. Thus the rigid, rodlike structure of the native viroid is based on a defective rather than a homogeneous RNA helix (18), in confirmation of conclusions arrived at earlier by different techniques (10, 51). It is difficult to believe that this unique structure of viroids does not have important biological significance.

SUBCELLULAR LOCATION

Bioassays of subcellular fractions from PSTV-infected tomato demonstrated that only the tissue debris and nuclear fractions contain appreciable infectivity (8). Chloroplast, mitochondrial, ribosomal, and "soluble" fractions contain only traces of infectivity. Most infectivity is chromatin-associated and can be extracted as free RNA with phosphate buffer (8). The citrus exocortis viroid (CEV) is also located primarily in the nuclear fraction in close association with chromatin (40a), but with CEV in *Gynura aurantiaca,* a significant portion of the viroid has been reported to be associated with a plasma-membranelike component of the endomembrane system (47). This may well be the case but, as discussed earlier (14), evidence presented for this association was not conclusive and further work is required to substantiate this claim.

VIROID-HOST CELL INTERACTIONS

When viroids are introduced into susceptible cells, they are replicated autonomously, that is, without the requirement of a helper virus (9). This basic biological fact raises a number of intriguing questions. Above all, one wonders by what mechanisms such small RNAs induce their own synthesis in susceptible cells and how they incite diseases in certain hosts, yet are replicated in many other susceptible plant species without discernible damage.

Possible Mechanisms of Replication

The fact that infectious PSTV is located primarily in the nuclei of infected cells does not prove that it is synthesized there, but experiments with an in vitro RNA-synthesizing system in which purified cell nuclei from infected plants were used as an enzyme source indicated that this is the case (52). It appears, therefore, that the infecting viroid migrates to the nucleus (by an unknown mechanism) and is replicated there. The absence of significant viroid amounts in cytoplasmic fractions of infected cells (8) suggests that most of the progeny viroid remains in the nucleus.

QUESTION OF VIROID TRANSLATION Viroids are of sufficient chain length to code for a polypeptide of about 10,000 daltons, although, with circular PSTV, the uneven number of 359 nucleotides theoretically permits three rounds of translation with a frame shift each time.

Testing for in vitro messenger function of PSTV and CEV in a variety of cell-free protein-synthesizing systems indicated that neither viroid functions in this capacity (5, 26). Also, CEV is not translated in *Xenopus laevis* oocytes, even after polyadenylation (45).

Although viroids do not act as mRNAs in these in vitro systems, they might be translated in vivo from a complementary RNA strand synthesized by preexisting host enzymes with the infecting viroid serving as a template. RNA sequences complementary to viroids have been identified in infected tissue (21), but whether these act as mRNAs is not known, but if they do, novel, viroid-specific proteins should be detectable in protein preparations from infected host tissue.

Comparisons of protein species in healthy and PSTV-infected tomato (54) and healthy and CEV-infected *Gynura aurantiaca* (4) did not, however, reveal qualitative differences between healthy and infected plants. In both studies, synthesis of at least two proteins was enhanced in infected as compared with healthy tissue (4, 54), but recent studies indicate that these proteins are host- and not viroid-specific (see below).

Although more sensitive methods of analysis may yet disclose the presence of viroid-specified polypeptides in infected cells, in the light of present knowledge one must conclude that viroids do not act as mRNAs. If so, the complementary RNA sequences found in infected tissue must be synthesized entirely by preexisting (but possibly activated) host enzymes.

RNA- OR DNA-DIRECTED REPLICATION? Whether the mechanism of viroid replication involves transcription from RNA or DNA templates is not known. Theoretically, an RNA-directed mechanism requires the presence of RNA sequences complementary to the entire viroid in infected tissue, as well as a preexisting host enzyme with the specificity of an RNA-directed RNA polymerase.

A DNA-directed mechanism requires the presence of DNA sequences complementary to the entire viroid. These DNA sequences might already be present in repressed form in uninfected hosts, or they might be synthesized as a consequence of infection with viroids. In the latter case, a preexisting host enzyme with the specificity of an RNA-directed DNA polymerase (reverse transcriptase) would also be required.

Inhibitor studies To distinguish between RNA- or DNA-directed replication, the effects of certain antibiotic compounds on viroid replication have been investigated. In an in vivo system in which leaf strips from healthy and PSTV-infected plants were treated with water or actinomycin D, viroid replication was found to be sensitive to this specific inhibitor of DNA-directed RNA synthesis (17). Similar results were obtained with an in vitro RNA synthesizing system in which purified cell nuclei from healthy or PSTV-infected tomato leaves were used as an enzyme source (52).

Sensitivity of viroid replication to actinomycin D has been confirmed in a study of cucumber pale fruit viroid (CPFV) synthesis in protoplasts isolated from tomato leaves (32). In the same study, the effect of α-amanitine on viroid synthesis has been investigated. Intracellular concentrations ($10^{-8}M$) of α-amanitine sufficient to inhibit tomato DNA-directed RNA polymerase II but not RNA polymerase III inhibited CPFV replication (32). In contrast to the studies with actinomycin D, in which the effect of the compound on viroid synthesis could conceivably be the result of a general toxic effect on cellular metabolism, rather than of a specific inhibition of DNA-directed RNA synthesis, the inhibitory effect of α-amanitine is not likely to be due to nonspecific, secondary effects of the compound on cell metabolism. This conclusion is strengthened by the demonstration that at an intracellular α-amanitine concentration sufficient to inhibit viroid replication by about 75%, the biosynthesis of tobacco mosaic virus RNA or that of prominent cellular RNA species (tRNA, 5S RNA, 7S RNA, and ribosomal RNA) was not appreciably affected (32).

All of these studies suggest that DNA-directed RNA synthesis is involved in viroid replication. This is most clearly evident in the experiments with α-amanitine, the results of which not only confirm those obtained with actinomycin D but, in addition, implicate one specific enzyme, namely RNA polymerase II.

Contradictory results, however, have been reported recently. Grill & Semancik (22), on the basis of infiltration of CEV-infected *G. aurantiaca* foliar tissue or of PSTV-infected potato tuber sprouts with varying concentrations of actinomycin D, concluded that the antibiotic had no specific inhibitory effect on viroid replication and that inhibitory effects reported earlier were due to a general toxic effect of actinomycin D on cell metabolism.

For two reasons these results must be regarded with caution. First, in both the CEV and PSTV experiments, total viroid concentration was conspicuously higher in nucleic acid preparations from tissue that had been infiltrated with actinomcyin D than in preparations from noninfiltrated tissue [see Figures 2a and 3a of (22)]. It appears that either nucleic acids isolated from unequal amounts of tissue were electrophoresed (which would make the reported data meaningless) or viroid extraction was more efficient from actinomycin D-treated than from untreated tissue. In the latter case, newly synthesized viroids must also have been extracted more efficiently from tissue samples that had been infiltrated with actinomycin D than from noninfiltrated samples, and any inhibitory effect of the antibiotic on viroid replication might well have gone unnoticed because of the larger amount of labeled viroid extracted [see Figures 2b and 3b of (22)]. Also, in the CEV experiment (22), the concentration of labeled 7S RNA appears not to have been affected by prior treatment with actinomycin D [see Figure 2b of (22)]. Thus, by the logic used regarding viroid synthesis, 7S RNA synthesis would also have to be regarded as insensitive to actinomycin D treatment. However, 7S RNA is a normal cellular RNA, and such a conclusion appears untenable.

Second, the authors leave unresolved (and unmentioned) the contradiction between their results and those reported earlier with α-amanitine (see above). If viroid replication is insensitive to actinomycin D, as claimed by Grill & Semancik (22), it should be equally insensitive to α-amanitine. It would thus appear that until the authors present evidence that this is the case and reasons why the results of the earlier, carefully executed study (32) are in error, the authors' results with actinomycin D appear to be questionable.

It is necessary to stress that sensitivity of viroid replication to actinomycin D or to α-amanitine does not prove that the viroid is transcribed from a DNA template. As has been pointed out earlier (12, 17, 32), it is possible that progeny viroids are transcribed from RNA templates, but that this process is dependent on the continued synthesis of a short-lived host RNA which, most plausibly, might serve as primer for RNA-directed viroid replication. Such a scheme would be analogous to that operative in influenza virus RNA replication, in which globin mRNA has been identified as primer for the transcription of viral RNA (1) and could explain actinomycin D sensitivity of viroid replication even though the latter was RNA directed.

Molecular probes Another approach to the study of the mechanisms of viroid replication consists in the development of viroid-specific molecular probes and identification of viroid-related RNA or DNA sequences in nucleic acid extracts from plants by molecular hybridization. Three types of molecular probes have been used: (*a*) purified viroids labeled in vitro with

^{125}I; (*b*) in vitro prepared, single-stranded viroid-complementary DNA (cDNA); (c) double-stranded viroid-related DNA obtained by recombinant DNA technology.

In experiments with ^{125}I-CEV, Semancik & Geelen (46) identified viroid-complementary DNA sequences in "DNA-rich preparations" from CEV-infected, but not from uninfected, plants. The authors speculated that this CEV-complementary DNA was either "synthesized de novo by the action of an induced RNA-directed DNA polymerase or exists in a limited number of copies and is amplified by normal DNA polymerase activity in the process of pathogenesis." Later, however, Grill and Semancik stated that what was reported as viroid-complementary DNA had subsequently been positively identified as complementary RNA, not DNA (22). Indeed, evidence presented by Grill & Semancik (21) indicates that the detected viroid-related sequences are RNA, not DNA. Most significant in this respect are the reported thermal denaturation properties of the complexes which are consistent with those of RNA·RNA duplexes, and not with those of RNA·DNA hybrid molecules (21). Unfortunately, the authors leave unexplained an earlier report in which the thermal denaturation properties of what was then believed to be an RNA·DNA hybrid was shown to be consistent with this assumption and not with that of an RNA·RNA duplex (44).

In another investigation of DNA sequences complementary to viroids (25), highly purified DNA preparations were used in molecular hybridization experiments with ^{125}I-PSTV, thus eliminating the possibility of mistaking viroid-complementary RNA-sequences with DNA sequences. These studies indicated that infrequent, if not unique, sequences complementary to PSTV occur in the DNAs of several uninfected, as well as PSTV-infected, host plants of the viroid (25). Although thermal denaturation of the presumed viroid DNA hybrids indicated that they were well-matched RNA·DNA duplex molecules, no competition experiments with unlabeled PSTV or host plant RNA were made to unambiguously rule out hybridization of the host DNA with cellular RNA contaminants in the ^{125}I-PSTV preparations (25). Recent efforts to confirm the existence of viroid-related sequences in DNA from either uninfected or viroid-infected host plants have failed.

Thus, in molecular hybridization experiments between [^{32}P]cDNA to the chrysanthemum stunt viroid (CSV) and the genomes of either healthy or CSV-infected hosts (chrysanthemum and *G. aurantiaca*), no sequence homologies could be detected (36). In this study, however, the large genome size of the chrysanthemum plants may have precluded detection of CSV sequences present only as single copies (or less) per haploid genome (36).

Similarly, in a study with PSTV, no viroid-complementary regions could be identified in DNA from either uninfected or viroid-infected tomato plants by conventional solution and filter hybridization techniques (55) or by Southern hybridization (2). Because, in the latter study, sensitivity of the experimental method was demonstrated to be adequate for the detection of less than one copy of viroid complement per haploid genome, these experiments appear to rule out the presence of even a single complete and contiguous complement of PSTV in host DNA. While the authors ascribed the earlier positive results (25) to hybridization of cellular RNA contaminants in the ^{125}I-PSTV preparations with host DNA, they did not rule out the possibility that PSTV-related sequences might be randomly located on host chromosomes or that host DNA contains only a portion of the PSTV genome. In the latter case, it is conceivable that short, viroid-complementary sequences might serve as recognition sites and be involved in viroid replication and/or pathogenesis. Present results seem to rule out, however, that any such DNA sequences could act as templates for the synthesis of progeny viroids; and it appears far more likely that viroids are replicated from RNA templates.

This conclusion is strengthened by observations indicating that the primary structure of viroids is faithfully maintained regardless of the host in which the viroid is replicated (7, 33, 35). Furthermore, the occurrence in viroid-infected plants of RNA sequences complementary to the viroid has been confirmed recently for PSTV-infected plants (34).

In efforts to develop a fully defined molecular probe, double-stranded cDNA has been synthesized from a polyadenylated PSTV template and cloned in the Pst I site of plasmid pBR322 using the oligo(dC) oligo(dG) tailing procedure (34). One recombinant clone contained a 460 base-pair insert and has been shown to represent almost the entire sequence of PSTV (34). Hybridization probes derived from this insert have allowed detection and preliminary characterization of RNA molecules having the same size as PSTV but the opposite polarity. This RNA is present during PSTV replication in infected cells (34).

In summary, indications that viroids are replicated by an RNA-directed mechanism have become increasingly convincing in recent years. Despite the plausibility of such a scheme, however, a more complete characterization of the putative RNA templates is required before their involvement in viroid replication can be regarded as a demonstrated fact.

Possible Mechanisms of Pathogenesis

Another intriguing question concerns viroid pathogenicity: By what mechanisms do viroids incite diseases in certain hosts yet replicate in other susceptible species without inflicting discernible damage? The nuclear location of viroids and their apparent inability to act as mRNAs suggest that viroid-induced disease symptoms may be caused by direct interaction of the viroid with the host genome; that is, by interference with gene regulation in the infected cells. If so, viroids might be regarded as abnormal regulatory molecules (6, 9, 11). No evidence exists, however, to substantiate such a scheme.

Many of the symptoms induced by viroid infection, such as stunting of plants, epinasty, curling, and deformation of leaves, suggest viroid-induced disturbances in the metabolism of growth substances. Indeed, in a comparison of the concentrations of some plant growth substances in healthy and CEV-infected *G. aurantiaca* plants, an auxin-like substance of unknown chemical nature was found to be formed as a consequence of viroid infection (39). Also, a significant decrease in the levels of endogenous gibberellins (probably GA_3 and/or GA_1) was observed in viroid-infected as compared with healthy plants, but no changes in the levels of abscisic or indoleacetic acids were detectable (39).

As discussed above, in both PSTV- and CEV-infected plants, certain alterations of protein synthesis have been noted (4, 54). Possibly these aberrations are related to the pathogenic properties of viroids and may constitute intermediate points in the causal chain leading from viroid infection to the appearance of macroscopic symptoms in infected plants. With CEV, two low molecular weight proteins, P_1, of 15,000 and P_2 of 18,000 daltons, accumulate in infected plants (20). Partial purification of P_1 showed that it is a slightly acidic protein with an isoelectric point of 5.21. Although ribonuclease activity was associated with some P_1 preparations, this was not consistent. No affinity was detected in in vitro or in vivo conditions between P_1 and gibberellin A1/A3 or indoleacetic acid (20). Convincing evidence that P_1 and P_2 are host proteins and not translation products of the viroid or its complementary strand has been reported by Conejero et al (3). These authors demonstrated that P_1 and P_2 are of significantly different size depending on the particular CEV-infected host species from which they are isolated, and that senescence in healthy *G. aurantiaca* plants induces the same low-molecular-weight proteins as does infection with CEV (3).

In addition to the macroscopic symptoms characteristic of certain viroid-infected plants (13), cytopathic effects of viroid infection have been observed. Thus, CEV infection of *G. aurantiaca* plants has been reported to result in the appearance of membranous structures of the cellular membrane, so-called "plasmalemmasomes" or "paramural bodies" (48). These bodies vary in size, internal structure, and shape, and their origin and function are not yet entirely understood. Appearance of these paramural bodies has been regarded as the primary cytopathic effect of viroid infection and, in view of the claimed association of CEV with plasma membranes (47), their appearance has been regarded as suggesting a direct causal relationship with the pathogenic RNA (48).

In another study of cytopathic effects of viroid infection, however, plasmalemmasomes were found to be present at equal frequency in CEV-infected and healthy *G. aurantiaca* plants (41), indicating that they could not be the primary cytopathic effect of viroid infection. Viroid infection, nevertheless, was shown to affect the structure of plasmalemmasomes. In healthy plants, plasmalemmasomes with vesicular or tubular internal structures were found, whereas in CEV-infected plants, the vesicular type exhibited pronounced irregularities in size and shape and contained malformed internal vesicles (41). In addition, pronounced cell wall irregularities were found with infected cells, consisting in large differences in thickness and width (41).

Hari (27), on the other hand, reported that PSTV-infected tomato leaf cells develop paramural bodies, as has been reported earlier for CEV-infected *G. aurantiaca* cells. Another, previously unreported cytopathic effect concerns chloroplasts of infected cells in which aberrations of the thylakoid membrane system and lack of development of grana were observed (27).

CONCLUSIONS

Evidence available in 1971, at the time when the viroid concept was advanced (9), clearly indicated that the prototype viroid PSTV differs basically from all known viruses. Viroids, nevertheless, could reasonably be regarded as relatives of conventional viruses, being either very primitive or else degenerate representatives of the latter. Knowledge accumulated since then has rendered this concept increasingly less likely. The apparent lack of messenger function of viroids (or their complements) and their novel molecular structure—that has no counterpart among viruses—imply a far greater phylogenetic distance from viruses than could be imagined previously. It can be predicted safely that future work with these unusual pathogens will yield further surprises and that the knowledge gained will be important not only for plant pathology but for plant physiology and molecular biology as well.

Literature Cited

1. Bouloy, M., Plotch, S. J., Krug, R. M. 1978. Globin mRNAs are primers for the transcription of influenza viral RNA *in vitro*. *Proc. Natl. Acad. Sci. USA* 75:4886–90
2. Branch, A. D., Dickson, E. 1980. Tomato DNA contains no detectable regions complementary to potato spindle tuber viroid as assayed by Southern hybridization. *Virology* 104:10–26
3. Conejero, V., Picazo, I., Segado, P. 1979. Citrus exocortis viroid (CEV): Protein alterations in different hosts following viroid infection. *Virology* 97:454–56
4. Conejero, V., Semancik, J. S. 1977. Exocortis viroid: Alteration in the proteins of *Gynura aurantiaca* accompanying viroid infection. *Virology* 77:221–32
5. Davies, J. W., Kaesberg, P., Diener, T. O. 1974. Potato spindle tuber viroid. XII. An investigation of viroid RNA as messenger for protein synthesis. *Virology* 61:281–86

6. Dickson, E. 1976. *Studies of plant viroid RNA and other RNA species of unusual function.* PhD thesis. Rockefeller Univ., New York. 134 pp.
7. Dickson, E., Diener, T. O., Robertson, H. D. 1978. Potato spindle tuber and citrus exocortis viroids undergo no major sequence changes during replication in two different hosts. *Proc. Natl. Acad. Sci. USA* 75:951–54
8. Diener, T. O. 1971. Potato spindle tuber virus: A plant virus with properties of a free nucleic acid. III. Subcellular location of PSTV-RNA and the question of whether virions exist in extracts or *in situ. Virology* 43:75–89
9. Diener, T. O. 1971. Potato spindle tuber "virus." IV. A replicating, low molecular weight RNA. *Virology* 45:411–28
10. Diener, T. O. 1972. Potato spindle tuber viroid. VIII. Correlation of infectivity with a UV-absorbing component and thermal denaturation properties of the RNA. *Virology* 50:606–9
11. Diener, T. O. 1977. Viroids: Autoinducing regulatory RNAs? In *Genetic Interaction and Gene Transfer,* ed. C. W. Anderson. *Brookhaven Symp. Biol.* 29:50–61
12. Diener, T. O. 1979. Viroids: Structure and function. *Science* 205:859–66
13. Diener, T. O. 1979. *Viroids and Viroid Diseases.* New York: Wiley Interscience. 252 pp.
14. Diener, T. O., Hadidi, A. 1977. Viroids. In *Comprehensive Virology,* ed. H. Fraenkel-Conrat, R. R. Wagner, 11:285–337. New York: Plenum
15. Diener, T. O., Lawson, R. H. 1973. Chrysanthemum stunt: A viroid disease. *Virology* 51:94–101
16. Diener, T. O., Raymer, W. B. 1967. Potato spindle tuber virus: A plant virus with properties of a free nucleic acid. *Science* 158:378–81
17. Diener, T. O., Smith, D. R. 1975. Potato spindle tuber viroid. XIII. Inhibition of replication by actinomycin D. *Virology* 63:421–27
18. Domdey, H., Jank, P., Sänger, H. L., Gross, H. J. 1978. Studies on the primary and secondary structure of potato spindle tuber viroid: Products of digestion with ribonuclease A and ribonuclease T_1, and modification with bisulfite. *Nucleic Acids Res.* 5:1221–36
19. Fernow, K. H., Peterson, L. C., Plaisted, R. L. 1970. Spindle tuber virus in seeds and pollen of infected potato plants. *Am. Potato J.* 47:75–80
20. Flores, R., Chroboczek, J., Semancik, J. S. 1978. Some properties of the CEV-P_1 protein from citrus exocortis viroid-infected *Gynura aurantiaca* DC. *Physiol. Plant Pathol.* 13:193–201
21. Grill, L. K., Semancik, J. S. 1978. RNA sequences complementary to citrus exocortis viroid in nucleic acid preparations from infected *Gynura aurantiaca. Proc. Natl. Acad. Sci. USA* 75:896–900
22. Grill, L. K., Semancik, J. S. 1980. Viroid synthesis: The question of inhibition by actinomycin D. *Nature* 283:399–400
23. Gross, H. J., Domdey, H., Lossow, C., Jank, P., Raba, M., Alberty, H., Sänger, H. L. 1978. Nucleotide sequence and secondary structure of potato spindle tuber viroid. *Nature* 273:203–8
24. Gross, H. J., Riesner, D. 1980. Viroids: A class of subviral pathogens. *Angew. Chem.* 19:231–43
25. Hadidi, A., Jones, D. M., Gillespie, D. H., Wong-Staal, F., Diener, T. O. 1976. Hybridization of potato spindle tuber viroid to cellular DNA of normal plants. *Proc. Natl. Acad. Sci. USA* 73:2453–57
26. Hall, T. C., Wepprich, R. K., Davies, J. W., Weathers, L. G., Semancik, J. S. 1974. Functional distinctions between the ribonucleic acids from citrus exocortis viroid and plant viruses: Cell-free translation and aminoacylation reactions. *Virology* 61:486–92
27. Hari, V. 1980. Ultrastructure of potato spindle tuber viroid-infected tomato leaf tissue. *Phytopathology* 70:385–87
28. Henco, K., Riesner, D., Sänger, H. L. 1977. Conformation of viroids. *Nucleic Acids Res.* 4:177–94
29. Langowski, J., Henco, K., Riesner, D., Sänger, H. L. 1978. Common structural features of different viroids: Serial arrangement of double helical sections and internal loops. *Nucleic Acids Res.* 5:1589–1610
30. McClements, W. L., Kaesberg, P. 1977. Size and secondary structure of potato spindle tuber viroid. *Virology* 76:477–84
31. Mohamed, N. A., Thomas, W. 1980. Viroid-like properties of an RNA species associated with the sunblotch disease of avocados. *J. Gen. Virol.* 46:157–67
32. Mühlbach, H.-P., Sänger, H. L. 1979. Viroid replication is inhibited by α-amanitin. *Nature* 278:185–88
33. Niblett, C. L., Dickson, E., Fernow, K. H., Horst, R. K., Zaitlin, M. 1978. Cross protection among four viroids. *Virology* 91:198–203

34. Owens, R. A., Cress, D. E. 1980. Molecular cloning and characterization of potato spindle tuber viroid cDNA sequences. *Proc. Natl. Acad. Sci. USA* 77:5302–6
35. Owens, R. A., Smith, D. R., Diener, T. O. 1978. Measurement of viroid sequence homology by hybridization with complementary DNA prepared *in vitro*. *Virology* 89:388–94
36. Palukaitis, P. 1980. *Molecular biology of viroids*. PhD thesis. Univ. Adelaide, Adelaide, South Australia. 138 pp.
37. Palukaitis, P., Hatta, T., Alexander, D. McE., Symons, R. H. 1979. Characterization of a viroid associated with avocado sunblotch disease. *Virology* 99:145–51
38. Randles, J. W. 1975. Association of two ribonucleic acid species with cadang-cadang disease of coconut palm. *Phytopathology* 65:163–67
39. Rodriguez, J. L., Garcia-Martinez, J. L., Flores, R. 1978. The relationship between plant growth substances content and infection of *Gynura aurantiaca* D.C. by citrus exocortis viroid. *Physiol. Plant Pathol.* 13:355–63
40. Romaine, C. P., Horst, R. K. 1975. Suggested viroid etiology for chrysanthemum chlorotic mottle disease. *Virology* 64:86–95
40a. Sänger, H. L. 1972. An infectious and replicating RNA of low molecular weight: The agent of exocortis disease of citrus. *Adv. Biosci.* 8:103–16
41. Sänger, H. L. 1979. Structure and function of viroids. In *Slow Transmissible Diseases of the Nervous System, Vol. 2: Pathogenesis, Immunology, Virology, and Molecular Biology of the Spongiform Encephalopathies*, ed. S. B. Prusiner, W. J. Hadlow, pp. 291–341. New York: Academic
42. Sänger, H. L., Klotz, G., Riesner, D., Gross, H. J., Kleinschmidt, A. K. 1976. Viroids are single-stranded covalently closed circular RNA molecules existing as highly base-paired rod-like structures. *Proc. Natl. Acad. Sci. USA* 73:3852–56
43. Sasaki, M., Shikata, E. 1977. On some properties of hop stunt disease agent, a viroid. *Proc. Japan Acad. Ser. B* 53: 109–12
44. Semancik, J. S. 1976. Structure and replication of plant viroids. In *Animal Virology*, ed. D. Baltimore, A. S. Huang, C. F. Fox, 4:529–45. ICN-UCLA Symp. Mol. Cell. Biol.
45. Semancik, J. S., Conejero, V., Gerhart, J. 1977. Citrus exocortis viroid: Survey of protein synthesis in *Xenopus laevis* oocytes following addition of viroid RNA. *Virology* 80:218–21
46. Semancik, J. S., Geelen, J. L. M. C. 1975. Detection of DNA complementary to pathogenic viroid RNA in exocortis disease. *Nature* 256:753–56
47. Semancik, J. S., Tsuruda, D., Zaner, L., Geelen, J. L. M. C., Weathers, L. G. 1976. Exocortis disease: Subcellular distribution of pathogenic (viroid) RNA. *Virology* 69:669–76
48. Semancik, J. S., Vanderwoude, W. J. 1976. Exocortis viroid: Cytopathic effects at the plasma membrane in association with pathogenic RNA. *Virology* 69:719–26
49. Semancik, J. S., Weathers, L. G. 1972. Exocortis disease: Evidence for a new species of "infectious" low molecular weight RNA in plants. *Nature New Biol.* 237:242–44
50. Singh, R. P. 1970. Seed transmission of potato spindle tuber virus in tomato and potato. *Am. Potato J.* 47:225–27
51. Sogo, J. M., Koller, T., Diener, T. O. 1973. Potato spindle tuber viroid. X. Visualization and size determination by electron microscopy. *Virology* 55:70–80
52. Takahashi, T., Diener, T. O. 1975. Potato spindle tuber viroid. XIV. Replication in nuclei isolated from infected leaves. *Virology* 64:106–14
53. Van Dorst, H. J. M., Peters, D. 1974. Some biological observations on pale fruit, a viroid-incited disease of cucumber. *Neth. J. Plant Pathol.* 80:85–96
54. Zaitlin, M., Hariharasubramanian, V. 1972. A gel electrophoretic analysis of proteins from plants infected with tobacco mosaic and potato spindle tuber viruses. *Virology* 47:296–305
55. Zaitlin, M., Niblett, C. L., Dickson, E., Goldberg, R. B. 1980. Tomato DNA contains no detectable regions complementary to potato spindle tuber viroid as assayed by solution and filter hybridization. *Virology* 104:1–9

Ann. Rev. Plant Physiol. 1981. 32:327–47

PHYCOBILISOMES[1,2]

♦7715

E. Gantt

Radiation Biology Laboratory, Smithsonian Institution, Rockville, Maryland 20852

CONTENTS

INTRODUCTION

Phycobilisomes are supramolecular pigment aggregates which serve as the primary light-gathering antennae in red algae and in cyanobacteria (blue-green algae). These aggregates, composed primarily of phycobiliproteins, greatly extend the range in which light is absorbed and are particularly important under light-limiting conditions (44a). In fact, organisms can adapt to light-limiting conditions by increasing the total phycobiliprotein content (65) or by specifically producing the phycobiliprotein type capable of absorbing the prevalent available wavelengths. Phycobiliproteins sometimes comprise up to 60% of the total soluble cell protein; thus their production commands a major part of the cell's metabolic resources (4).

[2]Abbreviations: APC, allophycocyanin; PBS, phycobilisome(s); PC, phycocyanin; PCB, phycocyanobilin; PE, phycoerythrin; PEB, phycoerythrobilin; PEC, phycoerythrocyanin; PSI, photosystem I; PSII, photosystem II; PUB, phycourobilin; SDS-PAGE, sodium dodecyl sulfate polyacrylamide gel electrophoresis.

The prefixes C-, R-, and B-(b) are respective designations for cyanophytan, rhodophytan, and bangiophycean algal origins.

Functionally, PBS can be considered to be analogous to the light-harvesting complexes containing chlorophyll *a* and *b* in green plants. Both are light-gathering aggregates which function in passing the trapped energy to the photosynthetic reaction centers in the photosynthetic membrane. They differ significantly, however, in their structure and location. Phycobilisomes are directly attached to the photosynthetic membrane, but are not major membrane constituents as are the chlorophyll light-harvesting complexes. The chromophores in the phycobiliproteins are tightly linked to the apoproteins, a property which has been advantageous in the characterization of phycobiliproteins and PBS. Energy transfer from PBS to the reaction centers is considered to be as efficient as transfer from the chlorophyll *a* and *b* complexes to the reaction centers in green plants. Under physiological conditions, energy absorbed by PBS is distributed to both photosystems. For example, in the cyanobacterium *Anacystis nidulans,* 40% of the quanta absorbed by phycocyanin were contributed to photosystem I and 60% to photosystem II (80, 81). In *Porphyridium cruentum* (synonymous with *P. purpureum*) 95% of the energy absorbed by phycoerythrin was transferred initially to PSII (46). Energy transfer from quanta absorbed by the peripheral PBS pigment is: PE (or PEC) $\rightarrow$ PC $\rightarrow$ APC $\rightarrow$ PSII. A very close physical relationship can be expected to exist between the PBS and PSII, because the transfer efficiencies are high and the number of chlorophyll molecules associated with PSII in these algae is very small (67, 77, 80).

The aim of this review is to discuss the more recent attempts to elucidate the phycobilisome structure. It will be concerned chiefly with recent contributions to the characterization and composition of PBS. Especially relevant to the topic is the ever increasing literature on phycobiliproteins, but since this area has been covered in several recent reviews, only the most pertinent and recent studies are cited. The last comprehensive coverage of phycobiliproteins in this series appeared in 1975 (4), at which time PBS had only recently been isolated. In addition to dealing generally with phycobiliprotein characteristics, several reviews can be consulted for special emphasis on phycobiliprotein structure (19, 20, 26, 27, 88); phycobilisome structure (9, 18, 19, 21); chromatic adaptation (4, 9); phycobiliprotein synthesis (78); chromophore structure (34, 58, 70, 70a); and evolutionary relationships of red algae and cyanobacteria (74). Phycobiliproteins may also serve as a nitrogen sink for cyanobacteria and red algae (21, 85), but this topic is beyond the scope of this review.

COMPOSITION

Phycobiliproteins are the primary constituents of PBS; they account for most of the total stainable protein on SDS-PAGE (76, 86). There are three main classes of phycobiliproteins: phycocyanins, phycoerythrins, and allophycocyanins. Another class, phycoerythrocyanin, is only found in a few cyanobacteria (6, 9). Phycobiliprotein spectral properties and molecular characteristics are detailed in Table I: several spectral forms can be found within each type. The PE's absorbance maxima range from ca 498 nm to 568 nm. Typical PC maxima range around 625 nm, except for R-PC which

Table 1 Phycobilisome components and their spectral and molecular characteristics[a]

Type[b]	Absorption λ max (nm)	Emission λ max (nm)	M_r[c]	Structure subunit	Chromophores per polypeptide[d]
Phycobiliproteins:					
R-PE	498, 542, 565	578	260,000	$(\alpha\beta)_6\gamma$	—
B-PE	498, 545, 563	575	260,000	$(\alpha\beta)_6\gamma$	α 2PEB, β 4PEB, γ 2PEB, 2 PUB
b-PE	545, 563	575	—	$(\alpha\beta)_n$	α 2PEB, β 4PEB
C-PE	565	578	230,000	$(\alpha\beta)_6$	α 2PEB, β 4PEB
C-PEI	555	578	—	—	—
C-PEII	540, 568	578	—	—	—
R-PC	553, 615	640	130,000	$(\alpha\beta)_3$	α 1PCB, β 1 PCB, 1 PEB
C-PC	620	650	120,000	$(\alpha\beta)_3$	α 1PCB, β 2PCB
PEC	568, 590	610	103,000	$(\alpha\beta)_3$	α 1PXB, β 2PCB
APC B	618, 673	680	98,000	$(\alpha\beta)_3$	α 1PCB, β 1PCB
APC I	654	680	145,000	$(\alpha\beta)_3\gamma$	—
APC II	650	660	105,000	$(\alpha\beta)_3$	α 1PCB, β 1PCB
APC III	650	660	105,000	$(\alpha\beta)_3$	—

Uncolored polypeptides M_r	*Synechococcus* 6301 Phycobilisomes	*Porphyridium cruentum* Phycobilisomes	PE-PC complex[e]
95,000	—	Present	—
75,000	Present	—	—
75–45,000	—	Present	—
45–35,000	Present	Present	—
35–30,000	Present	Present	Present
30–25,000	Present	Present	—
25–10,000	Present	Present	—

[a] Data compiled from tables in reviews (4, 9, 27, 70a) and other references (7, 31, 55a, 56, 66, 76, 79, 86, 87a). Absorption and emission maxima give general range for the phycobiliprotein types, but can vary with buffer conditions and species source.

[b] Prefix designations: C-cyanophycean, R-rhodophytan, b- and B-bangiophycean origins.

[c] Apparent molecular weights of lowest stable assembly form.

[d] Phycobilin abbreviations: PEB, phycoerythrobilin; PCB, phycocyanobilin; PUB, phycourobilin; PXB, chromophores of undetermined structure.

[e] From *Porphyridium sordidum* PBS.

has an additional green absorbing chromophore at 553 nm. The APCs have several wavelength-absorbing forms ranging from 618 to 673 nm, with the 650 nm being the most common. It should be noted that the absorption maxima (23, 45) and emission maxima of phycobiliproteins in vivo often occur at slightly longer wavelengths than in solution.

Each phycobiliprotein is made up of polypeptides α and β, usually occurring in equal amounts. Their molecular weights differ with the organism, but the α polypeptides (12,000–20,000) are generally smaller and the β polypeptides (15,000–22,000) larger. The phycobiliproteins exist as oligomers with the trimer $(\alpha\beta)_3$ usually being the smallest stable aggregate except for a C-PE from *Pseudoanabaena,* where monomers and dimers were the stable species (86a). The next highest aggregate, and the basic building block of the PBS, is a hexamer. In fact, by electron microscopy (6, 51, 55a) and X-ray crystallography (14) each hexamer can be resolved into two discs, each disc being equivalent to a trimer. In certain phycobiliproteins such as in B-PE and APC I, a larger polypeptide (30,000–35,000 mol wt) is present. Such larger polypeptides may be involved in stabilizing the larger aggregates (7, 55a). It is interesting that the amino acid sequence of phycobiliproteins is highly conserved (9, 15, 19, 27, 34, 78). Conservation of the amino acid sequence is higher within each α or β polypeptide group than between the groups, suggesting an early evolutionary divergence between the α and β polypeptides (28).

From 1–4 chromophores (open-chain tetrapyrroles) are covalently bound to each polypeptide (Table I), with APCs having the lowest number and PEs the highest. In most phycobiliproteins one covalent bond occurs between the chromophore and the apoprotein. This occurs by a thioether linkage between cysteine and ring A of the tetrapyrrole (5, 38, 44). Some evidence also exists for a second linkage to ring C through an ester bond to serine (57, 58, 70). It is remarkable that two main chromophore types, phycoerythrobilin in PE and phycocyanobilin in PC and APC, account for most of the spectroscopic variation (4, 57). When the chromophores are cleaved from the apoprotein and extracted, they exist as transisomers (16). In the past, considerable controversy had been engendered because the various methods used for preparations of the chromophores often resulted in destruction of some of the less common chromophores. Recently, milder cleavage conditions have lead to greater agreement on the ubiquity of PEB and PCB (34, 70a). Spectral evidence for phycourobilin in rhodophytan PE and an unidentified bilin (PXB) in PEC (27) have been reported, but their chemical structures have not been fully verified.

The spectral variations among the phycobiliproteins is largely dependent on the interaction of the linear tetrapyrrole chromophore with the apoprotein. Scheer and Kufer (43, 70b) analyzed PC and very convincingly showed stepwise changes in extinction coefficients and absorbance shifts which occurred with the unfolding of the apoprotein structure. Also in APC spectral shifts from 652 → 616 nm resulted when the APC trimers $(\alpha\beta)_3$ dissociated into monomers $(\alpha\beta)$ (51). These examples illustrate that aggregation state and conformation state of the apoproteins do play an important role in the absorption properties and function of phycobiliproteins.

After analysis of PBS on SDS-PAGE, Tandeau de Marsac & Cohen-Bazire (76) and Yamanaka et al (86) showed that the phycobiliproteins accounted for 85% of the Coomassie stained protein, whereas the colorless polypeptides accounted for 15%. These determinations were made by assuming that the Coomassie dye binding is proportional to the amount of protein present in all polypeptides. Uncolored polypeptides are present in PBS of cyanobacteria and red algae and range in molecular weight from 10,000 to 95,000. That the uncolored polypeptides occur as PBS components rather than as contaminants is supported by their presence in a purified PE-PC complex (50), their comigration with specific PBS fractions on a gradient (76, 86), and from effects observed during chromatic adaptation where specific polypeptide bands decreased or increased with specific phycobiliproteins (76). The role of these uncolored polypeptides will be discussed below.

MORPHOLOGY AND STRUCTURE

Phycobilisomes are structures somewhat larger than ribosomes and variable in shape and size with the species. They are attached to the photosynthetic lamellae, often forming a highly regular two-dimensional crystalline-like array. The number per unit of lamellar area differs with the organism. In *P. cruentum,* which has large PBS (82), there are ca 400 PBS per μm^2 (12). In *Oscillatoria brevis,* which has thin PBS, there are ca 1200 per μm^2 (49).

To isolate PBS, cells are suspended in 0.5–0.75 M phosphate buffer (pH 7), and after disruption in a French press or by sonication, a detergent (Triton X-100 or Deriphat) is added to release the PBS from the membrane. The PBS remain in the supernatant when cell fragments are removed by centrifugation. Subsequent centrifugation on a sucrose gradient (0.25–2.0 M) concentrates the PBS in the 1 M sucrose layer while chlorophyll and free phycobiliprotein remain on top (22, 30, 42, 76). In some filamentous cyanobacteria, PBS have been isolated by treating the cells with Triton X-100, followed by repeated rinses in a polyethylene glycol-Triton- 0.75M phosphate mixture (68, 69).

Many species of cyanobacteria and red algae have PBS that are broad and thin, and are thus said to be disc-shaped or hemidiscoidal in shape. Several red algae possess PBS that are hemispherical-prolate in shape, being almost as thick and high as they are wide (21). Among the various species, isolated PBS that were negatively stained with uranyl acetate or phosphotungstic acid have diameters ranging from 32–70 nm, heights of ca 25–45 nm, and thicknesses of ca 12–40 nm (6a, 21, 42, 69). The height and thickness are best determined on electron micrographs of sectioned cells because the orientation of the PBS to the thylakoid on these is unequivocal, but at the same time a correction for shrinkage caused by dehydration must be made (21). Sedimentation velocity analysis of hemidiscoidal PBS produced values of 61S for *Synechococcus* 6301 (synonymous with *Anacystis nidulans*), and 45–50S for LPP 7409 (*Lyngbya-Plectonema-Phormidium*) (6, 30). The smallest ones at 35–40S were obtained from *Synechocystis* 6701 cells grown in red light. Calculated molecular weight values of hemidiscoidal shaped PBS are $4.5–8 \times 10^3$, but PBS with the largest volume exist in several red algae having estimated molecular weights of $1.5–2.0 \times 10^7$ (12, 18, 23). These larger structures can possess as many as 2600 chromophores in one PBS.

The arrangement of the phycobiliproteins within the PBS (Figure 1) has been determined from spectral analysis of intact and selectively dissociated PBS and by electron microscopic examination (6a, 18, 23, 41, 55, 83). Phycobilisomes are judged to be intact by two criteria: a characteristic uniformity of size and shape and, more importantly, by their fluorescence emission (22). The characteristic high fluorescence emission from APC, the terminal pigment in the transfer chain, occurs only when PBS are intact and energetically well coupled. Lowering the ionic strength, usually from 0.75 M to 0.1 M phosphate buffer or even less, dissociates PBS, and the APC emission drops while emissions from the other phycobiliproteins emerge at their characteristic wavelengths (1, 23, 40, 55, 68).

The first determination of the phycobiliprotein arrangement in the PBS was made in *Porphyridium cruentum* by following the progressive release of the phycobiliproteins during dissociation (23). This was an advantageous choice because the hemispherical-prolate shape of these PBS allowed for an ordered dissociation rather than random PBS fragmentation. The observed order of release was PE > R-PC > APC, which lead to the proposal of model structures as seen in Figure 1 where APC exists as a centrally located core surrounded by PC and PE on the periphery. Such an arrangement is energetically consistent, and has been substantiated by morphological evidence and antibody labeling. In *Rhodella violaceae,* Koller et al (40, 41) have provided evidence for the peripheral location of PC and PE by isolating elongated rods from dissociated PBS. The rods or "tripartite units" were morphologically similar to those in PBS. Furthermore, by cross-linking experiments they showed that these rods contained a molar ratio of 2 B-PE and 1 PC as occurs in PBS. Allophycocyanin was not associated with the rods.

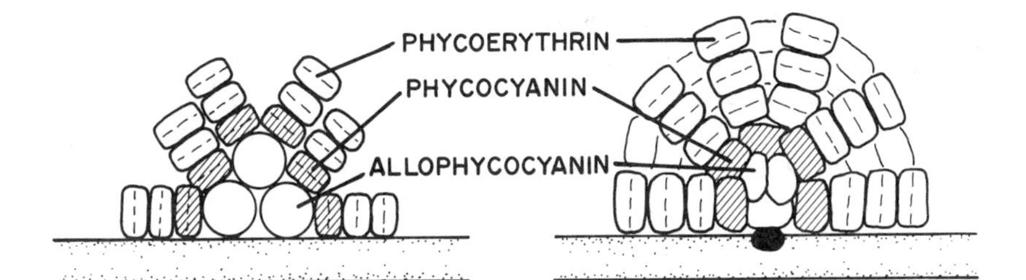

Figure 1 Phycobilisome models with the same fundamental phycobiliprotein arrangement. Stacked rods of phycobiliproteins are peripherally attached to a central allophycocyanin core. In organisms lacking phycoerythrin the stacked rods would be composed of phycocyanin. The model on left represents thin hemidiscoidally shaped phycobilisomes found in *Rhodella violaceae* (40, 55) and *Pseudoanabaena* (LLP)-7409 (6a). The three core units, assumed to be allophycocyanin, have been resolved by electron microscopy. The phycobilisome model of *Porphyridium cruentum* on right (23) is drawn so as to reveal the internal arrangement. Such a phycobilisome is thick, hemispherical-oblate in shape, with the peripheral rods extending in all directions. A protein (black) common to both phycobilisomes and thylakoids (66) is presumed to be involved in anchoring the phycobilisome to the thylakoid.

In the model (Figure 1) proposed for the red alga *Rhodella* (55) and for several cyanobacteria (6a), the PBS is seen to consist of six stacked rods which fan out from a morphologically distinctive core. Distinctive core structures have only been seen in hemidiscoidal PBS. By electron microscopy the peripheral rods are seen to be attached at right angles to the core units, i.e. the edge of a core unit is directly adjacent to the flat surface of a rod disc. The rod components, however, are attached to one another along their broad surface; this arrangement suggests dissimilar binding domains. Each stacked rod consists of 3–4 double discs, each double disc corresponding to a phycobiliprotein hexamer. The double discs have been clearly resolved in all cyanobacterial rods (6a, 30, 69, 83). Each disc has a dimension of ca 3 nm by 10 nm and is considered as a trimer with a molecular weight of 100,000–130,000 (6, 6a, 54). In their essential structure the phycobiliproteins of the C type are comparable with the B type, which exist at the hexamer level (2 disc equivalents). A hexamer of C-PC with a molecular weight of ca 220,000 is comparable to a B-PE molecule (5 X 10 nm) with a weight of ca 260,000 (4, 21, 27). In hemidiscoidal PBS the substructure of the stacked rod can be resolved along the entire length of the rod, but in the thicker hemispherical-oblate PBS the hexameric structures are resolvable only at the periphery. It seems valid to assume that the latter are similarly composed of stacked rods as well, even though the superposition of stain layers obscures much of the fine structure.

Among species, variation in length of the stacked rods is frequently observed. Two possible reasons for variation are that the rod length is specific for the species and light condition (intensity and/or wavelength), or that random dissociation and perhaps some reassociation occurs during PBS isolation. In *Synechococcus* 6301 (30), a cyanobacterium with PC and APC, the rods contain between 4–6 double discs. This may be a natural variation within the organism, or may reflect partial dissociation. A striking example exists in *Synechocystis* 6701. Responding to a shift from green to red light, with a reduction of PE synthesis, the rod length decreased from ca 4 double discs to ca 2 double discs, but the core structure remained unchanged (6a, 83). Some very interesting results have been obtained from mutants produced from *Synechocystis* 6701 (83). The wild type PBS (white light grown) has the typical structure of 6 rods, each consisting of ca 4 double discs. In the mutant NTG 31, however, the core structure remained constant, but only 1 or 2 peripheral rods were usually attached to the basal core units.

In most hemidiscoidal PBS the central core appears as three round units, two at the base and one on top in the groove between the basal units. However, in *Synechococcus* 6301, only two units were found per core (30). In *Rhodella* the core is assumed to consist of three APC hexamers (40, 55), but from studies on several cyanobacterial PBS Bryant et al (6a) concluded that the core consisted of six "hexameric equivalents" of APC. This interpretation was directly consistent with their observations from electron microscopy and with the observed molar ratios of the phycobiliproteins in PBS.

The observation of a morphologically identifiable core structure, as in hemiscoidal PBS, leads to the consideration of the composition of the core in all PBS types. Location of APC at the base of each PBS is expected because it is inherently consistent with the most direct pathway of energy transfer; this location has been verified in *P. cruentum* by selective reactions with ferritin-labeled antibodies to phycobiliproteins (23, 24). Phycobilisomes attached to thylakoids reacted positively with anti-APC only when most of the PE and PC were removed, but when the PBS was intact it reacted only with anti-PE. By analogy it has been assumed that the morphologically identifiable core in hemidiscoidal PBS consists of APC. A further assumption is that the core units in these hemidiscoidal PBS are near the photosynthetic membranes. These assumptions are reasonable, and direct verification should be obtainable by electron microscopic examination of PBS vesicles and from labeled studies with anti-APC.

If the core consists of APC as assumed, which of the core unit(s) contains the long wavelength emitting APC forms (F680)? Since the F680 APC forms (discussed more fully later) have been suggested as the probable bridging pigment between PBS and the photosynthetic membrane (7, 29, 48), they must be in or at the bottom of the core and near or in the lamella. One problem concerns the low concentration of APCB ($< 1\%$ of total

APC) (7, 48). If APC B is the only F680 form present, then at least one molecule (trimer, ca 100,000) should be present per PBS. In fact, it has been suggested by Cohen-Bazire & Bryant (9) that an APC B trimer could be accommodated in the space between the two basal hexamers according to their model. In *Rhodella* (40) and in the cyanobacterium *Synechocystis* 6701 (83), where such a space does not exist, one must assume that one or all of the basal core units consist partly or entirely of APC B. If APC B exists as a core trimer in vivo, it differs from PE and PC because they normally exist as hexamers (6, 40, 50). It is also possible that an APC B trimer is aggregated with a trimer of APC II (short wavelength emitting form).

In *Gloeobacter violaceus,* an unusual cyanobacterium lacking typical internal photosynthetic membranes, the PBS structure is different from those described above. Each PBS consists of 6 rods (50–70 nm long) which are not attached to a triangular core but probably to a flat plate of APC (33). Furthermore, the spectra of isolated PBS indicated that long wavelength-emitting APC seems to be absent.

BINDING FACTORS: DISASSEMBLY AND REASSEMBLY OF PHYCOBILISOMES

Purified phycobiliproteins can readily aggregate into specific PE, PC, or APC crystals. Even in a mixture of phycobiliproteins, under crystallizing conditions only homogeneous crystals form (6). Yet within PBS, phycobiliproteins exist as orderly specific arrays composed of several pigment types. It thus appears that some additional components may be involved in forming heterogeneous aggregates. Identification of possible binding components has been an active area of investigation since Tandeau de Marsac & Cohen-Bazire (76) reported the presence of uncolored proteins in the PBS of several cyanobacteria. They suggested that certain low molecular weight forms (30,000–70,000) may be involved as binding components among phycobiliproteins. They were the first to adapt a highly resolving SDS-PAGE system to the analysis of PBS composition. With this improved gel system, they were able to resolve the polypeptides of the phycobiliproteins and about 4–9 uncolored proteins. Indeed, it seems that the number of uncolored polypeptides increases with the complexity of the pigmentation. In *Synechoccus* 6301 where the PBS composition is simple (C-PC, APC, APC B), 10 polypeptides were resolved (86). Half of these were colored and half were uncolored. In *P. cruentum,* with more complex pigmentation (B-PE, b-PE, R-PC, APC, APC B), at least 20 polypeptides were resolved, of which half were colored and the other half uncolored (66).

Glazer's laboratory has isolated several uncolored polypeptides and found that they have isoelectric points more basic than those of phycobiliproteins, and that their amino acid compositions are distinct from those of phycobiliproteins [as quoted in (9)]. These features, together with the absence of chromophores, prove that these peptides are distinct from the phycobiliprotein polypeptides.

Uncolored polypeptides are definitely associated with peripheral rods of PBS (6a, 50, 86). They may serve either as sticky proteins like glue, or in stabilizing certain conformational states of specific phycobiliproteins, which then permits aggregation with like or unlike phycobiliprotein molecules. Their function(s) in aggregation is likely to occur at the hexameric to the oligomeric level. Aggregation in C-PC and APC (hexamers and trimers) can thus far be satisfactorily accounted for by assuming direct interactions among protomers (1a, 51, 52), apparently without binding proteins.

A PE-PC complex, isolated from the PBS of *P. sordidum* and dissociable and reassociable in vitro, has provided the most direct evidence for the involvement of a small uncolored polypeptide in phycobiliprotein binding (50). The complex contains B-PE and C-PC with a molar ratio of one. On SDS-PAGE, the uncolored 30,000 molecular weight polypeptide comprises 3–5% of the protein content (by Coomassie staining). The uncolored polypeptide appears to exist in conjunction with C-PC because it co-migrated with C-PC but not with B-PE when the components were separated. Phycocyanin fractions exhibiting the greatest reassociation (75% of total) possessed the highest concentration of uncolored polypeptide, whereas those with low recombination (15%) had only a very small amount of this polypeptide. This is strong evidence that this uncolored polypeptide plays a role in binding. It is notable that the recombined complex has the same spectral and sedimentation characteristics and quantum yield for fluorescence as the original complex, strongly demonstrating its functional integrity. Involvement of two uncolored polypeptides (30,500 and 31,500 mol wt) with PE has been inferred in *Synechocystis* 6701 (83). These polypeptides disappeared along with PE depletion during chromatic adaptation when cells grown in white light were changed to red light conditions. Possible involvement of a 30,000 uncolored polypeptide in binding between PC rods and APC cores has been suggested because this polypeptide was lost along with peripheral PC in *Synechoccocus* 6301 grown in nutrient-deficient conditions (85).

Hydrophobic and ionic interactions among phycobiliproteins (1a) are considered to be fundamental to the maintenance of the PBS structure. The stability of PBS at high phosphate buffer concentrations (0.5–0.75 M) and their dissociation at lower concentrations (< 0.5 M) support this assumption. Disruption of APC binding in PBS at lower temperature (22) can also be attributed to weakened hydrophobic interaction (22, 68). When PBS of

Fremyella and *Tolypothrix* were placed in dissociating conditions, it was shown by absorption spectroscopy and CD that the earliest changes occurred first with APC and later with PC and PE. These spectral changes could be directly correlated with uncoupling of energy transfer from PE to APC (68).

Reassembly of PBS has been a major goal ever since PBS were first isolated. The first step in the complete reassembly of PBS has been possible with APC and PE-PC complexes (8). Exploiting the fact that binding between APC and PC was weaker than between PE and PC, two separate fractions were isolated from *Nostoc* PBS. One fraction consisted of the APC complement (APC I-III, and –B), while another fraction consisted of C-PE and C-PC in a functional complex. From these fractions PBS have been reassembled essentially by reversing the dissociating conditions. Such reassembled PBS are virtually identical to native PBS by absorption, fluorescence emission, and sedimentation.

It has been generally assumed that PBS are directly linked, or at least in very close proximity to, PSII in the thylakoid membrane. A direct link would be through APC; therefore, certain long wavelength-emitting APC forms have been suggested for this role (7, 29, 48, 87). A PBS-thylakoid anchor component could be expected to be partly embedded in the membrane and partly in the PBS, and would thus be in close proximity to APC as well as with PSII chlorophyll. In fact, a large molecular weight polypeptide (95,000) which appears to exist in both PBS and thylakoids was recently isolated and partially characterized from *P. cruentum* and has been proposed as an anchor protein (66) (Figure 1). This polypeptide is blue, with an emission characteristic of APC when isolated from PBS on SDS-PAGE. Furthermore, this polypeptide when isolated from thylakoids occurs in a fraction with emission characteristics of PSII chlorophyll. Since the 95,000 mol wt polypeptide has identical peptide maps when isolated separately from PBS and from thylakoid membranes, it appears to be identical from the two sources.

Some other large molecular weight polypeptides (70,000–100,000) have also been suggested as possible linkers between the PBS and thylakoids because they were present in PBS and thylakoid fractions analyzed on SDS-PAGE (9, 76). Such large polypeptides have been found in many PBS of cyanobacteria (6a, 76, 83, 85, 86) and in a red alga (41). Particularly interesting is the observation that brief tryptic digestion of PBS caused degradation of a 75,000 mol wt polypeptide in *Synechoccus* 6301 (86). These results suggest that this polypeptide is located on an exposed surface, as would be expected if the thylakoid had been removed from the PBS. Thus far APC absorption has only been found in the 95,000 mol wt polypeptides of *P. cruentum* and *P. sordidum* (observations from our laboratory). If it were not for this difference, these large molecular weight polypeptides in the various PBS could be considered as functionally equivalent. Indeed, they may well be involved in anchoring the PBS to the thylakoid without participating in the energy transfer pathway.

ENERGY TRANSFER

Phycobiliprotein aggregates are particularly suited for efficient energy transfer. Recent picosecond studies on C-PC show that the larger aggregates have better chromophore-chromophore positioning for excitation energy transfer than monomers (39). The PBS is the ultimate aggregate and exists as one coherent unit. Phycobilisomes can be excited through any of their component pigments with the result that energy is transferred to APC and hence to photosystem II. Energy migration to APC from other phycobiliproteins in isolated PBS is evident from the fluorescence emission maximum of F675-680 nm (23°C), with a quantum fluorescence yield between 0.60 and 0.685 similar to that of isolated APC. The high efficiency of migration within the PBS is reflected by the low degree of polarized fluorescence (± 0.02) in intact PBS and transfer efficiencies of 95–99% (32, 72).

Time-resolved energy transfer in cells of *P. cruentum* were made by Porter et al (64) using laser-generated picosecond pulses. In line with the expected energy transfer, they showed a progressive increase in the mean fluorescence lifetimes of 70 ps for B-PE to 118 ps for APC, whereas that of chlorophyll was 175 ps (72). Direct energy transfer from PE to APC, or to chlorophyll, is considered to be unfavorable because the overlap between the emission and absorption spectra is small.

A major assumption inherently subscribed to is that the α and β polypeptide binding sites are critical to the chromophore orientation. From the completed amino acid sequence of APC and PC, and taking into account the predicted folding of the polypeptide chains, Zuber (88) has proposed a model for an energy transfer system in a linearly arranged hexamer. In his model for APC monomers the chromophore of each β-chain, although being firmly bound in an inflexible portion, nevertheless can interact directly with the chromophore of the flexible α-chain. The chromophores could thus orient for maximum transfer in the monomer as well as in higher aggregation states that are formed. He proposes that such a construct at the PBS level could result in chromophore distances of 1.5–4.0 nm in or between any one pigment.

Energy migration by inductive resonance, a case of very weak interaction, has been the assumed mechanism of transfer. It is based on the photophysical properties of the phycobiliproteins (31, 32, and references therein), particularly their overlapping PE→PC→APC absorption and emission spectra and sufficiently long lifetimes to make the transfer possible, providing the donor and acceptor chromophores are within ca 10 nm or less of one another.

In the last year, two transfer mechanisms requiring closer chromophore interaction have been proposed to occur in addition to resonance (Förster) transfer. From a spectroscopic study on a F660 nm APC form (APC II), MacColl et al (52) have proposed a model for intermediate coupling which is faster than resonance (Förster) transfer, and where chromophores of neighboring monomers could be in closer proximity to one another than in the monomer itself. This model is consistent with the low degree of polarization of fluorescence and with the observed electronic transitions. An even faster transfer occurs among some of the chromophores of the F680 nm APC forms (APC I and APC B). A delocalized exciton model was proposed for these long wavelength-absorbing and long wavelength-emitting APC forms, but not the short wavelength-absorbing forms (7). Strong exciton coupling between phycocyanobilin chromophores implies a unique structural feature where some of the chromophores in APC I and also in APC B are in close range of each other and, therefore, probably at the contact region between subunits. From maximum and minimum CD bands, an exciton split corresponding to a distance of 1.1 nm between chromophores was calculated for APC B. The exciton mechanism provides for the possibility of even more efficient energy transfer to chlorophyll than previously expected. A similar type of interaction has been proposed for a cryptomonad PC (36). In the cryptomonad algae PBS do not exist, and APC seems to be lacking (20). Since they have the same requirement for efficient energy transfer to chlorophyll, it is possible that cryptomonad PC serves the same function and therefore has the same properties as long wavelength-emitting APC.

Which pigment really is the terminal pigment in PBS? Is it possible that there is more than one terminal pigment? The possibility of chlorophyll being the final emitter in clean isolated PBS is low. APC B, on the basis of its long wavelength emission (F680) and its excellent overlap with chlorophyll absorption, fills that role. Furthermore, it has been isolated from a number of red algae and cyanobacteria (29, 48). The following pathway has been proposed:

(*a*) PE (or PEC)→ PC→ APC_{F660}→ $APC_{F675-680}$→ PSII chlorophyll

By the same criteria, APC I must also be considered as a terminal emitter. In both APC B and APC I the F680 arises from a common absorption form, recently resolved directly by low temperature spectroscopy (–196°C) (7), as well as by Gaussian curve-fitting analysis (87). APC I has been isolated from *Nostoc* sp. (7, 87a) and from *Fremyella* (observations, our laboratory) and probably will be found in other organisms as well. Because of the existence of the APC I and APC B, the transfer possibilities increase and could be one of the following, all of which are compatible with existing data:

(*b*) PE → PC → APC_{F660} → APC I_{F680} → PSII chlorophyll

(*c*) PE → PC → APC_{F660} → APC B_{F680} → PSII chlorophyll

(*d*) PE → PC → APC_{F660} ↗ APC I_{F680} ⇅ ↘ APC B_{F680} → PSII chlorophyll

Alternate transfer pathways may be advantageous to the organism. When the membrane composition becomes known as well as the detailed PBS structures, then the need for what now seem to be optional pathways may become clear. Whereas the above schemes indicate transfer to PSII, it must be pointed out that it may well be an oversimplification. Direct transfer to PSI chlorophyll must remain an open possibility because it cannot be ruled out by the present data. For example, in spheroplasts of some cyanobacteria, half of the energy absorbed by PC was found associated with PSI (77). Another unexplained phenomenon was noted in *Cyanidium* wild-type cells where only half of the PSII centers seemed to be connected with PBS (13).

Transfer from PC to chlorophyll cannot yet be ruled out. From a comparison of the fluorescence emission of PC-rich and PC-deficient cells, obtained by bleaching at high light intensities, Csatorday and colleagues (10) implied that in PC-rich cells there was considerable direct excitation transfer between PC and chlorophyll bypassing the path via APC and APC B. The amount of such direct PC-chlorophyll transfer remains to be determined.

It is possible that pigments other than APC B and APC I could be the terminal pigments. For example, in *Anabaena cylindrica,* the F685 nm (–196°C) PBS emission, according to Mimuro & Fujita (53), is not attributable to either APC I or APC B. In ammonium sulfate fractions of crude cell extracts, they were unable to detect the F685 nm emission, but since they were able to find an F681 nm peak, they postulated that F685 nm in PBS must have originated from a different source. Isolation of such a pigment from this alga would provide proof of its existence. Two exceptions exist where isolated PBS do not have the characteristic long wavelength emission at F670–F675 nm (23°C). Phycobilisomes of the red alga *Rhodella* and the cyanobacterium *Gloeobacter* have emission peaks at 665 nm (42) and 663 nm (9) respectively. Either the cells lack the long wavelength-emitting APCs or these pigments are lost during the isolation.

PHYCOBILIPROTEIN SYNTHESIS

The biosynthetic pathway of the phycobilin chromophores has been postulated to occur from ALA through magnesium protoporphyrin IX (78). However, it was recently demonstrated that phycocyanobilin synthesis in *Cyanidium caldarium* occurs directly via haem (5a). Nothing is known directly about where phycobiliproteins are produced and how PBS form. Are the phycobiliproteins synthesized on or near predetermined PBS sites on the thylakoid membrane, starting with APC followed by PC and perhaps PE, or are PBS formed in the stroma region and then attached to a thylakoid site? Fluorescence emission spectra of whole cells would argue against the presence of a large pool of free phycobiliproteins in the stroma region (24, 48) because there is relatively little emission from free phycobiliproteins as compared to the total present.

Light is known to regulate phycobiliprotein synthesis by enhancing their overall synthesis in all organisms, and in certain ones by determining the synthesis of a specific type of phycobiliprotein. Synthesis of PC has also been reported to occur in the dark in several cyanobacteria (61) and in at least one red alga (73), but synthesis in the dark is much lower in comparison to synthesis in the light. Direct photoregulation of phycobiliproteins is particularly notable in certain cyanobacteria, such as *Fremyella* and *Tolypothrix,* which can undergo complementary chromatic adaptation. For example, when exposed to green light they synthesize the complementary pigment PE, while in red light they synthesize PC. Chromatic adaptation was fully reviewed by Bogorad in 1975 in this series (4), and thus an updating will suffice. Cyanobacteria have been placed into three groups according to their adaptation capacity by Tandeau de Marsac (75). Species in group I exhibit no adaptation of their phycobiliprotein composition either in response to red or green light. In group II PE synthesis is enhanced by green light; PC synthesis is not affected. In group III both PC and PE synthesis are photoregulated. Green light induces while red light suppresses PE synthesis; conversely, green light suppresses, although incompletely, and red light stimulates PC synthesis in organisms of this class. Bryant et al (6a) in *Synechocystis* 6701 (group II) showed a significant increase in PBS size with PE synthesis, whereas in LPP 7409 (group III) the PBS were similar with or without PE synthesis. But in *Fremyella diplosiphon,* another example of group III, the PE-deficient PBS in red light were smaller by lacking one peripheral hexamer (69).

Chromatic adaptation in red algae is rare. However, in *Porphyra* and *Chondrus,* pigment changes with depth have been reported to occur (65, 79). Their whole cell spectra indicated changes in the pigment ratios as well as in the total amount. It will be interesting to ascertain the PBS composition from such plants. In red algae and many cyanobacteria, counter-chromatic adaptation operates. Here phycobiliprotein ratios show no significant change with different wavelengths, but instead changes occur in the spectrally variable chlorophyll forms (47). This type of adaptation is more prevalent than generally recognized.

What factors mediate the photoregulation of phycobiliprotein synthesis? For this role adaptochromes were proposed by Bogorad (4). According to action spectra, adaptochromes vary with the organism. In *Cyanidium caldarium* with maxima at 440 (or 460), 575, and 645 nm (70c), the adaptochrome may be a hemoprotein (4). In *Tolypothrix tenuis,* with major action peaks at 541 nm for PE and 641 nm for PC, a bile pigment appears to be involved (11, 17). In *Fremyella diplosiphon,* which is capable of chromatic adaptation like *Tolypothrix,* the peak for PE production is at 387 nm and 550 nm (with a 7:1 effectiveness ratio) and for PC at 463 nm and 641 nm (35). In *Fremyella,* metalloporphyrin complexes as well as bile pigments may be involved in the adaptation. Recently Gendel et al (25) provided evidence that PE synthesis in *Fremyella* is under translational control.

Photoreversible pigments isolated from phycobiliprotein fractions, designated phycochromes (2, 3), have been considered for the role of adaptochromes. Allophycocyanin fractions have looked particularly promising for this role, and their photoreversibility has been studied most extensively (4, 59, 60, 62). However, certain cogent factors need to be clarified: a direct physiological effect has not yet been demonstrated; APC reversible changes, similar to those caused by light, are also inducible by chaotropic agents, and the changes in normal cells only occur after photobleaching (62, 63) but not under normal conditions. Furthermore, photoreversion such as in APC also occurs in PC, thus placing the significance of the APC absorption changes in question.

FUTURE DIRECTIONS

Considerable progress has been made in elucidating the PBS structures since they were discovered in 1966 (21). Their major constituents have now been characterized and their overall structure is known. The main directions to be pursued are the involvement of the uncolored polypeptides in the stabilization of the PBS structure. Studies with phycobiliprotein-deficient mutants (83) will be particularly profitable in this area. Furthermore, the use of recombinable heteromer complexes as exemplified by the PE-PC complex of *P. sordidum* are promising (50). By reassociation studies of PBS, as begun with *Nostoc* sp. (8), it will be possible to ascertain the type of interaction between the components and to determine if they are species-specific.

Another exciting area involves the full elucidation of the PBS core, and one needs to know the location of the long wavelength-emitting APC forms in relation to the reaction centers. Also, the PBS-thylakoid anchor components (66) and the connection with the photosystems must be fully established. Whole cell studies on *A. nidulans* indicate that the attachment of PBS to thylakoid membranes is controllable by temperature (71). This opens the possibility of performing such experiments with PBS vesicles using a fluorescence assay for functional coupling (37) by which PBS binding sites can be directly explored. If appropriate conditions can be worked out, then it should also be possible to assess the binding requirements on PBS-free membranes. The mutant of *Cyanidium* which has a high PSII activity and lacks PBS (84) will be extremely useful in establishing such a relationship. Experiments on chlorophyll and PBS mixtures by Frackowiak et al (14a) are of direct interest because they showed that excitation of PE (545 nm) resulted in energy transfer to chlorophyll. Thus it appears that in vitro transfer conditions can be established. The site of phycobiliprotein synthesis, the photoreceptor(s) responsible for chromatic adaptation, and the control of phycobiliprotein production remain among the most important areas for future explorations.

ACKNOWLEDGMENTS

I appreciate the helpful suggestions for improvement of the manuscript made by D. Bryant, O. Canaani, R. Khanna, C. Lipschultz, M. Mooney, T. Redlinger, and W. Shropshire; my thanks also to the following authors for providing manuscripts in print: A. Glazer, G. Cohen-Bazire, R. MacColl, D. Berns, J. Rosinski, W. Siegelman, R. Troxler. Research in the author's laboratory on phycobilisome structure has been supported in part by DOE contract No. AS5-76ER04310A012.

Literature Cited

1. Bekasova, O. D., Shubin, L. M., Estignev, V. B. 1979. Phycobilisomes of blue-green algae *Aphanizomenon flos-aqua* and *Anabaena variabilis. Izv. Akad. Nauk SSSR Ser. Biol.* 2:198–207

1a. Berns, D. S. 1971. Phycocyanin and deuterated proteins. In *Biological Macromolecules,* ed. S. N. Timasheff, G. Fasman, pp. 105–48. New York: Dekker

2. Björn, G. S. 1978. Phycochrome D, a new photochromic pigment from the blue-green alga, *Tolypothrix distorta. Physiol. Plant.* 42:321–23

3. Björn, G. S., Björn, L. O. 1976. Photochromic pigments from blue-green algae: phycochromes A, B, and C. *Physiol. Plant.* 36:297–304

4. Bogorad, L. 1975. Phycobiliproteins and complementary chromatic adaptation. *Ann. Rev. Plant Physiol.* 26:369–401

5. Brown, A. S., Offner, G. D., Ehrhardt, M. M., Troxler, R. F. 1979. Phycobilin-apoprotein linkages in the α and β subunits of phycocyanin from the unicellular rhodophyte, *Cyanidium caldarium.* Amino acid sequences of ^{35}S-labelled chromopeptides. *J. Biol. Chem.* 254:7803–11

5a. Brown, B. B., Holroyd, J. A., Troxler, R. F., Offner, G. D. 1981. Bile pigment synthesis in plants: Incorporation of

haem into phycocyanobilin and phycobiliproteins in *Cyanidium caldarium. Biochem. J.* 191:137–47
6. Bryant, D. A., Glazer, A. N., Eiserling, F. 1976. Characterization and structural properties of the major biliproteins of *Anabaena* sp. *Arch. Microbiol.* 110:61–75
6a. Bryant, D. A., Guglielmi, G., Tandeau de Marsac, N., Castets, A.-M., Cohen-Bazire, G. 1979. The structure of cyanobacterial phycobilisomes: a model. *Arch. Microbiol.* 123:113–27
7. Canaani, O., Gantt, E. 1980. Circular dichroism and polarized fluorescence characteristics of blue-green algal allophycocyanins. *Biochemistry* 19: 2950–60
8. Canaani, O., Lipschultz, C. A., Gantt, E. 1980. Reassembly of phycobilisomes from allophycocyanin and a phycocyanin-phycoerythrin complex. *FEBS Lett.* 115:225–29
9. Cohen-Bazire, G., Bryant, D. A. 1981. Phycobilisomes: composition and structure. In *The Biology of the Cyanobacteria,* ed. N. Carr, B. Whitton. New York: Blackwell. In press
10. Csatorday, K., Kleinen Hammans, J. W., Goedheer, J. C. 1978. Excitation energy transfer in *Anacystis nidulans. Biochem. Biophys. Res. Commun.* 81: 571–75
11. Diakoff, S., Scheibe, J. 1973. Action spectra for chromatic adaptation in *Tolypothrix tenuis. Plant Physiol.* 51: 382–85
12. Dilworth, M., Gantt, E. 1981. Phycobilisome thylakoid topography on photosynthetically active vesicles of *Porphyridium cruentum. Plant Physiol.* 67: In press
13. Diner, B. 1979. Energy transfer from the phycobilisomes to photosystem II reaction centers in wild type *Cyanidium caldarium. Plant Physiol.* 63:30–34
14. Fisher, R. G., Woods, N. E., Fuchs, H. E., Sweet, R. M. 1980. Three-dimensional structure of C-phycocyanin and B-phycoerythrin at 5 Å resolution. *J. Biol. Chem.* 255:5082–89
14a. Frackowiak, D., Erokhina, L. G., Fiksinski, K. 1979. The influence of aggregation on the excitation energy transfer between phycobiliproteins and chlorophyllin. *Photosynthetica* 13:245–53
15. Frank, G., Sidler, W., Widmer, H., Zuber, H. 1978. The complete amino acid sequence of both subunits of C-phycocyanin from the cyanobacterium *Mastigocladus laminosus. Hoppe-Seylers Z. Physiol. Chem.* 359:1491–1507
16. Fu, E., Friedman, L., Siegelman, H. W. 1979. Mass spectral identification and purification of phycoerythrobilin and phycocyanobilin. *Biochem. J.* 179:1–6
17. Fujita, Y., Hattori, A. 1962. Photochemical interconversion between precursors of phycobilin chromoproteids in *Tolypothrix tenuis. Plant Cell Physiol.* 3:209–20
18. Gantt, E. 1975. Phycobilisomes: light harvesting pigment complexes. *BioScience* 25:781–87
19. Gantt, E. 1977. Recent contributions in phycobiliproteins and phycobilisomes. *Photochem. Photobiol.* 26:685–89
20. Gantt, E. 1979. Phycobiliproteins of cryptophytes. In *Biochemistry and Physiology of Protozoa,* ed. M. Levandowsky, S. A. Hutner, 1:121–37. New York: Academic. 462pp. 2nd ed.
21. Gantt, E. 1980. Structure and function of phycobilisomes: light harvesting pigment complexes in red and blue-green algae. *Int. Rev. Cytol.* 66:45–80
21a. Gantt, E., Lipschultz, C. A. 1980. Structure and phycobiliprotein composition of phycobilisomes from *Griffithsia pacifica* (Rhodophyceae). *J. Phycol.* 16:394–98
22. Gantt, E., Lipschultz, C. A., Grabowski, J., Zimmerman, B. K. 1979. Phycobilisomes from blue-green and red algae. Isolation criteria and dissociation characteristics. *Plant Physiol.* 63:615–20
23. Gantt, E., Lipschultz, C. A., Zilinskas, B. 1976. Further evidence for a phycobilisome model from selective dissociation, fluorescence emission, immunoprecipitation, and electron microscopy. *Biochim. Biophys. Acta* 430:375–88
24. Gantt, E., Lipschultz, C. A., Zilinskas, B. 1976. Phycobilisomes in relation to the thylakoid membranes. *Brookhaven Symp. Biol.* 28:347–57
25. Gendel, S., Ohad, I., Bogorad, L. 1979. Control of phycoerythrin synthesis during chromatic adaptation. *Plant Physiol.* 64:786–90
26. Glazer, A. N. 1976. Phycocyanins: structure and function. *Photochem. Photobiol.* 1:71–115
27. Glazer, A. N. 1977. Structure and molecular organization of the photosynthetic accessory pigments of cyanobacteria and red algae. *Mol. Cell. Biochem.* 18:125–40
28. Glazer, A. N., Apell, G. S. 1977. A common evolutionary origin for the biliproteins of cyanobacteria, rhodo-

phyta and cryptophyta. *FEMS Microbiol. Lett.* 1(2):113–16
29. Glazer, A. N., Bryant, D. A. 1975. Allophycocyanin B (λ_{max} 671, 618 nm). A new cyanobacterial phycobiliprotein. *Arch. Microbiol.* 104:15–22
30. Glazer, A. N., Williams, R. C., Yamanaka, G., Schachman, H. K. 1979. Characterization of cyanobacterial phycobilisomes in zwitterionic detergents. *Proc. Natl. Acad. Sci. USA* 76:6162–66
31. Grabowski, J., Gantt, E. 1978. Photophysical properties of phycobilisomes: fluorescence lifetimes, quantum yields, and polarization spectra. *Photochem. Photobiol.* 28:39–45
32. Grabowski, J., Gantt, E. 1978. Excitation energy migration in phycobilisomes: comparison of experimental results and theoretical predictions. *Photochem. Photobiol.* 28:47–54
33. Guglielmi, G., Cohen-Bazire, G., Bryant, D. A. 1981. The structure of *Gleobacter violaceus* and its phycobilisomes. *Arch. Microbiol.* 129: In press
34. Gysi, J. R., Chapman, D. J. 1980. Phycobiliproteins and phycobilins. In *CRC Handbook on Biosolar Resources,* Vol. 1 Fundamental Principle, ed. C. C. Black, A. Mitsui. In press
35. Haury, J., Bogorad, L. 1977. Action spectra of phycobiliprotein synthesis in a chromatically adapting cyanophyte, *Fremyella diplosiphon. Plant Physiol.* 60:835–39
36. Jung, J., Song, P.-S., Paxton, R. J., Edelstein, M. S., Swanson, R., Hazen, E. E. Jr. 1980. Molecular topography of the phycocyanin photoreceptor from *Chroomonas* species. *Biochemistry* 19: 24–32
37. Katoh, T., Gantt, E. 1979. Photosynthetic vesicles with bound phycobilisomes from *Anabaena variabilis. Biochim. Biophys. Acta* 546:383–93
38. Killilea, S. D., O'Carra, P., Murphy, R. F. 1980. Structures and apoprotein linkages of phycoerythrobilin and phycocyanobilin. *Biochem. J.* 187:311–20
39. Kobayashi, T., Degenkolb, E. O., Bersohn, R., Rentzepis, P. M., MacColl, R., Berns, D. S. 1979. Energy transfer among the chromophores in phycocyanins measured by picosecond kinetics. *Biochemistry* 18:5073–78
40. Koller, K.-P., Wehrmeyer, W. 1979. Biliproteid-Komposition und Pigmentverteilung in scheibchenförmigen Phycobilisomen von *Rhodella violacea. Ber. Dtsch. Bot. Ges.* 92:403–11
41. Koller, K.-P., Wehrmeyer, W., Mörschel, E. 1978. Biliprotein assembly in the disc-shaped phycobilisomes of *Rhodella violacea:* on the molecular composition of energy-transfer complexes (tripartite units) forming the periphery of the phycobilisome. *Eur. J. Biochem.* 91:57–63
42. Koller, K.-P., Wehrmeyer, W., Schneider, H. 1977. Isolation and characterization of disc-shaped phycobilisomes from the red alga *Rhodella violacea. Arch. Microbiol.* 112:61–67
43. Kufer, W., Scheer, H. 1979. Studies on plant bile pigments. VII. Preparation and characterization of phycobiliproteins with chromophores chemically modified by reduction. *Hoppe-Seylers Z. Physiol. Chem.* 360:935–56
44. Lagarias, J. C., Glazer, A. N., Rapoport, H. 1979. Chromopeptides from C-phycocyanin. Structure and linkage of a phycocyanobilin bound to the β subunit. *J. Am. Chem. Soc.* 101:5030–37
44a. Larkum, A. W. D., Weyrauch, S. K. 1977. Photosynthetic action spectra and light-harvesting in *Griffithsia monilis* (Rhodophyta). *Photochem. Photobiol.* 25:65–72
45. Leclerc, J. C., Hoarau, J., Remy, R. 1979. Analysis of absorption spectra changes induced by temperature lowering on phycobilisomes, thylakoids, and chlorophyll-protein complexes. *Biochim. Biophys. Acta* 547:398–409
46. Ley, A. C., Butler, W. L. 1977. The distribution of excitation energy between photosystem I and photosystem II in *Porphyridium cruentum.* In *Photosynthetic Organelles.* Spec. Issue *Plant Cell Physiol.* 3:33–46
47. Ley, A. C., Butler, W. L. 1980. Effects of chromatic adaptation on the photochemical apparatus of photosynthesis in *Porphyridium cruentum. Plant Physiol.* 65:714–22
48. Ley, A. C., Butler, W. L., Bryant, D. A., Glazer, A. N. 1977. Isolation and function of allophycocyanin B of *Porphyridium cruentum. Plant Physiol.* 59:974–80
49. Lichtle, C., Thomas, J. C. 1976. Étude ultrastructorale des thylacoides des algues à phycobiliproteines, comparaison des résultats obtenus par fixation classique et cryodećapage. *Phycologia* 15:393–404
50. Lipschultz, C. A., Gantt, E. 1981. Association of phycoerythrin and phycocyanin: *in vitro* formation of a functional energy transferring physobilisome com-

plex of *Porphyridium sordidum. Biochemistry.* 20: In press
51. MacColl, R., Csatorday, K., Berns, D. S., Traeger, E. 1980. The relationship of the quaternary structure of allophycocyanin to its spectrum. *Arch. Biochem. Biophys.* 206: In press
52. MacColl, R., Csatorday, K., Berns, D. S., Traeger, E. 1980. Chromophore interactions in allophycocyanin. *Biochemistry* 19:2817–20
53. Mimuro, M., Fujita, Y. 1980. Comparison of main emissions at –196°C from phycobilisomes of two blue-green algae *Anabaena cylindrica* and *Anacystis nidulans. Plant Cell Physiol.* 21:37–45
54. Mörschel, E., Koller, K.-P., Wehrmeyer, W. 1980. Biliprotein assembly in the disc-shaped phycobilisomes of *Rhodella violacea* electron microscopical and biochemical analyses of C-phycocyanin and allophycocyanin aggregates. *Arch. Microbiol.* 125:43–45
55. Mörschel, E., Koller, K.-P., Wehrmeyer, W., Schneider, H. 1977. Biliprotein assembly in the disc-shaped phycobilisomes of *Rhodella violacea* I. Electron microscopy of phycobilisomes *in situ* and analysis of their architecture after isolation and negative staining. *Cytobiologie* 16:118–29
55a. Mörschel, E., Wehrmeyer, W., Koller, K.-P. 1980. Biliprotein assembly in the disc-shaped phycobilisomes of *Rhodella violacea.* Electron microscopical and biochemical analysis of B-phycoerythrin and C-phycocyanin aggregates. *Eur. J. Cell Biol.* 4:319–27
56. Muckle, G., Rüdiger, W. 1977. Chromophore content from various cyanobacteria. *Z. Naturforsch. Teil C* 32:957–62
57. O'Carra, P., Murphy, R. F., Killilea, S. D. 1980. The native forms of the phycobilin chromophores of algal biliproteins. *Biochem. J.* 187:303–9
58. O'Carra, P., O'Eocha, C. 1976. In *Chemistry and Biochemistry of Plant Pigments,* ed. T. W. Goodwin, 1:328. New York: Academic
59. Ohad, I., Clayton, R. K., Bogorad, L. 1979. Photoreversible absorbance changes in solutions of allophycocyanin purified from *Fremyella diplosiphon:* temperature dependence and quantum efficiency. *Proc. Natl. Acad. Sci USA* 76:5655–59
60. Ohad, I., Schneider, H.-J., Gendel, S., Bogorad, L. 1980. Light-induced changes in allophycocyanin. *Plant Physiol.* 65:6–12
61. Ohki, K., Fujita, Y. 1978. Photocontrol of phycoerythrin formation in the blue-green alga *Tolypothrix tenuis* growing in the dark. *Plant Cell Physiol.* 19:7–15
62. Ohki, K., Fujita, Y. 1979. Photoreversible absorption changes of guanidine-HCl-treated phycocyanin and allophycocyanin isolated from the blue-green alga *Tolypothrix tenuis. Plant Cell Physiol.* 20:483–90
63. Ohki, K., Fujita, Y. 1979. *In vivo* transformation of phycobiliproteins during photobleaching of *Tolypothrix tenuis* to forms active in photoreversible absorption changes. *Plant Cell Physiol.* 20:1341–47
64. Porter, G., Tredwell, C. J., Searle, G. F. W., Barber, J. 1978. Picosecond time-resolved energy transfer in *Porphyridium cruentum.* Part I. In the intact alga. *Biochim. Biophys. Acta* 501:232–45
65. Ramus, J., Beale, S. I., Mauzerall, D., Howard, K. L. 1976. Changes in photosynthetic pigment concentration in sea weeds as a function of water depth. *Mar. Biol.* 37:223–29
66. Redlinger, T., Gantt, E. 1980. Identification of a 95,000 mw polypeptide common to both phycobilisomes and thylakoid membranes in *Porphyridium cruentum. Proc. 5th Int. Congr. Photosynth., Kassandra-Halkidiki, Greece.* In press
67. Ried, A., Hessenberg, B., Metzler, H., Ziegler, R. 1977. Distribution of excitation energy among photosystem I and photosystem II in red algae. I. Action spectra of light reactions I and II. *Biochim. Biophys. Acta* 459:175–86
68. Rigbi, M., Rosinski, J., Siegelman, H. W., Sutherland, J. C. 1980. Cyanobacterial phycobilisomes: selective dissociation monitored by fluorescence and circular dichroism. *Proc. Natl. Acad. Sci. USA* 77:1961–65
69. Rosinski, J., Hainfeld, J. E., Rigbi, M., Siegelman, H. W. 1980. Phycobilisome ultrastructure and chromatic adaptation in *Fremyella diplosiphon. Ann. Bot.* 47:1–12
70. Rüdiger, W. 1979. Struktur und Spektraleigenschaft von Phycobilinen und Biliproteiden. *Ber. Dtsch. Bot. Ges.* 92:413–26
70a. Scheer, H. 1980. Phycobiliproteins: Molecular aspects of photosynthetic antenna systems. In *Primary Processes in Photosynthesis,* ed. F. K. Fong. New York/Berlin/Heidelberg: Springer
70b. Scheer, H., Kufer, W. 1977. Studies on plant bile pigments. IV. Conforma-

tional studies on C-phycocyanin from *Spirulina platensis. Z. Naturforsch. Teil C* 32:513–19
70c. Schneider, H. A. W., Bogorad, L. 1979. Spectral response curves for the formation of phycobiliproteins, chlorophyll and δ-aminolevulinic acid in *Cyanidium caldarium. Z. Pflanzenphysiol.* 94:449–59
71. Schreiber, U., Rijgersberg, C. P., Amesz, J. 1979. Temperature-dependent reversible changes in phycobilisome-thylakoid membrane attachment in *Anacystis nidulans. FEBS Lett.* 104:327–31
72. Searle, G. F. W., Barber, J., Porter, G., Tredwell, C. J. 1978. Picosecond time-resolved energy transfer in *Porphyridium cruentum.* Part II. In the isolated light harvesting complex (phycobilisomes). *Biochim. Biophys. Acta* 501:246–56
73. Sheath, R. G., Hellebust, J. A., Sawa, T. 1977. Changes in plastid structure, pigmentation, and photosynthesis of the conchocelis stage of *Porphyra leucosticta* (Rhodophyta, Bangiophyceae) in response to low light and darkness. *Phycologia* 16:265–76
74. Stanier, R. Y. 1974. The origins of photosynthesis in eukaryotes. *Symp. Soc. Gen. Microbiol.* 24:219–40
75. Tandeau de Marsac, N. T. 1977. Occurrence and nature of chromatic adaptation in cyanobacteria. *J. Bacteriol.* 130:82–91
76. Tandeau de Marsac, N. T., Cohen-Bazire, G. 1977. Molecular composition of cyanobacterial phycobilisomes. *Proc. Natl. Acad. Sci. USA* 74:1635–39
77. Tel-Or, E., Malkin, S. 1977. The photochemical and fluorescence properties of whole cells, spheroplasts and spheroplast particles from the blue-green alga *Phormidium luridum. Biochim. Biophys. Acta* 459:157–74
78. Troxler, R. F. 1977. Synthesis of bile pigments in plants. In *Chemistry and Physiology of Bile Pigments,* ed. P. D. Beck, H. I. Berlin, pp. 431–54. Fogarty Int. Cent. Proc. No. 35, DHEW Publ. No (NIH) 77-1100
79. Troxler, R. F., Greenwald, L. S., Zilinskas, B. 1980. Allophycocyanin from *Nostoc* sp. phycobilisomes. Properties and amino acid sequence at the amino terminus of the α and β subunits of allophycocyanin I, II, and III. *J. Biol. Chem.* 255:9380–87
80. Wang, R. T., Myers, J. 1976. Simultaneous measurement of action spectra for photoreactions I and II of photosynthesis. *Photochem. Photobiol.* 23: 411–14
81. Wang, R. T., Stevens, C. L. R., Myers, J. 1977. Action spectra for photoreactions I and II of photosynthesis in the blue-green alga *Anacystis nidulans. Photochem. Photobiol.* 25:103–8
82. Wanner, G., Köst, H.-P. 1980. Investigations on the arrangement and fine structure of *Porphyridium cruentum* phycobilisomes. *Protoplasma* 102:97–109
83. Williams, R. C., Gingrich, J. C., Glazer, A. N. 1980. Cyanobacterial phycobilisomes. Particles from *Synechocystis* 6701 and two pigment mutants. *J. Cell Biol.* 85:558–66
84. Wollman, F.-A. 1979. Ultrastructural comparison of *Cyanidium caldarium* wild type and III-C mutant lacking phycobilisomes. *Plant Physiol.* 63: 375–81
85. Yamanaka, G., Glazer, A. N. 1980. Dynamic aspects of phycobilisome structure. Phycobilisome turnover during nitrogen starvation in *Synechococcus* sp. *Arch. Microbiol.* 124:39–47
86. Yamanaka, G., Glazer, A. N., Williams, R. C. 1978. Cyanobacterial phycobilisomes. Characterization of the phycobilisomes of *Synechococcus* sp. 6301. *J. Biol. Chem.* 253:8303–10
86a. Zickendraht-Wendelstadt, B., Friedenreich, J., Rüdiger, W. 1980. Spectral characterization of monomeric C-phycoerythrin from *Pseudoanabaena* W 1173 and its α and β subunits; energy transfer in isolated subunits and C-phycoerythrin. *Photochem. Photobiol.* 31:367–76
87. Zilinskas, B. A., Greenwald, L. S., Bailey, C. L., Kahn, P. C. 1980. Spectral analysis of allophycocyanin I, II, III, and B from *Nostoc* sp. phycobilisomes. *Biochim. Biophys. Acta* 592:267–76
87a. Zilinskas, B. A., Zimmerman, B. K., Gantt, E. 1978. Allophycocyanin forms isolated from *Nostoc* sp. phycobilisomes. *Photochem. Photobiol.* 27: 587–95
88. Zuber, H. 1978. Studies on the structure of the light-harvesting pigment-protein complexes from cyanobacteria and red algae. *Ber. Dtsch. Bot. Ges.* 91:459–75

Ann. Rev. Plant Physiol. 1981. 32:349–83

THE CARBOXYLATION AND OXYGENATION OF RIBULOSE 1,5-BISPHOSPHATE: The Primary Events in Photosynthesis and Photorespiration[1,2]

◆7716

George H. Lorimer

Central Research and Development Department, E. I. du Pont de Nemours and Company Experimental Station, Wilmington, Delaware 19801

CONTENTS

[1]Dedicated to the memory of Bessel Kok.

[2]Abbreviations used: $^{A}CO_2$, activator CO_2; CABP, 2-carboxyarabinitol 1,5-bisphosphate; M, divalent metal ion, Mg^{2+} or Mn^{2+}; PCA, photosynthetic carbon assimilation; PCO, photorespiratory carbon oxidation; PCR, photosynthetic carbon reduction; PPBP, 2-peroxypentitol 1,5-bisphosphate; 6-P-gluconate, gluconate 6-phosphate; 3-P-glycerate, glycerate 3-phosphate; P-glycolate, glycolate 2-phosphate; Rubisco, ribulose 1,5-bisphosphate carboxylase-oxygenase; RuBP, ribulose 1,5-bisphosphate; $^{S}CO_2$, substrate CO_2.

0066-4294/81/0601-0349$01.00

> If someone tells me that in making these conclusions I have gone beyond the facts I reply: "This is true, that I have put myself freely among ideas which cannot be rigorously proved. That is my way of looking at things. Every time a chemist concerns himself with these mysterious phenomena and every time he has the luck to make an important step forward he will be led instinctively to attribute their prime cause to a class of reactions in harmony with the general results of his own researches. That is the logical course of the human mind, in all controversial matters."
>
> Louis Pasteur, 1857

INTRODUCTION

In the mid-1950s Calvin and his colleagues mapped out the pathway for the photosynthetic conversion of CO_2 to carbohydrate (17). This pathway, familiar now to several generations of students, has withstood both sniper fire (73) and the opening salvos of the C-4 onslaught (83, 99) without need for serious revision. The C-3 PCR[2] cycle remains the master cycle to which other cycles are either subservient (as in the case of the C-4 PCA cycle) or dependent (as in the case of the C-2 PCO cycle).

Having recognized the central role of the C-3 PCR cycle, we must now concede that it is no longer an adequate description of the path of carbon in photosynthesis *under natural atmospheric conditions.* Under these conditions photosynthetic carbon metabolism can best be described as the integrated sum of the activities of two mutually opposing and interlocking cycles, the classic C-3 PCR cycle and the C-2 PCO cycle, as outlined in Figure 1. At the heart of this unified picture of photosynthetic carbon metabolism stands the bifunctional enzyme ribulose-1,5-bisphosphate carboxylase-oxygenase. The flow of carbon to either the C-3 PCR cycle (carboxylation) or to the C-2 PCO cycle (oxygenation) is dictated by the prevailing concentrations of CO_2 and O_2 and by the kinetic properties of this enzyme.

Rather than attempt an exhaustive review of all aspects of photosynthetic-photorespiratory carbon metabolism, I have elected to concentrate on the two primary reactions, the carboxylation and oxygenation of RuBP. Several reviews (4, 23, 43, 68, 77, 84, 94, 102, 103, 119, 127, 160, 173, 194) and books (35, 74, 178) dealing with various aspects of photorespiration have appeared recently. For historical perspective, earlier reviews (75, 88, 193, 198, 216) should be consulted.

A Philosophical Diversion

It is appropriate to begin with the question "What is the nature of the evidence which indicates that the carboxylation and oxygenation of RuBP are the primary reactions of the C-3 PCR and C-2 PCO cycles respectively?" After reviewing the evidence it must be emphasized that there is no single item of evidence which unequivocally and indisputably establishes this hypothesis. Each item of evidence can be subjected to an alternative

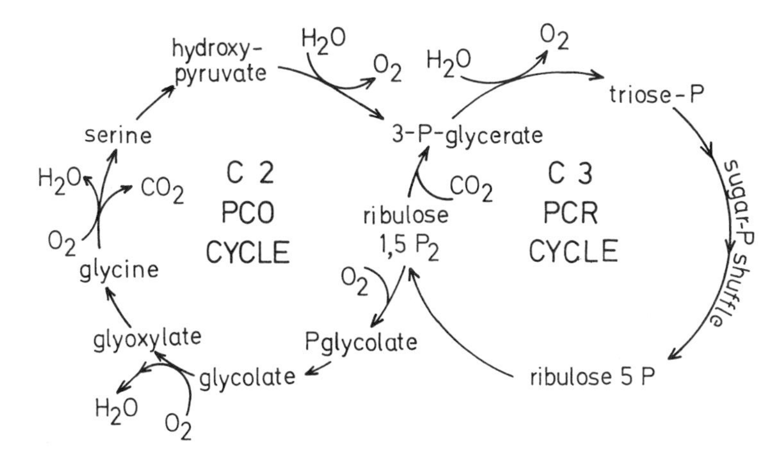

Figure 1 **Integration of the C-3 PCR and C-2 PCO cycles is accomplished by the carboxylation or oxygenation of RuBP. The C-3 PCR cycle is capable of independent operation. However, the C-2 PCO cycle is parasitic, being dependent upon the C-3 PCR cycle to regenerate RuBP from 3-P-glycerate. At the CO_2 compensation point this system constitutes what may be the first recorded instance of a futile bicycle. Under natural atmospheres, it more closely resembles a penny-farthing.**

explanation. The fact that the experimental evidence in support of such alternative hypotheses is often weak or nonexistent does not of itself establish the veracity of the most commonly accepted explanation. Instead, there are a number of necessary but insufficient criteria to be satisfied. These criteria should be considered in toto, as a series of mechanistic constraints and not as single items for which this, that, or the next alternative explanation can be found. Ockham's razor is still sharp and can serve us well.

The purpose of the next two sections is to summarize the evidence which indicates that the carboxylation and oxygenation of RuBP are the primary reactions respectively of the C-3 PCR and C-2 PCO cycles. Some of the data are not new. But a short reiteration of the evidence will refresh our memories and illustrate the almost complete harmony that exists between the physiological evidence on the one hand and the biochemical properties of Rubisco[3] on the other.

THE CARBOXYLATION OF RIBULOSE BISPHOSPHATE AS THE PRIMARY REACTION OF THE C-3 PHOTOSYNTHETIC CARBON REDUCTION CYCLE

3-Phosphoglycerate as the First Detectable Product

After very short periods of photosynthesis in $^{14}CO_2$, 3-P-glycerate was the compound most heavily labeled (37, 38). In addition, when the percentage of the total ^{14}C fixed into the various intermediates was plotted versus time, the slope of the plot for 3-P-glycerate was negative and extrapolated back to values close to 100% at time zero (37). When the distribution of ^{14}C within the 3-P-glycerate was examined after a short pulse of $^{14}CO_2$, the ^{14}C was confined predominantly to the C-1 position, an important mechanistic constraint (16, 39). While such results clearly establish 3-P-glycerate as the first detectable product of photosynthetic CO_2 fixation, they do not prove that 3-P-glycerate is derived from RuBP. For that we must consider evidence of a different nature.

Ribulose Bisphosphate as the Immediate Precursor of 3-Phosphoglycerate

Compelling evidence for this relationship was obtained by observing reciprocal changes in the pool sizes of RuBP and 3-P-glycerate during light-dark transients or as a result of suddenly lowering the CO_2 concentration (20,

[3]What's in a name! Despite sounding like a breakfast cereal, the acronym Rubisco (91) has the distinct advantage that it rolls off the tongue much more easily than ribulose-1,5-bisphosphate carboxylase-oxygenase or than 3-phospho-D-glycerate carboxylase (dimerizing).

39, 212). The changes in these pools were precisely what one would expect if 3-P-glycerate were derived from RuBP.

Early kinetic measurements (18) lent support to the notion that in vivo the products of the carboxylation of RuBP may not be two molecules of 3-P-glycerate as in vitro, but rather one molecule each of 3-P-glycerate and triose phosphate. A careful kinetic study was performed later and was specifically designed to distinguish between these two possibilities (69). The results clearly indicated that carboxylation in vivo led to the formation of two molecules of 3-P-glycerate.

The results of an enormous number of experiments performed since then (and usually for quite different reasons) support the view that RuBP is the CO_2 acceptor, although it has become distinctly passé to point this out. For example, Krause et al (104) demonstrated that $^{14}CO_2$ fixation by intact, isolated chloroplasts with sugar monophosphates as substrates was sensitive to inhibitors of photophosphorylation. Since, with the exception of RuBP, all of the intermediates of the C-3 PCR cycle can be formed from triose phosphates without ATP, the above result clearly implicates RuBP as the CO_2 acceptor.

$\delta^{13}C$ Values as Indicators of Carboxylation Reactions

Atmospheric CO_2 has a $\delta^{13}C$ value of about –6.7‰. During photosynthesis considerable discrimination against $^{13}CO_2$ occurs, such that the mean $\delta^{13}C$ value for C-3 plants is -27.8 ± 2.8‰ (196). In vitro measurements of the fractionation which accompanies the carboxylation of RuBP have yielded similar values (46, 215). That this agreement between the in vivo and in vitro values is not simply fortuitous can be shown by reference to *Chlorobium thiosulfatophilum.* This is a green sulfur bacterium capable of photosynthetic growth on CO_2. It has been reported to lack RuBP carboxylase (34) and to fix CO_2 by a pathway other than the C-3 PCR cycle (33), although this report is disputed (187). However, it is most significant that the $\delta^{13}C$ values for this organism differ significantly from the $\delta^{13}C$ values for organisms known to fix CO_2 via Rubisco (156, 181).

O_2 as a Competitive Inhibitor of CO_2 Fixation

In 1943 at a meeting of the Japanese Botanical Society, Tamiya & Huzisige (189) outlined their studies of the Warburg O_2 effect, the inhibition of photosynthesis by O_2 (204). In a remarkably prescient conclusion (the modern version of Calvinism had not appeared yet), they proposed that O_2 and CO_2 competed for a common site on the carboxylating enzyme. Lacking any knowledge of the biochemistry of CO_2 fixation, Tamiya and Huzisige were unable to put their hypothesis to experimental test. Curiously, the idea seems to have been forgotten. Although some speculation occurred (19), no one seems to have considered the idea worth testing experimentally, even when the existence of photorespiration and the Warburg O_2 effect were established and when the chemistry of the carboxylation reaction was known. This is surprising, for it clearly represents the simplest and most direct explanation of the inhibition of CO_2 fixation by O_2. In 1971, after reexamining some data on the effects of O_2 on photosynthesis and photorespiration, Ogren & Bowes (142) came to the same conclusion reached by Tamiya and Huzisige 28 years previously. More importantly, they were able to put the idea to experimental test in vitro (29, 30). Their successful demonstration that O_2 was not only a competitive inhibitor of carboxylation (with respect to CO_2) but also a substrate for the formation of P-glycolate from RuBP, was the necessary impetus for a series of studies which lead us to our present understanding of photosynthetic-photorespiratory carbon metabolism (Figure 1).

Data gathered in a number of laboratories using a variety of organisms (28, 57, 67, 92, 105, 106, 110, 111, 147, 148) have demonstrated the competitive effect of O_2 upon CO_2 fixation. This phenomenon, the inhibition of CO_2 fixation by O_2, is perhaps the criterion which most readily lends itself to alternative explanations. At least in vitro, O_2 can be demonstrated to inhibit a number of the partial reactions of photosynthesis. For example, the ability of O_2 to capture electrons from the photosynthetic electron transport system is well known and could conceivably account for the inhibition of CO_2 fixation. However, it is not sufficient to demonstrate the inhibitory effect of O_2 on this or that isolated reaction. For the inhibition to be significant in vivo that reaction must be close to rate-limiting for the overall process. There is evidence that the rate of light-saturated photosynthesis at atmospheric CO_2 is limited by the activity of Rubisco (26, 27, 159, 180, 205). Therefore, the inhibition of carboxylation by O_2 is highly significant. The effects of O_2 on in vivo photosynthesis have been described with some degree of success on the basis of the competitive interactions of CO_2 and O_2 with Rubisco (62, 63, 76, 110, 146, 190–192). This success would suggest that the major contributor to the inhibitory effect of O_2 upon photosynthesis is its effect upon Rubisco and that other inhibitory effects are of lesser importance.

For the case to be complete the above in vivo evidence needs only to be complimented by the demonstration that there exists an enzyme capable of carboxylating RuBP to give two molecules of 3-P-glycerate at rates equal to or greater than the in vivo rates of photosynthesis. The first studies of the carboxylation reaction established that it involved the incorporation of $^{14}CO_2$ into the carboxyl group of 3-P-glycerate (89). This result provided a compelling explanation for the labeling pattern observed in vivo after short periods of photosynthesis in $^{14}CO_2$. However, the low rates of carboxylation in vitro (149) and the seemingly low affinity of the enzyme for CO_2 (206) raised some doubts about its ability to fulfil its function as the primary carboxylation of the PCR cycle. These uncertainties have been largely resolved by improvements in assay methods. Both the affinity of Rubisco for CO_2 and the in vitro rates of carboxylation are now sufficiently high to account easily for the observed in vivo rates (8, 55, 114, 122, 202). Having established that the carboxylation of RuBP is the primary event in the C-3 PCR cycle, let us now turn to the evidence which indicates that the oxygenation of RuBP is the primary reaction of the C-2 PCO cycle.

THE OXYGENATION OF RIBULOSE BISPHOSPHATE AS THE PRIMARY REACTION OF THE C-2 PHOTORESPIRATORY CARBON OXIDATION CYCLE

2-Phosphoglycolate as the Immediate Precursor of Glycolate

If the oxygenation of RuBP is the primary and causal reaction of the C-2 PCO cycle, P-glycolate must be the immediate precursor of glycolate. A phosphatase, located within the chloroplast and highly specific for its substrate P-glycolate, catalyzes the conversion to glycolate (47, 161). But this alone does not establish P-glycolate as the sole precursor of glycolate. Experiments with intact isolated chloroplasts which demonstrate the accumulation of ^{14}C (from $^{14}CO_2$) in P-glycolate rather than in glycolate in response to fluoride (an inhibitor of phosphatases) are suggestive of a precursor-product relationship (112). However, such experiments are subject to the usual caveats associated with the use of inhibitors. Experiments with mutants of *Arabidopsis* deficient in P-glycolate phosphatase (182) provide much more compelling evidence. When such plants were allowed to fix $^{14}CO_2$ in the presence of a suicide inhibitor of glycolate oxidase, ^{14}C accumulated in P-glycolate and not in glycolate as it did in the wild type. This result can only have come about if glycolate is more or less exclusively derived from P-glycolate.

Glycolate Synthesis from Sugar Monophosphates Needs ATP

Kirk & Heber (96) were the first to exploit the fact that, with the exception of RuBP, all of the other intermediates of the C-3 PCR cycle can be synthesized readily from sugar monophosphates without ATP. Therefore, they reasoned, if glycolate is synthesized by the oxygenation of RuBP, its formation from triose-P should be blocked by the inhibitor of photophosphorylation, carbonylcyanide-4-trifluoromethoxyphenylhydrazone (FCCP). Using intact, isolated chloroplasts, they were able to demonstrate this point (96).

"Specific inhibitors are like beautiful virgins. Rare!"
Martin Gibbs, 1977

In replying to this witticism, Krause et al (104) adopted the same experimental strategy as Kirk & Heber (96), but extended their observations to include the inhibitors Dio-9 and arsenate. FCCP, Dio-9, and arsenate each inhibit photophosphorylation by a different mechanism. Yet all three inhibited glycolate synthesis from triose-P by intact, isolated chloroplasts (104). It would be most remarkable if these three inhibitors were to elicit this response via some nonspecific effect unrelated to their common ability to inhibit photophosphorylation. Inhibition of RuBP synthesis because of the lack of ATP, as first proposed by Kirk & Heber (96), is surely a more straightforward explanation.

Incorporation of an Atom of Molecular Oxygen

Shortly after its discovery, the in vitro oxygenation of RuBP was shown to occur with the incorporation of an atom of molecular O_2 into the carboxyl group of P-glycolate. (120). Thus, the in vivo synthesis of glycolate and those metabolites derived from it can be investigated using $^{18}O_2$ (5). However, there are two inherent difficulties associated with such experiments. The first is not unique to the use of isotopic oxygen. It is simply the complication which arises if there are inactive pools of the intermediate in which one is interested. Upon extraction, these inactive, unlabeled pools are combined with the active pool. One therefore underestimates the enrichment of the isotope in the active pool. The second complication is unique to photosynthetic systems evolving O_2. Ideally one wishes to compare the specific enrichments in precursor and product. Owing to the dilution of the precursor $^{18}O_2$ by photosynthetically produced $^{16}O_2$ at the site of glycolate synthesis, the externally measured specific enrichment of $^{18}O_2$ is an overestimate of the desired internal value. Consequently, the relative enrichment found in the product glycolate (and those metabolites derived from it) represents a minimum value.

In the case of leaves, these problems are especially acute, and the results one obtains with them are often indecisive for these reasons (5, 25). For example, take the observation that the relative enrichment of ^{18}O in glycolate at equilibrium was between 50 and 70% for spinach leaves at the CO_2 compensation point (25). Several interpretations of this result are permissible. One can invoke isotopic dilution or the presence of inactive pools to account for the failure to observe 100% relative enrichment. Or one can invoke mechanisms which do not bring about the incorporation of isotopic oxygen. In truth one cannot distinguish between these explanations, at least not with so complicated an object as a leaf. It was for this very reason that we turned to simpler systems in which some of the above problems could be minimized (25, 126). Intact, isolated chloroplasts are excellent objects for this purpose, since they can easily be shown to contain no inactive pools of glycolate. In addition, they permit the use of the Hoch-Kok mass spectrometer inlet system (85), which enables one to measure the specific enrichment of the isotopic oxygen in the solution phase where glycolate synthesis occurs. Thus, the problem of isotopic dilution by $^{16}O_2$ is reduced, although not entirely eliminated. The results obtained with intact, isolated chloroplasts are more decisive for these reasons. The observed relative enrichment of ^{18}O in glycolate was 92% or more, regardless of the carbon source with which the glycolate was synthesized (25). Since there are some small (~5%) losses of ^{18}O associated with preparing the glycolate for isotopic analysis, this result indicates that practically all the glycolate has been formed by a mechanism (or mechanisms) which brings about the incorporation of an atom of molecular oxygen.

Similar results have been obtained with *Chromatium* and *Chlorella,* both of which can be persuaded to excrete the product glycolate into the medium (126). Dimon & Gerster (56), using cultures of *Euglena,* observed a relative enrichment of ^{18}O in glycolate of 75%. This value was based upon an integration of the specific enrichment of the isotopic oxygen over the entire duration of the experiment, which does not necessarily correspond to the period during which the glycolate was synthesized. Indeed, during the first 4 min of the experiment, there was a sharp decline in the specific enrichment of the isotopic oxygen due to the evolution of $^{16}O_2$. Little or no $^{18}O_2$ was consumed during this period, suggesting that no glycolate was synthesized then. If this period is discounted and the specific enrichment of the isotopic oxygen calculated over the remaining time, the relative enrichment of ^{18}O in glycolate approaches 90%. The results of these in vivo $^{18}O_2$-labeling experiments together with those performed with intact, isolated chloroplasts indicate that the synthesis of glycolate occurs with the incorporation of an atom of molecular oxygen. This criterion represents a severe constraint upon the mechanism of glycolate synthesis. Mechanisms which fail to satisfy this criterion can be dismissed or relegated to a very minor role.

Gerster and his colleagues (71) have reported that ^{18}O-labeled glycolate continues to be synthesized by *Chlorella* treated with sufficient cyanide to inhibit the carboxylation of RuBP and, presumably, also the oxygenation of RuBP. They interpreted this result as an indication that there are additional pathways for the incorporation of ^{18}O from oxygen into glycolate other than via the oxygenation of RuBP. This interpretation involves involves the assumption that what takes place in the presence of cyanide also occurs in its absence. This is a dangerous assumption. By poisoning the carboxylase one automatically prevents the reduction of the normal electron acceptor CO_2. Under these conditions electrons will inevitably flow to the only available acceptor, oxygen. Under normal circumstances the reduced forms of oxygen, the superoxide radical and hydrogen peroxide, are rendered harmless by dismutation or reduction to water. But cyanide is an effective inhibitor of the enzymes which catalyze these reactions, superoxide dismutase, catalase, and ascorbate peroxidase. Therefore, in the presence of cyanide considerable increases in the steady state concentrations of the superoxide radical, hydrogen peroxide, and possibly also of the very reactive hydroxyl free radical, are to be expected. Under such artificial conditions it would not be surprising if some nonenzymatic free radical oxidation of sugars leading to the synthesis of ^{18}O-labeled glycolate occurred. [Glycolate is among the products formed during the pulse-radiolysis of oxygenated solutions of glucose and ribose 5-P (201). Pulse radiolysis is probably the best method for generating defined solutions of the free radicals of oxygen.] However, it does not necessarily follow that such reactions occur in the absence of cyanide.

Gerster & Tournier (72) have also reported that RuBP and fructose 6-P are readily oxidized by hydrogen peroxide and have suggested that this reaction could account for the incorporation of ^{18}O (from peroxide but ultimately from molecular O_2) into glycolate. Detailed information on the kinetics and on what precautions, if any, were taken to exclude redox metal ions, have not yet appeared. However, recent work in other laboratories involving the reaction of hydrogen peroxide with RuBP (9) and other keto sugar monophosphates (49) indicates that these reactions are much too slow in the absence of redox metal ions to account for the observed rates of glycolate synthesis in vivo.

Tournier et al (195) have demonstrated that the oxidative decarboxylation of hydroxypyruvate by hydrogen peroxide takes place with the incorporation of an atom of oxygen (from the peroxide) into the carboxyl group of glycolate. Since both hydroxypyruvate and hydrogen peroxide are generated within the peroxisomes during the normal operation of the C-2 PCO cycle, the implication is that this reaction might occur in vivo. As such this reaction not only represents a source of ^{18}O-labeled glycolate but also of photorespiratory CO_2. However, it should be noted that such a reaction would not account for the de novo synthesis of glycolate, since both hydroxypyruvate and hydrogen peroxide are themselves the products of the metabolism of glycolate. Therefore, we would anticipate that inhibitors of peroxisomal hydrogen peroxide and hydroxypyruvate formation (e.g. an inhibitor of glycolate oxidase) would inhibit glycolate synthesis. Clearly this is not the case.

Beck (21) has proposed a four-electron mechanism for the synthesis of two sugar acids (one of them presumably glycolate or P-glycolate) by the reaction of a keto sugar phosphate with not one but two molecules of hydrogen peroxide. This somewhat improbable reaction is depicted as bringing about the incorporation of an atom of oxygen from the peroxide into the carboxyl group of both products. This is quite untenable since the sugar acids 3-P-glycerate and glycerate remain unlabeled during exposures to $^{18}O_2$ in vivo long enough to label glycolate substantially (5, 25).

Carbon Labeling Patterns

The uniform ^{14}C labeling of glycolate from $^{14}CO_2$ (174) is the criterion most readily satisfied by all of the various mechanisms proposed for glycolate synthesis (217). But it is also the most ambiguous. During photosynthesis with $^{14}CO_2$ carbon atoms 1 and 2 of all the sugar phosphates of the C-3 PCR cycle, including RuBP, become and remain uniformly labeled with ^{14}C. The oxygenation of RuBP has been shown recently to occur with carbon-carbon bond cleavage between carbon atoms 2 and 3 of RuBP, thus satisfying the requirement of uniform labeling (151). However, this result alone does not exclude other mechanisms.

Competition between CO_2 and O_2

One of the characteristics of photorespiration which distinguishes it from "dark" respiration is the competition between CO_2 and O_2 (75, 88, 216). The influence of CO_2 and O_2 upon glycolate synthesis is most convincingly demonstrated in systems which permit glycolate to accumulate as an end product. There is a common belief (58, 163, 175, 176) that one can reach definitive conclusions by plotting the percentage of ^{14}C in glycolate (or any other metabolite for that matter) as a function of some environmental parameter such as the CO_2 concentration or the O_2 concentration or pH. This is a very questionable practice for it completely ignores the possibility of glycolate synthesis from nonradioactive sources. But specific activity measurements have established that glycolate can by synthesized from sources other than CO_2 (66, 162). Thus, while the conclusions based upon such plots may support (175, 176) or question (58) one's particular dogma, the truth of the matter is that they are indecisive and equivocal. The desired parameter, the rate of synthesis (change in mass per unit time), is frequently a good deal more difficult to measure. There are relatively few studies of glycolate synthesis which satisfy these criteria. Two which do (24, 95) demonstrate quite unequivocally the competition between CO_2 and O_2. In both the chemosynthetic system (*Alcaligenes*) and the photosynthetic system (*Chlamydomonas*), glycolate synthesis and CO_2 fixation were shown to be competitive processes. This particular criterion can be satisfied by a number of the proposed mechanisms of glycolate synthesis. But none offers a more simple and direct explanation for this phenomenon than the mutually competitive inhibition of RuBP carboxylation by O_2 and of RuBP oxygenation by CO_2 (8, 110, 142). The fact that H_2O_2 is an inhibitor of both carboxylation and oxygenation of RuBP (9) complicates but in no way invalidates the conclusion just reached.

Double Labeling Experiments (^3H_2O and $^{14}CO_2$)

In short-term double labeling experiments using ^3H_2O and $^{14}CO_2$, the observation has been made (150, 153, 179) that the $^3H/^{14}C$ ratio is greater in glycolate than in P-glycerate. From this result it has been argued (150, 218) that an appreciable source of glycolate must exist aside from the C-3 PCR cycle. However, there is a much simpler and more likely explanation. This is based upon the demonstration (131) that the reaction catalyzed by ribose-5-P isomerase, the interconversion of ribose 5-P and ribulose 5-P, occurs with the incorporation of an atom of hydrogen from water into the carbon-1 position. In double labeling experiments one would expect that RuBP will become labeled with ^3H at carbon-1 sooner than with ^{14}C at carbon-1. The glycolate that is formed from this RuBP will therefore have a higher $^3H/^{14}C$ ratio than 3-P-glycerate, as is observed. Thus, far from demonstrating the existence of multiple pathways of glycolate synthesis, the results of these double labeling experiments are precisely what one would expect if glycolate is indeed synthesized by the oxygenation of RuBP.

For the case to be complete, the above in vivo evidence needs to be complimented by the demonstration of an enzyme capable of converting RuBP to P-glycolate with the incorporation of an atom of molecular oxygen in a CO_2-sensitive reaction at rates equal to or greater than the in vivo rates of glycolate synthesis. Under normal photosynthetic conditions, the rate of photorespiration is about 15 to 20% of the rate of apparent photosynthesis (40). Thus, for a typical C-3 plant photosynthesizing at about 100 μmols/mg chlorophyll/h, the rates of glycolate synthesis needed to sustain the observed rates of photorespiration are between 30 and 40 μmol/mg chlorophyll/h. Zelitch (217) has measured similar rates of glycolate accumulation in tobacco leaf discs. Typically, the ratio of maximum rates (V_{max} ratio) carboxylase/oxygenase is between three and four to one (8, 45, 93). With maximum rates of carboxylation in excess of 250 μmols/mg chlorophyll/h now fairly commonplace (8, 55, 114, 136, 159) it is clear that there is sufficient oxygenase activity present, even when the rates under air are calculated (8).

A number of alternative mechanisms for glycolate synthesis have been proposed (217). While these mechanisms might satisfy some of the criteria established above, none of them meet all of the criteria. Only one mechanism does so. That is the oxygenation of RuBP. Let us now look more closely at the enzyme which catalyzes this reaction.

RIBULOSE-1,5-BISPHOSPHATE CARBOXYLASE-OXYGENASE

Since it catalyzes the primary reactions of both the C-3 PCR and C-2 PCO cycles, Rubisco is currently attracting considerable attention. Several reviews (1, 2, 4, 60, 80, 90, 119, 124, 133, 178, 208) and the proceedings of a symposium dealing with various aspects of Rubisco have recently appeared.

Structural Aspects

Considerable progress has been made toward elucidating the primary structure of both subunits of the higher plant-type enzyme (8 large and 8 small subunits). In addition, the amino acid sequence of the simpler dimeric form of the enzyme found in *Rhodospirillum rubrum* is currently being determined.

The complete amino acid sequence of the small subunit of spinach Rubisco has been determined by traditional sequencing techniques (128). Very recently the complete sequence of the pea small subunit has been elucidated with recombinant DNA technology (22). Unfortunately, these sequences do not provide us with any clues as to the function of the small subunit. This remains a mystery.

In another demonstration of the power of recombinant DNA technology, Bogorad and his colleagues (135) have recently announced the complete amino acid sequence of the large (catalytic) subunit of the maize Rubisco. This is shown in Figure 2 together with those segments of the spinach and barley large subunits which have been determined by traditional methods (81, 117, 155). Substantial homology is evident, particularly in those parts thought to constitute the catalytic site. In a series of elegant studies using the affinity labeling approach, Hartman and his colleagues have pinpointed some of the residues within the active-site domain (81, 172). Two lysyl residues, lysines 176 and 334, which are spatially close to one another in the native enzyme, and two cysteinyl residues, cysteines 173 and 458, have been identified in this manner. Lysine 176 is also derivatized by pyridoxal phosphate (144, 183). An additional lysyl residue, lysine 202, has been identified as the site of carbamate formation during activation of the enzyme with CO_2 and Mg^{2+} (117).

On the principle that residues and sequences essential to catalysis are likely to have been conserved throughout evolution, comparative studies of the primary structures of enzymes from evolutionarily distant species are often useful. With this in mind, Hartman, Tabita and their colleagues have begun to examine the primary structure of the *R. rubrum* Rubisco (80, 166). Hartman et al (80) have sequenced the tryptic peptides containing all five cysteinyl residues. These are listed below.

1. NH_2-Met-Asp-Gln-Ser-Ser-Arg-Tyr-Val-Asn-Leu-Ala-Leu-Lys-Glu-Glu-Asp-Leu-Ile-Ala- Gly-Gly-Glx-His-Val-Leu-Cys-Ala-Tyr-
2. -Ala-Cys-Thr-Pro-Ile-Ile-Ser-Gly-Gly-Met-Asn-Ala-Leu-Arg-
3. -Gly-Tyr-Thr-Ala-Phe-Val-His-Cys-Lys-

```
                                      10                                      20                                      30                                      40
M  Met Ser Pro Gln Thr Glu Thr Lys Ala Ser Val Gly Phe Lys Ala Gly Val Lys Asp Tyr Lys Leu Thr Tyr Tyr Thr Pro Glu Tyr Glu Thr Lys Asp Thr Asp Ile Leu Ala Ala Phe
B   -   -   -   -   -   -   -   -   -   -   -   -   -   -  Ala Gly Val Lys Asp Tyr Lys Leu Thr Tyr Tyr Thr Pro Glu Tyr Glu Thr Lys Asp Thr Asp Phe Leu Ala Ala Phe
S   -   -   -   -   -   -   -   -   -   -   -   -   -   -   -   -   -   -   -   -   -   -   -   -   -   -   -   -   -   -   -   -   -   -   -   -   -   -   -   -

                                      50                                      60                                      70                                      80
   Arg Val Thr Pro Gln Leu Gly Val Pro Pro Glu Glu Ala Gly Ala Ala Val Ala Ala Glu Ser Ser Ala Ala Gly Thr Trp Thr Thr Val Trp Thr Asp Gly Leu Thr Ser Leu Asp Arg
   Arg Val Ser Pro Gln Pro Gly Val Pro Pro Glu Glu Ala Gly Ala Ala Val Ala Ala Glu  -   -   -   -   -   -   -   -   -   -   -   -   -   -   -   -   -   -   -   -
    -   -   -   -   -   -   -   -   -   -   -   -   -   -   -   -   -   -   -   -   -   -   -   -   -   -   -   -   -   -   -   -   -   -   -   -   -   -   -   -

                                      90                                     100                                     110                                     120
   Tyr Lys Gly Arg Cys Tyr His Ile Glu Pro Val Pro Gly Asp Pro Asp Gln Tyr Ile Cys Tyr Val Ala Tyr Pro Leu Asp Leu Phe Glu Glu Gly Ser Val Thr Asn Met Phe Thr Ser
    -   -   -   -   -   -   -   -   -   -   -   -   -   -   -   -   -   -   -   -   -   -   -   -   -   -   -   -   -   -   -   -   -   -   -   -   -  Phe Thr Ser
    -   -   -   -   -   -   -   -   -   -   -   -   -   -   -   -   -   -   -   -   -   -   -   -   -   -   -   -   -   -   -   -   -   -   -   -   -   -   -   -

                                     130                                     140                                     150                                     160
   Ile Val Gly Asn Val Phe Gly Phe Lys Ala Leu Arg Ala Leu Arg Leu Glu Asp Leu Arg Ile Pro Pro Ala Tyr Ser Lys Thr Phe Gln Gly Pro Pro Arg Gly Met Gln Val Glu Arg
   Ile Val Gly Asn Val Phe Gly Phe Lys Ala Leu  -   -   -   -   -   -   -   -   -   -   -   -   -   -   -   -   -   -   -   -   -   -   -   -   -   -   -   -   -
    -   -   -   -   -   -   -   -   -   -   -   -   -   -   -   -   -   -   -   -   -   -   -   -   -   -   -   -   -   -   -   -   -   -   -   -   -   -   -   -

                                     170                                     180                                     190                                     200
   Asp Lys Leu Asn Lys Tyr Gly Arg Pro Leu Leu Gly Cys Thr Ile Lys Pro Lys Leu Gly Leu Ser Ala Lys Asn Tyr Gly Arg Ala Cys Tyr Glu Cys Leu Arg Gly Gly Leu Asp Phe
    -   -   -   -   -   -   -   -   -   -   -   -   -   -   -   -   -   -   -   -   -   -   -   -   -   -   -   -   -   -   -   -   -   -   -   -   -   -   -   -
    -   -   -   -   -  Tyr Gly Arg Pro Leu Leu Gly Cys Thr Ile Lys Pro Lys  -   -   -   -   -   -   -   -   -   -   -   -   -   -   -   -   -   -  Gly Gly Leu Asp Phe

                                     210                                     220                                     230                                     240
   Thr Lys Asp Asp Glu Asn Val Asn Ser Gln Pro Phe Met Arg Trp Arg Asp Arg Phe Val Phe Cys Ala Glu Ala Ile Tyr Lys Ser Gln Ala Glu Thr Gly Glu Ile Lys Gly His Tyr
    -   -   -   -   -   -   -   -   -   -   -   -  Met Arg Ile Arg Asp Arg Phe Val Phe Cys Ala Glu Ala Ile Tyr Lys Ser Gln Ala Glu Thr Gly Glu Ile Lys Gly His Tyr
   Thr Lys Asp Asp Glu Asn Val Asn Ser Gln Pro Phe  -   -   -   -   -   -   -   -   -   -   -   -   -   -   -   -   -   -   -   -   -   -   -   -   -   -   -   -

                                     250                                     260                                     270                                     280
   Leu Asn Ala Thr Ala Gly Thr Cys Asp Glu Met Ile Lys Gly Ala Val Phe Ala Arg Gln Leu Gly Val Pro Ile Val Met His Asp Tyr Leu Thr Gly Gly Phe Thr Ala Asn Thr Thr
   Leu Asn Ala  -   -   -   -   -   -   -  Met Ile Lys Gly Ala Val Phe Ala Arg Gln Leu Gly Val Pro  -   -  Met His Asp Tyr Leu Thr Gly Gly Phe Thr Ala Asn Thr Thr
    -   -   -   -   -   -   -   -   -   -   -   -   -   -   -   -   -   -   -   -   -   -   -   -   -   -   -   -   -   -   -   -   -   -   -   -   -   -   -   -

                                     290                                     300                                     310                                     320
   Leu Ser His Tyr Cys Arg Asp Asn Gly Leu Leu His Ile His Arg Ala Met His Ala Val Ile Asp Arg Gln Lys Asn His Gly Met His Phe Arg Val Leu Ala Lys Ala Leu Arg Met
   Leu Ala His Tyr Cys Arg Asp  -   -   -   -   -   -   -   -   -   -   -  Ala Val Ile Asp Arg Gln  -   -   -   -   -   -   -   -   -   -   -   -   -   -   -   -
    -   -   -   -   -   -   -   -   -   -   -   -   -   -   -   -   -   -   -   -   -   -   -   -   -   -   -   -   -   -   -   -   -   -   -   -   -   -   -   -

                                     330                                     340                                     350                                     360
   Ser Gly Gly Asp His Ile His Ser Gly Thr Val Val Gly Lys Leu Glu Gly Glu Arg Glu Ile Thr Leu Gly Phe Val Asp Leu Leu Arg Asp Asp Phe Ile Glu Lys Asp Arg Ser Arg
   Ser Gly Gly Asp His Ile His Ser Gly Thr Val Val Gly Lys Leu Glu Gly Glu Arg Glu  -   -   -   -   -   -   -   -   -   -   -   -   -   -   -   -   -   -   -   -
   Ser Gly Gly Asp His Ile His Ser Gly Thr Val Val Gly Lys Leu Glu Gly Glu Arg  -   -   -   -   -   -   -   -   -   -   -   -   -   -   -   -   -   -   -   -   -

                                     370                                     380                                     390                                     400
   Gly Ile Phe Phe Thr Gln Asp Trp Val Ser Met Pro Gly Val Ile Pro Val Ala Ser Gly Gly Ile His Val Trp His Met Pro Ala Leu Thr Glu Ile Leu Gly Asp Asp Ser Val Leu
    -   -   -   -   -   -   -   -   -   -  Met Pro Gly Val Ile Pro Val Ala Ser Gly Gly Ile His Val Trp His Met Pro Ala Leu Thr Glu Ile Val Gly Asp Asp Ser Val Leu
    -   -   -   -   -   -   -   -   -   -   -   -   -   -   -   -   -   -   -   -   -   -   -   -   -   -   -   -   -   -   -   -   -   -   -   -   -   -   -   -

                                     410                                     420                                     430                                     440
   Gln Phe Gly Gly Gly Thr Leu Gly His Pro Trp Gly Asn Ala His Gly Ala Ala Ala Asn Arg Val Ala Leu Glu Ala Cys Val Gln Ala Arg Asn Glu Gly Arg Asp Leu Ala Arg Glu
   Gln Phe Gly Gly Gly Thr Leu Gly His Pro Trp Gly Asn Ala Pro Gly Ala Ala Ala Asn Arg Val Ala Leu Glu Ala Cys Val Gln Ala Arg Asn Glu Gly Arg Asp Leu Ala Arg Glu
    -   -   -   -   -   -   -   -   -   -   -   -   -   -   -   -   -   -   -   -   -   -   -   -   -   -   -   -   -   -   -   -   -   -   -   -   -   -   -   -

                                     450                                     460                                     470
   Val Gln Ile Ile Lys Ala Ala Cys Lys Trp Ser Ala Glu Leu Ala Ala Ala Cys Glu Ile Trp Lys Glu Ile Lys Phe Asp Gly Phe Lys Ala Met Asp Thr Ile
    -   -   -   -   -   -   -   -   -   -   -   -   -   -   -   -   -   -   -   -   -   -   -   -   -   -   -   -   -   -   -   -   -   -   -
    -   -   -   -   -   -   -   -   -  Trp Ser Ala Glu Leu Ala Ala Ala Cys Glu Val Trp Lys  -   -   -   -   -   -   -   -   -   -   -   -   -
```

Figure 2 The complete amino acid sequence of the large subunit of maize (M) Rubisco, determined by DNA cloning and sequencing techniques (135). Also shown are those segments of the spinach (S) and barley (B) Rubisco-large subunits, determined by more classical methods (81, 117, 155).

4. -Pro-Phe-Ala-Glu-Ala-Cys-His-Ala-Phe-Trp-Leu-Gly-Gly-Asn-Phe-Ile-Lys-
5. -Ala-Gly-Tyr-Gly-Tyr-Val-Ala-Thr-Ala-Ala-His-Phe-Ala-Ala-Glu-Ser-Ser-Thr-Gly-Thr-Asp- Val-Glu-Val-Cys-Thr-Thr-Asx-Asx-Phe-Thr-Arg-

It is instructive to compare these sequences with the cystein-containing sequences of the spinach or maize Rubisco. The most striking feature is the apparent absence of any homology between these peptides and their cysteine-containing counterparts in the plant enzyme. This apparent absence of homology extends to the peptide bearing the lysine residue which interacts with pyridoxal phosphate, lysine 176 of the plant enzyme. The *R. rubrum* peptide contains a Gly-Arg-Pro sequence in common with the plant peptide, but the two sequences are otherwise not homologously aligned (166).

R. rubrum:-Ala-Leu-Gly-Arg-Pro-Glu-Val-Asp-Lys*-Gly-Thr-Leu-Val-Ile

Plant : Tyr-Gly-Arg-Pro-Leu-Leu-Gly-Cys-Thr-Ile-Lys*-Pro-Lys

The *R. rubrum* Rubisco has both carboxylase and oxygenase activities (132, 168) and is activated by CO_2 and Mg^{2+} (44, 143, 207) in much the same manner as the plant enzyme (117, 121, 125). Thus, the absence of homology (thus far !) is a surprising result. However, unlike the plant enzyme which is competitively inhibited by 6-phosphogluconate (with respect to RuBP) (12, 53, 169), *R. rubrum* Rubisco is either not inhibited (185, 186) or only weakly inhibited (171) by 6-phosphogluconate. This result might reflect differences in the active site topology.

Additionally, Takabe & Akazawa (188) have analyzed the amino acid compositions of a number of Rubiscos, and have pointed out that the *R. rubrum* Rubisco has a strikingly different composition. These facts, together with the lack of homology raise an intriguing possibility. Did the *R. rubrum* Rubisco evolve independently (F. C. Hartman, personal communication)? If the independent evolution of the *R. rubrum* Rubisco can be firmly established, considerable support would be provided for the view (4, 118) that the carboxylation of RuBP obligatorily involves the formation of an intermediate (the enediol ?) which is susceptible to attack by O_2.

Crystals of Rubisco suitable for X-ray analysis have been obtained from tobacco (59), potato (91), and the bacteria *Alcaligenes* (31, 145) and *R. rubrum* (171). Work is in progress in at least three laboratories on the crystal structure. These studies will surely yield valuable information concerning the mechanisms of carboxylation and oxygenation.

Activation of Rubisco by CO_2 and M

MECHANISM OF ACTIVATION Although it was long known that the activity of Rubisco could be enhanced by preincubation with CO_2 and Mg^{2+} (154), it was not until the mid-70s that this phenomenon received detailed kinetic (11, 44, 107, 121, 122) and physical (138) study. The impetus for this was a series of papers describing a form of the enzyme with a high affinity for CO_2 (3, 8, 13, 14). This form of the enzyme was subsequently shown to be the activated form (109, 121). The important features of the activation reaction are summarized below.

Activation is a readily reversible equilibrium process, the final position of equilibrium being dependent upon $[CO_2]$, [M] and $[H^+]$ (121). A number of chloroplast metabolites, notably 6-phosphogluconate and NADPH, also influence the final position of equilibrium, but their effect appears to be secondary (12, 130) (see also below).

Activation applies to both carboxylase and oxygenase activities. The kinetics of activation and inactivation are the same for both activities (11, 121, 122). CO_2, rather than HCO_3^- or CO_3^{2-}, is the active species involved in both activation (121) and catalysis (54). This has important mechanistic implications for both processes. Carbamate formation is known to involve CO_2 (64).

Activation involves the ordered addition of $^{A}CO_2$ to the enzyme in a rate-determining step followed by the rapid addition of M to form the activated ternary complex (121).

$$\underset{\text{(inactive)}}{E + {}^{A}CO_2} \underset{}{\overset{\text{slow}}{\rightleftharpoons}} E\cdot{}^{A}CO_2 + M \overset{\text{fast}}{\rightleftharpoons} \underset{\text{(active)}}{E\cdot{}^{A}CO_2\cdot M}$$

An earlier spectroscopic study had shown that the tight binding of one Mn^{2+} per protomer occurs only in the presence of CO_2 (138). Thus kinetic and physical data are in accord as to the order of addition of CO_2 and M to the enzyme.

The molecule of CO_2 involved in the activation reaction $^{A}CO_2$ is distinct from the substrate ($^{S}CO_2$) which becomes fixed during carboxylation (121). Kinetic and spectroscopic measurements permit a distinction to be made between two exchange classes of CO_2, fast and slow. Kinetic measurements (12, 107, 121, 122) of the rate of dissociation of the enzyme $\cdot{}^{A}CO_2$ complex yielded a value for k_{off} in the order of 10^{-2} s^{-1}, a rather slow event. Spectroscopic studies (138), on the other hand, have demonstrated the existence of a fast exchanging species with a k_{off} of 2×10^4 s^{-1}. The 10^6 difference in rates clearly indicates that the fast exchanging species is not $^{A}CO_2$. Is it $^{S}CO_2$? CABP is an analog of 2-carboxy 3-keto arabinitol 1,5-bisphosphate, the putative six-carbon intermediate of the carboxylase reaction (152, 177). It contains the structural elements of both $^{S}CO_2$ and RuBP, and it is reasonable to suppose that the carboxyl group of CABP occupies the same site on the enzyme as $^{S}CO_2$. The ^{13}C NMR data (138) is consistent with the view that CABP displaces the fast exchanging species of CO_2, as would be expected if it and $^{S}CO_2$ were one and the same (124).

Additional evidence that $^{A}CO_2$ and $^{S}CO_2$ are discrete entities comes from the response of the enzyme $\cdot{}^{A}CO_2\cdot M$ complex to RuBP (116). By permitting catalysis to occur in the presence of a vast molar excess of unlabeled CO_2, the stability of enzyme $\cdot[^{14}C]^{A}CO_2\cdot M$ could be measured. If $^{A}CO_2$ and $^{S}CO_2$ are the same, then, following catalytic turnover, the ^{14}C radiospecific activity of the enzyme-bound CO_2 should equal that of the bulk CO_2. If $^{A}CO_2$ and $^{S}CO_2$ are different, $^{A}CO_2$ will not necessarily come into isotopic equilibrium with the CO_2 in the medium. When the enzyme$\cdot{}^{A}CO_2\cdot M$ complex was reisolated following catalytic turnover, its ^{14}C radiospecific activity was 40 times that of the unbound CO_2 (116). This can only have come about if $^{A}CO_2$ and $^{S}CO_2$ are bound at physically distinct sites.

By challenging the enzyme $\cdot{}^{A}CO_2\cdot M$ complex with CABP, Miziorko (137) has reached a similar conclusion. CABP failed to displace $^{A}CO_2$ from the enzyme, indicating that the carboxyl group of CABP and $^{A}CO_2$ are bound at different sites. Addition of CABP to the ternary complex leads to the formation of an exceptionally stable quaternary complex of enzyme $\cdot{}^{A}CO_2\cdot M\cdot CABP$ with a protomer-based stoichiometry of 1:1:1:1 (137, 139). This complex is so stable that neither $^{A}CO_2$ nor M readily exchange with unbound ligand (137, 139). This has proven to be a very useful property since it has enabled $^{A}CO_2$ to be specifically trapped on the enzyme (124, 125). The stoichiometry of this quaternary complex (1:1:1:1) is of considerable mechanistic significance (see below).

Activation involves the reaction of $^{A}CO_2$ with the ϵ-amino group of lysine 202 of the large subunit to form a carbamate (117, 121, 125). ^{13}C NMR evidence for the formation of a carbamate during the activation of the *R. rubrum* Rubisco has been obtained also (143).

$$\text{Lys-NH}_3^+ \underset{}{\overset{\pm H^+}{\rightleftharpoons}} \text{Lys-NH}_2 + {}^{A}CO_2 \overset{\pm H^+}{\rightleftharpoons} \text{Lys-NH-}{}^{A}COO^-$$

The cleavage of carbamates of aliphatic amines has been shown recently to proceed by a mechanism involving N-protonation in a fast equilibrium step followed by CO_2 explusion in the rate limiting step (61). The coordination of a divalent metal ion M with the carbamate would be expected to retard this rate-determining step further. However, there is no hard and fast evidence that M coordinates with and stabilizes the carbamate at lysine 202. But the requirement for CO_2 to be present for tight Mn^{2+} binding (138,

139), the influence of [M] and [H^+] on the equilibrium position of the activation reaction (121), and the fact that CABP "locks in" not only $^ACO_2$ but also M into positions which prevent their ready exchange with unbound ligand (137, 139) are observations most simply explained by invoking a direct interaction between the carbamate and M. The structure of the peptide containing the lysine responsible for binding $^ACO_2$ also encourages this belief (117). The presence of acidic residues adjacent to lysine 202 is likely to influence the equilibrium associated with activation in two distinct ways. First, their presence is likely to increase the pK of the ϵ-amino group above its free solution value. Second, their presence is likely to shift the equlibrium of carbamate formation per se to the left, since carbamate formation introduces additional negative charge into an already anionic region. Thus the presence of these acidic residues will tend to promote the dissociation of the carbamate unless a mechanism exists for neutralizing this negative charge. Coordination of M with this anionic site would represent such a mechanism. The binding of M would effectively stablize the carbamate. Consistent with this proposal are the kinetics of activation (121) which indicate that M displaces the equilibrium for activation to the right.

THE ROLE OF SUGAR PHOSPHATES If one subjects Rubisco in vitro to the conditions believed to apply in vivo (about 10μM CO_2, 5 to 10mM Mg^{2+}, pH about 8.0), it remains substantially in the inactivated state. Yet in vivo or within the intact isolated chloroplast, the enzyme seems to be maintained in a substantially activated state (15, 123, 164). In seeking to determine what factors might be responsible for maintaining Rubisco in this state in vivo, a number of laboratories have reported that various chloroplast metabolites stimulate or inhibit the carboxylase activity (32, 41, 50–53, 82, 107, 113, 169, 170, 185, 199, 200, 207). Unfortunately, failure to distinguish between effects upon the activation reaction and effects upon catalysis has created a degree of confusion. The binding of the effector at one or more allosteric sites distinct from the catalytic site is frequently invoked to account for the observed phenomena. However, the results of two more recent studies (9, 130) indicate that the various effectors interact at a single site, the catalytic site for RuBP. Physical binding studies (9) established that there was only one binding site per protomer for 6-P-gluconate. When this site was first occupied with CABP, no binding of either 6-P-gluconate or NADPH could be detected. The product of the carboxylation reaction 3-P-glycerate also displaced the 6-P-gluconate in a competitive manner.

Most of the effectors reported in the literature are also linearly competitive inhibitors of catalysis with respect to RuBP. When combined with the competitive binding results, these competitive kinetics clearly show that the effectors interact with the enzyme at the catalytic site for RuBP. Additional kinetic and physical studies (9, 130) show that the responses to these effectors are secondary to the basic activation reaction outlined above. The effectors elicit their response by binding to the inactivated form and/or to the activated form of the enzyme. They promote the activation reaction by reversibly stabilizing the binding of $^{A}CO_2$ and M in much the same manner that CABP does. Unlike CABP, which remains bound, the effectors readily dissociate upon dilution into the assay solution to yield the activated ternary enzyme·$^{A}CO_2$· M complex. If the effectors bind more tightly to the inactivated form of the enzyme than to the activated form, they promote inactivation. However, it is the ratio of the dissociation constants describing the binding of effector to the activated and inactivated forms which determines whether a given effector promotes or inhibits activation and the degree to which it does so.

Effectors such as NADPH and 6-P-gluconate stabilize the activated form of the enzyme, and it is possible that they function in vivo in this manner. However, to that statement must be added a restraining caveat (9). Namely, an enzyme molecule cannot be simultaneously catalytically competent (capable of binding and carboxylating or oxygenating RuBP) and activated by an effector such as 6-P-gluconate or NADPH since the latter involves occupancy of the RuBP binding site. In short, you can't keep your cake and eat it too!

Mechanisms of Catalysis

AN ATOMIC AUDIT AND ITS MECHANISTIC CONSEQUENCES This being an annual report, it is appropriate to include an audit, an atomic one. This will define for us whence the atoms come and whither they go in the reactions catalyzed by Rubisco. It will provide some important mechanistic constraints. The two reactions are shown in Figure 3 in such a manner that the fate of the various atoms can be easily traced.

1. The stoichiometry of carboxylation (89, 206) and of oxygenation (6, 79, 93, 134) is as indicated.
2. CO_2 (not HCO_3^-) is the reactive species in carboxylation (7, 54). This requires that C-2 of RuBP be converted from an electrophilic to a nucleophilic center.
3. Carbon-carbon bond cleavage occurs between C-2 and C-3 in both reactions (140, 151). $^{S}CO_2$ becomes the carboxyl group of *top* 3-P-glycerate.[4]
4. The hydrogen atom on C-3 of RuBP is lost to the medium during carboxylation (65).

[4]The product formed from carbon atoms 1 and 2 of RuBP and $^{S}Co_2$ is referred to as *top* 3-P-glycerate. That formed from carbon atoms 3, 4, and 5 of RuBP is *bottom* 3-P-glycerate.

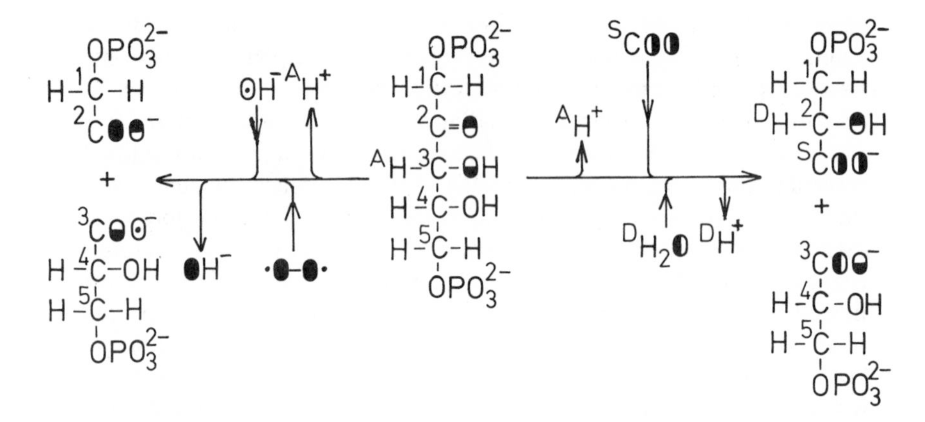

Figure 3 The reactions catalyzed by Rubisco, presented so that the fate of the atoms can be followed (see text).

5. The hydrogen atom on C-2 of *top* 3-P-glycerate is derived from water (87, 140).
6. The oxygen atom on C-2 of RuBP is retained during both reactions; it becomes the oxygen atom on C-2 of *top* 3-P-glycerate (carboxylation) (115, 184) or one of the two carboxyl oxygen atoms of P-glycolate (oxygenation) (151). These results eliminate from consideration mechanisms involving ketimine or thiohemiketal intermediates at C-2 of RuBP (157, 158, 197, 209).
7. The oxygen atom on C-3 of RuBP is retained during carboxylation to become one of the carboxyl oxygens of *bottom* 3-P-glycerate (115, 184). Again ketimine or thiohemiketal intermediates at C-3 can be discounted.
8. Only one atom of molecular oxygen is incorporated into the carboxyl group of P-glycolate during oxygenation. 3-P-glycerate remains unlabeled (120). This result eliminates mechanisms involving the formation of a dioxetane ring during oxygenation, since that would bring about the labeling of both products.

KINETIC CONSTRAINTS By far the most important kinetic constraint to be considered is the linearly competitive inhibition of carboxylation by O_2 (with respect to CO_2) and of oxygenation by CO_2 (with respect to O_2) (8, 110). This implies that O_2 and CO_2 compete with one another for a common enzyme-bound species of RuBP (or for different species in rapid equilibrium with one another). Therefore, mechanistic considerations of one reaction must take into account the existence of the other reaction. CO_2 and O_2 are linearly noncompetitive inhibitors with respect to RuBP (10).

Two studies (9, 108) dealing with the kinetic order of reaction have been reported recently. Using carbon oxysulfide (COS), an analog of CO_2, as a dead-end inhibitor of carboxylation, Laing & Christeller (108) showed that it behaved as a linearly competitive inhibitor with respect to CO_2 and as a linearly noncompetitive inhibitor with respect to RuBP. This kinetic behavior is consistent with a random mechanism of carboxylation. (This conclusion is valid only if COS is not a substrate.) The effect of COS on the oxygenase reaction was not reported.

Badger et al (9) have examined the effects of hydrogen peroxide on both carboxylation and oxygenation. As an inhibitor of carboxylation, hydrogen peroxide was linearly uncompetitive with respect to RuBP and noncompetitive (mixed) with respect to CO_2. Again this behavior indicates a random mechanism. In contrast, the oxygenase responded to peroxide in a linearly uncompetitive manner with respect to RuBP and in a linearly competitive manner with respect to O_2. This result implies that the oxygenase reaction is ordered with RuBP binding first. However, the interpretation of these results is complicated by the possibility that the carbonyl group of RuBP may be susceptible to nucleophilic attack by hydrogen peroxide. The product, an epimeric mixture of 2-peroxypentitol 1,5-bisphosphate (PPBP) would presumably exist in equilibrium with unreacted RuBP and hydrogen peroxide. PPBP would be an analog of the putative transition state peroxide intermediate of the oxygenase reaction in exactly the same manner as CABP is an analog of the putative 6-carbon transition state intermediate of the carboxylase reaction. Therefore, we might expect that PPBP would be a rather effective inhibitor of both reactions as is CABP, and to bind to the enzyme more tighly than RuBP. The observation that the combination of RuBP and hydrogen peroxide increased the stability of the enzyme $\cdot {}^{A}CO_2 \cdot M$ complex in a manner reminiscent of the behavior of CABP (137) strengthens the suspicion that the real inhibitor is PPBP and not hydrogen peroxide. Until this matter is resolved, the kinetic order of the oxygenase reaction must remain in doubt.

THE ROLE OF THE METAL ION (M) Until recently the role of the divalent metal ion (M) in catalysis had received little attention. Indeed, Laing & Christeller (107) have suggested that the role of M is restricted to activation and that it plays no role in catalysis. However, this conclusion is based upon the assumption that the dissociation of M from the enzyme$\cdot {}^{A}CO_2 \cdot$ M complex (or more precisely from the enzyme$\cdot {}^{A}CO_2 \cdot M \cdot$RuBP complex) is fast relative to the rate of catalysis. In view of the ability of compounds which occupy the active site, such as CABP (139) and 6-P-gluconate (12), to stabilize the binding of M to the enzyme, this assumption is probably incorrect. It certainly does not agree with the reports (210, 211) that the nature of M differentially influences the carboxylase and oxygenase activities. This result suggests a rather direct role for M in catalysis.

Detailed kinetic analyses of the effects of substituting Mg^{2-} with Mn^{2+} or Co^{2+} have been reported (45, 93, 165). The principal effect is to change the partitioning between carboxylation and oxygenation mainly by increasing the affinity of the enzyme for O_2. While this could be attributed to conformational effects, a case can be made for the direct participation of M in catalysis (124). Miziorko has measured the distance between the carbon atom of the fast exchanging species of CO_2, which has been identified as $^SCO_2$, and the enzyme-bound Mn^{2+} (138). The value 5.4 ± 0.1 Å was obtained. This is close enough to suggest some role for M in catalysis. In this respect the stoichiometry of the quaternary complex of enzyme·$^ACO_2$·M·CABP of 1 : 1 : 1 : 1 (137, 139) is especially interesting since it defines the composition of the transition state (214). The argument has been made above that M plays a role in both activation and catalysis. But there is only one metal ion in the quaternary complex, so it would seem that it must play a dual role; i.e. the carbamate-metal ion complex is formed within the catalytic site.

PROBLEMS OF STEREOCHEMISTRY Siegel & Lane (177) have presented evidence for the formation of 2-carboxy 3-ketopentitol 1,5-bisphosphate during the carboxylation reaction. Unfortunately, they used an epimeric mixture and so were unable to distinguish between the ribitol and arabinitol derivative. By separating CABP from carboxyribitol bisphosphate, Pierce et al (152) were able to show that CABP is the epimer which binds so tenaciously (K_d 10^{-11}M). Wolfenden (214) has outlined the theory whereby the very tight binding of an inhibitor to an enzyme can be used to infer structural similarity to the transition state of the enzyme-catalyzed reaction. Therefore, it is reasonable to suppose that the intermediate in the carboxylation reaction is 2-carboxy 3-ketoarabinitol 1,5-bisphosphate. The nonenzymatic hydrolysis of this intermediate would lead to the formation of a mixture of D- and L- *top* 3-P-glycerates. Since the enzymatic reaction produces only 3-P-glycerate of the D-configuration (89; J. Pierce and N. E. Tolbert, personal communication), the enzyme must exercise stereochemical control over this step.

OTHER PERTINENT POINTS Calvin (36) recognized that in order for CO_2 to add to the C-2 of RuBP, a rearrangement of RuBP was necessary so as to create a nucleophilic center at C-2. He suggested the formation of an enediol as the simplest means of accomplishing this. The formation of the enediol involves the removal of a proton from the C-3 position. However, early attempts to demonstrate that the enzyme could catalyze the exchange of [3C-^{3}H] RuBP with water in the absence of CO_2 failed (65). As the authors were careful to point out, this negative result could be attributed to the possible inactivation of the enzyme in the absence of CO_2. Preliminary evidence has been obtained confirming this conclusion. When fully activated *R. rubrum* Rubisco was used, this exchange reaction could be demonstrated (B. Saver and J. R. Knowles, personal communication).

Additional evidence for the formation of an enediol(ate) of RuBP has been provided with the preliminary report that the active site of *Pseudomonas facilis* Rubisco is amenable to paracatalytic modification with hexacyanoferrate (86). Christen (48) has used this approach successfully with several enzymes thought to form substrate carbanions during catalysis.

A MECHANISM FOR CARBOXYLATION A mechanism (Figure 4) which accommodates the constraints discussed above has been proposed (124, 152). While rather speculative, it provides an explanation for a number of phenomena. Aspects of it are amenable to experimental test.

The essence of the mechanism is the use of the carbamate-divalent metal ion complex to stabilize the transition state intermediate, 2-carboxy 3-ketoarabinitol 1,5-bisphosphate and the 3 carbon aci-acid which is to form the *top* 3-P-glycerate. The reaction begins with the removal of a proton from C-3 of RuBP to form the enediol I (step 1). The nature of the base B which brings this about is unknown. An attractive candidate for this role might be one of the lysyl residues, #176 or #334, identified by Hartman et al (81) as being within the active-site domain. Step 2, the binding of $^{S}CO_2$, may actually precede the formation of the enediol, since the kinetics indicate a random order of substrate addition. In any case, the quinternary complex II of enzyme·RuBP·$^{S}CO_2$·M·$^{A}CO_2$ will result. Whether $^{S}CO_2$ then adds to the *re* or to the *si* face of the enediol depends on whether the latter is in the *cis* or the *trans* configuration. This is unknown. It is conceivable that M contributes to the polarization of the carbon-oxygen bonds of $^{S}CO_2$ so as to facilitate the electrophilic attack on the C-2 of the enediol (step 3), leading to the formation of the transition state complex III. It is this complex which is mimicked by the very stable quaternary complex of enzyme·$^{A}CO_2$·M·CABP. This type of structure may account for the exceptional stability of the enzyme·$^{A}CO_2$·M·CABP complex and for why neither the $^{A}CO_2$ nor the M in this complex readily exchange with unbound ligand (137, 139).

Formation of *top* 3-P-glycerate of the D-configuration requires that the hydrolysis of the transition state complex III be stereochemically directed. Pierce et al (152) have suggested a carbanion inversion mechanism for achieving this. The expulsion of *bottom* 3-P-glycerate (step 4) is proposed to leave behind the aci-acid derivative of *top* 3-P-glycerate (IV), with the

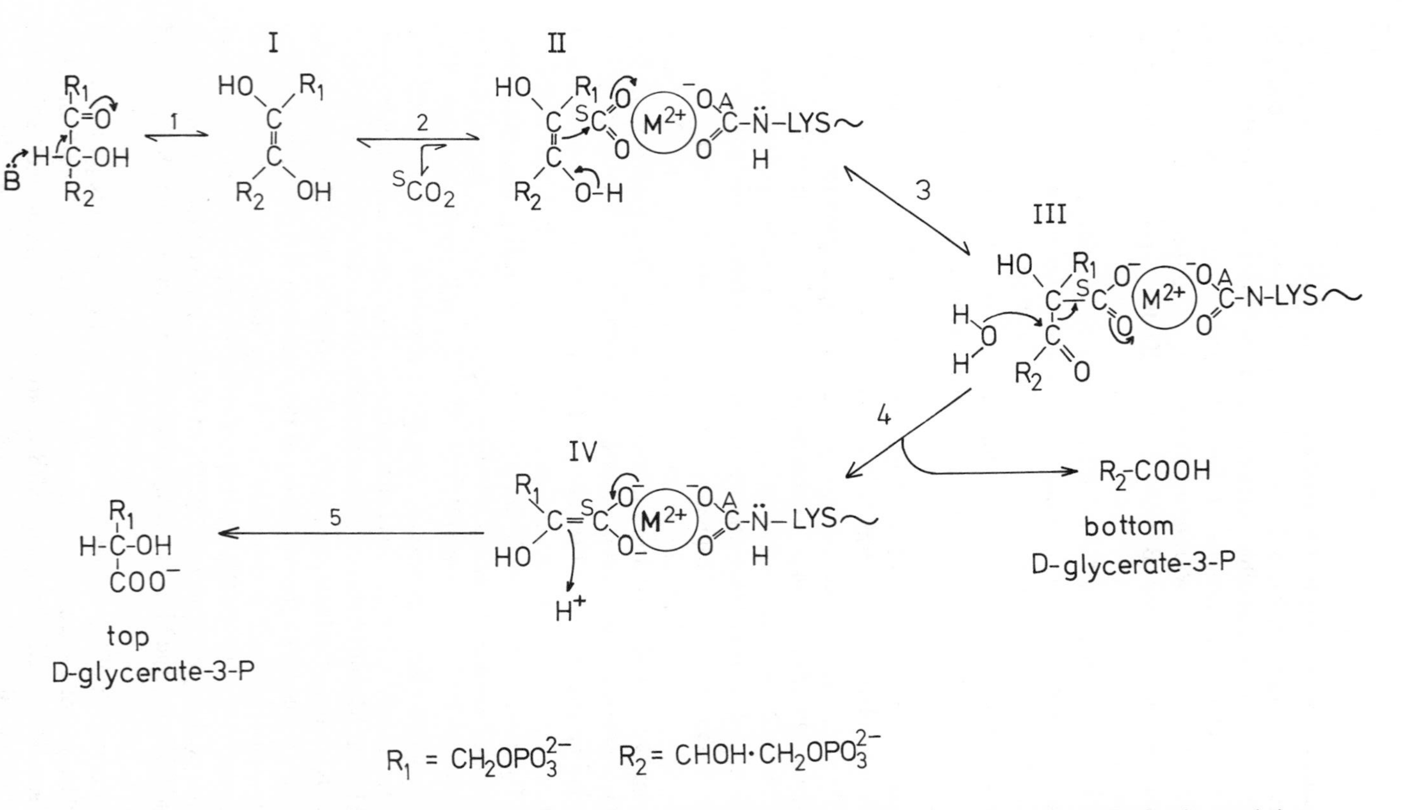

Figure 4 **A plausible mechanism for the carboxylation of RuBP (124, 152). While consistent with what is known, this mechanism is speculative.**

carbamate-divalent metal ion again providing stability. Addition of a proton to the top face of the aci-acid would then yield *top* 3-P-glycerate of the D-configuration.

The intriguing aspect of Rubisco is the competition between CO_2 and O_2, and the skeptical reader is entitled to ask "Can this mechanism account for the differential effects of M upon carboxylation and oxygenation?" The answer to that question is a tentative yes. For the moment let us accept the proposition that step 1 is common to both reactions and that the competition between CO_2 and O_2 is over the prostrate enediol(ate). Steps 2 and 3 of the proposed mechanism might well be reversible. It is conceivable therefore that the substitution of Mg^{2+} with Mn^{2+} could alter the equilibrium so as to cause the accumulation of the enediol(ate). The nature of M would then influence the partitioning between carboxylation and oxygenation. The decrease in the $K_m(O_2)$ associated with the substitution of Mg^{2+} with Mn^{2+} (42, 93) is consistent with this hypothesis.

THE OXYGENASE REACTION AND THE PROBLEM OF SPIN INVERSION One of the major unsolved problems in photorespiration is the mechanism of oxygenation of RuBP. Very little progress has been made toward the elucidation of the molecular mechanism, apart from the atomic auditing and the kinetic analyses already mentioned. Although a number of mechanistically plausible reaction-paths can be conceived (e.g. 119), experimentally established facts are few in number. The reaction involves spin inversion, the formation of singlet products from a mixture of singlet (RuBP) and triplet (O_2) reactants. Such reactions are generally considered to be spin forbidden since the lifetime of collisional complexes is usually too short for inversion to occur (78, 97, 100). As a result, the collision complex (a radical pair) either reverts to reactants or dissociates to independent radical species.

$$R\overset{\uparrow\downarrow}{^{-}} + \overset{\uparrow}{\bullet}O\text{-}O\overset{\uparrow}{\bullet} \rightleftharpoons [R\overset{\uparrow}{\bullet}\ \overset{\uparrow}{\bullet}O\text{-}O\overset{\uparrow\downarrow}{^{-}}] \not\longrightarrow R\overset{\uparrow}{\bullet}\ \overset{\downarrow}{\bullet}O\text{-}O\overset{\downarrow\uparrow}{^{-}}$$

$$\downarrow$$

$$R\overset{\uparrow}{\bullet} + \overset{\uparrow}{\bullet}O\text{-}O\overset{\uparrow\downarrow}{^{-}}$$

However, the lifetime of the radical pair and the probability of spin inversion can be enhanced by so-called cage effects (98, 141). The cage effect generally refers to the restraining influence that the solvent molecules surrounding the radical pair exact upon the motion of the latter. While they are rarely invoked in enzymatic reactions, it is clear that catalytic site of an enzyme could constitute such a cage.

The problem of spin inversion in reactions involving oxygen is most commonly solved in biology by the presence of spin delocalizing devices, transition metal ions such as copper or iron or organic cofactors (flavin, pterin, heme) capable of resonance stabilization. Interest was aroused by the report (213) of the presence of one mole of copper mole of Rubisco. However, in view of the presence of 8 catalytic sites per mole of Rubisco, it would seem unlikely that this copper could participate in catalysis. In addition, results from four other laboratories (42, 91, 101, 120) indicate the presence of only trace quantities of copper and iron. In the absence of redox metal ions or cofactors it is almost certain that the oxygenase reaction proceeds via a free radical mechanism. A highly pertinent review of free radical reactions of carbohydrates has recently appeared (201).

Figure 5 shows a plausible mechanism for the oxygenase reaction (100). It is consistent with the few facts at hand. Given the propensity for carbanions to undergo reaction with oxygen (70, 167) and from what we know about the carboxylase reaction (see discussion above), the most probable first step in the oxygenase reaction is the formation of the enediol(ate) (step 1). Addition of O_2 to the enediol(ate) and electron transfer (steps 2 and 3) creates the radical pair, the radical at C-2 of RuBP and the superoxide radical anion. Since CO_2 and O_2 are viewed as reacting with the same

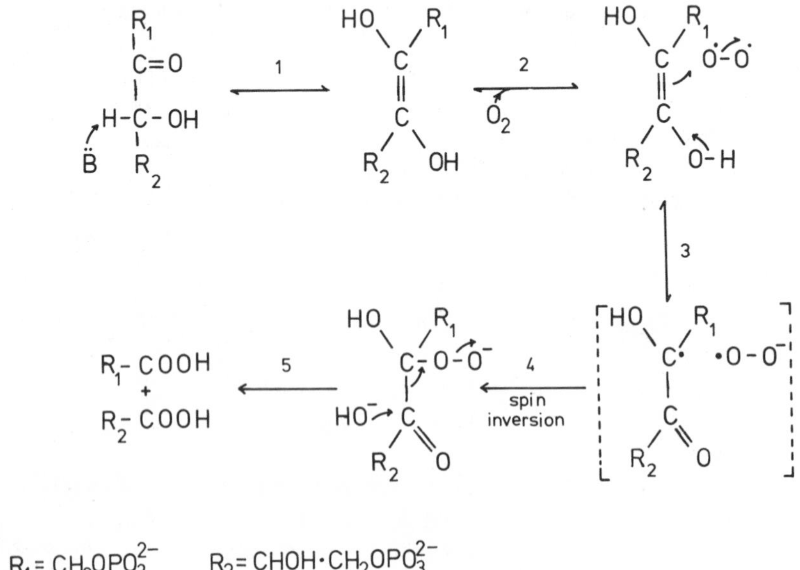

$R_1 = CH_2OPO_3^{2-}$ $R_2 = CHOH \cdot CH_2OPO_3^{2-}$

Figure 5 A plausible mechanism for the oxygenation of RuBP (100). While it too is consistent with what is known, it must be stressed that not very much is known.

species of enzyme-bound RuBP, the constraint imposed by the competitive kinetics is satisfied. By invoking reaction of O_2 with the enzyme·RuBP, kinetic constraint is imposed. Namely, that the oxygenase reaction be strictly ordered with RuBP binding first. Some tentative evidence for this has been obtained (9). In order for spin inversion to occur (step 4) it becomes necessary to postulate that the radical pair are held in place by the "cage" effect of the catalytic site. The peroxide formed at C-2 of RuBP then undergoes degradation to the final products by the addition and elimination of a hydroxyl ion. In this way only the carboxyl group of P-glycolate becomes labeled with an oxygen atom from O_2 (120).

This reaction is not without enzymatic precedent. The primary reactions of photorespiration and bioluminescence appear to have much in common mechanistically. Like Rubisco, the luciferin system lacks redox-metal ions and organic cofactors. The initial steps of the reaction involve the formation of a carbanion by the removal of a proton from luciferin, followed by reaction with oxygen. Formation of a radical pair within some form of cage, spin inversion, and recombination of the radicals is considered the most probable mechanism (129, 203).

CONCLUDING REMARKS

Throughout this article I have attempted to demonstrate *how* the primary reactions of photosynthesis and photorespiration occur. The evidence in favor of Rubisco is quite compelling. Further progress, especially in the area of the enzymatic mechanism, will surely demand that some of the views expressed here be abandoned. But perhaps the most intriguing question is *why* photorespiration occurs at all. Some ascribe a function to photorespiration (84). Others regard photorespiration as a consequence of the active-site chemistry of Rubisco (4, 118, 119). Both views depend heavily on teleological arguments. They are unlikely to be resolved unequivocally until the detailed mechanism of Rubisco is elucidated.

Literature Cited

1. Akazawa, T. 1978. Structure and function of ribulose bisphosphate carboxylase. In *Photosynthesis '77: Proc. 4th Int. Congr. Photosynth.*, ed. D. O. Hall, J. Coombs, T. W. Goodwin, pp. 447–56, London: Biochem. Soc. 827 pp.
2. Akazawa, T. 1979. Ribulose-1,5-bisphosphate carboxylase. See Ref. 74, pp. 208–29
3. Andrews, T. J., Badger, M. R., Lorimer, G. H. 1975. Factors affecting interconversion of kinetic forms of ribulose diphosphate carboxylase-oxygenase from spinach. *Arch. Biochem. Biophys.* 171:93–103
4. Andrews, T. J., Lorimer, G. H. 1978. Photorespiration—still unavoidable? *FEBS Lett.* 90:1–9
5. Andrews, T. J., Lorimer, G. H., Tolbert, N. E. 1971. Incorporation of molecular oxygen into glycine and serine during photorespiration in spinach leaves. *Biochemistry* 10:4777–82
6. Andrews, T. J., Lorimer, G. H., Tolbert, N. E. 1973. Ribulose diphosphate oxygenase. I. Synthesis of phospho-

glycolate by fraction-1 protein of leaves. *Biochemistry* 12:11–18
7. Badger, M. R. 1980. Kinetic properties of ribulose 1,5 bisphosphate carboxylase/oxygenase from *Anabaena variabilis. Arch. Biochem. Biophys.* 201: 247–54
8. Badger, M. R., Andrews, T. J. 1974. Effects of CO_2, O_2 and temperature on a high-affinity form of ribulose diphosphate carboxylase-oxygenase from spinach. *Biochem. Biophys. Res. Commun.* 60:204–10
9. Badger, M. R., Andrews, T. J., Canvin, D. T., Lorimer, G. H. 1980. Interactions of hydrogen peroxide with ribulose bisphosphate carboxylase-oxygenase. *J. Biol. Chem.* 255:7870–75
10. Badger, M. R., Collatz, G. J. 1977. Studies on the kinetic mechanism of ribulose-1,5-bisphosphate carboxylase and oxygenase reactions, with particular reference to the effects of temperature on kinetic parameters. *Carnegie Inst. Washington Yearbk.* 76:355–61
11. Badger, M. R., Lorimer, G. H. 1976. Activation of ribulose-1,5-bisphosphate oxygenase. The role of Mg^{2+}, CO_2 and pH. *Arch. Biochem. Biophys.* 175: 723–29
12. Badger, M. R., Lorimer, G. H. 1981. The interaction of sugar phosphates with the catalytic site of ribulose-1,5-bisphosphate carboxylase. *Biochemistry.* In press
13. Bahr, J. T., Jensen, R. G. 1974. Ribulose bisphosphate oxygenase activity from freshly ruptured spinach chloroplasts. *Arch. Biochem. Biophys.* 164: 408–13
14. Bahr, J. T., Jensen, R. G. 1974. Ribulose diphosphate carboxylase from freshly ruptured spinach chloroplasts having an in vivo K_m (CO_2). *Plant Physiol.* 53:39–44
15. Bahr, J. T., Jensen, R. G. 1978. Activation of ribulose bisphosphate carboxylase in intact chloroplasts by CO_2 and light. *Arch. Biochem. Biophys.* 185: 39–48
16. Bassham, J. A., Benson, A. A., Calvin, M. 1950. The path of carbon in photosynthesis. VIII. The role of malic acid. *J. Biol. Chem.* 185:781–87
17. Bassham, J. A., Calvin, M. 1957. *The Path of Carbon in Photosynthesis.* Englewood Cliffs, NJ: Prentice-Hall. 107 pp.
18. Bassham, J. A., Kirk, M. 1960. Dynamics of the photosynthesis of carbon compounds. I. Carboxylation reactions. *Biochim. Biophys. Acta* 43:447–64
19. Bassham, J. A., Kirk, M. 1962. The effect of oxygen on the reduction of CO_2 to glycolic acid and other products during photosynthesis by *Chlorella. Biochem. Biophys. Res. Commun.* 9:376–80
20. Bassham, J. A., Shibata, K., Steenberg, K., Bourdon, J., Calvin, M. 1956. The photosynthetic cycle and respiration: Light-dark transients. *J. Am. Chem. Soc.* 78:4120–24
21. Beck, E. 1979. Glycolate synthesis. See Ref. 74, pp. 327–37
22. Bedbrook, J. R., Smith, S. M., Ellis, R. J. 1980. Molecular cloning and sequencing of cDNA encoding the precursor to the small subunit of chloroplast ribulose-1,5-bisphosphate carboxylase. *Nature* 237:692–97
23. Berry, J. A., Björkman, O. 1980. Photosynthetic response and adaptation to temperature in higher plants. *Ann. Rev. Plant Physiol.* 31:491–543
24. Berry, J. A., Boynton, J., Kaplan, A., Badger, M. R. 1976. Growth and photosynthesis of *Chlamydomonas reinhardtii* as a function of CO_2 concentration. *Carnegie Inst. Washington Yearbk.* 75:423–32
25. Berry, J. A., Osmond, C. B., Lorimer, G. H. 1978. Fixation of $^{18}O_2$ during photorespiration. *Plant Physiol.* 62: 954–67
26. Björkman, O. 1968. Further studies of photosynthetic properties in sun and shade ecotypes of *Solidago virgaurea. Physiol. Plant.* 21:84–99
27. Björkman, O., Pearcy, R. W. 1971. Effect of growth temperature dependence of photosynthesis in vivo and on CO_2 fixation by carboxydismutase in vivo in C_3 and C_4 species. *Carnegie Inst. Washington Yearbk.* 70:511–20
28. Bowes, G., Berry, J. A. 1972. The effect of oxygen on photosynthesis and glycolate excretion in *Chlamydomonas reinhardtii. Carnegie Inst. Washington Yearbk.* 71:148–56
29. Bowes, G., Ogren, W. L. 1972. Oxygen inhibition and other properties of soybean ribulose 1,5-diphosphate carboxylase. *J. Biol. Chem.* 247:2171–76
30. Bowes, G., Ogren, W. L., Hageman, R. H. 1971. Phosphoglycolate production catalyzed by ribulose diphosphate carboxylase. *Biochem. Biophys. Res. Commun.* 45:716–22
31. Bowien, B., Mayer, F., Spiess, E., Pähler, A., Englisch, U., Saenger, W. 1980. On the structure of crystalline ribulosebisphosphate carboxylase from

Alcaligenes eutrophus. Eur. J. Biochem. 106:405–10

32. Buchanan, B. B., Schürmann, P. 1973. Regulation of ribulose 1,5-diphosphate carboxylase in the photosynthetic assimilation of carbon dioxide. *J. Biol. Chem.* 248:4956–64
33. **Buchanan, B. B., Schürmann, P., Shanmugam, K. T.** 1972. Role of the reductive carboxylic acid cycle in a photosynthetic bacterium lacking ribulose 1,5-diphosphate carboxylase. *Biochim. Biophys. Acta* 283:136–45
34. Buchanan, B. B., Sirevag, R. 1976. Ribulose 1,5-diphosphate carboxylase and *Chlorobium thiosulfatophilum. Arch. Microbiol.* 109:15–19
35. Burris, R. H., Black, C. C., eds. 1976. *CO_2 Metabolism and Plant Productivity.* Baltimore: Univ. Park Press. 431 pp.
36. Calvin, M. 1954. Chemical and photochemical reactions of thioctic acid and related disulfides. *Fed. Proc.* 13:697–708
37. Calvin, M., Bassham, J. A., Benson, A. A., Lynch, V., Ouellet, C., Schou, L., Stepka, W., Tolbert, N. E. 1951. Carbon dioxide assimilation in plants. *Symp. Soc. Exp. Biol.* 5:284–305
38. Calvin, M., Benson, A. A. 1948. The path of carbon in photosynthesis. *Science* 107:476–80
39. Calvin, M., Massini, P. 1952. The path of carbon in photosynthesis. XX. The steady state. *Experientia* 8:445–57
40. Canvin, D. T. 1979. Photorespiration: Comparisons between C_3 and C_4 plants. See Ref. 74, pp. 368–96
41. Chollet, R., Anderson, L. L. 1976. Regulation of ribulose-1,5-bisphosphate carboxylase-oxygenase activities by temperature pretreatment and chloroplast metabolites. *Arch. Biochem. Biophys.* 176:344–51
42. Chollet, R., Anderson, L. L., Hovespian, L. C. 1975. The absence of tightly bound copper, iron and flavin nucleotide in crystalline ribulose 1,5-bisphosphate carboxylase-oxygenase from tobacco. *Biochem. Biophys. Res. Commun.* 64:97–107
43. Chollet, R., Ogren, W. L. 1975. Regulation of photorespiration in C_3 and C_4 species. *Bot. Rev.* 41:137–79
44. Christeller, J. T., Laing, W. A. 1978. A kinetic study of ribulose bisphosphate carboxylase from the photosynthetic bacterium *Rhodospirillum rubrum. Biochem. J.* 173:467–73
45. Christeller, J. T., Laing, W. A. 1979. Effects of manganese ions and magnesium ions on the activity of soya-bean ribulose bisphosphate carboxylase/oxygenase. *Biochem. J.* 183:747–50
46. Christeller, J. T., Laing, W. A., Troughton, J. H. 1976. Isotopic discrimination by ribulose-1,5-bisphosphate carboxylase. No effect of temperature or HCO^-_3 concentration. *Plant Physiol.* 57:580–82
47. Christeller, J. T., Tolbert, N. E. 1978. Phosphoglycolate phosphatase. *J. Biol. Chem.* 253:1780–98
48. Christen, P .1970. Chemical approaches to intermediates of enzyme catalysis. *Experientia* 26:337–47
49. Christen, P., Gasser, A. 1980. Production of glycolate by oxidation of the 1,2-dihydroxyethylthiamine diphosphate intermediate of transketolase with hexacyanoferrate (III) or H_2O_2. *Eur. J. Biochem.* 107:73–77
50. Chu, D. K., Bassham, J. A. 1972. Inhibition of ribulose 1,5-diphosphate carboxylase by 6-phosphogluconate. *Plant Physiol.* 50:224–27
51. Chu, D. K., Bassham, J. A. 1973. Activation and inhibition of ribulose 1,5-diphosphate carboxylase by 6-phosphogluconate. *Plant Physiol.* 52:373–79
52. Chu, D. K., Bassham, J. A. 1974. Activation of ribulose 1,5-diphosphate carboxylase by nicotinamide adenine dinucleotide phosphate and other chloroplast metabolites. *Plant Physiol.* 54: 556–69
53. Chu, D. K., Bassham, J. A. 1975. Regulation of ribulose 1,5-diphosphate carboxylase by substrates and other metabolites. *Plant Physiol.* 55:720–26
54. Cooper, T. G., Filmer, D., Wishnick, M., Lane, M. D. 1969. The active species of "CO_2" utilized by ribulose diphosphate carboxylase. *J. Biol. Chem.* 244:1081–83
55. Delaney, M. A., Walker, D. A. 1978. Comparison of the kinetic properties of ribulose bisphosphate carboxylase in chloroplast extracts of spinach, sunflower and four other reductive pentose phosphate-pathway species. *Biochem. J.* 171:477–82
56. Dimon, B., Gerster, R. 1976. Incorporation d'oxygene dans le glycolate excrete a la lumiere par *Euglena gracilis. C. R. Acad. Sci. Ser. D* 283:507–10
57. Ehleringer, J., Björkman, O. 1977. Quantum yields for CO_2 uptake in C_3 and C_4 plants. *Plant Physiol.* 59:86–90
58. Eickenbusch, J. D., Beck, E. 1973. Evidence for the involvement of two types of reactions in glycolate formation during photosynthesis in isolated spinach chloroplasts. *FEBS Lett.* 31:225–28

59. Eisenberg, D., Baker, T. S., Suh, S. W., Smith, W. W. 1978. Structural studies of ribulose 1,5-bisphosphate carboxylase/oxygenase. See Ref. 178, pp. 271–81
60. Ellis, J. R. 1979. The most abundant protein in the world. *Trends Biochem. Sci.* 4:241–44
61. Ewing, S. P., Lockshon, D., Jencks, W. P. 1980. Mechanism of cleavage of carbamate anions. *J. Am. Chem. Soc.* 102:3072–84
62. Farquhar, G. D. 1979. Models describing the kinetics of ribulose bisphosphate carboxylase-oxygenase. *Arch. Biochem. Biophys.* 193:456–68
63. Farquhar, G. D., von Caemmerer, S., Berry, J. A. 1980. A biochemical model of photosynthetic CO_2 assimilation in leaves of C_3 plants. *Planta* 149:78–90
64. Faurholt, C. 1925. Etude sur les aqueues de carbamates et de carbonates. *J. Chim. Phys.* 22:1–44
65. Fiedler, F., Mullhofer, G., Trebst, A., Rose, I. A. 1967. Mechanism of ribulose diphosphate carboxydismutase reaction. *Eur. J. Biochem.* 1:395–99
66. Fock, H., Bate, G. C., Egle, K. 1974. On the formation of glycolate in photosynthesizing *Chlorella* using a new gas-liquid chromatography method. *Planta* 121:9–16
67. Forrester, M. L., Krotkov, G., Nelson, C. D. 1966. Effect of oxygen on photosynthesis, photorespiration and respiration in detached leaves. I. Soybean. *Plant Physiol.* 41:422–27
68. Foyer, C. H., Hall, D. O. 1980. Oxygen metabolism in the active chloroplast. *Trends Biochem. Sci.* 5:188–91
69. Galmiche, J. M. 1973. Studies on the mechanism of glycerate-3-phosphate synthesis in tomato and maize leaves. *Plant Physiol.* 51:512–19
70. Garst, J. F. 1973. Electron transfer reactions of organic anions. See Ref. 97, pp. 503–46
71. Gerster, R., Dimon, B., Tournier, P., Peybernes, A. 1977. Oxygen-18 as a tool for studying photorespiration: oxygen uptake and incorporation into glycolate, glycine and serine. In *Stable Isotopes in the Life Sciences,* pp. 293–301. Vienna: Int. Atomic Energy Assoc. Publ. 442 (In French)
72. Gerster, R., Tournier, P. 1977. Metabolic pathway of oxygen during photorespiration incorporation of ^{18}O into glycolate. *Abstr. 4th Int. Congr. Photosynth.,* pp. 129–30
73. Gibbs, M., Kandler, O. 1957. Assymetric distribution of C-14 in sugar formed during photosynthesis. *Proc. Natl. Acad. Sci. USA* 43:441–51
74. Gibbs, M., Latzko, E., eds. 1979. Photosynthesis II: Photosynthetic carbon metabolism and related processes. In *Encyclopedia of Plant Physiology,* New Ser., Vol. 6. Berlin: Springer. 578 pp.
75. Goldsworthy, A. 1970. Photorespiration. *Bot. Rev.* 36:321–40
76. Hall, A. E., Björkman, O. 1975. Model of leaf photosynthesis and respiration. In *Ecological Studies,* ed. D. M. Gates, R. B. Schmerl. Berlin: Springer
77. Halliwell, B. 1978. The chloroplast at work. *Prog. Biophys. Mol. Biol.* 33:1–54
78. Hamilton, G. A. 1974. Chemical models and mechanisms for oxygenases. In *Molecular Mechanisms of Oxygen Activation,* ed. O. Hayaishi, pp. 405–51. New York: Academic
79. Harris, G. C., Stern, A. I. 1978. Stoichiometry of the ribulose-1,5-bisphosphate oxygenase reaction. *J. Exp. Bot.* 29:561–66
80. Hartman, F. C., Fraij, B., Norton, I. L., Stringer, C. D. 1980. Ribulose bisphosphate carboxylase/oxygenase: Active-site characterization and partial sequence determination. *Proc. 5th Int. Congr. Photosynth.* In press
81. Hartman, F. C., Norton, I. L., Stringer, C. D., Schloss, J. V. 1978. Attempts to apply affinity labeling techniques to ribulose bisphosphate carboxylase/oxygenase. See Ref. 178, pp. 245–69
82. Hatch, A. L., Jensen, R. G. 1980. Regulation of ribulose-1,5-bisphosphate carboxylase from tobacco: Changes in pH response and affinity for CO_2 and Mg^{2+} induced by chloroplast intermediates. *Arch. Biochem. Biophys.* In press
83. Hatch, M. D., Slack, C. R. 1966. Photosynthesis by sugar-cane leaves. A new carboxylation reaction and the pathway of sugar formation. *Biochem. J.* 106: 103–11
84. Heber, U., Krause, G. H. 1980. What is the physiological role of photorespiration. *Trends Biochem. Sci.* 5:32–35
85. Hoch, G., Kok, B. 1963. A mass spectrometer inlet system for sampling gases dissolved in liquid phases. *Arch. Biochem. Biophys.* 101:160–70
86. Hsu, T.-C., Kuehn, G. D. 1978. Active-site directed, paracatalytic modification of ribulose 1,5-bisphosphate carboxylase-oxygenase by activation of the carbanionic enzyme-substrate intermediate with the oxidant hexocyanoferrate. *Fed. Proc.* 37:1511

87. Hurwitz, J., Jakoby, W. B., Horecker, B. L. 1956. On the mechanism of CO_2 fixation leading to phosphoglyceric acid. *Biochim. Biophys. Acta* 22:194–95
88. Jackson, W. A., Volk, R. J. 1970. Photorespiration. *Ann. Rev. Plant Physiol.* 21:385–432
89. Jakoby, W. B., Brummond, D. O., Ochoa, S. 1956. Formation of 3-phosphoglyceric acid by carbon dioxide fixation with spinach leaf enzymes. *J. Biol. Chem.* 218:811–22
90. Jensen, R. G., Bahr, J. T. 1977. Ribulose 1,5-bisphosphate carboxylase-oxygenase. *Ann. Rev. Plant Physiol.* 28:379–400
91. Johal, S., Bourque, D. P., Smith, W. W., Suh, S. W., Eisenberg, D. 1980. Crystallization and characterization of ribulose 1,5-bisphosphate carboxylase/oxygenase from eight plant species. *J. Biol. Chem.* 255:8873–80
92. Jolliffe, P. A., Tregunna, E. B. 1973. Environmental regulation of the oxygen effect on apparent photosynthesis in wheat. *Can. J. Bot.* 51:841–52
93. Jordan, D. B., Ogren, W. L. 1980. A sensitive assay procedure for simultaneous determination of ribulose 1,5-bisphosphate carboxylase and oxygenase activities. *Plant Physiol.* In press
94. Keys, A. J. 1980. Synthesis and interconversion of glycine and serine. In *The Biochemistry of Plants,* ed. B. J. Miflin, 5:359–74. New York: Academic
95. King, W. R., Andersen, K. 1980. Efficiency of CO_2 fixation in a glycolate oxidase mutant of *Alcaligenes eutrophus* which exports fixed carbon as glycolate. *Arch. Microbiol.* 128:84–90
96. Kirk, M., Heber, U. 1976. Rates of synthesis and source of glycolate by intact chloroplasts. *Planta* 132:131–41
97. Kochi, J. K., ed. 1973. *Free Radicals.* New York: Wiley. 713 pp.
98. Koenig, T., Fischer, H. 1973. "Cage" effects. See Ref. 97, pp. 157–190
99. Kortschak, H. P., Hartt, C. E., Burr, G. O. 1965. Carbon dioxide fixation in sugar cane leaves. *Plant Physiol.* 40:209–13
100. Kosman, D. J. 1978. Carbanions as substrates in biological oxidation reactions. In *Bioorganic Chemistry,* ed. E. E. van Tamelen, 2:175–95, New York: Academic. 371 pp.
101. Kosman, D. J., Ettinger, M. J., Bereman, R. D., Giordano, R. S. 1977. Role of tryptophan in the spectral and catalytic properties of the copper enzyme, galactose oxidase. *Biochemistry* 16: 1597–1601 (footnote 3)
102. Kowallik, W. 1982. Photocontrol of respiration. *Ann. Rev. Plant Physiol.* 33
103. Krause, G. H., Lorimer, G. H., Heber, U., Kirk, M. 1978. Photorespiratory energy dissipation in leaves and chloroplasts. See Ref. 1, pp. 299–310
104. Krause, G. H., Thorne, S. W., Lorimer, G. H. 1977. Glycolate synthesis by intact chloroplasts. *Arch. Biochem. Biophys.* 183:471–79
105. Ku, S., Edwards, G. E. 1977. Oxygen inhibition of photosynthesis. *Plant Physiol.* 59:986–90
106. Ku, S., Edwards, G. E. 1977. Oxygen inhibition of photosynthesis. *Plant Physiol.* 59:991–99
107. Laing, W. A., Christeller, J. T. 1976. A model for the kinetics of activation and catalysis of ribulose-1,5-bisphosphate carboxylase. *Biochem. J.* 159:563–70
108. Laing, W. A., Christeller, J. T. 1980. A steady state kinetic study on the catalytic mechanism of ribulose bisphosphate carboxylase from soybean. *Arch. Biochem. Biophys.* 202:592–600
109. Laing, W. A., Ogren, W. L., Hageman, R. H. 1975. Bicarbonate stabilization of ribulose-1,5-bisphosphate carboxylase. *Biochemistry* 14:2269–75
110. Laing, W. A., Ogren, W. L., Hageman, R. H. 1974. Regulation of soybean net photosynthetic CO_2 fixation by the interaction of CO_2, O_2 and ribulose 1,5-diphosphate carboxylase. *Plant Physiol.* 54:678–85
111. Laisk, A., Oja, V. 1972. A mathematical model of photosynthesis and photorespiration. II. Experimental verification. In *Theoretical Foundations of Photosynthetic Productivity,* pp. 362–68. Moscow: Nauka
112. Larsson, C. 1974. Photosynthetic glycolate formation via phosphoglycolate in isolated chloroplasts. *Proc. 3rd Int. Congr. Photosynth.,* ed. M. Avron, pp. 1321–28. Amsterdam: Elsevier
113. Lendzian, K. J. 1978. Activation of ribulose-1,5-bisphosphate carboxylase by chloroplast metabolites in a reconstituted spinach chloroplast system. *Planta* 143:291–96
114. Lilley, R. McC., Walker, D. A. 1975. Carbon dioxide assimilation by leaves, isolated chloroplasts, and ribulose bisphosphate carboxylase from spinach. *Plant Physiol.* 55:1087–92
115. Lorimer, G. H. 1978. Retention of the oxygen atoms at carbon-2 and carbon-3 during carboxylation of ribulose 1,5-bisphosphate. *Eur. J. Biochem.* 89:43–50
116. Lorimer, G. H. 1979. Evidence for the existence of discrete activator and sub-

strate sites for CO_2 on ribulose-1,5-bisphosphate carboxylase. *J. Biol. Chem.* 254:5599–5601

117. Lorimer, G. H. 1981. Ribulose bisphosphate carboxylase: Amino acid sequence of a peptide bearing the activator carbon dioxide. *Biochemistry.* In press
118. Lorimer, G. H., Andrews, T. J. 1973. Plant photorespiration—an inevitable consequence of the existence of atmospheric oxygen. *Nature* 248:359–60
119. Lorimer, G. H., Andrews, T. J. 1980. The C-2 photo- and chemo-respiratory carbon oxidation cycle. In *The Biochemistry of Plants,* ed. M. D. Hatch, N. K. Boardman, 8:329–74. New York: Academic. In press
120. Lorimer, G. H., Andrews, T. J., Tolbert, N. E. 1973. Ribulose diphosphate oxygenase. II. Further proof of reaction products and mechanism of action. *Biochemistry* 12:18–23
121. Lorimer, G. H., Badger, M. R., Andrews, T. J. 1976. The activation of ribulose-1,5-bisphosphate carboxylase by carbon dioxide and magnesium ions. Equilibria, kinetics, a suggested mechanism and physiological implications. *Biochemistry* 15:529–36
122. Lorimer, G. H., Badger, M. R., Andrews, T. J. 1977. D-ribulose-1,5-bisphosphate carboxylase-oxygenase. Improved methods for activation and assay of catalytic activities. *Anal. Biochem.* 78:66–75
123. Lorimer, G. H., Badger, M. R., Heldt, H. W. 1978. The activation of ribulose-1,5-bisphosphate carboxylase/oxygenase. See Ref. 178, pp. 283–306
124. Lorimer, G. H., Miziorko, H. M. 1980. RuBP carboxylase: The mechanism of activation and its relation to catalysis. *Proc. 5th Int. Congr. Photosynth.* In press
125. Lorimer, G. H., Miziorko, H. M. 1980. Carbamate formation on the ε-amino group of a lysyl residue as the basis for the activation of ribulosebisphosphate carboxylase by CO_2 and Mg^{2+}. *Biochemistry* 19:5321–28
126. Lorimer, G. H., Osmond, C. B., Akazawa, T., Asami, S. 1978. On the mechanism of glycolate synthesis by *Chromatium. Arch. Biochem. Biophys.* 185:49–56
127. Lorimer, G. H., Woo, K. C., Berry, J. A., Osmond, C. B. 1978. The C_2 photorespiratory carbon oxidation cycle in leaves of higher plants: pathway and consequences. See Ref. 1, pp. 311–22
128. Martin, P. G. 1979. Amino acid sequence of the small subunit of ribulose-1,5-bisphosphate carboxylase from spinach. *Aust. J. Plant Physiol.* 6:401–8
129. McCapra, F. 1976. Chemical mechanisms of bioluminescence. *Acc. Chem. Res.* 9:201–8
130. McCurry, S. D., Pierce, J., Tolbert, N. E., Orme-Johnson, W. H. 1981. Effect or molecules stimulate ribulose bisphosphate carboxylase through interaction at the catalytic site. *J. Biol. Chem.* In press
131. McDonough, M. W., Wood, W. A. 1961. The mechanism of pentose phosphate isomerization and epimerization studied with T_2O and $H_2{}^{18}O$. *J. Biol. Chem.* 236:1220–24
132. McFadden, B. A. 1974. The oxygenase activity of ribulose diphosphate carboxylase from *Rhodospirillum rubrum. Biochem. Biophys. Res. Commun.* 60:312–17
133. McFadden, B. A. 1980. A perspective of ribulose bisphosphate carboxylase/oxygenase, the key catalyst in photosynthesis and photorespiration. *Acc. Chem. Res.* 13:394–99
134. McFadden, B. A., Lord, J. M., Rowe, A., Dilks, S. 1975. Composition, quaternary structure and catalytic properties of D-ribulose-1,5-bisphosphate carboxylase from *Euglena gracilis. Eur. J. Biochem.* 54:195–206
135. McIntosh, L., Poulsen, C., Bogorad, L. 1980. The DNA sequence of the gene encoding the large subunit of maize ribulose bisphosphate carboxylase. *Nature* 288:556–60
136. McNeil, P., Foyer, C. 1980. Similarity of ribulose 1,5-bisphosphate carboxylase of isogenic diploid and teraploid ryegrass (*Lolium perenne*). *ARC Res. Group Photosynth., Ann. Rep.* 1:48–52
137. Miziorko, H. M. 1979. Ribulose-1,5-bisphosphate carboxylase. Evidence in support of the existence of discrete CO_2 activator and CO_2 substrate sites. *J. Biol. Chem.* 254:270–72
138. Miziorko, H. M., Mildvan, A. S. 1974. Electron paramagnetic resonance, ^1H and ^{13}C nuclear magnetic resonance studies of the interaction of manganese and bicarbonate with ribulose 1,5-bisphosphate carboxylase. *J. Biol. Chem.* 249:2743–50
139. Miziorko, H. M., Sealy, R. C. 1980. Characterization of ribulose bisphosphate carboxylase-carbon dioxide-divalent cation-carboxypentitol bisphosphate complex. *Biochemistry* 19: 1167–72

140. Mullhofer, G., Rose, I. A. 1965. The position of carbon-carbon bond cleavage in the ribulose diphosphate carboxydismutase reaction. *J. Biol. Chem.* 240:1341–46
141. Nelson, S. F., Bartlett, P. D. 1966. Azocumene. II. Cage effects and the question of spin coupling in radical pairs. *J. Am. Chem. Soc.* 88:143–49
142. Ogren, W. L., Bowes, G. 1971. Ribulose diphosphate carboxylase regulates soybean photorespiration. *Nature New Biol.* 230:159–60
143. O'Leary, M. H., Jaworski, R. J., Hartman, F. C. 1979. ^{13}C Nuclear magnetic resonance of the CO_2 activation of ribulosebisphosphate carboxylase from *Rhodospirillum rubrum. Proc. Natl. Acad. Sci. USA* 76:673–75
144. Paech, C., Tolbert, N. E. 1978. Active site studies of ribulose-1,5-bisphosphate carboxylase/oxygenase with pyridoxal 5'-phosphate. *J. Biol. Chem.* 253: 7864–73
145. Pähler, A., Saenger, W., Bowien, B. 1979. Crystallization of D-ribulose-1,5-bisphosphate carboxylase. *Acta Crystallogr. A* 34:S55
146. Peisker, M. 1974. A model describing the influence of oxygen on photosynthetic carboxylation. *Photosynthetica* 8:47–50
147. Peisker, M., Apel, P. 1976. Influence of oxygen on photosynthesis and photorespiration in leaves of *Triticum aestivium* L. *Photosynthetica* 10:140–48
148. Peisker, M., Apel, P. 1977. Influence of oxygen on photosynthesis and photorespiration in leaves of *Triticum aestivium* L. *Photosynthetica* 11:29–37
149. Peterkofsky, A., Racker, E. 1961. The reductive pentose phosphate cycle. III. Enzyme activities in cell-free extracts of photosynthetic organisms. *Plant Physiol.* 36:409–14
150. Peterson, R. B. 1980. Evidence for a non-Calvin cycle precursor of glycolate in photosynthesizing tobacco leaf tissue and inhibition of photorespiration by glyoxylate. *Plant Physiol. Suppl.* 65: Absr. No. 351
151. Pierce, J., Tolbert, N. E., Barker, R. 1980. The position of carbon-carbon bond cleavage in the reactions of ribulosebisphosphate carboxylase/oxygenase and the retention of the oxygen at C-2 of D-ribulose 1,5-bisphosphate in the oxygenase reaction. *J. Biol. Chem.* 255:509–11
152. Pierce, J., Tolbert, N. E., Barker, R. 1980. Interaction of ribulose bisphosphate carboxylase/oxygenase with transition state analogues. *Biochemistry* 19:934–42
153. Plamondon, J. E., Bassham, J. A. 1966. Glycolic acid labeling during photosynthesis with $^{14}CO_2$ and tritiated water. *Plant Physiol.* 41:1272–75
154. Pon, N. G., Rabin, B. R., Calvin, M. 1963. Mechanism of the carboxydismutase reaction. I. The effect of preliminary incubation of substrates, metal ion and enzyme on activity. *Biochem. Z.* 338:7–19
155. Poulsen, C., Martin, B., Svendsen, I. 1979. Partial amino acid sequence of the large subunit of ribulosebisphosphate carboxylase from barley. *Carlsberg Res. Commun.* 44:191–99
156. Quandt, L., Gottschalk, G., Ziegler, H., Stichler, W. 1977. Isotope discrimination by photosynthetic bacteria. *FEMS Microbiol. Lett.* 1:125–28
157. Rabin, B. R., Trown, P. W. 1964. Inhibition of carboxydismutase by iodioacetamide. *Proc. Natl. Acad. Sci. USA* 51:497–501
158. Rabin, B. R., Trown, P. W. 1964. Mechanism of action of carboxydismutase. *Nature* 202:1290–93
159. Randall, D. D., Nelson, C. J., Asay, K. H. 1977. Ribulose bisphosphate carboxylase. Altered genetic expression in tall fescue. *Plant Physiol.* 59:38–41
160. Rathnam, C. K. M., Chollet, R. 1980. Regulation of photorespiration. *Curr. Adv. Plant. Sci.* In press
161. Richardson, K. E., Tolbert, N. E. 1961. Phosphoglycolic acid phosphatase. *J. Biol. Chem.* 236:1285–90
162. Robinson, J. M., Gibbs, M. 1974. Photosynthetic intermediates, the Warburg effect and glycolate synthesis in isolated spinach chloroplasts. *Plant Physiol.* 53:790–97
163. Robinson, J. M., Gibbs, M., Cotler, D. N. 1977. Influence of pH upon the Warburg effect in isolated intact spinach chloroplasts. *Plant Physiol.* 59:530–34
164. Robinson, S. P., McNeil, P. H., Walker, D. A. 1979. Ribulose bisphosphate carboxylase-lack of dark inactivation of the enzyme in experiments with protoplasts. *FEBS Lett.* 97:296–300
165. Robison, P. D., Martin, M. N., Tabita, F. R. 1979. Differential effects of metal ions on *Rhodospirillum rubrum* ribulose bisphosphate carboxylase/oxygenase and stoichiometric incorporation of HCO_3 into a cobalt (III)-enzyme complex. *Biochemistry* 18:4453–58
166. Robison, P. D., Whitman, W. B., Waddill, F., Riggs, A. F., Tabita, F. R. 1980. Isolation and sequence of the pyridoxal

5'-phosphate active site peptide from *Rhodospirillum rubrum* ribulose-1,5-bisphosphate carboxylase/oxygenase. *Biochemistry* 19:4848–53

167. Russell, G. A., Bemis, A. G., Geels, E. J., Janzen, E. G., Moye, A. J. 1967. Oxidation of carbanions. *Adv. Chem. Ser.* 75:174–201
168. Ryan, F. J., Jolly, S. O., Tolbert, N. E. 1974. Ribulose diphosphate oxygenase V. Presence in ribulose diphosphate carboxylase from *Rhodospirillum rubrum*. *Biochem. Biophys. Res. Commun.* 59:1233–41
169. Ryan, F. J., Tolbert, N. E. 1975. Ribulose diphosphate carboxylase/oxygenase. IV. Regulation by phosphate esters. *J. Biol. Chem.* 250:4234–38
170. Salujah, A. K., McFadden, B. A. 1978. Inhibition of ribulose bisphosphate carboxylase/oxygenase by sedoheptulose 1,7-bisphosphate. *FEBS Lett.* 96: 361–63
171. Schloss, J. V., Phares, E. F., Long, M. V., Norton, I. L., Stringer, C. D., Hartman, F. C. 1979. Isolation, characterization and crystallization of ribulosebisphosphate carboxylase from autotrophically grown *Rhodospirillum rubrum*. *J. Bacteriol.* 137:490–501
172. Schloss, J. V., Stringer, C. D., Hartman, F. C. 1978. Identification of essential lysine and cysteinyl residues in spinach ribulosebisphosphate carboxylase/oxygenase modified by the affinity label N-bromoacetylethanolamine phosphate. *J. Biol. Chem.* 253:5705–11
173. Schnarrenberger, C., Fock, H. 1976. Interactions among organelles involved in photorespiration. In *Transport in Plants. III. Intracellular interactions and transport processes*, ed. C. R. Stocking, U. Heber, pp. 185–223. *Encyclopedia of Plant Physiology*, New Ser. Vol. 3. Berlin: Springer 517 pp.
174. Schou, L., Benson, A. A., Bassham, J. A., Calvin, M. 1950. The path of carbon in photosynthesis. XI. The role of glycolic acid. *Physiol. Plant.* 3:487–95
175. Servaites, J. C., Ogren, W. L. 1977. pH dependence of photosynthesis and photorespiration in soybean leaf cells. *Plant Physiol.* 60:693–96
176. Servaites, J. C., Ogren, W. L. 1978. Oxygen inhibition of photosynthesis and stimulation of photorespiration in soybean leaf cells. *Plant Physiol.* 61:62–67
177. Siegel, M., Lane, M. D. 1973. Chemical and enzymatic evidence for the participation of a 2-carboxy-3-ketoribitol-2,5-bisphosphate intermediate in the carboxylation of ribulose 1,5-bisphosphate. *J. Biol. Chem.* 248:5486–98
178. Siegelman, H. W., Hind, G., eds. 1978. *Photosynthetic Carbon Assimilation*. New York: Plenum. 445 pp.
179. Simon, H., Dorrer, H. D., Trebst, A. 1964. Photosynthese-versuche in Tritiumwasser mit *Chlorella*. *Z. Naturforsch. Teil B* 19:734–44
180. Singh, M., Ogren, W. L., Widholm, J. M. 1974. Photosynthetic characteristics of several C_3 and C_4 plant species grown under different light intensities. *Crop. Sci.* 14:563–66
181. Sirevag, R., Buchanan, B. B., Berry, J. A., Troughton, J. H. 1977. Mechanisms of CO_2 fixation in bacterial photosynthesis studied by the carbon isotope fractionation technique. *Arch. Microbiol.* 112:35–38
182. Somerville, C., Ogren, W. L. 1979. A phosphoglycolate phosphatase mutant of *Arabidopsis*. *Nature* 280:833–35
183. Spellman, M., Tolbert, N. E., Hartman, F. C. 1979. Isolation of a peptide from the active-site of ribulose bisphosphate carboxylase/oxygenase. *Abstr. 178th Natl. Meet. Am. Chem. Soc., Washington DC,* BIOL 3
184. Sue, J. M., Knowles, J. R. 1978. Retention of the oxygens at C-2 and C-3 of D-ribulose 1,5-bisphosphate in the reaction catalyzed by ribulose-1,5-bisphosphate carboxylase. *Biochemistry* 17:4041–44
185. Tabita, F. R., McFadden, B. A. 1972. Regulation of ribulose-1,5-diphosphate carboxylase by 6-phosphogluconate. *Biochem. Biophys. Res. Commun.* 48:1153–58
186. Tabita, F. R., McFadden, B. A. 1974. D-Ribulose-1,5-diphosphate carboxylase from *Rhodospirillum rubrum*. *J. Biol. Chem.* 249:3459–64
187. Tabita, F. R., McFadden, B. A., Pfennig, N. 1974. D-Ribulose-1,5-diphosphate carboxylase from *Chlorobium thisulfatophilum* Tassajara. *Biochim. Biophys. Acta* 341:187–94
188. Takabe, T., Akazawa, T. 1975. Molecular evolution of ribulose-1,5-bisphosphate carboxylase. *Plant Cell. Physiol.* 16:1049–60
189. Tamiya, H., Huzisige, H. 1949. Effect of oxygen on the dark reaction of photosynthesis. *Acta Phytochim.* 15:83–104
190. Tenhunen, J. D., Hesketh, J. D., Gates, D. M. 1980. Leaf photosynthesis models. In *Predicting Photosynthesis for Ecosystem Models*, pp. 123–81. West Palm Beach, Fla: CRC Press

191. Tenhunen, J. D., Hesketh, J. D., Harley, P. C. 1980. Modeling C_3 leaf respiration in the light. See Ref. 190, pp. 17–47
192. Tenhunen, J. D., Weber, J. A., Yocum, C. S., Gates, D. M. 1979. Solubility of gases and the temperature dependence of whole leaf affinities for carbon dioxide and oxygen. *Plant Physiol.* 63: 916–23
193. Tolbert, N. E. 1971. Microbodies—peroxisomes and glyoxysomes. *Ann. Rev. Plant Physiol.* 22:45–74
194. Tolbert, N. E. 1980. Photorespiration. In *The Biochemistry of Plants,* ed. D. D. Davies, Vol. 2. New York: Academic. In press
195. Tournier, P., Espinasse, A., Gerster, R. 1978. Decarboxylation par H_2O_2 de ceto-acides en relation avec le metabolisme de la photorespiration. *C. R. Acad. Sci. Ser. D* 287:729–32
196. Troughton, J. H. 1979. ^{13}C as an indicator of carboxylation reactions. See Ref. 74, pp. 140–49
197. Trown, P. W., Rabin, B. R. 1964. The mechanism of action of carboxydismutase. *Proc. Natl. Acad. Sci. USA* 52:88–93
198. Turner, J. S., Brittain, E. G. 1962. Oxygen as a factor in photosynthesis. *Biol. Rev.* 37:130–70
199. Vater, J., Gaudszun, T., Schnarnow, H., Salnikow, J. 1980. Competition of pyridoxal 5'-phosphate with ribulose 1,5-bisphosphate and effector sugar phosphates at reaction centres of the spinach ribulose 1,5-bisphosphate carboxylase/oxygenase. *Z. Naturforsch. Teil C* 35:416–22
200. Vater, J., Salnikow, J. 1979. Identification of two binding sites of D-ribulose 1,5-bisphosphate carboxylase/oxygenase from spinach for D-ribulose 1,5-bisphosphate and effectors of the carboxylation reaction. *Arch. Biochem. Biophys.* 194:190–97
201. von Sonntag, C. 1980. Free-radical reactions of carbohydrates as studied by radiation techniques. *Adv. Carbohydr. Chem. Biochem.* 37:7–77
202. Walker, D. A. 1976. Regulatory mechanisms in photosynthetic carbon metabolism. *Curr. Top. Cell. Regul.* 11:203–41
203. Walsh, C. T. 1979. *Enzymatic Reaction Mechanisms,* pp. 421–22. San Francisco: Freeman
204. Warburg, O. 1920. Über die Geschwindigkeit der photochemischen Kohlensäurezersetung in lebenden Zellen. II. *Biochem. Z.* 100:188–217
205. Wareing, P. F., Khalifa, M. M., Treharne, K. J. 1968. Rate-limiting processes in photosynthesis at saturating light intensities. *Nature* 220:453–57
206. Weissbach, A., Horecker, B. L., Hurwitz, J. 1956. The enzymatic formation of phosphoglyceric acid from ribulose diphosphate and carbon dioxide. *J. Biol. Chem.* 218:795–810
207. Whitman, W. B., Martin, M. N., Tabita, F. R. 1979. Activation and regulation of ribulose bisphosphate carboxylase-oxygenase in the absence of small subunits. *J. Biol. Chem.* 254:10184–89
208. Wildman, S. G. 1979. Aspects of fraction-1 protein evolution. *Arch. Biochem. Biophys.* 196:598–610
209. Wildner, G. F. 1976. The role of ribulose-1,5-bisphosphate carboxylase and its oxygenase activity in the events of photorespiration. *Ber. Dtsch. Bot. Ges.* 89:349–60
210. Wildner, G. F., Henkel, J. 1978. Differential reactivation of ribulose-1,5-bisphosphate oxygenase with low carboxylase activity by Mn^{2+}. *FEBS Lett.* 91:99–103
211. Wildner, G. F., Henkel, J. 1979. The effect of divalent metal ions on the activity of Mg^{2+} depleted ribulose-1,5-bisphosphate carboxylase-oxygenase. *Planta* 146:223–28
212. Wilson, A. T., Calvin, M. 1955. The photosynthetic cycle. CO_2 dependent transients. *J. Am. Chem. Soc.* 77: 5948–57
213. Wishnick, M., Lane, M. D., Scrutton, M. C., Mildvan, A. S. 1969. The presence of tightly bound copper in ribulose diphosphate carboxylase from spinach. *J. Biol. Chem.* 244:5761–63
214. Wolfenden, R. 1972. Analog approaches to the structure of transition states in enzyme reactions. *Acc. Chem. Res.* 5:10–18
215. Wong, W. W., Benedict, C. R., Kohel, R. J. 1979. Enzymatic fractionation of the stable carbon isotopes of carbon dioxide by ribulose-1,5-bisphosphate carboxylase. *Plant Physiol.* 63:852–56
216. Zelitch, I. 1971. *Photosynthesis, Photorespiration and Plant Productivity.* New York: Academic. 347 pp.
217. Zelitch, I. 1976. Biochemical and genetic control of photorespiration. See Ref. 35, pp. 343–58
218. Zelitch, I. 1980. Regulation of photorespiration. *Proc. 5th Int. Congr. Photosynth.* In press

Ann. Rev. Plant Physiol. 1981. 32:385–406

CELL WALL TURNOVER IN PLANT DEVELOPMENT

◆7717

John M. Labavitch

Pomology Department, University of California, Davis, California 95616

CONTENTS

INTRODUCTION

A rigid, polysaccharide-rich cell wall generally surrounds each plant cell. This wall provides physical support while the cell is growing, and after differentiation is over, it may offer protection against invasion by plant pathogens (5). The strength of this wall, which makes it structurally essential to plant tissues, constitutes a physical barrier to the cell elongation and shape changes which are a requirement for differentiation. Unlike the arthropod which periodically sheds its undersized, but otherwise intact, exoskeleton, the plant cell must carry its "exoskeleton" for the duration of its existence. As a result, the differentiating plant cell must modify the structural integrity of its wall while maintaining, to a varying degree, the wall's physical presence.

Many plants carry out developmentally related turnover of their walls. The turnover of proteins in biological systems is thought of as a steady-state

0066-4294/81/0601-0385$01.00

situation in which degradation of protein is accompanied by synthesis (129). Cell wall turnover events fall into two fairly distinct classes. There are situations in which only a transient weakening of the wall is required for development. Synthesis and wall degradation are integrated in these situations. In contrast, some instances of wall turnover precede cell or tissue senescence. There is wholesale dissolution of cell walls during fruit ripening (18, 105) and mobilization of endosperm reserves as seeds germinate (58, 134). Whether wall synthesis occurs, and what its role might be in these situations, is unclear.

While cell wall turnover could occur through a variety of chemical modifications of wall components (74), the most thoroughly researched instances of turnover appear to involve the cleavage (hydrolysis) of wall polysaccharides. Wall polymer breakdown has been associated with cell division in bacteria (130), with growth (139) and the differentiation of fruiting structures (147) in fungi, and with growth and cell division in yeast (131). A variety of developmental events in higher plants apparently involves wall hydrolysis. These include xylem differentiation (102), the abscission of various plant organs (2), and pollen germination and pollen tube growth (72, 122), as well as the subjects of this review: vegetative cell elongation, fruit ripening, and storage tissue breakdown during seed germination.

With a few exceptions, the suggestion that cell wall turnover has occurred in relation to a developmental event has been based on the measurement of a decrease in some wall monosaccharide component or the detection of carbohydrate-degrading enzyme(s) in tissue extracts. This kind of evidence of turnover is at best incomplete and may be misleading if the aim of research is to explain the control of cell wall metabolism.

Over the years carbohydrate chemists have provided structural descriptions of a variety of wall polysaccharide types. Recent concentrated efforts using improved analytical approaches have led to the publication of cell wall models which describe polymer structures and the interpolymer relationships in the intact primary cell wall of dicots (62, 89, 92). Although these models do not completely account for the structural positions and roles of cell wall components, they do provide a framework by which some data on wall turnover can be assessed (3, 76, 134). As more data are generated to extend the models to cell walls as they occur in plant tissues, the models take on more importance. In spite of the present shortcomings, one point is clear. Each of the cell wall monosaccharide types can be found in a variety of polysaccharide species and in different glycosidic linkages. Thus data which indicate, for example, that glucosyl residues are lost from the *Avena* coleoptile wall as tissue growth is promoted by indoleacetic acid (79) is incomplete. If we are to suggest a mechanism by which glucose is removed from the cell wall during growth, we must know if the missing glucose was in a homo- or heteropolysaccharide and in what way the glucose residues were linked to one another. Until recently wall turnover research has not included sufficient structural information.

The inference of wall turnover based on measurement of carbohydrate-degrading enzymes in tissue extracts can be misleading, especially in the absence of cell wall data describing specific cell wall changes. The β-1,3-glucanases are generally assayed by incubating extracts with algal β-1,3-glucan laminarin. Such glucanases are found in many tissues (36, 47, 106), yet clear evidence for structural β-1,3-glucans in the walls from these tissues is lacking. If these proteins play a role in cell wall turnover, they must be hydrolizing another polysaccharide type. Cellulases (β-1,4-glucanases) are often found in plant extracts (9, 31, 48, 106). Because these glucanases can degrade accessible β-1,4-glucans of cellulose as well as two distinct cell wall hemicelluloses [xyloglucan (17) and β-1,4-xylan (59)], it would be erroneous to automatically assign a specific role in turnover to cellulase. Glycosidases are enzymes which hydrolyze oligosaccharides to monomers. For convenience glycosidase activities in plant extracts are routinely assayed by incubation with artificial, *p*-nitrophenyl glycoside substrates. Pharr et al (104) have shown tomato β-glucosidase to be more active on the artificial substrate than on natural β-glucosides. Purified glycosidases from animal (23) and plant tissues (32) show activity on artificial substrates containing a multiplicity of sugar moieties. Because many different glycosidase activities are present in tissues where wall catabolism is occurring (57, 106), one must be careful to determine which, if any, of these enzymes is playing a role.

Ideally, descriptions of development-related cell wall turnover will contain compositional and structural descriptions of cell wall change and measurements of enzyme activity against those wall polymers which are metabolized. Fortunately, work in the last 10 years on wall turnover in a limited number of systems comes very close to satisfying that ideal. The remainder of this review will discuss these efforts.

WALL TURNOVER AND CELL ELONGATION

Plant cell growth is the result of turgor-dependent extension of the cell wall. Heyn (45, 46) demonstrated early that one of the primary growth-promoting actions of auxin was a weakening of the wall which allowed it to stretch. Subsequent studies (25, 84, 154, 156), including involved testing and mathematical modeling of the rheological properties of cell walls from hormone-treated tissues, have confirmed Heyn's observations. The parallel suggestion that auxin promotes tissue water uptake, thus increasing turgor pressure and hence cell growth, is apparently not justified (103).

The idea that weakened cell walls were the consequence of turnover of stress-bearing wall components followed. This suggestion led to a number of reports of decreases in cell wall sugars, generally glucose, accompanying growth of dicots (82, 87) and monocots (60, 78, 143). Initially structural information about "turned-over" components was lacking. These observations were followed by the identification of polysaccharide-degrading enzymes in extracts of growing tissues (31, 36, 45, 153). Most often enzymes were identified as cellulase (31) and β-1,3-glucanase (36, 47), although data indicating purity of these protein preparations or that these were the only carbohydrases present were not provided. Subsequently, Tanimoto & Masuda (136) and Katz & Ordin (61) reported auxin-related promotions in tissue glucanase activity that roughly paralleled elongation. Addition of fungal β-1,3-glucanase (of unknown purity) to the medium bathing oat coleoptile segments induced a decrease in cell wall tensile strength and a transient increase in growth rate (143). Although Ruesink (123) was unable to duplicate the latter result, these efforts led to the proposal that auxin promoted growth by enhancing the activity of glucanases which hydrolyzed glucans and weakened the cell wall.

Physiological and Biochemical Considerations

The observation that increased RNA and protein synthesis accompanied hormone-promoted tissue elongation led to the proposal that enhancement of de novo synthesis of hypothetical wall-loosening enzymes was the primary action of auxin (27). That this is probably not the case is discussed elsewhere (108); suffice it to say that the short latent period between presentation of IAA and a growth response (13) makes it unlikely that new proteins are responsible for accelerated growth. Nevertheless, continued synthesis of as yet unidentified, short-lived proteins is required for the continuation of enhanced growth rates (63).

A possible role for growth-limiting proteins is suggested by the observation that IAA promotes the synthesis of cell wall polysaccharides in monocots (10, 11) and dicots (1). The autoradiographic study of Ray (107) has shown that wall deposition in growing cells occurs throughout the wall's thickness although cellulose incorporation occurs chiefly at its inner surface. This wall synthesis could play a role in growth promotion. The introduction of new polymers could increase wall surface area, and Maclachlan (81) has proposed that cellulose synthesis could proceed via the intercalation of new β-1,4-glucan segments into preexisting cellulose chains. Either of these operations could confer an increment of growth potential on the wall. While promotion of extension could occur via wall synthesis, kinetic studies utilizing radioactive cell wall precursors have shown that hormonal promotion of wall synthesis (10, 11) follows increases in growth rate. Baker

& Ray (10, 11, 109), studying IAA-promoted incorporation of labeled precursors into *Avena* coleoptile walls and the effects of Ca^{2+} and other inhibitors of elongation on that incorporation, demonstrated that some of the hormone's effect on incorporation was not dependent on cell growth. They therefore argued that a small component of total wall synthesis ["intensile synthesis" (11)] could precede auxin effects on growth and thus be responsible for growth promotion. To date no such component of polymer synthesis has been identified. Nevertheless, because IAA-promoted growth is accompanied by the synthesis of wall material which is deposited largely at the wall-plasmalemma interface (and, therefore, probably becoming the stress-bearing, extension-controlling portion of the wall; see below), wall synthesis must be understood as an integral part of growth-related wall metabolism.

Turgor pressure constitutes the driving force for cell extension and apparently plays an important role in cell wall metabolism. When tissue segments are bathed in solutions containing turgor-reducing concentrations of mannitol, auxin-promoted cell wall synthesis is not seen (11). Under these conditions IAA-induced growth is blocked while hormone-mediated cell wall weakening continues (26), and the recovery of "lost" growth upon removal of mannitol inhibition has been reported for oat coleoptile segments. (28).

Although enzymes have been presumed to participate in cell wall weakening (85), how auxin might promote enzyme activity has remained unclear. Oat coleoptile (110) and pea stem (55, 83) segments reduce the pH of the bathing medium when treated with IAA. Medium acidification is inhibited by treatments which inhibit auxin-induced growth (110, 112). Furthermore, decreasing the medium pH generates an almost instantaneous increase in *Avena* coleoptile elongation rate (112). Thus Rayle & Cleland (112) and Hager et al (40) have proposed that the primary growth-promoting action of auxin is the triggering of hydrogen ion secretion into the tissue free space. The action of hydrogen ions is presumed to be directly on unidentified, acid-labile wall bonds or on wall-degrading enzymes with acidic pH optima (111). In any case, the minimum stress-relaxation time [T_o, an indicator of the strength of wall polymer interactions (154, 156)] for walls from an assortment of coleoptile tissues is reduced in the presence of acidic buffers (84, 127). T_o is similarly reduced when these coleoptiles are treated with auxin (126, 156, 157). The relationship of proton secretion to cell extension is currently undergoing intensive investigation.

If any instance of wall turnover is to be identified as a wall-weakening, growth-producing event, it should comply with the kinetic and other criteria described above. The work of Vanderhoef & Stahl (141) suggests one way to evaluate whether or not an auxin-mediated event is primary with regard to growth promotion. When the elongation rate of IAA-treated soybean hypocotyl segments is plotted against time, a biphasic curve is generated. The first rise begins 12 min after auxin is presented; the second rise is superimposed on the first, beginning after 30 min. Cytokinin added with IAA eliminates long-term auxin enhancement of growth yet does not prevent the first rise and fall in growth rate. Presumably the mechanisms involved in initiating more rapid growth operate in the presence of cytokinin, but there is a failure of the mechanisms responsible for maintaining the higher growth rate. If this presumption is correct, combinations of auxin and cytokinin can be used to test whether auxin effects on wall metabolism, or for that matter, hydrogen ion secretion and wall synthesis, are of primary importance to the stimulation of growth.

THE STRESS-BEARING PORTION OF THE CELL WALL The giant cells of the alga *Nitella* are often used in studies of the relationship of cell wall parameters and growth. Their size makes them especially well suited to these studies (measurements can be made on individual cells), and there are no complications arising from the presence of a middle lamella or neighboring cells. The cell environment can be changed rapidly to meet a variety of experimental regimes. In 1958, Green (37) showed that the alga added new cell wall material largely at the wall's inner surface and that increments of growth were uniformly distributed along the cell's length. As for higher plant cells, hydrogen ions promote *Nitella* growth (90, 91). Recently Richmond et al (121) have followed the orientation of cellulose microfibrils throughout the wall and at its inner surface and concluded that the innermost portion of the wall controls growth. Microfibril orientation appears to control growth in higher plants also (33, 132). Fibril orientation in elongating cells is transverse at the inner wall surface and random (or longitudinal) in older portions of the walls of some growing cells (49). Wall deposition in *Avena* is largely at the inner surface of the wall (107) and growth appears to be uniform along the length of the cell (22, 146). While there is no direct evidence that the inner wall of higher plants bears the burden of stress (the complexity of multicellular systems makes analysis difficult), the similarities to the situation in *Nitella* suggest that this must be so.

If the analogy between giant alga cells and cells of higher plants is valid, then the cell wall biochemist should focus attention on the inner wall to detect crucial turnover events. Rapid, hormone-mediated effects on recently formed wall substances must be measured and correlated with elongation. Instances of recent, growth-controlling cell wall events may be quantitatively insignificant against the background of older cell wall material which constitutes a "geologic" record of past development. Nevertheless, good evidence for growth-related turnover and its control exists. The cell walls of monocots and dicots differ in their content of various kinds of polysaccharides (21, 89). Therefore, wall turnover in each will be treated separately.

Turnover of Monocot Cell Wall Components

In 1972, Loescher & Nevins (79) showed an auxin-dependent loss of glucose from the noncellulosic portion of the walls of *Avena* coleoptile segments. The loss was seen beginning 4 hr after auxin presentation when glucose was provided in the medium. However, it could be seen within 1 hr if substrate was not supplied. Presumably, IAA-dependent synthesis of glucan had made detection of glucan turnover difficult; reduced synthesis in the absence of supplied sugar made detection more facile. Subsequently they demonstrated that glucan turnover occurred in the presence of growth-inhibiting (i.e. turgor-reducing) concentrations of mannitol (80).

The sugar analogue nojirimycin (5-amino-5-deoxy-D-glucopyranose) is often used as an inhibitor of carbohydrases (115). Nevins (95) used nojirimycin to test the proposition that glucan-degrading enzymes play a role in auxin-promoted coleoptile growth. After a 30 min lag period, nojirimycin inhibited IAA's effect on both cell elongation and loss of wall glucose. Treatment of coleoptile segments with a glucanase-containing preparation from beans mimicked auxin's promotion of wall glucose change (96).

Masuda and his co-workers extended these results by demonstrating glucan turnover in coleoptiles from a number of grasses (124, 125, 157), and showed a correlation between decreased wall glucose content and the resistance of isolated cell walls to uniaxial extension. Auxin-promoted growth and cell wall rheological properties and glucan content were measured in segments cut from 4-, 5-, and 6-day-old barley coleoptile segments (124). As the coleoptile tissue became older, it contained proportionally less noncellulosic glucose and gave less of a growth response to auxin. Moreover, IAA-promoted shortening of cell wall minimum stress relaxation times decreased in parallel with the decreased growth response. Following growth the average molecular weight of wall glucan decreased (128). From these results, they concluded, as have others (19, 120), that hemicellulosic glucan is the wall substrate whose metabolism permits cell extension; as glucan level falls so does coleoptile growth potential.

Although exogenously supplied glucanases had been shown to promote solubilization of coleoptile wall glucan (96), the nature of endogenous enzymes that might be responsible for this turnover was unclear, in large part because the structure of the wall glucan involved was not known. In 1967, Lee et al (77) had shown that corn coleoptile cell walls, prepared so as to minimize cytoplasmic contamination, would autolyse when incubated in buffer at 37°C. The autolytically solubilized material proved to be a glucan with a mixture of β-1,3- and β-1,4- linkages (64). Mixed linkage glucans are common to the wall hemicellulose of many monocots (19, 51, 97, 98, 155). Huber & Nevins (51) used gel filtration to examine corn wall autolysis products more closely. The fragments of glucan which were solubilized initially had an average molecular weight of 10^5 daltons; with extended incubation the only product was monosaccharide glucose. Walls incubated in the presence of $HgCl_2$ underwent autolysis, but while polymeric glucan was solubilized no monosaccharide was generated. As the time of incubation was extended, this polymeric glucan was reduced in size, finally stabilizing at an average molecular weight of ~10^4 daltons (~60 glucosyl residues). The polymeric glucan was demonstrated to be a mixed-linkage β-glucan (52). When $HgCl_2$ was removed, monomer glucose was generated. From these results the authors suggested that at least two enzyme activities were required for glucan autolysis—an endoglucanase to initiate digestion and a $HgCl_2$-sensitive exoglucanase to generate glucose monomer. Because the endoglucanase was limited in its ability to digest wall glucan, they proposed that its action was limited to specific sites (52).

Huber & Nevins (unpublished) then took advantage of the fact that 3M LiCl removed autolytic capacity from walls. They have identified two glucanases in LiCl extracts of corn coleoptile walls. One is an endoglucanase that is active on corn wall mixed-linkage β-glucan and on the algal β-1,3-glucan, laminarin. It is probably the same enzyme, described as a β-1,3-glucanase, that was shown to increase in activity following auxin treatment of *Avena* coleoptiles (61). The second enzyme is an exo-β-1,3-glucanase which produces monomer glucose when incubated with laminarin. Presumably autolysis also involves action of an exo-β-1,4-glucanase because the ultimate product is glucose and the glucan substrate contains β-1,4- linkages.

Although auxin-treated coleoptiles yield walls that are no more autolytically active than those from untreated coleoptiles (51), it is reasonable to propose that the enzyme systems responsible for autolysis also function in wall glucan turnover. This hypothesis would be especially attractive if the cell wall-bound glucanases of corn coleoptiles have pH optima at the free space pH attained during auxin-induced proton secretion. This hypothesis must be reconciled with the observation that nojirimycin inhibits auxin-induced extension but not growth promoted by low pH (127). It is possible that IAA and protons stimulate growth via different mechanisms. However, nojirimycin inhibits corn exoglucanase (D. J. Huber and D. J. Nevins, unpublished) and $HgCl_2$ inhibits low pH-induced growth of *Avena* coleoptiles (116).

This apparent difference between proton- and auxin-induced growth is but one of the questions that needs resolution before acceptance of a connection between wall glucan turnover, cell wall weakening, and growth promotion can be unequivocal. Does IAA-induced glucan turnover precede the hormone's effect on growth rate? If so, glucan turnover and effects of auxin on wall glucanase activity should occur more rapidly than 1 hr. Does wall synthesis make earlier detection of glucan turnover difficult? Pulse-chase studies, which allow a focus on recently synthesized wall polymers (75), could be of use in fine-tuning the kinetic analysis of glucan turnover. What is the role of wall synthesis in the control of elongation, and what structural role do mixed-linkage glucans play in strengthening the wall? In spite of these questions, the data provide substantial evidence for wall glucan turnover in elongating coleoptiles and for the existence of an enzymic mechanism to carry out that turnover.

Turnover of Dicot Cell Wall Components

The dicot cell wall model of Albersheim and his co-workers provides ever-increasing detail about polysaccharide structure and interpolymer relationships (62, 89). This is extremely valuable in the context of this review in that it provides a structural framework by which the significance of measured turnover events can be assessed. As it is currently formulated, the model, which is based on analysis of the wall of suspension-cultured sycamore cells, describes no role for mixed-linkage β-glucans. This may be an oversight because the *Bacillus subtilis* α-amylase preparation used to remove starch from sycamore wall preparations contains a potent β-glucanase (50). Nevertheless, the literature contains reports of growth-related wall turnover in dicots which gain in significance when interpreted according to the wall model.

Labavitch & Ray (75) used a pulse-chase protocol to study wall turnover in epicotyl segments from dark-grown peas. Wall constituents were labeled during incubation in ^{14}C-glucose. Soluble substrate pools were then flushed by bathing radioactive segments for 3 hr in unlabeled glucose. IAA was then provided and changes in labeled wall sugars (i.e. sugar residues that had been incorporated into the wall a few hours before) were monitored. They detected an auxin-dependent loss of ^{14}C-xylose and ^{14}C-glucose from the wall. Further study (76) showed that the liberated wall xylose and glucose could be recovered in ethanol-precipitable (i.e. polymeric) form from tissue homogenates. This led to the demonstration that the turnover could be measured within 15 min of auxin presentation and continued for as long as enhanced growth rate was in evidence. Furthermore, turnover occurred in the presence of 0.2 M mannitol which inhibited IAA-promoted growth [but presumably not wall weakening (26)], but could not be measured in the presence of concentrations of Ca^{2+} which inhibit both elongation and wall weakening. The authors did not suggest a mechanism by which the turnover was accomplished, although Jacobs & Ray (54) subsequently showed that it occurred in the presence of acidic buffers which stimulated pea segment growth. Structural studies later demonstrated that the wall component involved in this turnover was xyloglucan (35).

Xyloglucans have been subjected to intensive structural analysis and identified as significant hemicellulosic polysaccharides in walls of a variety of dicots (7, 8, 17, 73). The cell wall model described by Keegstra et al (62) places some, but not necessarily all, xyloglucan in what appears to be a key position: hydrogen-bonded to cellulose and covalently linked to pectic polysaccharides. Because its backbone is a β-1,4-glucan, xyloglucan can be degraded by cellulase (17). Thus, wall xyloglucan could be converted to soluble form (i.e. turned over) as a result of hydrogen-bond breakage, cleavage of pectic galactans or arabinans (135), or degradation by cellulase. Cellulase is present in pea stem tissues and increases following auxin treatment (31). Measured increases in extracted cellulase are seen only after 24 hr of IAA (31); however, it is possible that in situ cellulase activity would increase rapidly in response to decreased free space pH. To date there is no evidence for the presence of endogalactanase or endoarabanase in pea tissues. It is possible that xyloglucan is liberated by a nonenzymatic mechanism. However, hydrogen ion concentrations like those found in the tissue-free space following auxin treatment are not sufficient to alter xyloglucan binding to cellulose in vitro (140). Moreover, this turnover cannot be due to loosening of the wall during elongation because xyloglucan solubilization occurs in the presence of mannitol when elongation is inhibited (76).

Important clues about the mechanism by which wall xyloglucan is made soluble could come from structural analysis of the solubilized polymer itself. Is its size different from the average size [7–10 $\times$ 10^3 daltons (17)] of native wall xyloglucan? Does it bear remnants of pectic polymers through which it might have been bonded to the rest of the wall? Studies to date (6) have been inconclusive.

Recently Nishitani & Masuda have examined wall metabolism in azuki bean epicotyls (100). They have shown that relative wall galactose content decreases as one moves further from the apex of the epicotyl. This decrease correlates with a decreasing growth rate and increased wall rigidity (increased minimum stress relaxation time). The growth of intact plants or epicotyl segments supplied with sucrose was accompanied by increases in wall galactose in apical regions. However, wall galactose content of starved segments fell during growth over 20 hr (100). Beginning after 4 hr, this decrease appeared to be accentuated by auxin treatment. As judged by gel filtration, the average molecular weight of galactose-containing polymers appeared to decrease during growth (101). The work on azuki bean measured total cell wall sugars. Labavitch & Ray (75) measured turnover of labeled, recently synthesized wall components and reported an extensive, auxin-*independent* turnover of pea epicotyl cell wall galactose. If the two groups were measuring the same kind of wall metabolism in different dicots, it is reasonable to conclude that galactose turnover in the azuki bean does not involve recently synthesized wall material. This, combined with the lengthy auxin treatment required to bring azuki bean galactose turnover into evidence, suggests that while it is quantitatively significant, galactan turnover probably does not bear directly on cell wall changes that precede growth promotion. Nevertheless, it could be of interest in terms of wall metabolism which permits prolongation of cell growth.

Wong & Maclachlan (148, 149) have purified two pea epicotyl endoglucanases which show in vitro activity on β-1,3-glucans. Glucanase I predominates in apical tissue while glucanase II activity is located more basally. A 2-day auxin treatment promotes the development of glucanase II activity in apical tissues (149). Pea epicotyls stained with aniline blue exhibit fluorescence in the apex (stele tissue) and base (cortical parenchyma). Although it is not a specific agent, aniline blue induces fluorescence of β-1,3-glucans (29). Pretreatment of epicotyl tissues with the appropriate glucanase removed fluorescing material (149). Presumably this is the role of the endogenous glucanases. The authors argue that these enzymes are probably not involved in growth initiation. Nevertheless, their work provides an example of the sort of approach, involving the study of endogenous enzymes and wall components, required for an understanding of wall turnover.

Auxin-promoted turnover of xyloglucan in the pea appears to occur rapidly enough to precede elongation (76); other instances of turnover discussed in this section do not. Kinetic studies of turnover may be hampered by the presence of large amounts of nonresponsive cell wall. In view of the fact that epidermal tissues appear to control IAA-induced growth in peas (86, 138, 151), a focus on turnover in the epidermis may be warranted. Such a study would require clean separation of the epidermis from underlying tissues.

Currently there are no good candidates for enzymes responsible for the wall turnover that has been identified in auxin-responsive pea tissue. Other than cellulase, the carbohydrate-degrading enzymes that have been identified in extracts of dicot tissues are unlikely to play a role in xyloglucan metabolism (36, 94). Terry & Bonner (137) have used low-speed centrifugation to collect a pea tissue free space solution that contains minimal cytoplasmic contamination and have repeated the demonstration of auxin-promoted xyloglucan turnover. Wall-degrading enzymes found in these "cell wall solutions" ought to be good candidates for a role in wall metabolism.

WALL TURNOVER DURING TERMINAL DEVELOPMENT

Seed Germination

During the early stages of seed germination, plant embryonic tissues carry out metabolism at the expense of reserves established during seed development. In some cases wall dissolution has been observed in conjunction with germination and mobilization of stored material.

The primary reserve in seeds of barley is starch. Gibberellic acid (GA_3) treatment of aleurone layers isolated from barley seeds promotes the de novo synthesis of α-amylase which acts to digest starch reserves (24). Taiz & Jones (134) noted that the walls of aleurone cells underwent extensive breakdown following GA_3 treatment and that this dissolution was well under way prior to the onset of α-amylase introduction into the endosperm. They reasoned that wall dissolution facilitated transport between the degenerating endosperm and developing embryonic tissues. Aniline blue treatment of plant tissues followed by ultraviolet irradiation induces fluorescence if β-1,3-glucans are present (29), although a considerable amount of work has indicated that the fluorescence is not a certain indicator of β-1,3-glucans (142). Aniline blue-treated barley aleurone walls exhibited this fluorescence, and barley half-seeds incubated in GA_3-containing medium began secretion of β-1,3-glucanase within 4 hr of exposure to the hormone (134); over time the aleurone wall, including fluorescent material, underwent dissolution. Taiz & Jones (134) concluded that aleurone wall breakdown was accomplished through the action of the glucanase whose secretion, like that of α-amylase, was under gibberellin control.

Subsequent analysis of wall material prepared from aleurone tissues indicated the presence of cellulose and large amounts of arabinoxylan (a polysaccharide composed of a β-1,4-xylan backbone which bears arabinosyl substituents), but only a little noncellulosic glucose, none of which was β-1,3-linked (88). Study of aleurone wall breakdown was reinitiated, and the demonstration of aleurone enzymes able to completely digest arabinoxylan (i.e. β-1,4-xylanase, α-arabinosidase, and β-xylosidase) followed (30, 133). Some questions about the control of aleurone wall turnover remained. Dashek & Chrispeels (30) showed that GA_3-dependent release of wall pentose sugars (arabinose and xylose) into the medium in which isolated aleurone layers were incubated began prior to increases in xylanase activity. Although this could be an indication of initiation of wall turnover via a nonenzymic mechanism, it is possible that earlier increases in cell wall-bound xylanase, arabinosidase, and xylosidase were missed because tissue extractions were not in buffers of sufficiently high ionic strength to disrupt protein-cell wall interactions (38). Furthermore, a role for glucanase in aleurone wall digestion need not be ruled out. Aleurone walls which were subjected to structural analysis had been pretreated with α-amylase to remove starch (88). The *Bacillus subtilis* amylase used contained substantial activity of a β-glucanase which degrades mixed-linkage glucans (50). Thus such glucans could have been missed in the analysis. Recent reports have indicated the presence of mixed-linkage β-glucans in barley endosperm walls (34), and have shown that such glucans exhibit an intense aniline blue-induced fluorescence (142). Furthermore, some enzymes described as β-1,3-glucanases because they degrade laminarin are also active on mixed-linkage β-glucans (D. J. Huber & D. J. Nevins, unpublished). If this is true for the glucanase described by Taiz and Jones, it could act in aleurone wall turnover. Alternatively, such an enzyme could serve to digest β-glucans which may play a storage role in cereal grains (93, 150). The utilization of galactomannan, a storage polysaccharide which accumulates in the endosperm of some developing legume seeds (118), is accomplished through the action of carbohydrases which originate within the aleurone layer (117, 119).

Ikuma & Thimann (53) proposed that cell wall breakdown was an integral part of red light-promoted lettuce seed germination, and they detected low levels of pectinase and cellulase activity in extracts of lettuce cotyledons. Light microscopic examination of the lettuce endosperm confirmed that cell wall breakdown and extension of the radicle occurred in concert (58). Halmer et al (41–43) showed that an enzyme active on β-1,4-linked mannan appeared in lettuce seeds as they were induced to germinate by treatment with red light or GA_3. Analysis of wall material from lettuce endosperm showed it to have a composition rich in mannose (58% of the wall dry weight), quite different from the walls of other lettuce embryonic tissues (42). Although they provided no structural evidence to confirm that the endosperm wall would be labile to β-1,4-mannanase, they are probably justified in concluding that the mannanase acts in endosperm wall turnover and that walls of other seed tissues are "spared" because they contain little mannan. Cycloheximide treatment inhibits mannanase production although 40% of the seeds still germinate (41). Perhaps the enzyme merely enhances the ability of the radicle to penetrate the endosperm. In the normal course of germination, endosperm wall dissolution precedes utilization of reserves (44, 78).

Fruit Ripening

Ripening is the final stage of fruit development. It is marked by a number of biochemical changes which result in conversion of the fruit into an attractive seed container. Included among these changes for most fruits is a dramatic softening which is a consequence, at least in part, of cell wall breakdown (105). Histological studies confirm that ripening-related fruit softening involves extensive cell wall disruption (18).

WALL STRUCTURE IN POME FRUITS Structural anaylses of wall material from pear and apple fruit confirm the presence of many of the polysaccharide species described in the dicot wall model. Beyond this, there is evidence for covalent connections between cell wall polyuronide (pectin) and galactans and arabinans, as in the model (3, 70, 135); however, there is no indication of covalent interaction between pectin and hemicellulose. Treatment of apple walls with a fungal polygalacturonase (PG) solubilizes a substantial amount of uronic acid free of neutral sugars (70). This material was presumably derived from middle lamella, a portion of the wall of plant tissues that may not be described by Albersheim and his co-workers because they focused on the walls of cultured cells. Treatment of apple tissues with diazomethane, which converts uronide carboxyl groups to methyl esters, or with chelating agents solubilizes uronide material low in neutral sugar and causes loss of cell-to-cell cohesion (67). These results suggest a role for divalent cations and uronide carboxyl groups in stabilizing middle lamella structure (69, 113, 114).

RIPENING-ASSOCIATED CELL WALL CHANGES The clearest change in wall composition that accompanies fruit softening is a decrease in wall-bound uronic acids (pectin) that is closely matched by an increase in soluble uronide (105). Turnover of neutral sugar-containing cell wall components also occurs during ripening; generally galactose and arabininose are involved (56, 65, 71, 145). Turnover of cellulose, however, has not been demonstrated adequately in spite of the presence of β-1,4-glucanase in a variety of fruit tissues (106).

Fruit tissues contain an impressive array of carbohydrate-degrading enzymes (106). Many of these have been identified as cell wall-bound because they are solubilized from cell wall residue in high-ionic strength buffers (103). Total tissue activities often increase as ripening continues (106). A significant problem in understanding fruit wall metabolism is to know which of the many activities play an active role. Most fruits contain endo-PG, an enzyme that cleaves the α-1,4-linkages between nonesterified galacturonosyl residues at the interior of the pectin backbone (105). Endo-PG apparently plays a crucial role in converting wall-bound pectin to soluble form. Because PG is unable to act adjacent to an esterified galacturonosyl residue, it has been thought that PG could act in fruit wall metabolism only in conjunction with pectin esterase (PE). While this may be so, PE activity does not always change as fruits ripen (106), and PG is clearly the limiting enzyme of the two. *Rin* and *nor,* single-gene, nonripening mutants of tomato, contain considerable PE which increases as the fruits get older; however, the fruits contain no PG and do not soften (20, 99).

The means by which the turnover of neutral sugar-containing wall constituents is accomplished are not nearly as clear as the mechanism of pectin solubilization. The cell walls of the Japanese pear (*Pyrus serotina*) and of "Conference" and "Bartlett" pears (*Pyrus communis*) undergo considerable turnover of pectin and arabinose during ripening (3, 56, 153). While arabanase activity develops in the Japanese pear (152), no such enzyme is present in extracts of "Bartlett" pear fruits (4). Ahmed & Labavitch (3) studied polysaccharide metabolism in "Bartlett" pears by examining the structure of the polymers involved in wall turnover. They found that virtually all of the arabinose that was freed from the wall during ripening could be recovered in polymeric (i.e. ethanol-insoluble) form from the soluble fraction of tissue homogenates. Furthermore, this arabinose was apparently covalently linked with high molecular weight uronide. Because treatment of cell walls from unripe pears with a highly purified PG preparation solubilized a pectic arabinan like the soluble polymer identified in homogenates of ripe (soft) pears, they concluded that wall arabinose turnover, like pectin turnover, was a consequence of pear fruit PG activity (4).

Endo-PG activity is not responsible for the substantial turnover of galactose that occurs during apple and tomato ripening. Although PG appears to be the only identified tomato fruit enzyme that can directly digest a portion of tomato cell walls (145), there is considerable turnover of galactose in nonsoftening *rin* fruit which lack PG (39). Loss of galactose from the walls of mature apple fruits occurs in the absence of pectin solubilization (65, 66), and apples contain no endo-PG. Bartley (14, 15) has proposed that β-galactosidase in Cox's Orange Pippin apples is responsible for removal of galactose-containing side chains from the neutral sugar-rich galacturonan of apple fruit middle lamellae, thus converting that wall-bound pectin to the neutral sugar-deficient pectin characteristic of apple soluble pectin (68). Wallner (144) examined wall galactose metabolism in other apple varieties and found no correlation between galactose turnover and β-galactosidase activity.

Lackey et al (76a) have measured the synthesis of wall components in mature and ripening tomato fruits and have concluded that galactan turnover may occur throughout fruit development; it becomes apparent only after wall synthesis stops at the onset of ripening. In contrast, Knee & Bartley (69) have proposed a scheme whereby the wholesale wall turnover associated with fruit softening is triggered by wall synthesis. This proposal may be especially relevant for the apple which softens, albeit differently from most other fruits, in the absence of endo-PG. Knee (68) has measured incorporation from methyl-labeled ^{14}C-methionine and ^{3}H-inositol into the pectin of apple tissues. ^{14}C-Incorporation into pectin methyl esters continued after the onset of ripening although ^{3}H-incorporation stopped. Knee and Bartley suggest that tissue softening is initiated by methylation of pectin in the middle lamella. Esterified carboxyl groups are unable to participate in divalent cation-stabilized pectin junction zones and cell separation occurs. Knee & Casimir (unpublished) have shown that pears fail to soften properly if incorporation from methionine into pectin is inhibited. The combined action of PE and PG could then convert wall-bound pectin to soluble form, thus eliminating the possibility that the wall could be restabilized. In apples, which have exo- (16), but not endo-PG, less soluble pectin would be formed, but cell separation (tissue softening) could still occur. While this suggestion is attractive, it requires further testing. It is especially important that in situ pectin methyl ester synthesis be demonstrated.

CONCLUDING REMARKS

It is clear that turnover of specific cell wall components accompanies plant development. In many of the cases cited in this review, the structures of the polysaccharides involved in turnover are known. Enzymes capable of acting on these polymers are present in the tissues affected and presumed to be acting in wall metabolism. This latter point needs further elucidation. Ultrastructural localization of enzymes, perhaps using tagged antibodies for purified enzymes (12), and demonstrations of *in muro* activation of enzymes at the onset of turnover are important pieces of evidence that are currently lacking. Little is known about possible nonenzymic catalysts for polysaccharide metabolism. Perhaps the most crucial information relates to the elucidation of the role of synthesis in wall turnover. We must understand the way in which synthetic and degradative reactions are balanced to provide coordination of wall turnover with all of the other processes that are essential to plant cell development.

ACKNOWLEDGMENT

My thanks to the many colleagues who provided copies of papers discussed in this review. Thanks are also due to Bruce Bonner, Carl Greve, and Larry Strand for their reading of the manuscript and to Jackie DiClementine for her patience and cooperation in its typing.

Literature Cited

1. Abdul-Baki, A. A., Ray, P. M. 1971. Regulation of carbohydrate metabolism involved in cell wall synthesis by pea stem tissue. *Plant Physiol.* 47:537–44
2. Addicott, F. T., Wiatr, S. M. 1977. Hormonal controls of abscission: Biochemical and ultrastructural aspects. In *Plant Growth Regulation,* ed. P. E. Pilet, pp. 249–57. Berlin/Heidelberg/New York: Springer. 305 pp.
3. Ahmed, A. E. R., Labavitch, J. M. 1980. Cell wall metabolism in ripening fruit I. Cell wall changes in ripening 'Bartlett' pears. *Plant Physiol.* 65: 1009–13
4. Ahmed, A. E. R., Labavitch, J. M. 1980. Cell wall metabolism in ripening fruit II. Changes in carbohydrate-degrading enzymes in ripening 'Bartlett' pears. *Plant Physiol.* 65:1014–16
5. Albersheim, P., Anderson-Prouty, A. J. 1975. Carbohydrates, proteins, cell surfaces, and the biochemistry of plant pathogenesis. *Ann. Rev. Plant Physiol.* 26:31–52
6. Albersheim, P., McNeil, M., Labavitch, J. M. 1977. The molecular structure of the primary cell wall and elongation growth. See Ref. 2, pp. 1–12
7. Aspinall, G. D. 1969. Gums and mucilages. *Adv. Carbohydr. Chem. Biochem.* 24:333–79
8. Aspinall, G. D., Krishnamurthy, T. N., Rosell, K-G. 1977. A fucogalactoxyloglucan from rapeseed hulls. *Carbohydr. Res.* 55:11–19
9. Awad, J., Young, R. E. 1979. Postharvest variation in cellulase, polygalacturonase, and pectinmethylesterase in avocado (*Persea americana* Mill, cv. Fuerte) fruits in relation to respiration and ethylene production. *Plant Physiol.* 64:306–8
10. Baker, D. B., Ray, P. M. 1965. Direct and indirect effects of auxin on cell wall synthesis in oat coleoptile tissue. *Plant Physiol.* 40:345–52
11. Baker, D. B., Ray, P. M. 1965. Relationship between effects of auxin on cell wall synthesis and cell elongation. *Plant Physiol.* 40:360–68
12. Bal, A. K., Verma, D. P. S., Byrne, H., MacLachlan, G. A. 1976. Subcellular localization of cellulases in auxin-treated peas. *J. Cell Biol.* 69:97–105
13. Barkley, G. M., Evans, M. L. 1970. Timing of the auxin response in etiolated pea stem sections. *Plant Physiol.* 45:143–47
14. Bartley, I. M. 1974. β-galactosidase activity in ripening apples. *Phytochemistry* 13:2107–11
15. Bartley, I. M. 1977. A further study of β-galactosidase activity in apples ripening in store. *J. Exp. Bot.* 28:943–48
16. Bartley, I. M. 1978. Exo-polygalacturonase of apple. *Phytochemistry* 17: 213–16
17. Bauer, W. D., Talmadge, K. W., Keegstra, K., Albersheim, P. 1973. The structure of plant cell walls II. The hemicellulose of the walls of suspension-cultured sycamore cells. *Plant Physiol.* 51:174–87
18. Ben-Arie, R., Kislov, N., Frenkel, C. 1979. Ultrastructural changes in the cell walls of ripening apple and pear fruit. *Plant Physiol.* 64:197–202
19. Buchala, A. J., Wilkie, K. C. B. 1971. Ratio of β (1→3) to β (1→4) glucosidic linkages in non-endospermic hemicellulosic β-glucans from oat plant tissues at different stages of maturity. *Phytochemistry* 10:2287–91
20. Buescher, R. W., Tigchelaar, E. C. 1975. Pectinesterase, polygalacturonase, C_x-cellulase activities and softening of the *rin* tomato mutant. *HortScience* 10:624–25
21. Burke, D., Kaufman, P., McNeil, M., Albersheim, P. 1974. The structure of plant cell walls VI. A survey of the walls of suspension-cultured monocots. *Plant Physiol.* 54:109–15
22. Castle, E. S. 1955. The mode of growth of epidermal cells of the *Avena* coleoptile. *Proc. Natl. Acad. Sci. USA* 41:197–99
23. Chester, M. A., Hultberg, B., Ockerman, P. 1976. The common identity of five glycosidases in human liver. *Biochim. Biophys. Acta* 429:517–26
24. Chrispeels, M. J., Varner, J. E. 1967. Gibberellic acid enhanced synthesis and release of α-amylase and ribonuclease by isolated aleurone layers. *Plant Physiol.* 42:398–402
25. Cleland, R. 1967. Extensibility of isolated cell walls: Measurement and changes during cell elongation. *Planta* 74:197–209
26. Cleland, R. 1967. A dual role for turgor pressure in auxin-induced cell elongation in *Avena* coleoptiles. *Planta* 77:182–91
27. Cleland, R. 1971. Cell wall extension. *Ann. Rev. Plant Physiol.* 22:197–222
28. Cleland, R., Bonner, J. 1956. The residual effect of auxin on the cell wall. *Plant Physiol.* 31:350–54

29. Currier, H. B. 1957. Callose substance in plant cells. *Am. J. Bot.* 44:478–88
30. Dashek, W. V., Chrispeels, M. J. 1977. Gibberellic acid-induced synthesis and release of cell-wall-degrading endoxylanase by isolated aleurone layers of barley. *Planta* 134:251–56
31. Davies, E., Maclachlan, G. A. 1968. Effects of indoleacetic acid on intracellular distribution of β-glucanase activities in the pea epicotyl. *Arch. Biochem. Biophys.* 128:595–600
32. Dey, P. M. 1977. Polymorphism of some glycosidases from barley. *Phytochemistry* 16:323–25
33. Eisinger, W. R., Burg, S. P. 1972. Ethylene-induced pea internode swelling: Its relation to ribonucleic acid metabolism, wall protein synthesis, and cell wall structure. *Plant Physiol.* 50:510–17
34. Forrest, I. S., Wainwright, T. 1977. The mode of binding of β-glucans and pentosans in barley endosperm cell walls. *J. Inst. Brew.* 83:279–86
35. Gilkes, N. R., Hall, M. A. 1977. The hormonal control of cell wall turnover in *Pisum sativum* L. *New Phytol.* 78:1–15
36. Goldberg, R. 1977. On possible connections between auxin-induced growth and cell wall glucanase activities. *Plant Sci. Lett.* 8:233–42
37. Green, P. B. 1958. Concerning the site of the addition of new wall substances to the elongating *Nitella* cell wall. *Am. J. Bot.* 45:111–16
38. Greve, L. C., Ordin, L. 1977. Isolation and purification of an α-mannosidase from coleoptiles of *Avena sativa. Plant Physiol.* 60:478–81
39. Gross, K. C., Wallner, S. J. 1979. Degradation of cell wall polysaccharides during tomato fruit ripening. *Plant Physiol.* 63:117–20
40. Hager, A., Menzel, H., Krauss, A. 1971. Versuche und Hypothese zur Primärwirkung des Auxins beim Streckungswachstum. *Planta* 100:47–75
41. Halmer, P., Bewley, J. D. 1979. Mannanase production by the lettuce endosperm: Control by the embryo. *Planta* 144:333–40
42. Halmer, P., Bewley, J. D., Thorpe, T. A. 1975. Enzyme to break down lettuce endosperm cell wall during gibberellin- and light-induced germination. *Nature* 258:716–18
43. Halmer, P., Bewley, J. D., Thorpe, T. A. 1976. An enzyme to degrade lettuce endosperm cell walls. Appearance of a mannanase following phytochrome- and gibberellin-induced germination. *Planta* 130:189–96
44. Halmer, P., Bewley, J. D., Thorpe, T. A. 1978. Degradation of the endosperm cell walls of *Lactuca sativa* L. cv Grand Rapids. Timing of mobilisation of soluble sugars, lipid and phytate. *Planta* 139:1–8
45. Heyn, A. N. J. 1931. Der mechanismus der zellstreckung. *Rec. Trav. Bot. Neerl.* 28:113–244
46. Heyn, A. N. J. 1933. Further investigations on the mechanism of cell elongation and the properties of the cell wall in connection with elongation I. The load-extension relationship. *Protoplasma* 19:78–96
47. Heyn, A. N. J. 1969. Glucanase activity in coleoptiles of *Avena. Arch. Biochem. Biophys.* 132:442–49
48. Hobson, G. E. 1968. Cellulase activity during the maturation and ripening of tomato fruit. *J. Food Sci.* 33:588–92
49. Houwink, A. L., Roelofson, P. A. 1954. Fibrillar architecture of growing plant cell walls. *Acta Bot. Neerl.* 3:385–95
50. Huber, D. J., Nevins, D. J. 1977. Preparation and properties of a β-D-glucanase for the specific hydrolysis of β-D-glucans. *Plant Physiol.* 60:300–4
51. Huber, D. J., Nevins, D. J. 1979. Autolysis of cell wall β-D-glucan in corn coleoptiles. *Plant Cell Physiol.* 20:201–12
52. Huber, D. J., Nevins, D. J. 1980. β-D-glucan hydrolase activity in *Zea* coleoptile cell walls. *Plant Physiol.* 65:768–73
53. Ikuma, H., Thimann, K. V. 1963. The role of the seed coats in germination of photosensitive lettuce seeds. *Plant Cell Physiol.* 4:169–85
54. Jacobs, M., Ray, P. M. 1975. Promotion of xyloglucan metabolism by acid pH. *Plant Physiol.* 56:373–76
55. Jacobs, M., Ray, P. M. 1976. Rapid auxin-induced decrease in free space pH and its relationship to auxin-induced growth in maize and peas. *Plant Physiol.* 58:203–9
56. Jermyn, M. A., Isherwood, F. A. 1956. Changes in the cell walls of pear during ripening. *Biochem. J.* 64:123–32
57. Johnson, K. D., Daniels, D., Dowler, M. J., Rayle, D. L. 1974. Activation of *Avena* coleoptile cell wall glycosidases by hydrogen ions and auxin. *Plant Physiol.* 53:224–28
58. Jones, R. L. 1974. The structure of the lettuce endosperm. *Planta* 121:133–46
59. Kanda, T., Wakabayashi, K., Nisizawa, K. 1976. Xylanase activity of an endo-cellulase of carboxymethyl-cellulase

type from *Irpex lacteus* (*Polyporus tulipiferae*). *J. Biochem.* 79:989–95
60. Katz, M., Ordin, L. 1967. Metabolic turnover in cell wall constituents of *Avena sativa* L. coleoptile sections. *Biochim. Biophys. Acta* 141:118–25
61. Katz, M., Ordin, L. 1967. A cell wall polysaccharide-hydrolyzing enzyme system in *Avena sativa* L. coleoptiles. *Biochim. Biophys. Acta* 141:126–34
62. Keegstra, K., Talmadge, K. W., Bauer, W. D., Albersheim, P. 1973. The structure of plant cell walls III. A model of the walls of suspension-cultured sycamore cells based on the interconnections of the macromolecular components. *Plant Physiol.* 51:188–96
63. Key, J. L. 1969. Hormones and nucleic acid metabolism. *Ann. Rev. Plant Physiol.* 20:449–74
64. Kivilaan, A., Bandurski, R. S., Schulze, A. 1971. A partial characterization of an autolytically solubilized cell wall glucan. *Plant Physiol.* 48:389–93
65. Knee, M. 1973. Polysaccharide changes in cell walls of ripening apples. *Phytochemistry* 12:1543–46
66. Knee, M. 1975. Changes in structural polysaccharides of apples ripening during storage. In *Facteurs et Régulation de la Maturation des Fruits,* ed. R. Ullrich, pp. 341–45. Paris: C.N.R.S. 368 pp.
67. Knee, M. 1978. Properties of polygalacturonate and cell cohesion in apple fruit cortical tissue. *Phytochemistry* 17: 1257–60
68. Knee, M. 1978. Metabolism of polymethylgalacturonate in apple fruit cortical tissue during ripening. *Phytochemistry* 17:1261–64
69. Knee, M., Bartley, I. M. 1980. Composition and metabolism of cell wall polysaccharides in ripening fruits. In *The Biochemistry of Fruits and Vegetables.* In press
70. Knee, M., Fielding, A. H., Archer, S. A., Laborda, F. 1975. Enzymic analysis of cell wall structure in apple fruit cortical tissue. *Phytochemistry* 14:2213–22
71. Knee, M., Sargent, J. A., Osborne, D. J. 1977. Cell wall metabolism in developing strawberry fruits. *J. Exp. Bot.* 28:377–96
72. Konar, R. N., Stanley, R. G. 1969. Wall softening enzymes in the gynoecium and pollen of *Hemerocallis fulva. Planta* 84:304–10
73. Kooiman, P. 1967. The constitution of the amyloid from seeds of *Annona muricata* L. *Phytochemistry* 6:1665–73
74. Labavitch, J. 1972. *Auxin and the turnover of cell wall polysaccharides in elongating pea stem tissue.* PhD thesis. Stanford Univ., Stanford, Calif. 78 pp.
75. Labavitch, J. M., Ray, P. M. 1974. Turnover of cell wall polysaccharides in elongating pea stem segments. *Plant Physiol.* 53:669–73
76. Labavitch, J. M., Ray, P. M. 1974. Relationship between promotion of xyloglucan metabolism and induction of elongation by indoleacetic acid. *Plant Physiol.* 54:449–502
76a. Lackey, G. D., Gross, K. C., Wallner, S. C. 1980. Loss of tomato cell wall galactan may involve reduced rate of synthesis. *Plant Physiol.* 66:532–33
77. Lee, S., Kivilaan, A., Bandurski, R. S. 1967. *In vitro* autolysis of plant cell walls. *Plant Physiol.* 42:968–72
78. Leung, D. W. M., Reid, J. S. G., Bewley, J. D. 1979. Degradation of the endosperm cell walls of *Lactuca sativa* L., cv. Grand Rapids in relation to the mobilisation of proteins and the production of hydrolytic enzymes in the axis, cotyledons, and endosperm. *Planta* 146:335–41
79. Loescher, W., Nevins, D. J. 1972. Auxin-induced changes in *Avena* coleoptile cell wall composition. *Plant Physiol.* 50:556–63
80. Loescher, W., Nevins, D. J. 1973. Turgor-dependent changes in *Avena* coleoptile cell wall composition. *Plant Physiol.* 52:248–51
81. Maclachlan, G. A. 1976. A potential role for endo-cellulase in cellulose biosynthesis. *Appl. Polym. Symp.* 28: 645–58
82. Maclachlan, G. A., Duda, C. T. 1966. Changes in concentration of polymeric components in excised pea epicotyl tissue during growth. *Biochim. Biophys. Acta* 97:288–99
83. Marrè, E., Lado, P., Rasi-Caldogno, F., Colombo, R. 1973. Correlation between cell enlargement in pea internode segments and decrease in the pH of the medium of incubation I. Effects of fusicoccin, natural and synthetic auxins and mannitol. *Plant Sci. Lett.* 1:179–84
84. Masuda, Y. 1978. Auxin-induced cell wall loosening. *Bot. Mag.* Spec. Issue 1:103–23
85. Masuda, Y., Oi, S., Satumura, Y. 1970. Further studies on the role of cell wall-degrading enzymes in cell wall loosening in oat coleoptiles. *Plant Cell Physiol.* 11:631–38
86. Masuda, Y., Yamamoto, R. 1972. Control of auxin-induced stem elongation

by the epidermis. *Physiol. Plant* 27:109–15
87. Matchett, W. H., Nance, J. F. 1962. Cell wall breakdown and growth in pea seedling stems. *Am. J. Bot.* 49:311–19
88. McNeil, M., Albersheim, P., Taiz, L., Jones, R. L. 1975. The structure of plant cell walls VII. Barley aleurone cells. *Plant Physiol.* 55:64–68
89. McNeil, M., Darvill, A. G., Albersheim, P. 1980. The structural polymers of the primary cell walls of dicots. *Progress in the Chemistry of Organic Natural Products* 37:191–250
90. Métraux, J-P., Richmond, P. A., Taiz, L. 1980. Control of cell elongation in *Nitella* by endogenous cell wall pH gradients. *Plant Physiol.* 65:204–10
91. Métraux, J-P., Taiz, L. 1977. Cell wall extension in *Nitella* as influenced by acids and ions. *Proc. Natl. Acad. Sci. USA* 74:1565–69
92. Monro, J. A., Penny, D., Bailey, R. W. 1976. The organization and growth of primary cell walls of lupin hypocotyl. *Phytochemistry* 15:1193–98
93. Morall, P., Briggs, D. E. 1978. Changes in cell wall polysaccharides of germinating barley grains. *Phytochemistry* 17:1495–1502
94. Nevins, D. J. 1970. Relation of glycosidases to bean hypocotyl growth. *Plant Physiol.* 46:458–62
95. Nevins, D. J. 1975. The effect of nojirimycin on plant growth and its implications concerning a role for exo-β-glucanases in auxin-induced cell expansion. *Plant Cell Physiol.* 16:347–56
96. Nevins, D. J. 1975. The *in vitro* simulation of IAA-induced modification of *Avena* cell wall polysaccharides by an exo-glucanase. *Plant Cell Physiol.* 16:495–503
97. Nevins, D. J., Huber, D. J., Yamamoto, R., Loescher, W. H. 1977. β-D-glucan of *Avena* coleoptile cell walls. *Plant Physiol.* 60:617–21
98. Nevins, D. J., Yamamoto, R., Huber, D. J. 1978. Cell wall β-D-glucans of five grass species. *Phytochemistry* 17: 1503–5
99. Ng, T. J., Tigchelaar, E. C. 1977. Action of the non-ripening (nor) mutant on fruit ripening of tomato. *J. Am. Soc. Hortic. Sci.* 102:504–9
100. Nishitani, K., Masuda, Y. 1980. Modification of cell wall polysaccharides during auxin-induced growth in azuki bean epicotyl segments. *Plant Cell Physiol.* 21:169–81
101. Nishitani, K., Shibaoka, H., Masuda, Y. 1979. Growth and cell wall changes in azuki bean epicotyls II. Changes in wall polysaccharides during auxin-induced growth of excised segments. *Plant Cell Physiol.* 20:463–72
102. O'Brien, T. P. 1970. Further observations on hydrolysis of the cell wall in the xylem. *Protoplasma* 69:1–14
103. Ordin, L., Applewhite, T. H., Bonner, J. 1956. Auxin-induced water uptake by *Avena* coleoptile sections. *Plant Physiol.* 31:44–53
104. Pharr, D. M., Sox, H. N., Nesbitt, W. B. 1976. Cell wall bound nitrophenylglycosidases of tomato fruits. *J. Am. Soc. Hortic. Sci.* 101:397–400
105. Pilnik, W., Voragen, A. J. G. 1970. Pectic substances and other uronides. In *The Biochemistry of Fruits and Their Products,* ed. A. C. Hulme, 1:53–87. London: Academic. 620 pp.
106. Pressey, R. 1977. Enzymes involved in fruit softening. In *Enzymes in Food and Beverage Processing,* ed. R. L. Ory, A. J. St. Angelo, pp. 172–91. Washington DC: Am. Chem. Soc. 325 pp.
107. Ray, P. M. 1967. Radioautographic study of cell wall depositon in growing plant cells. *J. Cell Biol.* 35:659–74
108. Ray, P. M. 1969. The action of auxin on cell enlargement in plants. *Dev. Biol. Suppl.* 3:172–205
109. Ray, P. M., Baker, D. B. 1965. The effect of auxin on synthesis of oat coleoptile cell wall constituents. *Plant Physiol.* 40:353–60
110. Rayle, D. L. 1973. Auxin-induced hydrogen-ion secretion in *Avena* and its implications. *Planta* 114:63–73
111. Rayle, D. L., Cleland, R. 1970. Enhancement of wall loosening and elongation by acid solutions. *Plant Physiol.* 46:250–53
112. Rayle, D. L., Cleland, R. 1972. The *in vitro* acid growth response: Relation to *in vivo* growth responses and auxin action. *Planta* 104:282–96
113. Rees, D. A. 1969. Structure, conformation and mechanism in the formation of polysaccharide gels. *Adv. Carbohydr. Chem. Biochem.* 24:267–332
114. Rees, D. A. 1975. Stereochemistry and binding behaviour of carbohydrate chains. *MTP Int. Rev. Sci., Biochem.* Ser. 1, 5:1–42
115. Reese, E. T., Parrish, F. W., Ettlinger, M. 1971. Nojirimycin and D-glucono-1,5-lactone as inhibitors of carbohydrases. *Carbohydr. Res.* 18:381–88
116. Rehm, M. M., Cline, M. G. 1973. Inhibition of low pH-induced elongation in *Avena* coleoptiles by abscisic acid. *Plant Physiol.* 51:946–48

117. Reid, J. S. G., Davies, D., Meier, H. 1977. Endo-β-mannanase, the leguminous aleurone layer and the storage galactomannan in germinating seeds of *Trigonella foenum-groecum. L. Planta* 133:219–22
118. Reid, J. S. G., Meier, H. 1970. Formation of reserve galactomannan in the seeds of *Trigonella foenum-groecum. Phytochemistry* 9:513–20
119. Reid, J. S. G., Meier, H. 1972. The function of the aleurone layer during galactomannan mobilisation in germinating seeds of fenugreek (*Trigonella foenum-groecum* L.), crimson clover (*Trifolium incarnatum* L.), and lucerne (*Medicago sativa* L.): A correlative biochemical and ultrastructural study. *Planta* 106:44–60
120. Reid, J. S. G., Wilkie, K. C. B. 1969. Total hemicelluloses from oat plants at different stages of growth. *Phytochemistry* 8:2059–65
121. Richmond, P. A., Métraux, J-P., Taiz, L. 1980. Cell expansion patterns and directionality of wall mechanical properties in *Nitella. Plant Physiol.* 65:211–17
122. Roggen, H. P. J. R., Stanley, R. G. 1969. Cell wall hydrolysing enzymes in wall formation as measured by pollen tube extension. *Planta* 84:295–303
123. Ruesink, A. W. 1969. Polysaccharidases and the control of cell wall elongation. *Planta* 89:95–107
124. Sakurai, N., Masuda, Y. 1977. Effect of indole-3-acetic acid on cell wall loosening: Changes in mechanical properties and non-cellulosic glucose content of *Avena* coleoptile cell wall. *Plant Cell Physiol.* 18:587–94
125. Sakurai, N., Masuda, Y. 1978. Auxin-induced changes in barley coleoptile cell wall composition. *Plant Cell Physiol.* 19:1217–23
126. Sakurai, N., Masuda, Y. 1978. Auxin-induced extension, cell wall loosening and changes in the wall polysaccharide content of barley coleoptile segments. *Plant Cell Physiol.* 19:1225–33
127. Sakurai, N., Nevins, D. J., Masuda, Y. 1977. Auxin- and hydrogen ion-induced cell wall loosening and cell extension in *Avena* coleoptile segments. *Plant Cell Physiol.* 18:371–80
128. Sakurai, N., Nishitani, K., Masuda, Y. 1979. Auxin-induced changes in the molecular weight of hemicellulosic polysaccharides of the *Avena* coleoptile cell wall. *Plant Cell Physiol.* 20:1349–57
129. Schimke, R. T., Doyle, D. 1970. Control of enzyme levels in animal tissues. *Ann. Rev. Biochem.* 39:929–76
130. Schwarz, U., Asmus, A., Frank, H. 1969. Autolytic enzymes and cell division of *Escherichia coli. J. Mol. Biol.* 41:419–29
131. Shimoda, C., Yanagashima, N. 1971. Role of cell wall-degrading enzymes in auxin-induced cell wall expansion in yeast. *Physiol. Plant.* 24:46–51
132. Srivastava, L. M., Sawhney, V. K., Bonettemaker, M. 1977. Cell growth, wall deposition, and correlated fine structure of colchicine-treated lettuce hypocotyl cells. *Can. J. Bot.* 55:902–17
133. Taiz, L., Honigman, W. A. 1976. Production of cell wall hydrolyzing enzymes by barley aleurone layers in response to gibberellic acid. *Plant Physiol.* 58:380–86
134. Taiz, L., Jones, R. L. 1970. Gibberellic acid, β-1,3-glucanase, and the cell walls of barley aleurone layers. *Planta* 92:73–84
135. Talmadge, K. W., Keegstra, K., Bauer, W. D., Albersheim, P. 1973. The structure of plant cell walls I. The macromolecular components of the walls of suspension-cultured sycamore cells with a detailed analysis of the pectic polysaccharides. *Plant Physiol.* 51: 153–73
136. Tanimoto, E., Masuda, Y. 1968. Effect of auxin on cell wall degrading enzymes. *Physiol. Plant.* 21:820–26
137. Terry, M. E., Bonner, B. A. 1980. An examination of centrifugation as a method of extracting an extracellular solution from peas, and its use for the study of IAA-induced growth. *Plant Physiol.* 66:321–25
138. Thimann, K. V., Schneider, C. L. 1938. Differential growth in plant tissues. *Am. J. Bot.* 25:627–41
139. Thomas, D. S. 1970. Effects of water stress on induction and secretion of cellulase and on branching in the water mold *Achlya ambisexualis. Can. J. Bot.* 48:977–79
140. Valent, B. S., Albersheim, P. 1974. The structure of plant cell walls V. On the binding of xyloglucan to cellulose fibers. *Plant Physiol.* 54:105–8
141. Vanderhoef, L. N., Stahl, C. A. 1975. Separation of two responses to auxin by means of cytokinin inhibition. *Proc. Natl. Acad. Sci. USA* 72:1822–25
142. Vithinage, H. I. M. V., Gleeson, P. A., Clarke, A. E. 1980. The nature of callose produced during self-pollination in *Secale cereale. Planta* 148:498–509
143. Wada, S., Tanimoto, E., Masuda, Y. 1968. Cell elongation and metabolic turnover of the cell wall as affected by

auxin and cell wall degrading enzymes. *Plant Cell Physiol.* 9:369–76

144. Wallner, S. J. 1978. Apple fruit β-galactosidase and ripening in storage. *J. Am. Soc. Hortic. Sci.* 103:364–66
145. Wallner, S. J., Bloom, H. L. 1977. Characteristics of tomato cell wall degradation *in vitro:* Implications for the study of fruit-softening enzymes. *Plant Physiol.* 60:207–10
146. Wardrop, A. B. 1955. The mechanism of surface growth in parenchyma of *Avena* coleoptiles. *Aust. J. Bot.* 3:137–48
147. Wessells, J. G. H. 1969. A β-1,6-glucan glucanohydrolase involved in hydrolysis of cell wall glucan in *Schizophyllum commune. Biochim. Biophys. Acta* 178:191–92
148. Wong, Y. S., Maclachlan, G. A. 1979. β-1,3-glucanases from *Pisum sativum* seedlings I. Isolation and purification. *Biochim. Biophys. Acta* 571:244–55
149. Wong, Y. S., Maclachlan, G. A. 1980. 1,3-β-D-glucanases from *Pisum sativum* seedlings III. Development and distribution of endogenous substrates. *Plant Physiol.* 65:222–28
150. Woolard, G. R., Rathbone, E. B., Novellie, L. 1976. A hemicellulosic β-D-glucan from the endosperm of sorghum grain. *Carbohydr. Res.* 51:249–52
151. Yamagata, Y., Yamamoto, R., Masuda, Y. 1974. Auxin and hydrogen ion actions on light grown pea epicotyl segments II. Effect of hydrogen ions on extension of the isolated epidermis. *Plant Cell Physiol.* 15:833–41
152. Yamaki, S., Kakiuchi, N. 1979. Changes in hemicellulose-degrading enzymes during development and ripening of Japanese pear fruit. *Plant Cell Physiol.* 20:301–9
153. Yamaki, S., Kakiuchi, N. 1979. Changes in cell wall polysaccharides and monosaccharides during development and ripening of Japanese pear fruit. *Plant Cell Physiol.* 20:311–21
154. Yamamoto, R., Masuda, Y. 1971. Stress-relaxation properties of the *Avena* coleoptile cell wall. *Physiol. Plant* 25:330–35
155. Yamamoto, R., Nevins, D. J. 1978. Structural studies on the β-D-glucan of the *Avena* coleoptile cell wall. *Carbohydr. Res.* 67:275–80
156. Yamamoto, R., Shinozaka, K., Masuda, Y. 1970. Stress-relaxation properties of cell walls with special reference to auxin action. *Plant Cell Physiol.* 11:947–56
157. Zarra, I., Masuda, Y. 1979. Growth and cell wall changes in rice coleoptiles growing under different conditions II. Auxin-induced growth in coleoptile segments. *Plant Cell Physiol.* 20:1125–33

Ann. Rev. Plant Physiol. 1981. 32:407–49

INFECTION OF LEGUMES BY RHIZOBIA

◆7718

Wolfgang D. Bauer

C. F. Kettering Research Laboratory, Yellow Springs, Ohio 45387

CONTENTS

INTRODUCTION

The majority of biologically fixed nitrogen available for agriculture is formed by rhizobia in symbiotic association with legumes. The idea of making better use of rhizobia and other nitrogen-fixing microorganisms is most attractive, and recent support for biological nitrogen fixation research has generated a burst of new studies (72, 73, 89, 101, 146–48, 165, 189). The results from such studies have been sobering. A good deal has been learned, but there is little that can be used to practical advantage. It appears that

0066-4294/81/0601-0407$01.00

major practical benefits will not be available for some years, at least if it is first necessary to obtain a fairly sophisticated knowledge of the biology of the host plants, the rhizobia, and the means by which symbiotic associations between the two are established.

If socially useful results seem more distant now than in previous years, it is because we have gained rapidly in our appreciation of the complexity of symbiotic associations. Appreciation of these complexities is reflected here by a recurring emphasis on regulation, regulation of the interactions between host and microsymbiont, the developmental regulation of infectibility and infectivity, and the regulation of responses to changes in environmental conditions. Such regulation has now become an important area of study in itself. From this it would appear that our inquiries have reached an essential threshold of sophistication.

Establishment of the legume/*Rhizobium* symbiosis involves infection of the host root and the subsequent formation of nodular growths containing approximately equal weights of root and bacterial cells. Rhizobia are ordinary-appearing Gram-negative bacteria well adapted to survival in the soil. They proliferate in the rhizosphere of potential host plants and attach to the root surfaces. In the best studied legume hosts, compatible rhizobia induce curling and branching of root hairs and the subsequent formation of a tubular structure called the infection thread. The infection thread develops inward from its point of origin near the most acutely curled region of the hair. Rhizobia are carried within the thread, usually single file, as the tip of the thread follows the movement of the nucleus toward the base of the hair cell. The infection thread passes through the wall of the hair cell and the adjacent cortical cell and branches into many newly divided cortical cells. Rhizobia are released from the tips of the infection threads into the cytoplasm of host cells where they are surrounded by envelopes of host plasmamembrane. The enzyme nitrogenase is synthesized in the bacteria and converts dinitrogen to ammonia at the expense of host plant photosynthate.

This review is concerned with the initial phases of this remarkable sequence. It deals with the question of how the host is able to recognize its *Rhizobium* symbiont among the multitude of soil microorganisms, and it deals with the mechanics and control of the infection process. Other literature in addition to the previously cited collections is available for discussions of related topics, including the process of nodule development (169), the characteristics of free-living and bacteroid rhizobia (93, 117, 195, 198, 201), the soil ecology (80, 204) and genetics (17, 18, 150, 170, 171) of rhizobia, and associations with excised roots and root tissue cultures (1, 98). Readers should also consult the comprehensive reviews of *Rhizobium*/legume symbioses by Dart (54–56).

RECOGNITION

The term recognition is appropriate when one cell or organism shows a selective or specific response to another cell or organism. As such, recognition is a very common and fundamental biological phenomenon. Little is known of the nature and functioning of recognition mechanisms in either pathogenic or symbiotic plant/microorganism interactions. The most promising current hypothesis is that host plant lectins (carbohydrate binding proteins) interact selectively with microbial cell surface carbohydrates and serve as determinants of recognition or host specificity. Several reviews have appeared recently that deal with the question of recognition in plant/microorganism interactions and the possibility of lectin-mediated recognition (3, 12, 21, 34, 35, 38, 59, 60, 90, 143, 177, 178, 183, 184).

The basic hypothesis was first outlined by Albersheim & Anderson-Prouty (3). They suggested that the outcome of interactions between plants and pathogenic microorganisms was normally determined by the recognition capabilities of the plants. A potential host plant would be resistant to a particular microorganism if it were able to recognize the microorganism as a potential pathogen. The recognition mechanism would sense or detect the presence of the microorganism. Such detection would result in activation of inducible defense responses in the plant. According to this view, the range of plant species or cultivars that a particular pathogenic microorganism infects is dependent on two factors: whether or not a given type of plant can recognize the microbe as a potential pathogen, and whether or not the microbe can overcome the plant's constitutive defenses. If the microorganism is unable to overcome the plant's constitutive defenses, then it is a nonpathogen of that plant. If the microbe is able to overcome the constitutive defenses, but cannot avoid recognition and the consequent triggering of inducible defenses, then the microorganism is an incompatible pathogen of the given host. Only if the microorganism overcomes the constitutive defenses and also escapes recognition (or perhaps actively suppresses the induction of host defense responses) can it succeed in extensively colonizing the plant. In symbiotic associations, on the other hand, a successful microbial symbiont must be recognized by the host since the host must permit entry to the microsymbiont, but not to other microorganisms.

Based on their own work with elicitors of plant defense responses and on analogies with recognition mechanisms in other biological systems, Albersheim and Anderson-Prouty suggested that the most likely molecular determinants of plant-microorganism recognition would be host plant lectins and carbohydrate components of microbial cell surfaces (3). The experimental evidence indicating that host lectins and *Rhizobium* surface polysaccharides are involved in recognition is very briefly outlined in the following paragraphs. The original papers and the reviews cited earlier in this section should be consulted for detailed discussions.

Bohlool & Schmidt (28) provided the first experimental evidence demonstrating a strong correlation between the binding of a host plant lectin to microbial cell surfaces and the ability of such microorganisms to infect the plant. Fluorescent-labeled soybean seed lectin was reported to bind to the cell surfaces of bacteria in cultures of 22 different strains of *Rhizobium japonicum,* the *Rhizobium* species that infects soybean (28). No binding of the labeled lectin to bacteria could be detected with cultures of 23 different strains of heterologous rhizobia, or with 3 of the 25 *R. japonicum* strains tested (28). Subsequent studies provided independent confirmation of these initial observations (23). The correlation between lectin binding and infectivity was improved when the bacteria were cultured with host roots or host root exudates (22). Several strains of *R. japonicum* that had no lectin binding cells when cultured on artificial media did have lectin binding cells when cultured with host roots or root exudates (22).

In other studies, cell surface polysaccharides isolated from several different *Rhizobium* species were reported to interact specifically with lectins isolated from the seeds of the respective host legumes (209). The *Rhizobium* polysaccharides did not interact with lectins from nonhost legumes.

Excellent—indeed perfect—correlations between lectin binding and infectivity were also obtained for the clover/*R. trifolii* association (65). Common antigenic structures were reported to be present on the surfaces of both the host root and the bacterial symbiont (64, 65). Clover lectin, localized on the root (61, 69) as well as in the seed, was proposed to be a cross-linking agent binding to the common antigen structures on the root and the bacteria and in effect specifically gluing the bacterial symbiont to the root surface (65).

There are many gaps, inconsistencies, and experimental weaknesses that cloud the reliability and interpretation of the above mentioned studies and the numerous related investigations (12, 90). Moreover, several studies have been reported where no strong or simple correlations between lectin binding and nodulating ability were obtained (33, 44, 66, 120, 210). Although there are technical reasons for questioning some of the negative results (12, 59, 90), other results in these papers require further exploration and explanation. What seems clear at this time is that there are correlations between lectin binding and *Rhizobium* infectivity that are very difficult to explain away as coincidence. The evidence indicates that soybean, white clover, and perhaps other legumes are capable of synthesizing proteins (lectins) which can interact quite specifically with characteristic cell surface carbohydrates synthesized by their *Rhizobium* symbionts. This evidence does not by any means prove that the lectin recognition hypothesis is correct. However, the evidence is strong enough to suggest that it is now appropriate to study the mechanisms by which cell surface carbohydrates and lectins might be involved in recognition. Convincing proof for the role of cell surface carbohydrates and lectins in plant/microorganism recognition must come from evidence of how these components contribute to the infection process.

In their original paper on the subject of lectins in the white clover/*R. trifolii* symbiosis, Dazzo & Hubbell (65) proposed a simple model to suggest how the clover lectin and the *R. trifolii* cell surface receptor might function in determining host specificity. This model, illustrated schematically in Figure 1A, involves binding of the clover lectin to common antigen sites on the surface of both the root and the bacterial symbiont. In effect, the lectin is depicted as a specific glue, holding the bacterial symbiont to the root surface. Infection and nodulation follow. One would predict from this model that plants which do not have the lectin will not bind *R. trifolii* cells to the root surface and will not be infected or nodulated. Similarly, one would predict that rhizobia which lack the common antigen surface component to which the clover lectin binds will not be adsorbed to the root and will not infect or nodulate. Neither of these predictions is fully correct: nonhost pea roots rapidly bind large numbers of *R. trifolii* cells (44), and cells of the heterologous species *R. meliloti* are adsorbed in considerable numbers to white clover roots (67). The recognition model proposed by Dazzo & Hubbell (65) falls short because it does not indicate that there is any *functional* difference, in terms of infection and nodulation, between

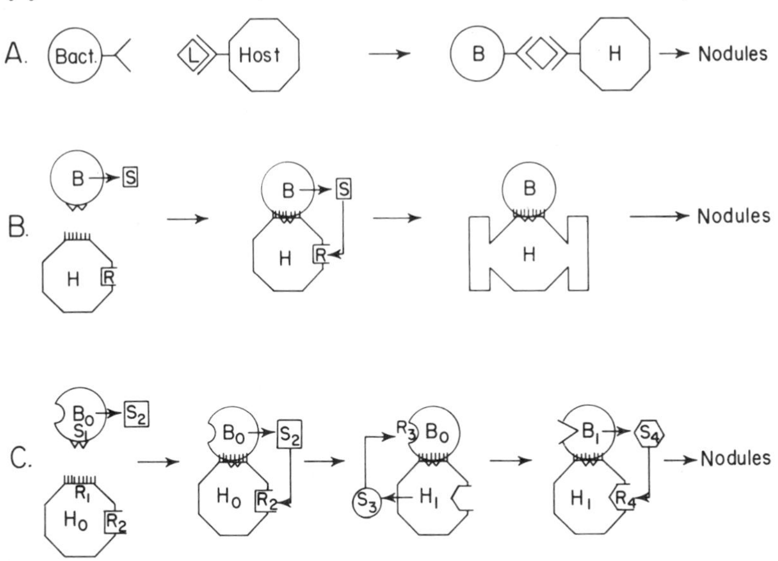

Figure 1. Recognition models for *Rhizobium*-legume associations.

clover lectin-mediated attachment and attachment that is not mediated by clover lectin.

An alternative model of recognition is illustrated in Figure 1B. This model reflects the approach taken previously by Albersheim & Anderson-Prouty (3), Bauer & Bhuvaneswari (12), and more recently by Graham (90). In this model, the bacterium attaches to a suitable host cell by some unspecified mechanism. An extracellular substance or surface component of the microorganism interacts with a receptor substance on the surface of the host cell. This interaction is coupled so that it triggers certain responses in the host that lead to infection and nodulation. As in the previous model, the "receptor" could be a host lectin and the "signal" substance could be a cell surface polysaccharide. And, as in the previous model, the interaction between signal and receptor could contribute to the physical attachment of the microorganism to the host and thereby enhance the probability of subsequent infection and nodulation. But the emphasis in this second model is on the *activation of host responses by the microorganism.*

This second model is also likely to be inadequate. It is probable that both the host and the rhizobia exchange signals and make appropriate responses, including the production of new signals, at each stage of the infection/nodulation process. If this is so, then it is arbitrary to consider only the first signal-and-response interaction as *the* recognition step. The first signal-and-response interaction is not a priori more special than the second or the third or the fourth. Perhaps after much investigation it will be established that the first signal-and-response interaction is by far the most selective one for a particular host/microbe combination. But this would provide no assurance that the same would be true for another host/microbe combination, or that the most selective signal-and-response interaction is the only significant one. In our present state of ignorance, the important thing is for all concerned to understand the same thing when using the term "recognition," and to have this common understanding based on a model that is sophisticated enough to at least approximately reflect the real biology of the associations. For this reason, a third recognition model is presented (Figure 1C). This model suggests that recognition is not a one-step process. It suggests that recognition is the cumulative effect of a series of signal-and-response interactions.

The initial step in the interaction sequence (Figure 1C) is arbitrarily depicted as attachment of the microbe to the target host cell. Attachment is considered here to be a required step in the infection process, although it is not necessarily the first step, or a host-specific one, and could occur at the same time as some other signal-and-response interaction. Following attachment, the model consists simply of an alternating sequence of signal production, signal reception, and reception-induced responses between the symbionts.

The overall recognition process in this model can be broken down into individual signal-and-response steps that can be studied and characterized independently. Characterization would include a testing of the biological specificity of signal production, of signal reception, and of reception-coupled responses. Since the overall specificity of the symbiotic association is the cumulative effect of several steps, there is no temptation to make any one step all-powerful in terms of explaining cross-inoculation specificity. Nor is there a temptation to discount any signal-and-response interaction as unimportant because it fails to account, by itself, for the full biological specificity of the symbiosis.

In concluding this section, it should be mentioned that Bowles has recently published a short theoretical paper of some significance (31). She has addressed the question of how the binding activity of cell surface lectins might be regulated and coupled to activities within the host cell. Her hypothesis suggests that cell surface lectins, localized in cell membranes, can be self-neutralized by binding to endogenous membrane (and perhaps transmembrane) receptors. According to the hypothesis, the formation, maintenance, and destruction of such self-neutralized complexes can serve to regulate lectin activity and can modulate any other biochemical activity of either member of the pair (31). Interaction of the lectin with exogenous carbohydrate receptors can, of course, affect the formation of the lectin complexes with endogenous membrane components, and vice versa. This model is useful beyond any reference to lectins, since it could apply equally well to almost any signal, signal reception, and reception-induced response mechanism, and need not be restricted to components of the cell membrane.

STRUCTURE AND ACTIVITY OF CELL SURFACE POLYSACCHARIDES

It seems highly probable that some kind of *Rhizobium* surface polysaccharide is important in establishing the symbiotic relationship (22, 23, 28, 63, 65, 68, 123, 124, 129, 140, 175, 209). It is now important to learn which surface polysaccharide is the important one and what it does. Several different kinds of *Rhizobium* cell surface polysaccharides have been described: exopolysaccharides (EPS), capsular polysaccharides (CPS), lipopolysaccharides (LPS), 2-linked glucans, 3-,6-,3,6-linked glucans, and β-4-linked glucans (cellulose). The recent review by Carlson (40) should be consulted for a comprehensive discussion of these *Rhizobium* polysaccharides.

The cellular location and morphology of surface polysaccharides can profoundly influence their biological activity. Therefore, I will attempt to maintain distinctions based on location or morphology wherever possible. The reader should be aware, however, that reports in the literature rarely specify or determine the cellular location or morphology of the polysaccharide material they describe.

Exopolysaccharides and Capsular Polysaccharides

Rhizobia in laboratory cultures generally produce abundant quantities of acidic heteroexopolysaccharides. Acidic heteroexopolysaccharides are polysaccharides containing several different sugars or sugar derivatives. Uronic acids are common acidic components. The term exopolysaccharide will be used here to indicate polysaccharides that are not attached to the bacteria. In general, the acidic heteroEPS of rhizobia are similar in sugar composition to those produced by other Gram-negative bacteria (190), although slow-growing rhizobia produce EPS that may contain one or more unusual sugars (8, 75, 115, 116). Carlson (40) summarizes the numerous studies of *Rhizobium* acidic heteroEPS in the following manner:

1. The acidic heteroEPS from various strains of *R. trifolii* and *R. leguminosarum* have identical or nearly identical structures (Figure 2A) (106, 168, 187).
2. Different strains of *R. meliloti* appear to synthesize an acidic heteroEPS (Figure 2B) that is very similar or identical to the EPS from *Agrobacterium tumefaciens,* but it is quite distinct from the EPS of other fast-growing rhizobia (27, 214).
3. Slow-growing rhizobia such as *R. japonicum* produce acidic heteroEPS that are quite different from those produced by the above mentioned fast-growing rhizobia. Moreover, different strains of slow-growing *Rhizobium* species may produce completely different EPS (Figure 2C and 2D) (75, 76; A. J. Mort and W. D. Bauer, unpublished).

The structures of the *Rhizobium* acidic heteroEPS that have been determined thus far are all complex, perhaps sufficiently complex that the proposed structures (Figure 2) must be regarded as somewhat tentative and subject to minor modification or elaboration. The determination of these difficult structures is a major and noteworthy advance. These structures provide the basis for further investigations of structure-function relationships *provided that biologically active polysaccharide materials were isolated for the structural analyses.*

Most unfortunately, it appears that the polysaccharide materials obtained for structural analyses from *R. trifolii* may not have been biologically active. Previous studies with *R. trifolii* 0403 indicated that broth cultures were reactive with anti-clover root antiserum during early exponential

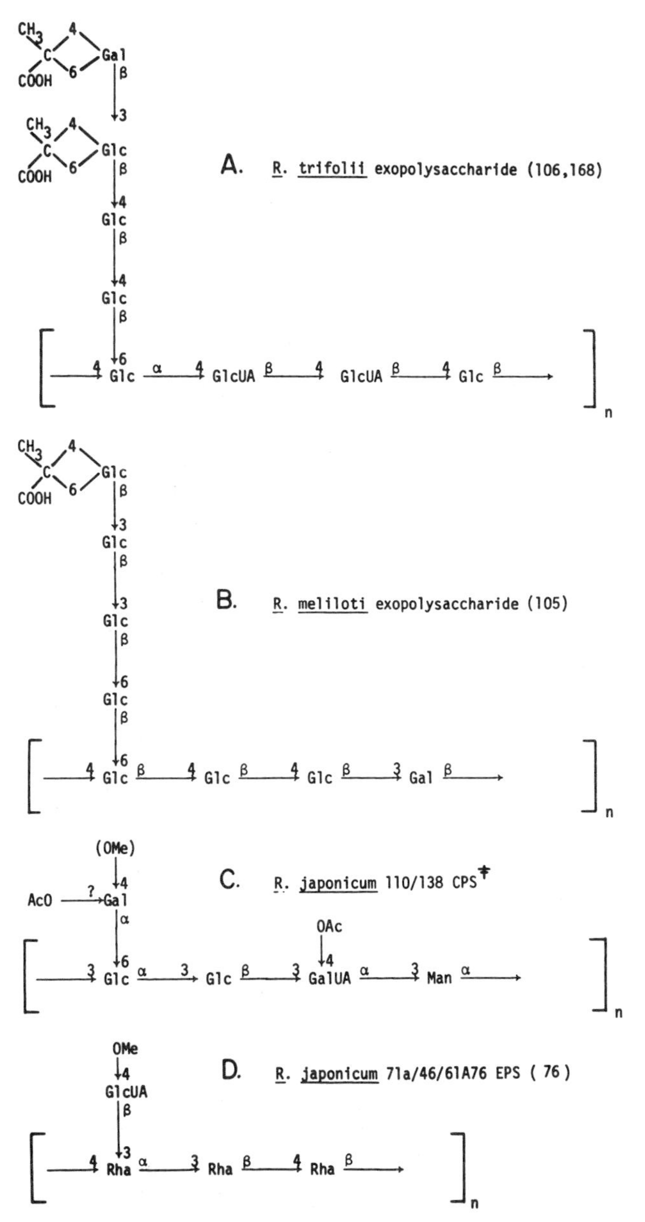

Figure 2. Primary structure of *Rhizobium* exopolysaccharides. Glc = D-glucose, Gal = D-galactose, Man = D-mannose, Rha = L-rhamnose, GlcUA = D-glucuronic acid, GalUA = D-galacturonic acid, Ac = acetate, Me = methyl, $CH_3C(4,6)COOH$ = acetal-linked pyruvate. The positions of acetate substituents on structures A and B are not shown.‡ A. J. Mort, W. D. Bauer, unpublished.

growth phase (26–30 h) and during early stationary phase (39–42 h), but not between these periods or later in stationary phase (68). Jansson et al (106) obtained their polysaccharide for structural analysis from culture filtrates of 4-day-old shake cultures. The growth phase of these *R. trifolii* cultures was not determined, but was quite possibly into late stationary phase. The polysaccharides used by Robertsen et al (168) were purified from culture filtrates obtained from cultures in late exponential growth phase. This apparent mismatch between the time of reactive polysaccharide production and the time of harvesting for analytical samples makes it necessary to suspend judgment on the relation between *R. trifolii* polysaccharide structure and function.

One striking feature of the proposed *R. trifolii* polysaccharide structure (Figure 2A) is the absence of 2-deoxyglucose. The lectin cross-bridging model of Dazzo & Hubbell (65) and the many supporting studies depend heavily on the specific inhibition of clover lectin binding and attachment to clover roots by 2-deoxyglucose. If 2-deoxyglucose or some close analog is not present in the capsular polysaccharide from *R. trifolii,* then either the model is on shaky ground or the wrong polysaccharide was analyzed. A recent preliminary report states that polysaccharides isolated from exponential and early stationary phase cultures of *R. trifolii* 0403 differ with respect to the relative amounts of 4 monosaccharide components and O-acetyl and O-pyruvate substituents (199). The variable sugars have been tentatively identified as deoxy- or dideoxyhexoses (F. B. Dazzo, personal communication). These observations suggest that a lectin-binding, deoxyhexose-containing polysaccharide may appear on cells in early exponential phase and then abruptly disappear. This needs to be demonstrated rather conclusively in order to resolve the present discrepancy between EPS structure and 2-deoxyglucose-related activity.

The uncertain relationship between biological activity and *R. trifolii* EPS structure is further complicated by the uncertain cellular localization of the active polysaccharide. None of the investigations have established yet whether or not reactive polysaccharide is present in *R. trifolii* culture filtrates, which were the source of polysaccharide for the structural analyses. Essentially all of the studies on *R. trifolii* by Dazzo and his colleagues have emphasized the capsular location of the reactive polysaccharide. This emphasis is especially clear in the study by Dazzo, Urbano & Brill (68) on the transient nature of the lectin receptors. The authors state that the reactive material is present on encapsulated cells of a particular age. Unencapsulated cells and encapsulated cells in late stationary growth phase did not show any lectin or antibody binding activity (68). However, in another relevant study, Dazzo & Brill (63) reported that at least two reactive acidic heteropolysaccharides were present in preparations of crude capsular material from *R. trifolii.* One of these reactive polysaccharides was reported to contain ketodeoxyoctonate (KDO), lipid A, and heptose, which are components characteristic of LPS molecules. But if *R. trifolii* has reactive LPS molecules, then why don't *unencapsulated* cells react with clover lectin or anti-clover root antiserum?

Since the complete structural analysis of any complex heteropolysaccharide is a major undertaking and a significant achievement when properly done, it is vital that such efforts be reserved for polysaccharides that have been isolated from the correct cellular location, that have been carefully purified to homogeneity, and that have been shown—after careful purification—to have some well-defined biological activity. Certainly it is unfortunate that neither Janssen et al (106) nor Robertsen et al (168) attempted to isolate their polysaccharide from the capsules of *R. trifolii* cells, and equally unfortunate that neither group attempted to assay their purified EPS for induction of polygalacturonase activity (123–125), root hair curling activity (102, 186, 212), increased nodulation (123), reactivity with clover lectin or anti-clover root antiserum (65), or inhibition of *R. trifolii* binding to clover root hairs by the EPS (67).

The structure of the acidic heteroEPS from *R. meliloti* has also been determined (Figure 2B). However, the relation between structure and function is even less satisfactorily established than for the EPS from *R. trifolii.* Partially purified EPS and sterile culture filtrates containing EPS have been reported to enhance polygalacturonase activity in alfalfa root bathing medium (123, 157). Sterile culture filtrates of *R. meliloti* have also been reported to induce root hair deformation (102, 212). There is no assurance, however, that these reported biological activities are actually due to the EPS since the active substances were not purified to homogeneity and characterized. Jansson et al (105) purified the acidic heteroEPS from *R. meliloti* to apparent homogeneity, but did not assay this material for biological activity, did not determine its cellular location, and did not report the growth phase of the cultures from which the EPS was obtained.

The situation with regard to the EPS of *R. japonicum* is somewhat more satisfactory, though still incomplete. The structures of the EPS from several strains of *R. japonicum* have recently been determined (Figure 2C and 2D). The acidic heteroEPS of two of these *R. japonicum* strains (110 and 138) have been shown to be reactive with soybean lectin in a biochemically specific manner (23, 39, 197). The lectin-binding polysaccharide is present on the bacterial surface under in vivo conditions (22).

Purified CPS from strain 138 was found to enhance the extent of nodulation above the SERH mark (see Figure 3) when used to pretreat soybean roots prior to inoculation (W. D. Bauer, T. V. Bhuvaneswari, B. G. Turgeon, and A. J. Mort, unpublished data). The optimal amount of purified, autoclaved CPS for this effect is between 5 and 50 nanograms per plant under the assay conditions used. This argues against the possibility that the observed activity is due to a minor contaminant of the CPS, as does the fact that purified oligosaccharides from the CPS are also active. In addition, *N*-acetylgalactosamine was found to be active in enhancing nodulation above the SERH mark, whereas *N*-acetylglucosamine had no effect. *N*-acetylgalactosamine is a potent monosaccharide hapten of soybean lectin binding, whereas *N*-acetylglucosamine does not bind to soybean lectin (23, 122). This suggests that the enhancement of nodulation is related to lectin binding activity.

Electron microscopic studies revealed that ferritin-labeled soybean lectin bound to the capsules of both *R. japonicum* 61A76 and 3I1b 138 cells (9, 39). Lectin-binding polysaccharide is also present in culture filtrates (23, 133, 197). The absence of labeled lectin on the outer membrane of the unencapsulated and the partially encapsulated cells examined by freeze etch and thin section techniques indicated that the outer membrane had no specific lectin binding components (9, 39). In particular, it seems unlikely that LPS from these *R. japonicum* strains can bind soybean lectin. This conclusion is in agreement with the Ouchterlony double-diffusion experiments of Kamberger (108). Precipitin bands were formed between soybean lectin and *R. japonicum* CB1809 EPS, but not between soybean lectin and *R. japonicum* CB1809 LPS (108). Previous results indicating a specific interaction between soybean lectin and *R. japonicum* LPS were probably due to EPS contaminants in the LPS preparations (209). The material identified as *R. japonicum* EPS was of very low molecular weight (209) and may have been 2-linked glucan.

The acidic heteropolysaccharides isolated from culture filtrates of *R. japonicum* strains 110 and 138 have been compared with the polysaccharides isolated from the capsules of these strains. The EPS and CPS are closely related, but not necessarily identical (133). The compositions of the CPS and EPS of these strains changed in a rather specific manner with culture age (133). The galactose content of the polysaccharide decreased as the cultures reached late exponential growth phase, and the amount of 4-O-methylgalactose increased correspondingly (133). These results suggest that galactosyl residues in the CPS may have been converted to 4-O-methylgalactosyl residues by a specific methylating enzyme. The relative amounts of the other monosaccharide components remained constant. Overall, the compositional analyses of the polysaccharides from these strains indicate that EPS in *R. japonicum* culture filtrates originates from CPS that is lost or released from the capsule.

The galactose side chains of the 110/138 type of polysaccharide (Figure 2C) are probably the principal determinants of lectin-binding activity, since soybean lectin binds specifically to galactose and related sugars [e.g. galactosamine, *N*-acetylgalactosamine, fucose (23, 122)]. The contribution of the sugar residues in the backbone of the polysaccharide to lectin-binding activity or nodulation enhancement activity is unknown. Soybean lectin binds to galactosyl residues but may not be able to bind to 4-O-methygalactosyl residues (96). The addition of methyl groups to the polysaccharide may thus be one means of regulating CPS activity or reactivity. Direct measurements of the binding and nodulation activity of 4-O-methylated vs unmethylated CPS are needed. A preliminary communication indicating the separation of lectin-binding *R. japonicum* polysaccharides by affinity chromatography on lectin-agarose columns is promising in this regard (194).

The synthesis of EPS with a very different structure (Figure 2D) by other *R. japonicum* strains (71a, 46, and 61A76) poses a difficult problem for the structure-function relationships suggested above. The second type of EPS structure contains no galactose or structural analogs of galactose that could serve as normal binding sites for soybean lectin. It is possible that *R. japonicum* strains such as 61A76, 46, and 71a also synthesize the galactose-containing type of polysaccharide (Figure 2C). Ultrastructural studies indicated that only a very small proportion of the cells in cultures of 61A76 were encapsulated (9). Only these encapsulated cells were reported to bind labeled soybean lectin. One to 5% of the cells bound lectin when this strain was cultured in association with soybean roots (22). Thus, *R. japonicum* strains such as 61A76 may produce the galactose side chain polysaccharide in small amounts as tight capsules, especially in the presence of the host. Further studies are needed to determine whether or not the synthesis of the galactose side chain polysaccharide is a characteristic of all *R. japonicum* strains.

Lipopolysaccharides

The LPS of rhizobia appear to be quite similar in most respects to the LPS of other Gram-negative bacteria (40, 41, 162, 208, 216). The sugar compositions of highly purified LPS from several strains and species of fast-growing rhizobia indicate that *LPS composition generally varies from strain to strain within a species almost as much as it varies between species* (40, 41, 216). Comparable data are not yet available for LPS from slow-growing *Rhizobium* species.

The biological significance of the great variability of LPS composition among strains of fast-growing rhizobia is not clear. On the one hand, the compositional diversity among different strains of the same species suggests that LPS does not contribute to host specificity. Host specificity requires components that are present in the LPS of all strains of a given species and that are absent from strains of other species. Carlson (40), on the other hand, does not believe that the compositional diversity and immunological specificity of LPS from different strains of fast-growing species necessarily rules out the possibility of LPS involvement in determining host specificity.

This issue will not be decided without the sugar linkage or linkage/sequence analyses of LPS from several *Rhizobium* strains and species. However, such structural analyses will not be very useful unless the homogeneity of the isolated LPS is well established and unless the cellular location and growth phase dynamics of the LPS are ascertained. It would be especially desirable if carefully purified LPS could be shown to have some specific, infection-related biological activity. The biological activity of *Rhizobium* LPS in the symbiosis is unknown. Carefully purified LPS does not seem to have been tested with any host legume for specific root hair curling activity, for polygalacturonase induction, nodulation enhancement activity, inhibition of *Rhizobium* attachment to roots, etc. LPS from several species of rhizobia have been reported to interact specifically with host seed lectins (20, 63, 108–110, 161, 209). These reports cannot be interpreted with any certainty as yet because of the questionable purity of the LPS preparations and because sugar hapten controls were not normally employed to eliminate possible artifacts due to biochemically nonspecific interactions. The most persuasive studies seem to be those of *R. meliloti* and *R. leguminosarum* LPS binding to alfalfa lectin, and to pea and lentil lectins, respectively, by Kamberger (108, 109). Nonetheless, these studies are subject to special difficulties. The first difficulty is that the lectins isolated from pea and lentil seeds are "specific" in their binding for glucose and mannose-containing polymers. As indicated below, rhizobia commonly produce cell surface glucans. These glucans can be extracted along with LPS when the standard hot phenol extraction method is used (161, 209, 216). It is not clear that glucan contaminants were fully removed from the LPS preparations (108, 109). The second difficulty is that the LPS from *R. meliloti* seems to be highly unusual. The very high proportion of KDO and the high cross-reactivity of antisera against LPS from different *R. meliloti* strains (40) suggest that LPS from *R. meliloti* may lack O-antigenic repeating units. LPS of this sort would be comparable to the rough and deep rough mutants characterized in *E. coli* (156). Whether the reported binding of alfalfa lectin to *R. meliloti* LPS (108) is an artifact of this unusual LPS structure or a biologically important interaction is uncertain.

It is quite relevant that similar questions and uncertainties about LPS and EPS function are of current interest among plant pathologists. Much of the pertinent literature on pathogenic associations has been discussed in the recent review by Sequeira (178). As yet no clear and general model of the role of either LPS or EPS in host/pathogen interactions has been developed. Indeed, no single pattern may hold for different species. LPS would seem to be required for the virulence of *Agrobacterium* (207) with EPS unimportant. However, EPS seems required for the virulence of *Pseudomonas solanacerum* (179) and *Erwinia amylovora* (16; D. J. Politis, R. N. Goodman, personal communication), with LPS being a factor that limits the infectivity of *P. solanacerum* (91) or a factor that is unimportant for *E. amylovora.*

Glucans

In addition to EPS, CPS, and LPS, rhizobia have been found to produce several glucans. A number of reports have shown that both *Rhizobium* and *Agrobacterium* species synthesize neutral β-1,2-linked glucans (70, 132, 164, 213, 215). Such glucans have not been reported in any other group of bacteria. It is biologically most intriguing that both *Rhizobium* and *Agrobacterium* species, but not other bacteria, share the ability to synthesize these unusual, low molecular weight (ca 3000 daltons) and probably cyclic (213, 215) polysaccharides. As yet no function has been indicated for these glucans in host plant/microbe interactions, although Zevenhuizen & Scholten-Koerselman (215) suggested that the 2-linked glucan was firmly attached to the bacterial cell wall and may serve to mask outer membrane determinants. It will be of considerable interest to identify the biological function of this class of polysaccharides. It should be possible to determine the cellular localization of these glucans with labeled antiserum against purified glucan, and to examine a wide variety of *Rhizobium* strains, species, and mutants with the labeled antiserum so as to test the generality and conditions of glucan synthesis. The neutral character, small size, and simple structure of the β-2-linked glucans should at least make them easy to purify to homogeneity—which is more than can be said for most other *Rhizobium* polysaccharides.

Another type of glucan has been discovered recently in culture filtrates of *R. japonicum* strain 71a (77). Two closely related glucans, each containing different proportions of terminal, 3-linked, 6-linked, and 3,6-linked glucosyl residues, were identified. These two glucans are therefore quite distinct from the 2-linked glucans described above, but are quite similar to the glucan isolated from *Phytopthora megasperma* var. *sojae,* a fungal pathogen of the same host, soybean (7, 77). Preliminary experiments indicated that one, but not both, of the 3- and 6-linked *R. japonicum* glucans acted as an elicitor of phytoalexin synthesis in soybean, as does the similar glucan from *P. megasperma* (7). The observation that a *Rhizobium* polysaccharide elicits an inducible defense response in the host rather clearly raises the question of how rhizobia are able to maintain intimate contact with their hosts in the face of both passive and inducible host defenses. This question should draw serious attention in the future, for it is basic to our understanding of how a symbiotic association can be achieved. If we hope some day to extend the host range of microbial symbionts or develop new plant/microbe symbioses, then certainly we must learn how microbial symbionts overcome, avoid, or suppress plant defense responses.

Rhizobia are reported to produce one other type of extracellular glucan: cellulose fibrils (71, 141). As yet the only known biological consequence of cellulose fibril production by rhizobia is flocculation of the bacterial cells (71). Fibril production may be involved in establishing firm attachment of the bacteria to surfaces.

EARLY EVENTS IN THE INFECTION PROCESS

Attachment

The available evidence indicates that *Rhizobium* cells can become attached to the host root surface within seconds or minutes after inoculation (44, 67). Thus, attachment of *Rhizobium* cells to legume roots is likely to be one of the first steps in the required sequence of interactions leading to infection and nodulation. The subject of *Rhizobium* attachment to roots has been reviewed recently (58) and will not be explored deeply here. Selective attachment has been proposed for both the white clover/*R. trifolii* association (65) and the soybean/*R. japonicum* association (188). Clover lectin is suggested to bind to cross-reactive antigen sites on *R. trifolii* and clover root surfaces, with consequent selective attachment of the bacterial symbiont to the roots (65). The lectin-mediated attachment mechanism is supported by direct microscopic observations of *Rhizobium* cells binding to root hair cells of seedlings in Fåhraeus slide assemblies (67). Microscopic, immunological, and hapten inhibition evidence for lectin-mediated attachment of *R. japonicum* cells to soybean roots has been described recently (188). Thus, lectin-mediated attachment may prove to be a reasonably common feature of legume/*Rhizobium* associations.

It is intriguing that the reverse attachment strategy, with microbial lectin and host carbohydrate binding sites, seems to be employed by a pathogenic fungus (100), a predatory fungus (149), and the intestinal bacterium *E. coli* (155). Perhaps associations favorable to the host are characterized by the host lectin attachment strategy and associations favorable to the microbe by the microbe lectin attachment strategy.

There is some danger of overemphasizing lectin-mediated attachment of rhizobia. It is important to keep in mind that rhizobia are very successful soil microorganisms even in the absence of their host. Attachment to any source of nutrients, or to any surface in a flow of nutrients, is a character that will be strongly selected for. Rhizobia readily attach to surfaces without the involvement of lectin binding. They clump together and attach to glass and plastic vessels and to the roots of host and nonhost plants (58, 181,

196). With regard to the establishment of symbiosis, it may not really matter whether *Rhizobium* cells attach by lectin-mediated or nonlectin-mediated mechanisms to infectible sites on the host root.

Although the ecological importance of microbial attachment to surfaces is now recognized (26), little is known yet about the nature of the surface components of the rhizobia and host root that are responsible for nonlectin-mediated attachment. Polar attachment of rod-shaped *Rhizobium* cells to various surfaces, including host roots, has been observed many times (29, 53, 57, 67, 119, 188). It seems likely that polar attachment is mediated by "polar bodies" (29, 196). The biochemical nature of the polar bodies and the environmental conditions that govern the formation and loss of polar bodies have not been ascertained. The capsules of rhizobia, which are probably distinct chemically as well as morphologically from polar bodies, may also be instrumental in the attachment of the bacteria to the host root (9). Comparisons need to be made of the host root attachment capabilities of cells with and without polar bodies, cells with and without lectin-binding capsules, and cells with and without nonlectin-binding capsules. Other bacterial structures, including cellulose fibrils and fimbriae, may also be involved in host root attachment.

There is little information available concerning the nature of the host root surface or the zonation and stratification of *Rhizobium* populations in the vicinity of the root surface. Investigations have pointed to the importance of the "mucigel" layer of the host root surface (57, 92). Dense populations of rhizobia were found within the body of the mucigel (57). A thin, electron-dense mucigel boundary layer has been observed on the outer surface of the mucigel in clover and alfalfa (57), pea (92), soybean (B. G. Turgeon and W. D. Bauer, unpublished data), and other plants (26). The mucigel layer has been reported to cover most of the root surface, quite thickly in some regions (57, 58, 92). Numerous rhizobia were observed in the clover and alfalfa mucigel just beneath the boundary layer 2 to 3 days after inoculation, whereas relatively few rhizobia were found above or attached to the surface of the boundary layer (57). These observations raise a number of questions that require further study. If host lectins on the root surface are responsible for selective binding of homologous rhizobia, are the lectins located on the epidermal cell walls, or within the mucigel, or on the surface of the mucigel boundary layer? If rhizobia initially attach to the outside of the mucigel boundary layer, how do they subsequently get inside this layer? How does attachment or lectin/polysaccharide binding on the surface of the boundary layer or within the mucigel induce responses in the host cell? [In this regard, K. E. Fjellheim and B. Solheim (personal communication) have recently observed that root extracts from pea and clover were able to selectively degrade radiolabeled capsular polysaccharide preparations from *R.*

leguminosarum and *R. trifolii,* respectively. Dialyzable oligosaccharide fragments of the CPS were formed. It has also been observed that oligosaccharide-containing factors induce cell differentiation in *Dictyostelium* (193). Thus, perhaps we should ask whether oligosaccharides derived from *Rhizobium* CPS migrate from the mucigel surface to the host cell plasma membrane and activate host responses.] We may also ask whether rhizobia multiply especially well in the mucigel layer because it traps high concentrations of nutrients. What are the components and permeabilities of the mucigel and the mucigel boundary layers? Are rhizobia motile within the mucigel? Do ordinary homogenization and plating techniques detect and accurately count mucigel populations? Is the partial pressure of oxygen in the mucigel environment sufficiently low (114) to induce nitrogenase synthesis in rhizobia? Is there significant nitrogen fixation by mucigel rhizobia?

Location of Infectible Root Cells

There are likely to be several intervening steps between attachment and root hair curling that cannot be detected at the light microscope level of resolution. One promising strategy for detecting and characterizing these early events is that of isolating and studying a variety of non-nodulating, noninfective mutants of the *Rhizobium* symbiont. This approach is just beginning to yield results (126, 170, 172, 203). Transposon mutagenesis has been a very effective way of generating non-nodulating mutants at high frequencies, at least for the fast-growing rhizobia (17, 126, 170).

A second, more general strategy is simply to look for changes in the rhizobia and in the host cells that occur between the time of attachment and the time that curling begins. Any number of techniques might be used to look for changes during this interval: ultrastructural, biochemical, physiological, immunological, and electrophysiological. However, such techniques cannot be used effectively unless it is reasonably certain that a particular *Rhizobium* cell and the host cell to which it is attached will interact to produce an infection. As a rule, the vast majority of rhizobia attached to host root cells never produce an infection. Infections are ordinarily generated in a very small percentage—less than 5%—of the root hair cells of temperate legumes such as pea, alfalfa, or clover, and many of these are abortive (54–56). Moreover, individual root hair cells and other epidermal cells may each have several attached rhizobia, perhaps as many as 20 or 30 (67). Thus, the few host/rhizobia cell pairs that do produce successful infections are ordinarily a fraction of 1% of the total cell pairs. It is extremely difficult to study changes occurring in the few productive cell pairs if the changes must be detected against a background of hundreds of unproductive pairs.

A useful contribution to the study of early events in the interaction sequence is provided by the recent development of a relatively simple method for determining the location of the infectible host cells on legume roots (24). The method for determining the location of infectible cells involves the growth of seedlings in plastic growth pouches [diSPo seed pouch, Scientific Products (206)]. The roots are visible in these pouches, and the position of the root tip and the zones of root hair development can be marked with a waterproof pen at the time of inoculation (Figure 3A). The marks are made on the plastic overlaying the root. Nodules develop on the primary roots to a readily visible size 4 to 8 days after inoculation. Their position can then be determined relative to the marks made on the pouch at the time of inoculation. In this way, the position of infectible root cells at the time of inoculation can be inferred from the position of subsequent nodule formation, provided of course that the effects of root elongation (Figure 3B) are taken into consideration.

The profile of nodulation frequency in Figure 4 illustrates the pattern of nodulation on the primary root obtained with soybean/*R. japonicum* (24). Nodules failed to develop in the zone where mature (fully elongated) root hairs were present at the time of inoculation. Nodules formed occasionally in the zone where developing root hairs were present at inoculation. The most frequent nodulation, however, occurred in the zone where *no* root hairs were present at inoculation. Bhuvaneswari et al (24) suggested that the infectible cells of soybean are developmentally restricted at any given time to the region of the root just below the smallest emergent root hairs and just above the zone of rapid root elongation.

Spot inoculations at specific points on the root surface have been used to confirm the developmentally restricted pattern of infectible host cells described above (24). More importantly, however, it appears that spot inoculation techniques can be used to greatly facilitate studies of the infection process. Inoculum droplets approximately 200 μ in diameter (5 nanoliters) have been applied to the surfaces of soybean roots at locations 70 to 80% of the RT-SERH distance above the root tip (Figures 3 and 4). Small ion exchange beads that adhere to the root surface were used to mark the inoculation sites. Nodules were observed to develop within a millimeter of the inoculation site on 80 to 85% of the seedlings (B. G. Turgeon and W. D. Bauer, unpublished). With such techniques it may be possible to make systematic studies of the early infection events at almost a single host cell level.

Delayed inoculation experiments have shown that the infectibility of soybean epidermal cells is a transient property, as expected from the developmentally restricted pattern of nodulation indicated above (24). Infectibility is acquired and lost acropetally over a period of about 5 h in soybean.

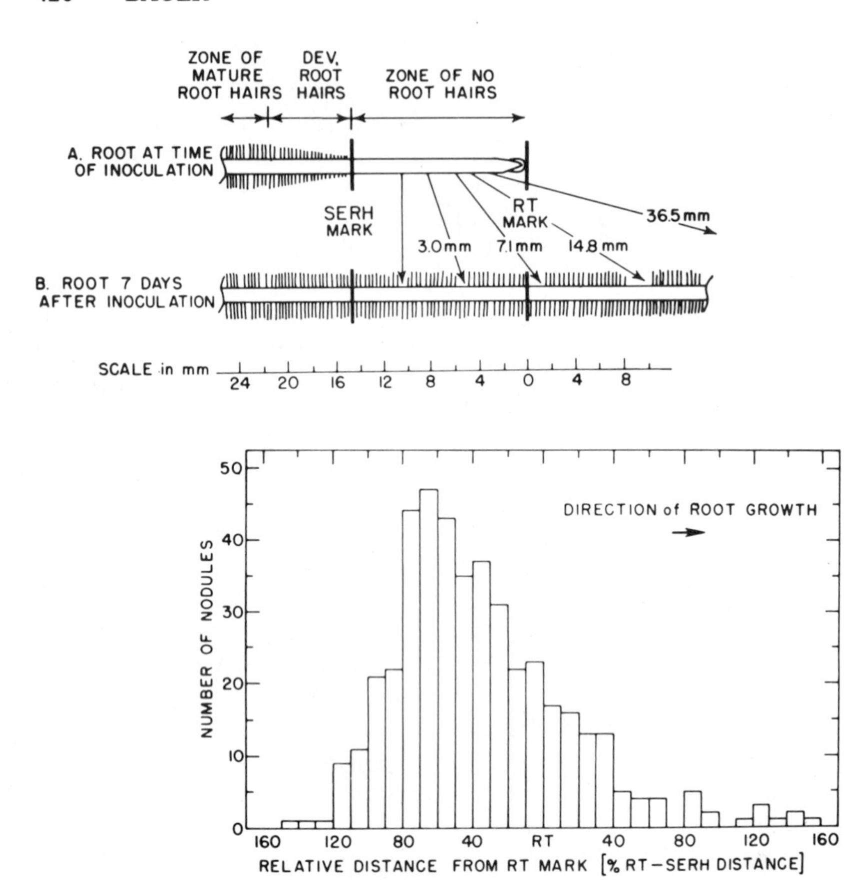

Figure 3. (*top*) Location and development of soybean root epidermal cells at A, the time of inoculation, and B, the time of nodule scoring. Arrows with associated numbers indicate the distances that epidermal cells are displaced by elongation relative to the root tip (RT) mark made at the time of inoculation. SERH = smallest emergent root hairs (From Ref. 24 with permission of *Plant Physiology.*)

Figure 4. (*bottom*) Profile of nodulation frequency at different positions on the primary root of soybean relative to the positions of the root tip (RT) and smallest emergent root hairs (SERH) at the time of inoculation. This figure is aligned with Figure 3 above and uses a common scale. (From Ref. 24 with permission of *Plant Physiology.*)

The average distance of the uppermost nodule on the primary root above the mark made at the root tip decreases rapidly in a linear manner as the interval between marking and inoculation is increased (24). Thus, the position of the uppermost nodule in soybean is a sensitive indicator of factors that affect the rate at which infections are initiated, and should be useful as such in future studies of the infection process.

Infections in soybean take place through infection threads formed in emergent root hair cells (25, 145, 166). Nonetheless, infections leading to nodule development appeared to be initiated most frequently in that zone of the root where no emergent root hairs were present at the time of inoculation (24). No certain explanation can be provided for this apparent discrepancy. Bhuvaneswari et al (24) speculated that perhaps "developing, preemergent hair cells must be activated by the bacteria, or by some substance from the bacteria, in order for these hair cells to develop as sites of subsequent infection and nodulation." The activation and development of preemergent hair cells into sites of infection appeared to require 2 to 3 h (24). It thus seems possible that the activation and development of preemergent hair cells is a very early step in the interaction sequence for soybean. Support for this possibility is provided by the observation that pretreatment of soybean roots with *R. japonicum* culture filtrates, purified *R. japonicum* capsular polysaccharide, or *N*-acetylgalactosamine has the effect of increasing the proportion of plants subsequently nodulated above the SERH mark (W. D. Bauer, T. V. Bhuvaneswari, B. G. Turgeon, and A. J. Mort, unpublished). These substances may be able to activate preemergent hair cells in the absence of the bacteria.

Studies of the localization and development of infectible root cells in other legumes have indicated both similarities and differences with regard to the pattern described for soybean. Cowpea, alfalfa, white clover, and peanut showed evidence of transient infectibility of developing epidermal cells (13). However, nodules on white clover also developed abundantly throughout the zone that contained mature root hairs at the time of inoculation (13, 81, 153). It is likely that infections in the mature root hair zone of white clover were "basal" infections, originating near the base of a fully elongated hair where a lateral branch of the hair had formed (52, 81). It appears that there is also a transient infectibility in white clover that is restricted to the developing root hair zone and the no-root hair zone (13, 52, 54, 81). In these zones, infections were probably "apical" (i.e. originating near the tip of a root hair) (13, 52, 81). The different patterns of nodulation observed for different legumes provide a valuable caution against projecting results from one legume/*Rhizobium* combination to other combinations that may infect by different mechanisms.

Infective Cell Type of Rhizobia

In the previous section considerable attention was given to the identification of the infectible cells on the host root. It was pointed out that the root cells capable of becoming infected so that nodules subsequently developed constitute only a small proportion of the total number of epidermal or hair cells. It seems pertinent to ask if some kind of developmental restriction also applies to rhizobia. Are all the cells in a *Rhizobium* culture equally infective? If all *Rhizobium* cells are not equally infective, which cells are the infective ones? Do the noninfective cells become infective after a brief period of contact with the host?

Some indirect evidence is available to suggest that only a small fraction of the cells in an ordinary laboratory culture of *Rhizobium* is actually infective at the time of inoculation. The effect of inoculum size on nodulation above the RT mark has been investigated. Extrapolation of the dose/response curves in such experiments to the inoculum dose axis consistently provided intercepts in the range of approximately 5 to 20 rhizobia per plant (24). This result suggested that perhaps only one cell in every 5 to 20 *Rhizobium* cells is infective (24). (Alternatively, perhaps a minimum of 5 to 20 *Rhizobium* cells must act in concert to initiate an infection.) Purchase & Nutman (163), using quite different methods, measures, strains, and hosts, have also reported that nodulation required a minimum of about 10 cells.

The question of which cell type of rhizobia is infective is a difficult one to investigate in any direct manner. Nonetheless, several indirect approaches to the identification of an infective cell type have provided interesting results. Two recent papers have examined the question of whether or not *Rhizobium* cells must be motile in order for them to cause infections (6, 139). *Rhizobium* cells are generally equipped with one or more flagella and are reported to have chemotactic responses (6, 49–51, 85). In the study by Ames et al (6), three types of mutants of the normally motile and chemotactic wild type of *R. meliloti* were isolated: nonmotile mutants lacking flagella; nonmotile mutants with inactive flagella; and motile but nonchemotactic mutants. The motility of *R. meliloti* cells was found to be strongly dependent on culture medium, growth phase, and temperature. Nodulation tests showed that the three types of mutants were able to nodulate alfalfa seedlings just as well as the wild type, demonstrating that motility and chemotaxis are not required for infection or nodule development (6). Similar results were reported for *R. trifolii* motility mutants on clover (139).

It is of interest that nonmotile mutants of a common bacterial plant pathogen, *Pseudomonas phaseolicola,* were also able to infect their host (158). While the inherent infectiveness of motile and nonmotile strains was apparently the same, motile strains of the pathogen were shown to have a significant quantitative advantage over nonmotile strains under certain limiting inoculation conditions (158). The same may be true of rhizobia. Motility, even though not required, may confer significant advantages under a number of conditions in vivo. Such advantages might not be detectable under saturating inoculum conditions in vitro.

Recent studies have indicated that the synthesis of polysaccharide capsules may be an important characteristic of infective *Rhizobium* cells. In an early study, Dudman (74) observed that several strains and species of rhizobia produced encapsulated cells, and found that *R. trifolii* TA1 could be recovered from nodules more frequently if seeds in field tests were inoculated with encapsulated cells than when seeds were inoculated with the normal mixture of encapsulated and unencapsulated cells. Not all cells in a given culture were encapsulated, and the proportion of encapsulated cells changed with culture age and the composition of the growth medium (74). This kind of variability has been borne out by subsequent studies. For example, although Dudman did not observe capsules on cells of the one strain of *R. japonicum* that he examined (74), there is evidence that most or all *R. japonicum* strains generate encapsulated cells under the right culture conditions (9, 22, 23, 28, 29, 39, 133, 180). Dazzo and his colleagues (68) have reported that clover lectin binds to encapsulated *R. trifolii* cells but not to unencapsulated cells. Several noninfective or nonnodulating mutants of *R. trifolii* did not bind clover lectin (65). Presumably these mutants lacked capsules or had altered capsules.

Sanders et al (175) have investigated the requirement for *Rhizobium* exopolysaccharide synthesis by selecting for mutants of *R. leguminosarum* which were unable to synthesize EPS. Several spontaneous mutants or variants with substantially diminished EPS synthesis were isolated. None of these mutants or variants were able to form nodules, indicating that EPS synthesis was required for nodulation (175). Subsequent studies have extended these observations. A mutant that produced only 10% as much EPS as the wild type strain, but which produced just as high a proportion of encapsulated cells as the wild type (80%), was found to induce just as many curled and infected root hairs as the wild type, although only a third as many nodules (140). A mutant that produced no encapsulated cells and only 1.5% as much EPS as the wild type induced no root hair curling, infection threads, or nodules (140). Two other mutants, intermediate with respect to EPS production and encapsulation, were also intermediate with respect to curling, infection thread formation, and nodulation.

Unpublished studies from our laboratory suggest that for *R. japonicum* 138 at least, capsules per se are not required for nodulation as long as the cells can make adequate amounts of the normal EPS. Mutants unable to form capsules or bind soybean lectin in either exponential or stationary growth phase were isolated by repetitive enrichment for cells without capsules by differential centrifugation (Y. Yamamoto, I. J. Law, A. J. Mort, and W. D. Bauer, unpublished). Those cap$^-$ mutants that synthesized reasonably normal amounts of normal EPS were able to nodulate soybeans approximately as well as the parental strain. However, cap$^-$ mutants which produced substantially reduced amounts of the normal EPS nodulated significantly less well than the parental strain. In another relevant study, Hubbell & Elkan (103) established that high EPS production was a characteristic of those *R. japonicum* strains that were able to nodulate a "nonnodulating" soybean mutant. These various studies indicate that capsule formation or EPS synthesis may be required for infection and nodulation, although the experiments are not definitive.

The isolation of mutant strains lacking "polar bodies" (196) and the comparison of relative rates of infection by cells with and without polar bodies may similarly prove to be valuable indirect approaches for establishing the importance of polar attachment for infectivity.

There has also been interest for many years concerning the role of the "swarmer" cell type of rhizobia. The swarmer form of *R. meliloti* has been characterized as cocoid cells 0.1 to 0.4 μ in diameter having as many as 50 flagella (57). Dart & Mercer (57) suggested that swarmers may be the infective cell type because their high motility and small cell size might allow them to pass between the cellulose microfibrils of the root hair wall. However, the genetic evidence just cited (6, 139) indicated that motility was not required for nodulation. Thus swarmers may have no direct function in the infection process, but serve primarily to extend the range of search for nutrients (4, 19). On the other hand, if swarmer formation is regulated by the presence or absence of particular nutrients, perhaps swarmer formation was suppressed under the assay conditions used to identify nonmotile mutants. It is possible that the presumed nonmotile mutants of *R. trifolii* and *R. meliloti* were able to form swarmers in the rhizosphere after inoculation.

ROOT HAIR CURLING AND INFECTION THREAD INITIATION

Root Hair Curling

The curling of root hairs after inoculation with rhizobia is the first obvious response of the host. Still, there is no certainty about what substances from the rhizobia induce curling, branching, and other root hair deformations—despite the fact that this response was described by Ward nearly 100 years ago (205). Moreover, it is not known how normal hair cell wall synthesis is altered to produce the characteristic deformations. Even the function of curling in the infection process is a matter of conjecture.

Several attempts have been made to isolate and characterize substances from rhizobia that induce root hair deformations. None of these attempts has lead to the clear identification of a deformation-inducing substance. For some time it was thought that IAA, produced by rhizobia, was responsible for the deformation of root hairs. However, IAA alone over a wide range of concentrations has no such effect: IAA causes deformations, but not the curling characteristic of *Rhizobium* interactions (52, 82, 173). Moreover, only infective rhizobia, and not other microorganisms that produce IAA, are able to induce the characteristic deformations (36, 111, 113). *Rhizobium* culture filtrates induce curling and branching responses in host root hairs (102, 186, 211, 212). These responses are less pronounced than the responses elicited by living *Rhizobium* cells (212). Culture filtrates of the bacteria will induce deformation of root hairs in nonhost plants, but this response is more modest than in the homologous interaction (212). A marked degree of curling (360° or greater) does not appear to be inducible by culture filtrates, and it is this most marked form of curling that is host specific—i.e. characteristic of the homologous interaction (94, 212). Although the marked curling response in clover is essentially restricted to the homologous interaction, there is no evidence to indicate that infections *require* marked curling, or even that infections occur more frequently in markedly curled hairs. Marked curling may be a symptom of a compatible association but not a requirement for establishing one.

Mechanism of Curling

There is no description in the literature of any mechanism by which the normal synthesis of root hair cell walls might be altered to produce curling or branching. In Figure 5, I have provided an explicitly speculative model of *Rhizobium*-induced curling and branching responses. The purpose of this model is to stimulate experimentation and discussion. Hopefully, it will be valuable even if proven wrong.

The model is based on what little is known of root hair structure and development. This may be summarized as follows: The cell walls of root hairs appear to be composed of two rather distinct layers, called α and β layers (15, 144). The α layer is considered to be flexible and the β layer relatively rigid. The α layer is continuous over the entire hair (see Figure 5B). It appears to consist primarily of cellulose microfibrils randomly oriented in a plane parallel to the hair plasma membrane, with pectic polysaccharides and perhaps hemicelluloses forming an amorphous matrix. The β layer consists primarily of cellulose microfibrils oriented in a plane parallel to the hair plasma membrane and oriented roughly parallel to the axis of the hair elongation. The β layer, in contrast to the α layer, does not extend over the entire hair (Figure 5B). It develops *inside* of the α layer, and extends as a cylinder almost to the hemispherical dome of the growing hair tip.

Root hair cells develop from short epidermal cells called trichoblasts in plants where these are formed by unequal epidermal cell divisions in the root meristem (45, 46). In other plants, where all the root epidermal cells are of roughly comparable length, root hairs form most frequently from the shorter epidermal cells, with the shortest epidermal cells generating the

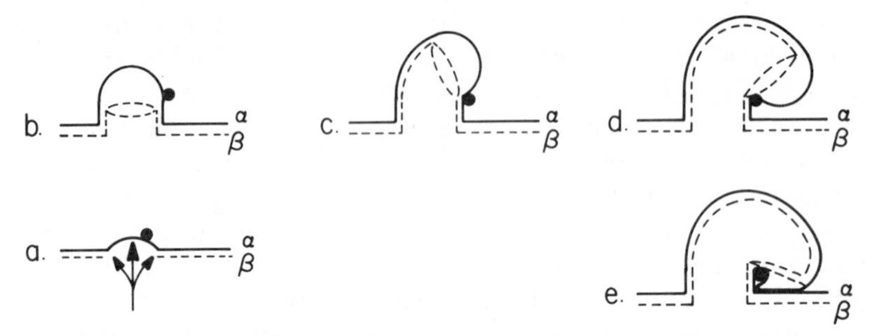

Figure 5. Model of root hair curling induction by rhizobia. See text for discussion.

longest hairs (45–48). Hairs develop only on epidermal cells that are still elongating and are initiated exclusively at the apical end of the epidermal cells (45, 46, 167).

In Figure 5A, a hair is emerging from the apical end of an epidermal cell. The flexible α layer tip of the hair bulges outward as the result of turgor pressure against an area of localized removal of the rigid β layer. The localized removal of the β layer is presumed to be a consequence of host-induced disintegration of the microfibrillar matrix or inhibition of β layer synthesis. It is also presumed that deposition of new α layer material is heaviest at the apex of the hemispherical swelling. As a consequence, any bacterial cell attached to the emerging tip will gradually be displaced from the apex to the edge of the hemisphere (Figure 5B). The model proposes that curling results from a localized inhibition of the β layer deposition in the emerging hair, induced by an attached *Rhizobium* cell. The rigid cylinder of β layer material thus does not develop past the attached *Rhizobium* cell, but continues to be deposited inside the α layer opposite the attached *Rhizobium* (Figure 5C). As the hair continues to elongate, the flexible hemispherical tip gradually pivots around the attached bacterium, resulting in a tight curl that envelops the *Rhizobium* (Figure 5D and E). Lateral branches are considered to develop in the same sequence, except that they are initiated by a localized disintegration of the hair wall β layer instead of the epidermal wall β layer. This disintegration is presumably *Rhizobium*-induced rather than host-induced.

The envelopment and enclosure of attached rhizobia in a pocket between cell walls is likely to be the biological function of root hair curling (82). The enclosing wall provides something solid to push against as the developing infection thread progresses inward against the turgor pressure of the hair cell. The formation of a pocket enclosing the bacteria may also serve to substantially increase the concentration of nutrients for the microsymbiont as well as the concentration of signals and effectors from the microsymbiont.

The β layer model and the results of Bhuvaneswari et al (13, 24) suggest that the first interactions critical to the infection process in plants such as alfalfa and soybean most probably take place at the apical ends of short epidermal cells, cells that have completed most of their axial elongation but have not produced a long hair. The data of Cormack (45–48) indicate that epidermal cells appear to have a limited capacity for elongation or cell wall formation. This capacity is divided between hair formation and elongation along with root growth axis. If attachment and β layer inhibition by *Rhizobium* cells occur too late, on well-developed hairs, then the cell wall forming capacity of the epidermal cell will be used up before the tip sliding (5A-B), hair curling (5C-D), and infection thread development stages are completed. On the other hand, attachment of rhizobia to epidermal cells that are still rapidly elongating is likely to be ineffectual because there appears to be a fast-acting regulatory mechanism that substantially diminishes the frequency of nodulation in the root zone occupied by these younger epidermal cells at the time of inoculation [Figure 4; see (24)]. These considerations lead to fairly precise predictions of the location of the infectible points on the epidermal surface, predictions suitable for direct experimental testing. They may also provide at least a partial explanation for the low percentage of infected hairs noted previously.

Infection Thread Formation

Infection threads are tubular structures that carry *Rhizobium* cells, often single file, from the root surface into the root cortex. Rhizobia are released from the ends of the infection threads in host membrane envelopes, establishing the bacterial symbiont in the host cortical cell cytoplasm (169). The initial formation of infection threads, their directional growth inward against host turgor pressure, their passage through root hair and cortical cell walls, and their timely dissolution to effect release of the rhizobia are still amazing and rather mysterious phenomena despite nearly a century of study.

Light microscopic studies of infection thread formation and development have been greatly facilitated by the development of the Fåhraeus slide technique (81). This technique enables one to make continuous microscopic observations of the growing roots of small seeded legumes under aseptic, hydroponic conditions. Fåhraeus reported that, prior to the formation of a visible infection thread, the hair cell cytoplasm and nucleus were concentrated near the site of infection and were intensely active (81). The site of infection thread initiation first appeared as a local swelling or refractile spot (81). The thread grew inward at about the same rate (5-8 μ/h) in clover as the hair previously elongated (52). Observations by several investigators indicated that the direction of growth of the infection thread tip was governed by the position and movement of the hair cell nucleus (52, 81, 94, 153). No one has suggested just how the movement of the nucleus might influence or direct the growth of the thread.

Until recently, studies with the electron microscope have added surprisingly little to our understanding of infection thread initiation and development. The earliest studies were made when techniques for electron microscopy were still in their infancy, and served primarily to strengthen the view that the thread wall and hair wall were one continuous structure (99, 174). A subsequent study by Napoli & Hubbell (142) was much more adequate. Sequential sections through an apically infected clover root hair were obtained. The thread wall appeared to be continuous with the hair wall (142). The thread tube, originating at the point of most acute curling, appeared to be an invagination of the hair wall (142).

Recent studies by Callaham (37) clearly surpass all previous attempts at this difficult problem. Callaham first established that infection threads developed in clover seedlings grown in Fåhraeus slide assemblies within 32-40 h. Threads reached the base of the hair cell within 48 h after inoculation (37). Sequential sections through 48-hour-old samples of both apically and laterally infected hairs were prepared. Callaham observed that the infection thread was not an invagination of the root hair tip, but rather was initiated at a point where two portions of the wall of the curled or branched hair formed an enclosure (37). A disintegration or degradation of the hair wall at the site of thread initiation in the enclosure was indicated. The thread wall appeared to be formed by the apposition of a new, fibrillar layer of wall material around the infection site containing the rhizobia (37). Infection threads that originated in apically curled and laterally branched root hairs appeared to be initiated in essentially the same way. Callaham concluded that "rhizobia directly penetrate the root hair wall by a process involving degradation of the cell wall polysaccharides at a localized site" (37). The apparent continuity of the thread wall with the original hair wall was due to the synthesis of a new wall layer. This new thread wall layer was not callose: it was fibrillar and did not react histochemically as callose does (37, 119).

Callaham also studied the single example he found of infection in an unbranched, uncurled hair (37). He suggested that this example showed that infection of hairs could take place by direct penetration at sites beneath a dense *Rhizobium* colony, and thus that enclosure of the bacteria between two walls was helpful but not required. Callaham, however, ignored the possibility that this exceptional infection was generated at the point of contact between two root hairs. Infections generated at points of contact between two hairs are well known (52, 54), and have been reported to account for approximately 4% of the infections in pea (94). Thus, Callaham is probably incorrect in this instance. Enclosure between two walls at the infection site is probably required for infection thread development.

Although many questions still remain regarding the initiation and development of infection threads, Callaham's results provide an excellent foundation of knowledge and a valuable assurance that the technical problems are tractable. Spot inoculation techniques and the growth pouch marking method should facilitate developmental EM studies of infection thread formation.

Alternate Modes of Infection

It is well established that the formation of infection thread structures in root hair cells provides the means of entry for rhizobia in several temperate crop and forage legumes, including pea, bean, clover, soybean, and alfafa (54). This mode of infection probably serves in a variety of other legume species as well since infection threads have been detected in sections of nodules from other species, even though the site of entry has not necessarily been established (54). Infection threads occasionally may be initiated in epidermal cells that have not formed root hairs (25, 54). This may be the exclusive mode of entry for a water plant such as *Neptunia oleracea,* which has no root hairs (176), but it is only a minor path of infection for example in soybean, which does have root hairs (25).

In at least two dozen legume species, sections through nodules have failed to reveal infection threads (54). It is possible that in some instances of this sort the infection thread structures disintegrated before examination, as reported for older nodules of soybean (88). However, Dart (54) points out that in nodules without infection threads, all of the host cells in the bacteroid zone have been invaded, whereas uninvaded host cells are common in the bacteroid zone of nodules that do have infection threads. It seems quite likely that rhizobia infect these apparently threadless legumes by a rather different mechanism than the root hair/infection thread mechanism that has been most thoroughly studied to date.

The recent study by Chandler (42) on the infection of peanut is a most valuable beginning to the characterization of alternate modes of infection. Root hairs and nodules develop on peanut only around the base of emergent lateral roots. No infection threads have been found in either the hairs or the nodules (5, 42, 43, 54). Chandler's observations indicate that infections are initiated at the basal junction between the axillary hair cells and adjacent epidermal cells, and then progress intercellularly through the middle lamellar region (42). Movement of the rhizobia from the middle lamella into the host cell cytoplasm appears to involve alteration or disintegration of the host wall and enclosure of the bacteria in membrane envelopes (42). The site of nodule initiation appears to be in the cortex of the lateral root, not the cortex of the primary root (42). A similar, but not identical, mode of infection and nodule initiation has been observed recently in *Stylosanthes* species (M. R. Chandler, R. J. Roughley, and R. N. Date, personal communication). Preliminary observations have indicated that the infectibility of peanut epidermal cells is developmentally restricted and transient (13).

These investigations are only the beginning of many studies that need to be made in order to clarify the mode of infection in legumes where infection thread formation does not seem to occur. The developmental and physiological requirements for the initiation of such infections need considerable clarification. It is noteworthy that the penetration mechanisms in both peanut and clover involve the enclosure of the rhizobia in a pocket between two cell walls and the subsequent disintegration or degradation of one of these walls (37, 42).

Role of Cell Wall Degrading Enzymes

Since the electron microscopic observations (37, 42) indicate that a region of the host cell wall is degraded when rhizobia enter the infection thread structure or the host cytoplasm, it seems likely that the cell wall is altered by hydrolytic enzymes, either from the host or from the rhizobia or both. The task of determining which enzymes are involved, if any, and of determining how their activity is induced, localized, and regulated, is an important but most difficult one. Enzymes with both cellulytic and pectolytic activities are normally present in roots and root exudates of legume seedlings (30, 78, 123, 125, 138, 202). Isolated and carefully washed root cell wall preparations from seedlings also contain autolytic enzymes, including pectinases, cellulases, glucan hydrolases, and various glycosidases (87, 112, 118; W. D. Bauer, unpublished). Recent evidence suggests that rhizobia may also produce pectinases, hemicellulases, and cellulases (104, 130).

Some years ago, Ljunggren and Fåhraeus reported that clover root polygalacturonase activity was specifically enhanced by homologous rhizobia (82, 123, 125). Partially purified exopolysaccharide preparations from homologous rhizobia also increased host polygalacturonase activity (123, 125). These authors hypothesized that rhizobia become entrapped in pockets between cell walls as a result of root hair deformation. The *Rhizobium*-induced host polygalacturonase was suggested to decrease the rigidity of the host wall in the pocket. The increasing amount of bacteria and bacterial substances was thought to produce a pressure in the pocket somewhat greater than the turgor in the hair. The pressure produced in the pocket, the depolymerization of cell wall material, and the deposition of new cell wall material as a defense response of the host to this invasion were suggested to act in concert to generate the infection thread structure (82). This hypothesis was most attractive, both as an explanation of host specificity and as a mechanism of root hair infection. Various attempts by other investigators to obtain similar induction of polygalacturonase, however, were not successful (30, 121, 128, 185). Neither the results nor the hypothesis advanced by Ljunggren and Fåhraeus had received much attention in recent years.

Perhaps a reassessment is in order. The polygalacturonase hypothesis is fully compatible with recent evidence indicating the specificity and importance of *Rhizobium* CPS/EPS. In addition, the polygalacturonase hypothesis describes the initial events in the formation of infection threads in a manner that is quite consistent with recent EM data (37, 42, 142): rhizobia in an enclosed pocket multiply and locally disrupt or degrade the hair wall, and a new layer of host cell wall-like material is deposited at the infection site. Whether the accumulation of this new wall-like material is a defense response by the host to the rhizobia at the infection site is by no means proved, but is nonetheless tenable. Certainly the localized deposition of wall-like materials such as papillae, collars, or sheaths at the point of penetration of microbial pathogens is well known (2, 32, 83). The infection thread is in essence a sheath. Moreover, recent reports indicate that fragments of pectic polysaccharides from the host cell wall can act as potent elicitors of defense responses in plants (95, 127). Thus, localized activation of host root polygalacturonase might result in both localized degradation of the hair cell wall and localized release of pectic elicitor substances that induce deposition of the infection thread/sheath material.

Of course, the central feature of the polygalacturonase hypothesis is the host polygalacturonase itself, and it is this feature that is least certain and most difficult to establish. The extensive observations of Ljunggren and Fåhreaus were primarily based on viscometric measurements of the polygalacturonase activity present in the liquid used to bathe approximately 250 seedlings for 2 to 4 days (123). The decrease in viscosity of added pectin was determined after 24–84 h of incubation with the solution to be assayed. The small viscosity differences obtained, generally 10 to 20%, are near the limits of reliable assay by this method. Any attempts to extend or confirm the results of Ljunggren and Fåhreaus must contend with this problem of very low activity, as well as with possible cultivar or seed lot differences, with the growth phase-dependent synthesis of specific *Rhizobium* polysaccharides (63, 133), and with a pronounced optimum in the concentration of inducing polysaccharide (123).

Rhizobia may be able to produce extracellular cellulases, hemicellulases, or pectinases. A recent attempt to detect extracellular *Rhizobium* proteins of any kind, however, gave negative results (86), and previous attempts to detect pectolytic activity in *Rhizobium* cultures have failed (123, 131, 182). Recent studies indicating that rhizobia may produce extracellular pectinases (104) involved the use of pectin-agar plate assays. A subsequent report (130) indicated that this assay technique produces ambiguous results when the microorganisms produce enzymes capable of hydrolyzing agar, as a number of *Rhizobium* strains were shown to do. Thus, it is not clear whether the agar, the pectin, or both were the substrates responsible for the observed activity on pectin-agar plates. Essentially no "hemicellulase" activity was reported for the *Rhizobium* strains tested (130), but this negative evidence has little value since the gum arabic used as a substrate is not a good analog of primary cell wall hemicelluloses (11, 14). The evidence for cellulase production by rhizobia is somewhat stronger, and includes an important attempt to localize cellulase histochemically at the ultrastructural level (130, 202). However, it should be noted that cellulases have little or no effect on plant cell walls by themselves (10, 191).

REGULATION OF INFECTION AND NODULATION

A broad range of evidence indicates that nodulation is self-regulated and optimized by the host. Ineffective nodules (i.e. nodules in which dinitrogen fixation does not take place) are usually formed in greater numbers than effective nodules (151), suggesting that nodule development is subject to negative feedback regulation by substances produced in effective nodules. Similarly, nodules are often found to occur in clusters, especially after a delayed inoculation (24, 151; see Figure 4), indicating that the frequency of subsequent nodulation below the cluster is diminished by self-regulation. Nodulation is increased by excision of previously formed nodules (152), again suggesting that nodules produce inhibitory substances that reduce further nodulation. Evidence for the existence of an inhibitor of nodulation in pea cotyledons has also been obtained (160). Cotyledon excision resulted in significantly increased nodulation. However, there was no effect of cotyledon removal on the number of infected root hairs, indicating that the cotyledonary inhibitor acted on a step between infection thread formation and macroscopic nodule development (160). Although it is not certain that this cotyledonary inhibitor is ordinarily involved in regulating nodule number, it is clear that the host plant has the means for blocking infections after they have been initiated.

It is well established that many or most infections abort before they can lead to nodule formation (54, 81, 136, 154). The high proportion of aborted infections may be a manifestation of self-regulation of nodule number. Infection thread abortion is clearly an aspect of nitrate inhibition of nodulation, since added nitrate substantially increases the proportion of aborted infections (136). Observations of infection thread formation in clover and *Vicia* indicated that the rate of infection thread formation was very high shortly after inoculation (154). The rate of thread formation then decreased abruptly and substantially at about the time that the first nodules had developed to a readily visible size. The most straightforward interpretation of this result is that the development of the first nodules inhibited subsequent infections at some stage prior to infection thread formation.

Evidence is provided in Figure 4 for a form of self-regulation of nodulation in soybean that is effective within a few hours after inoculation. Nodulation frequency reached a maximum about 9 mm above the RT mark, but this high frequency of nodulation was not maintained farther down the root. Nodulation was diminished to half maximal frequency or less just 7 to 12 mm below the peak of maximum nodulation frequency. The epidermal cells in this zone of diminished nodulation frequency were only about 3 to 5 h younger than those nodulated at maximal frequency. It appears that the events taking place in the most infectible cells soon after inoculation diminished the infectibility or the success of infections in these slightly younger cells. Similar profiles of nodulation frequency have been obtained with cowpea (13), indicating that quick-acting self-regulation is operative in this host plant as well.

In other relevant studies, the extent of nodulation has been found to vary in direct proportion to the logarithm of the inoculum size (24, 163). If the extent of nodulation depended solely on the probability of an infective *Rhizobium* cell coming in contact with a suitable site on a root hair cell, then doubling the inoculum size should approximately double the probability, and nodulation would increase linearly with the size of the inoculum. This does not happen. Nodulation increases much less than linearly with inoculum size, indicating that infections or nodule development are somehow inhibited when more than enough bacteria are present.

Inhibition of legume nodulation by exogenous nitrogen has been known and studied for a great many years (54, 192). It appears that inhibition can be expressed at several levels, from an inhibition of the initiation of infections during the first 24 h after inoculation to the induced senescence of mature, effective nodules (54, 136, 137, 159). Host plants are able to sever their association with rhizobia quite dramatically when external nitrogen is plentiful. The mechanisms by which nitrate and similar substances inhibit infection and nodulation are not known. Thornton (192) observed that exogenous nitrate caused fewer root hairs and fewer curled root hairs to be produced on alfalfa plants. He suggested that the reduced numbers of curled root hairs were responsible for the lower number of nodules formed in the presence of nitrate. Thornton's observations were later confirmed and extended in a classical series of papers by Munns (134–137). Munns concluded that the reduction in nodule number could not be attributed solely or even chiefly to interference by nitrate with any one phase of the nodulation process. However, as indicated previously, nitrate caused both a reduction in the number of infection threads and an increase in the proportion of abortive infection threads (136). Moreover, nitrate was able to delay nodulation by causing a delay in the formation of infection threads (153). The addition of IAA counteracted some of the effects of nitrate on infection and nodulation in alfalfa (137, 200). The amount of *R. trifolii*-binding lectin on clover roots and the amount of clover polygalacturonase inducible by *R. trifolii* polysaccharide were reportedly diminished when the plants were grown on nitrate (62, 82). There are sometimes large variations between *Rhizobium* strains with respect to the effect of ammonium and nitrate on nodule number (79, 84, 97, 159). This implies that the exogenous nitrogen has significant effects on the bacteria as well as the host plant, which considerably complicates investigations and interpretations.

Further study of self-regulatory and NO_3/NH_4-induced regulatory mechanisms will help us to understand how the symbiosis is maintained within useful limits, a subject that is central to the question of how to engineer new symbioses or improvements in known symbiotic associations.

CONCLUDING REMARKS

The foregoing pages have provided a brief summary of our current knowledge of the *Rhizobium* infection process. If we knew more, then the description of the infection process would be shorter, as greater understanding has a way of swallowing loose odds and ends and dissolving complicated conflicts. Several new discoveries in the past few years have added substantially to our understanding: the discovery of correlations between host lectin binding and infectivity; the discovery of selective *Rhizobium* attachment to host roots; the identification and localization of infectible root cells; the discovery that cell surface components and infectibility are transient; the structural analyses of several *Rhizobium* surface polysaccharides; and the demonstration that infection threads originate by localized degradation of the hair wall and synthesis of a new layer. The contribution of these discoveries to our specific knowledge of the infection process is important. Even more important is our growing awareness of the magnitude and intricacy of what remains to be learned.

There is also a growing awareness that the social investment in biological nitrogen fixation research needs to be repaid. The pressure of population against the limits of natural resources is generating major changes in our civilization. If these major changes are to be based on new knowledge and understanding, rather than on brutal readjustments in the ratio of population to resources, then the acquisition and extension of new knowledge must be directed to reach towards the fulfillment of practical needs.

ACKNOWLEDGMENTS

I would like to thank the many colleagues who so generously provided descriptions of unpublished data and comments on important developments for this review. I am most grateful to T. V. Bhuvaneswari, F. B. Dazzo, M. Lalonde, A. J. Mort, and B. G. Turgeon for critical discussions of the manuscript.

Grants from the National Science Foundation (PCM 7922947 & PFR 7727269) and the USDA (CRGO 5901-0410-8-0042-0) provided essential support for the studies cited as unpublished data from our laboratory. The patient and skillful assistance of Ms. Debbie Patten with preparation of the manuscript is much appreciated.

Literature Cited

1. Abe, M., Higashi, S. 1979. The infectivity of *Rhizobium trifolii* into a minute excised root of white clover. *Plant Soil* 53:81–88
2. Aist, J. R. 1977. Mechanically induced wall appositions of plant cells can prevent penetration by a parasitic fungus. *Science* 197:568–71
3. Albersheim, P., Anderson-Prouty, A. J. 1975. Carbohydrates, proteins, cell surfaces, and the biochemistry of pathogenesis. *Ann. Rev. Plant Physiol.* 26:31–52
4. Allen, E. K., Allen, O. N. 1958. Biological aspects of symbiotic nitrogen fixation. In *Encyclopedia of Plant Physiology*, ed. W. Rhuland, 8:64–118. Berlin: Springer
5. Allen, O. N., Allen, E. K. 1940. Response of the peanut plant to inoculation with rhizobia, with special reference to morphological development of the nodules. *Bot. Gaz.* 102:121–42
6. Ames, P., Schluederberg, S. A., Bergman, K. 1980. Behavioral mutants of *Rhizobium meliloti. J. Bacteriol.* 141: 722–27
7. Ayers, A. R., Ebel, J., Finelli, F., Berger, N., Albersheim, P. 1976. Host-pathogen interactions. IX. Quantitative assays of elicitor activity and characterization of the elicitor present in the extracellular medium of cultures of *Phytophthora megasperma* var. *sojae. Plant Physiol.* 57:751–59
8. Bailey, R. W., Greenwood, R. M., Craig, A. 1971. Extracellular polysaccharides associated with Lotus species. *J. Gen. Microbiol.* 65:315–24
9. Bal, A. K., Shantharam, S., Ratnam, S. 1978. Ultrastructure of *Rhizobium japonicum* in relation to its attachment to root hairs. *J. Bacteriol.* 133:1393–1400
10. Bateman, D. F., Basham, H. G. 1976. Degradation of plant cell walls and membranes by microbial enzymes. In *Physiological Plant Pathology*, ed. R. Heitefuss, P. H. Williams, pp. 316–55. Berlin: Springer. 890 pp.
11. Bauer, W. D. 1977. Plant cell walls. In *The Molecular Biology of Plant Cells*, ed. H. Smith, pp. 6–23. Oxford: Blackwell. 496 pp.
12. Bauer, W. D., Bhuvaneswari, T. V. 1979. The possible role of lectins in legume/*Rhizobium* symbiosis and other plant/microorganism interactions. See Ref. 189, pp. 344–79
13. Bauer, W. D., Bhuvaneswari, T. V., Bhagwat, A. A. 1980. Transient susceptibility of root cells in five common legumes to infection by rhizobia. *Plant Physiol.* Abstr. 748, p. 136
14. Bauer, W. D., Talmadge, K. W., Keegstra, K., Albersheim, P. 1973. The structure of plant cell walls. II. The hemicellulose of the walls of suspension-cultured sycamore cells. *Plant Physiol.* 51:174–87
15. Belford, D. S., Preston, R. D. 1961. The structure and growth of root hairs. *J. Exp. Bot.* 12:157–68
16. Bennett, R. A., Billing, E. 1978. Capsulation and virulence in *Erwinia amylovora. Ann. Appl. Biol.* 89:41–45
17. Beringer, J. E., Beyman, J. L., Buchanan-Wollaston, A. V., Johnston, A. W. B. 1978. Transfer of the drug-resistance transposon Tn5 to *Rhizobium. Nature* 276:633–34
18. Beringer, J. E., Brewin, N., Johnston, A. W. B., Schulman, H. M., Hopwood, D. A. 1979. The *Rhizobium*-legume

symbiosis. *Proc. R. Soc. London Ser. B.* 204:219–33

19. Bewley, W. F., Hutchinson, H. B. 1920. On the changes through which the nodule organism (*Ps. radicicola*) passes under cultural conditions. *J. Agric. Sci.* 10:144–62
20. Bhagwat, A. A., Thomas, J. 1980. Dual binding sites for peanut lectin on rhizobia. *J. Gen. Microbiol.* 117:119–25
21. Bhuvaneswari, T. V. 1981. The recognition mechanism and the infection process in legumes. *Econ. Bot.* In press
22. Bhuvaneswari, T. V., Bauer, W. D. 1978. Role of lectins in plant-microorganism interactions. III. Binding of soybean lectin to root cultured rhizobia. *Plant Physiol.* 62:71–74
23. Bhuvaneswari, T. V., Pueppke, S. G., Bauer, W. D. 1977. Role of lectins in plant-microorganism interactions. I. Binding of soybean lectin to rhizobia. *Plant Physiol.* 60:486–91
24. Bhuvaneswari, T. V., Turgeon, G., Bauer, W. D. 1980. Early stages in the infection of soybean (*Glycine max.* L. Merr.) by *Rhizobium japonicum.* I. Localization of infectible root cells. *Plant Physiol.* 66:1027–31
25. Bieberdorf, F. W. 1938. The cytology and histology of root nodules of some leguminosae. *J. Am. Soc. Agron.* 30: 375–89
26. Bitton, G., Marshall, K. C., eds. 1980. *Adsorption of Microorganisms to Surfaces.* New York: Wiley. 439 pp.
27. Bjorndal, H., Erbing, C., Lindberg, B., Fåhraeus, G., Ljunggren, H. 1971. Studies on an extracellular polysaccharide from *Rhizobium meliloti. Acta Chem. Scand.* 25:1281–86
28. Bohlool, B. B., Schmidt, E. L. 1974. Lectins: A possible basis for specificity in the *Rhizobium*-legume root nodule symbiosis. *Science* 185:269–71
29. Bohlool, B. B., Schmidt, E. L. 1976. Immunofluorescent polar tips of *Rhizobium japonicum:* possible site of attachment or lectin binding. *J. Bacteriol.* 125:1188–94
30. Bonnish, P. M. 1973. Pectolytic enzymes in inoculated and uninoculated red clover seedlings. *Plant Soil* 39: 319–28
31. Bowles, D. J. 1979. Lectins as membrane components: implications of lectin-receptor interaction. *FEBS Lett.* 102:1–3
32. Bracker, C. E., Littlefield, L. J. 1973. Structural concepts of host-pathogen interfaces. In *Fungal Pathogenicity and the Plant's Response,* ed. R. J. W. Byrde, C. V. Cutting, pp. 159–313. London: Academic. 499 pp.
33. Brethauer, T. S. 1977. *Soybean lectin binds to rhizobia unable to nodulate soybean. Lectins may not determine host specificity.* MS thesis. Univ. Ill., Urbana.
34. Brethauer, T. S., Paxton, J. 1977. The role of lectin in soybean—*Rhizobium japonicum* interactions. In *Cell Wall Biochemistry Related to Specificity in Host - Plant Pathogen Interactions,* ed. B. Solheim, J. Raa, pp. 381–88. New York: Columbia Univ. Press. 487 pp.
35. Broughton, W. J. 1978. Control of specificity in legume - *Rhizobium* associations. *J. Appl. Bacteriol.* 45:165–94
36. Brown, M. E. 1972. Plant growth substances produced by microorganisms of soil rhizosphere. *J. Appl. Bacteriol.* 19:195–99
37. Callaham, D. A. 1979. *A structural basis for infection of root hairs of* Trifolium Repens *by* Rhizobium Trifolii. MS thesis. Univ. Mass., Amherst, 41 pp.
38. Callow, J. A. 1977. Recognition, resistance and the role of plant lectins in host-parasite interactions. *Adv. Bot. Sci.* 4:1–49
39. Calvert, H. E., Lalonde, M., Bhuvaneswari, T. V., Bauer, W. D. 1978. Role of lectins in plant-microorganism interactions. IV. Ultrastructural localization of soybean lectin binding sites of *Rhizobium japonicum. Can. J. Microbiol.* 24:785–93
40. Carlson, R. W. 1980. Surface chemistry of *Rhizobium.* In *Ecology of Nitrogen Fixation,* ed. W. J. Broughton, Vol. 2. Oxford: Oxford Univ. Press. In press
41. Carlson, R. W., Sanders, R. E., Napoli, C., Albersheim, P. 1978. Host-symbiont interactions. III. Purification and partial characterization of *Rhizobium* lipopolysaccharides. *Plant Physiol.* 62: 912–17
42. Chandler, M. R. 1978. Some observations on infection of *Arachis hypogaea* L. by *Rhizobium. J. Exp. Bot.* 29: 749–55
43. Chandler, M. R., Dart, P. J. 1973. Structure of nodules of peanut and *Vigna* spp. *Rothamsted Exp. Stn. Rep.* (1972) 1:85
44. Chen, A. T., Phillips, D. A. 1976. Attachment of *Rhizobium* to legume roots as the basis for specific interactions. *Physiol. Plant.* 38:83–88
45. Cormack, R. G. H. 1935. Investigations on the development of root hairs. *New Phytol.* 34:30–54

46. Cormack, R. G. H. 1945. Cell elongation and development of root hairs in tomato roots. *Am. J. Bot.* 32:490–96
47. Cormack, R. G. H. 1949. The development of root hairs in angiosperms. *Bot. Rev.* 15:583–612
48. Cormack, R. G. H. 1962. The development of root hairs in angiosperms II. *Bot. Rev.* 28:446–64
49. Currier, W. W., Strobel, G. A. 1976. Chemotaxis of *Rhizobium* spp. to plant root exudates. *Plant Physiol.* 57:820–23
50. Currier, W. W., Strobel, G. A. 1977. Chemotaxis of *Rhizobium* spp. to a glycoprotein produced by birdsfoot trefoil roots. *Science* 196:434–36
51. Currier, W. W., Strobel, G. A. 1977. The chemotactic behavior of trefoil *Rhizobium. FEMS Microbiol. Lett.* 5: 243–46
52. Darbyshire, J. F. 1964. *A study of the initial stages of infection of clover by nodule bacteria.* PhD thesis. London: Univ. London Press. 146 pp.
53. Dart, P. J. 1971. Scanning electron microscopy of plant roots. *J. Exp. Bot.* 22:163–68
54. Dart, P. J. 1974. Development of root nodule symbioses. The infection process. See Ref. 165, pp. 381–429
55. Dart, P. J. 1975. Legume root nodule initiation and development. In *The Development and Function of Roots,* ed. J. G. Torrey, D. T. Clarkson, pp. 468–506. London: Academic. 618 pp.
56. Dart, P. J. 1977. Infection and development of leguminous nodules. In *A Treatise on Dinitrogen Fixation,* ed. R. W. F. Hardy, W. S. Silver, 3:367–72. New York: Wiley. 675 pp.
57. Dart, P. J. Mercer, F. V. 1964. The legume rhizosphere. *Arch. Mikrobiol.* 47:344–78
58. Dazzo, F. B. 1979. Adsorption of microorganisms to roots and other plant surfaces. See Ref. 26, pp. 253–316
59. Dazzo, F. B. 1980. Determinants of host specificity in the *Rhizobium* clover symbiosis. See Ref. 147, 2:165–87
60. Dazzo, F. B. 1980. Lectins and their saccharide receptors as determinants of specificity in the *Rhizobium*-legume symbiosis. In *The Cell Surface: Mediator of Developmental Processes,* ed. S. Subtleney, N. K. Wesselles, pp. 277–304
61. Dazzo, F. B., Brill, W. J. 1977. Receptor site on clover and alfalfa roots for *Rhizobium. Appl. Environ. Microbiol.* 33:132–36
62. Dazzo, F. B., Brill, W. J. 1978. Regulation by fixed nitrogen of host-symbiont recognition in the *Rhizobium*-clover symbiosis. *Plant. Physiol.* 62:18–21
63. Dazzo, F. B., Brill, W. J. 1979. Bacterial polysaccharide which binds *Rhizobium trifolii* to clover root hairs. *J. Bacteriol.* 137:1362–73
64. Dazzo, F. B., Hubbell, D. H. 1975. Antigenic differences between infective and noninfective strains of *Rhizobium trifolii. Appl. Microbiol.* 30:172–77
65. Dazzo, F. B., Hubbell, D. H. 1975. Cross-reactive antigens and lectins as determinants of symbiotic specificity in *Rhizobium trifolii*-clover association. *Appl. Microbiol.* 30:1017–33
66. Dazzo, F. B., Hubbell, D. H. 1975. Concanavalin A: lack of correlation between binding to *Rhizobium* and specificity in *Rhizobium*-legume symbiosis. *Plant Soil* 43:713–17
67. Dazzo, F. B., Napoli, C., Hubbell, D. H. 1976. Adsorption of bacteria to roots as related to host specificity in the *Rhizobium*-clover symbiosis. *Appl. Environ. Microbiol.* 32:166–71
68. Dazzo, F. B., Urbano, M. R., Brill, W. J. 1979. Transient appearance of lectin receptors on *Rhizobium trifolii. Curr. Microbiol.* 2:15–20
69. Dazzo, F. B., Yanke, W. E., Brill, W. J. 1978. Trifoliin: A *Rhizobium* recognition protein from white clover. *Biochim. Biophys. Acta* 539:276–86
70. Dedonder, R. A., Hassid, W. Z. 1964. The enzymatic synthesis of a (β-1,2-)-linked glucan by an extract of *Rhizobium japonicum. Biochim. Biophys. Acta.* 90:239–48
71. Deinema, M. H., Zevenhuizen, L. P. T. M. 1971. Formation of cellulose fibrils by Gram-negative bacteria and their role in bacterial flocculation. *Arch. Mikrobiol.* 78:42–57
72. Döbereiner, J., Burris, R. H., Hollaender, A., Franco, A. A., Neyra, C. A., Scott, D. B., eds. 1978. *Limitations and Potentials for Biological Nitrogen Fixation in the Tropics.* New York: Plenum. 398 pp.
73. Dommergues, Y. R., Krupa, S. V., eds. 1978. *Interactions Between Non-Pathogenic Soil Microorganisms and Plants.* Amsterdam: Elsevier. 475 pp.
74. Dudman, W. F. 1968. Capsulation in *Rhizobium* species. *J. Bacteriol.* 95: 1200–1
75. Dudman, W. F. 1976. The extracellular polysaccharides of *Rhizobium japonicum:* compositional studies. *Carbohydr. Res.* 46:97–110
76. Dudman, W. F. 1978. Structural studies of the extracellular polysaccharides of

Rhizobium japonicum strains 71a, CC708 and CB1795. *Carbohydr. Res.* 66:9–23
77. Dudman, W. F., Jones, A. J. 1980. The extracellular glucans of *Rhizobium japonicum* strain 3I1b 71a. *Carbohydr. Res.* 84:358–64
78. Ekdahl, I. 1953. Studies on the growth and osmotic conditions of root hairs. *Symb. Bot. Ups.* 11:6–13
79. El-Sherbeeny, M. H., Mytton, L. R., Lawes, D. A. 1977. Symbiotic variability in *Vica faba.* I. Genetic variation in the *Rhizobium leguminosarum* population. *Euphytica* 26:149–56
80. Evans, J., Barnet, Y. M., Vincent, J. M. 1979. Effect of a bacteriophage on the colonization and nodulation of clover roots by a strain of *Rhizobium trifolii. Can. J. Microbiol.* 25:968–73
81. Fåhraeus, G. 1957. The infection of clover root hairs by nodule bacteria studied by a simple glass slide technique. *J. Gen. Microbiol.* 16:374–81
82. Fåhraeus, G., Ljunggren, H. 1967. Preinfection phases of the legume symbiosis. In *The Ecology of Soil Bacteria.* Liverpool: Liverpool Univ. Press
83. Fullerton, R. A. 1970. An electron microscope study of the intracellular hyphae of some smut fungi (Ustilaginales). *Aust. J. Bot.* 18:285–92
84. Gibson, A. H. 1974. Consideration of the growing legume as a symbiotic association. *Proc. Indian Natl. Acad. Sci.* 40:741–67
85. Gitte, R. R., Vital Rai, P., Patil, R. B. 1978. Chemotaxis of *Rhizobium* sp. towards root exudate of *Cicer arietinum L. Plant Soil* 50:553–66
86. Glenn, A. R., Dilworth, M. J. 1979. An examination of *Rhizobium leguminosarum* for the production of extracellular and periplasmic proteins. *J. Gen. Microbiol.* 112:405–9
87. Goldberg, R. 1977. On possible connections between auxin-induced growth and cell wall glucanase activities. *Plant Sci. Lett.* 8:233–42
88. Goodchild, D. J., Bergersen, F. J. 1966. Electron microscopy of the infection and subsequent development of soybean nodule cells. *J. Bacteriol.* 92:204–13
89. Gordon, J. C., Wheeler, C. T., Perry, D. A., eds. 1979. *Symbiotic Nitrogen Fixation in the Management of Temperate Forests.* Workshop Proc., Oregon State Univ., Corvalis. 501 pp.
90. Graham, T. L. 1980. Recognition in the *Rhizobium*-legume symbiosis. In *Biology of the Rhizobiaceae,* ed. A. Atherly, K. Giles. New York: Academic. In press
91. Graham, T. L., Sequeira, L., Huang, T. S. R. 1977. Bacterial lipopolysaccharides as inducers of disease resistance in tobacco. *Appl. Environ. Microbiol.* 34:424–32
92. Greaves, M. P., Darbyshire, J. F. 1972. The ultrastructure of the mucilaginous layer on plant roots. *Soil Biol. Biochem.* 4:443–49
93. Gresshoff, P. M., Skotnicki, M. L., Eadie, J. F., Rolfe, B. G. 1977. Viability of *Rhizobium trifolii* bacteroids from clover root nodules. *Plant Sci. Lett.* 10: 299–304
94. Haak, A. 1964. Über den einfluss der Knöllchenbakterien auf die wurzelhaare von leguminosen und nicht leguminosen. *Zentralbl. Bakteriol. Parasitenkd. Infektionskr. Abt. 2* 117: 343–61
95. Hahn, M. G., Darvill, A. G., Albersheim, P. 1980. Polysaccharide fragments from the walls of soybean cells elicit phytoalexin (antibiotic) accumulation in soybean cells. *Plant Physiol.* 65: Abstr. 750, p. 136
96. Hammarström, S., Murphy, L. A., Goldstein, I. J., Etzler, M. E. 1977. Carbohydrate binding specificity of four N-acetyl-D-galactosamine-"specific" lectins: *Helix pomatia* A hemagglutinin. soybean agglutinin, lima bean lectin and *Dolichos biflorus* lectin. *Biochemistry* 16:2750–55
97. Heichel, G. H., Vance, C. P. 1979. Nitrate-N and *Rhizobium* strain roles in alfalfa seedling nodulation and growth. *Crop. Sci.* 19:512–18
98. Hermina, N., Reporter, M. 1977. Root hair cell enhancement in tissue cultures from soybean roots: a useful model system. *Plant Physiol.* 59:97–102
99. Higashi, S. 1966. Electron microscopic studies on the infection thread developing in the root hair of *Trifolium repens* L. infected with *Rhizobium trifolii. J. Gen. Appl. Microbiol.* 12:147–56
100. Hinch, J. M., Clarke, A. E. 1980. Adhesion of fungal zoospores to root surfaces is mediated by carbohydrate determinants of root slime. *Physiol. Plant Pathol.* 16:303–7
101. Hollaender, A., Burris, R. H., Day, P. R., Hardy, R. W. F., Helinski, D. R., Lamborg, M. R., Owens, L., Valentine, R. C., eds. 1977. *Genetic Engineering for Nitrogen Fixation.* New York: Plenum. 538 pp.
102. Hubbell, D. H. 1970. Studies on the

root hair "curling factor" of *Rhizobium*. *Bot. Gaz.* 131:337–42
103. Hubbell, D. H., Elkan, G. H. 1967. Correlation of physiological characteristics with nodulating ability in *Rhizobium japonicum*. *Can. J. Microbiol.* 13: 235–41
104. Hubbell, D. H., Morales, V. M., Umali-Garcia, M. 1978. Pectolytic enzymes in *Rhizobium*. *Appl. Environ. Microbiol.* 35:210–13
105. Jansson, P. E., Kenne, L., Lindberg, B., Ljunggren, H., Lönngren, J., Ruden, U., Svensson, S. 1977. Demonstration of an octasaccharide repeating unit in the extracellular polysaccharide of *Rhizobium meliloti* by sequential degradation. *J. Am. Chem. Soc.* 99:3812–15
106. Jansson, P. E., Lindberg, B., Ljunggren, H. 1979. Structural studies of the *Rhizobium trifolii* extracellular polysaccharide. *Carbohydr. Res.* 75:207–20
107. Johnston, A. W. B., Beynon, J. L., Buchanon-Wollaston, A. V., Setchell, S. M., Hirsch, P. R., Beringer, J. E. 1978. High frequency transfer of nodulating ability between strains and species of *Rhizobium*. *Nature* 276:634–36
108. Kamberger, W. 1979. An Ouchterlony double diffusion study on the interaction between legume lectins and rhizobial cell surface antigens. *Arch. Microbiol.* 121:83–90
109. Kamberger, W. 1979. Role of cell surface polysaccharides in *Rhizobium*-pea symbiosis. *FEMS Microbiol Lett.* 6: 361–65
110. Kato, G., Maruyama, Y., Nakamura, M. 1979. Role of lectins and lipopolysaccharides in the recognition process of specific legume-*Rhizobium* symbiosis. *Agric. Biol. Chem.* 43:1085–92
111. Katznelson, H., Sirois, J. C. 1961. Auxin production by species of *Arthrobacter*. *Nature* 191:1323
112. Keegstra, K., Albersheim, P. 1970. The involvement of glycosidases in the cell wall metabolism of suspension-cultured *Acer Pseudoplatanus* cells. *Plant Physiol.* 675–78
113. Kefford, N. P., Brockwell, J., Zwar, J. A. 1960. The symbiotic synthesis of auxin by legumes and nodule bacteria and its role in nodule development. *Aust. J. Biol. Sci.* 13:456–63
114. Keister, D. L., Evans, W. R. 1976. Oxygen requirement for acetylene reduction by pure cultures of rhizobia. *J. Bacteriol.* 129:149–53
115. Kennedy, L. D. 1978. Extracellular polysaccharides of *Rhizobium:* Identification of monosaccharides from strain CB756. *Carbohydr. Res.* 61:217–21
116. Kennedy, L. D., Bailey, R. W. 1976. Monomethyl sugars in extracellular polysaccharides from slow-growing rhizobia. *Carbohydr. Res.* 49:451–54
117. Kinje, J. W. 1975. The fine structure of pea root nodules. 2. Senescence and disintegration of the bacteriod tissue. *Physiol. Plant Pathol.* 7:17–21
118. Kivilaan, A., Bandurski, R. S., Schulze, A. 1971. A partial characterization of an autolytically solubilized cell wall glucan. *Plant Physiol.* 48: 968–72
119. Kumarashinge, R. M. K., Nutman, P. S. 1977. *Rhizobium*-stimulated callose formation in clover root hairs and its relation to infection. *J. Exp. Bot.* 28:961–76
120. Law, I. J., Strijdom, B. W. 1977. Some observations on plant lectins and *Rhizobium* specificity. *Soil. Biol. Biochem.* 9:79–84
121. Lillich, T. T., Elkan, G. H. 1968. Evidence countering the role of polygalacturonase in invasion of root hairs of leguminous plants by *Rhizobium* spp. *Can. J. Microbiol.* 14:618–25
122. Lis, H., Sela, B., Sachs, L., Sharon, N. 1970. Specific inhibition by N-acetylgalactosamine of the interaction between soybean agglutinin and animal cell surfaces. *Biochim. Biophys. Acta* 211:582–85
123. Ljunggren, H. 1969. Mechanism and pattern of *Rhizobium* invasion into leguminous root hairs. *Physiol. Plant. Suppl.* 5. 84 pp.
124. Ljunggren, H., Fåhraeus, G. 1959. Effect of *Rhizobium* polysaccharide on the formation of polygalacturonase in lucerne and clover. *Nature* 183: 1578–79
125. Ljunggren, H., Fåhraeus, G. 1961. The role of polygalacturonase in root-hair invasion by nodule bacteria. *J. Gen. Microbiol.* 26:521–28
126. Long, S. R., Meade, H. M., Ausubel, F. M. 1981. The use of transposon mutagenesis in the molecular genetic analysis of symbiotic nitrogen fixation. *Microbiology*. In press
127. Lyon, G., Albersheim, P. 1980. The nature of the phytoalexin elicitor of *Erwinia carotovora*. *Plant Physiol.* 65: Abstr. 752, p. 137
128. Macmillan, J. D., Cooke, R. C. 1969. Evidence against involvement of pectic enzymes in the invasion of root hairs by *Rhizobium trifolii*. *Can. J. Microbiol.* 15:643–45

129. Maier, R. J., Brill, W. J. 1978. Involvement of *Rhizobium japonicum* O-antigen in soybean nodulation. *J. Bacteriol.* 133:1295–99
130. Martinez-Molina, E., Morales, V. M., Hubbell, D. H. 1979. Hydrolytic enzyme production by *Rhizobium*. *Appl. Environ. Microbiol.* 38:1186–88
131. McCoy, E. 1932. Infection by *Bact. Radicicola* in relation to the microchemistry of the host's cell walls. *Proc. R. Soc. London Ser. B* 110:514–33
132. McIntire, F. C., Peterson, W. H., Riker, A. J. 1942. A polysaccharide product by the crown-gall organism. *J. Biol. Chem.* 143:491–96
133. Mort, A. J., Bauer, W. D. 1980. Composition of the capsular and extracellular polysaccharides of *Rhizobium japonicum*. Changes with culture age and correlations with binding of soybean seed lectin to the bacteria. *Plant Physiol.* 66:158–63
134. Munns, D. N. 1968. Nodulation of *Medicago sativa* in solution culture. I. Acid sensitive steps. *Plant Soil* 28: 129–46
135. Munns, D. N. 1968. Nodulation of *Medicago sativa* in solution culture. II. Compensating effects of nitrate and prior nodulation. *Plant Soil* 28:246–57
136. Munns, D. N. 1968. Nodulation of *Medicago sativa* in solution culture. III. Effects of nitrate on root hairs and infection. *Plant Soil* 29:33–47
137. Munns, D. N. 1968 Nodulation of *Medicago sativa* in solution culture. IV. Effects of indole-3-acetate in relation to acidity and nitrate. *Plant Soil* 29: 257–61
138. Munns, D. N. 1969. Enzymic breakdown of pectin and acid inhibition of the infection of *Medicago* roots by *Rhizobium*. *Plant Soil* 30:117–20
139. Napoli, C., Albersheim, P. 1980. Infection and nodulation of clover by nonmotile *Rhizobium trifolii*. *J. Bacteriol.* 141:979–80
140. Napoli, C., Albersheim, P. 1980. *Rhizobium leguminosarum* mutants incapable of normal extracellular polysaccharide production. *J. Bacteriol.* 141: 1454–56
141. Napoli, C., Dazzo, F. B., Hubbell, D. 1975. Production of cellulose microfibrils by *Rhizobium*. *Appl. Microbiol.* 30:123–32
142. Napoli, C., Hubbell, D. H. 1975. Ultrastructure of *Rhizobium*-induced infection threads in clover root hairs. *Appl. Microbiol.* 30:1003–9
143. Napoli, C., Sanders, R. E., Carlson, R. W., Albersheim, P. 1980. Host-symbiont interactions: recognizing *Rhizobium*. See Ref. 147, 2:189–204
144. Newcomb, E. H., Bonnett, H. T. 1965. Cytoplasmic microtubule and wall microfibril orientation in root hairs of radish. *J. Cell. Biol.* 27:575–88
145. Newcomb, W., Sippell, D., Peterson, R. L. 1979. The early morphogenesis of *Glycine max* and *Pisum sativum* root nodules. *Can. J. Bot.* 57:2603–16
146. Newton, W. E., Nyman, C. J., eds. 1975. *Nitrogen Fixation*, Vols. 1, 2. Pullman: Washington State Univ. Press. 312 pp., 381 pp.
147. Newton, W. E., Orme-Johnson, W. H. eds. 1980. *Nitrogen Fixation*, Vols. 1, 2. Baltimore: Univ. Park. 394 pp., 323 pp.
148. Newton, W. E., Posgtate, J. R., Rodriguez-Barreuco, C., eds. 1977. *Recent Developments in Nitrogen Fixation*. New York: Academic. 622 pp.
149. Nordbring-Hertz, B., Mattiasson, B. 1979. Action of a nematode-trapping fungus shows lectin-mediated host-microorganism interaction. *Nature* 281: 477–79
150. Nuti, M. P., Lepidi, A. A., Prakash, R. K., Schilperoot, R. A., Cannon, F. C. 1979. Evidence for nitrogen fixation (nif) genes on indigenous *Rhizobium* plasmids. *Nature* 282:533–35
151. Nutman, P. S. 1949. Physiological studies on nodule formation. II. The influence of delayed inoculation on the rate of nodulation in red clover. *Ann. Bot.* 13:261–83
152. Nutman, P. S. 1952. Studies on the physiology of nodule formation. III. Experiments on the excision of root tips and nodules. *Ann. Bot.* 16:79–101
153. Nutman, P. S. 1959. Some observations on root-hair infection by nodule bacteria. *J. Exp. Bot.* 10:250–62
154. Nutman, P. S. 1962. The relation between root hair infection by *Rhizobium* and nodulation in *Trifolium* and *Vicia*. *Proc. R. Soc. London Ser. B* 156:122–37
155. Ofek, I., Mirelman, D., Sharon, N. 1977. Adherence of *Escherichia coli* to human mucosal cells mediated by mannose receptors. *Nature* 265:623–25
156. Ørskov, I., Ørskov, F., Jann, B., Jann, K. 1977. Serology, chemistry and genetics of O and K antigens of *Escherichia coli*. *Bacteriol. Rev.* 41:667–710
157. Palomares, A., Montoya, E., Olivares, J. 1979. Quality and rate of extracellular polysaccharides produced by *Rhizobium meliloti* and their inducing effect on polygalacturonase production in

legume roots as derived from the presence of extrachromosomal DNA. *Microbios* 22:7–13

158. Panopoulos, N. J., Schroth, M. N. 1974. Role of flagellar motility in the invasion of bean leaves by *Pseudomonas phaseolicola. Phytopathology* 64: 1389–97
159. Pate, J. S., Dart, P. J. 1961. Nodulation studies in legumes. IV. The influence of inoculum strain and time of application of ammonium nitrate on symbiotic response. *Plant Soil* 15:329–46
160. Phillips, D. A. 1971. A cotyledonary inhibitor of root nodulation in *Pisum sativum. Physiol. Plant.* 25:482–87
161. Planqué, K., Kijne, J. W. 1977. Binding of pea lectins to a glucan-type polysaccharide in the cell walls of *Rhizobium leguminosarum. FEBS Lett.* 73:64–66
162. Planqué, K., van Nierop, J. J., Burgers, A. 1979. The lipopolysaccharide of free-living and bacteroid forms of *Rhizobium leguminosarum. J. Gen. Microbiol.* 110:151–59
163. Purchase, H. F., Nutman, P. S. 1957. Studies on the physiology of nodule formation. VI. The influence of bacterial numbers in the rhizosphere on nodule initiation. *Ann. Bot.* 21:439–54
164. Putman, E. W., Potter, A. L., Hodgson, R., Hassid, W. Z. 1950. The structure of crown-gall polysaccharide. *J. Am. Chem. Soc.* 72:5024–26
165. Quispel, A., ed. 1974. *The Biology of Nitrogen Fixation.* Amsterdam: North Holland. 769 pp.
166. Rao, R. V., Keister, D. L. 1978. Infection threads in root-hairs of soybean (*Glycine max.*) plants inoculated with *Rhizobium japonicum. Protoplasma* 97: 311–16
167. Roberts, E. A. 1916. The epidermal cells of roots. *Bot. Gaz.* 62:488–506
168. Robertsen, B., Åman, P., Darvill, A. G., McNeil, M., Albersheim, P. 1981. Host-symbiont interactions. V. The structure of the acidic extracellular polysaccharides secreted by *Rhizobium leguminosarum* and *Rhizobium trifolii. Plant Physiol.* In press
169. Robertson, J. G., Farnden, K. J. F. 1980. The ultrastructure and metabolism of the developing legume root nodule. In *Amino Acids and Derivatives,* ed. B. J. Miflin. Vol. 5 of *The Biochemistry of Plants; A Comprehensive Treatise,* ed. P. K. Stumpf, E. E. Conn. New York: Academic. In press
170. Rolfe, B. G., Gresshoff, P. M., Shine, J., Vincent, J. M. 1980. Interaction between a non-nodulating and an ineffective mutant of *Rhizobium trifolii* resulting in effective (nitrogen fixing) nodulation. *Appl. Environ. Microbiol.* 39: 449–52
171. Rolfe, B. G., Shine, J., Gresshoff, P. M. 1980. Will a molecular biological analysis of the *Rhizobium* legume symbiosis help agriculture? *Proc. Aust. NZ Microbiol. Soc.* In press
172. Ruvkun, G. B., Long, S. R., Meade, H. M., Ausubel, F. M. 1980. Molecular genetics of symbiotic nitrogen fixation. *Cold Spring Harbor Symp. Quant. Biol.* 45: In press
173. Sahlman, K., Fåhraeus, G. 1962. Microscopic observations on the effect of indole-3-acetic acid upon root hairs of *Trifolium repens. K. Lantbrukshoegsk. Annlr.* 28:261–68
174. Sahlman, K., Fåhraeus, G. 1963. An electron microscope study of root-hair infection by *Rhizobium. J. Gen. Microbiol.* 33:425–31
175. Sanders, R. E., Carlson, R. W., Albersheim, P. 1978. A *Rhizobium* mutant incapable of nodulation and normal polysaccharide secretion. *Nature* 271: 240–42
176. Schaede, R. 1940. Die knöllchen der adventiven wasserwurtzeln von *Neptunia oleracea* und ihre bakteriensymbiose. *Planta* 31:1–21
177. Schmidt, E. L. 1979. Initiation of plant root-microbe interactions. *Ann. Rev. Microbiol.* 33:355–76
178. Sequeira, L. 1978. Lectins and their role in host-pathogen specificity. *Ann. Rev. Phytopathol.* 16:453–81
179. Sequeira, L., Graham, T. L. 1977. Agglutination of avirulent strains of *Pseudomonas solanacearum* by potato lectin. *Physiol. Plant Pathol.* 11:43–54
180. Shantharam, S., Gow, J. A., Bal, A. K. 1980. Fractionation and characterization of two morphologically distinct types of cells in *Rhizobium japonicum* broth culture. *Can. J. Microbiol.* 26:107–14
181. Shimshick, E. J., Herbert, R. R. 1978. Adsorption of rhizobia to cereal roots. *Biochem. Biophys. Res. Commun.* 84:736–42
182. Smith, W. K. 1958. A survey of the production of pectic enzymes by plant pathogenic and other bacteria. *J. Gen. Microbiol.* 18:33–41
183. Smittle, D. A. 1979. A synopsis of symbiotic nitrogen fixation and other apparently similar host-microorganism interactions. *J. Plant Nutr.* 1:377–95
184. Solheim, B., Paxton, J. 1980. Recognition in *Rhizobium*-legume systems.

Proc. Rockefeller Found. Conf., Lake Como, Italy. In press
185. Solheim, B., Raa, J. 1971. Evidence countering the theory of specific induction of pectin degrading enzymes as a basis for specificity in *Rhizobium-Leguminosae* associations. *Plant Soil* 35:275–80
186. Solheim, B., Raa, J. 1973. Characterization of the substances causing deformation of root-hairs of *Trifolium repens* when inoculated with *Rhizobium trifolii. J. Gen. Microbiol.* 77:241–47
187. Sømme, R. 1980. Pyruvic acid-containing mono- and oligo-saccharides from *Rhizobium trifolii* Bart A. *Carbohydr. Res.* 80:325–32
188. Stacey, G., Paau, A. S., Brill, W. J. 1980. Host recognition in the *Rhizobium*-soybean symbiosis. *Plant Physiol.* 66:609–14
189. Subba Rao, N. S., ed. 1979. *Recent Advances in Biological Nitrogen Fixation.* New Delhi: Oxford and IBH. 486 pp.
190. Sutherland, I. W. 1977. Bacterial exopolysaccharides: their nature and production. In *Surface Carbohydrates of the Prokaryotic Cell,* ed. I. W. Sutherland, pp. 27–96. London: Academic. 472 pp.
191. Talmadge, K. W., Keegstra, K., Bauer, W. D., Albersheim, P. 1973. The structure of plant cell walls. I. The macromolecular components of the walls of suspension-cultured sycamore cells with a detailed analysis of the pectic polysaccharides. *Plant Physiol.* 51: 158–73
192. Thornton, H. G. 1936. Action of sodium nitrate on infection of lucerne root hairs by nodule bacteria. *Proc. R. Soc. London Ser. B* 119:47–92
193. Town, C., Stanford, E. 1979. An oligosaccharide-containing factor that induces cell differentiation in *Dictyostelium discoideum. Proc. Natl. Acad. Sci. USA* 76:308–12
194. Tsien, H. C., Anderson, J. S., Schmidt, E. L. 1980. Purification and compositional studies of soybean lectin binding polysaccharide of *Rhizobium japonicum. Abstr. Ann. Meet. Am. Soc. Microbiol.* p. 155
195. Tsien, H. C., Cain, P. S., Schmidt, E. L. 1977. Viability of *Rhizobium* bacteroids. *Appl. Environ. Microbiol.* 34: 854–56
196. Tsien, H. C., Schmidt, E. L. 1977. Polarity in the exponential phase *Rhizobium japonicum* cells. *Can. J. Microbiol.* 23:1274–84
197. Tsien, H. C., Schmidt, E. L. 1980. Accumulation of soybean lectin-binding polysaccharide during growth of *Rhizobium japonicum* as determined by hemagglutination assay. *Appl. Environ. Microbiol.* 39:1100–4
198. Urban, J. E. 1979. Nondividing, bacteroid-like *Rhizobium trifolii:* In vitro induction via nutrient enrichment. *Appl. Environ. Microbiol.* 38:1173–78
199. Urbano, M. R., Dazzo, F. B. 1980. Age dependent determinants of bacterial adhesion in the *Rhizobium*-clover symbiosis. *Abstr. Ann. Meet. Am. Soc. Microbiol.,* p. 163
200. Valera, C. L., Alexander, M. 1965. Reversal of nitrate inhibition of nodulation by indole-3-acetic acid. *Nature* 206: 326–31
201. Van Brussel, A. A. N., Costerton, J. W. 1979. Nitrogen fixation by *Rhizobium* 32H1. A morphological and ultrastructural comparison of asymbiotic and symbiotic nitrogen-fixing forms. *Can. J. Microbiol.* 25:352–61
202. Verma, D. P. S., Zogbi, V., Bal, A. K. 1978. A cooperative action of plant and *Rhizobium* to dissolve the host cell wall during development of root nodule symbiosis. *Plant Sci. Lett.* 13:137–42
203. Vincent, J. M. 1980. Factors controlling the legume-*Rhizobium* symbiosis. See Ref. 147, 2:103–30
204. Walker, N., ed. 1975. *Soil Microbiology.* London: Butterworth. 262 pp.
205. Ward, H. M. 1887. On the tubercular swellings on the roots of *Vicia faba. Philos. Trans. R. Soc. London Ser. B* 178:539–62
206. Weaver, R. W., Frederick, L. R. 1972. A new technique for most probable number counts of rhizobia. *Plant Soil* 36:219–22
207. Whatley, M. H., Bodwin, J. S., Lippincott, B. B., Lippincott, J. A. 1976. Role for *Agrobacterium* cell envelope lipopolysaccharide in infection site attachment. *Infect. Immun.* 13:1080–83
208. Wilkinson, S. G. 1977. Composition and structure of bacterial lipopolysaccharides. See Ref. 190, pp. 97–175
209. Wolpert, J. S., Albersheim, P. 1976. Host-symbiont interactions. I. The lectins of legumes interact with the O-antigen containing lipopolysaccharides of their symbiont rhizobia. *Biochem. Biophys. Res. Commun.* 70:729–37
210. Wong, P. P. 1980. Interactions between rhizobia and lectins of lentil, pea, broad bean and jackbean. *Plant Physiol.* 65: 1049–52

211. Yao, P. Y., Vincent, J. M. 1969. Host specificity in root hair "curling factor" of *Rhizobium* spp. *Aust. J. Biol. Sci.* 22:413–23
212. Yao, P. Y., Vincent, J. M. 1976. Factors responsible for the curling and branching of clover root hairs by *Rhizobium. Plant Soil.* 45:1–16
213. York, W. S., McNeil, M., Darvill, A. G., Albersheim, P. 1980. β-2-linked glucans secreted by fast-growing species of *Rhizobium. J. Bacteriol.* 142:243–48
214. Zevenhuizen, L. P. T. M. 1973. Methylation analysis of acidic exopolysaccharides of *Rhizobium* and *Agrobacterium. Carbohydr. Res.* 26:409–19
215. Zevenhuizen, L. P. T. M., Scholten-Koerselman, H. J. 1979. Surface carbohydrates of *Rhizobium.* I. β-1,2-glucans. *Antonie Van Leeuwenhoek J. Microbiol. Serol.* 45:165–75
216. Zevenhuizen, L. P. T. M., Scholten-Koerselman, H. J., Postmus, M. A. 1980. Lipopolysaccharides of *Rhizobium. Arch. Microbiol.* 125:1–8

Ann Rev. Plant Physiol. 1981. 32:451–63

THE PHYSICAL STATE OF PROTOCHLOROPHYLL(IDE) IN PLANTS

◆7719

Hemming I. Virgin

Department of Plant Physiology, University of Göteborg, S-413 19 Göteborg, Sweden

CONTENTS

INTRODUCTION

Protochlorophyll (PChl) is the immediate precursor to chlorophyll (Chl) *a* in the green plant and is like the Chls localized to the chloroplasts in the irradiated plant and to their precursors, the pro/etioplasts in the nonirradiated plants. The pigment exists in both esterified (PChl in true sense) and nonesterified (PChlide) forms (55). The PChl molecule (the chromophore) differs from Chl *a* by the lack of two hydrogen atoms at carbon atoms 7 and 8 of the porphyrin ring (7). Higher plants contain at least four different molecular species, namely mono- and divinylprotochlorophyllide (PChlide), together with possibly two unknown fluorescent compounds (5). The major component consists of divinylprotochlorophyllide. The mono- and divinylprotochlorophyllides show similar emission maxima but differ in the position of their excitation maxima in the Soret region.

0066-4294/81/0601-0451$01.00

In the following, the term PChl will be used as a comprehensive term for the pigment irrespective of whether it is esterified or not. When dealing with the different kinds of molecules, this will be particularly specified. This review deals with the physical state and localization of PChl. The "physical state" can be interpreted in various ways. My intention here is to discuss and describe data available on localization and molecular pattern of the PChl(ide) complexes found primarily in dark-grown plants. Some emphasis will also be given to the complex events taking place in connection with the photoconversion of PChl(ide) into Chl(ide). The information extractable from such studies can in many instances contribute to the understanding of how changes in absorption and fluorescence properties can be brought back to structural changes in the etio-chloroplast, which are intimately connected with the physical state of the pigments present.

The subject has been partly treated in other connections (e.g. 3, 7, 9, 51, 52, 63). This review is a survey of pertinent studies on the subject and will cover results from recent years' research. The article deals mainly with the situation in higher plants. Studies on extracts of PChl-protein complexes ("holochromes") will be dealt with only insofar as they have any bearing on the in vivo system.

SPECTROSCOPIC FORMS OF PROTOCHLOROPHYLL IN VIVO

Absorption and fluorescence spectra of leaves in vivo are mainly determined by the pigments present. Differences in their aggregation and in the protein moieties to which they are bound can contribute to differences in absorption properties. It is generally agreed that etioplasts of dark-grown leaves contain three spectroscopically distinguishable PChl complexes (13). They are characterized by their main absorption maxima in the red part of the spectrum and are here called $PChl_{628}$, $PChl_{636}$, and $PChl_{650}$, respectively. In vivo these forms show fluorescence maxima at 632, 657, and 657 nm, respectively. This means that all energy absorbed by $PChl_{636}$ is internally transferred to $PChl_{650}$, indicating a closeness in space between the last two species. The suggestion that $PChl_{636}$ and $PChl_{650}$ could be a single species (47) has been rejected (21).

LOCALIZATION OF THE PROTOCHLOROPHYLL

The PChl found in dark-grown material has been thought mainly to be localized to the prolamellar bodies of the etioplasts. Action spectra for light effects on this structure speak in favor of such a localization (40, 41, 80). Fluorescence microscopy (8) of etioplasts from initially dark-grown leaves of bean indicate that the PChl of the etioplasts is confined to small discrete optically dense bodies, 0.7–1.3 μm in diameter, called "1 μ centers" (8). Normally the single etioplast contains one or two such centers. These centers obtained from plants irradiated for 6 hours are more strongly fluorescent than at the beginning of illumination, indicating that new Chl is probably deposited within these regions as well as in other regions of the plastid. From (8) it seems probable that the "1 μ centers" are the prolamellar bodies seen in electron micrographs and that they or their immediate surroundings are the site for the PChl found in dark-grown plants. Recent studies seem to indicate that the main bulk of PChl in dark-grown material is located just outside the prolamellar body—primarily to the primary thylakoid membranes adjacent to the prolamellar body and not to the body itself. The bodies proper are thought to consist mainly of a mixture of proteins and saponins (49, 50, 57). A localization of the PChl outside the prolamellar bodies would speak against the conclusion reached from action spectra for the breaking up of the prolamellar bodies (38, 80)—spectra which are very similar to the action spectrum for PChl-Chl *a* transformation (54). It does not, however, exclude the possibility that smaller portions of PChl could exist within the prolamellar structure.

Protein denaturants are the only agents which inhibit the conversion of PChlide to Chlide. From this and other findings the conclusion has been drawn (6) that both active PChlide and the source for the hydrogen atoms and electrons required for the photoconversion are located in a hydrophobic region of the PChl-protein complex which is not readily accessible to aqueous reagents (6).

Upon irradiation the tubules of the prolamellar body are broken up into smaller tubule fractions simultaneously with the photochemical reduction of PChl to Chl (30, 80, 81). If it is assumed that the PChl-holochrome is a component of the walls of the tubules (37) and remains associated with this membraneous material after conversion to Chl-holochrome, the latter would reside in or on the primary lamellar layers after the dispersal process is completed.

As we do not yet know for sure the size of the smallest holochrome unit obtainable (see below), it is fruitless to speculate about the way in which the membranes are formed. Several models have been proposed (12), but it is important to stress that there might be differences in respect to membrane composition between the original PChl-containing system in, or in close vicinity to, the prolamellar body and the system formed at the building up of the thylakoids.

CORRELATION BETWEEN STRUCTURE AND ABSORPTION PROPERTIES

Pigment Molecule Aggregates

The sequence of spectral changes taking place after the phototransformation of PChl is accompanied by morphological changes taking place when the etioplast gradually develops into a "normal" chloroplast with well-defined grana and stroma regions (10). These changes include the aforementioned breaking up of the crystalline prolamellar body lattice into small tubules, their dispersal, and the gradual formation of thylakoids containing grana regions as described in a series of papers on the macromolecular physiology of plastids (30, 39–41, 81). Summarizing the results from these earlier studies, it must be said that they all give rather good evidence for postulating that the PChl in the dark-grown plant is localized in (or is in close proximity to) the prolamellar body.

The relationship between etioplast or chloroplast structure and absorption properties in vivo of the leaves has been the object of many studies. By treating the material with various agents which presumably also act on the structural architecture of the pigment-protein complexes, changes in absorption values often take place which are sometimes reversible. At such treatments the 650 nm peak of the in vivo PChl shifts to around 635 nm or to 628 nm (21, 26, 27, 32). The main cause for the change from the 650 form to the 636 form is probably the change from a dimer to a monomer. Circular dichroism (CD) measurements of homogenate of etiolated bean leaves (59) suggest that the PChlide molecules are initially in an aggregated, probably dimeric state in the etioplasts and dissociate to a monomeric configuration after irradiation (66). On the basis of the different bleaching properties of dimers vs monomers, it has been concluded that the long wavelength forms are more aggregated than the short wavelength species (45), conclusions which are in agreement with the general ideas in this respect.

The extent to which the PChl molecules form aggregates has been investigated in various ways (78). The polarization of Chl fluorescence was found to decrease during the course of photoconversion before 50% conversion was reached (experiments on PChl holochrome from beans), suggesting that absorbed energy is transferred among pigments before emission (78). From quantitative study of the relation between fluorescence polarization and percentage photoconversion the conclusion could be drawn that there could not be less than three pigment molecules forming an aggregate. Probably the group size is five or larger, thus comprising a mixture of PChlide and Chlide molecules. Studies of polarization of fluorescence of Chl-protein holochromes in 2 M sucrose (66) containing a more purified holochrome than described above showed less depolarization, probably the result of the presence of smaller protein aggregates. It was suggested (23) that the aggregation of pigment molecules in the structures of the plastids is the cause for the shifts to longer wavelengths and that consequently the spectral shifts to shorter wavelengths would be due to disaggregation following the disruption of plastid structure. A support for this idea is the fact that PChl, dissolved in solvents of low dielectric constants, slowly shows a red shift with time, a shift which is reversed upon addition of trace amounts of a more polar solvent (67). The aggregates presumedly consist of polymers made up of a large number of pigment molecules. $PChl_{650}$ dissolved in dry nonpolar solvents tends to form a dimer predominently (15, 16, 18, 19, 62), while the PChlide $_{628}$ and $_{636}$ are monomers in nonpolar solvents (15). In binary dioxane-water mixtures, PChl forms aggregates with absorption spectra in the red which slowly (after 3-5 hours) shift a few nanometers toward longer wavelengths (83). This change indicates a change to other types of aggregates with another type of molecular packing in the solution. Polarization and CD spectra of such aggregates indicate a definite orientation of the molecules joined in such molecular packets.

It is always tempting to use results from in vitro systems to explain the conditions found in vivo. What we can say for the present is that at least some of the absorption and fluorescence properties of the in vivo form of the pigments are a function of different aggregation states. Unfortunately, however, the present data do not allow us to get a clear picture of the orientation of the molecules nor of the way in which they are incorporated in a protein structure. CD spectra can inform us to a certain extent of the degree of order, but as our knowledge of the composition of the membranes in which they are located is still very scanty, only few reliable conclusions can be drawn. As the PChl molecules with all probability are associated in vivo with lipoproteins in a highly specific way in the membranes, the aggregation state plays a great role in providing an interpretation of the mechanism of the photochemical reactions (28, 69–72). Absorption properties of the PChl forms in dark-grown leaves and the spectral shifts taking place in the newly formed Chlide after irradiations indicate (77) that PChl is present in the leaf (beans) in molecular groups, the size of which should be around 20 or possibly multiples of this number, and as such comprise a fundamental unit for the photoconversion. This value is in contrast to other findings indicating that PChl is present in the leaf as single molecules (56). The value of 20 tallies well with the calculated value for the size of the energy transfer unit—for which the smallest unit for the photoconversion of PChl to Chl should consist of about 18 pigment molecules (69).

To summarize our knowledge about in vivo aggregates of PChl(ide) molecules one could say that the earlier studies on this subject indicated a low number (one or two) while later works indicate a considerably higher value for this number (~ 20). It is, however, important to have in mind the possibility of the existence of smaller subunits with only a few pigment molecules because these subunits may form units of tighter structure, thus containing a considerably higher proportion of pigment molecules to protein (42).

Protein Complexes

It is now generally believed that the molecular weight of the PChl holochrome complex is around 550,000 (65) to 600,000 (24, 46) with possible subunit molecular weights of about half of these values (65). Lower values have been found for bean [100,000 (42), 45,000 (24)], and for barley [63,000 (42)]. The 600,000 complex, containing at least 4 PChlide molecules, has been shown to be composed of one type of polypeptide with a molecular weight of 45,000 when analyzed before and immediately after the photoconversion reaction. The PChlide and Chlide pigment molecules are supposed to be associated with this polypeptide. The values found for the number of PChl molecules attached to each protein unit show great variation (46, 65, 66). Most of the values presented are obtained for holochrome preparations. Determinations on holochrome preparations seem in general to give lower values for the units than those made on materials less maltreated. The actual size for the complex in vivo might be greater and approach the highest values found (~ 20) (77). It is also important to distinguish between the different absorption forms of the PChl present since, as mentioned above, the various spectral forms of the PChl complexes in vivo in all probability differ in respect to aggregation. There are indications of the presence, in holochromes, of complexes with a molecular weight of around 63,000 with only one PChl molecule attached to the protein (42). It is obvious that holochrome preparations can contain holochrome units with different sizes of the pigment aggregates. In the hitherto reported studies on complexes of PChl-protein, a distinction has not been made between the different forms of optically discernable PChl forms in vivo. By means of labeling experiments with ^{3}H-δ-aminolevulinic acid followed by acrylamide gel electrophoresis it has indeed been possible to distinguish two peptide chains with molecular weights of 21,000 and 29,000, respectively (36) (cf the holochrome with a molecular weight of 45,000 (24)). It would be tempting to relate these two forms of holochromes to the two main forms of PChl complexes existing in the dark ($PChl_{636}$ and $PChl_{650}$), but the results presented are not conclusive enough to allow this conclusion.

TRANSFER UNITS

In a series of papers (20, 22, 29, 60, 69, 72), certain formal relations for the PChl(ide) → Chl *a* photoconversion have been developed dealing with efficiency of energy transfer, intensity of fluorescence, and degree of reduction of PChl in the etiolated leaf. These studies have resulted in the coining of the idea of the existence of so-called "transfer units" (70), consisting of a number of PChl(ide) units existing before irradiation and containing PChl(ide) plus Chl(ide) molecules after partial photoreduction. These "transfer units" themselves consist of subunits, each of which consists of a polypeptide chain to which PChl(ides) are attached. The theory of the transfer unit considers pigments distributed among two lipoprotein subunits classes, each comprised of reduced pigments and nonreduced pigments. According to the theory, there could be dark reductions of pigment molecules in addition to that predicted by photochemistry alone and a light regulation of the concentration of the availability of some reductant. It is beyond the scope of this article to make a deeper analysis of the theories, but it is interesting that in many respects they tally with experimentally found data (61, 72, 74). The discussion leading to the concept of transfer units is purely theoretical, and the microscopic mechanisms involved are not considered.

ORIENTATION OF THE PROTOCHLOROPHYLL IN THE MEMBRANES

The literature has little data on the orientation of PChl species in the membranes. Measurements of the fluorescence polarization ratio in magnetically oriented plastids from etiolated corn and cotyledons from cucumber (31) show that the chromophores are probably oriented at random with respect to the plane of the thylakoid membranes. This random orientation in respect to PChl molecules which is found for both long and short wavelength-absorbing species could possibly be an effect of the compartmentalization of the PChl synthesis (31). A localization of PChl molecules to the three-dimensional prolamellar body would give practically no change in polarization ratio with respect to different directions of the plane, while a localization to lamellae would give a more pronounced effect, as in such a case the molecules could be attached to the membranes in an orderly fashion—as are the chlorophylls in the thylakoids. Some results (14) speak in favor of PChl—at least in dark-grown leaves—being localized to the prolamellar bodies.

It is now well established (2, 4, 53, 58, 79, 82) that only the nonesterified pigment, i.e. PChlide, is able to be phototransformed and the esterification of the Chlide formed takes place in a series of steps (55) more or less synchronously with the Shibata shift (73), even if the two processes probably are not connected with each other (1, 34, 43). Failure of PChl to be phototransformed could possibly be the result of a steric hindrance from the phytyl residue (33). The inability of PChl to act as a precursor of Chl would be because of the interference by the bulky phytyl group of a membrane-located photochemically active $PChl_{650}$ complex formed in darkness (34). It was shown (35) that PChlide added to green deactivated membranes in the presence of NADPH forms phototransformable $PChlide_{650}$, whereas no such complex is formed from added PChl.

PHOTOCHLOROPHYLL IN SEED COATS OF CUCURBITACEAE SPECIES

In addition to being found in dark-grown seedlings, PChl is found in large amounts in the inner seed coats of certain species belonging to the family of Cucurbitaceae (68). Here the pigment is present in such high concentrations as to give the tissue a greyish-green color. The pigment is nonphototransformable, and it is not certain whether or not it is protein bound. Analyses of different species of Cucurbitaceae have shown that in addition to PChl, PChlide, Mg-2,4,divinylpheoporphyrin a_5 and Mg-free vinyl pheoporphyrin a_5 are also present (44, 45). The pigment is localized to plastids which are especially rich in osmophilic globuli in *Cucurbita* (45, 75). The spectra of the PChl moieties differ to a certain extent between the species. Thus, the in vivo spectra of seed coats of *C. pepo* and *C. maxima* are very similar to those of leaves of dark-grown wheat and bean, with maxima at 635 nm and 650 nm with a fluorescence emission peak at 655 nm, while the spectrum of seed coats of *C. moschata* also show small peaks at 680 nm. The latter species also contains Mg-free vinylpheoporphyrin a_5 (76). In *Cyclanthera* the maximum in absorption lies at 670 nm with a shoulder at 640 nm and a fluorescence maximum at 691 nm (76).

In seed coats of mature *Cyclanthera,* the concentration of PChl is amazingly high (14,000 μg/g as compared to 500 μg/g in mature *Cucurbita* seeds). According to CD studies, it is evident that at least one of the three forms existing in vivo is present in a crystalline form (76). Electron microscopy revealed that the plastids contain high amounts of crystalloids (64). The mature seed is very rich in lipids, at least in the cell layer just underneath the PChl-containing layer. It is highly probable that crystalline PChl is nonprotein bound. CD spectra of seed coats of *Cyclanthera* in vivo give a very strong and characteristic signal in the red region with negative asymmetrical Cotton effects at 644(+), 669(0), and 687(–) nm (76). This result indicates a strong interaction between the PChl molecules, suggesting the presence of water molecules in the aggregates, probably between the porphyrin rings (76). Such asymmetrical and negative Cotton effects have also been found in the CD spectrum of crystalline Chl a_{745} prepared with iso-octane (25). In model systems (PChlide in liquid paraffin oil) the pigment shows an infrared spectrum indicating that the PChl molecules in these aggregates are connected through water molecules (15, 17), thus showing the same physical state as Chl a_{745} which is thought to be a "watered aggregate" in vivo (48). The similarity between the CD spectra of seed coats from pumpkin seeds and that of solid forms of PChl (in both cases a negative signal at 638 nm and a positive signal at 652 nm) suggests that the 650 nm PChl in the seed is present as a molecular aggregate (11).

CONCLUDING REMARKS

While our knowledge is still rather scanty concerning the kind of pigment aggregates present in holochrome preparations, it can be said that what we know about the conditions in live plants is still more fragmentary. In this connection it is worthwhile to stress that the different spectroscopic in vivo forms of PChl(ide) extensively described by a number of authors and characterized by their absorption and fluorescence properties have not been successfully isolated. All efforts made hitherto to retain the absorption properties of the live, intact leaf have met with no success. There are several reasons for this failure. An important one is obvious. The disintegration of the cells and the plastids in solutions with low water potentials—necessary to retain the photoactivity—must have a profound effect on the original macro as well as microstructure of the system. Considerable knowledge of the submicroscopic structures and development of the etio/chloroplasts has been obtained during the last decade. We still lack, however, that piece of knowledge which would make it possible to find the missing link between spectrophotometric observations and characterization of the pigment species and their spatial localization in the membrane systems observable in the electron micrograph.

Acknowledgments

I would like to thank Hans Ryberg for fruitful discussions and Carin Thysell for great help with the manuscript.

Literature Cited

1. Akoyunoglou, G., Michalopoulos, G. 1971. The relation between the phytylation and the 682–672 nm shift in vivo of chlorophyll *a*. *Physiol. Plant.* 25: 324–29
2. Akoyunoglou, G., Siegelman, H. 1968. Protochlorophyllide resynthesis in dark grown bean leaves. *Plant Physiol.* 51:66–68
3. Anderson, J. M. 1975. The molecular organization of chloroplast thylakoids. *Biochim. Biophys. Acta* 416:191–235
4. Augustinessen, E., Madsen, A. 1965. Regeneration of protochlorophyllide in etiolated barley seedlings following different light treatments. *Physiol. Plant.* 18:828–37
5. Belanger, F. C., Rebeiz, C. A. 1980. Chloroplast biogenesis. Detection of divinyl protochlorophyllide in higher plants. *J. Biol. Chem.* 225:1266–72
6. Boardman, N. K. 1962. Studies on protochlorophyll-protein complex. II. The photo-conversion of protochlorophyll to chlorophyll *a* in the isolated complex. *Biochim. Biophys. Acta* 64:279–93
7. Boardman, N. K. 1966. Protochlorophyll. In *The Chlorophylls,* ed. L. P. Vernon, G. R. Seely, pp. 437–80. New York/London: Academic. 679 pp.
8. Boardman, N. K., Anderson, J. M. 1964. Studies on the greening of dark-grown bean plants. I. Formation of chloroplasts from proplastids. *Aust. J. Biol. Sci.* 17:86–92
9. Boardman, N. K., Anderson, J. M., Goodchild, D. J. 1978. Chlorophyll-protein complexes and structure of mature and developing chloroplasts. *Curr. Top. Bioenerg.* 8(Pt.B):35–109
10. Boardman, N. K., Anderson, J. M., Kahn, A., Thorne, S. W., Treffrey, T. E. 1970. Formation of photosynthetic membranes during chloroplast development. In *Autonomy and Biogenesis of Mitochondria and Chloroplasts,* ed. N. K. Boardman, A. W. Linnane, R. S. Smillie, pp. 70–84. Amsterdam: North Holland. 511 pp.
11. Böddi, B., Lang, F. 1979. A study of 650 nm protochlorophyll from pumpkin seed coat. *Plant Sci. Lett.* 16:75–79
12. Bogorad, L., Laber, L., Gassman, M. 1968. Aspects of chloroplast development: Transitory pigment-protein complexes and protochlorophyllide regeneration. In *Comparative Biochemistry and Biophysics of Photosynthesis,* ed. K. Shibata, A. Takamiya, A. T. Jagendorf, R. C. Fuller, pp. 299–312. State College, Pa: Univ. Park Press. 445 pp.
13. Bovey, F., Ogawa, T., Shibata, K. 1974. Photoconvertible and nonphotoconvertible forms of protochlorophyll(ide) in etiolated bean leaves. *Plant Cell Physiol.* 15:1133–37
14. Breton, J., Michel-Villaz, M., Paillotin, G. 1973. Orientation of pigments and structural proteins in the photosynthetic membrane of spinach chloroplasts: A linear dichroism study. *Biochim. Biophys. Acta* 314:42–56
15. Brouers, M. 1972. Optical properties of in vivo aggregates of protochlorophyllide in non-polar solvents. I. Visible absorption and fluorescence spectra. *Photosynthetica* 6:415–23
16. Brouers, M. 1975. Optical properties of in vitro aggregates of protochlorophyllide in non-polar solvents. II. Fluorescence polarization, delayed fluorescence and circular dichroism spectra. *Photosynthetica* 9:304–10
17. Brouers, M. 1976. *L'aggregation de la protochlorophyllide dans de systems models et dans la feuille.* PhD thesis. Univ. Liège, Belgium
18. Brouers, M. 1977. Electron spin resonance of protochlorophyllide aggregates. *Plant Sci. Lett.* 10:13–17
19. Brouers, M. 1979. Optical properties of in vitro aggregates of protochlorophyllide in non-polar solvents. III. Infra-red spectra, visible absorption and fluorescence of fractions obtained by differential centrifugation. *Photosynthetica* 13:9–14
20. Brouers, M., Sironval, C. 1974. Evidence for energy transfer from protochlorophyllide to chlorophyllide in leaves treated with δ-aminolevulinic acid. *Plant Sci. Lett.* 2:67–72
21. Brouers, M., Sironval, C. 1975. Restoration of a $P_{557-647}$ form from $P_{645-638}$ in extracts of etiolated primary bean leaves. *Plant Sci. Lett.* 4:175–181
22. Brouers, M., Sironval, C. 1978. The reduction of protochlorophyllide into chlorophyllide. VII. Relations between energy transfer, 690 fluorescence emission and reduction; a theory. *Photosynthetica* 12:399–405
23. Butler, W. L., Briggs, W. R. 1966. The relation between structure and pigments during the first stages of proplastid greening. *Biochim. Biophys. Acta* 112:45–53
24. Canaani, O. D., Sauer, K. 1976. The subunit structure of protochlorophyll-holochrome. In *Chlorophyll-proteins,*

Reaction Centers and Photosynthetic Membranes, ed. J. M. Olson, G. Hind, pp. 360–61. Brookhaven Symp. Biol. No. 28
25. Dratz, E. A. 1966. *The geometry and electronic structure of biologically significant molecules as observed by natural and magnetic optical activity.* PhD thesis. Lawrence Radiation Lab., Rep. UCRL-20202, Univ. Calif. Berkeley
26. Dujardin, E. 1973. Pigment-lipoprotein complexes in the lyophilized etiolated leaf. *Photosynthetica* 7:121–31
27. Dujardin, E. 1976. Reversible transformation of the $P_{657-650}$ form into $P_{633-628}$ in etiolated bean leaves. *Plant Sci. Lett.* 7:91–94
28. Dujardin, E., Sironval, C. 1970. The reduction of protochlorophyllide into chlorophyllide. III. The phototransformability of the forms of the protochlorophyllide-lipoprotein complex found in darkness. *Photosynthetica* 4:129–38
29. Dujardin, E., Sironval, C. 1977. Transitory pigment-protein complexes similar to photosynthesis active centres during protochlorophyll(ide) photoreduction. *Plant Sci. Lett.* 10:347–55
30. Eriksson, G., Kahn, A., Walles, B., von Wettstein, D. 1961. Zur makromolekularen Physiologie der Chloroplasten. III. *Ber. Dtsch. Bot. Ges.* 75: 221–32
31. Garab, G. I., Sundqvist, C., Mustardy, L. A., Faludi-Daniel, A. 1980. Orientation of short wavelength and long wavelength protochlorophyll species in greening chloroplasts. *Photochem. Photobiol.* 31:491–94
32. Gassman, M. L. 1973. A reversible conversion of phototransformable protochlorophyll(ide)$_{650}$ to photoinactive protochlorophyll(ide)$_{633}$ by hydrogen sulfide in etiolated bean leaves. *Plant Physiol.* 51:139–45
33. Godnev, T. N., Galaktionov, S. G., Raskin, V. I. 1968. On the problem of steric conditions of the reaction of hydration of the C_7 and C_8 atoms of the 4th pyrolic ring of protochlorophyll pigments. *Dokl. Akad. Nauk. SSSR* 181:237–40. In Russian
34. Griffiths, W. T. 1974. Protochlorophyll and protochlorophyllide as precursors for chlorophyll synthesis in vitro. *FEBS Lett.* 49:196–200
35. Griffiths, W. T. 1975. Some observations on chlorophyll(ide) synthesis by isolated etioplasts. *Biochem. J.* 146: 17–24
36. Guignery, G., Luzzati, A., Duranton, J. 1974. On the specific binding of protochlorophyllide and chlorophyll to different peptide chains. *Planta* 115: 227–43
37. Gunning, B. E. S. 1965. The greening process in plastids. I. The structure of the prolamellar body. *Protoplasma* 60:111–30
38. Henningsen, K. W. 1968. Spectral shifts of chlorophyllous pigments associated with rearrangements of plastid membranes. *Ann. Acad. Sci. Fenn.* 128: 39–40
39. Henningsen, K. W. 1970. Macromolecular physiology of plastids. VI. Changes in membrane structure associated with shifts in the absorption maxima of the chlorophyllous pigments. *J. Cell Sci.* 7:587–621
40. Henningsen, K. W., Boynton, J. E. 1969. Macromolecular physiology of plastids. VII. The effect of a brief illumination on plastids of dark-grown barley leaves. *J. Cell Sci.* 5:757–93
41. Henningsen, K. W., Boynton, J. E. 1970. Macromolecular physiology of plastids. VIII. Pigment and membrane formation in plastids of barley greening under low light intensity. *J. Cell Biol.* 44:290–304
42. Henningsen, K. W., Kahn, A. 1971. Photoactive subunits of protochlorophyll(ide) holochrome. *Plant Physiol.* 47:685–90
43. Henningsen, K. W., Thorne, S. W. 1974. Esterification and spectral shifts of chlorophyll(ide) in wild type and mutant seedlings developed in darkness. *Physiol. Plant.* 30:82–89
44. Hossier, C., Sauer, K. 1969. Optical properties of the protochlorophyll pigments. I. Isolation, characterization, and infrared spectra. *Biochim. Biophys. Acta* 172:476–91
45. Jones, O. T. G. 1966. A protein-protochlorophyll complex obtained from inner seed coats of *Cucurbita pepo. Biochem. J.* 101:153–60
46. Kahn, A., Boardman, N. K., Thorne, S. W. 1970. Energy transfer between protochlorophyllide molecules. Evidence for multiple chromophores in the photoactive protochlorophyllide-protein complex in vivo and in vitro. *J. Mol. Biol.* 48:85–101
47. Kahn, A., Nielsen, O. F. 1974. Photoconvertible protochlorophyll(ide)$_{635/650}$ in vivo: A single species or two species in dynamic equilibrium? *Biochim. Biophys. Acta* 333:409–14

48. Katz, J. J. 1972. Chlorophyll function in photosynthesis. In *The Chemistry of Plant Pigments,* ed. C. O. Chichester, pp. 103–22. New York/London: Academic. 218 pp.
49. Kesselmeier, J., Budzikiewicz, H. 1979. Identification of saponins as structural building units in isolated prolamellar bodies from etioplasts of *Avena sativa* (L.) *Z. Pflanzenphysiol.* 91:333–44
50. Kesselmeier, J., Ruppel, H. G. 1979. Relation between saponin concentration and prolamellar body structure in etioplasts of *Avena sativa* during greening and re-etiolating and in etioplasts of *Hordeum vulgare* and *Pisum sativum. Z. Pflanzenphysiol.* 93:171–84
51. Kirk, J. T. O. 1970. Biochemical aspects of chloroplast development. *Ann. Rev. Plant Physiol.* 21:11–42
52. Kirk, J. T. O., Tilney-Bassett, R. A. E. 1978. *The Plastids. Their Chemistry, Structure, Growth and Inheritance.* Amsterdam/New York/Oxford: Elsevier/North-Holland. 960 pp. 2nd ed
53. Klein, S. 1962. Phytylation of chlorophyllide and formation of lamellae in chloroplasts. *Nature* 196:992–93
54. Koski, V. M., French, C. S., Smith, J. H. C. 1951. The action spectrum for the transformation of protochlorophyll to chlorophyll *a* in normal and albino corn seedlings. *Arch. Biochem. Biophys.* 31:1–17
55. Liljenberg, C. 1977. Chlorophyll formation: The phytylation step. In *Lipids and Lipid Polymers in Higher Plants,* ed. M. Tevini, H. K. Lichtenthaler, pp. 259–70. Berlin/Heidelberg: Springer. 306 pp.
56. Litvin, F. F., Belyaeva, O. B. 1971. Characteristics of individual reactions and a general scheme of the biosynthesis of native forms of chlorophyll in etiolated plant leaves. *Biochimija* 36: 615–22
57. Lütz, C., Klein, S. 1979. Biochemical and cytological observations on chloroplast development. VI. Chlorophylls and saponins in prolamellar bodies and protylakoids separated from etioplasts of etiolated *Avena sativa* L. leaves. *Z. Pflanzenphysiol.* 95:227–37
58. Madsen, A. 1962. Protochlorophyll/chlorophyll conversion and regeneration of protochlorophyll in etiolated leaves. *Physiol. Plant.* 15:815–20
59. Mathis, P., Sauer, K. 1972. Circular dichroism studies on the structure and the photochemistry of protochlorophyllide and chlorophyllide holochrome. *Biochim. Biophys. Acta* 267: 498–511
60. Michel, J.-M., Sironval, C. 1977. Shifts to $C_{675-670}$ and to $C_{696-684}$ in etiolated leaves illuminated with series of brief flashes. *Plant Cell Physiol.* 18:1223–34
61. Nielsen, O. F., Kahn, A. 1973. Kinetics and quantum yield of photoconversion of protochlorophyll(ide) to chlorophyll(ide) *a. Biochim. Biophys. Acta* 292: 117–29
62. Rasquain, A., Hossier, C., Sironval, C. 1977. The dimerization of protochlorophyll pigments in non-polar solvents. *Biochim. Biophys. Acta* 462:622–41
63. Rosinski, J., Rosen, W. G. 1972. Chloroplast development: Fine structure and chloroplast synthesis. *Q. Rev. Biol.* 47:160–91
64. Ryberg, H. 1980. *Structural development and pigment formation in photoactive and non-photoactive green plastids.* PhD thesis. Univ. Göteborg, Sweden. 32 pp.
65. Schopfer, P., Siegelman, H. W. 1968. Purification of protochlorophyllide holochrome. *Plant Physiol.* 43:990
66. Schultz, A., Sauer, K. 1972. Circular dichroism and fluorescence changes accompanying the protochlorophyllide to chlorophyllide transformation in greening leaves and holochrome preparations. *Biochim. Biophys. Acta* 267: 320–40
67. Seliskar, C. J., Ke, B. 1968. Protochlorophyllide aggregation in solution and associated spectral changes. *Biochim. Biophys. Acta* 153:685–91
68. Singh, B. 1953. Studies on the structure and development of seeds of Cucurbitaceae. *Phytomorphology* 3:224–39
69. Sironval, C. 1972. The reduction of protochlorophyllide into chlorophyllide. VI. Calculation of the size of the transfer unit and the initial quantum yield of the reduction in vivo. *Photosynthetica* 6:375–80
70. Sironval, C., Brouers, M. 1980. The reduction of protochlorophyllide. VIII. The theory of transfer units. *Photosynthetica* 14:213–21
71. Sironval, C., Brouers, M., Michel, J.-M., Kuiper, Y. 1968. The reduction of protochlorophyllide into chlorophyllide. I. The kinetics of the $P_{657-647} \rightarrow P_{688-676}$ phototransformation. *Photosynthetica* 2:268–87
72. Sironval, C., Kuiper, Y. 1972. The reduction of protochlorophyllide into chlorophyllide. IV. The nature of the intermediate $P_{688-676}$ species. *Photosynthetica* 6:254–75

73. Sironval, C., Michel-Wolwertz, M. R., Madsen, A. 1965. On the nature and possible functions of the 673– and 684–mμ forms in vivo of chlorophyll. *Biochim. Biophys. Acta* 94:344–54
74. Stetler, D. A. 1973. Nonphotoconvertible protochlorophyllide in etiolated tissue lacking prolamellar bodies. *Bot. Gaz.* 134:290–95
75. Sundqvist, C., Ryberg, H. 1979. Structure of protochlorophyll-containing plastids in the inner seed coat of pumpkin seed (*Cucurbita pepo*). *Physiol. Plant.* 47:124–28
76. Sundqvist, C., Ryberg, H., Böddi, B., Lang, F. 1980. Spectral properties of a long-wavelength absorbing form of protochlorophyll in seeds of *Cyclanthera explodens. Physiol. Plant.* 48:297–301
77. Thorne, S. W. 1971. The greening of etiolated bean leaves. I. The initial photoconversion process. *Biochim. Biophys. Acta* 226:113–27
78. Vaughan, G. D., Sauer, K. 1974. Energy transfer from protochlorophyllide to chlorophyllide during photoconversion of etiolated bean holochrome. *Biochim. Biophys. Acta* 347:383–94
79. Virgin, H. I. 1958. Studies on the formation of protochlorophyll and chlorophyll *a* under varying light treatments. *Physiol. Plant.* 11:347–62
80. Virgin, H. I., Kahn, A., von Wettstein, D. 1963. The physiology of chlorophyll formation in relation to structural changes in chloroplasts. *Photochem. Photobiol.* 2:83–91
81. von Wettstein, D., Kahn, A. 1960. Macromolecular physiology of plastids. *Eur. Reg. Conf. Electron Microsc., Delft,* 2:1051–54
82. Wolff, J. B., Price, L. 1957. Terminal steps of chlorophyll *a* biosynthesis in higher plants. *Arch. Biochem. Biophys.* 72:293–301
83. Zen'kevich, E. I., Kochubeev, G. A., Losev, A. P., Gurinovich, G. P. 1978. Spectral and luminescence properties of detached aggregated forms of pigments in solution. *Mol. Biol.* 12:1002–11. English transl. (1979) 12:767–75, Plenum Publ.

Ann. Rev. Plant Physiol. 1981. 32:465–84

PHLOEM STRUCTURE AND FUNCTION

♦7720

James Cronshaw

Department of Biological Sciences, University of California, Santa Barbara, California 93106

CONTENTS

INTRODUCTION

The phloem tissue of vascular plants transports carbohydrates produced as a result of photosynthesis, and other substances, to meristems, developing fruits, storage organs, and other sites of carbohydrate utilization. Elucidation of the mechanism of this long-distance transport process has been one of the most challenging and continuing problems in plant physiology and a major goal of research into the structure and function of phloem since the discovery of the sieve tube by Hartig in 1860. Although this mechanism has

0066-4294/81/0601-0465$01.00

not been determined unequivocally, tremendous progress has been made in recent years as can be seen by comparing the proceedings of two major international conferences, one in 1974 held at Banff in Canada (5) and the second, 5 years later at Badgrund in Germany (14). The reader is referred to the proceedings of these conferences and to other extensive reviews and texts for the older literature and for detailed considerations of the physiology and biophysics of phloem transport (17, 25, 48, 59, 64, 125, 129, 131, 150, 184, 185, 191, 206).

For some years now there has been agreement on most of the major physiological parameters of phloem transport; for many plants it has been determined what substances move in the phloem and at what rate these substances move. It is generally agreed that translocation takes place on a source-to-sink pathway, that the loading of nutrients into the phloem controls the rate of movement, and that the direction of movement is controlled by unloading at the sink. In the species analyzed, sugars or sugar alcohols together with water form the bulk of the transported substances. Most species transport sucrose as the predominant sugar, together with small amounts of oligosaccharides of the raffinose type (205). Other plants transport considerable amounts of raffinose-type oligosaccharides, as well as sucrose, and yet others transport considerable quantities of sugar alcohols in addition to sucrose and raffinose-type oligosaccharides. The concentration of those carbohydrates in the phloem varies between species but is of the order of 10 to 30% sucrose or sucrose equivalent (207). There is also agreement on the speed of movement of those substances in the phloem. In most species the speed is of the order of 50–100 cm/h, although there is variation between species.

For many species, the specific mass transfer or mass of substance transported per unit time per square centimeter of phloem has been determined (19). The rates of movement are greater than can be accounted for by diffusion, and several physiological mechanisms have been proposed that would be compatible with them. These mechanisms have been reviewed recently at length (18, 76, 128, 165).

The mechanisms of translocation must be compatible with both the physiological characteristics of the transport process and also with the structure of the sieve element. Often structural data form a basis for the determination of physiological mechanisms, but in the case of phloem tissue, determining normal sieve element structure has been difficult because of the extreme sensitivity of the sieve element protoplast to manipulation and chemical fixation for microscopical investigation. Because of the sugar concentration in the sieve element protoplast, the contents are under high turgor pressure. When a sieve element is cut at the time of sampling, or the semipermeable properties of its membrane are destroyed by chemical fixatives, the contents surge toward the site of pressure release. The surge causes a displacement of the normal arrangement of the cytoplasmic components and even a rupturing of some organelles. Cytoplasmic contents accumulate against the sieve plates, forming so-called "slime plugs" on the sides of pressure release. These surge artifacts have created controversy as to the structure of mature sieve elements. The two most controversial structural features of mature sieve elements are precisely those that would help in determining the mechanism of translocation, namely, the distribution of P-protein in the mature sieve element and the nature of the contents of sieve-area pores. However, although as yet the picture is not precise enough to enable an unequivocable correlation of structure and function to be made and consequently determination of the translocation mechanism, great strides have been made in the past 10 years in understanding the unique structural features of phloem sieve elements and their associated cells.

THE SIEVE ELEMENT

Developmental Anatomy

Many attempts have been made to prevent surging artifacts and structural changes in mature sieve elements in preparing samples for microscopic examination (164). One of the approaches that we and others have taken is to investigate structural changes during differentiation because young sieve elements at early stages of differentiation have not developed a high hydrostatic pressure and are amenable to study by normal cytological techniques. These changes are essentially similar for all species examined (15, 26, 27, 33–35, 73, 135, 141, 167, 194). Differentiation involves both synthetic events and selective autophagy, resulting in a reorganization of some and a disappearance of other cellular components.

Young sieve elements contain the normal complement of cell organelles typical of plant meristematic cells. They are first recognizable by a thickening of the wall and by small deposits of electron-transparent callose associated with the plasmodesmata. During intermediate stages of differentiation, the endoplasmic reticulum is reorganized; changes occur in the plastids and mitochondria; the nucleus, vacuolar system, dictyosomes, and ribsomes are disassembled and disappear; and in the dicotyledons and some monocotyledons a new component P-protein is fabricated. At the same time changes occur in the cell wall and the specialized sieve area pores are developed.

P-proteins

For many years it has been known that proteinaceous accumulations occur in phloem cells. The term P-protein (phloem protein) was introduced to describe proteinaceous substances in the phloem which were sufficiently characteristic when observed with the electron microscope to warrant the special term (54). P-protein has been encountered in the sieve elements of all dicotyledenous species examined and in some monocotyledons (27, 201). It is, however, lacking from a great many monocotyledons, including many palms (140, 141, 142, 144, 145), *Lemna minor* (127), *Hordeum vulgare* (69), *Triticum aestivum* (122), and *Zea mays* (182). P-protein is also absent from the gymnosperms and lower vascular plants (16, 46, 47, 68, 119, 124, 132, 151, 187–189, 197).

In species where P-protein does occur, it first appears in the cytoplasm as small aggregates which later enlarge to form P-protein bodies (27). As the sieve elements mature, these P-protein bodies usually disperse with a disaggregation of the P-protein components. P-protein occurs in several morphological forms that vary both from species to species and often within one species. It is often tubular, as in *Nicotiana tabacum* and *Coleus blumeii,* but may be fibrillar, as in *Cucurbita maxima* and *Ricinus communis,* or granular or crystalline, as in *Phaseolus vulgare* and *Glycine max* (27, 37, 139. 194). Conformational changes from one morphological type of P-protein to another may take place within a differentiating sieve element (27, 33, 37, 139, 146, 168, 194), and it has been suggested that the subunits of P-protein may be interconvertible into the various morphological forms. The concept of P-protein interconvertibility has been questioned by Sabnis & Hart (153, 154).

P-proteins, particularly those of the Cucurbitaceae, have been characterized chemically and shown to consist of proteins or polypeptides of molecular weights 14,000–158,000 (62, 112, 114, 115, 180, 181, 192, 193, 199). Phosphatases have been localized cytochemically in association with P-proteins. Acid phosphatase has been localized on P-protein in both immature and mature sieve elements of *Phaseolus vulgaris* by Esau & Charvat (52). Catesson (20), Catesson & Czaninski (22, 23), Bentwood & Cronshaw (12), and Cronshaw (29) found activity associated only with the dispersed P-protein in mature sieve elements of *Acer pseudoplatanus, Nicotiana tabacum,* and *Pisum sativum* respectively. Other workers found no acid phosphatase associated with the P-proteins of *Pisum sativum* (203) or *Impatiens holstii* (78).

Nucleoside phosphatase activity has been found associated with the P-proteins of mature sieve elements of *Nicotiana tabacum* (36, 93, 94) and *Cucurbita maxima* (92). Activity was not found, however, associated with the P-protein in P-protein bodies of young sieve elements of these species. Nucleoside phosphatase activity could not be detected in association with P-proteins in *Pisum sativum, Phaseolus vulgaris* and *Ricinus communis*

(29), in *Tetragonia expansa* (200), or in *Acer pseudoplatanus* and *Robina pseudoacacia* (20).

Because P-proteins occur exclusively in sieve elements and associated parenchyma cells and because of a superficial morphological similarity to microtubules and microfilaments, it has been postulated that they may be directly or indirectly involved in assimilate transport and possibly contraction (1, 4, 75, 77, 93, 109, 123, 125, 152, 191). Further research has indicated, however, that P-protein cannot be equated with actin, tubulin, microfilaments, or microtubules (137). P-protein is dissimilar in biochemistry, morphology, and interaction with drugs, and it is now thought that P-protein does not play a role in generating the force for translocation.

The distribution of P-protein in mature sieve elements has not been determined equivocally. Some workers have described a more or less even distribution of P-protein filaments in the sieve element lumen (3, 27, 196); others have described a mostly or entirely parietal distribution (70, 74, 81, 90, 141, 177). Most workers now agree that P-protein does not normally exist in the form of transcellular strands or as a component of such strands as had been envisaged by some authors (41, 43, 75, 77, 108, 111, 152, 174, 175). However, in sieve elements where there have been obvious changes due to the release of hydrostatic pressure, strands of P-protein have been observed on the downstream side of the sieve plate pores (27, 111).

The amount of P-protein in sieve elements in different regions of the plant does vary. Sieve elements in roots, both during differentiation and when mature, contain much less P-protein than do sieve elements in stems (27). No differences have been found in the distribution of P-protein in mature sieve elements as plants age, and it is unlikely that P-protein normally moves in the assimilate stream (27). Possibly P-proteins are anchored to the membranes of sieve elements by lectin bridges (114, 154).

Membrane Systems and Ribosomes

During the development of sieve elements, the electron-opaque regions of the plasma membrane may be thicker than their counterparts in adjacent cells (26) and may be asymmetric with the outer electron-opaque line more prominent than the inner (15).

Nucleoside phosphatase has been found associated with the sieve element plasma membranes of *Cucurbita maxima* (92), *Nicotiana tabacum* (93, 94), *Robinia pseudacacia* (20), *Acer pseudoplatanus* (20), *Tetragonia expansa* (199), and *Pinus nigra* (155). In contrast to these species, no nucleoside phosphatase activity was found associated with the sieve element plasma membrane of *Pisum sativum* (13, 29, 30). In some studies acid phosphatase has been reported localized on the sieve element plasma membrane of

Phaseolus vulgare (52) and *Pisum sativum* (203), but in other studies no association of acid phosphatase with the plasma membrane of *Nicotiana tabacum* (12) or *Pisum sativum* (29) was found. Peroxidase activity is especially characteristic of phloem cells and has been localized on the plasma membrane of *Robinia pseudo-acacia* (38), *Acer pseudoplatanus* (38), *Dianthus caryophyllus* (21) and *Triticum aestivum* (21).

Immature sieve elements usually contain one or more vacuoles. During differentiation the vacuolar membrane apparently changes its permeability properties and subsequently breaks down. The majority of workers have recorded an absence of vacuoles in mature sieve elements (26, 33, 48). Evert and co-workers described intact vacuoles in mature sieve elements of some species but now regard their previous reports as anomalies or instances in which the vacuolar membrane was late in disappearing from the cell (64).

Endoplasmic reticulum, mostly of the rough form, and dictyosomes are normal components of immature sieve elements. Early in the development of the sieve element, some cisternae of endoplasmic reticulum become associated with the sites of developing sieve plate pores (55). As the sieve element matures, other cisternae of rough endoplasmic reticulum that were evenly distributed in the cells lose their ribosomes and begin to move to the peripheral cytoplasm. Cisternae may form stacks, some of which are associated with the nuclear envelope (3, 15, 49, 57, 141, 202). Between the cisternae of endoplasmic reticulum is often a fibrous, dark-staining component which resembles fibrils of P-protein (27). The mature sieve element has smooth endoplasmic reticulum in a parietal position as a more or less continuous reticulum along the inner surface of the plasma membrane and as regions of flattened or convoluted stacks of cisternae also close to the plasma membrane (15, 49–51, 55, 67, 142, 182). In the sieve cells of gymnosperms the endoplasmic reticulum exists as a parietal network of smooth tubules, which at the sieve areas form massive aggregates (11, 65, 133, 176, 197).

Nucleoside phosphatases (13, 29, 30, 92–94), acid phosphatase (12, 20, 52, 203), and peroxidase (J. Cronshaw, unpublished observations) are associated with both the reticular and stacked endoplasmic reticulum systems. In some species peroxidase activity is associated with the Golgi cisternae and the ribosomes of the rough endoplasmic reticulum in differentiating sieve elements (21). The endoplasmic reticulum system with its phosphatases may be an important source of enzymes involved on the autophagic events of sieve element maturation and in the mature cell may represent a sequestering of the membrane system in an inactive form. It has also been suggested that the endoplasmic reticulum may be a specific cytoplasmic differentiation related to the conduction function of the cell (52). Esau & Hoefert (58) have described endoplasmic reticulum tubules containing microtubules surrounded by electron-opaque material in the sieve elements of *Beta vulgaris* and *Spinacia oleracea.*

Plastids and Mitochondria

Mitochondria are normal components of young sieve elements and remain apparently unmodified throughout the differentiation of the cell (15, 26, 67, 141, 182). Mitochondria have been shown to stain with 3-3'-diaminobenzidine in mature sieve elements of *Oryza sativa* (136), *Acer pseudoplatanus* and *Polypodum vulare* (24), and *Dianthus caryophyllus* (21). We have obtained similar results for mitochondria in mature sieve elements of *Nicotiana tabacum.* This staining indicates that mature sieve element mitochondria possess at least the terminal part of the respiratory chain and may be able to carry out some oxidative reactions. Nucleoside phosphatases (93, 94) and acid phosphatase (52) have been localized in mitochondria of mature and differentiating sieve elements.

The plastids of young sieve elements have little differentiation of their internal membrane systems. During differentiation of the sieve elements, the plastids enlarge to some extent but the internal membrane systems remain poorly developed. Two basic types of plastids have been recognized; those that store starch (S-type plastids) and those that elaborate protein inclusions and often also contain starch (P-type plastids) (9, 10). The starch of sieve element plastids has a characteristic reddish-brown staining reaction with iodine, and cytochemical studies have shown that the sieve element starch is composed of highly branched molecules with numerous –1,6 linkages (138).

Microtubules and Microfilaments

Microtubules are abundant in the peripheral cytoplasm of young sieve elements but disappear from the cells at intermediate stages of differentiation (26). Microfilament bundles are frequently seen in differentiating sieve elements (147) and have been reported in mature sieve elements (33).

Cell Wall

Sieve elements typically have nonlignified cellulosic walls which are often thickened. This thickening is described as nacreous because of its refractive properties and characteristic luster when viewed with the light microscope. The basic structure of the wall has been reviewed by Cronshaw (28) and Evert (64). Nucleoside phosphatases (92), acid phosphatase (52), and peroxidase (21, 38–40; J. Cronshaw, unpublished information) have been localized in the walls of sieve elements.

The Sieve Plate

Because of its importance in relation to the proposed mechanisms of phloem transport and the controversy surrounding the nature of the contents of the pores, the sieve plate has received great attention from phloem workers. Sieve plate development in the angiosperms has been thoroughly documented, and a fairly uniform picture has emerged for most species examined (26, 28, 32, 44, 45, 48, 183).

The sieve areas of young sieve elements are penetrated by a variable number of plasmodesmata, each of which is associated with cisternae of the endoplasmic reticulum. These plasmodesmata are the sites of future pores and become recognizable as such by the deposition of electron-translucent callose in the form of platelets or cones on either sides of the wall. Endoplasmic reticulum cisternae remain closely appressed to the plasma membrane at the region of the platelets throughout pore development but are removed as the pores attain their full size. The callose deposits undergo rapid enlargement and appear to replace previously deposited wall material. Eventually the opposing callose masses fuse to form cylinders in the wall surrounding the plasmodesmata. The plasmodesmata then enlarge at the middle lamella region to form median nodules. The median nodules enlarge with dissolution of callose, resulting in the formation of a pore with the plasma membrane continuous through it from cell to cell. Residual callose is nearly always observed as a thin walled cylinder around the pore. Pore formation thus involves synthetic events and the formation of callose and degradation phenomena involving the removal of wall material. However, little is known about the control of this process, how enzymes perform localized events, or how their activity is controlled.

The amount of callose observed surrounding sieve plate pores varies according to preparation procedures and the degree of injury to the sieve elements (31, 60). Callose deposition in response to wounding, and in some sieve elements prior to dormancy, are well-known phenomena, and the literature on the properties of callose and the factors influencing its deposition have been reviewed extensively (25, 48, 60).

Much less is known about the development of sieve area pores in gymnosperms. As with the angiosperms, the sieve area pores develop from plasmodesmata, and widening of the plasmodesmatal canals results in the formation of pores (11, 48, 66, 133, 134). Pore formation and the formation of median cavities occur more or less simultaneously, and neither endoplasmic reticulum cisternae nor callose platelets are involved.

Sieve Area Pore Contents

The question of the nature of the contents of the sieve area pores has not been answered unequivocally. Some investigators believe that the plugged condition of the pores observed in many published electron micrographs (130, 159, 162, 163, 166) represents the normal state in mature functioning sieve elements; others have suggested that when plugged pores are observed, this condition is an artifact created by the sudden release of the sieve element hydrostatic pressure (2, 3, 8, 25, 31, 70, 142). The question, however, is not simply whether the pores are plugged, but also one of the degree of occlusion. The sieve plates resist the mass flow of a solution, and any plugging of the sieve plate pores will further increase this resistance. As Weatherly & Johnson (191) have calculated, if there are more than just a few strands of P-protein in the pores, then the resistance would be too high for a pressure-driven mass flow mechanism to operate.

Sieve plate pores have been investigated by electron microscopy of sections of both chemically fixed and of freeze-substituted material, and of freeze fracture replicas. With chemical fixation, not only have the results varied with the type of manipulation of the tissue prior to fixation, but also with the type of chemical fixatives used. Early work with osmium tetroxide and potassium permanganate showed plugged and unplugged pores respectively (53, 98, 113, 156), but the results can be discounted because of the obviously unsatisfactory fixation. With the introduction of glutaraldehyde and acrolein for the fixation of phloem tissue (15), plugged pores were described in material which according to most criteria was well fixed. Continued use of aldehyde fixatives has led to many published electron micrographs, and most workers agree that sieve plate pores in careful preparations are predominantly occluded (28, 164).

Although conventionally but carefully fixed material nearly always shows occluded pores, if steps are taken which are calculated to reduce surge artifacts, there is a shift in the observations toward more open pores. This has been found after fixation with the rapidly penetrating fixative acrolein (31, 158); when whole plants were rapidly frozen in liquid nitrogen and then transferred to chemical fixative (31); when isolated sieve elements induced to differentiate in tobacco pith cultures are fixed (28); or when starved or wilted plants are fixed (3, 70). When these results are coupled with those showing that some exceptional plants such as *Zea* (182), *Lemna* (183), and some palms (142) do have unplugged pores; that carbon black particles (7, 169) and mycoplasm-like bodies (107, 143, 198) have been shown to pass through sieve plate pores; and that virus particles can replace the P-protein pore plugs in some specimens (56), a strong argument can be made that the pores are unoccluded in vivo.

The conclusion is substantiated by studies of freeze-fractured and freeze-substituted specimens. Johnson (109, 110) studied sieve area pores in phloem cells prepared for electron microscopy by freeze fracturing and found that there was a uniform network of P-protein continuous from cell to cell through the pores with no dense plugging. Unplugged pores have also been observed in material prepared by freeze substitution (42, 43, 80, 81, 90, 91). Spanner (164) has questioned these results and documented numerous objections such as dissection damage and volume changes, and pointed out that the quality of micrographs produced by freeze substitution is much inferior to that given in good chemical fixation.

In gymnosperms, the sieve areas of the sieve elements all have the same degree of specialization. The sieve areas, however, are commonly more numerous on the overlapping ends of the cells, and the endoplasmic reticulum forms massive aggregates at these regions (11, 66, 133, 176, 197). The tubular elements of endoplasmic reticulum are continuous from cell to cell, occluding the pores. The speed of translocation in the gymnosperms, however, is similar to that of the angiosperms (17, 195).

In some species of the lower vascular plants such as *Botrychium, Ophioglossum, Angiopteris, Marattia,* and *Osmunda psilotum,* the sieve element pores are occluded with endoplasmic reticulum as in the gymnosperms (63, 151, 186). In others such as *Selaginella, Isoetes,* and *Lycopodium,* the sieve area pores are normally unoccluded by any cytoplasmic material except for an occasional tubule of endoplasmic reticulum (16, 119–121, 189). Sieve area pores unoccluded but transversed by variable numbers of endoplasmic reticulum membranes are found in heterosporous ferns (16, 190), homosporous leptosporangiate ferns (186) and *Equisetum* (46, 47).

MINOR VEINS AND PHLOEM LOADING

Minor Veins

In the minor veins of angiosperms are small sieve elements and larger companion and parenchyma cells. Companion cells are characterized by prominent plasmodesmatal connections with the sieve elements, dense cytoplasm, and usually an absence of starch; however, companion cells and other parenchyma cells intergrade with one another and are not always distinguishable (48). Most of the parenchyma cells of minor veins have dense cytoplasm, rich in organelles that Fischer (79) called "intermediary cells." Some of these cells develop wall ingrowths and have been termed "transfer cells" (95, 148, 149). A-type transfer cells have the dense protoplasts of companion cells and wall ingrowths on all cell walls; B-type transfer cells are phloem parenchyma cells with wall ingrowths best developed on walls next to the sieve elements or their companion cells. It has been proposed that transfer cells in the phloem of minor veins collect and pass on photosynthates and also retrieve and recycle solutes that enter the leaf apoplast in the transpiration stream (96).

Photosynthate Pathway and Loading

The pathway of metabolites from the mesophyll cells to the sieve elements may be symplastic, apoplastic, or a combination of the two. A symplastic pathway depends on the availability of plasmodesmata, but it is conceivable that photosynthates could diffuse from the mesophyll cells to the sieve tubes along concentration gradients (204). Increasing evidence indicates, however, that at some place on the pathway from the mesophyll cells, sucrose enters the apoplast from which it is actively loaded into the sieve element-companion cell complex of the minor veins (72, 83, 86–88, 161). In an extensive series of experiments using sugar beet leaves, Geiger and his co-workers have demonstrated that in this species sucrose is loaded into the companion cell-sieve element complex from the apoplast (82–85).

Phloem-loading from the apoplast has also been shown for *Zea mays* (100, 101). This species has two types of sieve elements, thin-walled sieve elements that have companion cells and thick-walled sieve elements that lack companion cells. The thin-walled sieve elements have abundant connections with their companion cells, and sugar is actively accumulated from the apoplast by the companion cell-sieve element complex. The thick-walled sieve elements abut the xylem, lack companion cells, but have abundant connections with contiguous vascular parenchyma cells. The significance of the thick-walled sieve elements remains to be determined.

Tubular extensions of the plasma membrane of vascular parenchyma and some other cells of the *Zea mays* leaf have been described which extend the plasma membrane surface and may be functionally similar to the wall and plasma membrane ingrowths of transfer cells (71, 72).

Movement of photosynthetic intermediates between mesophyll and the bundle sheath cells in *Zea mays* probably occurs via a symplastic pathway, as the suberin lamella of the bundle sheath cells is fairly impermeable to water (72). The numerous plasmodesmata between the mesophyll cells are characterized by electron-dense "sphincters" which may represent plasmodesmatal valves and be capable of controlling the direction and rates of transport of substances throughout the desmotubules of plasmodesmata (71; see also 103).

ATPase Localization

As phloem loading is an active process, determination of the precise sites at which enzymes such as ATPase are located has been undertaken by several workers. ATPase has been localized on the plasma membrane of sieve elements and companion cells of several species (20, 29, 30, 36, 92–94, 155, 200). In tobacco similar localization patterns were obtained with the nucleoside triphosphates ATP, CTP, GTP, ITP, and UTP. Activity was also localized on the plasma membranes with ADP and 5'-AMP but not with the competitive inhibitors 2'– and 3'–AMP. Deposition of reaction product was strongly inhibited by fluoride and by the sulfhydryl blocking agents NEM and PCMB (94). In *Pisum sativum,* however, the sieve elements show no ATPase associated with the plasma membrane at any stage of differentiation (13). In the minor veins of this species, ATPase is localized on the transfer cell plasma membrane but appears only at the stage of differentiation when the wall ingrowths are developing and is strongest in the mature transfer cell.

Sucrose-Proton Co-transport

The cytochemical localization of ATPase activity associated with phloem cell plasma membranes supports the hypothesis put forward in recent years that the driving force for solute accumulation may be provided for by a proton pump energized by the hydrolysis of ATP at the plasma membrane and involving a sucrose-proton co-transport system (6, 88, 89, 99, 104–106, 116, 117, 126, 170). Earlier it was shown that electrochemical gradients of protons are the immediate energy source for the accumulation of sugars and amino acids into bacteria, fungi, and unicellular algae (118, 157, 160, 171–173). Although the phloem loading system of vascular plants has not been analyzed in such detail, it appears to be very similar (170). In the loading of sugar and amino acids from the apoplast an electrochemical potential gradient of protons is established by the activity of the plasma membrane ATPase. Solutes from the apoplast may then be loaded into the phloem by a co-transport with protons via specific carriers (6, 89, 99, 117, 170).

Phloem Unloading

Less work has been done on the unloading of assimilates from the phloem. In recent work with tomato fruits, Ho and co-workers have shown that the import rate is related to fruit metabolism and thus to the developmental stage of the fruit (102, 178); that the import rate is also affected by fruit temperature (179), fruit sucrose concentration, and to some extent to fruit starch content. It has been suggested that the limiting step in the unloading process is the activity of an invertase (102). Eschrich (61) has also stressed the importance of invertase in the unloading process and has suggested that free space invertase may act as a "reflux valve" to prevent a reloading of sucrose that was unloaded by a concentration gradient. The free space invertase is an acid invertase and may in turn be controlled by pH. Eschrich considers that the other factor regulating unloading is the active transport of hexoses into growing cells.

Plasma membrane ATPase activity has been demonstrated for phloem cells of the roots of *Nicotiana tabacum* (29), but nothing is known of its possible involvement in the unloading process.

CONCLUDING REMARKS

At the present time there appears to be a consensus developing concerning the translocation process. A good deal of evidence is accumulating that photosynthates are loaded into the sieve element-companion cell complex by a sucrose-proton co-transport system. The energy for loading is provided by the hydrolysis of ATP at the plasma membranes of these cells. In species with transfer cells or tubular ingrowths, the area of contact between the plasma membrane and the cytoplasm is increased, allowing for increased transport by exposure of extra potential "carrier" sites. Sucrose uptake would cause an increased osmotic pressure in the sieve elements which, in turn, would produce a pressure-driven mass flow out of the leaf. The direction of flow would be determined by the various sinks, and the strength of the sinks could be regulated by an acid invertase. Along the phloem conduits there could be efflux or influx of sucrose, depending on local conditions. Most workers agree that there is a mass flow of solution through unplugged sieve plate pores. However, the evidence in favor of unplugged pores is as yet inconclusive.

Literature Cited

1. Aikman, D. P., Anderson, W. P. 1971. A quantitative investigation of a peristaltic model for phloem translocation. *Ann. Bot.* 35:761–72
2. Anderson, R., Cronshaw, J. 1969. The effects of pressure release on the sieve plate pores of *Nicotiana. J. Ultrastruct. Res.* 29:50–59
3. Anderson, R., Cronshaw, J. 1970. Sieve-plate pores in tobacco and bean. *Planta* 91:173–80
4. Anderson, W. P. 1973. The mechanisms of phloem translocation. *Symp. Soc. Exp. Biol.* No. 28, "Transport at the Cellular Level," ed. M. A. Sleigh, D. H. Jennings
5. Aronoff, S., Dainty, J., Gorham, P. R., Srivastava, L. M., Swanson, C. A., eds. 1975. *Phloem Transport.* New York: Plenum. 363 pp.
6. Baker, D. A., Malek, F., Dehvar, F. D. 1980. Phloem loading of amino acids from the petioles of *Ricinus* leaves. *Ber. Dtsch. Bot. Ges.* 93:203–9
7. Barclay, G. F., Fensom, D. S. 1973. Passage of carbon black through sieve plates of unexcised *Heracleum sphondylium* after micro-injection. *Acta Bot. Neerl.* 22:228–32
8. Behnke, H.-D. 1971. The contents of sieve plate pores in *Aristolochia. J. Ultrastruct. Res.* 36:493–98
9. Behnke, H.-D. 1972. Sieve-tube plastids in relation to angiosperm systematics —an attempt towards a classification by ultrastructural analysis. *Bot. Rev.* 38:155–97
10. Behnke, H.-D. 1975. P-type sieve-element plastids: a correlative ultrastructural and ultrahistochemical study on the diversity and uniformity of a new reliable character in seed plant systematics. *Protoplasma* 83:91–101
11. Behnke, H.-D., Paliwal, D. S. 1973. Ultrastructure of phloem and its development in *Gnetum gnemom,* with some observations on *Ephedra campylopoda. Protoplasma* 78:305–19
12. Bentwood, B. J., Cronshaw, J. 1976. Biochemistry and cytochemical localization of acid phosphatase in the phloem of *Nicotiana tabacum. Planta* 130:97–104
13. Bentwood, B. J., Cronshaw, J. 1978. Cytochemical localization of adenosine triphosphatase in the phloem of *Pisum sativum* and its relation to the function of transfer cells. *Planta* 140:111–20
14. *Ber. Dtsch. Bot. Ges.* 1980. 93:1–378
15. Bouck, G. B., Cronshaw, J. 1965. The fine structure of differentiating sieve tube elements. *J. Cell Biol.* 25:79–96
16. Burr, F. A., Evert, R. F. 1973. Some aspects of sieve- element structure and

development in *Selaginella kraussiana. Protoplasma* 78:81–97
17. Canny, M. J. P. 1973. *Phloem Translocation.* Cambridge:Univ. Press. 301 pp.
18. Canny, M. J. P. 1975. Protoplasmic streaming. See Ref. 206, pp. 289–300
19. Canny, M. J. P. 1975. Mass transfer. See Ref. 206, pp. 139–53
20. Catesson, A.-M. 1973. Observations cytochimiques sur les tubes criblés de quelques angiospermes. *J. Microsc. (Paris)* 16:95–104
21. Catesson, A.-M. 1980. Localization of phloem oxidases. *Ber. Dtsch. Bot. Ges.* 93:141–52
22. Catesson, A.-M., Czaninski, Y. 1967. Mise en evidence d'une activité phosphatasique acide dans le reticulum endoplasmique des tissus conducteurs de Robinier et de Sycomore. *J. Microsc.* 6:509–14
23. Catesson, A.-M., Czaninski, Y. 1968. Localization ultrastructurale de la phosophatase acide et cycle saisonnier dans les tissus conducteurs de quelques arbres. *Bull. Soc. Fr. Physiol. Vég.* 14:165–73
24. Catesson, A.-M., Liberman-Maxe, M. 1974. Les mitochondries des cellules criblées: réactions avec la 3,3' - diaminobenzidine. *C. R. Acad. Sci. Ser. D* 278:2771–73
25. Crafts, A. S., Crisp, C. E. 1971. *Phloem Transport in Plants.* San Francisco: Freeman. 481 pp.
26. Cronshaw, J. 1974. Phloem differentiation and development. In *Dynamic Aspects of Plant Ultrastructure.* ed. A. W. Robards, 391–415. Berkshire, Engl: McGraw-Hill. 546 pp.
27. Cronshaw, J. 1975. P-proteins. See Ref. 5, pp. 79–115
28. Cronshaw, J. 1975. Sieve element cell walls. See Ref. 5, pp. 129–47
29. Cronshaw, J. 1980. Histochemical localization of enzymes in the phloem. *Ber. Dtsch. Bot. Ges.* 93:123–39
30. Cronshaw, J. 1980. ATPase in mature and differentiating phloem and xylem. *J. Histochem. Cytochem.* 28:375–77
31. Cronshaw, J., Anderson, R. 1969. Sieve plate pores of *Nicotiana. J. Ultrastruct. Res.* 27:134–48
32. Cronshaw, J., Anderson, R. 1971. Phloem differentiation in tobacco pith culture. *J. Ultrastruct. Res.* 34:244–59
33. Cronshaw, J., Esau, K. 1967. Tubular and fibrillar components of mature and differentiating sieve elements. *J. Cell Biol.* 34:801–16
34. Cronshaw, J., Esau, K. 1968. P-protein in the phloem *Cucurbita.* I. The development of P-protein bodies. *J. Cell Biol.* 38:25–39
35. Cronshaw, J., Esau, K. 1968. P-protein in the phloem of *Cucurbita.* II. The P-protein of mature sieve elements. *J. Cell Biol.* 38:292–303
36. Cronshaw, J., Gilder, J. 1972. Localization of adenosine triphosphatase activity in the phloem of *Nicotiana tabacum.* 30th Ann. Proc. Electron Microscopy Soc. Am., Los Angeles. ed. C. J. Arceneau, pp. 230–31. Baton Rouge: Claitor's
37. Cronshaw, J., Gilder, J., Stone, D. 1973. Fine structural studies of P-protein in *Cucurbita, Cucumis* and *Nicotiana. J. Ultrastruct. Res.* 45:192–205
38. Czaninski, Y., Catesson, A.-M. 1969. Localisation ultrastructurale d'activités peroxydasiques dans les tissus conducteurs végétaux au cours du cycle annuel. *J. Microsc.* 8:875–88
39. Czaninski, Y., Catesson, A.-M. 1970. Activités peroxydasiques d'origines diverses dans les cellules d' *Acer pseudoplatanus* (tissus conducteurs et cellules en culture). *J. Microsc.* 9:1089–1101
40. de Jong, D. W. 1967. An investigation of the role of plant peroxidase in cell wall development by the histochemical method. *J. Histochem. Cytochem.* 15:335–46
41. DeMaria, M. E., Thaine, R. 1974. Strands in sieve tubes in longitudinal cryostat sections of *Cucurbita pepo* stems, *J. Exp. Bot.* 25:871–75
42. Dempsey, G. P., Bullivant, S., Bieleski, R. L. 1975. The distribution of P-protein in mature sieve elements of celery. *Planta* 126:45–59
43. Dempsey, G. P., Bullivant, S., Bieleski, R. L. 1976. The distribution of P-protein in mature sieve tube elements. In *Transport and Transfer Processes in Plants.* eds. I. F. Wardlaw, J. B. Passioura, 247–51. New York: Academic. 484 pp.
44. Deshpande, B. P. 1974. Development of the sieve plate in *Saxifraga sarmentosa.* L. *Ann. Bot.* 38:151–58
45. Deshpande, B. P. 1975. Differentiation of the sieve plate of *Cucurbita:* A further view. *Ann. Bot.* 39:1015–22
46. Dute, R. R., Evert, R. F. 1977. Sieve-element ontogeny in the root of *Equisetum hyemale, Am. J. Bot.* 64:421–38
47. Dute, R. R., Evert, R. F. 1978. Sieve-element ontogeny in the aerial shoot of *Equisetum hyemale* L., *Ann. Bot.* 42:23–32

48. Esau, K. 1969. The phloem. In *Encyclopedia of Plant Anatomy,* eds. W. Zimmermann, P. Ozenda, H. D. Wulff. Vol. 5, Pt. 2. Stuggart: Gerbruder Borntraeger. 505 pp.
49. Esau, K. 1972. Changes in the nucleus and the endoplasmic reticulum during differentiation of a sieve element in *Mimosa pudica* L. *Ann. Bot.* 36:703–10
50. Esau, K. 1975. The phloem of *Nelumbo nucifera* Gaertn., *Ann. Bot.* 39:901–13
51. Esau, K. 1978. Developmental features of the primary phloem in *Phaseolus vulgaris* L., *Ann. Bot.* 42:1–13
52. Esau, K., Charvat, I. D. 1975. An ultrastructural study of acid phosphatase localization in cells of *Phaseolus vulgaris* phloem by the use of the azo dye method. *Tissue Cell.* 7:619–30
53. Esau, K., Cheadle, V. I. 1965. Cytologic studies on phloem *Univ. Calif. Publ. Bot.* 36:253–344
54. Esau, K., Cronshaw, J. 1967. Tubular components in cells of healthy and tobacco mosaic virus infected *Nicotiana. Virology* 33:26–35
55. Esau, K., Cronshaw, J. 1968. Endoplasmic reticulum in the sieve elements of *Cucurbita. J. Ultrastruct. Res.* 23:1–14
56. Esau, K., Cronshaw, J., Hoefert, L. L. 1967. Relation of beet yellow virus to the phloem and to translocation in the sieve tube. *J. Cell Biol.* 32:71–87
57. Esau, K., Gill, R. H. 1971. Aggregation of endoplasmic reticulum and its relation to the nucleus in a differentiating sieve element. *J. Ultrastruct. Res.* 34:144–58
58. Esau, K., Hoefert, L. 1980. Endoplasmic reticulum and its relation to microtubules in sieve elements of sugarbeet and spinach. *J. Ultrastruct. Res.* 71:249–57
59. Eschrich, W. 1970. Biochemistry and fine structure of phloem in relation to transport. *Ann. Rev. Plant Physiol.* 21:193–214
60. Eschrich, W. 1975. Sealing systems in phloem. See Ref. 206, pp. 36–56
61. Eschrich, W. 1980. Free Space Invertase, its possible role in phloem unloading. *Ber. Dtsch. Bot. Ges.* 93:363–78
62. Eschrich, W., Evert, R. F., Heyser, W. 1971. Proteins of the sieve tube exudate of *Cucurbita maxima. Planta* 100: 208–21
63. Evert, R. F. 1976. Some aspects of sieve-element structure and development in *Botrychium virginianum. Isr. J. Bot.* 25:101–26
64. Evert, R. F. 1977. Phloem structure and histochemistry. *Ann. Rev. Plant Physiol.* 28:199–222
65. Evert, R. F., Bornman, C. H., Butler, V., Gilliland, M. G. 1973. Structure and development of the sieve-cell protoplast in leaf veins of *Welwitschia. Protoplasma* 76:1–21
66. Evert, R. F., Bornman, C. H., Butler, V., Gilliland, M. G. 1973. Structure and development of sieve areas in leaf veins of *Welwitschia. Protoplasma* 76:23–34
67. Evert, R. F., Deshpande, B. P. 1969. Electron microscope investigation of sieve-element ontogeny and structure in *Ulmus americana. Protoplasma* 68:403–32
68. Evert, R. F., Eichhorn, S. E. 1976. Sieve-element ultrastructure in *Platycerium bifurcatum* and some other polypodiaceous ferns: the nacreous wall thickening and maturation of the protoplast. *Am. J. Bot.* 63:30–48
69. Evert, R. F., Eschrich, W., Eichhorn, S. E. 1971. Sieve-plate pores in leaf veins of *Hordeum vulgare. Planta* 100: 262–67
70. Evert, R. F., Eschrich, W., Eichhorn, S. E. 1973. P-protein distribution in mature sieve elements of *Cucurbita maxima. Planta* 109:193–210
71. Evert, R. F., Eschrich, W., Heyser, W. 1977. Distribution and structure of the plasmodesmata in mesophyll and bundle-sheath cells of *Zea mays* L. *Planta* 136:77–89
72. Evert, R. F., Eschrich, W., Heyser, W. 1978. Leaf structure in relation to solute transport and phloem loading in *Zea mays* L. *Planta* 138:279–94
73. Evert, R. F., Murmanis, L., Sachs, I. B. 1966. Another view of the ultrastructure of *Cucurbita* phloem. *Ann. Bot. N.S.* 30:563–81
74. Fellows, R. J., Geiger, D. R. 1974. Structural and physiological changes in sugar beet leaves during sink to source conversion. *Plant Physiol.* 54:877–85
75. Fensom, D. S. 1972. A theory of translocation in phloem of *Heracleum* by contractile protein microfibrillar material. *Can. J. Bot.* 50:479–97
76. Fensom, D. S. 1975. Other possible Mechanisms. See Ref. 206, pp. 354–66
77. Fensom, D. S., Williams, E. J. 1974. A note on Allen's suggestion for long-distance translocation in the phloem of plants. *Nature* 250:490–92
78. Figier, J. 1972. Localisation infrastructurale de la phosphatase acide dans les glandes pétiolaires d' *Impatiens holstii. Planta* 108:215–26

79. Fischer, A. 1884. Untersuchungen uber das Siebrohren-System der Cucurbitaceen. *Gebruder Borntraeger.* 109 pp.
80. Fischer, R. A., MacAlister, T. J. 1975. An evaluation of chemical and freeze-substituting techniques in determining the *in vivo* conditions of the plasmodesmata in *Chara coraleina. Can. J. Bot.* 53:555–69
81. Fisher, D. B. 1975. Structure of functional soybean sieve elements. *Plant Physiol.* 56:555–69
82. Geiger, D. R. 1975. Phloem loading. See Ref. 206, pp. 395–431
83. Geiger, D. R. 1975. Phloem loading and associated processes. See Ref. 5, pp. 251–81
84. Geiger, D. R. 1976. Phloem loading in source leaves. In *Transport and Transfer Processes in Plants,* ed. I. F. Wardlaw, J. B. Passioura, pp. 167–83. New York/San Francisco/London: Academic. 484 pp.
85. Geiger, D. R., Fondy, B. R. 1980. Response of phloem loading and export to rapid changes in sink demand. *Ber. Dtsch. Bot. Ges.* 93:177–86
86. Geiger, D. R., Giaquinta, R. T., Sovonick, S. A., Fellows, R. J. 1973. Solute distribution in sugar beet leaves in relation to phloem loading and translocation. *Plant Physiol.* 52:585–89
87. Giaquinta, R. 1976. Evidence for phloem loading from the apoplast. Chemical modification of membrane sulfhydryl groups. *Plant Physiol.* 57: 872–75
88. Giaquinta, R. 1977. Phloem loading of sucrose. pH dependence and selectivity. *Plant Physiol.* 59:750–55
89. Giaquinta, R. 1980. Mechanism and control of phloem loading of sucrose. *Ber. Dtsch. Bot. Ges.* 93:187–201
90. Giaquinta, R. T., Geiger, D. R. 1973. Mechanism of inhibition of translocation by localized chilling. *Plant Physiol.* 51:372–77
91. Giaquinta, R. T., Geiger, D. R. 1977. Mechanism of cyanide inhibition of phloem translocation. *Plant Physiol.* 59:178–80
92. Gilder, J., Cronshaw, J. 1973. Adenosine triphosphatase in the phloem of *Cucurbita. Planta* 110:189–204
93. Gilder, J., Cronshaw, J. 1973. The distribution of adenosine triphosphatase activity in differentiating and mature phloem cells of *Nicotiana tabacum* and its relationship in phloem transport. *J. Ultrastruct. Res.* 44:388–404
94. Gilder, J., Cronshaw, J. 1974. A biochemical and cytochemical study of adenosine triphosphatase activity in the phloem of *Nicotiana tabacum. J. Cell Biol.* 60:221–35
95. Gunning, B. E. S., Pate, J. S., Briarty, L. 1968. Specialized "transfer cells" in minor veins of leaves and their possible significance in phloem translocation. *J. Cell Biol.* 37:C7–C12
96. Gunning, B. E. S., Pate, J. S., Minchin, F. R., Marks, I. 1974. Quantitative aspects of transfer cell structure in relation to vein loading in leaves and solute transport in legume nodules; In *Transport at the the Cellular Level,* eds. M. A. Sleigh, D. H. Jennings, 87–126. Cambridge: Univ. Press
97. Hendrix, J. E. 1977. Phloem loading in squash. *Plant Physiol.* 60:567–69
98. Hepton, C. E. L., Preston, R. D., Ripley, G. W. 1955. Electron microscope observations on the structure of sieve plates in *Cucurbita. Nature* 176:868–70
99. Heyser, W. 1980. Phloem loading in the maize leaf. *Ber. Dtsch. Bot. Ges.* 93:221–28
100. Heyser, W., Evert, R. F., Fritz, E., Eschrich, W. 1978. Sucrose in the free space of translocating maize leaf bundles. *Plant Physiol.* 62:491–94
101. Heyser, W., Heyser, R., Eschrich, W., Leonard, O. A., Rautenberg, M. 1976. The influence of externally applied organic substances on phloem translocation in detached maize leaves. *Planta* 132:269–77
102. Ho, L. C. 1980. Control of import into tomato fruits. *Ber. Dtsch. Bot. Ges.* 93:315–25
103. Hughes, J. E., Gunning, B. E. S. 1980. Glutarldehyde-induced deposition of callose. *Can. J. Bot.* 58:250–58
104. Humphreys, T. 1978. A model for sucrose transport in the maize scutellum. *Phytochemistry* 17:679–84
105. Hutchings, V. M. 1978. Sucrose and proton cotransport in *Ricinus* cotyledons. I. H+ influx associated with sucrose uptake. *Planta* 138:229–35
106. Hutchings, V. M. 1978. Sucrose and proton cotransport in *Ricinus* cotyledons. II. H+ efflux and associated K+ uptake. *Planta* 138:237–41
107. Jacoli, G. G. 1974. Translocation of mycoplasma-like bodies through sieve pores in plant tissue cultures infected with aster yellows. *Can. J. Bot.* 52:2085–88
108. Jarvis, P., Thaine, R., Leonard, J. W. 1973. Structures in sieve elements cut with a cryostat following different rates of freezing. *J. Exp. Bot.* 24:905–19

109. Johnson, R. P. C. 1968. Microfilaments in pores between frozen-etched sieve elements. *Planta* 81:314–32
110. Johnson, R. P. C. 1973. Filaments but no membranous transcellular strands in sieve pores in freeze-etched translocating phloem. *Nature* 244:464–66
111. Johnson, R. P. C., Freundlich, A., Barclay, G. F. 1976. Transcellular strands in sieve tubes; What are they? *J. Exp. Bot.* 27:1117–36
112. Kleinig, H., Dorr, I., Weber, C., Kollmann, R. 1971. Filamentous proteins from plant sieve tubes. *Nature New Biol.* 229:152–53
113. Kollmann, R. 1960. Untersuchungen uber das protoplasma der siebrohren von *Passiflora coerulea.* II. Elektronenoptische Untersuchungen. *Planta* 55:67–107
114. Kollmann, R. 1980. Fine structural and biochemical characterization of phloem proteins. *Can. J. Bot.* 58:802–6
115. Kollmann, R., Dorr, I., Kleinig, H. 1970. Protein filaments—structural components of the phloem exudate. *Planta* 95:86–94
116. Komor, E., Rotter, M., Tanner, W. 1977. A proton cotransport system in a higher plant: sucrose transport in *Ricinus communis.* *Plant Sci. Lett.* 9:153–62
117. Komor, E., Rotter, M., Waldhauser, J., Martin, E., Cho, B. H. 1980. Sucrose proton symport for phloem loading in the *Ricinus* seedling. *Ber. Dtsch. Bot. Ges.* 93:211–19
118. Komor, E., Tanner, W. 1980. In *Plant Membrane Transport* (workshop, Toronto 1979), ed. J. Dainty. Amsterdam: Elsevier/North Hollywood Biomed. Press. In press
119. Kruatrachue, M., Evert, R. F. 1974. Structure and development of sieve elements in the leaf of *Isoetes muricata.* *Am. J. Bot.* 61:253–66
120. Kruatrachue, M., Evert, R. F. 1977. The lateral meristem and its derivatives in the corm of *Isoetes muricata.* *Am. J. Bot.* 64:310–25
121. Kruatrachue, M., Evert, R. F. 1978. Structure and development of sieve elements in the root of *Isoetes muricata* Dur. *Ann. Bot.* 42:15–21
122. Kuo, J., O'Brien, T. P., Zee, S.-Y. 1972. The transverse veins of the wheat leaf. *Aust. J. Biol. Sci.* 25:721–37
123. Lee, D. R., Arnold, D. C., Fensom, D. S. 1971. Some microscopial observations of functioning sieve tubes of *Heracleum* using Nomarski optics. *J. Exp. Bot.* 22:35–38
124. Liberman-Maxe, M. 1971. Etude cytologique de la différenciation des cellules criblées de *Polypodium vulgare* (Polypodiacée). *J. Microsc.* 12:271–88
125. MacRobbie, E. A. C. 1971. Phloem translocation. Facts and mechanisms: a comparative survey. *Biol. Rev. Cambridge Philos. Soc.* 46:429–81
126. Malek, F., Baker, D. A. 1977. Proton co-transport of sugars in phloem loading. *Planta* 135:297–99
127. Melaragno, J. E., Walsh, M. A. 1976. Ultrastructural features of developing sieve elements in *Lemna minor* L.—the protoplast. *Am. J. Bot.* 63:1145–57
128. Milburn, J. A. 1975. Pressure flow. See Ref. 206, pp. 328–53
129. Minchin, P. E. H., Troughton, J. H. 1980. Quantitative interpretation of phloem translocation data. *Ann. Rev. Plant. Physiol.* 31:191–215
130. Mishra, U., Spanner, D. C. 1970. The fine structure of sieve tubes of *Salix caprea* L and its relation to the electroosmotic theory. *Planta* 90:43–56
131. Moorby, J. 1977. Integration and regulation of translocation within the whole plant. *Symp. Soc. Exp. Bot.* 31:425–54
132. Neuberger, D. S., Evert, R. F. 1974. Structure and development of the sieve-element protoplast in the hypocotyl of *Pinus resinosa.* *Am. J. Bot.* 61:360–74
133. Neuberger, D. S., Evert, R. F. 1975. Structure and development of sieve areas in the hypocotyl of *Pinus resinosa.* *Protoplasma* 84:109–125
134. Neuberger, D. S., Evert, R. F. 1976. Structure and development of sieve cells in the primary phloem of *Pinus resinosa.* *Protoplasma* 87:27–37
135. Northcote, D. H., Wooding, F. B. P. 1966. Development of sieve tubes in *Acer pseudoplatanus.* *Proc. R. Soc. London Ser. B* 163:524–37
136. Opik, H. 1975. Staining of sieve tube mitochondria in coleoptiles of rice (*Oryza sativa* L.) with diaminobenzidine. *Planta* 122:269–71
137. Palevitz, B. A., Hepler, P. K. 1975. Is P-protein actin-like?—Not yet. *Planta* 125:261–71
138. Palevitz, B. A., Newcomb, E. H. 1970. A study of sieve element starch using sequential enzymatic digestion and electron microscopy. *J. Cell Biol.* 45:383–98
139. Palevitz, B. A., Newcomb, E. H. 1971. The ultrastructure and development of tubular and crystalline P-protein in the sieve elements of certain papilionaceous legumes. *Protoplasma* 72:399–426

140. Parthasarathy, M. V. 1974. Ultrastructure of phloem in palms. I. Immature sieve elements and parenchymatic elements. *Protoplasma* 79:59–91
141. Parthasarathy, M. V. 1974. Ultrastructure of phloem in palms. II. Structural changes, and fate of the organelles in differentiating sieve elements. *Protoplasma* 79:93–125
142. Parthasarathy, M. V. 1974. Ultrastructure of phloem in palms. III. Mature phloem. *Protoplasma* 79:265–315
143. Parthasarathy, M. V. 1974. Mycoplasmalike organisms associated with lethal yellowing disease of palms. *Phytopathology* 64:667–74
144. Parthasarathy, M. V., Klotz, L. H. 1976. Palm "wood". II. Ultrastructural aspects of sieve elements, tracheary elements and fibers. *Wood Sci. Technol.* 10:247–71
145. Parthasarathy, M. V., Klotz, L. H. 1976. Palm "wood". I. Anatomical aspects. *Wood Sci. Technol.* 10:215–29
146. Parthasarathy, M. V., Muhlethaler, K. 1969. Ultrastructure of protein tubules in differentiating sieve elements. *Cytobiology* 1:17–36
147. Parthasarathy, M. V., Pesacreta, T. C. 1980. Microfilaments in plant vascular cells. *Can. J. Bot.* 58:807–15
148. Pate, J. S., Gunning, B. E. S. 1969. Vascular transfer cells in angiosperm leaves. A taxonomic and morphological survey. *Protoplasma* 68:135–56
149. Pate, J. S., Gunning, B. E. S. 1972. Transfer cells. *Ann. Rev. Plant Physiol.* 23:173–96
150. Peel, A. J. 1974. *Transport of Nutrients in Plants.* London: Butterworths, 258 pp.
151. Perry, J. W., Evert, R. F. 1975. Structure and development of the sieve elements in *Psilotum nudum. Am. J. Bot.* 62:1038–52
152. Robidoux, J., Sandborn, E. B., Fensom, D. S., Cameron, M. L. 1973. Plasmatic filaments and particles in mature sieve elements of *Heracleum sphondylium* under the electron microscope. *J. Exp. Bot.* 24:349–59
153. Sabnis, D. D., Hart, J. W. 1976. A comparative analysis of phloem exudate proteins from *Cucumis melo, Cucumis sativus* and *Cucurbita maxima* by polyacrylamide gel electrophoresis and iso-electric focusing. *Planta* 130:211–18
154. Sabnis, D. D., Hart, J. W. 1979. Heterogeneity in phloem protein complements from different species. *Planta* 145:459–66
155. Sauter, J. J. 1977. Electron microscopical localization of adenosine triphosphatase and B-glycerophosphatase in sieve cells of *Pinus nigra* var. *austriaca* (Hoess) *Badoux. Z. Pflanzenphysiol.* 81:438–58
156. Schumacher, W., Kollmann, R. 1959. Zur Anatomie des Siebrobrenplasmas bei *Passiflora coerulea. Ber. Dtsch. Bot. Ges.* 72:176–79
157. Seaston, A., Inkson, C., Eddy, A. A. 1973. The absorption of protons with specific amino acids and carbohydrates by yeast. *Biochem. J.* 134:1031–43
158. Shih, C. Y., Currier, H. B. 1969. Fine structure of phloem cells in relation to translocation in the cotton seedling. *Am. J. Bot.* 56:464–72
159. Siddiqui, A. W., Spanner, D. C. 1970. The state of the pores in functioning sieve plates. *Planta* 91:181–89
160. Slayman, C. L., Slayman, C. W. 1974. Depolarization of the plasma membrane of *Neurospora* during active transport of glucose: evidence for a proton-dependent co-transport system *Proc. Natl. Acad. Sci. USA* 71:1935–39
161. Sovonick, S. A., Geiger, D. R., Fellows, R. J. 1974. Evidence for active phloem loading in the minor veins of sugar beet. *Plant Physiol.* 54:886–91
162. Spanner, D. C. 1970. The electroosmotic theory of phloem transport in the light of recent measurements on *Heracleum* phloem. *J. Exp. Bot.* 21:325–34
163. Spanner, D. C. 1978. The Munch hypothesis, freeze substitution and the structure of sieve-plate pores. *Ann. Bot.* 42:485–88
164. Spanner, D. C. 1978. Sieve-plate pores, open or occluded? A critical review. *Plant Cell Environ.* 1:7–20
165. Spanner, D. C. 1979. The electroosmotic theory of phloem transport: a final restatement. *Plant Cell Environ.* 2:107–21
166. Spanner, D. C., Jones, R. L. 1970. The sieve tube wall and its relation to translocation. *Planta* 92:64–72
167. Steer, M. W., Newcomb, E. H. 1969. Development and dispersal of P-protein in the phloem of *Coleus blumei* Benth. *J. Cell Sci.* 4:155–69
168. Stone, D. L., Cronshaw, J. 1973. Fine structure of P-protein filaments from *Ricinus communis. Planta* 113:193–206
169. Tammes, P. M. L., Ie, T. S. 1971. Studies on phloem exudation from *Yucca flaccida* Haw. IX. Passage of carbon black particles through sieve plate pores. *Acta Bot. Neerl.* 20:309–17

170. Tanner, W. 1980. Proton sugar co-transport in lower and higher plants. *Ber. Dtsch. Bot. Ges.* 93:167–76
171. Tanner, W., Komor, E. 1975. In *Biomembranes: Structure and Function.* eds. G. Gardos, I. Szasz, 145–54. New York: Elsevier
172. Tanner, W., Komor, E. 1979. In *Function and Molecular Aspects of Biomembrane Transport.* eds. E. Quagliariello, F. Palmieri, S. Papa, M. Klingenberg, pp. 259–69. Amsterdam: Elsevier
173. Tanner, W., Komor, E., Fenzl, F., Decker, M. 1977. In *Regulation of Cell Membrane Activities in Plants.* ed. E. Marre, O. Ciferrie, pp. 79–90. Amsterdam: Elsevier Biomed. Press
174. Thaine, R. 1969. Movement of sugars through plants by cytoplasmic pumping. *Nature* 222:873–75
175. Thaine, R., DeMaria, M. E., Sarisalo, H. I. M. 1975. Evidence of transcellular strands in transverse cryostat sections of *Cucurbita pepo* sieve tubes. *J. Exp. Bot.* 26:91–101
176. Timell, T. E. 1973. Ultrastructure of the dormant and active cambial zones and the dormant phloem associated with formation of normal and compression woods in *Picea abies* (L.) Karst, Tech. Publ. 96. Syracuse, New York: SUNY, Coll. Environ. Sci. For.
177. Turgeon, R., Webb, J. A., Evert, R. F. 1975. Ultrastructure of minor veins in *Cucurbita pepo* leaves. *Protoplasma* 83:217–32
178. Walker, A. J., Ho, L. C. 1977. Carbon translocation in the tomato: carbon import and fruit growth. *Ann. Bot.* 41:813–23
179. Walker, A. J., Ho, L. C. 1977. Carbon translocation in the tomato: effects of fruit temperature on carbon metabolism and the rate of translocation. *Ann. Bot.* 41:825–32
180. Walker, T. S. 1972. The purification and some properties of a protein causing gelling in phloem sieve tubes exudate from *Cucurbita pepo. Biochim. Biophys. Acta* 257:433–44
181. Walker, T. S., Thaine, R. 1971. Proteins and fine structural components in exudate from sieve tubes from *Cucurbita pepo* stems. *Ann. Bot.* 35:773–90
182. Walsh, M. A., Evert, R. F. 1975. Ultrastructure of metaphloem sieve elements in *Zea mays. Protoplasma* 83:365–88
183. Walsh, M. A., Melaragno, J. E. 1976. Ultrastructural features of developing sieve elements in *Lemna minor* L.-sieve plate and lateral sieve areas. *Am. J. Bot.* 63:1174–83
184. Wardlaw, I. F. 1974. Phloem transport: physical, chemical or impossible. *Ann. Rev. Plant Physiol.* 25:515–39
185. Wardlaw, I. F., Passioura, J. B. 1976. *Transport and Transfer Processes in Plants.* New York: Academic. 484 pp.
186. Warmbrodt, R. D. 1978. The comparative leaf structure of ferns. PhD thesis. Univ. Wisc., Madison
187. Warmbrodt, R. D. 1980. Characteristics of structure and differentiation in the sieve element of lower vascular plants. *Ber. Dtsch. Bot. Ges.* 93:13–28
188. Warmbrodt, R. D., Evert, R. F. 1974. Structure of the vascular parenchyma in the stem of *Lycopodium lucidulum. Am. J. Bot.* 61:437–43
189. Warmbrodt, R. D., Evert, R. F. 1974. Structure and development of the sieve element in the stem of *Lycopodium lucidulum. Am. J. Bot.* 61:267–77
190. Warmbrodt, R. D., Evert, R. F. 1978. Comparative leaf structure of six species of heterosporous ferns. *Bot. Gaz.* 139:393–429
191. Weatherley, P. E., Johnson, R. P. C. 1968. The form and function of the sieve tube: a problem in reconciliation. *Int. Rev. Cytol.* 24:149–92
192. Weber, C. W., Franke, W., Kartenbeck, J. 1974. Structure and biochemistry of phloem-proteins isolated from *Cucurbita maxima. Exp. Cell Res.* 87:97–106
193. Weber, C., Kleinig, H. 1971. Molecular weight of *Cucurbita* sieve tube proteins. *Planta* 99:179–82
194. Wergin, W. P., Newcomb, E. H. 1970. Formation and dispersal of crystalline P-proteins in sieve elements of soybean *Glycine max L. Protoplasma* 71:365–88
195. Willenbrink, J., Kollmann, R. 1966. Uber den assimilattransport in phloem von *Metasequoia. Z. Pflanzenphysiol.* 55:42–53
196. Wooding, F. B. P. 1969. P-protein and microtubular systems in *Nicotiana* callus phloem. *Planta* 85:284–98
197. Wooding, F. B. P. 1974. Development and fine structure of angiosperm and gymnosperm sieve tubes. *Symp. Soc. Exp. Biol.* 28:27–41
198. Worley, J. F. 1973. Evidence in support of "open" sieve tube pores. *Protoplasma* 76:129–32
199. Yapa, P. A. J., Spanner, D. C. 1972. Isoelectric focusing of sieve tube protein. *Planta* 106:369–73
200. Yapa, P. A. J., Spanner, D. C. 1974. Localization of adenosine triphospha-

tase activity in mature sieve elements of *Tetragonia. Planta* 117:321–28

201. Zahur, M. S. 1959. Comparative study of secondary phloem of 423 species of woody dicotyledons belonging to 85 families. *Cornell Univ. Agric. Exp. Sta. Mem.* 358 pp.
202. Zee, S.-Y. 1969. Fine structure of the differentiating sieve elements of *Vicia faba. Aust. J. Bot.* 17:441–56
203. Zee, S.-Y. 1969. The localization of acid phosphatase in the sieve element of *Pisum. Aust. J. Biol. Sci.* 22:1051–54
204. Ziegler, H. 1974. Biochemical aspects of phloem transport. In *Transport at the Cellular Level,* ed. M. A. Sleigh, D. H. Jennings, pp. 43–62. Cambridge: Univ. Press
205. Ziegler, H. 1975. Nature of transported substances. See Ref. 206, pp. 59–100
206. Zimmermann, M. H., Milburn, J. A., eds. 1975. *Encyclopedia of Plant Physiology, Vol. 1. Transport in Plants 1. Phloem Transport.* Berlin/Heidelberg: Springer. 535 pp.
207. Zimmermann, M. H., Ziegler, H. 1975. List of sugars and sugar alcohols in sieve-tube exudates. See Ref. 206, pp. 480–503

Ann. Rev. Plant Physiol. 1981. 32:485-509

PHOTOSYNTHESIS, CARBON PARTITIONING, AND YIELD

❖7721

Roger M. Gifford and L. T. Evans

Division of Plant Industry, CSIRO, Canberra, A.C.T. 2601, Australia

CONTENTS

Increase in the yield of crop plants can come from many quarters, such as better adaptation to environmental conditions, greater resistance to pests and diseases, improved agronomic practices, increased genetic yield potential, and interactions between these. In this review we are concerned with the physiological basis of increased genetic yield potential, and in particular with the control and improvement in distribution and storage of photosynthetic assimilates.

Photosynthesis and translocation in relation to crop yield have been reviewed previously (e.g. 112, 138). To comprehend the development of

0066-4294/81/0601-0485$01.00

yield we need to treat photosynthesis, translocation, growth, and storage as an integrated whole, since these processes are linked by numerous interactions. We start here by considering the physiological basis of man's achievements in crop breeding over past millenia before examining the details of photosynthesis and carbon partitioning.

LESSONS FROM CROP PLANT EVOLUTION

Changes that have occurred between wild species and modern cultivar may be grouped into four categories (43): changes conferring adaptedness to cultivation and systematic harvesting; modification of daylength and vernalization requirements; quality improvements by selection against undesirable, and for desirable, components; and increase in yield potential. So far, increase in yield potential has been achieved by direct selection for ability to yield the organs of interest under progressively improved systems of agronomic inputs; changes in individual physiological characteristics have resulted only indirectly from such selection. Recently, however, there has been increased interest in the possibility of raising yield by direct selection for important physiological attributes.

Marked increase in the size of seeds, pods, fruit, or inflorescences (e.g. in maize) has occurred in many crops during their evolution. The diameter of sugarcane stems and of sugar beet roots has likewise increased. In some cases, as Charles Darwin observed, the increase in size has been confined to the organs or tissues harvested or eaten by man (e.g. in brussels sprouts and some stone fruits), but in others there has been a parallel increase in the size of several organs. In wheat there has been parallel increase in leaf and seed size (44). In sugarcane, both stem diameter and the area of the upper leaves have increased over a twentyfold range from *Saccharum spontaneum* through *S. robustum* to modern forms of *S. officinarum* (14, 15). Greater cell size, with increasing ploidy, has contributed to the parallel increases in organ size during the domestication of wheat (40) and sugarcane (14), but a rise in cell numbers was also involved.

Despite these changes, the relative growth rate (RGR) of young plants does not appear to have increased in the course of domestication, even when comparisons are made with seedlings of the same size (118). This has been shown for wheat (44,100), maize (38), tomato (196), and cowpea (118).

The relative leaf area growth rate is an important determinant of the time taken for the crop canopy to close. Before the canopy achieves full interception of the light, variation in leaf area is a much more powerful determinant of variation in crop growth rate than is variation in photosynthesis rate per unit leaf area (64). After canopy closure, photosynthetic CO_2 exchange per unit leaf area may become an important determinant of canopy photosynthesis. Yet there is no evidence to date of any indirect selection for increase in the maximum light-saturated CO_2 exchange rate per unit leaf area (CER) during the domestication and improvement of wheat (41, 44, 99), maize (38), sorghum (37), pearl millet (106), sugarcane (16), cotton (42), or cowpea (117). Indeed, the highest CERs recorded for wheat, sorghum, pearl millet, and cotton have been found in the wild relatives, not in the modern cultivars. In wheat, the higher CER of the more primitive species were associated with higher stomatal and residual conductances (41) and with higher Hill-reaction activity per unit of chlorophyll (206).

The lack of advance in leaf CER during crop evolution may have been due in part to negative relationships between CER and leaf area and the persistence of photosynthetic activity as discussed in the next section. Also the need for increase in the genetic potential of CER may have been mitigated in some cropping environments by the use of nitrogenous fertilizers.

In the absence of genetic increases in photosynthesis and growth, past improvements in yield potential have derived largely from increase in the proportion of accumulated dry weight which is invested in the organs harvested by man, i.e. in the harvest index (32). Both the size and the number of these organs have been increased, as have the rate and the duration of their growth in many species. Increases in duration have been associated with greater longevity of leaves, or with the storage phase making up a larger proportion of the crop life cycle. Increased rate of storage, on the other hand, has been shown, at least in wheat, to involve a greater allocation of photosynthetic assimilates to the grain during grain filling (44). Associated with this greater allocation has been a parallel increase in the cross-sectional area of phloem in the peduncle (45). In sugarcane also, the phloem cross-sectional area serving the storage tissues has increased in parallel with the latter (14).

Clearly, many coordinated changes have occured in the course of evolution from wild plant to modern cultivar. In general, developmental processes—such as those concerned with germination, flowering, plant form and composition, dehiscence, and shedding etc—appear to have been more readily modified by empirical selection than has photosynthesis.

Two lines of argument may be followed from the observation of improved carbon partitioning during crop evolution. One is that any remaining scope for further improvement in carbon allocation must be small, so it would be better now to aim at increasing photosynthetic and growth rates. The other is that since partitioning is where the flexibility has been in the past, it is better to aim for further increases in harvest index. We briefly examine the first argument before returning to the second one.

IMPROVING PHOTOSYNTHETIC RATE

Genotypic differences in components of photosynthesis have been found and used as selection criteria. The great genetic variation in chlorophyll content per unit leaf area seems generally to have little impact on variation in CER or productivity, as for example in barley (47), where even chlorophyll-deficient mutants have near-normal CER (123). The reduction in CER of soybean lines having abnormally low leaf chlorophyll (< 35 μg cm^{-2}) (18) may be associated with low specific leaf weight (see below) in low chlorophyll genotypes (114). Similarly, there is no indication that variation in Hill activity or photophosphorylation *per unit area* is reflected in variation in CER (73, 187). Determinations of Hill activity per unit chlorophyll cannot be interpreted in relation to CER without information on chlorophyll per unit leaf area.

In contrast to the light reactions, variation in ribulose bisphosphate carboxylase (RuBPC) in C_3 species has often been found to correlate with leaf CER, as Randall et al (159) found when comparing a high CER mutant of tall fescue with normal tall fescues. Since RuBPC represents 30–50% of the soluble protein of leaves, there is a risk that breeding for its high activity might be at the expense of other important enzymes. Selection for high specific activity forms of the enzyme might be better, but none have been found yet, although a ryegrass tetraploid with a lower than normal K_m (CO_2) for RuBPC has been described (52, 160). Also, the oxygenase activity of the carboxylase protein (113) yields a substrate for photorespiration, for which a function has not yet been established. The substantial effort to breed out photorespiration by several methods including screening seedlings for low CO_2 compensation point (e.g. 1, 104, 131) has proved fruitless so far. There is no unequivocal evidence of C_3 plant genotypes with low oxygenase and normal carboxylase activity.

A morphological character which often (e.g. 5, 36), but not always (e.g. 12, 25, 41), correlates with CER is specific leaf weight (SLW = leaf dry weight per unit area) or, more simply, leaf thickness (21). Intergenotypic variation in SLW at a chosen ontogenetic stage can show stability of ranking from season to season (114) and is heritable (179). Whether or not it is a good breeding strategy to select for high SLW depends on its relationship to leaf area development (95). Expansion of leaf area and thickening of leaves can be inversely related (132). A sound strategy might be to produce plants which expand large thin leaves early in the season and then thick leaves after the canopy intercepts all the light.

Selection for stomatal frequency and length has not met with success in changing CER through stomatal conductance, since frequency and length are inversely related (125, 145) and other countering compensations in leaf area and longevity can also occur (91). Further, stomatal aperture is a much more powerful determinant of stomatal conductance, and behaves as if it were continuously controlled to resolve optimally the conflict between the carbon needs of growth and the environmental demand on plant water through transpiration (24). So even if stomatal frequency were increased, it is likely that the water loss/carbon gain optimization process would lead to the same resolution in terms of stomatal conductance.

The complexity of the internal controls over CER, together with the little success at improving it by selection for component processes, suggests that direct selection for CER itself might be more effective. There is much evidence of heritable variation in leaf CER, and high and low lines have been selected in several species (e.g. 26, 191, 198, 204). Unfortunately, the relationship between CER and growth and yield is tenuous (e.g. 12, 44, 76, 96, 136), so much so that it was possible to select tall fescues with high growth rate but low CER and vice versa (198). The fact that CO_2 enrichment of the atmosphere will enhance growth and yield of C_3 crops (e.g. 62, 63, 122) shows that growth is photosynthetically limited in some environments. So selection for high CER might be failing because of counterproductive associations.

Part of the problem is that CER is determined to some extent by sink growth rate in relation to leaf area, as will be discussed later. Ontogenetic changes in the partitioning of leaf dry weight between leaf area and leaf thickness, together with sink demand effects, can result in poor correlation between CER at one ontogenetic stage and at another (41, 107, 117). In wheat the persistence of flag leaf photosynthesis was greatest for the modern hexaploids, so that toward grain maturity they have higher CER than the primitive types despite a reverse ranking early in grain filling (44). An inverse relationship between area per leaf and maximum CER among wheat species and within triticales (44, 51, 156), and in maize (72), alfalfa (27), and soybean (17), is partly a reflection of the inverse association between leaf thickness (SLW) and leaf area. It is also reflected in a negative relation between mesophyll cell size and CER (40, 205).

It is likely that any manipulable physiological imbalances in crops which have, through long selection, become well adapted to their agronomic environment, will be subtle and small. Wilson (201) appears to have identified one such imbalance in perennial ryegrass (*Lolium perenne*). He has selected genotypes with low rates of dark respiration in fully expanded leaves. The low respiration lines have about a 10% growth rate advantage in both simulated swards and field plots (167, 203). There is indication that there may also be wasteful respiration, amenable to elimination, in maize (78) and tall fescue (93). Given the disappointments in breeding for high CER and for low photorespiration, it is remarkable that a few cycles of selection were sufficient to eliminate completely an apparent inefficiency which nature or agricultural practice had not already overcome (202).

CARBON PARTITIONING BETWEEN ORGANS

Several stages exist in the partitioning of photosynthetic products. For a developing leaf, photosynthetic assimilate is partitioned within the leaf between exported material and further leaf growth or temporary storage. Assimilate exported is partitioned between different sinks, and within sinks the incoming carbon is partitioned between different chemical constituents. We are concerned here with the partitioning of net assimilate (often approximated by dry weight) between organs which are separated by vascular transport channels. Sinks may be meristematic, elongation, or storage sinks, and the characteristics of storage sinks depend on the type of product stored —sucrose, starch, proteins, or lipids. There may also be respiratory sinks in some plants, involving cyanide-resistant respiration which is not an integral part of tissue growth and maintenance (178).

The data discussed here relate mainly to sucrose or starch storage sinks such as sugarcane internodes and cereal grains. Although the biological yield of cereal cultivars has not been increased, grain yield continues to increase because partitioning of assimilate to grains (as measured by harvest index = grain yield/shoot yield) has increased over many years for oats (e.g. 174, 176), barley (33), and wheat (2, 49, 190). The same is true for lipid- and protein-storing peanuts (39). Although there must be a limit to how high harvest index can go, [Austin et al (2) suggest about 62% for wheat], progress still seems possible, the best wheats now having harvest indices around 50%.

The inadequacy of our understanding of the control of partitioning of photosynthetic assimilate is indicated by the empirical ways in which modelers of crop productivity handle it (80, 112). It is sometimes handled by rules based on descriptive allometry (128, 149, 171), sometimes by "priority" concepts for sinks (112), sometimes by a "nutritional control" approach based on "functional equilibrium" between organs (29), and sometimes by transport resistance/substrate gradient/growth response curve models (184). The most appropriate viewpoint will depend on the purpose of the model and the type of sink of central interest. We next review the major characteristics of the movement of photosynthetic assimilate in source-path-sink systems to arrive at a simple conceptual framework which guides our further examination of the regulation of partitioning in whole plants.

CHARACTERISTICS OF ESTABLISHED SOURCE-PATH-SINK SYSTEMS

Photosynthesis in Relation to Sink Demand

The dominant primary sources of assimilate are leaves, although green stems and floral organs (46, 144) can sometimes make substantial contributions. While environmental factors have strong direct influence on photosynthesis, the demand by sinks for assimilate can also determine photosynthetic supply. Unfolding leaves, before photosynthetic autonomy, must compete with other sinks for assimilate, and it is a common observation that heavily fruiting or tuberizing plants experience reduced growth of vegetative parts including leaves [see (130, 162) for wheat] though less essential growth—like increase of stem diameter in apple—may suffer more than leaf development (79) by heavy fruiting. Countering this tendency, CER may be enhanced for a leaf which developed and unfolded in a regime of intense competition from other sinks, whether the intense competition is generated by the onset of fruit growth as in soybean (35, 59) or by removal of other leaves (65, 143).

After full leaf expansion, CER can still be increased or decreased by a natural or manipulated change in sink demand. Responses occurring within hours (101) may represent stomatal responses (65, 109). Longer term responses over several days can be due to parallel change in both stomatal and internal mesophyll control of CER such that intercellular space CO_2 concentration is almost constant (69). But also senescent leaves can be rejuvenated to full photosynthetic performance when the sink:source ratio is increased substantially (68, 84). By contrast, when sink growth is competing with leaves for remobilizable nitrogen, senescence and the fall in CER may be accelerated, as postulated by Sinclair & de Wit (175). Also, removal of fruits (127, 155) or meristematic sinks may arrest senescence of older leaves for unknown but possibly hormonal reasons (168).

Thus, the quantity of assimilate available for partitioning to sinks is itself dependent on the presence and properties of sinks in a complex way, involving opposing responses which cannot yet be predicted. Evidence for the simplest explanation of photosynthetic stimulation by sink demand—that assimilate level in the leaf controls photosynthesis by end-product inhibition (140)—remains equivocal [e.g. contrast (74, 137, 139, 168) with (54, 66, 111, 127)]. This uncertainty indicates that it is not a simple mechanism, and involvement of growth regulators, perhaps emanating from sinks, seems likely in many situations (168).

Phloem Loading

Movement of photosynthetic assimilate toward veins is considered to be in the symplast (53). A downhill gradient of tissue sucrose concentration from the mid-interveinal region to the veins was found in wheat (108), but an active cross-mesophyll transport could not be ruled out. However, mesophyll cells have the capacity to release sucrose and hexoses into the free space (85), although this capacity may not be expressed except by mesophyll cells close to the phloem. There sugars are believed to be transferred to the apoplast close to the minor veins (53, 55). There is no direct evidence for this, but such a pathway would be advantageous in maintaining the large differences in osmotic and turgor potentials and in discriminating between compounds during transfer from mesophyll to phloem (194). This unloading into the vascular free space is stimulated by K^+ ions in sugar beet, suggesting that there may be an active proton cotransport mechanism involved (31). Uptake into sieve tubes from the vascular free space is probably via phloem parenchyma and/or companion cells and is an active, carrier-mediated process having saturation kinetics, with respect to the donor side of the membrane, in C_3 species at least. Transfer from the companion cells to the sieve tubes is probably symplastic. In the sieve tubes, sugar (in most species almost solely sucrose) reaches very high levels, usually in the range 200–800 mM (50, 192, 208). The weight of evidence suggests that sucrose is the molecule loaded rather than hexoses in most species (53, 60), though evidence for prior inversion in some species does exist (13). This active accumulation of sucrose in the sieve-tube–companion cell complex leads to osmotic uptake of water and hence to a considerable hydrostatic pressure in these cells.

Interorgan Transport

That assimilate transport in sieve tubes occurs down the sucrose gradient, as originally found by Mason & Maskell (121), or at least down the osmotic pressure gradient—which for most species (208) amounts to almost the same thing—still seems to be generally accepted (20, 53, 148, 192). Fisher (50) found in soybeans that there was a sufficient gradient in sieve-tube sucrose between source leaf and sink to create a hydrostatic pressure gradient to drive Münch-type mass flow at the observed rate. Assuming open sieve-plate pores, it would not take a large osmotic pressure gradient, relative to the absolute osmotic pressure in the tubes, to generate enough differential in hydrostatic pressure to drive translocation over the distances involved in crop species, but such small gradients are difficult to detect (56, 146).

Minor vein sieve tubes extend into stems as bundles and as isolated vascular strands. Numerous bridging strands occur between and within bundles in lamina, petiole, and stem (115, 183), and in grasses many anastomoses also occur at the nodes (142, 150). This continuous network of sieve tubes allows a molecule entering at any point to follow several routes to any sink organ, perhaps even without leaving the sieve-tube lumen. The pathway lacks directional valves. A severed vein is quickly by-passed, presumably via bridging strands, without the surrounding tissue being flooded with unloaded translocate (115). So there is the possibility of continuous traffic around the plant as seen with the reciprocal interchange of carbon between young tillers and main shoot of grass plants (120), and the simultaneous influx and efflux of phosphorus in the shoot apex (165). Nevertheless, labeled assimilate can follow specific routes to particular sinks, especially in the short term; for example, ^{14}C is translocated specifically to younger leaves in the same orthostichy in tobacco (92, 173), cotton (188), apple (4), and other species (102). This does not necessarily mean that there are no linking connections to other sinks along the path from source to favored sink, for in soybean such sink specificity was found even though functional cross-links were known to occur en route (186), and transport routes quickly adjust to changes in the pattern of supply (10, 101). In terms of our assumptions about the mechanism of translocation, it simply means that for the configuration of competing sources and sinks any gradients in solute concentration across the links were insufficient to drive detectable flow through them in the period of the experiment (often a few hours). Over a long period some label can reach all parts of the plant, this being particularly evident in grasses which have many cross-links between bundles (22, 34). Even where there is usually little net C translocation from one organ to another, as between well-developed grass tillers, the vascular channel is still functional for long periods, as shown by repeatedly removing the source leaves from the tillers (65). Study of the short-term distribution of carbon from a particular source may tell us something of the preferred routes and the pattern of gradients, but it does not establish the control mechanisms underlying the long-term distribution of dry weight.

Phloem Unloading

Sieve-tube unloading has received little attention. Phloem retains its solutes with minimal radial exchanges except where remobilizable stores are being laid down or in growing regions where unloading occurs. Autoradiographs of ^{14}C assimilate moving from a mature tobacco leaf to recipient expanding leaves have shown that major veins in the exporting leaf can retain labeled assimilate without leakage detectable on the autoradiographic film, whereas major veins of the importing leaves are the first parts to show up, indicating some leakage or unloading before translocate reaches the minor veins (158). One way that leakage out of the vascular bundle may be prevented en route is for a suberized lamella to form in cell walls of the layer of mesophyll cells which ensheath the vascular bundle in some species, as found by O'Brien & Carr (141) in the mestome sheath of temperate cereals. But this would not prevent apoplastic leakage into xylem vessels. The transpiration stream rarely contains appreciable sugars (147) where transport between well-developed sources and sinks is involved. But for some woody perennials like grapevine (75) and sugar maple (182), xylem sucrose can be very high before leaf development in spring, and in apple some xylem transport of carbon was evident for trees with developed leaves (4). Where the source leaf is immature, carbon export can be partly through the xylem (6), presumably before the leaf's phloem has acquired full loading capability. There may be temporary storage depots en route between source and final sink, but transfer to them does not represent leakage, depots being sinks too. It is possible, therefore, that the sieve-tube/companion cell/phloem parenchyma complex is countering leakage by actively loading sucrose along almost its entire length against the radial concentration gradient; the pH gradient from alkaline sieve tube to nearby acid xylem contents would assist that if proton cotransport were the loading mechanism [(3) and see below]. If so, unloading could occur by leakage wherever active uptake by the phloem ceases. A simple possible corollary would be that sinks in some way locally inhibit the phloem-loading mechanism. This could take several forms according to the type of sink. In apical meristems it could simply be that the phloem is not developmentally mature enough to have acquired the sucrose-loading mechanism, or that the absence of xylem vessels adjacent to phloem termini leads to an inadequate proton gradient for loading. In storage tissues, such as sugarcane stem internode (67) and maize endosperm (170), it may be that sink control over the activity of invertase, which is closely associated with the phloem or sieve-tube surfaces, prevents sucrose being reloaded by inverting it to hexoses. Whether or not the phloem actively unloads with a carrier-mediated process in any situation is unknown (98). It probably varies according to the type of sink, but for some sinks the necessity to unload water may constrain options for sugar unloading; apoplastic sucrose inversion could set up an osmotic gradient in a direction assisting water removal from the sieve tubes. Whether unloading into the apoplast be active or passive, there must be some feedback control over the process related to the sugar concentration in the recipient free space and to the "needs" of sink growth.

Sink Properties

Despite much work on prolonged tissue culture in vitro, only a restricted range of organs has been examined in terms of short-term sugar uptake. The latter is more relevant to defining properties of plant parts as sinks. Organs so studied include sugarcane internode segments (e.g. 8, 67), cotton hypocotyl segments (70, 71), wheat kernels (86, 89), expanding cotyledons of *Ricinus* seedlings (103, 105), and developing barley embryos (19). Generally sucrose is found to be the preferred substrate for uptake and short-term growth or storage by these tissues, although the cotton hypocotyl system could also utilize fructose (70), and in sugarcane stem, glucose is preferred even though it is sucrose which accumulates in the stem cells (8). The ready uptake of sugars from the bathing media by these tissues is suggestive that in vivo they may take up assimilate delivered by the phloem from the free space, and this apoplastic route is most commonly favored. In wheat kernels, the endosperm is separated from the vascular strand by a cell-free endosperm cavity containing sucrose and little hexose (87), suggesting uptake from the apoplast. In roots, where water flow through the free space occurs in the opposite direction, symplastic transport of sucrose from stele to cortex and apical cells seems more likely and is supported experimentally (30).

In all the systems cultured in sucrose in vitro, except perhaps the wheat kernel (87, 88), sucrose uptake by the cells of the tissue has saturation kinetics and can occur against a concentration gradient from the outside to the inside of the cells.This indicates that a membrane carrier mechanism is involved. In wheat kernels, Jenner found no evidence for active transport of sucrose into the cells, and he argues that the regulation of sugar uptake by grains lies not in the kinetics of uptake by the cells but in the concentration of sucrose in the vascular bundle running along the length of the groove in the grain (89). By contrast, in maize kernels the interpretation is that sucrose is cleaved to hexoses in the free space of the placento-chalazal region of the kernel as it diffuses from phloem to endosperm cells (170), and the invertase reaction is a control point in sugar uptake, and hence in starch synthesis, by the maize endosperm. Similarly, apoplastic invertase plays a regulatory role in the passage of sugar into sugarcane internode cells; the phloem delivers and the cells store sucrose, but the sugar enters the cells as hexose (11, 67).

The apparent Michaelis-Menten constant for sugar uptake (K_m) by the tissues varies widely from 0.9 mM for glucose uptake by cane internodes (8) to 36 mM for sucrose uptake by barley embryos (19). In cotton hypocotyl, K_m varied threefold, this probably being related to the season during which the plants were grown (70). The specificity of carriers varied between systems, although in general the sucrose carriers were not interfered with by other sugars. In both cotton hypocotyls (71) and wheat kernels (87), the endosperm free-space sucrose concentration in vivo was low enough for the rate of uptake and storage by the cells to be considered a linear function of free-space sucrose concentration—a conclusion which simplifies the approach taken to modeling assimilate distribution. At free-space sugar concentrations higher than are generally found in vivo, several of the systems showing carrier-mediated uptake revealed a diffusional component of sugar influx.

Other phenomena observed in short-term sugar uptake studies add awkward complexity to the task of constructing source-path-sink models to interpret carbon distribution. First, in *Ricinus* cotyledon (103), cotton hypocotyl (71), and barley embryo (19) systems, the rate of carrier-mediated sucrose uptake was not only determined by substrate concentration in the free space but also by the concentration of transported moiety inside the cells. The higher the intracellular concentration, the less the transport. So although transport operates against the gradient in these systems, the rate is not independent of the steepness of the gradient across the cell membrane. Thus the rate of removal of cytoplasmic sucrose by accumulation in the vacuole or by conversion is not only determined by, but also influences the rate of sucrose transport into the cells. Second, in cotton hypocotyls, incorporation of sucrose into insoluble fractions was dependent on current uptake of sucrose from the medium (70); sucrose taken up previously by the tissue was apparently unavailable for synthesis. Perhaps this is due to sequestering in the vacuole. When this occurs it is not meaningful to relate tissue growth rate simply to *average* cellular sucrose concentration. The third feature found for cotton hypocotyl (71) which creates difficulties in interpreting in vitro tissue culture studies in vivo was that the maximum rate of sucrose uptake exceeded the maximum rate at which insoluble compounds could be synthesized. This suggests that in vivo some kind of control over sucrose uptake might operate to prevent its concentration from rising to the very high levels possible in vitro.

The variety of strategies which the above sinks use to control their growth makes universal description of sink capabilities, in-so-far as they determine assimilate partitioning, difficult. But one variable which can be a powerful determinant of sink growth is the apoplastic sugar concentration in the free space of sink tissues.

REGULATION OF CARBON FLOW THROUGH SOURCE-PATH-SINK SYSTEMS

A Conceptual Framework

The following paragraph is a generalization of the above, intended as a working hypothesis onto which further details of the control of carbon partitioning can be built.

The amount of photosynthetic substrate available for distribution is in part determined by sinks themselves through feedback control of photosynthesis. The same may be true of substrate from remobilizable reserves (81, 82, 163). The sieve tubes are separated from both source cells and sink cells by the free space of sources and sinks although for some sinks, such as growing roots, symplastic connections between phloem and sink cells may be more important than apoplastic ones. Phloem can actively take up sucrose from a low concentration solution in the source apoplast, accumulating high concentrations in the sieve tubes, the rate of accumulation being a function, among other things, of free-space sucrose concentration. The phloem may be actively loading against a steep sucrose concentration gradient along its entire length, except where unloading is somehow stimulated by the presence of sinks. Sinks can actively accumulate sugars from the free space of unloading phloem. In excised tissues, sink growth saturates with respect to free-space sugar at around 60-200 mM sucrose and typically operates in apoplastic solution less concentrated than that. Since sieve-tube contents can range up to 800 mM sucrose or higher, it is easy to envision that gradients could develop in these tubes between sources and sinks, creating hydrostatic pressure gradients sufficient to drive whatever mass flow is needed to equate supply and demand provided sieve-plate pores are not plugged in vivo. In this simple picture the average sucrose concentrations and the gradient within the sieve tube would automatically rise and fall, balancing supply and demand. Temporary storage in stems can also help reconcile inputs and outputs at each end of the sieve tubes. The observation that localized stem cooling in species not damaged by chilling causes no change in the rate of translocation after an hour or two of adjustment supports this passive, physical interpretation of sieve-tube transport (57, 192). Passioura & Ashford (146) have shown that rates of translocation (specific mass transfer) ten times greater than the highest previously recorded in the literature can be induced by restriction of the sieve-tube cross-section available. They conclude that rate of transport in sieve tubes is unlikely to limit flow from source to sink.

Additional Complexities

Features which complicate the above picture include: the role of energy and light; evidence that the phloem exerts more directional and quantitative control over C flow than is credited to it above, including hormone-directed transport; buffering of flow between primary sources and ultimate sinks by temporary stores; claims that source leaves can exert control over destination; and the problem of whether the movement of carbon down sucrose gradients in the sieve tubes necessarily implies that it flows from high sucrose source organs to low sucrose sink organs.

ENERGY AND LIGHT The energy requirements for the above model are primarily in the source, for loading sucrose into the phloem parenchyma (83), and in the sink, for loading the sink tissue cells. The only energy needed en route would be for tissue maintenance and for keeping the sieve tube loaded (leak prevention) (57). The mechanism may be for a membrane ATPase to pump protons into the sugar donor side of the membrane, thereby maintaining a proton gradient necessary to allow a sugar-proton cotransport mechanism to operate across the membrane (60, 119, 157).

Besides driving the photosynthetic supply of sucrose, light may play indirect roles in controlling translocation and partitioning at several steps.

Light has frequently been noted to foster active transport of solutes across membranes of living systems (161, 189), though the mechanism is unknown (157). Light might be involved in controlling export of photosynthetic metabolites from the chloroplasts (94) or generally from leaves (77, 83, 169, 193). Similarly, there is evidence of light stimulation of uptake of sugars by sink tissues like apical meristems (183) or cotton hypocotyls (71, 166), but apparently not in young bean leaves (181). Such evidence is not strong, but if the processes of source supply, phloem loading, short-term storage, and sink uptake were influenced or atuned by irradiance, one might obtain reduced sink growth in low light without any change in free-space sugar concentration in source and sink. This would confound interpretations based on simple transport resistance/substrate concentration gradient models. Jenner (90) tested that possibility by examining grain growth and grain endosperm sucrose concentration in shaded and well-illuminated wheat; in fact, both parameters declined with shading consistent with the simple picture.

LIMITATIONS ON TRANSLOCATION CAPACITY Milthorpe & Moorby (124) concluded that vascular transport does not usually exert any control over sink growth. Experiments involving incisions in the culm of wheat (195) and sorghum (133) have shown that dry-weight accumulation in the grain was not reduced by restricting phloem cross-section. In terms of our conceptual framework, phloem restriction would create an increased sucrose gradient in the sieve tubes across the restriction, causing flow to bypass through interconnecting strands. This change of gradient could reduce sink growth indirectly by reducing the sucrose concentration to which the sink is exposed while increasing the concentration that the source must work against.

One way that the transport mechanism could be inadequate to accommodate an enhanced rate of photosynthesis is if the carrier-mediated loading of sucrose at the phloem parenchyma plasmalemma reached the saturation point with respect to mesophyll apoplastic sucrose concentration. Geiger (53) believes this to be unlikely, saturation not occurring until mesophyll sucrose reaches about 100 mM. Moreover, for sucrose applied exogenously to the mesophyll free space of sugar beet leaves, a second uptake curve is superimposed which saturates at about 400 mM sucrose (180). For C_4 grasses, it has been suggested (116) that when leaf photosynthesis is driven high enough, the bundle sheath mesophyll may reach sucrose concentrations exceeding the sieve-tube contents such that loading occurs *down* a gradient (cf 185). But even then the proportion of assimilate retained by the leaf did not increase, indicating that an upper limit to translocation capacity out of the leaf had not been reached.

So although the sieve tubes undoubtedly impose a resistance to mass flow, and they can channel the flow from particular sources to particular sinks, there is enough system flexibility for translocation not to limit flow from sources to sinks in most situations. The flexibility derives from the existence of alternative routes not necessarily having much higher resistances, the possibility of adjusting solute gradients at a small expense of energy, the potential for cycling of materials between organs having both source and sink attributes, occasionally the transport of carbon compounds in the xylem, and the possibility of turgor-dependent loading (177) which might provide a mechanism, involving K^+ pumping, for matching loading and transport-to-sink demand independent of any apparent impediments in the phloem channels.

Another reason why the vascular system may rarely exert appreciable control over flow of material from source to sink is that the vasculature may develop in such a way as to meet the "anticipated" needs for C flow as it does in the wheat peduncle, where the number and size of the vascular bundles correlates with the number of spikelets (45). During the most rapid phase of wheat stem elongation, vascular connections are being continually broken, but rate of differentiation of new sieve tubes exceeds the rate of destruction, and specific mass transfer of carbon compounds (per unit area of sieve-tube lumen) never rises to particularly high levels during stem extension (151).

Phloem develops in cultured tissues in response to IAA and sucrose in the medium (172). Primordia produce auxins and the extending end of a vascular bundle releases sucrose, so it is easy to envisage how vascular connections may extend to wherever they are needed. Indeed, complex venation patterns can be simulated by a mathematical model based on auxin diffusion from growing tissues (126). Moreover, callus cultured in media containing auxin can develop vascular tissues, and the ratio of phloem to xylem is a function of sucrose concentration in the medium and of no other sugar (172).

HORMONE-DIRECTED TRANSLOCATION Growth regulators have long been implicated in assisting translocation in established source-path-sink systems (129, 197). Although most discussion relates to growth regulators released from sinks, hormones from sources have also been considered (28, 207). When IAA (9) and other regulators like cytokinins, ethylene, and gibberellic acid (135) are applied to a cut stem surface, or ABA is dissolved in the rooting solution of *Phaseolus* plants (97), assimilates accumulate in the region of application. Combinations of growth regulators can have additive, synergistic (152), or inhibitory effects (110, 135). The idea that such IAA-directed transport (the most studied hormone in this regard) is due solely to stimulation of growth rate of tissue at the site of application was argued against from the outset (9), and there is evidence for direct effects on transport (153). However, the systems studied usually involve very low translocation rates and may not reflect the most important controls for intact plants. In bean seedlings, which were regenerating root and shoot apices following their excision, the main control over the distribution of sucrose between root and shoot sinks was attributed to hormonal influences (IAA and cytokinin) on sink activity per se, although some superimposed influence on relative sucrose availability to the respective sinks could not be excluded (58). By contrast, the enhanced accumulation of ^{14}C assimilate in peapods caused by pod warming was due both to an effect on ovule growth directly and to an influence on transport outside the warmed zone (200); this is suggestive of a hormonal effect emanating from the sink.

Each growth regulator might act primarily on different parts of source-path-sink systems, and most steps in the overall sequence have been examined: phloem loading (110, 119), long-distance transport in the sieve tubes (9, 153), sieve-tube unloading (154), and sugar uptake by sink tissues (134, 207). Such hormonal influences are yet too ill-defined to be able to generalize in relation to assimilate partitioning.

DESTINATION CONTROL BY LEAVES It has been suggested occasionally that leaves can determine the destination of their assimilate (7). We know of no clear evidence on this.

DIRECTION OF INTERORGAN ASSIMILATE GRADIENTS Although there is little doubt that phloem transport occurs down pressure potential and solute gradients in the sieve tubes, this does not mean that assimilate distribution necessarily occurs from organs of high to organs of low average soluble carbohydrate. Phloem loading and uptake by sink cells and vacuoles (199) against the gradient mitigate against that. If a sink stores sucrose in its symplasm and vacuoles and has a capacity to accumulate sugars against a steep gradient, then the sink apoplast could have a lower sugar concentration than the sieve-tube contents in the source leaves despite a higher sucrose concentration in the sink as a whole. This presumably is what happens in sugar beet, sugarcane, and sugary fruits. Another mechanism of maintaining a downhill solute gradient in the phloem, despite an uphill sucrose gradient between source and sink organs, is to convert soluble carbohydrate from one form to another. Sucrose is inverted in cane-stem apoplast and taken into the cells as hexoses; *Rosaceae* translocate sorbitol but store sucrose in the fruits (164). The possibility of loading being controlled more by sieve-tube turgor than by sieve-tube or apoplastic sucrose concentrations (177) can, in the presence of K^+ pumping, provide a further mechanism.

SINKS AS DETERMINANTS OF CARBON PARTITIONING

It is difficult therefore to avoid concluding that it is phenomena in the sinks which largely determine the distribution of assimilate between them. Although in the very short term label moves from certain sources to certain sinks, in the longer term it is the ability of sinks to take up sugars from the sink free space which determines partitioning, not the relative adequacy of vascular connections between sources and sinks nor the relative activity of various sources. From studies on competition between two wheat ears containing different numbers of grains but receiving assimilate from a common source leaf equidistant from the two sinks (23), a similar conclusion could be drawn; dry weight growth of the ears was more or less proportional to the number of grains in the ear, but in the short term the ear with the larger number of grains drew a disproportionately large share of its growth from the source leaf equidistant from both ears. These findings are consistent with the notion that a rapidly growing sink generates a steeper gradient in sieve-tube assimilate concentration leading to flow from more distant sources than does a weakly growing competing sink.

If sinks control partitioning of dry weight, then a corollary is that it is factors determining the setting up and relative activity of sinks which determine the pattern of assimilate distribution. This implicates the entire fields of environmental and developmental plant physiology. Among the many factors determining the existence and activity of sinks is the supply of photosynthetic assimilate at an earlier ontogenetic stage. For example, the development of lateral meristems in grasses is strongly influenced by the light and also by the carbon dioxide level (61). The number of grains set in wheat is determined by the adequacy of the photosynthetic environment before anthesis (63), especially during the phase of ear development between 10 days after terminal spikelet formation and ear emergence (48).

SUMMARY AND CONCLUSION

Improvement in potential crop yield by breeding has largely been by selecting plants for high yield of the sink of economic interest. Clearly there must be a limit to how far this can go. We know from CO_2 and light enrichment studies that crop yield is frequently, perhaps usually, photosynthetically limited. Yet genetic improvement in growth rate or photosynthetic rate potential has not occurred so far. Continuation of improvement in the ratio of economic sink to total plant still seems to be the most effective route, until it is at the expense of light-intercepting leaf surface or robustness of the crop. Improvement in the maximum rate of leaf photosynthesis may then become essential to further increase in yield potential. In the meantime, the key to understanding the distribution of photosynthetic assimilate to particular organs lies not so much in the leaf mesophyll or the phloem loading system, nor in the translocation system, but more in the determination of the properties of the sinks themselves. It lies in understanding what determines the establishment and premature abortion of sinks, what determines the duration of sink growth, what it is about some sinks which enables them to stimulate leaf photosynthesis, how a sink controls unloading of the phloem, and what determines the response of sink growth to the sucrose concentration in its free space.

ACKNOWLEDGMENTS

We thank I. F. Wardlaw, R. W. King, and J. W. Patrick for comments on the manuscript and V. Viviani and F. Kelleher for typing.

Literature Cited

1. Apel, P. 1979. Dark and photorespiration. In *Crop Physiology and Cereal Breeding: Proceedings of a Eucarpia Workshop,* ed. J. H. J. Spiertz, Th. Kramer, pp. 102–5. Wageningen: Centre Agric. Publ. Doc. 186 pp.
2. Austin, R. B., Bingham, J., Blackwell, R. D., Evans, L. T., Ford, M. A., Morgan, C. L., Taylor, M. 1980. A comparison of the yields of old and new varieties of winter wheat at two levels of soil fertility. *J. Agric. Sci* 94:675–89
3. Baker, D. A. 1978. Proton co-transport of organic solutes by plant cells. *New Phytol.* 81:485–97
4. Barlow, H. W. B. 1979. Sectorial patterns in leaves on fruit tree shoots produced by radioactive assimilates and solutions. *Ann. Bot.* 43:593–602
5. Barnes, D. K., Pearce, R. B., Carlson, G. E., Hart, R. H., Hanson, C. H. 1969. Specific leaf weight differences in alfalfa associated with variety and plant age. *Crop Sci.* 9:421–23
6. Biddulph, O., Cory, R. 1965. Translocation of ^{14}C-metabolites in the phloem of the bean plant. *Plant Physiol.* 40:119–29
7. Bidwell, R. G. S. 1974. *Plant Physiology.* New York: MacMillan 643 pp.
8. Bieleski, R. L. 1962. The physiology of surgarcane V. Kinetics of sugar accumulation. *Aust. J. Biol. Sci.* 15:429–44
9. Booth, A., Moorby, J., Davies, C. R., Jones, H., Wareing, P. F. 1962. Effects of indolyl-3-acetic acid on the movement of nutrients within plants. *Nature* 194:204–5
10. Borchers, C., Swanson, C. A. 1977. The kinetics of compensated translocation in *Phaseolus vulgaris* L. *Plant Physiol.* 59:125 (Suppl)
11. Bowen, J. E., Hunter, J. E. 1972. Sugar transport in immature internodal tissue of sugarcane II. Mechanism of sucrose transport. *Plant Physiol.* 49:789–93
12. Brinkman, M. A., Frey, K. J. 1978. Flag leaf physiological analysis of oat isolines that differ in grain yield from their recurrent parents. *Crop Sci.* 18:67–73
13. Brovchenko, M. I. 1965. On the movement of sugars from the mesophyll to the conducting bundles in sugar beet leaves. *Sov. Plant Physiol.* 12:230–37
14. Bull, T. A. 1965. *Taxonomic and physiological relationships in the Saccharum complex.* PhD thesis. Univ. Queensland, Australia
15. Bull, T. A. 1967. The taxonomic significance of quantitative morphological characters and physiological studies of *Saccharum. Proc. Int. Soc. Sugar-Cane Technol., 12th Congr.* pp. 985–94
16. Bull, T. A. 1971. The C_4 pathway related to growth rates in sugarcane. In *Photosynthesis and Photorespiration,* ed. M. D. Hatch, C. B. Osmond, R. O. Slatyer, pp. 68–75, New York: Wiley. 565 pp.
17. Burris, J. S., Edje, O. T., Wahab, A. H. 1973. Effects of seed size on seedling performance in soybeans. II. Seedling growth and photosynthesis and field performance. *Crop Sci.* 13:207–10
18. Buttery, B. R., Buzzell, R. I. 1977. The relationship between chlorophyll con-

tent and rate of photosynthesis in soybeans. *Can. J. Plant Sci.* 57:1–5
19. Cameron-Mills, V., Duffus, C. M. 1979. Sucrose transport in isolated immature barley embryos. *Ann. Bot.* 43:559–69
20. Canny, M. J. 1973. *Phloem Translocation.* London: Cambridge Univ. Press. 301 pp.
21. Charles-Edwards, D. A. 1978. An analysis of the photosynthesis and productivity of vegetative crops in U.K. *Ann. Bot.* 42:717–32
22. Chonan, N., Kawahara, H., Matsuda, T. 1974. Morphology of vascular bundles of leaves in graminaceous crops. *Proc. Crop Sci. Soc. Jpn.* 43:425–32 (In Japanese)
23. Cook, M. G., Evans, L. T. 1978. Effect of relative size and distance of competing sinks on the distribution of photosynthetic assimilates in wheat. *Aust. J. Plant Physiol.* 5:495–509
24. Cowan, I. R. 1977. Stomatal behaviour and environment. *Adv. Bot. Res.* 5:117–27
25. Crosbie, T. M., Mock, J. J., Pearce, R. B. 1977. Variability and selection advance for photosynthesis in Iowa stiff stalk synthetic maize population. *Crop Sci.* 17:511–14
26. Crosbie, T. M., Mock, J. J., Pearce, R. B. 1978. Inheritance of photosynthesis in a diallel among eight maize inbred lines from Iowa stiff stalk synthetic. *Euphytica* 27:657–64
27. Delaney, R. H., Dobrenz, A. K. 1974. Morphological and anatomical features of alfalfa leaves as related to CO_2 exchange. *Crop Sci.* 14:444–47
28. de Stigter, H. C. M. 1961. Translocation of ^{14}C-photosynthates in the graft muskmelon/*Cucurbita ficifolia.* *Acta Bot. Neerl.* 10:466–73
29. de Wit, C. T., Brouwer, R., Penning de Vries, F. W. T. 1970. The simulation of photosynthetic systems. In *Prediction and Measurement of Photosynthetic Productivity,* ed. I. Setlik, pp. 47–70. Wageningen: Pudoc. 632 pp.
30. Dick, P. A., ap Rees, T. 1975. The pathway of sugar transport in roots of *Pisum sativum.* *J. Exp. Bot.* 26:305–14
31. Doman, D. C., Geiger, D. R. 1979. Effect of exogenously supplied foliar potassium on phloem loading in *Beta vulgaris* L. *Plant Physiol.* 64:528–33
32. Donald, C. M. 1962. In search of yield. *J. Aust. Inst. Agric. Sci.* 28:171–78
33. Donald, C. M., Hamblin, J. 1976. The biological yield and harvest index of cereals as agronomic and plant breeding criteria. *Adv. Agron.* 28:361–405
34. Doodson, J. K., Manners, J. G., Myers, A. 1964. The distribution of ^{14}C assimilated by the third leaf of wheat. *J. Exp. Bot.* 15:96–103
35. Dornhoff, G. M., Shibles, R. M. 1970. Varietal differences in net photosynthesis of soybean leaves. *Crop Sci.* 10:42–45
36. Dornhoff, G. M., Shibles, R. M. 1976. Leaf morphology and anatomy in relation to CO_2 exchange rate of soybean leaves. *Crop Sci.* 16:377–81
37. Downes, R. W. 1971. Relationship between evolutionary adaptation and gas exchange characteristics of diverse *Sorghum* taxa. *Aust. J. Biol. Sci.* 24:843–52
38. Duncan, W. G., Hesketh, J. D. 1968. Net photosynthetic rates, relative growth rates and leaf numbers of 22 races of maize grown at eight temperatures. *Crop Sci.* 8:670–74
39. Duncan, W. G., McCloud, R. L., McGraw, R. L., Boote, K. J. 1978. Physiological aspects of peanut yield improvement. *Crop Sci.* 18:1015–20
40. Dunstone, R. L., Evans, L. T. 1974. Role of changes in cell size in the evolution of wheat. *Aust. J. Plant Physiol.* 1:157–65
41. Dunstone, R. L., Gifford, R. M., Evans, L. T. 1973. Photosynthetic characteristics of modern and primitive wheat species in relation to ontogeny and adaptation to light. *Aust. J. Biol. Sci.* 26:295–307
42. El-Sharkawy, M., Hesketh, J. D., Muramoto, H. 1965. Leaf photosynthetic rates and other growth characteristics among 26 species of *Gossypium.* *Crop Sci.* 5:173–75
43. Evans, L. T. 1976. Physiological adaptation to performance as crop plants. *Philos. Trans. Soc. London Ser. B* 275:71–83
44. Evans, L. T., Dunstone, R. L. 1970. Some physiological aspects of evolution in wheat. *Aust. J. Biol. Sci.* 23:725–41
45. Evans, L. T., Dunstone, R. L., Rawson, H. M., Williams, R. F. 1970. The phloem of the wheat stem in relation to requirements for assimilate by the ear. *Aust. J. Biol. Sci.* 23:743–52
46. Evans, L. T., Rawson, H. M. 1970. Photosynthesis and respiration by the flag leaf and components of the ear during grain development in wheat. *Aust. J. Biol. Sci.* 23:245–54
47. Ferguson, H., Eslick, R. F., Aase, J. K. 1973. Canopy temperatures of barley as influenced by morphological characteristics. *Agron. J.* 65:425–28

48. Fischer, R. A., Aquilar, I. 1976. Yield potential in a semi-dwarf spring wheat and the effect of carbon dioxide fertilization. *Agron. J.* 68:749–52
49. Fischer, R. A., Kertesz, Z. 1976. Harvest index in spaced populations and grain weight in micro plots as indicators of yielding ability in spring wheat. *Crop Sci.* 16:55–59
50. Fisher, D. B. 1978. The estimation of sugar concentration in individual sieve-tube elements by negative staining. *Planta* 139:19–24
51. Gale, W. D., Edrich, J., Lupton, F. G. H. 1974. Photosynthetic rates and the effects of applied gibberellin in some dwarf, semi-dwarf and tall wheat varieties (*Triticum aestivum*.) *J. Agric. Sci.* 83:43–46
52. Garrett, M. K. 1978. Control of photorespiration at RuBP carboxylase-oxygenase level in ryegrass cultivars. *Nature* 274:913–15
53. Geiger, D. R. 1975. Phloem loading. In *Encyclopaedia of Plant Physiology 1 Transport in Plants. I. Phloem Transport,* ed. M. H. Zimmermann, J. A. Milburn, 17:395–431. Berlin: Springer. 535 pp.
54. Geiger, D. R. 1976. Effects of translocation and assimilate demand on photosynthesis. *Can. J. Bot.* 54:2337–45
55. Geiger, D. R. 1979. Control of partitioning and export of carbon in leaves of higher plants. *Bot. Gaz.* 140:241–48
56. Geiger, D. R., Giaquinta, R. T., Sovonick, S. A., Fellows, R. J. 1973. Solute distribution in sugar beet leaves in relation to phloem loading and translocation. *Plant Physiol.* 52:585–89
57. Geiger, D. R., Sovonick, S. A. 1975. Effects of temperature, anoxia and other metabolic inhibitors on translocation. See Ref. 53, pp. 256–86
58. Gersani, M., Lips, S. H., Sachs, T. 1980. The influence of shoots, roots and hormones on sucrose distribution. *J. Exp. Bot.* 31:177–84
59. Ghorashy, S. R., Pendleton, J. W., Peters, D. B., Boyer, J. F., Beuerlein, J. E. 1971. Internal water stress and apparent photosynthesis with soybeans differing in pubescence. *Agron. J.* 63:674–76
60. Giaquinta, R. T. 1977. Phloem loading of sucrose: pH dependence and selectivity. *Plant Physiol.* 59:750–55
61. Gifford, R. M. 1977. Growth pattern, carbon dioxide exchange and dry weight distribution in wheat growing under differing photosynthetic environments. *Aust. J. Plant Physiol.* 4:99–110
62. Gifford, R. M. 1979. Growth and yield of CO_2-enriched wheat under water-limited conditions. *Aust. J. Plant Physiol.* 6:367–68
63. Gifford, R. M., Bremner, P. M., Jones, D. B. 1973. Assessing photosynthetic limitation to grain yield in a field crop. *Aust. J. Agric. Res.* 24:297–307
64. Gifford, R. M., Jenkins, C. L. 1981. Prospects of applying knowledge of photosynthesis toward improving crop production. In *Photosynthesis: CO_2 Assimilation and Plant Productivity,* ed. Govindjee, Vol. 2. New York: Academic. In press
65. Gifford, R. M., Marshall, C. 1973. Photosynthesis and assimilate distribution in ***Lolium multiflorum*** Lam. following differential tiller defoliation. *Aust. J. Agric. Res.* 24:297–307
66. Glasziou, K. T., Bull, T. A. 1971. Feedback control of photosynthesis in sugar cane. See Ref. 16, pp. 82–88
67. Glasziou, K. T., Gayler, K. R. 1972. Storage of sugars in stalks of sugarcane. *Bot. Rev.* 38:471–90
68. Hall, A. J., Brady, C. J. 1977. Assimilate source-sink relationships in *Capsicum annuum* L. II. *Aust. J. Plant Physiol.* 4:771–83
69. Hall, A. J., Milthorpe, F. L. 1978. Assimilate source sink relationships in *Capsicum annuum* L. III. *Aust. J. Plant Physiol.* 5:1–13
70. Hampson, S. E., Loomis, R. S., Rains, D. W. 1978. Characteristics of sugar uptake in hypocotyls of cotton. *Plant Physiol.* 62:846–50
71. Hampson, S. E., Loomis, R. S., Rains, D. W. 1978. Regulation of sugar uptake in hypocotyls of cotton. *Plant Physiol.* 62:851–55
72. Hanson, W. D. 1971. Selection for differential productivity among juvenile maize plants: associated net photosynthetic rate and leaf area changes. *Crop Sci.* 11:334–39
73. Hanson, W. D., Grier, R. E. 1973. Rates of electron transfer and of non-cyclic photophosphorylation for chloroplasts isolated from maize populations selected for juvenile productivity and in leaf widths. *Genetics* 75:247–57
74. Hanson, W. D., Yeh, R. Y. 1979. Genotypic differences in reduction in carbon dioxide exchange rates as associated with assimilate accumulation in soybean leaves. *Crop Sci.* 19:54–58
75. Hardy, P. J., Possingham, J. V. 1969. Studies on translocation of metabolites in the xylem of grapevine shoots. *J. Exp. Bot.* 20:325–35

76. Hart, R. H., Pearce, R. B., Chatterton, N. J., Carlson, G. E., Barnes, D. K., Hanson, C. H. 1978. Alfalfa yield, specific leaf weight, CO_2 exchange rate and morphology. *Crop Sci.* 18:649–53
77. Hartt, C. E. 1965. Light and translocation of ^{14}C in detached blades of sugarcane. *Plant Physiol.* 40:718–24
78. Heichel, G. H. 1971. Confirming measurements of photosynthesis with dry matter accumulation. *Photosynthetica* 5:93–98
79. Heim, G., Landsberg, J. J., Watson, R. L., Brain, P. 1979. Ecophysiology of apple trees: dry matter production and partitioning by young Golden Delicious trees in France and England. *J. Appl. Ecol.* 16:179–94
80. Hesketh, J. D., Jones, J. W. 1976. Some comments on computer simulators for plant growth - 1975. *Ecol. Modelling* 2:235–47
81. Ho, L. C. 1976. The relation between the rates of carbon transport and of photosynthesis in tomato leaves. *J. Exp. Bot.* 27:87–97
82. Ho, L. C. 1979. Regulation of assimilate translocation between leaves and fruits in the tomato. *Ann. Bot.* 43:437–48
83. Ho, L. C., Thornley, J. H. M. 1978. Energy requirement for assimilate translocation from mature tomato leaves. *Ann. Bot.* 42:481–83
84. Hodgkinson, K. C. 1975. Influence of partial defoliation on photosynthesis, photorespiration and transpiration by lucerne leaves of different ages. *Aust. J. Plant Physiol.* 1:561–78
85. Huber, S. C., Moreland, D. E. 1980. Translocation: Efflux of sugars across the plasmalemma mesophyll protoplast. *Plant Physiol.* 65:560–62
86. Jenner, C. F. 1968. Synthesis of starch in detached ears of wheat. *Aust. J. Biol. Sci.* 21:597–608
87. Jenner, C. F. 1974. Factors in the grain regulating the accumulation of starch. In *Mechanisms of Regulation of Plant Growth,* ed. R. L. Bieleski, A. R. Ferguson, M. M. Cresswell, pp. 901–8. Wellington: R. Soc. N. Z. Bull. 12. 934 pp.
88. Jenner, C. F. 1974. An investigation of the association between hydrolysis of sucrose and its absorption by grains of wheat. *Aust. J. Plant Physiol.* 1:319–29
89. Jenner, C. F. 1976. Physiological investigations on restriction to transport of sucrose in ears of wheat. *Aust. J. Plant Physiol.* 3:337–47
90. Jenner, C. F. 1980. Effects of shading or removing spikelets in wheat: testing assumptions. *Aust. J. Plant Physiol.* 7:113–22
91. Jones, H. G. 1977. Transpiration in barley lines with differing stomatal frequency. *J. Exp. Bot.* 28:162–68
92. Jones, H., Martin, R. V., Porter, H. K. 1959. Translocation of ^{14}C in tobacco following assimilation of $^{14}CO_2$ by a single leaf. *Ann. Bot.* 23:493–508
93. Jones, R. J., Nelson, C. J. 1979. Respiration and concentration of water soluble carbohydrates in plant parts of contrasting tall fescue genotypes. *Crop Sci.* 19:367–72
94. Kaiser, W. M., Bassham, J. A. 1979. Light-dark regulation of starch metabolism in chloroplasts. *Plant Physiol.* 63:109–13
95. Kallis, A., Tooming, H. 1974. Estimation of the influence of leaf photosynthetic parameters, specific leaf weight and growth functions on yield. *Photosynthetica* 8:91–103
96. Kaplan, S. L., Koller, H. R. 1977. Leaf area and CO_2 exchange rate as determinants of the rate of vegetative growth in soybean plants. *Crop Sci.* 17:35–38
97. Karmoker, J. L., van Steveninck, R. F. M. 1979. The effect of abscisic acid on sugar levels in seedlings of *Phaseolus vulgaris* L. cv. Redland Pioneer. *Planta* 146:25–30
98. Keener, M. E., deMichele, D. W., Sharpe, P. J. H. 1979. Sink metabolism: a conceptual framework for analysis. *Ann. Bot.* 44:659–69
99. Khan, M. A., Tsunoda, S. 1970. Evolutionary trends in leaf photosynthesis and related leaf characters among cultivated wheat species and its wild relatives. *Jpn. J. Breed.* 20:133–40
100. Khan, M. A., Tsunoda, S. 1970. Growth analysis of cultivated wheat species and their wild relatives with special reference to dry matter distribution among different plant organs and to leaf area expansion. *Tohoku J. Agric. Res.* 21:47–59
101. King, R. W., Wardlaw, I. F., Evans, L. T. 1967. Effect of assimilate utilization on photosynthetic rate in wheat. *Planta* 77:261–76
102. King, R. W., Zeevaart, J. A. D. 1973. Floral stimulus movement in *Perilla* and flower inhibition caused by non-induced leaves. *Plant Physiol.* 51: 727–38
103. Komor, E. 1977. Sucrose uptake by cotyledons of *Ricinus communis* L.: Characteristics, mechanism, and regulation. *Planta* 137:119–31

104. Krenzer, E. G. Jr., Moss, D. N., Crookston, R. K. 1975. Carbon dioxide compensation points of flowering plants. *Plant Physiol.* 56:194–206
105. Kriedemann, P., Beevers, H. 1967. Sugar uptake and translocation in the castor beans I. Characteristics of transfer in intact and excised seedlings. *Plant Physiol.* 42:161–73
106. Lavergne, D., Bismuth, E., Champigny, M. L. 1979. Physiological studies on two cultivars of *Pennisetum: P. americanum* 23DB, a cultivated species and *P. mollissimum,* a wild species. *Z. Pflanzenphysiol.* 91:291–303
107. Lawes, D. A., Treharne, K. J. 1971. Variation in photosynthetic activity in cereals and its implications in a plant breeding programme. *Euphytica* 20: 86–92
108. Lee, H. J., Ashley, D. A., Brown, R. H. 1980. Sucrose concentration gradients in wheat leaves. *Crop Sci.* 20:95–99
109. Lenz, F., Williams, C. N. 1973. Effect of fruit removal on net assimilation and gaseous diffusion resistance of soybean leaves. *Angew. Bot.* 47:57–63
110. Lepp, N. W., Peel, A. J. 1970. Effects of IAA and kinetin upon the movement of sugars in the phloem of willow. *Planta* 90:230–35
111. Little, C. H. A., Loach, K. 1973. Effect of changes in carbohydrate concentration on the rate of net photosynthesis in mature leaves of *Abies balsamea. Can. J. Bot.* 51:751–58
112. Loomis, R. S., Rabbinge, R., Ng, E. 1979. Explanatory models in crop physiology. *Ann. Rev. Plant Physiol.* 30:339–67
113. Lorimer, G. H., Andrews, T. J. 1973. Plant photorespiration—an inevitable consequence of the existence of atmospheric oxygen. *Nature* 243:359–60
114. Lugg, D. G., Sinclair, T. R. 1979. A survey of soybean cultivars for variability in specific leaf weight. *Crop Sci.* 19:887–92
115. Lush, W. M. 1976. Leaf structure and translocation of dry matter in a C_3 and a C_4 grass. *Planta* 130:235–44
116. Lush, W. M., Evans, L. T. 1974. Longitudinal translocation of ^{14}C-labelled assimilates in leaf blades of *Lolium temulentum. Aust. J. Plant Physiol.* 1:433–43
117. Lush, W. M., Rawson, H. M. 1979. Effects of domestication and region of origin on leaf gas exchange in cowpea (*Vigna unguiculata* L.) *Photosynthetica* 13:419–27
118. Lush, W. M., Wien, H. C. 1980. The importance of seed size in early growth of wild and domesticated cowpeas. *J. Agric. Sci.* 94:177–82
119. Malek, F., Baker, D. A. 1977. Proton co-transport of sugars in phloem loading. *Planta* 135:297–99
120. Marshall, C., Sagar, G. R. 1968. The interdependence of tillers in *Lolium multiflorum* Lam.—a quantitative assessment. *J. Exp. Bot.* 19:785–94
121. Mason, T. G., Maskell, E. J. 1928. Studies on the transport of carbohydrate in the cotton plant II. The factors determining the rate and direction of movement of sugars. *Ann. Bot.* 42:571–636
122. Mauney, J. R., Fry, K. E., Guinn, G. 1978. Relationship of photosynthetic rate to growth and fruiting of cotton, soybean, sorghum and sunflower. *Crop Sci.* 18:259–63
123. McCashin, B. G., Canvin, D. T. 1979. Photosynthetic and photorespiratory characteristics of mutants of *Hordeum vulgare L. Plant Physiol.* 64:354–60
124. Milthorpe, F. L., Moorby, J. 1969. Vascular transport and its significance in plant growth. *Ann. Rev. Plant Physiol.* 20:117–38
125. Miskin, K., Rasmussen, D. C. 1970. Frequency and distribution of stomata in barley. *Crop. Sci.* 10:575–78
126. Mitchison, G. J. 1980. A model for vein formation in higher plants. *Proc. R. Soc. London Ser. B.* 207:79–109
127. Mondal, M. H., Brun, W. A., Brenner, M. L. 1978. Effects of sink removal on photosynthesis and senescence in leaves of soybean (*Glycine max* L.) plants *Plant Physiol.* 61:394–97
128. Monsi, M., Murata, Y. 1970. Development of photosynthetic systems as influenced by distribution of dry matter. See Ref. 29, pp. 115–30
129. Moorby, J. 1968. The effect of growth substances on transport in plants. In *Transport of Plant Hormones,* ed. Y. Vardar, pp. 192–206. Amsterdam: New Holland. 457 pp.
130. Moorby, J., Milthorpe, F. L. M. 1975. Potato. In *Crop Physiology,* ed. L. T. Evans, 8:225–57. Cambridge Univ. Press. 374 pp.
131. Moss, D. N. 1975. Studies on increasing photosynthesis in crop plants. In *CO_2 Metabolism and Plant Productivity,* ed. R. H. Burris, C. C. Black, pp. 31–41. Baltimore: Univ. Park Press. 431 pp.
132. Motto, M., Soressi, G. P., Salamini, F. 1979. Growth analysis in a reduced leaf mutant of common bean (*Phaseolus vulgaris L.*). *Euphytica* 28:593–600

133. Muchow, R. C., Wilson, G. L. 1976. Photosynthetic and storage limitations to yield in *Sorghum bicolor* (L. Moench). *Aust. J. Agric. Res.* 27:489–500
134. Mulligan, D. R., Patrick, J. W. 1979. Gibberellic acid-promoted transport of assimilates in stems of *Phaseolus vulgaris L. Planta* 145:233–38
135. Mullins, M. G. 1970. Hormone-directed transport of assimilates in decapitated internodes of *Phaseolus vulgaris L. Ann. Bot.* 34:897–909
136. Murthy, K. K., Singh, M. 1979. Photosynthesis, chlorophyll content and ribulose diphosphate carboxylase in relation to yield in wheat genotypes. *J. Agric. Sci.* 93:7–11
137. Nafzier, E. D., Koller, H. R. 1976. Influence of leaf starch concentration on CO_2 assimilation in soybeans. *Plant Physiol.* 57:560–63
138. Nasyrov, Y. S. 1978. Genetic control of photosynthesis and improving of crop productivity. *Ann. Rev. Plant Physiol.* 29:215–37
139. Natr, L., Watson, B. T., Weatherley, P. E. 1974. Glucose absorption, carbohydrate accumulation, presence of starch and rate of photosynthesis in barley leaf segments. *Ann. Bot.* 38:589–93
140. Neales, T. F., Incoll, L. D. 1968. The control of leaf photosynthesis rate by level of assimilate concentration in the leaf: a review of the hypothesis. *Bot. Rev.* 34:107–25
141. O'Brien, T. P., Carr, D. J. 1970. A suberized layer in the cell walls of the bundle sheath of grasses. *Aust. J. Biol. Sci.* 23:275–87
142. O'Brien, T. P., Zee, S.-Y. 1971. Vascular transfer cells in the vegetative nodes of wheat. *Aust. J. Biol. Sci.* 24:207–17
143. Ollerenshaw, J. H., Incoll, L. D. 1979. Leaf photosynthesis in pure swards of two grasses (*Lolium perenne* and *Lolium multiflorum*) subjected to contrasting intensities of defoliation. *Ann. Appl. Biol.* 92:133–42
144. Ong, J. K., Colville, K. E., Marshall, C. 1978. Assimilation of $^{14}CO_2$ by the inflorescence of *Poa annua L.* and *Lolium perenne L. Ann. Bot.* 42:855–62
145. Pallardy, S. G., Kozlowski, T. T. 1979. Frequency and length of stomata of 21 *Populus* clones. *Can. J. Bot.* 57: 2519–23
146. Passioura, J. B., Ashford, A. E. 1974. Rapid translocation in the phloem of wheat roots. *Aust. J. Plant Physiol.* 1:521–27
147. Pate, J. S. 1975. Exchange of solutes between phloem and xylem and circulation in the whole plant. See Ref. 53, pp. 451–73
148. Pate, J. S., Atkins, C. A., Hamel, K., McNeil, D. L., Layzell, D. B. 1979. Transport of organic solutes in phloem and xylem of a nodulated legume. *Plant Physiol.* 63:1082–88
149. Patefield, W. M., Austin, R. B. 1971. A model for the simulation of the growth of *Beta vulgaris. Ann. Bot.* 35:1227–50
150. Patrick, J. W. 1972. The vascular system of wheat I. Mature state. *Aust. J. Bot.* 20:48–63
151. Patrick, J. W. 1972. Distribution of assimilates during stem elongation in wheat. *Aust. J. Biol. Sci.* 25:455–67
152. Patrick, J. W. 1976. Hormone-directed transport of metabolites. In *Transport and Transfer Processes in Plants,* ed. I. F. Wardlaw, J. B. Passioura, 37:433–46. New York: Academic. 484 pp.
153. Patrick, J. W. 1979. An assessment of auxin-promoted transport in decapitated stems and whole shoots of *Phaseolus vulgaris* L. *Planta* 146:107–112
154. Patrick, J. W., Wareing, P. F. 1976. Auxin-promoted transport of metabolites in stems of *Phaseolus vulgaris L.:* effects at the site of hormone application. *J. Exp. Bot.* 27:969–82
155. Patterson, T. G., Brun, W. A. 1980. Influence of sink removal on the senescence pattern of wheat. *Crop Sci.* 20:19–23
156. Planchon, C. 1979. Photosynthesis, transpiration, resistance to CO_2 transfer, and water efficiency of flag leaf of bread wheat, durum wheat and Triticale. *Euphytica* 28:403–8
157. Poole, R. J. 1978. Energy coupling for membrane transport. *Ann. Rev. Plant. Physiol.* 29:437–60
158. Porter, H. K. 1966. Leaves as collecting and distributing agents of carbon. *Aust. J. Sci.* 29:31–40
159. Randall, D. D., Nelson, C. J., Asay, K. H. 1977. Ribulose bisphosphate carboxylase: altered genetic expression in tall fescue. *Plant Physiol.* 59:38–41
160. Rathnam, C. K. M., Chollet, R. 1980. Photosynthetic and photorespiratory carbon metabolism in mesophyll protoplasts and chloroplasts isolated from isogenic diploid and tetraploid cultivars of ryegrass. *Plant Physiol.* 65:489–94
161. Raven, J. A. 1976. Transport in algal cells. In *Encyclopedia of Plant Physiology. New Series 2. Transport in Plants. II A: Cells,* ed. U. Luttge, M. G. Pitman, pp. 129–88. Berlin: Springer. 400 pp.

162. Rawson, H. M. 1970. Spikelet number, its control and relation to yield per ear in wheat. *Aust. J. Biol. Sci.* 23:1–15
163. Rawson, H. M., Evans, L. T. 1971. The contribution of stem reserves to grain development in a range of wheat cultivars of different height. *Aust. J. Agric. Res.* 22:851–63
164. Reid, M. S., Bieleski, R. L. 1974. Sugar changes during fruit ripening—whither sorbitol. See Ref. 87, pp. 823–30
165. Rijven, A. H. G. C., Evans, L. T. 1967. Inflorescence initiation in *Lolium temulentum* X. Changes in ^{32}P incorporation into nucleic acids of the shoot apex at induction. *Aust. J. Biol. Sci.* 20:13–24
166. Robbins, W. J., Maneval, W. E. 1924. Effect of light on growth of excised root tips under sterile conditions. *Bot. Gaz.* 78:424–32
167. Robson, M. J. 1979. Growth of simulated swards in controlled environments. *Grassl. Res. Inst. (Hurley) 1978 Ann. Rep.*, pp. 60–61
168. Satoh, M., Kriedemann, P. E., Loveys, B. R. 1977. Changes in photosynthetic activity and related processes following decapitation in mulberry trees. *Physiol. Plant.* 41:203–10
169. Shangina, Z. I. 1965. Influence of 2,4-D on the outflux of carbohydrates from tomato leaves. *Sov. Plant Physiol.* 12:912–17
170. Shannon, J. C., Dougherty, C. T. 1972. Movement of ^{14}C-labelled assimilates into kernels of *Zea mays* L. II. Invertase activity of the pedicel and placento-chalazal tissue. *Plant Physiol.* 49:203–6
171. Sheehy, J. E., Cobby, J. M., Ryle, G. J. A. 1979. The growth of perennial ryegrass: a model. *Ann. Bot.* 43:335–54
172. Shininger, T. L. 1979. The control of vascular development. *Ann. Rev. Plant Physiol.* 30:313–37
173. Shiroya, M., Lister, G. R., Nelson, C. D., Krotkov, G. 1961. Translocation of ^{14}C in tobacco at different stages of development following assimilation of $^{14}CO_2$ by a single leaf. *Can. J. Bot.* 39:855–64
174. Sims, H. J. 1963. Changes in hay production and the harvest index of Australian oat varieties. *Aust. J. Exp. Agric. Anim. Husb.* 3:198–202
175. Sinclair, T. R., de Wit, C. T. 1975. Photosynthate and nitrogen requirements for seed production by various crops. *Science* 189:565–67
176. Slootmaker, L. A. J. 1974. Aims and objectives in breeding cereals. *Outlook Agric.* 8:133–40
177. Smith, J. A. C., Milburn, J. A. 1980. Phloem turgor and the regulation of sucrose loading in *Ricinus communis* L. *Planta* 148:42–48
178. Solomos, T. 1977. Cyanide-resistant respiration in higher plants. *Ann. Rev. Plant Physiol.* 28:279–97
179. Song, S. P., Walton, P. D. 1975. Inheritance of leaflet size and specific leaf weight in alfalfa. *Crop Sci.* 15:649–52
180. Sovonick, S. A., Geiger, D. R., Fellows, R. J. 1974. Evidence for active phloem loading in the minor veins of sugar beet. *Plant Physiol.* 54:886–91
181. Swanson, C. A., Hoddinott, J. 1978. Effect of light and ontogenetic stage on sink strength in bean leaves. *Plant Physiol.* 62:454–57
182. Taylor, F. H. 1956. Variation in sugar content of maple sap. *Vt. Agric. Exp. Stn. Bull.* 587:1–39
183. Thaine, R., Ovenden, S. L., Turner, J. S. 1959. Translocation of labelled assimilates in the soybean. *Aust. J. Biol. Sci.* 12:349–72
184. Thornley, J. H. M. 1976. *Mathematical Models in Plant Physiology,* 8:152–71. London: Academic. 318 pp.
185. Thorpe, M. R., Minchin, P. E. H., Dye, E. A. 1979. Oxygen effects on phloem loading. *Plant Sci. Lett.* 15:345–50
186. Thrower, S. L. 1962. Translocation of labelled assimilates in the soybean II. The pattern of translocation in intact and defoliated plants. *Aust. J. Biol. Sci.* 15:629–49
187. Treharne, K. J. 1972. Biochemical limitation to photosynthetic rates. In *Crop Processes in Controlled Environments,* ed. A. R. Rees, pp. 285–303. New York: Academic. 391 pp.
188. Tsing, T. 1963. Translocation of assimilates of leaf of mainstem in relation to the phyllotaxis of cotton plants. *Acta Biol. Exp. Sinica* 8:656–63
189. van Bel, A. J. E., van Erven, A. J. 1979. A model for proton and potassium cotransport during the uptake of glutamine and sucrose by tomato internode disks. *Planta* 145:77–82
190. van Dobben, W. H. 1962. Influence of temperature and light conditions on dry matter distribution, rate of development and yield in arable crops. *Neth. J. Agric. Sci.* 10:377–89
191. Vietor, D. M., Musgrave, R. B. 1979. Photosynthetic selection of *Zea mays* L. II. The relationship between CO_2 exchange and dry matter accumulation of 2 hybrids. *Crop Sci.* 19:70–75
192. Wardlaw, I. F. 1974. Phloem transport:

physical, chemical or impossible. *Ann. Rev. Plant Physiol.* 25:515–39

193. Wardlaw, I. F. 1976. Assimilate movement in *Lolium* and *Sorghum* leaves. I. Irradiance effects on photosynthesis, export and the distribution of assimilates. *Aust. J. Plant Physiol.* 3:377–87
194. Wardlaw, I. F. 1980. Translocation and source-sink relationships. In *The Biology of Crop Productivity.* ed. P. S. Carlson, 8:297–339. New York: Academic. 471 pp.
195. Wardlaw, I. F., Moncur, L. 1976. Source, sink and hormonal control of translocation in wheat. *Planta* 128:93–100
196. Warren-Wilson, J. 1972. Control of crop processes. In *Crop Processes in Controlled Environments,* ed. A. R. Rees, I. E. Cockshull, D. W. Hand, R. G. Hurd, pp. 7–30. New York: Academic. 391 pp.
197. Went, F. W. 1939. Some experiments on bud growth. *Am. J. Bot.* 26:109–17
198. Wilhelm, W. W., Nelson, C. J. 1978. Irradiance response of tall fescue genotypes with contrasting levels of photosynthesis and yield. *Crop Sci.* 18:405–8
199. Willenbrink, J., Doll, S. 1979. Characteristics of the sucrose uptake system of vacuoles is dated from red beet tissue. *Planta* 147:159–62
200. Williams, A. M., Williams, K. R. 1978. Regulation of movement of assimilate into ovules of *Pisum sativum* cv. Greenfeast. *Aust. J. Plant Physiol.* 5:295–300
201. Wilson, D. 1975. Variation in leaf respiration in relation to growth and photosynthesis of *Lolium. Ann. Appl. Biol.* 80:323–38
202. Wilson, D. 1978. Dark respiration. *Welsh Plant Breed. Stn. (Aberystwyth) 1977 Ann. Rep.,* pp. 157–58
203. Wilson, D. 1979. Developmental genetics. *Welsh Plant Breed. Stn. (Aberystwyth) 1978 Ann. Rep.,* p. 152
204. Wilson, D., Cooper, J. P. 1969. Diallel analysis of photosynthetic rate and related leaf characters among contrasting genotypes of *Lolium perenne* L. *Heredity* 24:633–49
205. Wilson, D., Cooper, J. P. 1970. Effect of selection for mesophyll cell size on growth and assimilation in *Lolium perenne* L. *New Phytol.* 69:233–45
206. Zelenskii, M. I., Mogileva, G., Shitova, I., Fattakhova, F. 1978. Hill reaction of chloroplasts from some species, varieties and cultivars of wheat. *Photosynthetica* 12:428–35
207. Zimmermann, M. H. 1958. Translocation of organic substances in trees III. The removal of sugars from sieve tubes in the white ash. *Plant Physiol.* 33:213–17
208. Zimmermann, M. H., Ziegler, H. 1975. List of sugars and sugar alcohols in sieve-tube exudates. See Ref. 53, pp. 480–503

Ann. Rev. Plant Physiol. 1981. 32:511–38

MODERN METHODS FOR PLANT GROWTH SUBSTANCE ANALYSIS

❖7722

Mark L. Brenner

Department of Horticulture and Landscape Architecture, University of Minnesota, St. Paul, Minnesota 55108

CONTENTS

INTRODUCTION

During the past decade, physicochemical methods for the analysis of plant growth substances[1] (PGS) have been developed and widely accepted, en-

[1]In accordance with Trewavas (164), the term plant growth substance will be used in place of plant hormone.

0066-4294/81/0601-0511$01.00

abling their use for routine analysis. Bioassays have been and will continue to be invaluable for the detection of biologically active material possessing growth substance like activity, particularly when the compound of interest is not chemically characterized. However, when one is seeking to identify and quantify a PGS that has been chemically characterized, the physicochemical procedures are superior to bioassays.

In general, physicochemical methods provide greater specificity than bioassays, permitting identification of a specific compound; greater sensitivity, permitting detection of considerably smaller amounts of PGS; less variation of results, permitting resolution of smaller differences between samples; and shorter analysis times, permitting the analysis of more samples.

During the past 3 years, there have been a number of reviews written which deal with procedures used for PGS analysis (17, 18, 68, 95, 136, 159, 176). In addition, there are articles that focus on methods of analysis of each of the major classes of PGS including abscisic acid (ABA) (114, 143, 167), auxins (109, 122, 144), cytokinins (75, 94), ethylene (126, 169), and gibberellins (56, 58, 64, 135, 141). This article will attempt to integrate the current information on PGS analysis, with special attention devoted to the establishment of criteria necessary for acceptable PGS analysis.

TERMINOLOGY AND CRITERIA FOR ANALYTICAL METHODS

Accuracy and Precision

Physiologists should be concerned with the accuracy of PGS analysis. The accuracy of an analysis is a description of how close the estimated value is to the true value (level). Correct identification of the given compound is inherent in this determination. An accurate estimate of the level of a given compound must be based on the analysis of a standard of the compound.

Precision is a measurement of the reproducibility of an analysis, generally expressed in terms of variance. It is all too easy to have a precise estimate (minimal variance) of the level of a given PGS that is inaccurate. Adequate replication of the unknowns, standards and blanks is required for precision (32).

The accuracy of a PGS analysis is dependent upon the quality (selectivity and sensitivity) of the procedures used while the precision of analysis is dependent on error control and replication.

Sensitivity

PGSs generally occur, on a fresh weight basis, at levels of 1 to 100 ng/g for IAA (8, 144), 1 to 1000 ng/g for gibberellins (58), 1 to 1000 ng/g for cytokinins (34, 112, 154, 156, 157), and 10 to 4000 ng/g for ABA (114).

This means that trace analysis procedures are required for proper identification and quantitation of these substances.

Selectivity

The presence of an impurity in a PGS sample is one of many factors which will adversely affect the accuracy of quantitation. When one attempts to analyze a substance which occurs at trace levels and near the limit of detection, the probability of simultaneously detecting interfering substances is great. Near the limit of detection, the signal-to-noise ratio is low, making accurate detection and quantitation difficult. The use of selective separation and identification procedures is required to prevent analysis of the interfering substance(s) together with the PGS. Selectivity or specificity provides a means for distinguishing the substance of interest from most other substances present in the sample. Reeve & Crozier (136) have rigorously reviewed criteria for accurate PGS analysis. They make the point that accuracy may be improved by increasing the selectivity of the analysis procedure or by increasing the amount of the PGS relative to interfering substances (136). Correspondingly, it is essential to emphasize that as smaller samples are analyzed for trace levels of PGS, greater selectivity or purification is demanded if the same degree of accuracy is sought.

Qualitative and Quantitative Analysis

Qualitative analysis demonstrates the presence of a given compound in a sample. Bioassays help establish the presence of a given class of PGS, such as cytokinins, but they cannot be used to identify a specific compound. Physicochemical detection procedures, on the other hand, when coupled with adequate purification procedures, can provide specific identification of a compound. Quantitative analysis requires accurate information regarding the recovery efficiency of the compound in question, starting at the initial sampling through to the final step of quantitative analysis. Thus, for quantitative analysis, the use of internal standards is required and will be discussed in a later section of this article.

There are numerous published reports describing the occurrence and amounts of specific PGSs in plant tissues. Unfortunately, many of the reports lack rigorous proof of the accuracy of the analyses. Reeve & Crozier (136) described two approaches that may be used to establish the accuracy of an analysis. The first approach is called "successive approximations" although successive determinations may be used synonymously. The second approach is called the "information content."

SUCCESSIVE DETERMINATIONS The successive determinations procedure calls for a series of analytical purification steps in which the putative compound is quantified after each step (136). Each step of purification and/or identification should employ a different method (minimal correlation) from the previous steps. As the determination becomes more accurate, due to greater purity, the variance from the previous determination should decrease. Thus, further purification will yield little change in the analysis results. If any of the detection steps are destructive, then aliquots from the sample must be analyzed as it is further purified. Although not clearly pointed out by Reeve & Crozier (136), it is necessary to use an internal standard [specifically the isotope dilution procedure (7)] to ensure that the observed higher purity obtained by this procedure is the result of removal of interfering substances, not the loss of the putative compound.

Once a minimum variance (i.e. 5%) is determined for three successive determinations, the identification/quantitation can be accepted with confidence of accuracy. This procedure is analagous to drying a sample to a constant weight with the distinction that each successive step either incorporates a new dimension of purification or uses a different selective method of detection.

Proving that one analytical procedure for a PGS is accurate does not mean the procedure is valid for all types of plant samples. Errors may be introduced when the sample composition is changed by sampling plants of different physiological states or different plant materials. Different samples may have different types or amounts of substances which interfere with the selectivity of an analysis.

INFORMATION CONTENT The second procedure that Reeve & Crozier (136) describe for accurate identification of a PGS is based on the amount of information gathered to describe the compound. This is essentially a statistical method, based on the premise that if enough information is derived about the compound, its identification can be accepted with a high probability. The information content procedure does not dictate any particular sequence of methods for an accurate analysis. It requires that enough steps be taken to assure the desired accuracy. Utilizing steps of high selectivity decreases the number of steps required for an analysis. Reeve & Crozier have assigned numerical values (based on binary numerics) to various separation and detection procedures to generate mathematical criteria for accuracy. An acceptable identification/quantitation sequence must accumulate a minimum number of bits in order to be accurate. This approach is intriguing but remains untested. The critical question is, what is the minimal amount of information required for accurate identification of a PGS? Reeve & Crozier's (136) model suggests that one must obtain enough information to characterize the compound in relation to all of the other compounds that might be present. In order to accomplish this, we must first obtain a reasonable estimate of the population size of the compounds that occur in our plant samples in order to establish the resolution required for analyses. Secondly, we must fully understand the resolving power of the purification techniques used to accomplish the analysis. Finally, a numerical value must be assigned to the selectivity of the detection procedure(s) for each PGS.

The amount of information required for unequivocal identification of a PGS in a plant sample therefore requires that we know how many other substances are present in our sample and the relative amount of PGS present in the sample in relation to impurities. The analysis would be more difficult if the plant sample consisted of similar compounds that would not be distinguished by the purification and identification process than if a sample consisted of dissimilar compounds that could be resolved.

In addition, there is scant information on the actual resolving power achieved by the procedures available for plant extracts. The resolving power of a purification step has been numerically described (85, 136) as the fraction capacity. For a multistep purification process, where each step is independent (having a different chemical selectivity) of the other steps, the fraction capacity of the entire process is the product of all the steps. In practice, there is often overlap in the type of selectivity utilized from one step to the next. This reduces the degree of resolution from the theoretically anticipated separation (85). Therefore, a correction factor to account for correlative selectivities must be developed. Further, a data base must be established to compare the relative selectivities of the available detection procedures for each PGS. Reeve & Crozier's (136) specific assignment of bits or quality points to the various detectors seems somewhat arbitrary. For example, a gas chromatograph electron capture detector (GLC-EC) is equated to a flame ionization detector (GLC-FID; four bits each). Certainly this is not the case for GLC detection of the methyl ester of ABA (ABA-ME) (145).

Once the above information is acquired, then a series of experiments should be initiated in which the PGS analysis accuracy is established by successive determinations. At the same time, the information content of the overall analysis procedures should be calculated. If the information content model is correct, then the forecasted minimum number of bits should be accumulated at the point at which accuracy is established by successive determinations. Until such rigorous proof of the information content model is established, it is the author's recommendation that we rely on successive determinations. It should be emphasized that the concept of the information content model is interesting, but many additional facts are necessary before its use can be accepted.

APPROACHES USED FOR PGS ANALYSIS

Hundreds of articles have been written during the past decade describing methods for PGS analysis. In too many of these articles, little collaborative proof has been provided to justify selection of one procedure over another. The majority of these methods focus on a single class of PGS. As it has become apparent that many physiological processes are a result of the interplay of endogenous hormones, greater emphasis should be placed on the analysis of more than one class of PGS from a single plant sample.

Methods of Sampling

Dennis (37) has succinctly reviewed the problem of selecting a proper sampling technique in order to realize meaningful results from PGS analysis. The sampling technique must provide a sample with adequate amounts of PGS, yet it must also provide a sample that will help answer the question being examined. Most analytical procedures used in the past have required 0.1 to 50 μg of a given PGS to be present in a sample for adequate identification. Due to the trace levels of PGS, the amount of tissue necessary for analysis has been in the 0.2 to 4 kg range. An unfortunate trap has been the sampling of an entire shoot when just buds or even smaller portions of the shoot should have been sampled to obtain meaningful physiological data. The use of correct sampling procedures can represent a formidable task. For example, 2000 lateral buds of *Phaseolus* were required per sample for identification of IAA (69). Minimal data is available regarding endogenous PGS at the subcellular level; yet on a physiological basis, this is where they certainly are synthesized, metabolized, and where they function. ABA has been analyzed at the subcellular level and has been found to occur in the chloroplast (133), which has been described as a site of synthesis (98).

Extraction Procedures

Selection of an appropriate solvent for extraction of PGS from tissue is one of the more confusing aspects of PGS analysis. Although 80% aqueous methanol is most commonly selected, there are numerous reports of other solvent systems. Improvement of the existing practices has been limited for several reasons. First, absolute extraction efficiencies are difficult to obtain when internal standards are added to the extraction solvent rather than to the tissue before extraction. Secondly, extraction procedures have often been examined only for the free form of the respective PGS. It is thus incumbent upon physiologists to establish that improved recovery of a PGS from tissue is the result of the better extraction procedure and not the result of a net change in the homeostatic endogenous PGS balance. Extraction procedures must be examined to prevent alteration of the PGSs from their original state during extraction. Alteration may occur from hydrolysis, enzyme action, or chemical derivatization. In the case of IAA, it is now well established that the majority of IAA occurs as either sugar or amino acid conjugates (8). It is reasonable to suspect that some of the "improved" extraction procedures which yielded greater recovery of free IAA might have been the result of enhanced hydrolysis of the conjugated IAA.

Enzymatic activity in the extraction medium may interfere with PGS recovery. The use of an organic extracting solvent or rapid heating does not assure inactivation of enzymes such as phosphatase (75, 94). For recovery of cytokinin nucleotides, a special solvent mix is recommended to prevent phosphate hydrolysis (12, 13). Care must also be taken to minimize chemical derivatization of the PGS during extraction. When esters of ABA are extracted in nonacidic conditions in methanol, transesterification may occur yielding ABA-ME (115). Extraction with acetone or with acidified methanol in the presence of the antioxidant 2,6-di-*t*-butyl-4-methyl phenol (BHT) blocks transesterification of ABA (115).

The use of extraction solutions containing antioxidants, organic solvents, aqueous solvents, or Triton X-100 have been shown to improve recovery of PGS by several scientists. Mann & Jaworski (104) found that the addition of the antioxidant sodium diethyldithiocarbamate improved the recovery of IAA. Acetone has been used extensively for the recovery of the sugar esters of IAA (7, 8, 42–44), although methanol has been found to be equivalent to acetone, ethanol, diethyl ether, chloroform, or ethyl acetate for recovery of free IAA (110). For recovery of 4-Cl-IAA, l-butanol has been used (46). Although not documented as being superior to 80% methanol, organic solvent mixes have been used to extract ABA (70, 78, 149). However, Walton et al (168) did find extraction with water-saturated ethyl acetate to simplify ABA purification compared to extraction with 80% aqueous methanol. Water or aqueous buffers have also been used for extraction of ABA (121, 138, 140), auxins (72, 73), cytokinins (36), and gibberellins (99). None of these reports examined the impact of the extracting solvent on release of free PGS from conjugates.

Triton X-100 incorporated in the extraction solvent has been reported to yield a 1000-fold increase in gibberellin recovery from chloroplasts (20). It was suggested that the Triton X-100 reduced gibberellin binding to plastid membranes during the extraction, but this effect of Triton X-100 could not be confirmed by other workers (132) using a different tissue.

Careful selection of an extraction solvent is not only important for maximizing the recovery of PGS from plant tissue but also for minimizing the contribution of artifacts. It is now well established that solvents used for extraction and subsequent sample purification can contain numerous contaminants (105, 107). This problem can usually be averted by using high purity solvents and analyzing solvent "blanks" processed in the same manner as plant samples.

Sample Fractionation

Quantitation of PGS from plant extracts requires a high degree of sample purification when physicochemical analysis procedures are applied. The optimum sequence of steps for sample purification is dependent upon the number and diversity of PGS to be recovered. The best approach for the analysis of a single PGS is to use several high resolution steps (high fraction or peak capacity) in sequence. Each step should be based on a different type of chemical selectivity (i.e. size separation, partition, ion exchange, etc). When analyzing multiple PGSs, the initial steps should ideally yield groups of compounds, and these steps should then be followed by several high resolution steps (85). Initial high resolution steps would generate an inordinate number of samples, each of which must be processed a number of additional times.

PARTITIONING Often the initial purification step after extraction is the partitioning of the plant sample between immiscible solvents. The solvents chosen should yield optimum partition coefficients at the conditions selected (pH, etc) for recovery of acidic, neutral, or basic compounds. Partition coefficients for some of the more commonly used solvents have been reported for ABA (26, 168), gibberellins (GA_1-GA_{27}) (40), IAA (6), and cytokinins (66, 67, 75, 93). Solute interactions often confound the solvent partitioning steps and reduce PGS recovery. For example, phospholipids or other materials may form an emulsion at the solvent system interface (26, 36, 75), making it impossible to resolve the two phases.

It should be noted that water is slightly soluble in some organic solvents such as 1-butanol and ethyl acetate. After partitioning, care must be taken to remove traces of buffer before these organic solvents are taken to dryness. Otherwise, as the buffer is concentrated, the sample will be exposed to extremes in pH. The buffer can be neutralized before drying (M. L. Brenner, unpublished data) or removed by extracting the organic solvent with water (106).

CONVENTIONAL COLUMN CHROMATOGRAPHY Many PGS analysis procedures have incorporated at least one form of conventional column chromatography. Removal of phenolics and pigments are two of the main reasons for use of these columns. PGSs from standards and tissue extracts have been eluted from columns of insoluble polyvinylpyrrolidone (PVP) with aqueous buffers (10, 11, 57, 84, 116) to remove phenolics and pigments. Recovery of PGS from PVP columns was improved when the columns were eluted with methanol rather than aqueous buffer (123). PVP has also been used in a batch method in which the material was slurried with the plant extract and then removed by filtration (1, 4, 57, 149).

Another column packing, Sephadex LH-20, has often been used for cleanup for cytokinins (5, 23, 89, 97, 162, 165, 170), gibberellins (102), and 4-Cl-IAA (46). The mechanism of separation with this column packing, when 35% aqueous ethanol is used as the mobile phase, is reverse phase partitioning [(23, 75) polar compounds elute before nonpolar compounds].

McDougall & Hillman (110) found DEAE-cellulose superior to PVP for removal of plant pigments and recovery of IAA from plant tissue extracts. The DEAE-cellulose column also had greater sample capacity and yielded faster separations. They noted that the LH-20 columns, with 80% aqueous methanol or ethanol as the mobile phase, would not separate plant pigments from IAA.

Ion exchange columns have been incorporated into many PGS cleanup procedures. Dowex-50 has been used for recovery of cytokinins (116, 165, 166, 171) and sugar esters of IAA (42). It has been shown that 70% aqueous ethanol is required to elute the cytokinins from Dowex-50 columns, and even then, only 70% recovery is obtained (166). This indicates that the retention mechanism is partitioning rather than ion exchange. A milder anion exchange material like Duolite permits nearly total recovery of cytokinins (166). Cellulose phosphate also has been used for this purpose (112), and it has been reported to be better than Dowex-50 for cytokinin purification (36).

Adsorption chromatography using charcoal (1), Celite (89), or silica gel (41) has been demonstrated to help purify ABA, zeatin, and gibberellins respectively.

Size exclusion chromatography (SEC), employing columns packed with BioBeads, may be used to separate many of the PGSs from plant extracts. The PGSs are separated on the basis of their molecular size. It has been reported that high recovery efficiency is obtainable using SEC (134).

THIN-LAYER CHROMATOGRAPHY Until quite recently, most PGS analysis procedures had incorporated at least one step involving thin-layer chromatography (TLC). Due to the greater sample capacity and resolving power of high-performance liquid chromatography [HPLC (85)], it is the author's opinion that HPLC will totally replace TLC procedures except for economic constraints. A further disadvantage of TLC is that substantial losses of polar cytokinins (75) and gibberellins (58) occur when silica plates are used. Silica-coated TLC plates have also been reported to decrease recovery of IAA when compared to cellulose-coated plates (142).

HIGH-PERFORMANCE LIQUID CHROMATOGRAPHY High-performance liquid chromatography, when first reported for PGS analysis (22, 129, 130), offered greater resolving power and much faster separation than previous conventional column or TLC procedures. Technology has rapidly advanced to a point where column efficiencies in excess of 50,000 plates per meter can be expected (151). HPLC not only offers high resolving power but also has ample sample capacity for PGS recovery from crude plant extracts (18). Speed is another benefit of HPLC since most separations are accomplished in less than one hour.

A wide range of HPLC packing materials is available, representing all the modes of separation that are known for column chromatography (151). Reverse phase HPLC has proved to be the most adaptable and reliable method of separation for PGS. The most popular column packing material has been the octadecyl silane (ODS or C_{18}) type which consists of an octadecyl carbon chain bonded to silica microparticles. There are numerous papers describing conditions for separation of PGS standards on HPLC C_{18} columns (3, 18, 21, 39, 62, 76, 119, 120). It should be noted that there are extensive differences in selectivity among the commercially available C_{18} packings (76, 151).

Preparative C_{18} columns (6 to 10 mm ID X 0.25 to 2 m) have been used to recover fractions containing ABA (9, 26, 39, 147, 148, 174), auxin (39, 160, 174, 177), cytokinins (23, 120, 154, 155), and gibberellins (81, 113, 119, 178) from plant extracts. In a few cases (26, 147, 148) prior cleanup was eliminated. Most scientists have used relatively large particles (35 to 70 μm diameter) for these semipreparative HPLC purifications. It is the author's experience (18) that microparticle (7 or 10 μm diameter) preparative columns (10 mm ID X 25 cm) offer greater sample capacity than the larger particle-packed preparative columns. It is possible to apply a crude extract from as much as 5 to 10 grams fresh weight of tissue without overloading these microparticle columns. More significantly, a 20- to 100-fold increase in column efficiency (50,000 plates per meter) is obtained when spherical, uniformly sized particles are used (151).

There are a limited number of reports describing the use of other bonded phase packings. Some of the other packings used include: C_8 (74) and C_2 (76) for cytokinin standards; phenyl for cytokinins from plant extracts (154, 155) and ABA metabolites (9); amine for ABA (26) and ABA metabolites (9); nitrile for derivatized gibberellin standards (119); and a fatty acid analysis column for cytokinin standards (3).

Steric exclusion chromatography (SEC) has been used for the separation of PGS from crude plant extracts (31). In this case, most of the PGS standards eluted from the column on the basis of molecular size. The column did exhibit some adsorption characteristics even though tetrahydrofuran with 0.1 *M* acetic acid was used as the solvent. The separation occurred in 10 min with an efficiency of 4800 plates per meter and a peak capacity of 10 (31). Due to its high sample capacity (>100 mg), it has been suggested that SEC is more suitable than reverse phase HPLC for initial sample purification (31). It is the author's experience that the preparative microparticle C_{18} column described above has an equally high sample capacity. The peak capacity is far greater for reverse phase columns (>50), but a group separation as accomplished by SEC may be more practical for the first chromatographic separation step if more than one PGS is to be analyzed. A comparison of SEC and reverse phase separation techniques as initial fractionation steps has not appeared in the literature.

Ion exchange packing, one of the first types of HPLC packing available, has been used for separation of IAA (22, 38, 160), ABA (22, 38, 161), and cytokinins (22, 25, 129, 130). Unfortunately, ion exchange packing materials are not as stable as other types of HPLC packing (22).

Adsorption chromatography, using HPLC grade silica, has mainly been used for the analysis of ABA in highly purified plant samples (22, 26, 168, 174) and for cytokinin standards (76). Some separations of gibberellins have been achieved with silica packing that has been modified by silylation (178) or argentation [treatment of the silica with $AgNO_3$(65)]. Silica columns have also been used for semipreparative separations of benzyl methyl gibberellin esters (137).

GAS CHROMATOGRAPHY Providing that a volatile PGS derivative can be formed, gas chromatography (GC) is a powerful and fast separation technique (85). Due to its limited sample capacity, GC is often the last step in a multistep separation sequence. Until recently, packed column GC has been used for all PGS separations. Wall-coated open tube (WCOT) capillary columns are beginning to replace conventionally packed columns now that fused silica capillary columns are available (80). These new columns typically have efficiencies of 10,000 to 100,000 theoretical plates, depending upon their lengths. This represents a 10- to 100-fold increase in net efficiency over packed columns. In addition to this greater efficiency, faster separations are obtained with WCOT columns.

WCOT columns have been used for ABA and IAA analysis (4, 73, 140; E. Knegt, personal communication; B. Bray, M. Hein, and M. L. Brenner, unpublished data). Peaks that cochromatographed with ABA-ME on packed GC columns can be well resolved by using a more efficient WCOT column (B. Bray and M. L. Brenner, unpublished data).

Gas chromatography used in the separation of most PGS is called gas-liquid chromatography (GLC) since the support material (in the case of packed columns) or the capillary walls (in the case of WCOT columns) is coated with a nonvolatile liquid phase. The thermodynamics of the GLC separation process dictates that the liquid phase plays a less significant role than the bonded phases do in HPLC (85). Most often nonpolar phases such as SE-30, OV-1, OV-101, and intermediate polarities such as OV-17 are used for PGS separations (14, 34, 56, 77, 92, 109, 146).

Unfortunately, few reports on HPLC or GC separations of PGSs provide information on column performance parameters. In order to evaluate or duplicate a PGS separation, it is essential that reports contain information on the column performance parameters (80, 151), including the effective number of theoretical plates (N_{eff}), a measure of sample retention (K'), and the peak symmetry factor. Since all columns degrade with use, the range of performance during the column's use should also be reported.

Methods of PGS Identification and Quantification

BIOASSAYS Bioassays have played an essential role in the discovery and subsequent identification of PGSs. However, due to their limited sensitivity and selectivity, they cannot be used for unequivocal PGS analysis. Most bioassays have large response curves, often covering several orders of magnitude. Unfortunately, the variance associated with this response range is proportional to the response. This, coupled with the fact that bioassays have a log-linear concentration response curve, makes it impossible to determine small differences in amounts of PGSs between samples. Reeve & Crozier (136) have presented a succinct tabulation of the major PGS bioassays.

PHYSICOCHEMICAL DETECTION PROCEDURES As described above in regard to qualitative analysis, selective procedures are necessary when studying PGSs since they occur at trace levels. The selectivity necessary for PGS analysis comes from the combination of selective separation procedures and detection methods. Increasing the selectivity for part of an analytical procedure may reduce the selectivity requirements of the other parts. Thus the use of highly selective detectors should be sought; only those detectors that are sensitive to less than 100 ng of a PGS will be discussed.

High-performance liquid chromatographic detectors The UV detector is one of the most commonly used detectors for HPLC. Fixed wavelength detectors, usually operated at 254 or 280 nm, should be considered only partially selective for PGS analytical methods as they will detect all molecules which absorb UV light at these wavelengths. The detection limit for ABA at 254 nm has been reported to be in the range of 0.5 to 5.0 ng per injection (26, 161, 174). As little as 4 ng of cytokinins may be detected at 254 nm (84), and the reported limit for IAA at 280 nm is 5 to 20 ng (160,

177). Since the detector response is concentration-dependent (151), delivering a sample to the detector as a sharper peak (i.e. higher concentration) will extend these detection limits. The use of more efficient columns will thus extend the detection limits. Gibberellins absorb UV radiation only below 220 nm, but they have been derivatized to absorb at 254 nm (119, 135). The reported detection limit for the derivatized gibberellins is 5 ng (119). A problem associated with derivatization is that other compounds present in the sample may also react with the derivatizing agent and can interfere with the analysis.

Fluorescence detection of PGSs, especially indoles, offers greater selectivity and sensitivity than UV detection (30, 160). The natural fluorescence of IAA and other indoles allows detection without derivatization. By present HPLC-fluorescence detection techniques, IAA has a detection limit of 10 to 20 pg (30). It is also possible to prepare fluorescent derivatives of other PGSs to improve detection limits of these compounds. The use of methoxycoumaryl esters of a number of the gibberellin standards facilitated their detection to a lower limit of 5 pg (A. Crozier, J. B. Zaerr, and R. O. Morris, personal communication).

Electrochemical detectors operated in the oxidative mode are also both selective and sensitive to IAA (160) and other indole auxins, provided that the amino-nitrogen is not blocked (M. L. Brenner, unpublished results). When the HPLC column effluent is sequentially passed through a fluorescence detector and an electrochemical detector, two independent measurements can be obtained. The ratio of the two detector signals for unknown samples should match that of standards if the IAA samples are free of impurities (M. Hein and M.L. Brenner, unpublished results). A graphic representation of the procedure has been published (176, Figure 12), although it was incorrectly credited to Sweetser & Swartzfager. Zeatin but not zeatin riboside may be detected by an electrochemical detector in the oxidative mode (M. G. Carnes, personal communication).

Radioactivity monitors directly attached to HPLC systems (9, 137, 147) permit detection of radiolabeled compounds as they are eluted from an HPLC column. The first radioactivity detectors for HPLC systems required the addition of liquid scintillant to the column eluent (137). Quartz encapsulated scintillant now provides a nondestructive means of monitoring radiolabeled compounds. This system has been used successfully to examine the metabolism of ABA (9, 147).

Gas chromatographic detectors and mass spectrometry Flame-ionization detectors (FID) will detect any compound that ionizes upon combustion in a hydrogen flame. This detector is essentially nonselective and will detect as little as 1.0 ng of a compound. Although there are reports of identification of PGS on the basis of GLC with FID (61, 92, 175), without further proof of identification, the results must be questioned.

The electron capture (EC) detector responds only to those compounds that are electron negative (halogenated and certain aromatic hydrocarbon compounds). Not only is this detector highly selective, but it is also quite sensitive and has a large dynamic range. With pulsed-width amplifiers, the dynamic range of the EC detector is greater than 10^4 for ABA (26). ABA-ME is naturally electron negative (145), which facilitates selective detection within plant samples. GLC-EC is one of the most common and accepted means of ABA analysis (114, 167). With a packed column, as little as 5 to 10 pg per injection of ABA can be detected (145). Using WCOT columns, the detection limit per injection can be extended to at least 0.3 pg per injection (M. Hein, B. Bray, and M. L. Brenner, unpublished data). Phaseic acid and dihydrophaseic acid also can be detected by GLC-EC with slightly less sensitivity than for ABA (182). Even though GLC-EC analysis of ABA offers substantial selectivity, adequate purification is required for accurate quantitation. A single step purification, such as using microsilica cartridges (Sep-Pak) (78) before GLC-EC analysis, must be questioned. It is necessary to provide rigorous proof that the peak used for quantitation is free of all impurities. The use of successive determinations, previously described, is appropriate.

Other PGSs have been derivatized with halogen-containing functional groups so that they can be detected with GLC-EC. The main benefit of such derivatives is a 10- to 1000-fold increase in detection limits (10 pg) for the derivatized compounds. The following are some of the halogenated derivatives of PGS reported: trichloroethyl IAA ester (15), heptafluorobutyryl IAA methyl ester (72, 139, 146), indole-3-acetylpentafluorobenzyl ester (47), trifluoroacetyl esters, of GA_3 methyl ester (90, 146), and heptafluorobutyryl gibberellin methyl esters (146). Derivatizing these PGSs with halogenated material increases detection sensitivity but may not improve selectivity. Any other compounds present in the sample with functional groups reactive to the derivatizing agent will also become halogenated.

An alkali flame ionization detector (NPD) can be set to selectively detect nitrogen or phosphorus-containing molecules. This detector is also more sensitive than FID units and has been found to detect IAA-methyl ester (108, 158) and permethylated cytokinins (183). Limits of detection are reported to be in the 5 pg range for both types of compounds, although the data provided in the reports (108, 183) show sensitivity only to 1 ng per injection.

The mass spectrometer (MS) is the most selective detector that can be coupled to a GC. The coupling of the GC to the MS system combines the separation power of the GC with the selective detection qualities of the MS. Compounds are ionized within the mass spectrometer and then the fragmented ions are separated and analyzed on the basis of their mass to charge (m/z) ratio. The presence of specific fragments and their relative intensities provide a mass spectrum which is unique to a compound. A mass spectrum can be matched to reference spectra of known compounds for sample identification.

The mass spectrometer is a powerful analytical tool by itself, without being coupled to a GC. This direct probe procedure requires highly purified samples. Although samples must be volatile for GC-MS, direct probe MS may be used for a broader range of compounds. All PGS compounds except ethylene must be derivatized to make them volatile for GC-MS. The fragmentation pattern obtained from MS is characteristic of the derivative of the initial compound and the method of ionization. Until recently, all PGS samples have been ionized by bombarding them with high-energy electrons (electron impact, EI). Examples of some of the publications on MS and GC-MS analysis of PGS compounds using EI ionization include the following: derivatives of ABA and related compounds (1, 59, 71, 88, 101, 163, 182); IAA, derivatives of IAA, and IAA conjugates (43, 44, 46, 48, 49, 60, 103, 110); cytokinins and derivatives of cytokinins (34, 35, 82, 89, 100, 118, 154–157, 171, 181); and gibberellins, derivatives of gibberellins, and gibberellin conjugates (14, 50, 51, 64, 65, 106, 113, 119, 124, 154, 180). Chemical ionization and field desorption cause less compound fragmentation than EI ionization. These ionization techniques generally yield a more intense molecular ion peak and are more useful than EI ionization for the determination of molecular weights. Chemical ionization has been shown to be superior to EI ionization for studying several cytokinins (154, 156). Field desorption ionization has also been applied to cytokinin analysis (29).

When coupled to a GC, the mass spectrometer can be used to monitor a few unique major ion fragments (m/z) that are characteristic of the compound being studied. Since the absolute intensity of an ion is proportional to the amount of a compound applied to the mass spectrometer, selective ion current monitoring (SICM) can be used to quantify a compound. SICM provides a detection limit of 1 to 10 pg for most compounds. In contrast, a complete MS fragmentation scan of a compound requires at least 100 to 200 ng for analysis.

The GC-MS-SICM technique has been used for the analysis of many PGSs including ABA (3, 70, 96, 133, 138, 140), IAA and other auxins (19, 24, 63, 69, 110, 125, 139), cytokinins (34, 35, 154, 156, 157, 162, 170), and gibberellins (20, 50, 51, 79, 91, 179).

Although sensitivity is increased with SICM, selectivity is reduced. The validity of the compound identification depends upon the degree of prior sample purification and the uniqueness of the ion(s) monitored. For example, 190 m/z is the major fragment from ABA after electron impact ionization. Comparing detection of ABA by SICM and EC, Andersson et al (4) found 10% more ABA using SICM. They suggested that results from SICM were overestimated as a result of interfering substances. This example should caution researchers that it is essential to use multiple independent determinations for accurate analysis of PGSs.

Considerably more information is obtained when several major fragments are followed by SICM. The ratios of the intensities of several fragments monitored for the PGS compound should match those of the reference standards if the compound is free of impurities.

The use of GC-MS is especially important in the study of gibberellins. These compounds have few unique attributes that lend themselves to selective detection by the available physicochemical detectors. Since standards are not readily available for most of the gibberellins, the only accepted means of identification is the use of GC-MS and comparison of the observed mass spectra to reference spectria. Quantitative analysis requires access to gibberellin standards for calibration of the MS. Hedden (64) recently reviewed some of the more important aspects of GC-MS as applied to gibberellin analysis.

Fluorimetric assay for IAA A unique assay for IAA is based on its reaction with acetic anhydride to form indolo-α-pyrone (45, 83, 87, 122, 153). This product fluoresces at 490 nm when excited at 440 nm, allowing selective detection in crude plant extracts. The procedure was first applied to plant samples by Stoessl & Venis (153) and has been modified by a number of other scientists. By partially purifying the plant extract with solvent partitioning (87), the assay can be used to detect as little as 10 ng of IAA in a plant sample. Since indolo-α-pyrone is not stable, the assay product must be measured within a short time (1 min) after the reaction is stopped (45, 83, 87, 153). Eliasson et al exploited the short half-life (15 min) of the indolo-α-pyrone by remeasuring the fluorescence of each sample 4 h after the reaction. This later fluorescence measurement served as a blank to reduce errors from fluorescence contributed by impurities.

Kamisaka & Larsen (83) observed that the method of stopping the reaction can affect the stability of the indolo-α-pyrone. Instead of stopping the reaction with water (45, 87, 122, 153), they used an acetic acid solution. The indolo-α-pyrone is more stable in the acetic acid and the solution is less turbid, allowing the limit of detection for IAA to be extended to 3 ng. However, a separate blank must be used with this modified procedure.

The indolo-α-pyrone assay is specific for indoles with an acetic acid side chain at the 3 position (153). Thus, if the naturally occurring 5-hydroxy-indoleacetic acid (5-OH-IAA) or 4-chloro-indoleacetic acid (4-Cl-IAA) is present in a sample, an overestimate for IAA will be made (16). The only way to remove the possible interference of 5-OH-IAA and 4-Cl-IAA in the assay is prior chromatographic purification. A single HPLC column separation step will separate IAA from these two compounds (V. Sjut, personal communication).

IMMUNOLOGICAL ASSAYS Immunological assays have been widely applied for animal hormone assays. Due to their high selectivity, minimal sample purification is required and sensitivity is in the same range as the most sensitive physicochemical procedures. The procedure requires the formation of an antibody for the compound of interest. As with animal hormones, PGS molecules are too small to be recognized for formation of specific antibodies. It is thus necessary to covalently bind the PGS to a larger molecule, usually a protein. The PGS-protein molecule is then intravenously fed to an animal in order to stimulate antibody formation. Thus far, immunological studies have used only bovine [BSA (33, 52–55, 128, 168, 173)] or human (172) serum albumin PGS conjugates. In all cases to date, rabbits have been the source of antiserum for PGS analysis.

The position on the PGS molecule used for attachment to the protein is critical to the selectivity of the antibody formed. Fuchs et al (52, 53) prepared an IAA-BSA conjugate through the carboxyl group of IAA. The antibody formed had low sensitivity (required more than 1.0 μg of IAA) and would react to a number of IAA-related compounds (cross-reactivity). In contrast, Pengelly & Meins (128) bound IAA through the indole nitrogen to BSA. The resultant antibody was shown to cross-react only with naphthaleneacetic acid, and the sensitivity was much greater (0.2 ng IAA) than previously reported (128). Antibodies to ABA conjugated with BSA or human serum albumin (carbonyl linked) cross-react equally well with ABA-glucose ester tetra acetate (168, 172) as with ABA, and react to ABA-ME with a higher affinity (33, 168, 172).

In most immunological assays, quantitation of PGSs has been accomplished by competitive binding of a radiolabeled PGS (radioimmunological assay, RIA) with the PGS in plant samples. The amount of radiolabel bound to antibodies is inversely proportional to the amount of PGS in the sample. Recently, Daie & Wyse (33) used ABA bound to the enzyme alkaline phosphatase, rather than a radiolabeled ABA, for an immunological assay. In this assay, ABA from the plant sample and an aliquot of alkaline phosphatase-ABA are incubated with a limited amount of immobilized antibody. After a suitable incubation time, free ABA and alkaline phosphatase-ABA not complexed with the antibody are removed. Then the level of ABA in the plant sample can be determined by adding suitable phosphatase substrate. The amount of alkaline phosphatase activity detected is inversely proportional to the amount of ABA in the plant sample.

The response curves of the immunoassays are linear in the range of about one order of magnitude [i.e. 0.1 to 2.5 ng for ABA (33, 168, 172)]. The detection limits reported for the more sensitive immunological assays for PGS are as follows: ABA, 0.1 ng (33, 168, 172); IAA, 0.2 ng (128); isopentenyladenosine, 5 ng (86); and *t*-zeatin riboside, 0.02 ng (173). The coefficient of variance for the immunological assays is a percentage of the amount of PGS detected, as compared to an absolute level of variance for physicochemical measurements.

With the exception of the report by Walton et al (168), the RIA assays reported for PGS in plant extracts have relied on the theoretical high degree of the assays' selectivity, without providing collaborative proof of the amounts of PGS present. Pengelly et al (127) recently observed that the RIA assay for IAA released by alkaline hydrolysis is accurate only if samples are first purified to remove interfering material. In order to verify immunological assays, at least two different methods of analysis should be compared. When properly executed, immunological assays offer a rapid, sensitive method for the analysis of PGS compounds. Better immunological assays will result by preparing PGS-protein complexes between nonfunctional sites of the PGS molecule and the carrier protein.

Internal Standards—A Requirement for Quantitative Analysis

Essentially all PGS analysis procedures require multiple purification procedures before the compound(s) of interest can be quantified. As the number of steps increases, the probability of losing some of the compound(s) of interest becomes greater. Losses of PGS may result from incomplete partitioning between solvents, entrapment in other components [phospholipids or proteins (26, 36)], chemical reaction, or adsorption onto glassware or onto solid chromatographic supports. The net recovery of an internal standard added at the initial point of tissue extraction is now the accepted procedure for estimating recovery efficiency. Absolute losses of PGS due to oxidation, incomplete partitioning or entrapment will likely be proportional to the amount of PGS in a sample over a wide range of concentrations. However, absolute losses due to adsorption are proportional to the amount of PGS in a sample only at low concentrations. Once all the adsorption sites are filled, no additional PGS will be lost to adsorption. Internal standards must be added at low concentrations, similar to the levels of PGSs in the sample, so that losses due to adsorption may be accurately accounted for.

Physiologists have often added internal standards at the microgram level to estimate the recovery of substances occurring at the nanogram level within their samples. They most likely masked all of the adsorption sites with the excess standard and recorded an overestimate of sample recovery. An example of how adsorption can distort PGS analysis can be found by examining the report by Horgan & Neill (77) on the loss of ABA-ME on a GC column. ABA could be quantified when injected in microgram amounts, but it was not detected when injected in nanogram amounts. When the column was treated to deactivate adsorption sites, the nanogram injections of ABA-ME were detected.

Choice of an internal standard is equally critical for an accurate estimate of sample recovery. PGS analogs such as *t*-ABA for ABA (92), indole-3-butyric acid for IAA (125), 5-methyl-IAA for IAA (73, 139), and ^{14}C-kinetin for zeatin (36) have been used in the past. Since such analogs will fractionate differently in a high-resolution purification procedure, they are not the best choices for an internal standard to estimate PGS sample recovery. Bandurski & Schultze (7) first suggested adding ^{14}C-IAA to estimate recovery of IAA using an isotope dilution procedure. This procedure requires the addition of an isotope of the compound of interest. By knowing the specific activity of the internal standard at the beginning of the analysis and calculating its dilution at the end of the analysis, recovery efficiency can be accurately determined. A number of other scientists have adopted this method for determining recovery efficiency using ^{14}C or ^{3}H-labeled ABA, IAA, GA_3, and zeatin (26, 69, 79, 83, 87, 96, 104, 110, 111, 117, 150, 155, 160, 161).

Deuterium-labeled PGSs can be used as internal standards when GC-MS-SICM is selected for PGS analysis. The peak intensity of the deuterium-labeled fragments is measured along with the naturally occurring fragments. Monodeuterated IAA (2, 24) and GA (152) have been used. However, polydeuterated standards are preferred since ions with masses five or six above the natural PGS fragments will be distinct from the naturally occurring heavy ions which are one and two mass units above the normal PGS fragments. Polydeuterated ABA (138, 140), IAA (103), and several naturally occurring cytokinins (154, 156) have been used as internal standards. An advantage of deuterated internal standards is that recovery efficiency is determined at the quantitation step rather than in a subsequent radiochemical determination.

Recovery efficiencies for different PGSs are independent. This is certainly the case for IAA and ABA (18, 96) where the loss and coefficient of variation for IAA is much greater. Substantial loss of IAA may be attributed to oxidation (104); however, the IAA conjugates are not affected by peroxidase-catalyzed oxidation (27) and are probably recovered at higher efficiencies than IAA. This means separate internal standards are required for the IAA conjugates. It appears that for multiple PGS analyses from a single sample, multiple internal standards are required. If radiolabeled internal standards are used, then care must be taken to separate the compounds completely before determining recovery with a scintillation spectrophotometer.

As pointed out by J. D. Cohen and A. Schulze (submitted to *Anal. Biochem.*), the purity of an internal standard must be verified before it is added to a plant sample. When a radiolabeled internal standard is used, its specific activity should also be verified by the same procedures that will be used for the PGS analysis.

The above discussion has emphasized the necessity of using internal standards to determine recovery efficiency of PGS while preparing samples for analysis. When a GC-MS is available, accurate recovery efficiency can be determined at the time of sample quantitation with polydeuterated internal standards. However, when other chromatographic methods of quantitation are used (GC or HPLC), then a second internal standard might be considered to account for errors due to inaccurate estimation of sample volume (due to evaporation), errors in injection volume, variable or incomplete sample derivatization, variation in chromatographic performance (temperature or mobile phase flow rate), and nonlinear detector response. The secondary internal standard is added to the sample before chromatographic analysis. Several of the secondary internal standards used include ^{14}C-IAA for IAA (28) and ABA ethyl ester for ABA (131), both for GC, and androsta-1,4-dien-3,17 dione for ABA (21) for HPLC. J. D. Cohen and A. Schulze (manuscript submitted to *Anal. Biochem.*) recommend the use of a second radiolabeled internal standard to account for chromatographic sampling errors.

CONCLUDING REMARKS

At the present time, it is inappropriate to identify one particular method as the best way to analyze PGSs. Since PGS analysis involves determinations at the trace level, the procedures used should have high resolving capability to remove all possible interfering compounds. Detection and quantitation should be a highly selective method capable of distinguishing the compound of interest from other compounds.

Regardless of the procedure selected for PGS analysis, it is absolutely necessary to prove that the method is accurate for the specific plant material being analyzed. This is best done by determining the quantity of the PGS with several independent analytical methods on aliquots from the same sample. Internal standards are required to account for variations in recovery efficiency for each sample analyzed. The internal standard should resemble the PGS as closely as possible in structure and quantity.

Physicochemical and immunological procedures offer potential for improving analytical methods for PGS. However, if the analyst cannot substantiate the validity of a method, then the method should not be used. Bioassays, with all of their limitations, are probably still better than unsubstantiated physicochemical methods. At least the bioassay will indicate biological activity "like-the-substance" being studied, while an incomplete physicochemical analysis might only detect unknown impurities.

ACKNOWLEDGMENTS

I wish to thank Beth Bray, Jim Elvecrog, and Mich Hein for their thoughtful discussions and review of the manuscript. The assistance of Marilyn Clement and Angie Klidzejs in the preparation of the manuscript was invaluable. Research support in the author's laboratory came from the Minnesota Agricultural Experiment Station, NSF PCM-78-13255; USDA-SEA 5901-0410-0183-0; and the Minnesota Soybean Research and Promotion Council.

Literature Cited

1. Adesomoju, A. A., Okogun, J. I., Ekong, D. E. U. 1980. GC-MS identification of abscisic acid and abscisic acid metabolites in seeds of *Vigna unguiculata*. *Phytochemistry* 19:223–25
2. Allen, J. R. F., Baker, D. A. 1980. Free tryptophan and indole-3-acetic acid levels in the leaves and vascular pathways of *Ricinus communis* L. *Planta* 148: 69–74
3. Andersen, R. A., Kemp, T. R. 1979. Reversed-phase high-performance liquid chromatography of several plant cell division factors (cytokinins) and their *cis* and *trans* isomers. *J. Chromatogr.* 172:509–12
4. Andersson, B., Haggstrom, N., Andersson, K. 1978. Identification of abscisic-acid in shoots of *Picea abies* and *Pinus sylvestris* by combined gas chromatography mass spectrometry. A versatile method for cleanup and quantification. *J. Chromatogr.* 157:303–10
5. Armstrong, D. J., Burrows, W. J., Evans, P. K., Skoog, F. 1969. Isolation of cytokinins from tRNA. *Biochem. Biophys. Res. Commun.* 37:451–56
6. Atsumi, S., Kuraishi, S., Hayashi, T. 1976. An improvement of auxin extraction procedure and its application to cultured plant cells. *Planta* 129:245–47
7. Bandurski, R. S., Schulze, A. 1974. Concentrations of indole-3-acetic acid and its esters in *Avena* and *Zea*. *Plant Physiol.* 54:257–62
8. Bandurski, R. S., Schulze, A. 1977. Concentration of indole-3-acetic acid and its derivatives in plants. *Plant Physiol.* 60:211–13
9. Belke, C. J., Sjut, V., Brenner, M. L. 1980. Metabolism and phloem transport of [2-^{14}C]-ABA in tomato leaves. *Plant Physiol.* 65:94 (Suppl)
10. Biddington, N. L., Thomas, T. H. 1973. Chromatography of five cytokinins on an insoluble polyvinylpyrrolidone column. *J. Chromatogr.* 75:122–23
11. Biddington, N. L., Thomas, T. H. 1976. Effect of pH on the elution of cytokinins from polyvinylpyrrolidone columns. *J. Chromatogr.* 121:107–9
12. Bieleski, R. L. 1964. The problem of halting enzyme action when extracting plant tissues. *Anal. Biochem.* 9:431–42
13. Bieleski, R. L. 1968. Levels of phosphate esters in spirodela. *Plant Physiol.* 43:1297–1308
14. Binks, R., MacMillan, J., Pryce, R. J. 1969. Combined gas chromatography-mass spectrometry of the methyl esters of gibberellins A_1 to A_{24} and their trimethylsilyl ethers. *Phytochemistry* 8: 271–84
15. Bittner, S., Even-Chen, Z. 1975. A GLC procedure for determining sub-nanogram levels of indol-3-yl acetic acid. *Phytochemistry* 14:2455–57
16. Böttger, M., Engvild, K. C., Kaiser, P. 1978. Response of substituted indoleacetic acids in the indole-α-pyrone

fluorescence determination. *Physiol. Plant.* 43:62–64
17. Brenner, M. L. 1978. Use of modern forms of chromatography for plant growth substance analysis - II. *Proc. 5th Ann. Meet. Plant Growth Regul. Work. Group,* pp. 20–32
18. Brenner, M. L. 1979. Advances in analytical methods for plant growth substance analysis. In *Plant Growth Substances,* ed. N. B. Mandava, 111:215–44. Washington DC: Am. Chem. Soc. 310 pp.
19. Bridges, I. G., Hillman, J. R., Wilkins, M. B. 1973. Identification and localization of auxin in primary roots of *Zea mays* by mass spectrometry. *Planta* 115:189–92
20. Browning, G., Saunders, P. F. 1977. Membrane localised gibberellins A_9 and A_4 in wheat chloroplasts. *Nature* 265: 375–77
21. Cargile, N. L., Borchert, R., McChesney, J. D. 1979. Analysis of abscisic acid by high performance liquid chromatography. *Anal. Biochem.* 97:331–39
22. Carnes, M. G., Brenner, M. L., Andersen, C. R. 1974. Rapid separation and identification of plant growth substances using high pressure liquid chromatography. In *Plant Growth Substances,* ed. S. Tamura, pp. 99–110. Tokyo: Hirokawa
23. Carnes, M. G., Brenner, M. L., Andersen, C. R. 1975. Comparison of reversed-phase high-pressure liquid chromatography with Sephadex LH-20 for cytokinin analysis of tomato root pressure exudate. *J. Chromatogr.* 108: 95–106
24. Caruso, J. L., Smith, R. G., Smith, L. M., Cheng, T., Daves, G. D. Jr. 1978. Determination of indole-3-acetic acid in douglas fir using a deuterated analog and selected ion monitoring. Comparison of microquantities in seedling and adult tree. *Plant Physiol.* 62:841–45
25. Challice, J. S. 1975. Separation of cytokinins by high pressure liquid chromatography. *Planta* 122:203–7
26. Ciha, A. J., Brenner, M. L., Brun, W. A. 1977. Rapid separation and quantification of abscisic acid from plant tissues using high performance liquid chromatography. *Plant Physiol.* 59:821–26
27. Cohen, J. D., Bandurski, R. S. 1978. The bound auxins: Protection of indole-3-acetic acid from peroxidase-catalyzed oxidation. *Planta* 139:203–8
28. Cohen, J. D., Schulze, A., Bandurski, R. S. 1978. A double-standard method for isotope dilution analysis of IAA by gas liquid chromatography. *Plant Physiol.* 61:63 (Suppl)
29. Cole, D. L., Leonard, N. J., Cook, J. Jr. 1974. A system for separation and identification of the naturally occurring cytokinins: High performance liquid chromatography-field desorption mass spectrometry. In *Recent Developments in Oligonucleotide Synthesis and Chemistry of Minor Bases of tRNA,* pp. 153–74. Poznan-Kiekrz, Poland: Uniw. Adama Mickiewicza Press
30. Crozier, A., Loferski, K., Zaerr, J. B., Morris, R. O. 1980. Analysis of picogram quantities of indole-3-acetic acid by high performance liquid chromatography-fluorescence procedures. *Planta.* In press
31. Crozier, A., Zaerr, J. B., Morris, R. O. 1980. High performance steric exclusion chromatography of plant hormones. *J. Chromatogr.* In press
32. Currie, L. A. 1979. Sources of error and the approach to accuracy in analytical chemistry. In *Treatise on Analytical Chemistry,* ed. I. M. Kolthoff, P. J. Elving, 1:156–75, 230–33. New York: Wiley. 2nd ed.
33. Daie, J., Wyse, R. 1980. Elisa: A new technique for the quantification of abscisic acid. *Plant Physiol.* 65:94 (Suppl)
34. Dauphin, B., Teller, G., Durand, B. 1977. Mise au point d'une nouvelle méthode de purification, caractérisation et dosage de cytokinines endogènes, extraites de bourgeons de Mercuriales annuelles. *Physiol. Veg.* 15:747–62
35. Dauphin, B., Teller, G., Durand, B. 1979. Identification and quantitative analysis of cytokinins from shoot apices of *Mercurialis ambiqua* by gas chromatography-mass spectrometry computer system. *Planta* 144:113–19
36. Dekhuijzen, H. M., Gevers, E. C. T. 1975. The recovery of cytokinins during extraction and purification of clubroot tissue. *Physiol. Plant.* 35:297–302
37. Dennis, F. G. 1977. Growth hormones: Pool size, diffusion or metabolism? *HortScience* 12:217–20
38. Düring, H. 1977. Analysis of abscisic acid and indole-3-acetic acid from fruits of *Vitis vinifera* L. by high pressure liquid chromatography. *Experientia* 33: 1666–67
39. Durley, R. C., Crozier, A., Pharis, R. P., McLaughlin, G. E. 1972. Chromatography of 33 gibberellins on a gradient eluted silica gel partition column. *Phytochemistry* 11:3029–33
40. Durley, R. C., Kannangara, T., Simpson, G. M. 1978. Analysis of abscisins

and 3-indolylacetic acid in leaves of *Sorghum bicolor* by high performance liquid chromatography. *Can. J. Bot.* 56:157–61
41. Durley, R. C., Pharis, R. P. 1972. Partition coefficients of 27 gibberellins. *Phytochemistry* 11:317–26
42. Ehmann, A. 1974. Identification of 2-*O*-(indole-3-acetyl)-D-glucopyranose, 4-*O*-(indole-3-acetyl)-D-glucopyranose and 6-*O*-(indole-3-acetyl)-D-glucopyranose from kernels of *Zea mays* by gas-liquid chromatography-mass spectrometry. *Carbohydr. Res.* 34:99–114
43. Ehmann, A., Bandurski, R. S. 1972. Purification of indole-3-acetic acid myoinositol esters on polystyrene-divinylbenzene resins. *J. Chromatogr.* 72:61–70
44. Ehmann, A., Bandurski, R. S. 1974. The isolation of di-*O*-(indole-3-acetyl)-*myo*-inositol and tri-*O*-(indole-3-acetyl)-*myo*-inositol from mature kernels of *Zea mays. Carbohydr. Res.* 36:1–12
45. Eliasson, L., Stromquist, L. H., Tillberg, E. 1975. Reliability of the indolo-α-pyrone fluorescence method for indole-3-acetic acid determination in crude plant extracts. *Physiol. Plant.* 56:16–19
46. Engvild, K. C., Egsgaard, H., Larsen, E. 1978. Gas chromatographic-mass spectrometric identification of 4-chloroindole-3-acetic acid methyl ester in immature green peas. *Physiol. Plant.* 42:365–68
47. Epstein, E., Cohen, J. D. 1980. Microscale preparation of indole-3-acetylpentafluorotoluene for electron capture detection-gas chromatography. *Plant Physiol.* 65:76 (Suppl)
48. Feung, C-S., Hamilton, R. H., Mumma, R. O. 1975. Indole-3-acetic acid. Mass spectra and chromatographic properties of amino acid conjugates. *J. Agric. Food Chem.* 23:1120–24
49. Feung, C-S., Hamilton, R. H., Mumma, R. O. 1976. Metabolism of indole-3-acetic acid. III. Identification of metabolites isolated from crown gall callus tissue. *Plant Physiol.* 58:666–69
50. Frydman, V. M., Gaskin, P., MacMillan, J. 1974. Qualitative and quantitative analyses of gibberellins throughout seed maturation in *Pisum sativum* cv Progress no. 9. *Planta* 118:123–32
51. Frydman, V. M., MacMillan, J. 1973. Identification of gibberellins A_{20} and A_{29} in seed of *Pisum sativum* cv. Progress no. 9 by combined gas chromatography-mass spectrometry. *Planta* 115:11–15
52. Fuchs, S., Fuchs, Y. 1969. Immunological assay for plant hormones using specific antibodies to indoleacetic acid and gibberellic acid. *Biochim. Biophys. Acta* 192:528–30
53. Fuchs, S., Haimovich, J., Fuchs, Y. 1971. Immunological studies of plant hormones: Detection and estimation by immunological assays. *Eur. J. Biochem.* 18:384–90
54. Fuchs, Y., Gertman, E. 1974. Insoluble antibody column for isolation and quantitative determination of gibberellins. *Plant Cell Physiol.* 16:629–33
55. Fuchs, Y., Mayak, S., Fuchs, S. 1972. Detection and quantitative determination of abscisic acid by immunological assay. *Planta* 103:117–25
56. Gaskin, P., MacMillan, J. 1978. GC and GC-MS techniques for gibberellins. See Ref. 68, pp. 79–95
57. Glenn, J. L., Kuo, C. C., Durley, R. C., Pharis, R. P. 1972. Use of insoluble polyvinylpyrrolidone for purification of plant extracts and chromatography of plant hormones. *Phytochemistry* 11: 345–51
58. Graebe, J. E., Ropers, H. J. 1978. Gibberellins. See Ref. 95, pp. 107–204
59. Gray, R. T., Mallaby, R., Ryback, G., Williams, V. P. 1974. Mass spectra of methyl abscisate and isotopically labelled analogues. *J. Chem. Soc. Perkin Trans.* 2:919–24
60. Greenwood, M. S., Shaw, S., Hillman, J. R., Ritchie, A., Wilkins, M. B. 1972. Identification of auxin from *Zea* coleoptile tips by mass spectrometry. *Planta* 108:179–83
61. Hahn, H. 1975. Cytokinins: A rapid extraction and purification method. *Physiol. Plant.* 34:204–7
62. Hahn, H. 1976. High performance liquid chromatography and its use in cytokinin determination in *Agrobacterium tumefaciens* B6. *Plant Cell Physiol.* 17:1053–8
63. Hall, S. M., Medlow, G. C. 1974. Identification of IAA in phloem and root pressure saps of *Ricinus communis* L. by mass spectrometry. *Planta* 119:257–261
64. Hedden, P. 1979. Recent developments in the identification and quantification of gibberellins. See Ref. 17, pp. 33–44
65. Heftmann, E., Saunders, G. A., Haddon, W. F. 1978. Argentation high pressure liquid chromatography and mass spectrometry of gibberellin esters. *J. Chromatogr.* 156:71–77

66. Hemberg, T. 1974. Partitioning of cytokinins between ethyl acetate and acid water phases. *Physiol. Plant.* 32:191–92
67. Hemberg, T., Westlin, P. E. 1973. The quantitative yield in purification of cytokinins. Model-experiments with kinetin, 6-furfuryl-amino-purine. *Physiol. Plant.* 28:228–31
68. Hillman, J. R., ed. 1978. *Isolation of Plant Growth Substances.* Cambridge: Cambridge Univ. Press. 157 pp.
69. Hillman, J. R., Math, V. B., Medlow, G. C. 1977. Apical dominance and the levels of indole acetic acid in *Phaseolus* lateral buds. *Planta* 134:191–93
70. Hillman, J. R., Young, I., Knights, B. A. 1974. Abscisic acid in leaves of *Hedera helix* L. *Planta* 119:263–66
71. Hirai, N., Fukui, H., Koshimizu, K. 1978. A novel abscisic acid metabolite from seeds of *Robinia pseudacacia. Phytochemistry* 17:1625–27
72. Hofinger, M. 1980. A method for the quantitation of indole auxins in the picogram range by high performance gas chromatography of their N-heptafluorobutyrylmethyl esters. *Phytochemistry* 19:219–22
73. Hofinger, M., Böttger, M. 1979. Identification by GC-MS of 4-chloroindolylacetic acid and its methyl ester in immature *Vicia faba* seeds. *Phytochemistry* 18:653–54
74. Holland, J. A., McKerrell, E. H., Fuell, K. J., Burrows, W. J. 1978. Separation of cytokinins by reversed-phase high-performance liquid chromatography. *J. Chromatogr.* 166:545–53
75. Horgan, R. 1978. Analytical procedures for cytokinins. See Ref. 68, pp. 97–114
76. Horgan, R., Kramers, M. R. 1979. High-performance liquid chromatography of cytokinins. *J. Chromatogr.* 173:263–70
77. Horgan, R., Neill, S. 1979. Column conditioning for trace analysis of abscisic acid by gas chromatography. *J. Chromatogr.* 177:116–17
78. Hubick, K. T., Reid, D. M. 1980. A rapid method for the extraction and analysis of abscisic acid from plant tissue. *Plant Physiol.* 65:523–25
79. Ingram, T. J., Browning, G. 1979. Influence of photoperiod on seed development in the genetic line of peas G2 and its relation to changes in endogenous gibberellins measured by combined gas chromatography mass spectrometry. *Planta* 146:423–32
80. Jennings, W. 1980. *Gas Chromatography with Glass Capillary Columns.* New York: Academic. 320 pp. 2nd ed.
81. Jones, M. G., Metzger, J. D., Zeevaart, J. A. D. 1980. Fractionation of gibberellins in plant extracts by reverse phase high performance liquid chromatography. *Plant Physiol.* 65:218–21
82. Kaiss-Chapman, R. W., Morris, R. O. 1977. Trans-zeatin in culture filtrates of *Agrobacterium tumefaciens. Biochem. Biophys. Res. Commun.* 76:453–59
83. Kamisaka, S., Larsen, P. 1977. Improvement of the indolo-α-pyrone fluorescence method for quantitative determination of endogenous indole-3-acetic acid in lettuce seedlings. *Plant Cell Physiol.* 18:595–602
84. Kannangara, T., Durley, R. C., Simpson, G. M. 1978. High performance liquid chromatographic analysis of cytokinins in *Sorghum bicolor* leaves. *Physiol. Plant.* 44:295–99
85. Karger, B. L., Snyder, L. R., Horvath, C. 1973. *An Introduction to Separation Science.* New York: Wiley. 586 pp.
86. Khan, S. A., Humayun, M. Z., Jacob, T. M. 1977. A sensitive radioimmunoassay for isopentenyladenosine. *Anal. Biochem.* 83:632–35
87. Knegt, E., Bruinsma, J. 1973. A rapid, sensitive accurate determination of indolyl-3-acetic acid. *Phytochemistry* 12: 753–56
88. Koshimizu, K., Inui, M., Fukui, H., Mitsui, T. 1968. Isolation of (+)-abscisyl-β-D-glucopyranoside from immature fruit of *Lupinus luteus. Agric. Biol. Chem.* 32:789–91
89. Koyama, S., Kawai, H., Kumazawa, Z., Ogawa, Y., Iwamura, H. 1978. Isolation and identification of trans-zeatin from the roots of *Raphanus sativus* L. cv Sakurajima. *Agric. Biol. Chem.* 42:1997–2001
90. Küllertz, G., Eckert, H., Schilling, G. 1978. Quantitative gas chromatographic determination of gibberellic acid (GA_3) in nanogram quantities by electron capture detection. *Biochem. Physiol. Pflanz.* 173:186–87
91. Kurogochi, S., Murofushi, N., Ota, Y., Takahashi, N. 1979. Identification of gibberellins in the rice plant and quantitative changes of gibberellin A_{19} throughout its life cycle. *Planta* 146:185–92
92. Lenton, J. R., Perry, V. M., Saunders, P. F. 1971. The identification and quantitative analysis of abscisic acid in plant extracts by gas-liquid chromatography. *Planta* 96:271–80

93. Letham, D. S. 1974. Regulators of cell division in plant tissues. XXI. Distribution coefficients for cytokinins. *Planta* 118:361–64
94. Letham, D. S. 1978. Cytokinins. See Ref. 95, pp. 205–63
95. Letham, D. S., Goodwin, P. B., Higgins, T. J. V., eds. 1978. *Phytohormones and Related Compounds: A Comprehensive Treatise, Volume I. The Biochemistry of Phytohormones and Related Compounds.* Amsterdam: Elsevier/North Holland. 641 pp.
96. Little, C. H. A., Heald, J. K., Browning, G. 1978. Identification and measurement of indoleacetic and abscisic acids in the cambial region of *Picae sitchensis* (Bong.) Carr. by combined gas chromatography-mass spectrometry. *Planta* 139:133–38
97. Lorenzi, R., Horgan, R., Wareing, P. F. 1975. Cytokinins in *Picea sitchensis* Carriere: Identification and relation of growth. *Biochem. Physiol. Pflanz.* 168:333–39
98. Loveys, B. R. 1977. The intracellular location of abscisic acid in stressed and non-stressed leaf tissue. *Physiol. Plant.* 40:6–10
99. Luke, N., Chin, C., Eck, P. 1977. Dialysis extraction of gibberellin-like substances from cranberry tissue. *HortScience* 12:245–46
100. MacLeod, J. K., Summons, R. E., Letham, D. S. 1976. Mass spectrometry of cytokinin metabolites. Per(trimethylsilyl) and permethyl derivatives of glucosides of zeatin and 6-benzylaminopurine. *J. Org. Chem.* 41:3959–67
101. MacMillan, J., Pryce, R. J. 1968. Phaseic acid, a putative relative of abscisic acid, from seed of *Phaseolus multiflorus. Chem. Commun.* 124–26
102. MacMillan, J., Wels, C. M. 1973. Partition chromatography of gibberellins and related diterpenes on columns of Sephadex LH-20. *J. Chromatogr.* 87: 271–76
103. Magnus, V., Bandurski, R. S., Schulze, A. 1980. Synthesis of 4,5,6,7 and 2,4,-5,6,7 deuterium labeled indole-3-acetic acid for use in mass spectrometric assays. *Plant Physiol.* 66:775–81
104. Mann, J. D., Jaworski, E. G. 1970. Minimizing loss of indoleacetic acid during purification of plant extracts. *Planta* 92:285–91
105. Martin, G. C., Dennis, F. G. Jr., Gaskin, P., MacMillan, J. 1975. Contaminants present in materials commonly used to purify plant extracts for hormone analysis. *HortScience* 10:598–99
106. Martin, G. C., Dennis, F. G. Jr., Gaskin, P., MacMillan, J. 1977. Identification of gibberellins A_{17}, A_{25}, A_{45}, abscisic acid, phaseic acid, and dihydrophaseic acid in seeds of *Pyrus communis. Phytochemistry* 16:605–7
107. Martin, G. C., Nishijima, C. 1977. Contaminants in solvents and chromatographic materials. *HortScience* 12: 212–16
108. Martin, G. C., Nishijima, C., Labavitch, J. M. 1980. Analysis of indoleacetic acid by the nitrogen-phosphorus detector gas chromatograph. *J. Am. Soc. Hortic. Sci.* 105:46–50
109. McDougall, J., Hillman, J. R. 1978. Analysis of indole-3-acetic acid using GC-MS techniques. See Ref. 68, pp. 1–25
110. McDougall, J., Hillman, J. R. 1978. Purification of IAA from shoot tissues of *Phaseolus vulgaris* and its analysis by GC-MS. *J. Exp. Bot.* 29:375–86
111. McWha, J. A., Hillman, J. R. 1974. Endogenous abscisic acid in lettuce fruits. *Z. Pflanzenphysiol.* 74:292–97
112. Menhenett, R., Carr, D. J. 1973. Cytokinins in etiolated barley leaves. *Aust. J. Biol. Sci.* 26:1073–80
113. Metzger, J. D., Zeevaart, J. A. D. 1980. Identification of six endogenous gibberellins in spinach shoots. *Plant Physiol.* 65:623–26
114. Milborrow, B. V. 1978. Abscisic acid. See Ref. 95, pp. 295–347
115. Milborrow, B. V., Mallaby, R. 1975. Occurrence of methyl(+)-abscisate as an artefact of extraction. *J. Exp. Bot.* 26:741–48
116. Miller, C. O. 1975. Revised methods for purification of ribosyl-*trans*-zeatin from *Vinca rosea* crown gall tumor tissue. *Plant Physiol.* 55:448–49
117. Monselise, S. P., Varga, A., Knegt, E., Bruinsma, J. 1978. Course of the zeatin content in tomato fruits and seeds developing on intact or partially defoliated plants. *Z. Pflanzenphysiol.* 90: 451–60
118. Morris, R. O. 1977. Mass spectroscopic identification of cytokinins: Glucosyl zeatin and glucosyl ribosylzeatin from *Vinca rosea* crown gall. *Plant Physiol.* 59:1029–33
119. Morris, R. O., Zaerr, J. B. 1978. 4-bromophenacyl esters of gibberellins, useful derivatives for high performance liquid chromatography. *Anal. Lett.* 11:73–83
120. Morris, R. O., Zaerr, J. B., Chapman, R. W. 1976. Trace enrichment of

cytokinins from Douglas fir xylem extrudate. *Planta* 131:271–74
121. Most, B. H. 1970. The occurrence of abscisic acid in inhibitors B_1 and C from immature fruit of *Ceratonia siliqua* L. (Carob) and in commercial carob syrup. *Planta* 92:41–49
122. Mousdale, D. M. A., Butcher, D. N., Powell, R. G. 1978. Spectrophotofluorimetric methods of determining indole-3-acetic acid. See Ref. 68, pp. 27–39
123. Mousdale, D. M. A., Knee, M. 1979. Poly-N-vinylpyrrolidone column chromatography of plant hormones with methanol as eluent. *J. Chromatogr.* 177:398–400
124. Murofushi, N., Sugimoto, M., Itoh, K., Takahashi, N. 1979. Three novel gibberellins produced by *Gibberella fujikuroi*. *Agric. Biol. Chem.* 43:2179–85
125. Nishio, M., Zushi, S., Ishii, T., Furuya, T., Syono, K. 1976. Mass fragmentographic determination of IAA in callus tissues of *Panax ginseng* and *Nicotiana tabacum*. *Chem. Pharm. Bull.* 24: 2038–42
126. Osborne, D. J. 1978. Ethylene. See Ref. 95, pp. 265–94
127. Pengelly, W. L., Bandurski, R. S., Schulze, A. 1980. Validation of a radioimmunoassay for indole-3-acetic acid using gas chromatography selected ion monitoring mass spectrometry. *Plant Physiol.* 65:157 (Suppl)
128. Pengelly, W., Meins, F. Jr. 1977. A specific radioimmunoassay for nanogram quantities of the auxin, indole-3-acetic acid. *Planta* 136:173–80
129. Pool, R. M., Powell, L. E. 1972. The use of pellicular ion-exchange resins to separate plant cytokinins by high-pressure liquid chromatography. *HortScience* 7:26
130. Pool, R. M., Powell, L. E. 1973. Cytokinin analysis by high pressure liquid chromatography. See Ref. 22, pp. 93–98
131. Quarrie, S. A. 1978. A rapid and sensitive assay for abscisic acid using ethyl abscisate as an internal standard. *Anal. Biochem.* 87:148–56
132. Railton, I. D., Rechav, M. 1979. Efficiency of extraction of gibberellin-like substances from chloroplasts of *Pisum sativum* L. *Plant Sci. Lett.* 14:75–78
133. Railton, I. D., Reid, D. M., Gaskin, P., MacMillan, J. 1974. Characterization of abscisic acid in chloroplasts of *Pisum sativum* L. cv. Alaska by combined gas-chromatography-mass spectrometry. *Planta* 117:179–82
134. Reeve, D. R., Crozier, A. 1976. Purification of plant hormone extracts by gel permeation chromatography. *Phytochemistry* 15:791–93
135. Reeve, D. R., Crozier, A. 1978. The analysis of gibberellins by high performance liquid chromatography. See Ref. 68, pp. 41–77
136. Reeve, D. R., Crozier, A. 1980. Quantitative analysis of plant hormones. In *Molecular and Sub-cellular Aspects of Hormonal Regulation in Plants,* ed. J. MacMillan, 1:1–108. Berlin: Springer
137. Reeve, D. R., Yokota, T., Nash, L. J., Crozier, A. 1976. The development of a high performance liquid chromatograph with a sensitive on-stream radioactivity monitor for the analysis of ^3H- and ^{14}C-labelled gibberellins. *J. Exp. Bot.* 27:1243–58
138. Rivier, L., Milon, H., Pilet, P. E. 1977. Gas chromatography-mass spectrometric determinations of abscisic acid levels in the cap and the apex of maize roots. *Planta* 134:23–27
139. Rivier, L., Pilet, P. E. 1974. Indolyl-3-acetic acid in cap and apex of maize roots: Identification and quantification by mass fragmentography. *Planta* 120: 107–12
140. Rivier, L., Pilet, P. E. 1980. Abscisic acid levels in the root tips of seven *Zea mays* varieties. *Phytochemistry.* In press
141. Russell, S. 1976. Extraction purification and chemistry of gibberellins. In *Gibberellins and Plant Growth,* ed. H. N. Krishnamoorthy, pp. 1–34. New York: Halsted, div. Wiley. 356 pp.
142. Sagi, F. 1969. Silica gel or cellulose for the thin-layer chromatography of indole-3-acetic acid? *J. Chromatogr.* 39: 334–35
143. Saunders, P. F. 1978. The identification and quantitative analysis of abscisic acid in plant extracts. See Ref. 68, pp. 115–34
144. Schneider, E. A., Wightman, F. 1978. Auxins. See Ref. 95, pp. 29–105
145. Seeley, S. D., Powell, L. E. 1970. Electron capture-gas chromatography for sensitive assay of abscisic acid. *Anal. Biochem.* 35:530–33
146. Seeley, S. D., Powell, L. E. 1974. Gas chromatography and detection of microquantities of gibberellins and indoleacetic acid as their fluorinated derivatives. *Anal. Biochem.* 58:39–46
147. Setter, T. L., Brenner, M. L., Brun, W. A. 1981. Abscisic acid translocation and metabolism in soybean following depodding and petiole girdling treatments. *Plant Physiol.* 67:In press

148. Setter, T. L., Brun, W. A., Brenner, M. L. 1980. Effect of obstructed translocation on leaf abscisic acid, and associated stomatal closure and photosynthesis decline. *Plant Physiol.* 65:1111–15
149. Shaybany, B., Martin, G. C. 1977. Abscisic acid identification and its quantitation in leaves of *Juglans* seedlings during waterlogging. *J. Am. Soc. Hortic. Sci.* 102:300–2
150. Singh, B. N., Galson, E., Dashek, W., Walton, D. C. 1979. Abscisic acid levels and metabolism in the leaf epidermal tissue of *Tulipa gesnariana* L. and *Commelina communis* L. *Planta* 146: 135–38
151. Snyder, L. R., Kirkland, J. J. 1979. *Introduction to Modern Liquid Chromatography.* New York: Wiley. 2nd ed.
152. Sponsel, V. M., MacMillan, J. 1978. Metabolism of gibberellin A_{29} in seeds of *Pisum sativun* cv. Progress no. 9: Use of [^{2}H] and [^{3}H] GAs, and the identification of a new GA catabolite. *Planta* 144:69–78
153. Stoessl, A., Venis, M. A. 1970. Determination of submicrogram levels of indole-3-acetic acid: A new, highly specific method. *Anal. Biochem.* 34:344–51
154. Summons, R. E., Duke, C. D., Eichholzer, J. V., Entsch, B., Letham, D. S., MacLeod, J. K., Parker, C. W. 1979. Mass spectrometric analysis of cytokinins in plant tissues. II. Quantitation of cytokinins in *Zea mays* kernels using deuterium labelled standards. *Biomed. Mass Spectrom.* 6:407–13
155. Summons, R. E., Entsch, B., Letham, D. S., Gollnow, B. I., MacLeod, J. K. 1980. Regulators of cell division in plant tissues. XXVIII. Metabolites of zeatin in sweet-corn kernels: Purifications and identifications using high-performance liquid chromatography and chemical-ionization mass spectrometry. *Planta* 147:422–34
156. Summons, R. E., Entsch, B., Parker, C. W., Letham, D. S. 1979. Mass spectrometric analysis of cytokinins in plant tissues. III. Quantitation of the cytokinin glycoside complex of lupin pods by stable isotope dilution. *FEBS Lett.* 107:21–25
157. Summons, R. E., MacLeod, J. K., Parker, C. W., Letham, D. S. 1977. The occurrence of raphanatin as an endogenous cytokinin in radish seed: Identification and quantitation by gas chromatographic mass spectrometric analysis using deuterium labeled standards. *FEBS Lett.* 82:211–14
158. Swartz, H. J., Powell, L. E. 1979. Determination of indoleacetic acid from plant samples by an alkali flame ionization detector. *Physiol. Plant.* 47:25–28
159. Sweetser, P. B. 1978. Use of modern forms of chromatography for plant growth substance analysis - I. See Ref. 17, pp. 1–19
160. Sweetser, P. B., Swartzfager, D. G. 1978. Indole-3-acetic acid levels of plant tissue as determined by a new high performance liquid chromatographic method. *Plant Physiol.* 61:254–58
161. Sweetser, P. B., Vatvars, A. 1976. High-performance liquid chromatographic analysis of abscisic acid in plant extracts. *Anal. Biochem.* 71:68–78
162. Thompson, A. G., Horgan, R., Heald, J. K. 1975. A quantitative analysis of cytokinin using single-ion-current-monitoring. *Planta* 124:207–10
163. Tietz, D., Dörffling, K., Wöhrle, D., Erxleben, I., Liemann, F. 1979. Identification by combined gas chromatography-mass spectrometry of phaseic acid and dihydrophaseic acid and characterization of further abscisic acid metabolites in pea seedlings. *Planta* 147: 168–73
164. Trewavas, A. J. 1979. Plant growth substances: What is the molecular basis of their action? *What's New In Plant Physiol.* 10:33–36
165. Van Staden, J. 1976. Extraction and recovery of cytokinin glucosides by means of a cation exchange resin. *Physiol. Plant.* 38:240–42
166. Vreman, H. J., Corse, J. 1975. Recovery of cytokinins from cation exchange resins. *Physiol. Plant.* 35:333–36
167. Walton, D. C. 1980. Biochemistry and physiology of abscisic acid. *Ann. Rev. Plant. Physiol.* 31:453–89
168. Walton, D. C., Dashek, W., Galson, E. 1979. A radioimmunoassay for abscisic acid. *Planta* 146:139–45
169. Ward, T. M., Wright, M., Roberts, J. A., Self, R., Osborne, D. J. 1978. Analytical procedures for the assay and identification of ethylene. See Ref. 68, pp. 135–51
170. Watanabe, N., Yokota, T., Takahashi, N. 1978. *cis*-Zeatin riboside: Its occurrence as a free nucleoside in cones of the hop plant. *Agric. Biol. Chem.* 42: 2415–16
171. Watanabe, N., Yokota, T., Takahashi, N. 1978. Identification of zeatin and zeatin riboside in cones of the hop plant and their possible role in cone growth. *Plant Cell Physiol.* 19:617–25

172. Weiler, E. W. 1979. Radioimmunoassay for the determination of free and conjugated abscisic acid. *Planta* 144:255–63
173. Weiler, E. W. 1980. Radioimmunoassays for trans-zeatin and related cytokinins. *Planta* 149:155–62
174. Wheaton, T. A., Bausher, M. G. 1977. Separation and identification of endogenous growth regulators in citrus. *Proc. Int. Soc. Citricult.* 2:673–76
175. Wightman, F. 1977. Gas chromatographic identification and quantitative estimation of natural auxins in developing plant organs. In *Plant Growth Regulation: Proc. 9th Int. Conf. Plant Growth Subst. Lausanne, Aug. 30-Sept. 4, 1976,* ed. P. E. Pilet, pp. 77–90. Berlin/Heidelberg: Springer. 305 pp.
176. Wightman, F. 1979. Modern chromatographic methods for the identification and quantification of plant growth regulators and their application to studies of the changes in hormonal substances in winter wheat during acclimation to cold stress conditions. In *Plant Regulation and World Agriculture,* ed. T. K. Scott, pp. 327–77. New York: Plenum
177. Wurst, M., Prikryl, Z., Vancura, V. 1980. High-performance liquid chromatography of plant hormones. I. Separation of plant hormones of the indole type. *J. Chromatogr.* 191:130–35
178. Yamaguchi, I., Yokota, T., Yoshida, S., Takahashi, N. 1979. High pressure liquid chromatography of conjugated gibberellins. *Phytochemistry* 18:1699–1702
179. Yamane, H., Takahashi, N., Takeno, K., Furuya, M. 1979. Identification of gibberellin A_9 methyl ester as a natural substance regulating formation of reproductive organs in *Lygodium japonicum. Planta* 147:251–56
180. Yokota, T., Hiraga, K., Yamane, H., Takahashi, N. 1975. Mass spectrometry of trimethylsilyl derivatives of gibberellin glucosides and glucosylesters. *Phytochemistry* 14:1569–74
181. Young, H. 1977. Identification of cytokinins from natural sources by gas-liquid chromatography mass spectrometry. *Anal. Biochem.* 79:226–33
182. Zeevaart, J. A. D., Milborrow, B. V. 1976. Metabolism of abscisic acid and the occurrence of *epi*-dihydrophaseic acid in *Phaseolus vulgaris. Phytochemistry* 15:493–500
183. Zelleke, A., Martin, G. C., Labavitch, J. M. 1980. Detection of cytokinins using a gas chromatograph equipped with a sensitive nitrogen-phosphorus detector. *J. Am. Soc. Hortic. Sci.* 105:50–53

Ann. Rev. Plant Physiol. 1981. 32:539–67

STRUCTURE, BIOSYNTHESIS, AND BIODEGRADATION OF CUTIN AND SUBERIN

◆7723

P. E. Kolattukudy

Institute of Biological Chemistry and Biochemistry/Biophysics Program, Washington State University, Pullman, Washington 99164

CONTENTS

0066-4294/81/0601-0539$01.00

INTRODUCTION

The chemistry of cutin and suberin has been studied for a century, but very little was known about these polymers when I reviewed biosynthesis of cuticular lipids in this series a decade ago (55). During the last decade, however, much progress has been made in our understanding of the composition of cutin mainly because of the availability of modern analytical techniques, especially combined gas-liquid chromatography/mass spectrometry. Although our understanding of the structure and composition of suberin is still rather incomplete, sufficient chemical information about this polymer became available to distinguish between cutin and suberin. Biosynthesis and biodegradation of cutin also entered the era of modern biochemistry during the last decade; biochemical pathways involved in the synthesis of the biopolyester were elucidated, and enzymes involved in the degradation of the polymer were isolated and characterized. Modest beginnings were also made in our understanding of the biosynthesis of suberin. In this paper I shall summarize these areas of progress in the chemistry and biochemistry of cutin and suberin. The terms "cutin" and "suberin" refer to the insoluble polymeric materials, and the soluble waxes, which are always physically associated with these polymers, are not covered in this review. Other comprehensive reviews on the polymers and waxes are available (62, 63, 83, 112).

LOCATION AND ULTRASTRUCTURE OF CUTIN AND SUBERIN

Cutin is the insoluble polymeric structural component of the cuticle of all aerial parts of plants except periderms. Some internal tissues such as the inner seed coats and juice sacs of citrus also contain cutin (32). Even primitive plants such as liverworts (14) and mosses and aquatic higher plants such as *Zoestra marina* (60) have cutin. Electron microscopy shows that in most cases cutin-containing layers have an amorphous appearance but in some cases they have lamellar appearances (85, 110). Suberin is the insoluble polymeric material attached to the cell walls of periderms, including wound periderms formed on aerial parts of plants (26), endodermis (Casparian bands) (35), bundle sheaths of grasses (36), and the seed coat cells in the area of attachment of the vascular bundle (32). Suberized walls have a lamellar appearance, probably because of the presence of layers of wax (101, 108). "Cutin or suberin" type layers of unknown composition have also been observed in the endodermis, epidermis, and hypodermis of roots (37, 87, 95), the pigment strand of cereal grains (121), lining of nectaries (91), around idioblasts containing calcium oxalate crystals (117), and in the basal wall of trichomes (111) and oil glands (82).

CHEMISTRY OF CUTIN AND SUBERIN

Methodology

ISOLATION The cuticle can be removed from plant tissues by chemical or enzymatic disruption of the pectinaceous layer which attaches the cuticle to the cell wall, and any residual carbohydrates should be removed by further treatment of the isolated cuticle (53, 67, 70). The waxes associated with the cuticle can be removed by extensive extraction with organic solvents. Suberin can be obtained only as an "enriched preparation" because this polymer is tightly attached to the cell wall. The barrier layers such as periderms, physically removed from the rest of the tissue, are treated as noted above for cuticle to yield a suberin-enriched preparation (66, 96).

DEPOLYMERIZATION Since cutin is a polyester, the commonly used depolymerization techniques are alkaline hydrolysis, methanolysis with $NaOCH_3$ or BF_3 as catalysts, and hydrogenolysis with $LiAlH_4$. The esterified aliphatic components of suberin and phenolic acids esterified to cutin and suberin can also be recovered by similar depolymerization techniques. However, the phenolic components which apparently constitute major components of suberin cannot be readily isolated because they are held together by linkages somewhat similar to those found in lignin. Therefore, lignin-cleavage methods such as alkaline nitrobenzene or CuO oxidation, treatment with aqueous or acidic dioxane under high temperature and pressure (18, 72) have to be used to get soluble phenolic products from suberin for structural studies.

ANALYTICAL TECHNIQUES Thin-layer chromatographic techniques and gas-liquid chromatography/mass spectrometry are the most useful tools for the determination of the structure and composition of the monomers derived from cutin and suberin. Depolymerization with $LiAlD_4$ results in specific labeling of the monomers, and mass spectrometry of such monomers reveals the presence of functional groups which might not survive other types of depolymerization methods (59, 115). The highly favored cleavages on either side of the trimethylsiloxy functions in an aliphatic chain make mass spectrometry of the trimethylsilyl derivatives of the aliphatic components of cutin a simple procedure even for those who are not experienced in mass spectrometry (30, 115).

Composition of Cutin

Cutin is composed of mainly two families of monomers: a C_{16} family and a C_{18} family. The most predominant component of the former group is 10,16-dihydroxypalmitic acid and/or its positional isomers in which the mid-chain hydroxyl group is at C-9, C-8, or C-7; in most cases one isomer predominates. The dihydroxy acid from papaya and tomato cutin showed a plain positive rotation and therefore, L-configuration has been assigned to this acid (34). Smaller quantities of 16-hydroxypalmitic acid and palmitic acid are also found in most cases. Major components of the C_{18} family of monomers are 18-hydroxyoleic acid, 18-hydroxy-9,10-epoxystearic acid and *threo*-9,10,18-trihydroxystearic acid together with their analogs containing an additional double bond at C-12.

Cutin of fast growing plants is usually composed of chiefly the C_{16} family of monomers whereas cutin of slower growing plants with thick cuticle contains both the C_{16} and C_{18} families of monomers. In apple, for instance, the fruit cutin contains both families of monomers and in leaves the proportion of the C_{18} monomers is much less than that in the fruits, and in flowers the C_{18} monomers are only minor components (33). In plants which contain C_{18} monomers as the major components, the hydroxy-epoxy acid is often the dominant one [e.g. *Senecio odoris* leaves and *Vitis vinifera* fruits (115)], whereas in cases where both C_{16} and C_{18} components are major constituents the trihydroxy C_{18} as well as the epoxy acids occur as major components.

In addition to the most common major components discussed above, several less common but major monomers and numerous minor components have been found in cutin. The less common major components appear to be derived by further oxidation or reduction of the common major components. Examples in the C_{16} family include 16-oxo-9 or 10-hydroxypalmitic acid in embryonic *Vicia faba* (59), 16-hydroxy-10-oxopalmitic acid in *Citrus* (28, 32), and hexadecane-1,8,16-triol in some lower plants (15). In the C_{18} family, examples of further oxidation of major components include 18-oxo-9,10-epoxy C_{18} acid in apple (58); 9,10,18-trihydroxy-12,13-epoxystearic acid, 9,10,12,13,18-pentahydroxystearic acid and derivatives of ricinoleic acid in *Rosmarinus officinalis* (12, 21). NaB^3H_4 treatment of apple cutin resulted in incorporation of tritium into all monomer fractions, indicating the presence of small quantities of the ω-oxo derivatives in all monomer types (W. Köller and P. E. Kolattukudy, unpublished). Hydroxy acids longer and shorter than the major components, mid-chain hydroxylated compounds which do not contain an ω-hydroxy group, dicarboxylic acids, and mid-chain hydroxylated dicarboxylic acids are also minor and rare components of cutin (14, 28, 50, 51).

Minor quantities of functionally significant ligands might be covalently attached to the polyester. *p*-Coumaric acid and smaller amounts of ferulic acid were found covalently attached to some leaf and fruit cutins (94), and reminiscent of this linkage were the 1,16-dioxo-,1-hydroxy-16-oxo- and 1,16-dihydroxyhexadecan-7-yl *p*-coumarates found in pine pollen (99). The amount of covalently attached coumarate and flavanoids in tomato fruit cutin increased during fruit development and climacteric and accounted for as much as 6% of the cutin membranes (52). The nature of the insoluble residue (10–30%) left after all of the depolymerization procedures on most cutins is not known.

Structure of Cutin

Since cutin is an insoluble polymer in which the monomers are held together largely by ester linkages, it is difficult to use selective chemical cleavage techniques to determine the precise nature of the intermolecular linkages. Amorphous nature of the polymer precluded the use of X-ray diffraction techniques. A limited number of indirect approaches have been used to probe into the nature of the functional groups involved in intermonomer linkages. Depolymerization with $LiBH_4$, which reduces esterified but not free carboxyl groups showed that only a few percent of the monomers has free carboxyl groups (R. E. Purdy and P. E. Kolattukudy, unpublished). Chemical studies indicated that the majority of the free hydroxyl groups present in tomato fruit cutin are secondary and that the ester linkages which hold the polymer together involve mainly primary hydroxyl groups (29, 60). This conclusion is also supported by the observation that pancreatic lipase, which is specific for primary alcohol esters, catalyzes extensive depolymerization of cutin (13). The presence of cross-links or branching involving secondary hydroxyl groups has been shown by both chemical methods and by the release of oligomers composed of secondary alcohol esters from cutin by pancreatic lipase (I. B. Maiti & P. E. Kolattukudy, unpublished). The degree of cross-linking and/or branching in tomato fruit cutin did not appear to change during development (29).

Composition of Suberin

The composition of the aliphatic monomers of suberin, which constitute 5-30% of suberin-enriched preparations from several sources, is known (11, 64, 68). The most common aliphatic components are fatty acids, fatty alcohols, ω-hydroxyfatty acids, and dicarboxylic acids. The fatty acid and fatty alcohol fractions from suberin are usually characterized by the presence of very long chain (C_{20} to C_{30}) components. In the ω-hydroxy acid and dicarboxylic acid fractions, Δ^9 monoenoic C_{18} and/or saturated C_{16} is usually the major component, but in some cases saturated C_{22} also is a dominant component. Very long chain ($>C_{20}$) homologs with an even number of carbon atoms are often significant components of such fractions.

The more polar acids, which contain epoxy, dihydroxy, and trihydroxy functions similar to those found in cutin, are usually but not always (36, 49) minor components of suberin. Dicarboxylic acids derived from further oxidation of the ω-hydroxyl group of such polar acids are also found in suberin.

The composition of phenolic components of suberin remains obscure. The view that suberin contains both phenolic components and aliphatic components has been expressed many times over the past several decades (61, 83, 114). Suberin-enriched preparations from the periderm of the storage organs of potato, sweet potato, turnip, rutabaga, carrot, and red beet contained small amounts (≃1% of the mass) of esterified ferulic acid (94), but the bulk of the phenolics contained in such preparations have not been examined. Recently we started to apply some of the techniques developed for lignin chemistry to suberin-enriched preparations from potato tubers. Alkaline nitrobenzene oxidation gave three major phenolic fractions: (*a*) an ether soluble fraction (≃15%) containing vanillin, *p*-hydroxybenzaldehyde, and other unidentified components (but little syringaldehyde was found in this fraction) (19); (*b*) an ethyl acetate soluble fraction (≃6%); (*c*) a condensed phenolics fraction (≃20%). These fractions showed UV spectra and NMR spectra characteristic of phenolics, but they have not been fully characterized. Similar analysis of suberized walls, biosynthetically labeled with exogenous labeled cinnamate in wound-healing potato tuber, showed that the relative amounts of label found in the three fractions were similar to the mass recoveries noted above (W. Cottle, K. Espelie, and P. E. Kolattukudy, unpublished). Thus, the approximate composition of the suberin-enriched fraction from potato tuber was the following: carbohydrates 50%, phenolics 40–45%, aliphatics 5–10%. The relative amounts of the three classes of materials present in suberin-enriched preparations depend upon the tissues examined and the methods used to isolate the preparation.

Deposition of polymeric materials containing phenolic substances occurs as a result of injury caused by fungal attack or mechanical wounding (5, 38, 93, 113). In such cases the polymeric material has been variously designated as lignin, induced lignin, lignin-like, wound lignin, ligno-suberin, suberin, etc, because of the lack of information concerning chemical composition of such materials. In cases where the phenolic components of the polymeric material were examined by nitrobenzene or CuO oxidation, the results obtained were similar to those obtained with potato tuber periderm. The wound periderm formed on bean pods, jade leaves, and tomato fruit contained aliphatic components typical of suberin (26). Unfortunately, both the aliphatics and the aromatics of the suberin of wound periderm have been examined only in the case of potato tuber tissue. Based on the limited amount of available information, it is my working hypothesis that suberization is the general response to mechanical damage and that the variously designated polymeric materials containing phenolic substances (and probably aliphatics) observed in injured tissues represent suberin.

Structure of Suberin

It is difficult to determine the structure of suberin without knowing the composition of the polymer. Based on the scanty evidence available, a working hypothesis concerning the structure of suberin was proposed (Figure 1). According to this hypothesis, a phenolic matrix somewhat similar to lignin is attached to the cell wall, and the aliphatic components are covalently attached to the phenolic matrix. The ω-hydroxy acids and dicar-

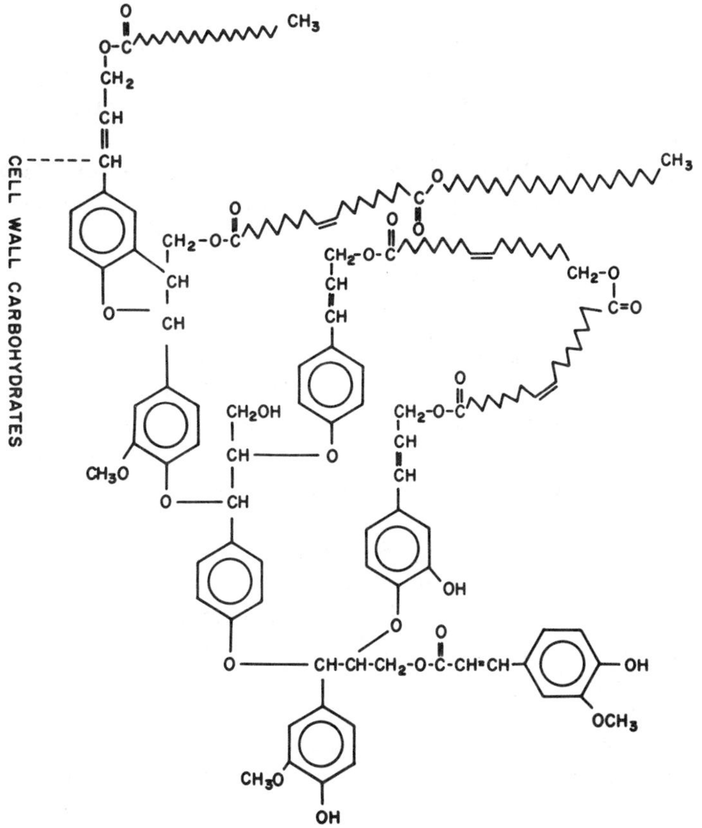

Figure 1 A tentative model for suberin structure.

boxylic acids cross-link the aromatic matrix, and ω-hydroxy acids could also form linear polyesters. In some cases where cutin-type polar acids constitute more than minor components, limited regions of cutin-like polyesters could also be present. The very long-chain fatty acids and alcohols are esterified obviously in terminal positions. The aliphatic domains provide the hydrophobicity for interacting with the waxes.

Observations which support the above working hypothesis concerning suberin structure are the following: (*a*) Suberized cell walls respond positively to staining tests for phenolics; relatively specific stains suggest that suberin contains monohydroxy aromatic rings but not many dihydroxy aromatic rings (97). In accordance with these observations, nitrobenzene oxidation of suberin generates vanillin and *p*-hydroxybenzaldehyde. There is evidence that the aromatic components of suberin are less substituted and contain fewer *O*-methoxy groups than those of lignin (5; K. E. Espelie and P. E. Kolattukudy, unpublished). Nitrobenzene oxidation also released other phenolic materials into the soluble phase probably representing condensed regions of the polymer. (*b*) Treatment of potato suberin with aqueous dioxane or dioxane-HCl, as well as milling by the Björkman method (6), released soluble phenolic materials in a manner somewhat similar to that observed with lignin. When milled for one week and extracted with dioxane, the aliphatic components released were only in proportion to the total mass released, strongly suggesting that the aliphatic components were covalently attached to insoluble polymeric matrix (K. E. Espelie and P. E. Kolattukudy, unpublished). (*c*) Depolymerization techniques which cleave ester bonds release the aliphatic components (11, 64, 68). The aliphatic components with their high proportion of dicarboxylic acids can hardly form an extensive polymer by themselves, although the hydroxyl groups of the ω-hydroxy acids in potato suberin are known not to be free. These experimental results are consistent with the cross-linking role suggested for the aliphatics. (*d*) Both phenolic acids (54) and fatty acids (27) are involved in the biosynthesis of suberin, and phenolic acids are not synthesized in tissue slices which do not suberize their cell walls (10, 54). Aminooxyphenylpropionic acid, a potent inhibitor of phenylalanine:ammonia lyase, inhibited suberization in potato tissue (C. L.Soliday and P. E. Kolattukudy, unpublished). (*e*) Time-course of deposition of the aromatic components (which generated *p*-hydroxybenzaldehyde and vanillin) indicated that the aromatic polymer was deposited prior to or simultaneously with the deposition of the aliphatic components of suberin in wound-healing potato tissue (19). Furthermore, the time-course of appearance of wall peroxidases, probably involved in suberization, correlated with the time-course of deposition of the aliphatic components of suberin (10).

BIOSYNTHESIS OF CUTIN

The finding that rapidly expanding plant organs incorporated exogenous labeled precursors into cutin (56) opened the way to a systematic study of the biosynthesis of cutin. Based on the incorporation of exogenous specifically labeled precursors and intermediates into cutin monomers, pathways were proposed for the biosynthesis of both the C_{16} and C_{18} family of acids. Subsequently all of the steps postulated to be involved in such pathways have been demonstrated in cell-free preparations although none of the enzymes have been purified.

Biosynthesis of C_{16} Family of Cutin Monomers

Labeled palmitate was incorporated into the three C_{16} monomers in rapidly expanding *Vicia faba* leaves (56). Exogenous 16-hydroxy[G-^{3}H]palmitate and 16-hydroxy[1-^{14}C]palmitate were directly incorporated into dihydroxypalmitate of cutin in *V. faba* leaves (69). The following experimental evidence strongly suggested that the conversion of ω-hydroxypalmitate to the dihyroxy acid involved direct hydroxylation by a mixed function oxidase: (*a*) neither [10-^{14}C]palmitoleic acid nor [10-^{14}C]palmitelaidate were incorporated into 10,16-dihydroxypalmitate, suggesting that a dehydrogenation followed by hydration is probably not involved in the introduction of mid-chain hydroxy groups; (*b*) only one of the four H atoms present at C-9 and C-10 of [10-^{14}C-9,10-^{3}H]palmitate was lost during the conversion of this doubly labeled acid into dihydroxypalmitate; (*c*) incorporation of exogenous labeled 16-hydroxypalmitate into dihydroxypalmitate required O_2, was inhibited by chelators such as phenanthroline, and this inhibition was reversed by Fe^{+2}. In some tissues the primary hydroxyl group of the dihydroxyacid is oxidized to an aldehyde, as was shown to be the case in germinating shoots and very young leaves of *V. faba* (59). In this case exogenous labeled dihydroxy acid was converted to the 16-oxo derivative in cutin. Similarly, the mid-chain keto derivatives found in *Citrus* (28, 32) must be produced by oxidation of the dihydroxy acid. Based on the above experimental evidence, a biosynthetic pathway was proposed for the C_{16} family of cutin monomers (Figure 2).

The two hydroxylation reactions postulated on the basis of the above results were demonstrated in cell-free preparations. Endoplasmic reticulum fraction from the shoots of germinating *V. faba* catalyzed ω-hydroxylation of palmitic acid with NADPH and O_2 as required cofactors (105). Free fatty acid appears to be the substrate because neither ATP and CoA stimulated the reaction nor was palmitoyl-CoA a preferred substrate. Stearate was ω-hydroxylated at very slow rates ($<10\%$), whereas myristate and oleate

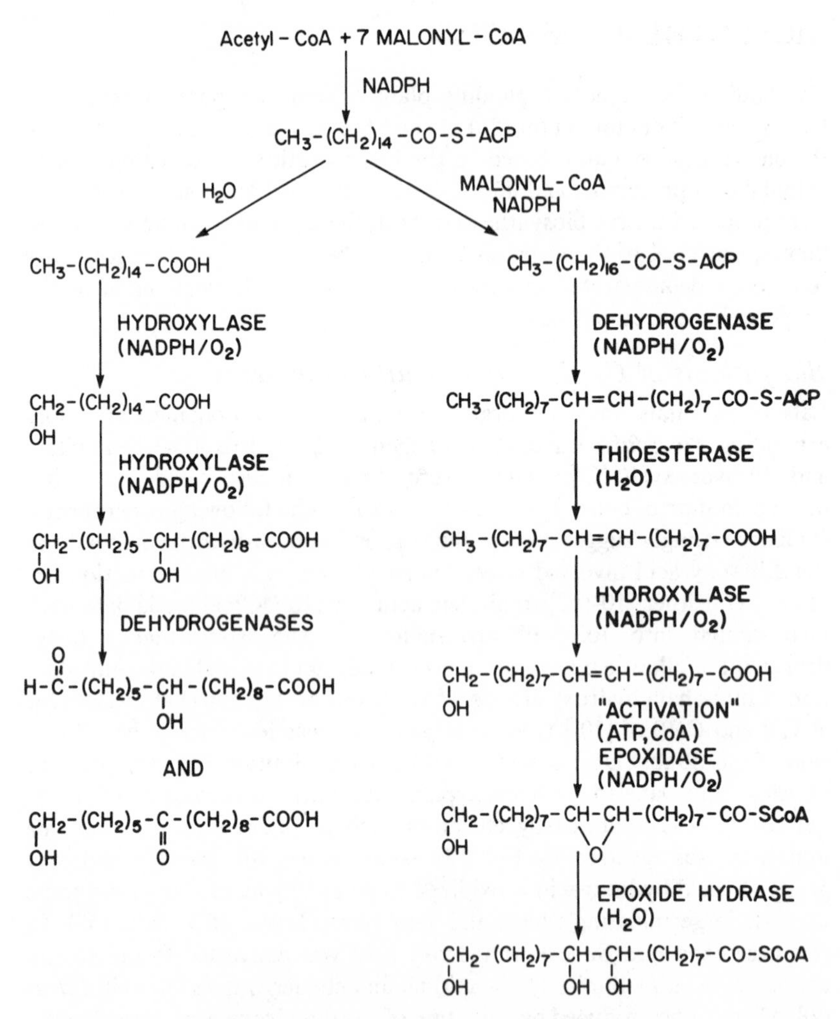

Figure 2 Biosynthetic pathway for cutin monomers.

were ω-hydroxylated at rates comparable to that observed with palmitate. Classical mixed function oxidase inhibitors such as metal ion chelators NaN_3 and thiol-directed reagents inhibited the ω-hydroxylation. Involvement of cytochrome P_{450} in this reaction remains in doubt because light at 420 to 460 nm did not reverse the inhibition caused by CO even when 10% CO (which caused only 30% inhibition) was used. This is the first example of ω-hydroxylation of a fatty acid with a known biological function.

Mid-chain hydroxylation of ω-hydroxypalmitate was demonstrated with cell-free preparations from excised epidermis of *V. faba* (116). More recently, this hydroxylation was characterized with a microsomal fraction from the shoots of germinating *V. faba* using a high performance liquid chromatographic assay (106). This hydroxylation required O_2 and NADPH and was inhibited by NaN_3, metal ion chelators and thiol-directed reagents. Photoreversible inhibition of the midchain hydroxylation by CO strongly suggested that cytochrome P_{450} was involved in this reaction. Although both ω-hydroxylation and mid-chain hydroxylation are catalyzed by the endoplasmic reticulum fraction, the drastic differences in the sensitivity of the two reactions to CO and in the photoreversibility of this inhibition suggest that two different enzymes catalyze the two hydroxylation reactions.

Biosynthesis of C_{18} Family of Cutin Monomers

Incorporation of exogenous labeled precursors into the C_{18} family of cutin monomers in skin slices of developing apple fruit and grape berries, and leaves of apple and *Senecio odoris* provided experimental evidence which suggested the pathway shown in Figure 2. [1-^{14}C]Acetate was incorporated into all monomers and [1-^{14}C]palmitate was incorporated exclusively into the C_{16} family of monomers (71). [1-^{14}C]Stearate was not incorporated into any mid-chain substituted C_{18} monomers, suggesting that direct mid-chain hydroxylation such as that demonstrated in the case of the C_{16} family of monomers was not involved in the introduction of the mid-chain functional groups found in this family of monomers. [1-^{14}C]Oleate, on the other hand, was readily incorporated into 18-hydroxyoleate, 18-hydroxy-9,10-epoxy-stearate, and 9,10,18-trihydroxystearate. Similarly, labeled linoleate and linolenate were incorporated specifically into the corresponding 18-hydroxy, 18-hydroxy-9,10-epoxy, and 9,10,18-trihydroxy acids. These results strongly suggested that ω-hydroxylation, epoxidation of the double bond at C-9, and subsequent hydration of the epoxide are the three major steps involved in the biosynthesis of this family of monomers. This hypothesis was strongly supported by the direct conversion of exogenous 18-hydroxy[18-^{3}H]oleate into 18-hydroxy-9,10-epoxystearate and 9,10,18-trihydroxystearate in plant tissue slices (22). The proposed sequence of reactions was further supported by the direct conversion of exogenous 18-hydroxy-9,10-epoxy[18-^{3}H]stearate to 9,10,18-trihydroxystearate in tissue slices. The epoxidation-hydration process is usually limited to the double bond at C-9, but in some cases these reactions also occur at the double bond at C-12, resulting in pentahydroxy acids. For example, 9,10,18-trihydroxy-12,13-epoxystearate and 9,10,12,13,18-pentahydroxystearate were found in *Rosmarinus officinalis* cutin (12, 21), and exogenous [1-^{14}C]linoleate was incorporated into these two acids in addition to the other more common components of the C_{18} family of acids (21).

Epoxidation and epoxide hydration, the two unique reactions postulated to be involved in the biosynthesis of the C_{18} family of acids, have been demonstrated in cell-free preparations. Epoxidation of 18-hydroxy[18-^{3}H]oleate to the corresponding *cis*-epoxyacid was catalyzed by a 3000 *g* particulate preparation from young spinach leaves in which this epoxyacid is a major component of cutin (24). This reaction required ATP, CoA, NADPH, and O_2 as cofactors, suggesting that this mixed function oxidase required an activated carboxyl group seven methylene groups away from the epoxidation site. The enzyme required *cis*-Δ^9 and a free ω-hydroxyl group in the substrate. This reaction was inhibited by chelators and this inhibition was reversed by Fe^{+2}. Involvement of cytochrome P_{450} in this epoxidation was strongly suggested by photoreversible inhibition of the reaction by CO.

Hydration of 18-hydroxy-*cis*-9,10-epoxystearate to *threo*-9,10,18,-trihydroxystearate was catalyzed by a 3000 *g* particulate fraction from the skin of rapidly growing apple fruit (25). This *trans* hydration of the epoxide is consistent with the known occurrence of *cis*-epoxyacid and *threo*-trihydroxy C_{18} acid in apple cutin. This epoxide hydrase was located exclusively in the 3000 *g* particulate preparation from the skin of the fruit (but not in the internal tissue). This biosynthetic epoxide hydrase, associated with cuticular membranes, showed a fairly stringent specificity for its substrate. Trichloropropene oxide, a potent inhibitor of the mammalian catabolic epoxide hydrase, was a poor inhibitor of the plant enzyme whereas thiol-directed reagents severely inhibited the enzyme involved in cutin synthesis. Epoxide hydrase activity similar to that found in the apple fruit preparation was also found in similar particulate preparations from excised epidermis of *S. odoris* leaves and spinach leaves.

Synthesis of Cutin From Monomers

Since cutin is an extracellular insoluble polymer, the enzyme(s) which catalyzes the formation of the polymer from the monomers would be expected to be at an extracellular location, probably in close association with the polymer. In fact, a 3000 *g* particulate fraction containing cuticular membrane fragments from excised epidermis of rapidly expanding *V. faba* leaves catalyzed incorporation of labeled hydroxy acids into an insoluble polymer in the presence of ATP and CoA (20). The following observations suggested that this incorporation into the insoluble material represented cutin synthesis: (*a*) Particulate preparation only from the epidermis but not that from the mesophyll tissue catalyzed the synthesis of the insoluble material. (*b*) Labeled hydroxyfatty acids from the labeled insoluble material was released only upon treatment with purified cutinase but not with other hydrolases. (*c*) Depolymerization of the labeled insoluble material by alkaline hydrolysis, transesterification with BF_3 in methanol and reductive cleavage with $LiAlH_4$, released the labeled monomers in a manner expected from a polyester.

Incorporation of the monomers into the polymer required ATP and CoA, suggesting that activation of the carboxyl group of the monomers is involved in the enzymatic esterification process. The enzyme preparations from leaves of *V. faba,* in which cutin is composed of chiefly the C_{16} family of monomers, showed preference for the C_{16} family of acids, but other acids could also be incorporated into the polymer by this enzyme (23). Particulate enzyme preparations from *V. faba* flower petals and epidermis of *Senecio odoris* leaves also catalyzed incorporation of labeled hydroxyfatty acids into cutin, suggesting that the synthesis of the polymer from the monomer occurs in a manner similar to that described for *V. faba.*

The particulate preparations which synthesized cutin from the monomers contained endogenous cutin primer, and therefore primer dependence could not be studied with such preparations. Ultrasonic treatment of the 3000 *g* particulate preparation from *V. faba* provided a soluble preparation which required exogenous cutin as primer for incorporation of labeled hydroxyfatty acids into the polymer (23). The most active primer for this preparation was cutin from very young leaves of *V. faba,* and the priming activity of cutin decreased with increasing age of the leaf; cutin from a variety of plants could also serve as the primer. Chemical treatments which increased the number of hydroxyl groups present in cutin increased the priming efficiency. Limited treatment of cutin with cutinase increased the priming efficiency, suggesting that this enzyme probably produced nicks and generated more esterification sites in the polymer. All of the available experimental evidence suggests that hydroxy and epoxy acyl groups are transferred from their thioester derivatives to the hydroxyl groups of the growing polymer.

BIOSYNTHESIS OF SUBERIN

Biosynthesis of Aliphatic Components of Suberin

The composition of the aliphatic components of suberin deposited in the wound periderm of potato tuber tissue is identical to that of the natural skin of intact potato tubers (66). This finding allowed the use of wound-healing potato tuber tissue slices to study suberin biosynthesis, and virtually all of the information presently available on the biosynthesis of suberin was obtained from this system. Since ω-hydroxyoleic acid and the corresponding dicarboxylic acid are the major aliphatic components of potato suberin, the amount of octadecene-1,18-diol generated by $LiAlH_4$ reduction of the polymer could be used as a measure of the aliphatic components of suberin in this tissue. With such an assay, it was found that deposition of the aliphatic components of suberin started after about 3 days of wound healing, the most rapid deposition occurred during the fifth and sixth days, and the process was completed after about 8 days. Tissue slices, which had wound-healed for 4 days, incorporated [1-^{14}C]oleic acid into ω-hydroxyoleic acid and the corresponding dicarboxylic acid of suberin polymer (27). Labeled acetate was incorporated into not only these unsaturated bifunctional C_{18} acids, but also into very long chain (up to C_{28}) fatty acids and alcohols. The time-course of development of the ability of the tissue to incorporate the exogenous labeled precursors into these suberin components was consistent with the observed time-course of deposition of the aliphatic components noted above.

On the basis of the chemical composition of the aliphatic components of suberin and the biosynthetic studies thus far conducted, the biosynthesis of these components can be depicted as shown in Figure 3. The more polar acids, which are usually minor components of suberin, are synthesized most probably by reactions analogous to those discussed under cutin. Synthesis of very long chain acids and alcohols, ω-hydroxylation, and conversion of the ω-hydroxyacids to the corresponding dicarboxylic acids are the major steps involved in the formation of the aliphatic components of suberin. Chain elongation is induced in wound-healing potato tissues as early as one day after wounding (8, 120), and this process can be demonstrated readily by incorporation of acetate into very long aliphatic chains in wound-healing potato tissue slices (27). Even though chain elongation has not been studied with cell-free preparations from suberizing tissues, on the basis of the results obtained with cell-free preparations from other plant tissues (17, 65, 80), it is most probable that a membranous elongating system catalyzes elongation with malonyl-CoA as the elongating agent and NADPH as the reductant. Acyl-CoA reductase, similar to that found in other plant tissues (57), probably catalyzes the reduction of the elongated products. ω-Hydroxylation of fatty acids has been shown to be catalyzed by microsomal preparations from *V. faba* roots (105), and the properties of this enzyme system were quite similar to those described under cutin synthesis.

Oxidation of ω-hydroxy acids to the corresponding dicarboxylic acids, a unique reaction involved in suberin synthesis, has been studied extensively (2–4). Extracts of suberizing potato tuber slices catalyzed oxidation of 16-hydroxypalmitate to 16-oxopalmitate, which was subsequently oxidized to the corresponding dicarboxylic acid. These two steps were catalyzed by two dehydrogenases which could be resolved by gel filtration with Sepharose 6B. The ω-hydroxyacid dehydrogenase, which catalyzes the first step in this process, was induced for suberization while the other dehydrogenase activity was present at all times in the potato tissue (3). This wound-

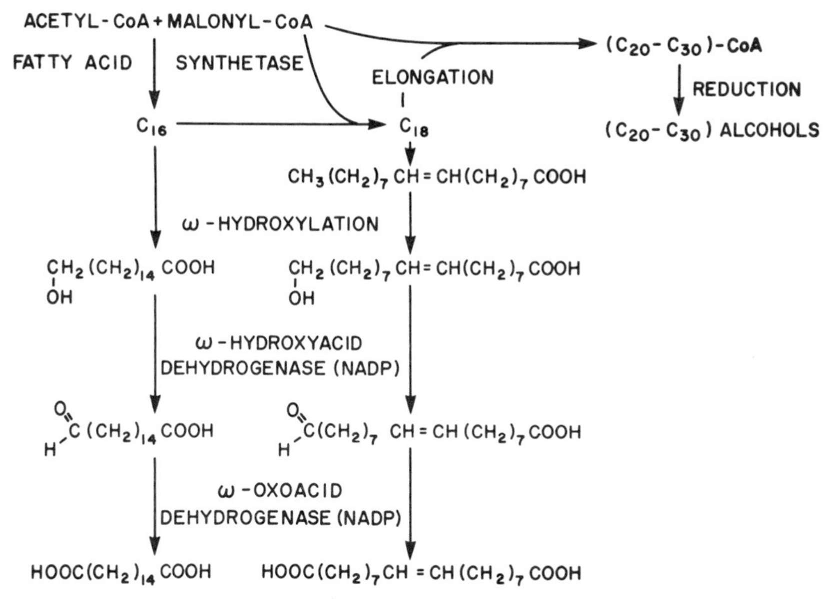

Figure 3 Biosynthesis of the aliphatic components of suberin.

induced ω-hydroxyacid dehydrogenase was purified to homogeneity from acetone powder extracts of suberizing potato slices using a combination of gel filtration, anion exchange, and hydroxyapatite chromatography, and NADP-Sepharose affinity chromatography (3). This enzyme, which consists of two 30 K dalton protomers, catalyzes the oxidation of ω-hydroxypalmitate (V_{max} 140 nmoles/min/mg) and the reduction of the corresponding ω-oxo acid (V_{max} 3200 nmoles/min/mg) with a pH optimum of 9.5 and 8.5, respectively. The equilibrium constant of the reaction at pH 9.5 and 30°C was 14×10^{-9} M. An unusual feature of this enzyme was that hydride from the A side of NADPH was transferred to 16-oxopalmitate at nearly the same rate as that from the B side.

Kinetic and chemical modification studies on the ω-hydroxyacid dehydrogenase provided information concerning the mechanism of action of this enzyme (4). Initial velocity and product inhibition studies suggested an ordered sequential mechanism where NADPH was bound to the enzyme first, followed by the oxoacid, and NADP was released after the hydroxy acid. Chemical modification with phenylglyoxal suggested that an arginine residue was involved in binding NADPH. Involvement of the ϵ-amino group of a lysine residue in binding the distal carboxyl group of the ω-oxoacid was suggested by chemical modification with pyridoxal phosphate and by the pH dependence of the K_m of the oxoacid but not of hexadecanal. Chemical modification with diethylpyrocarbonate strongly suggested that the imidazol group of a histidine residue was involved in the reaction. Kinetic studies with substrate analogs suggested that the binding of the substrates at the active site of this dehydrogenase involves a hydrophobic region and that aliphatic chains longer than C_{20} could not be accommodated in this active site. The latter finding explains why very long chain ($>C_{20}$) dicarboxylic acids are not found in potato suberin although it contains very long chain ω-hydroxyacids (11, 64). A comparison of the properties of ω-hydroxy acid dehydrogenase with those of other dehydrogenases suggests that this dehydrogenase has evolved a binding site specialized for the ω-hydroxyacids while retaining the same features as those of other dehydrogenases for catalytic purposes.

Biosynthesis of Aromatic Components of Suberin

Biosynthesis of the aromatic components of suberin has not been studied extensively although much information is available on the phenolic metabolism in wounded plant tissues (93). The major difficulty in relating such information to the suberization process arises from the fact that such biochemical studies, for the most part, dealt with the early phases of the wound-healing process. Even when the deposition of the soluble phenolic compounds into insoluble polymeric materials was observed, the nature of the insoluble material was not defined (39, 92, 109). However, it is probable that the active phenolic metabolism observed during the early phases of wound healing provides the monomers required for suberization. The sequence of reactions most probably involved in this process can be depicted as shown in Figure 4. The individual reactions involved in this process have been elucidated with cell-free preparations and at least partially purified enzymes. These studies are beyond the scope of the present review and have been described elsewhere (41, 46, 93, 122).

Recently we have initiated a study of the phenolic metabolism associated with suberization. The amount of *p*-hydroxybenzaldehyde and vanillin generated by alkaline nitrobenzene oxidation of suberin-enriched preparation from the wound periderm of potato tubers was used as a measure of deposition of the phenolic matrix during suberization. The time-course of deposition of such phenolics in the insoluble material showed a lag period of about 3 days, and during the subsequent several days there was a rapid increase in phenolic deposition which ceased in about 10–12 days (19). During the early phases, the amount of *p*-hydroxybenzaldehyde exceeded the amount of vanillin, but after 10 days of wound healing more vanillin than *p*-hydroxybenzaldehyde was obtained, possibly reflecting a change in the phenolic composition or degree of condensation of the polymer. Labeled L-phenylalanine was incorporated into insoluble material in wound-healing potato tuber disks. Exogenous labeled *trans*-cinnamate was a more efficient

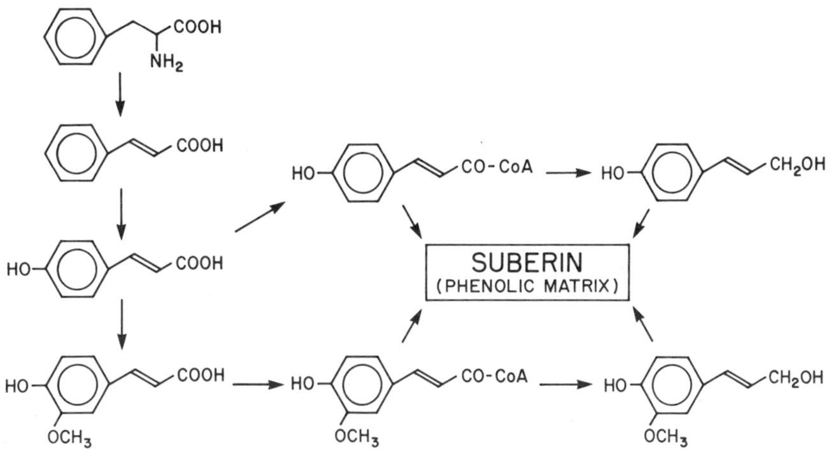

Figure 4 Biosynthesis of the aromatic components of suberin. Some phenolic acids and alcohols might first be esterified with fatty alcohols and fatty acids, respectively (via the CoA derivatives of the respective acids), before being incorporated into the polymer by peroxidase type enzymes.

precursor for the phenolic polymer than was phenylalanine. Nitrobenzene oxidation of the insoluble material derived from labeled cinnamic acid gave labeled *p*-hydroxybenzaldehyde and vanillin. To test whether chlorogenic acid, which is known to accumulate in wounded plant tissues, was used in suberization, the level of chlorogenic acid in wound-healing potato tuber tissue was measured by a high performance liquid chromatographic technique. The amount of chlorogenic acid increased to a maximum level in about 4 days of wound healing, prior to the most rapid deposition of the phenolic components of suberin. However, even after 8 days of suberization, the level of chlorogenic acid did not decrease drastically, and labeled chlorogenic acid generated from exogenous labeled cinnamate also did not disappear during suberization. Therefore, it remains uncertain whether the accumulated chlorogenic acid is in fact the precursor of the phenolic matrix of suberin.

Biosynthesis of Suberin From Monomers

Synthesis of the polymer from the monomers has not been studied in cell-free preparations. The aromatic monomers presumably undergo polymerization in a manner analogous to that suggested for lignin synthesis (40, 41). It is well known that peroxidase, present in cell walls, can catalyze polymerization of phenolic monomers and covalent bond formation between cell walls and the phenolic polymer (40, 41, 119). Among the various forms of this enzyme, the wall-bound forms appear to be more efficient in catalyzing the bond formation between carbohydrates and phenolic polymers, and the linkages generated by such forms of the enzyme appear to be resistant to base hydrolysis (118). The following observations suggest that the phenolic matrix of suberin is probably formed by a similar process involving peroxidases. (*a*) The polymeric materials generated by potato parenchyma from *p*-coumaric and ferulic acids were highly or partly condensed, resulting in low yields of aromatic aldehydes upon nitrobenzene oxidation (7). Similar treatment of potato suberin from the intact tuber and from wound periderm also gave low yields of *p*-hydroxybenzaldehyde and vanillin (as discussed above). (*b*) During the wound healing of potato tissue, specific peroxidases were found to be induced, and they were present specifically in the suberizing cells (9). (*c*) Both the time-course and spatial distribution of wound-induced activity of peroxidase in this tissue highly correlated with suberization (10). (*d*) A washing of the wound with cyanide and the use of 10% CO_2 in the gas phase suppressed suberization and prevented the induction of the peroxidases characteristic of suberizing cells but not that of the internal peroxidase (10). (*e*) Peroxidase in roots appears to be localized in the epidermal and endodermal walls (84, 102), the known areas of suberization.

The source of H_2O_2 needed for the polymerization process could be similar to that suggested to be involved in lignification (41). A wall-bound malate dehydrogenase similar to that found in horseradish (31, 43) and *Forsythia* (42) might produce sufficient amounts of NADH to allow generation of adequate amounts of H_2O_2 by the wall-bound peroxidase. The advantages of such a system for generating H_2O_2 at the extracellular site, where it is to be utilized, are obvious in the case of suberization, which occurs in a localized area of the cell wall.

The aliphatic components are presumably esterified to the functional groups in the phenolic polymer. Activated ω-hydroxyacids and dicarboxylic acids are most probably involved in this process as was shown for cutin synthesis. Fatty alcohols might be incorporated into the polymer by the peroxidase-catalyzed polymerization of phenylpropanoic acid esters of the alcohols, such as the feruloyl esters of C_{18}–C_{28} alcohols found in potato periderm (1).

Regulation of Suberin Biosynthesis

The finding that wounding triggers suberization even on tissues normally covered by cutin (26) suggests that this process is a general response to wounding of plants. It appears probable that some chemical generated by the wound initiates the chain of events which leads to suberization; in fact, thorough washing of the wound surface of potato tuber inhibited suberization (103). This washing removed abscisic acid and addition of abscisic acid to the washed tissue partially restored suberization. Furthermore, in potato tissue culture suberization was induced by abscisic acid in a dose-dependent manner. Although these observations suggested that abscisic acid plays a role in suberization, this role is presumably indirect because suberization could be inhibited by washing the tissue even 3 days after wounding, although abscisic acid was not removed by this washing. Experiments with actinomycin D and cycloheximide suggested that the transcriptional and translational processes directly related to the synthesis of the aliphatic components of suberin occurred between 72 and 96 hrs after wounding (2). Washing of the tissue any time prior to this period (72 hr) inhibited suberization. Thus, it appears that abscisic acid, generated during the first day after wounding, triggers a process which results in the production of the suberization-inducing factor which causes the induction of the enzymes involved in suberization.

BIODEGRADATION OF CUTIN

Degradation of Cutin by Plants

Since cutin is an extracellular polymer, it does not undergo much metabolic turnover. However, during the expansion of plant organs the polymer might undergo localized cleavages to generate new sites for extension of the polymer. A recent ultrastructural study raised the possibility that enzymatic hydrolysis of cutin creates an opening in the cuticle located above the newly differentiated sunken stomata (16). Another suspected case of degradation of cutin by a plant-generated enzyme is that catalyzed by a pollen enzyme during the penetration of stigmatic cuticle. The presence of such a cutinase was suggested by the finding that addition of cutin preparation to germinating pollen caused a slight increase in titratable acidity (48, 79). More recently, germinating pollen of nasturtium (*Tropaeolum majus*) was shown to excrete a cutinase which catalyzed the hydrolytic release of all types of monomers from labeled cutin (98). Actinomycin D and cycloheximide did not have any effect on the amount of cutinase released from pollen, suggesting that the enzyme was already present in the mature pollen. This cutinase was purified to homogeneity by Sephadex G-100 gel filtration, QAE-sephadex chromatography, and isoelectric focusing (81). It was a 40 K dalton peptide which contained about 7% carbohydrates and catalyzed hydrolysis of *p*-nitrophenyl esters of C_2 to C_{18} fatty acids with similar K_m and V_{max}. The pH optima for the hydrolysis of cutin and the *p*-nitrophenyl esters were 6.8 and 8.0, respectively. This enzyme showed a preference for primary alcohol esters, but triglycerides were not hydrolyzed at significant rates. It was severely inhibited by thiol-directed reagents and it was totally insensitive to active serine-directed reagents. The properties of the pollen enzyme are in sharp contrast to those of the fungal cutinases described below.

Degradation of Cutin by Microbes

Many microorganisms, including plant pathogens, can grow on cutin as their sole carbon source and produce extracellular cutin hydrolyzing enzymes (5a, 45, 47, 88, 100). Two isozymes of cutinase were isolated in homogeneous form from the growth medium of *Fusarium solani pisi* (89). The two isozymes were very similar in amino acid composition and immunological properties as well as in their catalytic properties (89, 90, 104). The only difference between the two proteins is that cutinase II contained what appeared to be a proteolytic nick near the middle of the polypeptide. More recently, cutinase was isolated from *Fusarium roseum culmorum* (104), *F. roseum sambucinum, Helminthosporum sativum, Ulocladium consortiale,* and *Streptomyces scabies* (78). The enzymes from all of these fungal sources are quite similar in size (25 K daltons) and amino acid composition. The two cutinases from *F. solani pisi* have no free SH groups and have one disulfide bridge which is essential for activity (T.-S. Lin and P. E. Kolattukudy, unpublished). In cutinase II the two halves of the molecule, which had undergone a proteolytic nick, are held together by this disulfide bridge. Since cutinases from the other sources (except that from *F. roseum culmorum*) also show two 1/2 cys and they are not affected by SH-directed reagents, it appears likely that they also have one disulfide bridge in them.

The catalytic properties of fungal cutinases are also quite similar. All of the fungal cutinases hydrolyze cutin at maximal rates at pH 10 and the enzyme from *F. solani pisi,* during short incubation periods, generated oligomers from cutin but further incubation resulted in the hydrolysis of the oligomers (90). All of the fungal cutinases also hydrolyze *p*-nitrophenyl esters of short chain fatty acids, such as butyrate, but not of long chain acids such as palmitate (90). Unlike the pollen cutinase, the fungal cutinases were completely insensitive to SH-directed reagents but were severely inhibited by active serine-directed reagents. Covalent attachment of one mole of diisopropylphosphoryl moiety per mole of cutinase caused complete inhibition of the enzyme.

The extracellular fungal cutinases thus far examined are glycoproteins containing 3.5 to 6% carbohydrates (73, 77). The carbohydrates are attached to the protein via *O*-glycosidic linkages in all cases except the cutinases from *S. scabies* and *H. sativum,* in which case the carbohydrates are presumably attached via asparaginyl residues (78). Cutinases, which contain *O*-glycosidic linkages, have been examined for the nature of the amino acid residues and carbohydrates involved in such linkages. β-Elimination of the carbohydrates by alkali, followed by reduction of the resulting dehydroaminoacyl residues and the released carbohydrates by NaB^3H_4, generated labeled protein and labeled carbohydrates. Analysis of the hydrolysate of the labeled protein thus generated revealed that the two cutinases from *F. solani pisi* contained *O*-glycosidic linkages involving serine and threonine, two amino acids previously known to be involved in such linkages in other glycoproteins. In addition, β-hydroxyphenylalanine and β-hydroxytyrosine, two amino acids heretofore not found in any other protein, were also found to be involved in *O*-glycosidic linkages in these proteins (73). The two novel hydroxyamino acids were isolated and identified from cutinase in the following manner (76): induction of cutinase synthesis in glucose-grown *F. solani pisi* by cutin hydrolysate (75) in the presence of labeled phenylalanine generated labeled cutinase. Removal of the *O*-glycosidically attached carbohydrates from this labeled enzyme by HF treatment followed by enzymatic hydrolysis gave four labeled amino acids which were chromatographically identified as phenylalanine, tyrosine, β-hydroxyphenylalanine and β-hydroxytyrosine.

Analyses of the products generated by alkaline NaB^3H_4 treatment of cutinases from several pathogenic fungi showed that *O*-glycosidic linkages involving all four hydroxyamino acids were not present in the enzyme from all sources. The following amino acids were found to be involved in the *O*-glycosidic linkages: serine, threonine and β-hydroxyphenylalanine in *U. consortiale,* serine and β-hydroxyphenylalanine in *F. roseum culmorum,* and only serine in *F. roseum sabucinum* (78). The carbohydrates attached to cutinase also showed novel features in that they were monosaccharides which had not been previously found to be *O*-glycosidically linked to proteins. Analyses of the labeled carbohydrates released by alkaline NaB^3H_4 treatment of the two cutinases from *F. solani pisi* showed that the monosaccharides attached to the protein were mannose, arabinose, glucosamine (possibly N-acetyl), and glucuronic acid (77). Cutinases from *F. roseum culmorum, F. roseum sambucinum,* and *U. consortiale* also contained the same *O*-glycosidically attached carbohydrates as above except the pentose (78).

Introduction of the *O*-glycosidically linked mannose into proteins via dolichol phosphoryl mannose was catalyzed by a microsomal preparation from glucose-grown *F. solani pisi* in which cutinase synthesis was induced with cutin hydrolysate (107). Such preparations catalyzed transfer of single mannosyl residues from GDP-mannose to endogenous dolicholphosphate and glycoproteins (107). The microsomal preparation also catalyzed mannosyl transfer from GDP-mannose to exogenous dolichol phosphate generating β-D-mannosyl phosphoryl dolichol. The endogenous mannosyl acceptor lipids from this fungus were isolated and identified to be phosphates of C_{95} and C_{100} dolichols. The microsomal preparation also catalyzed the transfer of mannose from exogenous mannosyl phosphoryl dolichol to glycoproteins. Analyses of the β-elimination products of the glycoproteins generated from both GDP-mannose and dolichol phosphoryl mannose showed that single mannosyl residues were transferred to hydroxyl groups of the endogenous proteins. Bacitracin and amphomycin but not tunicamycin inhibited the mannosyl transfer reactions. Sodium dodecyl sulfate electrophoresis indicated that cutinase and its possible precursors were among the in vitro glycosylation products. Since exogenous cutinase was not glycosylated even after denaturation, sulfitolysis, or removal of carbohydrates by HF treatment, it appears that exogenous cutinase cannot reach the glycosylation site which probably is the lumen of the endoplasmic reticulum. The mechanism of introduction of the other *O*-glycosidically linked carbohydrates into cutinase remains unknown.

The following experimental evidence showed that the *N*-terminal amino group of the cutinases from *F. solani pisi* was in amide linkage with glucuronic acid, an unusual blocking group: (*a*) The *N*-terminal amino group did not react with phenylisothiocyanate or dansyl chloride (74). (*b*) Upon treatment of the protein with neutral NaB^3H_4, gulonic acid attached to the protein became labeled and only gulonic acid was labeled when the protein was deglycosylated with HF prior to alkaline NaB^3H_4 treatment (77). (*c*) *N*-Gulonylglycine was isolated from the pronase digest of the enzyme after treatment with a neutral solution of NaB^3H_4 (74). All of the fungal cutinases, which contained *O*-glycosidically linked carbohydrates, also had glucuronic acid attached to the protein via an alkali stable linkage, presumably an amide linkage at the *N*-terminal amino group. Such a structural feature was found in cutinase from *F. solani pisi, F. roseum culmorum, Ulocladium consortiale,* and *F. roseum sambucinum* (78).

Induction of cutinase by cutin or cutin hydrolysate was always accompanied by the production of very small quantities of a nonspecific esterase which catalyzed hydrolysis of a variety of small esters but not hydrolysis of cutin or oligomers generated from cutin by cutinase (89, 90). This enzyme catalyzed hydrolysis of *p*-nitrophenyl esters of C_2 to C_{18} fatty acids. That this protein is biosynthetically related to cutinase, possibly as precursor, was suggested by the following experimental evidence (60, 62): (*a*) Induction of cutinase in all fungi under all conditions thus far examined resulted in the production of the esterase. (*b*) Like cutinase, the esterase contains an "active" serine, which is involved in catalysis (90). (*c*) Ouchterlony double diffusion analysis showed that the nonspecific esterase from *F. solani pisi* cross-reacted with rabbit anticutinase I, and immunoelectrophoresis showed that this cross-reactivity was not due to contamination of the nonspecific esterase by cutinase. (*d*) The amino acid composition of the esterase was fairly similar to that of cutinase I. (*e*) The nonspecific esterase, which has twice the molecular weight of cutinase, contains about 50% carbohydrates, and therefore it appears that the nonspecific esterase might be cutinase with an equal mass of carbohydrate attached to it. (*f*) Alkaline NaB^3H_4 treatment of this enzyme labeled carbohydrates much larger than those obtained from cutinase by a similar treatment. The large carbohydrates present in the "proenzyme" presumably prevent access of the macromolecular substrate (cutin) but not of small substrates to the active site of the enzyme. (*g*) In both cutinase and the nonspecific esterase, the carbohydrate and amino acid components involved in the *O*-glycosidic linkages are the same (T.-S. Lin and P. E. Kolattukudy, unpublished). (*h*) In NaB^3H_4 treated esterase, a labeled gulonyl residue was found to be attached to the peptide by an alkali-stable linkage, presumably by an amide linkage similar to that found in cutinase.

BIODEGRADATION OF SUBERIN

Ultrastructural studies indicate that fungal penetration of roots and barks probably involves enzymatic degradation of suberin (44, 86). However, the enzymes involved in this process have not been studied extensively. From the extracellular fluid of *Streptomyces scabies* and *Fusarium solani pisi,* grown on potato tuber suberin, enzymes, which catalyze the release of aliphatic monomers from suberin, were isolated (61). Since the assay used for purification was based on cutin hydrolysis, any enzymes which could release aromatic components of suberin would not have been detected. More recently, *Phytophthora cactorum* and *P. erythroseptica* were grown on suberin, and the extracellular fluid from these cultures released radioactivity from potato suberin biosynthetically derived from exogenous labeled cinnamate (C. Allan and P. E. Kolattukudy, unpublished). The enzymes involved in the degradation of the aromatic matrix are yet to be isolated.

CONCLUSION

Although much progress has been made in our knowledge of cutin and suberin during the past decade, we have only begun to understand the chemistry and biochemistry of these extremely complex phytopolymers. Major gaps remain in our knowledge of these polymers. For example, little is known about the intermolecular structure of cutin and the regulation of its biosynthesis, and none of the enzymes involved in the biosynthesis have been purified. The mechanics of synthesis of this insoluble polymer at an extracellular location is not understood. Even the chemical composition of suberin, particularly the composition of the aromatic components, is not known, and the vague description of the structure of this polymer shown in the proposed model is only a beginning working hypothesis. Since suberin presents the complexity of both lignin and cutin, a combination of techniques developed for studying such polymers should be applied to suberin. Biosynthesis of the aromatic components of suberin is also poorly understood, and the mechanisms by which the suberization process is triggered by wounding is unknown. It is quite possible that small quantities of ligands covalently attached to the polymers play an important role as biologically active components involved in the interaction of the plant with microorganisms and insects, but such aspects have not been investigated.

It is most likely that cutin and suberin contribute significantly to the making of coal from plant residues. Therefore, the new impetus in coal research, triggered by the recent awareness of the energy crisis, might reveal the structural features of the polymers left in coal and thus contribute to our understanding of the chemistry of these phytopolymers. Conversely, elucidation of the chemistry of these phytopolymers could help to understand the coalification process. In any case, considering the important functions these polymers play in the life of a plant [discussed in (62, 63, 67, 83)], and in view of the possibility of recovering useful chemicals from the polymeric residues present in agricultural wastes and in deposits such as peat and coal, chemical and biochemical studies of such polymers are likely to produce information relevant to the production and better utilization of plant products.

ACKNOWLEDGMENTS

I thank Dr. Karl Espelie for assistance in preparing this manuscript. Supported by NSF grant PCM 7700927 and a grant from Washington State Tree Fruit Commission.

Literature Cited

1. Adamovics, J. A., Johnson, G., Stermitz, F. R. 1977. Ferulates from cork layers of *Solanum tuberosum* and *Pseudotsuga menziesii*. *Phytochemistry* 16: 1089–90
2. Agrawal, V. P., Kolattukudy, P. E. 1977. Biochemistry of suberization; ω-hydroxyacid oxidation in enzyme preparations from suberizing potato tuber disks. *Plant Physiol.* 59:667–72
3. Agrawal, V. P., Kolattukudy, P. E. 1978. Purification and characterization of a wound-induced ω-hydroxyfatty acid:NADP oxidoreductase from potato tuber disks. *Arch. Biochem. Biophys.* 191:452–65
4. Agrawal, V. P., Kolattukudy, P. E. 1978. Mechanism of action of a wound-induced ω-hydroxyfatty acid:NADP oxidoreductase isolated from potato tubers (*Solanum tuberosum* L). *Arch. Biochem. Biophys.* 191:466–78
5. Asada, Y., Ohguchi, T., Matsumoto, I. 1976. Biosynthesis of lignin in Japanese radish root infected by downy mildew fungus. In *Biochemistry and Cytology of Plant-Parasite Interactions*, ed. K. Tomiyama, J. M. Daly, I. Uritani, H. Oku, S. Ouchi, pp. 200–12. Amsterdam: Elsevier. 256 pp.

5a. Baker, C. J., Bateman, D. F. 1978. Cutin degradation by plant pathogenic fungi. *Phytopathology* 68:1577–84

6. Björkman, A. 1956. Studies on finely divided wood. Part 1. Extraction of lignin with neutral solvents. *Sven. Papperstidn.* 59:477–85
7. Bland, D. E., Logan, A. F. 1965. The properties of syringyl, guaiacyl and *p*-hydroxyphenyl artificial lignins. *Biochem. J.* 95:515–20
8. Bolton, P., Harwood, J. L. 1976. Fatty acid synthesis in aged potato slices. *Phytochemistry* 15:1501–6
9. Borchert, R. 1974. Isoperoxidases as markers of the wound-induced differentiation pattern in potato tuber. *Dev. Biol.* 36:391–99
10. Borchert, R. 1978. Time course and spatial distribution of phenylalanine ammonia-lyase and peroxidase activity in wounded potato tuber tissue. *Plant Physiol.* 62:789–93
11. Brieskorn, C. H., Binneman, P. H. 1975. Carbonsaüren und Alkanole des Cutins und Suberins von *Solanum tuberosum. Phytochemistry* 14:1363–67
12. Brieskorn, C. H., Kabelitz, L. 1971. Hydroxyfettsäuren aus dem Cutin des Blattes von *Rosmarinus officinalis. Phytochemistry* 10:3195–204
13. Brown, A. J., Kolattukudy, P. E. 1978. Evidence that pancreatic lipase is responsible for the hydrolysis of cutin. *Arch. Biochem. Biophys.* 190:17–26
14. Caldicott, A. B., Eglington, G. 1976. Cutin acids from bryophytes: an ω-1 hydroxy alkanoic acid in two liverwort species. *Phytochemistry* 15:1139–43
15. Caldicott, A. B., Simoneit, B. R. T., Eglington, G. 1976. Alkane triols in Psilotophyte cutins. *Phytochemistry* 14:2223–28
16. Carr, D. J., Carr, S. G. M. 1978. Origin and development of stomatal microanatomy in two species of *Eucalyptus. Protoplasma* 96:127–48
17. Cassagne, C., Lessire, R. 1978. Biosynthesis of saturated very long chain fatty acids by purified membrane fractions from leek epidermal cells. *Arch. Biochem. Biophys.* 191:146–52
18. Chang, H-M., Allan, G. G. 1971. Oxidation. In *Lignins,* ed. K. V. Sarkanen, C. H. Ludwig, pp. 433–86. New York: Wiley-Interscience. 916 pp.
19. Cottle, W., Kolattukudy, P. E. 1980. Aromatic constituents of potato suberin and their synthesis. *Plant Physiol. Suppl.* 65:97
20. Croteau, R., Kolattukudy, P. E. 1973. Enzymatic synthesis of a hydroxy fatty acid polymer, cutin, by a particulate preparation from *Vicia faba* epidermis. *Biochem. Biophys. Res. Commun.* 52: 863–69
21. Croteau, R., Kolattukudy, P. E. 1974. Biosynthesis of pentahydroxystearic acid of cutin from linoleic acid in *Rosmarinus officinalis. Arch. Biochem. Biophys.* 162:458–70
22. Croteau, R., Kolattukudy, P. E. 1974. Direct evidence for the involvement of epoxide intermediates in the biosynthesis of the C_{18} family of cutin acids. *Arch. Biochem. Biophys.* 162:471–80
23. Croteau, R., Kolattukudy, P. E. 1974. Biosynthesis of hydroxy fatty acid polymers. Enzymatic synthesis of cutin from monomer acids by cell-free preparations from the epidermis of *Vicia faba* leaves. *Biochemistry* 13:3193–202
24. Croteau, R., Kolattukudy, P. E. 1975. Biosynthesis of hydroxy fatty acid polymers. Enzymatic epoxidation of 18-hydroxy oleic acid to 18-hydroxy-*cis*-9,10-epoxystearic acid by a particulate preparation from spinach (*Spinacia oleracea*). *Arch. Biochem. Biophys.* 170: 61–72
25. Croteau, R., Kolattukudy, P. E. 1975. Biosynthesis of hydroxy fatty acid polymers. Enzymatic hydration of 18-hydroxy-*cis*-9,10-epoxystearic acid to *threo*-9,10,18-trihydroxystearic acid by a particulate preparation from apple (*Malus pumila*). *Arch. Biochem. Biophys.* 170:73–81
26. Dean, B. B., Kolattukudy, P. E. 1976. Synthesis of suberin during wound-healing in jade leaves, tomato fruit, and bean pods. *Plant Physiol.* 58:411–16
27. Dean, B. B., Kolattukudy, P. E. 1977. Biochemistry of suberization. Incorporation of [1-^{14}C]oleic acid and [1-^{14}C]acetate into the aliphatic components of suberin in potato tuber disks (*Solanum tuberosum*). *Plant Physiol.* 59:48–54
28. Deas, A. H. B., Baker, E. A., Holloway, P. J. 1974. Identification of 16-hydroxyoxo-hexadecanoic acid monomers in plant cutins. *Phytochemistry* 13: 1901–5
29. Deas, A. H. B., Holloway, P. J. 1977. The intermolecular structure of some plant cutins. In *Lipids and Lipid Polymers in Higher Plants,* ed. M. Tevini, H. K. Lichtenthaler, pp. 293–300. Berlin: Springer. 306 pp.
30. Eglington, G., Hunneman, D. H., McCormick, A. 1968. Gas chromatographic-mass spectrometric studies of long chain hydroxy acids. *Org. Mass. Spectrom.* 1:593–611

31. Elstner, E. F., Heupel, A. 1976. Formation of hydrogen peroxide by isolated cell walls from horseradish (*Armoracia lapathifolia* Gilib.). *Planta* 130:175–80
32. Espelie, K. E., Davis, R. W., Kolattukudy, P. E. 1980. Composition, ultrastructure and function of the cutin- and suberin-containing layers in the leaf, fruit peel, juice-sac and inner seed coat of grapefruit (*Citrus paradisi* Macfed). *Planta* 149:498–511
33. Espelie, K. E., Dean, B. B., Kolattukudy, P. E. 1979. Composition of lipid-derived polymers from different anatomical regions of several plant species. *Plant Physiol.* 64:1089–93
34. Espelie, K. E., Kolattukudy, P. E. 1978. The optical rotation of a major component of plant cutin. *Lipids* 13:832–33
35. Espelie, K. E., Kolattukudy, P. E. 1979. Composition of the aliphatic components of suberin of the endodermal fraction from the first internode of etiolated *Sorghum* seedlings. *Plant Physiol.* 63: 433–35
36. Espelie, K. E., Kolattukudy, P. E. 1979. Composition of the aliphatic components of 'suberin' from the bundle sheaths of *Zea mays* leaves. *Plant Sci. Lett.* 15:225–30
37. Ferguson, I. B., Clarkson, D. T. 1976. Ion uptake in relation to the development of a root hypodermis. *New Phytol.* 77:11–14
38. Friend, J. 1975. Lignification in infected tissues. In *Biochemical Aspects of Plant-Parasite Relationships,* ed. J. Friend, D. R. Threfall, pp. 291–303. New York: Academic. 354 pp.
39. Gamborg, O. 1967. Aromatic metabolism in plants V. The biosynthesis of chlorogenic acid and lignin in potato cell cultures. *Can. J. Biochem.* 45: 1451–57
40. Grisebach, H. 1981. Lignins. In *The Biochemistry of Plants,* Vol. 7, ed. E. E. Conn. New York: Academic. In press
41. Gross, G. G. 1977. Biosynthesis of lignin and related monomers. *Recent Adv. Phytochem.* 11:141–84
42. Gross, G. G., Janse, C. 1977. Formation of NADH and hydrogen peroxide by cell wall-associated enzymes from *Forsythia* xylem. *Z. Pflanzenphysiol.* 84:447–52
43. Gross, G. G., Janse, C., Elstner, E. F. 1977. Involvement of malate, monophenols, and the superoxide radical in hydrogen peroxide formation by isolated cell walls from horseradish (*Armoracia lapathifolia* Gilib.). *Planta* 136:271–76
44. Grünwald, J., Seemüller, E. 1979. Zerstörung der Resistenzeigenschaften des Himbeerrutenperiderms als Folge des Abbaus von Suberin und Zellwandpolysacchariden durch die Himbeerrutengallmücke *Thomasiniana theobaldi* Barnes (Dipt., Cecidomyiidae). *Z. Pflanzenkr. Pflanzenschutz* 86:305–14
45. Hankin, L., Kolattukudy, P. E. 1971. Utilization of cutin by a Pseudomonad isolated from the soil. *Plant Soil* 34:525–29
46. Hanson, K. R., Havir, E. A. 1979. An introduction to the enzymology of phenylpropanoid biosynthesis. *Recent Adv. Phytochem.* 12:91–137
47. Heinen, W., de Vries, H. 1966. Stages during the breakdown of plant cutin by soil microorganisms. *Arch. Mikrobiol.* 54:331–38
48. Heinen, W., Linskens, H. F. 1961. Enzymic breakdown of stigmatic cuticula of flowers. *Nature* 191:1416
49. Holloway, P. J. 1972. The composition of suberin from the corks of *Quercus suber* L. and *Betula pendula* Roth. *Chem. Phys. Lipids* 9:158–70
50. Holloway, P. J., Deas, A. H. B., Kabarra, A. M. 1972. Composition of cutin from coffee leaves. *Phytochemistry* 11:1443–47
51. Hunneman, D. H., Eglington, G. 1972. The constituent acids of gymnosperm cutins. *Phytochemistry* 11:1989–2001
52. Hunt, G. M., Baker, E. A. 1980. Phenolic constituents of tomato fruit cuticles. *Phytochemistry* 19:1415–19
53. Jones, J. H. 1978. Chemical changes in cutin obtained from cuticles isolated by the zinc chloride-hydrochloric acid method. *Plant Physiol.* 62:831–32
54. Kahl, G. 1974. Metabolism in plant storage tissue slices. *Bot. Rev.* 40:263–314
55. Kolattukudy, P. E. 1970. Biosynthesis of cuticular lipids. *Ann. Rev. Plant Physiol.* 21:163–92
56. Kolattukudy, P. E. 1970. Biosynthesis of a lipid polymer, cutin: The structural component of plant cuticle. *Biochem. Biophys. Res. Commun.* 41:299–305
57. Kolattukudy, P. E. 1971. Enzymatic synthesis of fatty alcohols in *Brassica oleracea. Arch. Biochem. Biophys.* 142: 701–9
58. Kolattukudy, P. E. 1973. Identification of 18-oxo-9,10-epoxystearic acid, a novel compound in the cutin of young apple fruits. *Lipids* 8:90–92
59. Kolattukudy, P. E. 1974. Biosynthesis of a hydroxy fatty acid polymer, cutin. Identification and biosynthesis of 16-

oxo-9 or 10-hydroxy palmitic acid, a novel compound in *Vicia faba*. *Biochemistry* 13:1354–63

60. Kolattukudy, P. E. 1977. Lipid polymers and associated phenols, their chemistry, biosynthesis, and role in pathogenesis. In *The Structure, Biosynthesis, and Degradation of Wood*, ed. F. A. Loewus, V. C. Runeckles, pp. 185–246. New York: Plenum. 527 pp.
61. Kolattukudy, P. E. 1978. Chemistry and biochemistry of the aliphatic components of suberin. In *Biochemistry of Wounded Plant Tissues*, ed. G. Kahl, pp. 43–84. Berlin: de Gruyter. 680 pp.
62. Kolattukudy, P. E. 1980. Biopolyester membranes of plants: Cutin and suberin. *Science* 208:990–1000
63. Kolattukudy, P. E. 1980. Cutin, suberin and waxes. In *The Biochemistry of Plants*, ed. P. K. Stumpf, E. E. Conn, 4:571–645. New York: Academic. 685 pp.
64. Kolattukudy, P. E., Agrawal, V. P. 1974. Structure and composition of aliphatic constituents of potato tuber skin (suberin). *Lipids* 9:682–91
65. Kolattukudy, P. E., Buckner, J. S. 1972. Chain elongation of fatty acids by cell-free extracts of epidermis from pea leaves (*Pisum sativum*). *Biochem. Biophys. Res. Commun.* 46:801–7
66. Kolattukudy, P. E., Dean, B. B. 1974. Structure, gas chromatographic measurement, and function of suberin synthesized by potato tuber tissue slices. *Plant Physiol.* 54:116–21
67. Kolattukudy, P. E., Espelie, K. E., Soliday, C. L. 1981. Hydrophobic layers attached to cell walls: Cutin, suberin and associated waxes. In *Encyclopedia of Plant Physiology*, Vol. 14, ed. W. Tanner, F. A. Loewus. Berlin: Springer. In press
68. Kolattukudy, P. E., Kronman, K., Poulose, A. J. 1975. Determination of structure and composition of suberin from the roots of carrot, parsnip, rutabaga, turnip, red beet and sweet potato by combined gas-liquid chromatography and mass spectrometry. *Plant Physiol.* 55:567–73
69. Kolattukudy, P. E., Walton, T. J. 1972. Structure and biosynthesis of the hydroxy fatty acids of cutin in *Vicia faba* leaves. *Biochemistry* 11:1897–1907
70. Kolattukudy, P. E., Walton, T. J. 1973. The biochemistry of plant cuticular lipids. *Prog. Chem. Fats Other Lipids* 13:119–75
71. Kolattukudy, P. E., Walton, T. J., Kushwaha, R. P. S. 1973. Biosynthesis of the C_{18} family of cutin acids: ω-hydroxyoleic acid, ω-hydroxy-9,10-epoxystearic acid, 9,10,18-trihydroxystearic acid, and their Δ^{12}-unsaturated analogs. *Biochemistry* 12:4488–98
72. Lai, Y. Z., Sarkanen, K. V. 1971. Isolation and structural studies. See Ref. 18, pp. 165–240
73. Lin, T.-S., Kolattukudy, P. E. 1976. Evidence for novel linkages in a glycoprotein involving β-hydroxyphenylalanine and β-hydroxytyrosine. *Biochem. Biophys. Res. Commun.* 72:243–50
74. Lin, T.-S., Kolattukudy, P. E. 1977. Glucuronyl glycine, a novel N-terminus in a glycoprotein. *Biochem. Biophys. Res. Commun.* 75:87–93
75. Lin, T.-S., Kolattukudy, P. E. 1978. Induction of a biopolyester hydrolase (cutinase) by low levels of cutin monomers in *Fusarium solani f. sp. pisi. J. Bacteriol.* 133:942–51
76. Lin, T.-S., Kolattukudy, P. E. 1979. Direct evidence for the presence of β-hydroxyphenylalanine and β-hydroxytyrosine in cutinase from *Fusarium solani pisi. Arch. Biochem. Biophys.* 196:225–64
77. Lin, T.-S., Kolattukudy, P. E. 1980. Structural studies on cutinase, a glycoprotein containing novel amino acids and glucuronic acid amide at the N terminus. *Eur. J. Biochem.* 106:341–51
78. Lin, T.-S., Kolattukudy, P. E. 1980. Isolation and characterization of a cuticular polyester (cutin) hydrolyzing enzyme from phytopathogenic fungi. *Physiol. Plant Pathol.* 17:1–15
79. Linskens, H. F., Heinen, W. 1962. Cutinase-nachweis in pollen. *Z. Bot.* 50:338–47
80. Macey, M. J. K., Stumpf, P. K. 1968. Fat metabolism in higher plants XXXVI: Long chain fatty acid synthesis in germinating peas. *Plant Physiol.* 43:1637–47
81. Maiti, I. B., Kolattukudy, P. E., Shaykh, M. 1979. Purification and characterization of a novel cutinase from nasturtium (*Tropaeolum majus*) pollen. *Arch. Biochem. Biophys.* 196:412–23
82. Maron, R., Fahn, A. 1979. Ultrastructure and development of oil cells in *Laurus nobilis* L. leaves. *Bot. J. Linn. Soc.* 78:31–40
83. Martin, J. T., Juniper, B. E. 1970. *The Cuticles of Plants.* New York: St. Martins. 347 pp.
84. Mueller, W. C., Beckman, C. H. 1978. Ultrastructural localization of polyphenol oxidase and peroxidase in roots

and hypocotyls of cotton seedlings. *Can. J. Bot.* 56:1579–87

85. Olesen, P. 1979. Ultrastructural observations on the cuticular envelope in salt glands of *Frankenia pauciflora. Protoplasma* 99:1–9
86. Parameswaran, N., Wilhelm, G. E. 1979. Micromorphology of naturally degraded beech and spruce barks. *Eur. J. For. Pathol.* 9:103–12
87. Peterson, C. A., Peterson, R. L., Robards, A. W. 1978. A correlated histochemical and ultrastructural study of the epidermis and hypodermis of onion roots. *Protoplasma* 96:1–21
88. Purdy, R. E., Kolattukudy, P. E. 1973. Depolymerization of a hydroxy fatty acid biopolymer, cutin, by an extracellular enzyme from *Fusarium solani f. pisi:* Isolation and some properties of the enzyme. *Arch. Biochem. Biophys.* 159:61–69
89. Purdy, R. E., Kolattukudy, P. E. 1975. Hydrolysis of plant cuticle by plant pathogens. Purification, amino acid composition, and molecular weight of two isozymes of cutinase and a non-specific esterase from *Fusarium solani f. pisi. Biochemistry* 14:2824–31
90. Purdy, R. E., Kolattukudy, P. E. 1975. Hydrolysis of plant cuticle by plant pathogens. Properties of Cutinase I, Cutinase II, and a non-specific esterase isolated from *Fusarium solani pisi. Biochemistry* 14:2832–40
91. Rachmilevitz, T., Fahn, A. 1973. Ultrastructure of nectaries of *Vinca rosea* L., *Vinca major* L. and *Citrus sinensis* Osbeck cv. Valencia and its relation to the mechanism of nectar secretion. *Ann. Bot.* 37:1–9
92. Rhodes, M. J. C., Wooltorton, L. S. C. 1973. Stimulation of phenolic acid and lignin biosynthesis in swede root tissue by ethylene. *Phytochemistry* 12:107–18
93. Rhodes, M. J. C., Wooltorton, L. S. C. 1978. The biosynthesis of phenolic compounds in wounded plant storage tissues. See Ref. 61, pp. 243–86
94. Riley, R. G., Kolattukudy, P. E. 1975. Evidence for covalently attached *p*-coumaric and ferulic acid in cutins and suberins. *Plant Physiol.* 56:650–54
95. Robards, A. W., Jackson, S. M., Clarkson, D. T., Sanderson, J. 1973. The structure of barley roots in relation to the transport of ions into the stele. *Protoplasma* 77:291–311
96. Robards, A. W., Payne, H. L., Gunning, B. E. S. 1976. Isolation of the endodermis using wall-degrading enzymes. *Cytobiologie* 13:85–92
97. Scott, M. G., Peterson, R. L. 1979. The root endodermis in *Ranunculus acris.* II. Histochemistry of the endodermis and the synthesis of phenolic compounds in roots. *Can. J. Bot.* 57:1063–77
98. Shaykh, M., Kolattukudy, P. E., Davis, R. 1977. Production of a novel extracellular cutinase by the pollen and the chemical composition and ultrastructure of the stigma cuticle of nasturtium (*Tropaeolum maus*). *Plant Physiol.* 60:907–15
99. Shibuya, T., Funamiza, M., Kitahara, Y. 1978. Novel *p*-coumaric acid esters from *Pinus densiflora* pollen. *Phytochemistry* 17:979–81
100. Shishiyama, J., Araki, F., Akai, S. 1970. Studies on cutin-esterase II. Characteristics of cutin-esterase from *Botrytis cinerea* and its activity on tomato-cutin. *Plant Cell Physiol.* 11:937–45
101. Sitte, P. 1975. Die Bedeutung der molekularen Lamellenbauweise von Korkzellwanden. *Biochem. Biophys. Pflanz.* 168:287–97
102. Smith, M. M., O'Brien, T. P. 1979. Distribution of autoflourescence and esterase and peroxidase activities in the epidermis of wheat roots. *Aust. J. Plant Physiol.* 6:201–19
103. Soliday, C. L., Dean, B. B., Kolattukudy, P. E. 1978. Suberization: Inhibition by washing and stimulation by abscisic acid in potato disks and tissue culture. *Plant Physiol.* 61:170–74
104. Soliday, C. L., Kolattukudy, P. E. 1976. Isolation and characterization of a cutinase from *Fusarium roseum culmorum* and its immunological comparison with cutinases from *F. solani pisi. Arch. Biochem. Biophys.* 176:334–43
105. Soliday, C. L., Kolattukudy, P. E. 1977. Biosynthesis of cutin. ω-Hydroxylation of fatty acids by a microsomal preparation from germinating *Vicia faba. Plant Physiol.* 59:1116–21
106. Soliday, C. L., Kolattukudy, P. E. 1978. Midchain hydroxylation of 16-hydroxypalmitic acid by the endoplasmic reticulum fraction from germinating *Vicia faba. Arch. Biochem. Biophys.* 188:338–47
107. Soliday, C. L., Kolattukudy, P. E. 1979. Introduction of *O*-glycosidically linked mannose into proteins via mannosyl phosphoryl dolichol by microsomes from *Fusarium solani* f. *pisi. Arch. Biochem. Biophys.* 197:367–78
108. Soliday, C. L., Kolattukudy, P. E., Davis, R. W. 1979. Chemical and ultrastructural evidence that waxes asso-

ciated with the suberin polymer constitute the major diffusion barrier to water vapor in potato tuber (*Solanum tuberosum* L.) *Planta* 146:607–14

109. Taylor, A. O., Zucker, M. 1966. Turnover and metabolism of chlorogenic acid in *Xanthium* leaves and potato tubers. *Plant Physiol.* 41:1350–59
110. Thomson, W. W., Platt-Aloia, K. A., Endress, A. G. 1976. Ultrastructure of oil gland development in the leaf of *Citrus sinensis* L. *Bot. Gaz.* 137:330–40
111. Thomson, W. W., Platt-Aloia, K. A., Koller, D. 1979. Ultrastructure and development of the trichomes of *Larrea* (creosote bush). *Bot. Gaz.* 140:249–60
112. Tulloch, A. P. 1976. Chemistry of waxes of higher plants. In *Chemistry and Biochemistry of Natural Waxes,* ed. P. E. Kolattukudy, pp. 235–87. Amsterdam/New York: Elsevier. 459 pp.
113. Uritani, I., Oba, K. 1978. The tissue slice system as a model for studies of host-parasite relationships. See Ref. 61, pp. 287–308
114. Van Fleet, D. S. 1961. Histochemistry and function of the endodermis. *Bot. Rev.* 27:165–220
115. Walton, T. J., Kolattukudy, P. E. 1972. Determination of the structures of cutin monomers by a novel depolymerization procedure and combined gas chromatography and mass spectrometry. *Biochemistry* 11:1885–97
116. Walton, T. J., Kolattukudy, P. E. 1972. Enzymatic conversion of 16-hydroxypalmitic acid into 10,16-dihydroxypalmitic acid in *Vicia faba* epidermal extracts. *Biochem. Biophys. Res. Commun.* 46:16–21
117. Wattendorf, J. 1976. Ultrastructure of the suberized styloid cells in *Agave* leaves. *Planta* 128:163–65
118. Whitmore, F. W. 1976. Binding of ferulic acid to cell walls by peroxidases of *Pinus elliottii. Phytochemistry* 15: 375–78
119. Whitmore, F. W. 1978. Lignin-carbohydrate complex formed in isolated cell walls of callus. *Phytochemistry* 17:421–25
120. Willemot, C., Stumpf, P. K. 1967. Fat metabolism in higher plants. XXXIII. Development of fatty acid synthetase during the "aging" of storage tissue slices. *Can. J. Bot.* 45:579–84
121. Zee, S.-Y., O'Brien, T. P. 1970. Studies on the ontogeny of the pigment strand in the caryopsis of wheat. *Aust. J. Biol Sci.* 23:1153–71
122. Zenk, M. H. 1979. Recent work on cinnamoyl CoA derivatives. *Recent Adv. Phytochem.* 12:139–76

Ann. Rev. Plant Physiol. 1981. 32:569–96

AUXIN RECEPTORS ◆7724

Philip H. Rubery

Department of Biochemistry, University of Cambridge, Cambridge CB2 1QW, England

---

CONTENTS

INTRODUCTION

The term "auxin receptor" both expresses a hypothesis and suggests a methodology. The hypothesis is that modulation by auxin of the activity and spectrum of gene products will be explicable in terms of hormone recognition by specific receptors whose occupancy then initiates a chain of biochemical/biophysical events culminating in some observed hormone action. The methodology then follows heuristically and has derived inspiration from analogous fields dealing with the actions of drugs, animal hormones, and neurotransmitters. Thus one major approach involves in vitro binding studies where criteria for receptor identification include, as neces-

0066-4294/81/0601-0569$01.00

sary but insufficient conditions, the demonstration that hormone binding to a cellular component occurs with a specificity, speed, and affinity consistent with the hormonal dose-response characteristics of the biological starting material. The main limitation on further progress is the need for functional assays to filter such indirect evidence and distinguish auxin receptors from other auxin-binding components, including enzymes and transport proteins as well as fortuitous associations of questionable biological relevance. The common practice of referring to auxin binding sites as "receptors" (e.g. 101) is a convenient shorthand requiring qualification by the reader.

Three intensively studied aspects of auxin physiology have largely set the stage for most investigations and discussion of auxin-binding proteins and their possible functions. These are: (*a*) stimulation of cell elongation in shoot tissue about 10 min after auxin application, where the "acid growth hypothesis" has focused attention on the possibility of auxin-stimulated transmembrane proton pumping (17); (*b*) polar auxin transport which probably involves carriers (22, 26, 89, 90, 93a) and for which potent inhibitors such as 1-naphthylphthalamic acid (NPA) and 2,3,5-triiodobenzoic acid (TIBA) are available; (*c*) the ability of auxin to modulate or trigger changes in macromolecule biosynthesis (43, 97b) which is implicit in its manifold morphogenetic effects.

When hormone binding was first reviewed in this series in 1976 (51), one particularly promising approach to the receptor problem concerned membrane-located binding sites for auxins and transport inhibitors in maize coleoptiles. Substantial progress has indeed occurred in our knowledge of the variety and intracellular distribution of such proteins. Nevertheless, direct positive evidence as to their functions is still lacking. There has been little development of in vitro systems where "soluble" auxin-dependent transcription factors had been reported to stimulate DNA-dependent RNA synthesis (43, 51, 101). This review attempts to assess the current status and future prospects of auxin receptor research and to suggest some experimental systems and approaches that might profitably be applied to the elucidation of receptor structure and function.

AUXIN-BINDING COMPONENTS FROM MAIZE AND OTHER SOURCES

Binding Assays

The concept of a binding assay for putative receptors is a simple one, but careful experimental design and data analysis (68a) are needed to guard against possible artifacts. In principle, a radiolabeled ligand is equilibrated with the receptor preparation at a temperature where all components are stable and the concentrations of free and bound ligand are determined. The ligand must have a sufficiently high specific radioactivity and radiochemical purity (at least 90%) to allow accurate and reliable measurement of low concentrations. It is also essential to check that (bio)chemical transformation has not occurred during the assay. In particular, indole-3-acetic acid (IAA) may be converted by IAA oxidase to 3-methyleneoxindole which can form covalent adducts with -SH groups (3). The experimental procedure must allow binding to sites that may be physiologically relevant to be distinguished from weak nonspecific interactions of radioligand with biological material as well as from background counts due to the particular methodology used. The minimum discriminating criterion for identifying such "high affinity" sites has usually been similarity of dissociation constants (K_D)[1] and physiological ligand concentrations on the grounds of reflecting significant in vivo site occupancy. These values are often taken as micromolar or less, representing the lower range of hormonally effective concentrations of applied auxin (e.g. 81) or the level of IAA present in total tissue water (94). This may be too narrow a view since dose-response curves may span three or more concentration decades (51, 81), and also the concentration of auxin could vary widely between different tissue compartments (77, 90). In practice, physiological relevance is seldom attributed to sites with a K_D larger than 20 μM.

The standard method for resolving binding to the high affinity sites (37, 38) involves subtracting the bound radioactivity of a blank sample in which specific activity is lowered by a large excess of unlabeled ligand (~ 100 X the K_D and the radioligand concentration) from the bound radioactivity of a parallel incubation containing radioligand alone. The rationale is that unlabeled ligand will compete effectively for high affinity sites but not for sites where total binding increases linearly with concentration. Thus the radioactivity bound to these low affinity sites is not decreased. The high affinity sites are often referred to as "saturable" or "specific" sites to contrast them with the large number of low affinity "nonspecific sites." Although the term "specific binding" is convenient and widely used, it is, strictly speaking, a misnomer and specificity as such must be established by further experimentation.

A variety of methods are available for measuring bound radioactivity. Soluble ligand-receptor complexes may be analyzed by (equilibrium) dialysis (102, 104, 105, 110) or gel filtration (18–20, 93), or by adsorption of free ligand to dextran-coated charcoal (11, 40, 71, 88). They may also be precipitated at high ammonium sulfate concentration (68, 110, 111). For particulate sites, filtration (98) or, more usually, centrifugation techniques

[1]Throughout this chapter, K_D is used to describe equilibrium dissociation constants, *not* rate constants of dissociation.

have been used (e.g. 37, 38, 80). The nonspecific binding estimated from the blank incubation will include background counts retained, for example, in the free space of pellets or adsorbed to assay vehicles such as plastic centrifuge tubes or filtration media. The main constraint that limits reduction of nonspecific binding by washing is the value of the dissociation rate constant (k_{off}) which determines the half-life ($t_{½} = \ln 2/k_{off}$) of the high affinity receptor-ligand complex. In general, the separation time in which the complex is exposed to free ligand concentrations significantly below the equilibrium value should not exceed 0.15 $t_{½}$ to avoid losing more than about 10% of specifically bound ligand. Since $K_D = k_{off}/k_{on}$, the tighter and slower the binding, the longer is $t_{½}$. Most of the particulate auxin binding sites so far studied appear to dissociate too fast for washing to be useful, especially in centrifugation assays, and relatively high blanks may be unavoidable.

Binding data may be analyzed by linear transformations of the Michaelis saturation isotherm. For example, 1/[free ligand] may be plotted against 1/[bound ligand] as ordinate (the Klotz plot) and a straight line fitted by weighted linear regression (e.g. 5, 102). More often the concentration of specifically bound ligand corresponding to a particular free ligand concentration is plotted as abscissa against the ratio of bound to free ligand [Scatchard plot, e.g. (38)]. A line of best fit can be calculated by hyperbolic regression (e.g. 107). The slope is $-1/K_D$ and the abscissa intercept gives the total concentration of binding sites. The free ligand concentration has usually been measured explicitly, but may be approximated by the total ligand concentration if this greatly exceeds receptor concentration. Nonlinear Scatchard or Klotz plots where the gradient becomes less steep as the abscissa value increases may indicate site heterogeneity (especially if there is a sharp break in the curve), failure to correct adequately for nonspecific binding, or negative cooperativity within one class of site (68a). In practice it may be difficult to tell a straight line from a shallow curve (cf 23, 80). An index of cooperativity is provided by the Hill coefficient [see (52) for further explanation]. Data for plotting may be generated in two ways, both requiring correction for nonspecific binding. First, the concentration of radioactive ligand can be increased at constant specific activity. Second, increasing amounts of unlabeled ligand can be added to a series of tubes containing a fixed concentration of radioligand, and the amounts of total bound and free ligand can be calculated from the specific activities. Also, computer-aided model fitting to uncorrected data has been explored (68a).

The specificity of an auxin-binding site can be assessed from displacement curves by comparing the concentrations of different analogs that competitively inhibit high affinity binding of radiolabeled auxin by 50% (e.g. 81). These IC_{50} values approximate the corresponding analog K_Ds if both radioligand and the receptor sites are at fixed concentrations below about one-tenth of the dissociation constant for the radioligand (14, 42). Provided that reliable measurements can be made, Scatchard plots are independent of these restrictions (42).

Nonmembrane Auxin-binding Sites

In a careful study, Wardrop & Polya showed that the high capacity and high affinity auxin-binding component first demonstrated in bean leaf homogenates (109) can be accounted for by ribulose bisphosphate carboxylase (RBPC) (110, 111). This conclusion was adumbrated by the report of Fraction I protein as a repository of "bound auxin" (112). Scatchard plots at the pH optimum (pH 8) revealed high affinity IAA-binding sites on the purified RBPC with a K_D of 0.8 ± 0.3 μM. Assuming a M_r of 550,000 (that of RBPC comprising 8 large and 8 small subunits), a stoichiometry of 1.2 ± 0.5 sites mol^{-1} could be calculated. This may be an underestimate because of lability of auxin binding at some stages of the purification. The binding was rapid (within 10 min at 0-4°C), reversible, nondegradative, and inhibited by aliphatic and aromatic -SH reagents. As judged by inhibition of 0.2 μM [^{14}C]IAA binding by over 100 test compounds (111), the protein has a binding spectrum broadly similar to that of other auxin "receptors" (cf 81, 107). In discussing whether IAA binding to RBPC may be of general in vivo significance, it was pointed out that equilibrium dialysis failed to reveal high affinity binding in over half the preparations tested, which were nevertheless active in the physiologically less representative ammonium sulfate precipitation assay that was routinely used (110, 111), and which may favor particular auxin-binding conformers of the protein (68). Also, auxin-binding activity could be reversibly and irreversibly lost while RBPC activity remained unchanged. The absence of demonstrable auxin binding to RBPC preparations from species other than pea and bean, including maize (111), may reflect a need for separate optimization of protocols for different tissues and/or developmental stages. Since RBPC is a major chloroplast protein, a study of the equilibrium distribution of labeled auxin between incubation medium and isolated leaf cells could help to resolve this question. An auxin-binding protein located within chloroplasts at a concentration greatly in excess of its K_D for IAA is unlikely to play a direct receptor role in signal transduction, but it would be an important component in the synthesis/destruction/transport/sequestration network (2) determining free auxin levels, particularly if its auxin-binding capacity could be modulated.

The extent to which RBPC released from ruptured chloroplasts or etioplasts (15) may contribute to both soluble and membranous auxin-binding activity merits careful investigation. The logical difficulties in assessing the relevance of in vitro auxin binding are highlighted by Murphy's argument

(68) that bovine serum albumin would be regarded as a putative auxin receptor had it been isolated from a plant.

A soluble auxin receptor has been partially purified from tobacco pith explants (11, 72) and callus (71). This resembled the bean leaf protein in its K_D, M_r [about 300,000 (71), revised from an earlier value (56) of 75,000], pH optimum and specificity as judged by 13 test compounds. In contrast, binding was very slow, taking 30 min at 25°C, and was undetectable at 0°C. IAA binding was barely detectable in fresh pith (11) and achieved levels of only 10 fmol g^{-1} fresh weight after culture on nutrient agar in continuous light (11, 72). In established light-grown callus the auxin-binding capacity peaked (10-50 fmol g^{-1} fresh weight) about 10 days after subculturing and then declined (71). A role for this soluble protein in regulation of RNA transcription has been postulated (72), perhaps as part of some general wound response to pith excision (11).

In vivo labeling was attempted by Ihl (40), who incubated soybean cotyledons with [2–^{14}C]IAA $\pm$ excess unlabeled auxins for 1 h. The radioactivity associated with macromolecular material after homogenization was reduced if unlabeled auxin had been supplied, but the chemical nature of the bound radioactivity was not determined. The in vivo labeling approach has been successful in isolating high-affinity ethylene and fusicoccin receptor complexes which are sufficiently stable to survive manipulation of tissue extracts under nonequilibrium conditions (7,8).

No biological response to auxin binding, in a transducing sense, has been shown in the systems described above; nor have any of the putative particulate receptors been so characterized. This highlights approaches that seek to use changes in RNA transcription as a basis for functional receptor assay. The corpus of experiments that link auxin with regulation of RNA metabolism have been summarized critically in two recent reviews in this series (43, 51) and will not be rehearsed here. Recently there has been a limited pursuit of two themes discussed next; in both cases (85, 88) attractive candidates for auxin receptors are reported, but regrettably without strong coherent and direct evidence.

Further features of the auxin-responsive transcription system from immature coconut endosperm have been described (88). The earlier work on this tissue has been reviewed with some reserve (51, 101) because of insufficient published experimental detail to allow full evaluation and to lack of some critical controls. Two IAA binding proteins were obtained from, respectively, the nucleoplasmic fraction (designated n-IRP) and the nonhistone chromatin protein fraction (c-IRP) of isolated nuclei (88). Binding activity was detected (dextran-coated charcoal assay) in 1 M KSCN eluates from IAA-Sepharose columns. The c-IRP exhibited two IAA binding sites (K_Ds 58 nM and 8.2 μM) and had a M_r of 70,000 according to unpublished

SDS-PAGE data. Specificity was not investigated and overall yield was not given. When solubilized chromatin was incubated at 0°C with a [^{14}C]IAA-n-IRP complex that had been prepared at 25°C, radioactivity was retained on a filter. No radioactivity was bound to the filter when [^{14}C]IAA alone or [^{14}C]IAA and n-IRP (added separately) were included in the incubation. Binding of the [^{14}C]IAA-n-IRP to the chromatin was assumed to have occurred (although binding of the *complex* alone to the filter medium was not tested) since a doubling of transcription by *E. coli* RNA polymerase was found, in a separate experiment, apparently involving 9 - 12S RNA. A model was proposed, analogous to that for steroid receptors in chick oviduct (70), whereby the c-IRP in chromatin acts as an acceptor for the IAA-n-IRP complex to bring about specific gene transcription. While a model based on this paradigm is as good a working hypothesis as any in our current state of knowledge, and recognizes the ease of membrane penetration by auxin (89, 90), much more extensive experimental support is clearly needed. In particular the interaction between IAA, n-IRP, and c-IRP requires clarification.

Rizzo et al (85) have emulated Venis's earlier work on pea and corn shoots (99). They isolated a transcription factor from soybean hypocotyl by "affinity chromatography" using 2,4-dichlorophenoxyacetyl-ϵ-L-lysine coupled to Sepharose. A small fraction (0.04%) of the total protein was eluted with 2 mM KOH (pH 11.3). This factor could stimulate RNA transcription (X 2–7) in a heterologous system using native calf thymus DNA and *E. coli* RNA polymerase. When partially purified soybean RNA polymerases were substituted, the factor was inactive unless 0.1 μM 2,4-dichlorophenoxyacetic acid (2,4-D) was present. A small stimulation (25-80%) was obtained only with RNA *pol I* (the enzyme primarily concerned with rRNA synthesis). It was suggested (85) that the factor could be involved in the observed in vivo stimulation of soybean RNA *pol I* by 2,4-D. It probably differs from the ethanol- and water-soluble *pol II*-stimulating factor that is released by auxin from soybean (plasma)membrane fractions (34) and which may be glycolipid-like (66). The ability of the factor to bind 2,4-D was not investigated, and specificity or the effectiveness of other 2,4-D concentrations were not assessed. It is questionable whether true affinity chromatography is possible if 2,4-D or IAA (88) is linked to Sepharose by the carboxyl group; the presence of the free α-COOH group in the spacer lysine residue seems unlikely to compensate for lack of this structural feature which is essential for auxin activity (49). Various nonspecific weak interactions, such as may be broken by the chaotropic agent KSCN (88), could contribute to the observed retention. Venis (99) found that 50 mM 2,4-D eluted the pea factor in a larger volume than did 2 mM KOH, but was more effective than benzoic acid.

Auxin-binding Sites Located on Membranes

MAIZE COLEOPTILES High yields and reproducible results (68a) have favored the dark-grown maize coleoptile for auxin-binding studies. How representative the tissue is remains to be established. Kende & Gardner (51) have reviewed the pioneering work of Hertel and colleagues (38, 54, 95–97), which has since been refined to reveal distinct auxin-binding sites in different cellular membranes (4, 23, 35, 41, 69, 78, 80, 81). These developments have accompanied, and in part stimulated, a critical appraisal of the whole field of plant cell fractionation and its interpretation. The importance of recovery balance sheets and the desirability of combining morphological and biochemical markers when analyzing separations on, for example, continuous density gradients has been stressed. Also, because of flow differentiation of the endomembrane system, particular constituents may occur in membranes ontogenically related to their primary locations. Further, membranes fragment during homogenization, and a constituent originating from sparsely or asymmetrically distributed domains would not be fully representative of its parent membrane. This applies particularly to the highly differentiated plasmamembrane (37, 75, 83) and might be expected for receptors concerned with polar auxin transport (37, 77, 89). Differential pelleting is useful as a preliminary screen, but continuous isopycnic gradients give the greatest resolution and can guide the choice of density steps for preparative discontinuous gradients (e.g. 23). Step gradients alone can give a misleading impression of high resolution by compressing a range of variously dense components into single bands.

Careful optimization enabled the yield of specific 1-naphthylacetic acid (NAA) binding sites (80) to be improved tenfold over the original microsomal preparation from maize coleoptiles (hybrid WF9 X Bear 38)(38). The binding (pH optimum 5.5; compare pH 8 for RBPC) was cold-stable but was appreciably thermolabile with 40% inactivation occurring after 20 min at 37°C (80). In practice, membranous binding sites from corn shoots have been studied at 0-4°C, below the likely transition temperature of the membranes. Thiols such as mercaptoethanol, included in the early protocol (38), partially inactivated the binding without altering the affinity. NAA (0.1 mM) afforded some protection, and the inactivation could be partly reversed by oxidants. The optimization also revealed an inhibitory supernatant factor (SF) which could modulate the affinity of various ligands for the binding sites. Specific NAA binding to washed microsomes (133,000 *g* pellet) approached saturation over the range 5 nM–3.6 μM [^{14}C]NAA. A linear Scatchard plot with six points spanning 0.38 – 50 μM NAA indicated the presence of a single class of sites (K_D = 0.5–0.7 μM) at a tissue concentration of about 50 pmol g^{-1} fresh weight (80). The NAA binding was unaffected by a variety of molecules, including other growth substances (38, 80) and compounds present in experimental media such as sucrose, tris, ethanol and Cl^- ions (80).

Subsequently, Ray (78) clearly resolved the peaks of specific NAA and NPA binding on continuous isopycnic sucrose gradients. He concluded that most of the NAA sites were located on the endoplasmic reticulum (ER) since the peaks of a recognized marker [antimycin-insensitive NAD(P)H cytochrome *c* reductase] and NAA binding coincided whether ribosomes remained membrane-bound or had been stripped under low Mg^{2+} conditions. Step gradient fractionation lead to a similar conclusion (69). However, the presence of additional auxin-binding sites was suggested by pronounced penetration (35, 69, 78) of NAA binding, relative to ER markers, into the broad peaks at higher buoyant densities of both a putative Golgi membrane marker [glucan synthetase I(79)] and of glucan synthetase II (79) and NPA binding. The latter activities are regarded as likely plasmamembrane markers on two grounds. First, their presumed physiological functions in β(1-4)glucan synthesis (79) and polar auxin transport (54, 98) suggest a location at the plasmamembrane. Second, their distributions correlate (54, 75) with the presence of membrane fragments that can be selectively stained in vitro, under specified conditions, with a phosphotungstate/chromate reagent (PTC) which can show selectivity for plasmamembrane in vivo (87, 114). In a detailed discussion of the somewhat controversial PTC stain, Quail (75; see also 12, 83) emphasizes the lack of *independent* direct criteria for plasmamembrane identification.

Chemistry of supernatant factor There is evidence that maize SF activity is largely attributable to the glucoside of 6,7-dimethoxybenzoxazinone which occurs only in some Gramineae and therefore cannot be a general regulator (105). The corresponding benzoxazolinone, to which the glucoside is easily converted, also inhibits auxin binding to maize microsomes, and at comparable concentrations antagonized IAA-stimulated elongation of oat coleoptiles from which the parent compound is however absent (105); its effect on *maize* coleoptile elongation does not seem to have been reported.

Multiple sites Dohrmann et al (23) followed up the possibility of multiple receptors and found that the characteristics of auxin binding were not uniform throughout isopycnic sucrose gradients. They distinguished three distinct auxin binding sites on the following grounds:

1. The weak antiauxin 2-NAA was used as a probe. Binding assays of gradient fractions were set up using three tubes each containing 0.1 μM [^{14}C]1-NAA. No further additions were made to the first tube (α), 0.1 mM

unlabeled 1-NAA was added to the second tube (β), and 10 μM unlabeled 2-NAA was added to the third tube (γ). The total 1-NAA specific binding was given by (α - β), (γ - β) gave the specific binding component that was not saturated by (or was resisted by) 10 μM 2-NAA. This procedure revealed a population of sites (designated "site II") that had a relatively low affinity for 2-NAA and whose distribution on the gradient resembled that of glucan synthetase I. In contrast, the binding site designated as "site I" had a high affinity for both 1-NAA and 2-NAA and corresponded to the major ER-located binding activity that peaked with cytochrome *c* reductase.

2. Sites I and II can be similarly distinguished by their differing affinities for the weak auxin phenylacetic acid (PAA) which has a K_D of 3.2 μM for *unfractionated* microsomes (81). The distribution of 0.1 μM [^{14}C]PAA binding that was saturable by 0.1 mM 1-NAA coincided with cytochrome *c* reductase. If gradient fractions were first treated with dithioerythritol, the specific binding of 1-NAA which was resistant to 30 μM PAA peaked in the "site II region" at a higher density than ER markers: the pretreatment was considered selectively to inactivate site I. While consistent with partial inhibition of total NAA binding to microsomes by thiols (80), the claim was however based on only a single set of data obtained with "dense" or "light" particles from a step gradient. This same experiment also indicated that SF inhibited site I more than site II binding. The intracellular location of site II is not certain. Its distribution correlated well with that of sedimentable acid phosphatase during the *approach* to isopycnic equilibrium although *at* equilibrium it substantially overlapped the Golgi marker glucan synthetase I *and* the acid phosphatase. It was tentatively suggested (23), from circumstantial evidence and analogies with animal lysosomes, that site II may be on the tonoplast. Clearly, direct evidence from isolated vacuoles is desirable. It should be noted that site II was defined in the above experiment as specific 1-NAA binding that was resistant to 0.1 mM PAA plus 50 μM D-2,4-dichlorophenoxyisopropionic acid (D-2,4-DP). The rationale was to blot out site I with PAA, site III (see below) with D-2,4-DP, which does not compete strongly at sites I and II (41; cf 38).

3. The gradient regions where plasmamembrane markers peaked, at about 36% sucrose, preferentially bound IAA and 2,4-D rather than 1-NAA. Although complete data were not cited, the K_D for 2,4-D was estimated at 5 μM when measured in the presence of 5 μM 1-NAA. IAA was stated not to compete strongly with 2,4-D. The concentration of these 2,4-D sites (designated "site III") was estimated as 40 pmol g^{-1} fresh weight (23). In subsequent work (41), site III in unfractionated coleoptile membranes was also apparently manifested at pH 6.5 where specific IAA binding, although much smaller than at the pH 5.5 optimum, was *stimulated*

by low concentrations of the transport inhibitor TIBA but was not inhibited by 10 μM 1-NAA. As judged by a Scatchard plot, site III had a relatively low affinity for IAA ($K_D = 20$ μM in the presence of TIBA) but occurred at a higher tissue concentration (150 pmol g^{-1} fresh weight) than sites I and II. The sucrose gradient distribution of D-2,4-DP-sensitive IAA binding correlated well with NPA binding, and a plasmamembrane location of site III was inferred (41). The properties of site III in maize and in *Cucurbita pepo* (zucchini) will be discussed further in a later section.

The concentrations of sites I and II were estimated as 40 and 20 pmol g^{-1} fresh weight respectively from Scatchard plots of specific NAA binding to unfractionated microsomes without and with added 0.1 mM PAA; the quoted K_D for NAA of 1.3 μM for site II should be halved to take account of the level of PAA binding at site II [$K_D = 0.1$ mM; table 5 of (23)]. The NAA dissociation constant for site I was estimated as 0.63 ± 0.21 μM using pooled "light" gradient fractions (ER); the pooled "dense" fractions (predominantly site II-bearing membranes) gave a K_D of 2.7 ± 1.8 μM for NAA.

Also working with maize coleoptile membranes (cv "Kelvedon 33"), Batt & Venis (4) demonstrated major (32 pmol g^{-1} fresh weight; $K_D = 1.16$ μM, designated "site 2") and minor (24 pmol g^{-1} fresh weight; $K_D = 0.39$ μM, designated "site 1") NAA binding sites on respectively heavy and light bands from a one-step sucrose gradient fractionation of a 4 - 38,000 *g* pellet. This supported their deduction of two types of NAA binding site from biphasic narrow range Scatchard plots relating to unfractionated microsomes (5). However, corrections for nonspecific binding were not made (cf 68a, 80) so that the site concentration and K_D values are likely to have been overestimated. Subtraction of arbitrary values for nonspecific binding does not convert the biphasic plots for NAA to downward-sloping straight lines, but can do so for IAA [Figures 1, 2 of (5)]. The reason that linear plots were obtained with microsomes from "Bear hybrid" maize (23, 80) may be that in contrast to (4, 5), the higher affinity sites were the more numerous.

Site specificity An important indirect criterion for receptor identification is the extent to which ligand binding specificity parallels biological activity. This question has been approached as follows. The concentrations of 33 auxin analogs that gave half-maximal response in a 16–18 h maize coleoptile elongation test (C_{50}) were compared with the concentrations needed for 50% inhibition (IC_{50}) of specific NAA binding to unfractionated microsomes (predominantly site I) assayed plus and minus SF (81). The C_{50} values of 14 analogs were also compared with the corresponding IC_{50} values for inhibition of NAA binding to an enriched "dense" membrane fraction assayed in the presence and absence of 0.1 mM PAA to emphasise site II binding (23). The pIC_{50} values were taken to represent pK_Ds for sites I and II as appropriate (23, 81). For site II, a correction (+ 0.3) is needed to allow for PAA binding.

Supernatant factor tended to narrow the binding spectrum of the unfractionated membranes, both increasing and decreasing apparent pK_Ds, so that it resembled the specificity pattern uncovered for site II (23). It appeared as if SF had conferred site II affinity characteristics on site I, perhaps by some allosteric action. There was a stronger correlation between biological activity (pC_{50}) and binding affinity (pK_D) when the latter was measured in the presence of SF than in its absence. This suggests that the benzoxazinone glucosides which comprise SF (105) may have a physiological role, at least in maize, to control the chemical specificity of auxin action: site I has been suggested to be a precursor, at the ER, of site II (23). Given the uncertainties of penetration, metabolism (11), and in vivo pH vs binding assay pH, the correlation was remarkably good and in line with established structure activity relations (49). However, of the auxin analogs which stimulated the basal rate of coleoptile elongation, it was notable that all but 1-NAA had pC_{50} values greater than their pK_Ds when measured plus SF. This may be reconciled with a simple model where biological response is proportional to receptor occupancy if the auxin concentration in equilibrium with the receptor is larger than the concentration applied externally to the tissue at the bioassay pH of 6.3 (81). This may be the case if the cytoplasmic compartment acts as an anion trap to accumulate lipophilic, weakly acidic auxins (77).

The NAA binding sites I and II (23) may be equivalent to the less thoroughly characterized sites 1 and 2 of Venis et al (4, 5) in that the lower affinity sites (II and 2) which have the narrower binding specificity are present in denser membranes (23). The suggestion (4) that site 2 is associated with plasmamembrane vesicles and site 1 with Golgi and/or ER membranes arose from discontinuous step gradients that have a lower resolution than the continuous gradients which clearly differentiated site I (assigned to ER) from a Golgi marker and site II (perhaps on the tonoplast) from plasmamembrane markers (23). A comparison of the specificities reported by the two groups (4, 5; 23, 81) is limited to consideration of the *patterns* of displacement due to (*a*) use of different maize cultivars grown either in continuous darkness (4, 5) or with 2 h red light per day (23); (*b*) site heterogeneity; and (*c*) lack of correction for nonspecific binding when constructing Klotz plots (4, 5). The clearest qualitative difference is that benzoic acid appears to saturate site 1 (4, 5) but not site I of Dohrmann et al (23). Also, the K_D values for the compounds tested as displacers by Batt et al (4, 5) were usually considerably smaller than the corresponding values for "Bear hybrid" maize (23, 81).

In a recent reappraisal of NAA binding to maize membranes, Murphy (68a) has discussed the use of computer curve fitting according to models with varying numbers of binding parameters. He concluded that his own (68a) and some (4, 5, 19, 23) but not all (102) published data are consistent with the presence of only a single major class of NAA binding affinity, although multiple sites that differ in specificity (23, 41; cf 81) are not excluded.

OTHER TISSUES There are numerous reports of particulate auxin receptors from sources other than maize. Jacobs & Hertel (41) analyzed zucchini hypocotyl homogenates using a similar approach to (23). In contrast to maize, site I NAA binding was not detected although site II, again correlating with acid phosphatase, was present at about 15 pmol g^{-1} fresh weight. IAA binding to plasmamembrane-rich fractions was *stimulated* by noncompetitive inhibitors of polar transport (more prominently than in maize) with a K_D (measured in the presence of 3 μM TIBA) of 1.5 μM and a tissue concentration of 20 pmol g^{-1} fresh weight. This "site III" binding had a pH optimum around pH 5 that was established using a zwitterionic buffer since citrate (80) was found to inhibit.

An assay procedure likely to have eliminated most noncovalent IAA binding was the basis for claims of low affinity sites in *Avena* roots (10). Crude membrane fractions from pea roots and epicotyls had low levels of rapidly reversible specific IAA binding that could be displaced by physiological concentrations of phenoxyacetate herbicides; IAA itself did not displace labeled 2-methyl-4-chloro- or 4-chlorophenoxyacetate (24). This is suggestive of conformational change; similar behavior was reported but not documented for site III in maize (23).

The tobacco system (11, 71, 72) also contains particulate receptors (107) that resemble the maize sites (80) in pH optimum, concentration (87 and 28 pmol g^{-1} fresh weight in cultured and fresh pith respectively), and specificity pattern (11 analogs tested). The affinity for IAA ($K_D = 17$ μM) was again lower than for NAA ($K_D = 0.3$ μM). No auxin metabolism occurred during the assay as judged by one TLC system. In contrast to maize, NAA specific binding could not be detected at 0°C, and was very slow at 35°C, taking 30 min to reach equilibrium. From the time course and other cited data, an approximate value for $k_{on} = 2 \times 10^3$ $M^{-1}s^{-1}$ can be calculated (57, Equation 4). From the K_D, k_{off} is $6.7 \times 10^{-4}s^{-1}$ and corresponds to a half-life of about 17 min for the complex. These rates of loading and unloading seem improbably slow if rapid adjustments of receptor occupancy to auxin concentration are to be made. The 4–180,000 *g* pellet used in binding assays could not be fractionated on sucrose gradients (107) and membrane aggregation could have occluded NAA binding sites. In subsequent work (108), high affinity NAA sites that were present in crude membrane fractions from tobacco leaves could not be detected in freshly isolated leaf protoplasts. After 3 - 4 days culture, when cell division and presumably cell wall regeneration had started, NAA binding reappeared and rose to 40 pmol g^{-1} fresh weight after 10–12 days. The leaf receptor was stable to mechanical damage and plasmolysis and may have been removed from the plasmamembrane and/or cell wall during enzymatic isolation. This novel approach could provide valuable information about receptor location, biosynthesis, and function.

Quantitative electron microscope morphometry was applied (without complementary biochemical characterization) to step gradient fractions of soybean hypocotyl homogenates (113). IAA binding correlated best with plasmamembrane vesicles, as judged by the PTC stain. The assay conditions would have accentuated site III detection in maize membranes. The affinity, specificity, and total concentration of the sites were not determined. It was noted that "light" and "heavy" PTC-staining vesicles differed in relative auxin-binding activity. This may reflect lack of staining specificity or fragmentation of the plasmamembrane into domains or into oppositely oriented vesicles (cf 62). In an analogous study with mung bean (48), much of the IAA binding (which was not easily displaced by analogs) may have been irreversible. It resisted "careful washing." Also pretreating the tissue with 10 μM unlabeled IAA increased the already large retention of counts by particles centrifuged through sucrose (48, and Ref. 12 therein).

Solubilization of NAA Binding Activity from Maize Membranes

The characterization of solubilized NAA binding protein from maize microsomes (5, 80) has been undertaken by Cross et al (18–20) and by Venis (102–104), who consistently obtain M_rs of around 90,000 and 45,000 respectively. This discrepancy has not been resolved by exchange of procedures (104; J. W. Cross, personal communication), although different maize cultivars were used by the two groups. Venis achieved a 200-fold purification of a buffer extract of acetone-dried microsomes after successive chromatographic steps involving DEAE-Biogel, poly(U)-Sepharose, aminohexyl-Sepharose, and finally Sephacryl gel filtration. The purification was monitored by equilibrium dialysis assay using 0.15μM [^{14}C]NAA. M_rs of 40–50,000 were consistently found by calibrated gel filtration, irrespective of the presence of a potent protease inhibitor (phenylmethylsulfonylfluoride, routinely used by Cross et al) in the extraction medium (102, 104). As judged by a biphasic Klotz plot (102), the crude buffer extract appeared to contain the two binding sites previously shown by sucrose gradient fractionation (4). Complete binding isotherms have not been presented for the more purified preparations. NAA binding could be analytically (but not preparatively) resolved into two main peaks by isoelectric focusing (pH 3.5–5) of partially purified extracts and by shallow-gradient elution of DEAE-Biogel (104). The purest preparations showed two principal bands after PAGE (±) SDS, although binding activity could not be recovered from the nondenaturing gels. The apparent M_r values were 40,000 and 46,000. It is thus unclear whether two discrete binding species are present. However, the amount of NAA bound (a maximum of 4.2 nmol mg^{-1} protein under standard assay conditions) is close to the theoretical maximum specific activity (4.5 nmol mg^{-1}) for 40,000 M_r proteins with affinities and relative concentrations as calculated for the crude buffer extract (102, 104).

Cross et al (18, 20) showed that Triton X-100 completely solubilized NAA binding while only releasing a minority of microsomal ATPases. Gel filtration largely separated NAA binding material from basal and K-ATPase (20) as Venis also reported (104). Buffer extraction of acetone powders was used for further characterization because of the simpler binding assay procedure, with no need to correct for NAA binding to detergent micelles. Only one binding site was kinetically distinguishable (K_D = 46 nM; 32 pmol g^{-1} fresh weight) and the M_r was about 80,000, although the protein tended to aggregate at low ionic strength (18). The 80,000 M_r species itself may be a homo- or heterooligomer (104). Estimation of the M_r of *membrane-bound* auxin receptors by irradiation inactivation analysis, as applied to glucagon receptors in liver membranes (59), would help to elucidate this question. The solubilized NAA receptor is derived substantially from the ER (19). It has a specificity similar to unfractionated microsomes (81), although a roughly tenfold rise in affinity for 1- and 2-NAA accompanied solubilization; also, IAA and 2,4-D K_Ds were no longer altered by SF (19), suggesting that conformational change may have occurred. No IAA oxidase or auxin-conjugation activities have been associated with solubilized binding proteins (104).

APPROACHES TO RECEPTOR STRUCTURE AND FUNCTION

Some Quantitative Aspects

The tissue concentrations reported for auxin-binding proteins range (excluding RBPC) over 0.01–150 pmol g^{-1} fresh weight (11, 41). To give a scale, 1 pmol g^{-1} fresh weight represents about 1 nM in total cell water and 10 nM in cytoplasm, assuming receptor to be absent from the vacuole; for an orthorhombic cell of dimensions 10 X 10 X 100 μm, the corresponding values are about 6000 molecules per cell or 1.4 molecules μm^{-2} surface area. For comparison, the plasmamembrane proton-translocating ATPase (a likely major component) may occur at about 6000 molecules μm^{-2} (77) and insulin receptors in fat cells at about 90 μm^{-2} [(47); cf other values cited therein]. The K_Ds for IAA binding (generally 0.1–20 μM) are larger than for most animal hormones and neurotransmitters (47), commensurate with the higher likely IAA concentrations available to the receptors: the endogenous IAA level in nonseed organs is about 1 μM in total cell water (2, 94), but the cytoplasmic IAA concentration may exceed those in the vacuole and apoplast about tenfold, depending on transmembrane pH gradients (89, 90). Thus a substantial fraction of IAA binding proteins could be occupied in vivo. Rate constants have been determined for the NPA receptor (98) and gave a K_D in agreement with equilibrium data. The k_{on} value, 4.5×10^4 $M^{-1}s^{-1}$, is 10–100 times smaller than is usual for animal hormones (47) and may be atypical [although fusicoccin and ethylene behave similarly (8, 73)], but it is fast enough to account for the onset of transport inhibition within 1 min (98).

Auxin and Membranes: Dose-Response Curves

The ability of auxin to partition into lipid membranes probably does not account for their hormonal specificity [although see comments in (51, 97b)] but could nevertheless influence the ease and extent of protein receptor occupancy, depending on the relative accessibility of the binding site to auxin from within the membrane and from the aqueous phase. Auxins can alter the phase characteristics of lipids. IAA increases the microviscosity (or at least changes the environment of a fluorescent probe) of model membranes (53) and plasmamembrane-enriched vesicles (36). In the latter, the mean width of membranes fixed and stained after auxin treatment is significantly decreased relative to controls (65). This ultrastructural effect is specific for active auxins and PTC-staining vesicles, rapid ($t_{1/2}$ less than 1 min), temperature dependent, and reversible (65). A displacement of bound Ca^{2+} may be involved (64). High IAA concentrations (~ mM) alter the distribution of intraplasmamembrane particles seen in freeze-fractured protoplasts (86).

Particular attention has been drawn to the characteristic extended dose-response curves of plant hormones (51; cf 97b). For membrane receptors this could partly reflect negative cooperativity due to hormone-induced changes in the lipid bilayer causing concomitant changes in the properties of free and liganded receptors. However, such curves need not be incompatible with simple hyperbolic binding to independent sites. In general, the primary signal generated by a hormone binding to a receptor is likely to be a conformational change that is transduced and amplified into a detected response by an effector mechanism allowing modulation of the activity or level of an enzyme or membrane porter. The extent to which dose-response curves may parallel binding isotherms will depend on how far receptor binding is removed from the site of hormone application, the complexity of the coupling mechanism for effector activation, and the closeness of the measured response to effector activity (21). We are ignorant of any coupling mechanism for auxin, although there is evidence against direct activation of ATPase (20). Modulation of a catalyst whose concentration greatly exceeds that of auxin would require an amplification mechanism. There are precedents for direct coupling in the activation of ion gates by acetylcholine whose sigmoid concentration-effect curves have lead to multistate cooperative binding models (13, 47). A wide conceptual framework is suggested by the mobile receptor hypothesis concerning the modulation of adenylate cyclase by nonpenetrant animal hormones (47, 59). It envisages intramembrane interactions of hormone-receptor complexes with discrete effector and regulatory proteins with no necessarily simple relationship between "response" and the affinity of the initial hormone binding to receptor. Further interpretative difficulties arise if different auxin receptors could influence a particular monitored response somewhat removed from the molecular level (see 17). Also auxin conjugating and deconjugating enzymes could act as a buffer system, desensitizing the cell to external concentration changes (2, 11). Finally, it is of some importance whether plasmamembrane receptors face the symplast or apoplast since auxin can readily penetrate membranes as its protonated species and is accordingly accumulated by relatively alkaline compartments (89, 90); similar considerations apply to any receptors present on the tonoplast (23).

Structure-activity Relationships of Auxins

Attempts to map receptors by correlating auxin activity with chemical structure have tended to become outflanked by "exceptions." There has been renewed interest in this field. Kaethner (46) proposed a planar "recognition conformation" for initial auxin binding with the side chain atoms in the plane of the aromatic ring and the carboxyl oxygens lying astride the ring plane: a concerted conformational change to a perpendicular "modulation conformation" was then envisaged. Special arguments were made for arylcarboxylate auxins. This theory was challenged (30, 31) on the grounds that the planar conformation would be only a small proportion of the total molecular population. It was concluded that for IAA and other auxins there was a fractional positive charge on the ring separated by 5 Å from the carboxylate oxygens with the side chain in the thermodynamically favored perpendicular conformation. This new correlation embraces certain exceptions to the original 5.5 Å charge separation theory (74). Kateckar (49) also found it unnecessary to postulate conformational change in arriving at his map of the auxin receptor which features an electrophilic planar region, larger than the indole nucleus, with which aromatic rings and electron-rich substituents can interact because of their delocalized π electrons. The key idea is that increased coverage of this critical area by electron-rich groupings may correlate with auxin activity. Rakhaminova et al (76) included both conformational change and an "aromatic platform" in their auxin receptor model. While structure-activity correlations are perhaps the best surrogate for functional assay (81; cf 111), a problem may arise with multiple receptors; except for transport inhibitors (50), it is difficult to distinguish subsets of correlations.

Chemical Probes

The protein nature of high affinity auxin-binding sites has been confirmed (e.g. 18, 104). Chemical probes have provided some preliminary information to guide future studies of binding site topology. Although putative reduction of disulfide bridges was reported to inactivate site I (23, 80), other workers found no effect of dithiols (18, 100); rather the aromatic -SH reagent *p*-hydroxymercuribenzoate showed NAA-protected inhibition of particulate and solubilized auxin binding (18, 100). Aliphatic reagents were less potent, suggesting that essential cysteine residues could be located in a hydrophobic environment. Protection with NAA could aid receptor isolation by first treating with unlabeled *p*-hydroxymercuribenzoate plus protectant and then removing NAA and reacting with radioactive *p*-hydroxymercuribenzoate.

The pH-dependence of NAA binding and of the rates of inactivation by diazotized auxin analogs ("affinity labels") lead Venis (100) to suggest the presence of aspartate, glutamate, tyrosine, and lysine residues in the auxin binding domain, as would be consistent with charge separation theory. Structural interpretation of inferred pK values should be cautious in view of the widely ranging pKs of particular ionizing groups within proteins. The pH/binding profiles of auxins of differing acid strengths could indicate any preferential binding of the auxin anion. (Photo)affinity covalent labeling has obvious potential both for radioactive tagging and for probing the binding site (63). 4-Azido-2-chlorophenoxyacetate (55) and 4- and 5-azido-IAA (45) have been synthesized as possible photoaffinity reagents. They have moderate auxin activity in darkness whereas 3-azido-5-chlorophenoxyacetate (based on a nonauxin) was ineffective. No application of these chemicals has yet been reported although radioactive analogs are in preparation (L. Vanderhoef, personal communication). Future synthesis could include derivatives of transport inhibitors (50), benzoxazolinones (105), and of molecules suggested by structure-activity correlations (49). A disadvantage of affinity labels is that they tend to be unselective. At least with photoaffinity reagents, the investigator can trigger production of the active species (16): the tighter the "dark binding" and the more transient the reactive species, the more selective the reagent is likely to be (16).

The purification of auxin-binding proteins is an important, though not essential, step toward the production of powerful immunochemical probes including monoclonal antibodies (60). One useful approach could involve the use of antibodies to a chemical residue (e.g. dinitrophenyl-) or ligand that had first been specifically bound to receptors in a heterogeneous population (60).

Possible Receptor Functions of Auxin-binding Proteins

ION PUMPS The apparent incongruity of the acid growth hypothesis (17) with predominantly intracellular NAA binding sites in maize (23, 78) prompted the suggestion (78) that auxin might stimulate proton pumping into the lumen of the ER, which is then pinched off into vesicles that could discharge their contents into the wall by reversed pinocytosis. A similar proposal has been made for tonoplast receptors (23). These ideas are qualitatively attractive and offer an explanation of the 10 min lag that precedes detectable external acidification. But they raise questions about the accompanying plasmamembrane hyperpolarization, overall charge balance, and the required vesicle throughput and internal pH. Also, site I (ER) binding appears absent from zucchini hypocotyls (41) whereas plasmamembrane fractions from zucchini, maize, and soybean (23, 41, 113) may possess IAA binding sites which could concern proton pumping and/or IAA transport. Vanadate could be a useful probe since it inhibits both nonmitochondrial ATPase and auxin-stimulated proton secretion and elongation (40a, 94a).

Vectorially oriented ATPases or electron transport chains and transmembrane ion gradients are possible energy sources for "uphill" proton movements. Microsomal cytochromes and flavoproteins are well known (e.g. 84), and plasmamembrane vesicles contain an ATP-driven Ca^{2+} porter (32, 41) and a *b*-type cytochrome (37, 44). In vitro stimulation of membranous ATPase by auxin [or indeed by fusicoccin (6, 58)] has proved difficult to repeat (101, 107). However, when precautions were taken to inhibit phospholipase activity, a small but significant stimulation of soybean microsomal ATPase by 1 nM and 1 μM 2,4-D (relative to the effect of the nonauxin 2,3-D) was found (91). The cation and auxin sensitivity of a K^+, Ca^{2+}, ATPase from rice root microsomes was lost after acetone extraction but could be restored by phosphatidylcholine (28). A Ca^{2+}-dependent coupling was proposed to link auxin binding to ATPase activity in the native lipid environment. Such an indirect effect is consistent with lack of ATPase activity of solubilized maize ER auxin-binding protein, which does show competition of *trans*-cinnamate, *p*-coumarate, and azide for NAA binding (19). Auxin interaction with cinnamate 4-hydroxylase [an ER cytochrome P450 linked monooxygenase (35, 84)] could be investigated via enzyme kinetics and by looking for a characteristic "type I" substrate-binding difference spectrum (84). One useful tissue could be potato tuber slices which both synthesize cinnamate 4-hydroxylase and show auxin-stimulated fresh weight gain (33, 84). Any connection between ER redox systems and hormonal action on wall acidification is likely to involve auxin as an allosteric effector rather than an enzyme substrate.

Modulation of Ca^{2+} pumps or gates is an attractive possibility for auxin action (77) in view of the second messenger role of Ca^{2+} for many animal hormones and the occurrence of calmodulin, a widespread Ca^{2+}-dependent regulatory protein, in plants (1a, 61). Recently, phytochrome-mediated Ca^{2+} uptake has been shown in *Mougeotia* (25) and *Avena* coleoptiles (33a). But the tendency to hitch theories of auxin action to fashionable bandwagons is not necessarily helpful—witness the frustrating search for a role of 3',5' cyclic AMP in higher plants.

AUXIN TRANSPORT Auxin binding to transport proteins is probably closer to biological function than is the case for other putative receptors, and it offers correspondingly greater scope for in vitro manifestation of physiologically inferred properties. Thus "site III" may be concerned with polar auxin transport (41) because of its probable plasmamembrane localization, the unique stimulation of IAA binding by TIBA and NPA, and the degree of correlation between IAA transport inhibition by auxin analogs and their inhibition of specific IAA binding in the presence of TIBA (41). The site concentration [20 – 150 pmol g^{-1} fresh weight (23, 41)] should meet polar auxin transport needs (77). NAA binding to intracellular membranes is quite strongly inhibited by TIBA (4, 23, 81), and the basis of the seemingly paradoxical stimulation of IAA binding is unclear. Perturbation of net transport into membrane vesicles by TIBA, perhaps inhibiting IAA efflux, has not been excluded but is thought unlikely (41). Use of uncouplers to collapse any pH or electrical gradient set up during membrane preparation and investigation of solubilized binding activity could help to clarify this point. Conformational changes whereby TIBA and NPA increase the affinity or capacity of IAA binding could be involved; their inhibition of transport might reflect stabilization of a nontranslocatable form of IAA carrier complex. NPA binding sites on maize membranes (54, 95, 97, 98) have been regarded as separate from both TIBA and auxin receptors because of lack of appropriate competitive displacement and because NPA, unlike TIBA, is not subject to polar transport (96). However, solubilization renders the NPA receptor susceptible to IAA (but not benzoic acid) competition (93). It has been suggested that both site III and the NPA receptor may represent conformations of an IAA carrier (41, 93). NPA may be an analog of a natural regulator "Kartoffelfaktor" (82), of unknown structure. Concave Scatchard plots are found for specific NPA binding, but the kinetics of association and dissociation (98) fitted a simple noncooperative model reasonably well, although the large free energy of activation for binding (98) is consistent with a concomitant conformational change. The molecular mechanism of transmembrane IAA transport will certainly involve conformational changes, but whether distinct conformers could be isolated is debatable. Interpretation of binding data should take account of the possibility that two carrier systems differing in specificity toward auxins and transport inhibitors may exist in plasmamembranes (22, 89, 90, 93a).

GENE EXPRESSION The rapid development of eukaryotic molecular biology has revealed numerous stages at which auxin could influence gene expression. These include formation of the initial RNA transcript, posttranscriptional processing, turnover of mRNA, and translational and posttranslational control of catalyst level (43, 70). Systems where induced developmental changes are marked by substantial auxin-promoted synthesis of known proteins (e.g. 106; see also 9, 92) should be further explored for receptor participation. However, particular morphogenetic programs such as meristem activation may often be triggered by various environmental and nonspecific stimuli as well as by growth substances. Trewavas (97b) has argued that such developmental plasticity could reflect a common coupling mechanism, perhaps involving altered membrane permeability, and that any initial auxin intervention need not require mediation by a protein receptor. Also, in a balanced perspective it is important to recognize the hierarchical nature of development and that control of the activity and location of "preexisting" gene products may often be of more immediate significance than differential gene expression.

Prospects for Receptor Recognition by Functional Assay

A functional assay depends on monitoring some activity coupled to auxin binding, using either an isolated complex or a reconstituted system (cf 31a). Searches for auxin-stimulated ATPases (28, 91) illustrate the former approach which could be extended to ion transport or carrier-mediated auxin transport itself by closed vesicles. One difficulty may be slow translocation or poor receptor-effector coupling at low temperatures necessary for receptor stability; use of membranes from chilling-resistant plants such as pea (67) could help. For auxin transport, the high permeability of the protonated species restricts possible assay methods. Plasmamembrane vesicles of opposite orientations could perhaps be separated by polyethyleneglycol-supported sedimentation (62) or on affinity columns, for example by exploiting lectin binding (12, 97a).

Reconstitution has been a powerful tool in other areas of membrane biochemistry (e.g. 31a). Basic techniques include (29) the formation of proteoliposomes by sonication, from mixed detergent solutions of proteins and phospholipids by dialysis or dilution, and also by incorporation of proteins into preformed liposomes. Clearly, the difficulties multiply the greater the number of protein components involved so that auxin transport could be an easier prospect for reassembly than putative coupling of a receptor to an ion-translocating ATPase or redox array.

By extrapolation of a technique developed for animal cells (27), the feasibility could be explored of implanting auxin receptors into plasmamembranes of protoplasts under fusion conditions and so conferring some hormone-sensitive response that was initially absent or had been lost during protoplast isolation.

The further characterization of auxin-dependent transcription factors (43, 85, 101, 103) will demand painstaking and patient technical optimization of homologous systems [e.g. coconut endosperm; (88) and Ref. 3 therein], preferably relating to a clear hormonal response, to achieve the high specificity of RNA chain initiation that is probably not attainable with *E. coli* RNA polymerase and eukaryotic chromatin (43, 70). The methodology should aim at characterizing RNA products by hybridization back to native template and, where applicable, by in vitro translation to an identifiable protein.

USE OF MUTANTS Investigation of the properties of auxin-binding proteins in mutants with modified or lost auxin responses is a potentially powerful approach in which higher and lower plants, organ and tissue cultures, and protoplast fusion could all play a role. Attractive systems include duckweed (Lemnaceae) which can be rapidly cultured in large quantities; *Arabidopsis thaliana,* a crucifer with a generation time of only a few weeks (39); and tissue cultures that show hormone-dependent differentiation (e.g. 9). Selection procedures should allow detection of conditional mutants whose expression can be regulated by the investigator. A promising existing system is a category of morphological mutants of the moss *Physcomitrella patens* which are blocked, perhaps at a receptor site, between auxin synthesis and auxin-stimulated development of caulonemata (1; D. J. Cove, personal communication).

CONCLUDING REMARKS

The unraveling of auxin action and the recognition and characterization of receptors are obviously complementary objectives. Our present understanding emphasizes membrane permeability and gene expression, but specific molecular target processes have not yet been directly identified. Our deeper knowledge of animal hormones and neurotransmitters offers some attractive paradigms for working hypotheses. But the biochemical instinct for generalization across phylogenetic boundaries must be balanced by the assertion that multicellular plants are not green animals and that their sessile and autotrophic life style may reflect a distinctively evolved approach to communication between cells and from the environment. The extent of our ignorance of the occurrence, properties, and functions of auxin receptors is only too apparent from the following list of some fundamental questions whose answers need to be sought by future research. How many types of auxin-binding protein can be distinguished in terms of primary structure or subunit association, and to what extent are the known binding sites (membranous and soluble) molecularly heterogeneous? What events are coupled to auxin binding: can auxin act as a direct effector of a catalyst or a gene regulatory protein? do auxin-receptor complexes bind to and modulate other proteins, and if so, what are they? How do different tissues and organs vary in their complement of auxin receptors, and what is the molecular basis for interaction between different growth substances? I hope that some of the suggestions made in this review will contribute toward the ultimate goal of understanding the physiology and developmental biology of auxin action in molecular terms.

ACKNOWLEDGMENTS

I am grateful to Hugh Jones and Mary Astle for reading and commenting on the manuscript.

Literature Cited

1. Ashton, N. W., Grimsley, N. H., Cove, D. J. 1979. Analysis of gametophytic development in the moss, *Physcomitrella patens*, using auxin and cytokinin resistant mutants. *Planta* 144:427–35

1a. Anderson, J. M., Charbonneau, H., Jones, H. P., McCann, R. O., Cormier, M. J. 1980. Characterization of the plant nicotinamide adenine dinucleotide kinase activator protein as calmodulin. *Biochemistry* 19:3113–20

2. Bandurski, R. S. 1979. Chemistry and physiology of conjugates of indole-3-acetic acid. *ACS Symp. Ser.* No. 111, ed. N. B. Mandava, pp. 1–17

3. Basu, P. S., Tuli, V. 1972. The binding of indole-3-acetic acid and 3-methyleneoxindole to plant macromolecules. *Plant Physiol.* 50:507–9

4. Batt, S., Venis, M. A. 1976. Separation and localization of two classes of auxin binding sites in corn coleoptile membranes. *Planta* 130:15–21

5. Batt, S., Wilkins, M. B., Venis, M. A. 1976. Auxin binding to corn coleoptile membranes: kinetics and specificity. *Planta* 130:7–13

6. Beffagna, N., Cocucci, S., Marrè, E. 1977. Stimulating effect of fusicoccin on K-activated ATPase in plasmalemma preparations from higher plant tissues. *Plant Sci Lett.* 8:91–98

7. Beffagna, N., Pesci, P., Tognoli, L., Marrè, E. 1979. Distribution of fusicoccin bound in vivo among subcellular fractions from maize coleoptiles. *Plant Sci Lett.* 15:323–30

8. Bengochea, T., Dodds, J. H., Evans, D. E., Jerie, P. H., Niepel, B., Shaari, A. R., Hall, M. A. 1980. Studies on ethylene binding by cell-free preparations from cotyledons of *Phaseolus vulgaris* L. I. Separation and characterisation. *Planta* 148:397–406

9. Bevan, M., Northcote, D. H. 1979. The interaction of auxin and cytokinin in the induction of phenylalanine ammonia-

lyase in suspension cultures of *Phaseolus vulgaris*. *Planta* 147:77–81
10. Bhattacharyya, K., Biswas, B. B. 1978. Membrane-bound auxin receptors from *Avena* roots. *Ind. J. Biochem. Biophys.* 15:445–48
11. Bogers, R. J., Kulescha, Z., Quint, A., van Vliet, T. B., Libbenga, K. R. 1980. The presence of a soluble auxin receptor and the metabolism of 3-indoleacetic acid in tobacco-pith explants. *Plant. Sci. Lett.* 19:311–17
12. Boss, W. F., Ruesink, A. W. 1979. Isolation and characterisation of concanavalin A-labeled plasma membranes of carrot protoplasts. *Plant Physiol.* 64:1005–11
13. Cash, D. J., Hess, G. P. 1980. Molecular mechanism of acetylcholine receptor-controlled ion translocation across cell membranes. *Proc. Natl. Acad. Sci. USA* 77:842–46
14. Chang, K-J., Jacobs, S., Cuatrecasas, P. 1975. Quantitative aspects of hormone-receptor interactions of high affinity. Effect of receptor concentration and measurement of dissociation constants of labeled and unlabeled hormones. *Biochim. Biophys. Acta* 406:294–303
15. Chen, S., McMahon, D., Bogorad, L. 1967. Early effects of illumination on the activity of some photosynthetic enzymes. *Plant Physiol.* 42:1–5
16. Chowdhry, V., Westheimer, F. H. 1979. Photoaffinity labeling of biological systems. *Ann. Rev. Biochem.* 48:293–325
17. Cleland, R. E. 1977. The control of cell enlargement. In *Integration of Activity in the Higher Plant,* ed. D. H. Jennings. Soc. Exp. Biol. Symp. 31:101–15. Cambridge: Cambridge Univ. Press
18. Cross, J. W., Briggs, W. R. 1978. Properties of a solubilized microsomal auxin-binding protein from coleoptiles and primary leaves of *Zea mays*. *Plant Physiol.* 62:152–57
19. Cross, J. W., Briggs, W. R. 1979. Solubilized auxin-binding protein; subcellular localization and regulation by a soluble factor from homogenates of corn shoots. *Planta* 146:263–70
20. Cross, J. W., Briggs, W. R., Dohrmann, U. C., Ray, P. M. 1978. Auxin receptors of maize coleoptile membranes do not have ATPase activity. *Plant Physiol.* 61:581–84
21. Cuatrecasas, P., Hollenberg, M. D. 1976. Membrane receptors and hormone action. *Adv. Protein Chem.* 30: 251–451
22. Davies, P. J., Rubery, P. H. 1978. Components of auxin transport in stem segments of *Pisum sativum* L. *Planta* 142:211–19
23. Dohrmann, U. C., Hertel, R., Kowalik, H. 1978. Properties of auxin binding sites in different subcellular fractions from maize coleoptiles. *Planta* 140:97–106
24. Döllstädt, R., Hirschberg, K., Winkler, E., Hübner, G. 1976. Bindung von Indolylessigsäure an Fraktionen aus Epicotylen und Wurzeln von *Pisum sativum* L. *Planta* 130:105–11
25. Dreyer, E. M., Weisenseel, M. H. 1979. Phytochrome-mediated uptake of calcium in *Mougeotia* cells. *Planta* 146: 31–39
26. Edwards, K. L., Goldsmith, M. H. M. 1980. pH-Dependent accumulation of indoleacetic acid by corn coleoptile sections. *Planta* 147:457–66
27. Eimerl, S., Neufeld, G., Korner, M., Schramm, M. 1980. Functional implantation of a solubilised β-adrenergic receptor in the membrane of a cell. *Proc. Natl. Acad. Sci. USA* 71:760–64
28. Erdei, L., Tóth, I., Zsoldos, F. 1979. Hormonal regulation of Ca^{2+} stimulated K^+influx and Ca^{2+},K^+-ATPase in rice roots: in vivo and in vitro effects of auxins and reconstitution of the ATPase. *Physiol. Plant.* 45:448–52
29. Eytan, G. D., Kanner, B. I. 1978. Reconstitution of biological membranes. In *Receptors and Recognition* ed. P. Cuatrecasas, M. F. Greaves, 6A:65–105. London: Chapman & Hall
30. Farrimond, J. A., Elliott, M. C., Clack, D. W. 1978. Charge separation as a component of the structural requirements for hormone activity. *Nature* 274:401–2
31. Farrimond, J. A., Elliott, M. C., Clack, D. W. 1980. Auxin structure/activity relationships: benzoic acids and phenols. *Phytochemistry* 19:367–71
31a. Gonzalez-Ros, J. M., Paraschos, A., Martinez-Carrion, M. 1980. Reconstitution of functional membrane-bound acetylcholine receptor from isolated *Torpedo californica* receptor protein and electroplax lipids. *Proc. Natl. Acad. Sci. USA* 77:1796–1800
32. Gross, J., Marmé, D. 1978. ATP-dependent Ca^{2+} uptake into plant membrane vesicles. *Proc. Natl. Acad. Sci. USA* 75:1232–36
33. Hackett, D. P. 1952. The osmotic change during auxin-induced water uptake by potato tissue. *Plant Physiol.* 27:279–84
33a. Hale, C. C., Roux, S. J. 1980. Photoreversible calcium fluxes induced by

phytochrome in oat coleoptile cells. *Plant Physiol.* 65:658–62
34. Hardin, J. W., Cherry, J. H., Morré, D. J., Lembi, C. A. 1972. Enhancement of RNA polymerase activity by a factor released by auxin from plasma membrane. *Proc. Natl. Acad. Sci. USA* 69:3146–50
35. Hartmann-Bouillon, M-A., Benveniste, P. 1978. Sterol biosynthetic capability of purified membrane fractions from maize coleoptiles. *Phytochemistry* 17:1037–42
36. Helgerson, S. L., Cramer, W. A., Morré, D. J. 1976. Evidence for an increase in microviscosity of plasma membranes from soybean hypocotyls induced by the plant hormone, indole-3-acetic acid. *Plant Physiol.* 58:548–51
37. Hertel, R. 1979. Auxin binding sites: subcellular fractionation and specific binding assays. See Ref. 83, pp. 173–83
38. Hertel, R., Thomson, K-S., Russo, V. E. A. 1972. In-vitro auxin binding to particulate cell fractions from corn coleoptiles. *Planta* 107:325–40
39. Hess, D. 1968. *Biochemische Genetik.* Heidelberg: Springer. 353 pp.
40. Ihl, M. 1976. Indole-acetic acid binding proteins in soybean cotyledon. *Planta* 131:223–28
40a. Jacobs, M., Gepstein, S., Fichmann, J., Taiz, L. 1980. Vanadate as a tool for studying the role of H^+ secretion in plant development. *Plant Physiol.* 65:S720
41. Jacobs, M., Hertel, R. 1978. Auxin binding to subcellular fractions from *Cucurbita* hypocotyls: in vitro evidence for an auxin transport carrier. *Planta* 142:1–10
42. Jacobs, S., Chang, K-J., Cuatrecasas, P. 1975. Estimation of hormone receptor affinity by competitive displacement of labeled ligand: effect of concentration of receptor and of labeled ligand. *Biochem. Biophys. Res. Commun.* 66:687–92
43. Jacobsen, J. V. 1977. Regulation of ribonucleic acid metabolism by plant hormones. *Ann. Rev. Plant Physiol.* 28:537–64
44. Jesaitis, A. J., Heners, P. R., Hertel, R., Briggs, W. R. 1977. Characterization of a membrane fraction containing a *b*-type cytochrome. *Plant Physiol.* 59:941–47
45. Jones, A. M., Melhado, L. L., Lu, T-Y. S., Leonard, N. J., Vanderhoef, L. N. 1979. Fluorescent photoaffinity labeling: azido-auxin analogs. *Plant Physiol.* 63:S117
46. Kaethner, T. M. 1977. Conformational change theory for auxin structure-activity relationships. *Nature* 267:19–23
47. Kahn, C. R. 1976. Membrane receptors for hormones and neurotransmitters. *J. Cell Biol.* 70:261–86
48. Kasamo, K., Yamaki, T. 1976. In vitro binding of IAA to plasma membrane-rich fractions containing Mg^{++}–activated ATPase from mung bean hypocotyls. *Plant Cell Physiol.* 17:149–64
49. Katekar, G. F. 1979. Auxins: on the nature of the receptor site and molecular requirements for auxin activity. *Phytochemistry* 18:223–33
50. Katekar, G. F., Geissler, A. E. 1977. Auxin transport inhibitors: III. chemical requirements of a class of auxin transport inhibitors. *Plant Physiol.* 60:826–29
51. Kende, H., Gardner, G. M. 1976. Hormone binding in plants. *Ann. Rev. Plant Physiol.* 27:267–90
52. Lamb, C. J., Rubery, P. H. 1976. Inhibition of co-operative enzymes by substrate-analogs: possible implications for the physiological significance of negative co-operativity illustrated by phenylalanine metabolism in higher plants. *J. Theor. Biol.* 60:441–47
53. Lelkes, P. I., Bach, D., Miller, I. R. 1979. Interactions of auxin (IAA) with lipid membranes. In *Plant Membrane Transport: Current Conceptual Issues,* ed. R. M. Spanswick, W. J. Lucas, J. Dainty, pp. 539–40. Amsterdam: Elsevier
54. Lembi, C. A., Morré, D. J., Thomson, K-S., Hertel, R. 1971. N-1-Naphthylphthalamic-acid-binding activity of a plasma membrane-rich fraction from maize coleoptiles. *Planta* 99:37–45
55. Leonard, N. J., Greenfield, J. C., Schmitz, R. Y., Skoog, F. 1975. Photoaffinity-labeled auxins. Synthesis and biological activity. *Plant Physiol.* 55:1057–61
56. Libbenga, K. R. 1978. Hormone receptors in plants. In *Frontiers of Plant Tissue Culture 1978,* ed. T. A. Thorpe, pp. 325–33. Calgary: Int. Assoc. Plant Tissue Cult. 1978
57. Maelicke, A., Fulpius, B. W., Klett, R. P., Reich, E. 1977. Acetylcholine receptor. Responses to drug binding. *J. Biol. Chem.* 252:4811–30
58. Marrè, E. 1979. Fusicoccin: a tool in plant physiology. *Ann. Rev. Plant Physiol.* 30:273–88

59. Martin, B. R., Stein, J. M., Kennedy, E. L., Doberska, C. A., Metcalfe, J. C. 1979. Transient complexes. A new structural model for the activation of adenylate cyclase by hormone receptors (guanine nucleotides/irradiation inactivation). *Biochem. J.* 184:253–60
60. Mayer, R. J., Walker, J. H. 1980. Immunochemical methods in the biological sciences: enzymes and proteins. In *Biological Techniques, Series 3,* ed. J. E. Treherne, P. H. Rubery, p. 168. New York/London: Academic
61. Means, A. R., Dedman, J. R. 1980. Calmodulin—an intracellular calcium receptor. *Nature* 285:73–77
62. Michalke, W., Schmieder, B. 1979. Fractionation of particulate material from maize coleoptile homogenates with polyethylene glycol. *Planta* 145: 129–35
63. Mornet, R., Theiler, J. B., Leonard, N. J., Schmitz, R. Y., Moore, F. H., Skoog, F. 1979. Active cytokinins; Photoaffinity labeling agents to detect binding. *Plant Physiol.* 64:600–10
64. Morré, D. J. 1979. Response of isolated plasma membranes to auxins. *Abstr. Int. Conf. Plant Growth Subst., 10th, Madison, Rep. 021,* p. 5
65. Morré, D. J., Bracker, C. E. 1976. Ultrastructural alteration of plant plasma membranes induced by auxin and calcium ions. *Plant Physiol.* 58:544–47
66. Morré, D. J., Cherry, J. H. 1977. Auxin hormone-plasma membrane interactions. *Int. Conf. Plant Growth Subst., 9th, Lausanne,* ed. P. E. Pilet, pp. 35–43
67. Morris, D. A. 1979. The effect of temperature on the velocity of exogenous auxin transport in intact chilling-sensitive and chilling-resistant plants. *Planta* 146:603–5
68. Murphy, G. J. P. 1979. Plant hormone receptors: comparison of naphthaleneacetic acid binding by maize extracts and by a non-plant protein. *Plant Sci. Lett.* 15:183–91
68a. Murphy, G. J. P. 1980. A reassessment of the binding of naphthaleneacetic acid by membrane preparations from maize. *Planta* 149:417–26
69. Normand, G., Hartmann, M-A., Schuber, F., Benveniste, P. 1975. Caractérisation de membranes de coléoptiles de Maïs fixant l'auxine et l'acide N-naphtyl phtalamique. *Physiol Vég.* 13:743–61
70. O'Malley, B. W., Towle, H. C., Schwartz, R. J. 1977. Regulation of gene expression in eukaryotes. *Ann. Rev. Genet.* 11:239–75
71. Oostrom, H., Kulescha, Z., van Vliet, T. B., Libbenga, K. R. 1980. Characterisation of a cytoplasmic auxin receptor from tobacco-pith callus. *Planta* 149:44–47
72. Oostrom, H., van Loopik-Detmers, M. A., Libbenga, K. R. 1975. A high affinity receptor for indoleacetic acid in cultured tobacco pith explants. *FEBS Lett.* 59:194–97
73. Pesci, P., Tognoli, L., Beffagna, N., Marrè, E. 1979. Solubilisation and partial purification of a fusicoccin-receptor complex from maize microsomes. *Plant Sci. Lett.* 15:313–22
74. Porter, W. L., Thimann, K. V. 1965. Molecular requirements for auxin action: I. halogenated indoles and indoleacetic acid. *Phytochemistry* 4:229–43
75. Quail, P. H. 1979. Plant cell fractionation. *Ann. Rev. Plant Physiol.* 30: 425–84
76. Rakhaminova, A. B., Khavkin, É. E., Yaguzhinskii, L. S. 1978. Construction of a model of the auxin receptor. *Biokhimiya* 43:639–53
77. Raven, J. A., Rubery, P. H. 1981. Coordination of development: hormone receptors, hormone action, and hormone transport. In *Molecular Biology of Plant Development,* ed. H. Smith, D. Grierson. Oxford: Blackwells. In press
78. Ray, P. M. 1977. Auxin-binding sites of maize coleoptiles are localized on membranes of the endoplasmic reticulum. *Plant Physiol.* 59:594–99
79. Ray, P. M. 1979. Maize coleoptile cellular membranes bearing different types of glucan synthetase activity. See Ref. 83, pp. 135–46
80. Ray, P. M., Dohrmann, U. C., Hertel, R. 1977. Characterisation of naphthaleneacetic acid binding to receptor sites on cellular membranes of maize coleoptile tissue. *Plant Physiol.* 59: 357–64
81. Ray, P. M., Dohrmann, U. C., Hertel, R. 1977. Specificity of auxin-binding sites on maize coleoptile membranes as possible receptors sites for auxin action. *Plant Physiol.* 60:585–91
82. Reichert, T. L., Peterson, W., Hertel, R. 1978. Two plant substances which inhibit, respectively, auxin and naphthylphthalamic acid binding. *Plant Physiol.* 61:S358
83. Reid, E., ed. 1979. Plant organelles. *Methodological Surveys—subseries (B): Biochemistry,* Vol. 9. Chichester: Horwood. 232 pp.
84. Rich, P. R., Lamb, C. J. 1977. Biophysical and enzymological studies upon the

interaction of *trans*-cinnamic acid with higher plant microsomal cytochromes. *Eur. J. Biochem.* 72:353–60
85. Rizzo, P. J., Pedersen, K., Cherry, J. H. 1977. Stimulation of transcription by a soluble factor isolated from soybean hypocotyl by 2,4-D affinity chromatography. *Plant Sci. Lett.* 8:205–11
86. Robenek, H. 1979. Der Einfluss von Indolyl (3) essigsäure (IES) auf die Verteilung der intramembranosen Partikel des Plasmalemmas isolierter Sprosskallus-protoplasten von *Skimmia japonica.* Thunb. *Z. Pflanzenphysiol.* 93:317–24
87. Roland, J-C., Lembi, C. A., Morré, D. J. 1972. Phosphotungstic acid-chromic acid as a selective electron-dense stain for plasma membranes of plant cells. *Stain Technol.* 47:195–200
88. Roy, P., Biswas, B. B. 1977. A receptor protein for indoleacetic acid from plant chromatin and its role in transcription. *Biochem. Biophys. Res. Commun.* 74:1597–1606
89. Rubery, P. H. 1979. The effects of 2,4-dinitrophenol and chemical modifying reagents on auxin transport by suspension-cultured crown gall cells. *Planta* 144:173–78
90. Rubery, P. H. 1980. The mechanism of transmembrane auxin transport and its relation to the chemiosmotic hypothesis of the polar transport of auxin. *Int. Conf. Plant Growth Subst., 10th, Madison,* 1979, ed. F. Skoog, pp. 50–60. Heidelberg: Springer
91. Scherer, G. F. E., Morré, D. J. 1978. In vitro stimulation by 2,4-dichlorophenoxyacetic acid of an ATPase and inhibition of phosphatidate phosphatase of plant membranes. *Biochem. Biophys. Res. Commun.* 84:238–47
92. Schröder, J., Kreuzaler, F., Schäfer, E., Hahlbrock, K. 1980. Concomitant induction of phenylalanine ammonia-lyase and flavanone synthetase in irradiated plant cells. *J. Biol. Chem.* 254:57–65
93. Sussman, M. R., Gardner, G. M. 1980. Solubilization of the receptor for N-1-naphthylphthalamic acid. *Plant Physiol.* 66:1074–78
93a. Sussman, M. R., Goldsmith, M. H. M. 1981. Auxin uptake and action of N-1-naphthylphthalamic acid in corn coleoptiles. *Planta* 151:15–25
94. Sweetser, P. B., Swartzfager, D. G. 1978. Indole-3-acetic acid levels of plant tissue as determined by a new high performance liquid chromatographic method. *Plant Physiol.* 61:254–58
94a. Taiz, L., Jacobs, M. 1980. Vanadate inhibition of auxin-enhanced H^+ secretion and elongation in pea epicotyls and oat coleoptiles. *Plant Physiol.* 65:S719
95. Thomson, K-S. 1972. The binding of 1-N-naphthylphthalamic acid (NPA), an inhibitor of auxin transport, to particulate fractions of corn coleoptiles. In *Hormonal Regulation in Plant Growth and Development. Proc. Adv. Study Inst. Izmir 1971,* ed. H. Kaldewey, Y. Vardar, pp. 83–88. Weinheim: Verlag Chem.
96. Thomson, K-S, Hertel, R., Müller, S., Tavares, J. E. 1973. 1-N-Naphthylphthalamic acid and 2,3,5-triiodobenzoic acid: in-vitro binding to particulate cell fractions and action on auxin transport in corn coleoptiles. *Planta* 109:337–52
97. Thomson, K-S., Leopold, A. C. 1974. In-vitro binding of morphactins and 1-N-naphthylphthalamic acid in corn coleoptiles and their effects on auxin transport. *Planta* 115:259–70
97a. Travis, R. L., Berkowitz, R. L. 1980. Characterization of soybean plasmamembrane during development. Free sterol composition and concanavalin A binding studies. *Plant Physiol.* 65:871–79
97b. Trewavas, A. 1979. What is the molecular basis of plant hormone action? *Trends Biochem. Sci* 4:N199–202
98. Trillmich, K., Michalke, W. 1979. Kinetic characterisation of N-1-naphthylphthalamic acid binding sites from maize coleoptile homogenates. *Planta* 145:119–27
99. Venis, M. A. 1971. Stimulation of RNA transcription from pea and corn DNA by protein retained on Sepharose coupled to 2,4-dichlorophenoxyacetic acid. *Proc. Natl. Acad. Sci. USA* 68:1824–27
100. Venis, M. A. 1977. Affinity labels for auxin binding sites in corn coleoptile membranes. *Planta* 134:145–49
101. Venis, M. A. 1977. Receptors for plant hormones. *Adv. Bot. Res.* 5:53–88
102. Venis, M. A. 1977. Solubilization and partial purification of auxin-binding sites of corn membranes. *Nature* 266:268–69
103. Venis, M. A. 1979. Receptors for plant growth regulators. *Adv. Pestic. Sci.* ed. H. Geissbühler, Part. 3, pp. 487–93
104. Venis, M. A. 1979. Purification and properties of membrane-bound auxin receptors in corn. See Ref. 90, pp. 61–70
105. Venis, M. A., Watson, P. J. 1978. Naturally occurring modifiers of auxin-receptor interaction in corn: identification as benzoxazolinones. *Planta* 142:103–7

106. Verma, D. P. S., Maclachlan, G. A., Byrne, H., Ewings, D. 1975. Regulation and in vitro translation of messenger ribonucleic acid for cellulase from auxin-treated pea epicotyls. *J. Biol. Chem.* 250:1019–26
107. Vreugdenhil, D., Burgers, A., Libbenga, K. R. 1979. A particle-bound auxin receptor from tobacco pith callus. *Plant Sci. Lett.* 16:115–21
108. Vreugdenhil, D., Harkes, P. A. A., Libbenga, K. R. 1980. Auxin-binding by particulate fractions from tobacco leaf protoplasts. *Planta* 150:9–12
109. Wardrop, A. J., Polya, G. M. 1977. Properties of a soluble auxin-binding protein from dwarf bean seedlings. *Plant Sci. Lett.* 8:155–63
110. Wardrop, A. J., Polya, G. M. 1980. Copurification of pea and bean leaf soluble auxin-binding proteins with ribulose-1,5-bisphosphate carboxylase. *Plant Physiol.* 66:105–11
111. Wardrop, A. J., Polya, G. M. 1980. The ligand specificity of bean leaf soluble auxin-binding protein. *Plant Physiol.* 66:112–18
112. Wildman, S. G., Bonner, J. 1947. The proteins of green leaves. I. Isolation, enzymatic properties, and auxin content of spinach cytoplasmic proteins. *Arch. Biochem.* 14:381–413
113. Williamson, F. A., Morré, D. J., Hess, K. 1977. Auxin binding activities of subcellular fractions from soybean hypocotyls. *Cytobiologie* 16:63–71
114. Yunghans, W. N., Clark, J. E., Morré, D. J., Clegg, E. D. 1978. Nature of the phosphotungstic acid-chromic acid (PACP) stain for plasma membranes of plants and mammalian sperm. *Cytobiologie* 17:165–72

AUTHOR INDEX

(Names appearing in capital letters indicate authors of chapters in this volume.)

C

D

E

F

N

S

Y

Z

SUBJECT INDEX

N

CUMULATIVE INDEXES

CONTRIBUTING AUTHORS, VOLUMES 28–32

CHAPTER TITLES, VOLUMES 28–32